ALLE ZEIT WACH 1842

NICOLAS BOURBAKI

ELEMENTS OF MATHEMATICS

Lie Groups and Lie Algebras

Chapters 1–3

Springer-Verlag
Berlin Heidelberg New York
London Paris Tokyo

Originally published as
ÉLÉMENTS DE MATHÉMATIQUE,
GROUPES ET ALGÈBRES DE LIE 1et 2–3

Mathematics Subject Classification (1991): 17B05, 22E

Distribution rights worldwide:
Springer-Verlag Berlin Heidelberg New York London Paris Tokyo

ISBN 978-3-540-64242-8 Springer-Verlag Berlin Heidelberg New York
ISBN 978-3-540-50218-0 2nd printing Springer-Verlag Berlin Heidelberg New York

Softcover edition of the 2nd printing 1989

Library of Congress Cataloging-in-Publication Data
Bourbaki, Nicolas. [Groupes et algèbres de Lie. English] Lie groups and Lie algebras / Nicolas Bourbaki.
p. cm.-(Elements of mathematics / Nicolas Bourbaki) Translation of: Groupes et algèbres de Lie.
Bibliography: p. Includes index.
ISBN 978-0-387-50218-2 (U.S.)
1. Lie groups 2. Lie algebras. I. Title. II. Series: Bourbaki, Nicolas. Éléments de mathématique. English.
QA387.B6513 1988 512'.55-dc 19 88-31203

SPIN 11548713
41/3111-5 4 3 2 1
Printed on acid-free paper

TO THE READER

1. This series of volumes, a list of which is given on pages ix and x, takes up mathematics at the beginning, and gives complete proofs. In principle, it requires no particular knowledge of mathematics on the readers' part, but only a certain familiarity with mathematical reasoning and a certain capacity for abstract thought. Nevertheless, it is directed especially to those who have a good knowledge of at least the content of the first year or two of a university mathematics course.

2. The method of exposition we have chosen is axiomatic and abstract, and normally proceeds from the general to the particular. This choice has been dictated by the main purpose of the treatise, which is to provide a solid foundation for the whole body of modern mathematics. For this it is indispensable to become familiar with a rather large number of very general ideas and principles. Moreover, the demands of proof impose a rigorously fixed order on the subject matter. It follows that the utility of certain considerations will not be immediately apparent to the reader unless he has already a fairly extended knowledge of mathematics; otherwise he must have the patience to suspend judgment until the occasion arises.

3. In order to mitigate this disadvantage we have frequently inserted examples in the text which refer to facts the reader may already know but which have not yet been discussed in the series. Such examples are always placed between two asterisks: *...*. Most readers will undoubtedly find that these examples will help them to understand the text, and will prefer not to leave them out, even at a first reading. Their omission would of course have no disadvantage, from a purely logical point of view.

4. This series is divided into volumes (here called "Books"). The first six Books are numbered and, in general, every statement in the text assumes as known only those results which have already been discussed in the preceding

volumes. This rule holds good within each Book, but for convenience of exposition these Books are no longer arranged in a consecutive order. At the beginning of each of these Books (or of these chapters), the reader will find a precise indication of its logical relationship to the other Books and he will thus be able to satisfy himself of the absence of any vicious circle.

5. The logical framework of each chapter consists of the *definitions*, the *axioms*, and the *theorems* of the chapter. These are the parts that have mainly to be borne in mind for subsequent use. Less important results and those which can easily be deduced from the theorems are labelled as "propositions", "lemmas", "corollaries", "remarks", etc. Those which may be omitted at a first reading are printed in small type. A commentary on a particularly important theorem appears occasionally under the name of "scholium".

To avoid tedious repetitions it is sometimes convenient to introduce notations or abbreviations which are in force only within a certain chapter or a certain section of a chapter (for example, in a chapter which is concerned only with commutative rings, the word "ring" would always signify "commutative ring"). Such conventions are always explicitly mentioned, generally at the beginning of the chapter in which they occur.

6. Some passages in the text are designed to forewarn the reader against serious errors. These passages are signposted in the margin with the sign ☡ ("dangerous bend").

7. The Exercises are designed both to enable the reader to satisfy himself that he has digested the text and to bring to his notice results which have no place in the text but which are nonetheless of interest. The most difficult exercises bear the sign ¶.

8. In general, we have adhered to the commonly accepted terminology, except where there appeared to be good reasons for deviating from it.

9. We have made a particular effort always to use rigorously correct language, without sacrificing simplicity. As far as possible we have drawn attention in the text to *abuses of language*, without which any mathematical text runs the risk of pedantry, not to say unreadability.

10. Since in principle the text consists of the dogmatic exposition of a theory, it contains in general no references to the literature. Bibliographical references are gathered together in *Historical Notes*, usually at the end of each chapter. These notes also contain indications, where appropriate, of the unsolved problems of the theory.

The bibliography which follows each historical note contains in general only those books and original memoirs which have been of the greatest importance in the evolution of the theory under discussion. It makes no sort of pretence to completeness; in particular, references which serve only to determine questions of priority are almost always omitted.

As to the exercises, we have not thought it worthwhile in general to indicate their origins, since they have been taken from many different sources (original papers, textbooks, collections of exercises).

11. References to a part of this series are given as follows:

a) If reference is made to theorems, axioms, or definitions presented *in the same section*, they are quoted by their number.
b) If they occur *in another section of the same chapter*, this section is also quoted in the reference.
c) If they occur *in another chapter in the same Book*, the chapter and section are quoted.
d) If they occur *in another Book*, this Book is first quoted by its title.

The *Summaries of Results* are quoted by the letter R; thus *Set Theory*, R signifies "*Summary of Results of the Theory of Sets*".

CONTENTS
OF
THE ELEMENTS OF MATHEMATICS SERIES

CONTENTS OF THE ELEMENTS OF MATHEMATICS SERIES

Lie groups and lie algebras

1. Lie algebras. 2. Free Lie algebras. 3. Lie groups. 4. Coxeter groups and Tits systems. 5. Groups generated by reflections. 6. Root systems.

Commutative algebra

1. Flat modules. 2. Localization. 3. Graduations, filtrations, and topologies. 4. Associated prime ideals and primary decomposition. 5. Integers. 6. Valuations. 7. Divisors.

Spectral theories

1. Normed algebras. 2. Locally compact groups.

Differentiable and analytic manifolds

Summary of results.

CONTENTS

CHAPTER I

Lie Algebras

In paragraphs 1, 2 *and* 3, K *denotes a commutative ring with unit element. In paragraph* 4, K *denotes a commutative field. In paragraphs* 5, 6 *and* 7, K *denotes a commutative field of characteristic* 0.†

§ 1. DEFINITION OF LIE ALGEBRAS

1. ALGEBRAS

Let M be a unitary module over K with a *bilinear* mapping $(x, y) \mapsto xy$ of $M \times M$ into M. All the axioms for algebras are satisfied except associativity of multiplication. By an abuse of language, M is called a *not necessarily associative algebra* over K, or sometimes, when no confusion can arise, an *algebra* over K. In this no. we shall use the latter notation.

If the K-module M is given the multiplication $(x, y) \mapsto yx$ an algebra is obtained called the *opposite* of the above algebra.

A sub-K-module N of M which is stable under multiplication is given the structure of an algebra over K in an obvious way. N is called a *subalgebra* of M. N is called a *left* (resp. *right*) *ideal* of M if the conditions $x \in N$, $y \in M$ imply $yx \in N$ (resp. $xy \in N$). If N is both a left ideal and a right ideal of M, N is called a *two-sided ideal* of M. In this case the multiplication on M enables us to define, on passing to the quotient, a bilinear multiplication on the quotient module M/N such that M/N has an algebra structure. M/N is called the *quotient algebra* of M by N.

† The propositions proved in this chapter depend only on the properties established in Books I to VI and on certain results of *Commutative Algebra*, Chapter III, § 2.

Let M_1 and M_2 be two algebras over K and ϕ a mapping of M_1 into M_2. ϕ is called a *homomorphism* if ϕ is K-linear and $\phi(xy) = \phi(x)\phi(y)$ for $x \in M_1$, $y \in M_1$. The kernel N of ϕ is a two-sided ideal of M_1 and the image of ϕ is a subalgebra of M_2. On passing to the quotient, ϕ defines an isomorphism of the algebra M_1/N onto the algebra $\phi(M_1)$.

Let M be an algebra over K. A mapping D of M into M is called a *derivation* of M if it is K-linear and $D(xy) = (Dx)y + x(Dy)$ for all $x \in M$ and $y \in M$. This definition generalizes Definition 3 of *Algebra*, Chapter IV, § 4, no. 3. The kernel of a derivation of M is a subalgebra of M. If D_1 and D_2 are derivations of M, then $D_1D_2 - D_2D_1$ is a derivation of M (cf. *Algebra*, Chapter IV, § 4, no. 3, Proposition 5: the proof of this proposition does not use the associativity of the algebra).

Let M_1 and M_2 be two algebras over K. On the product K-module $M = M_1 \times M_2$ we define a multiplication by writing

$$(x_1, x_2)(y_1, y_2) = (x_1y_1, x_2y_2),$$

for all x_1, y_1 in M_1, x_2, y_2 in M_2. The algebra thus defined is called the *product algebra* of M_1 and M_2. The mapping $x_1 \mapsto (x_1, 0)$ (resp. $x_2 \mapsto (0, x_2)$) is an isomorphism of M_1 (resp. M_2) onto a two-sided ideal of M. Under these isomorphisms M_1 and M_2 are identified with two-sided ideals of M. The K-module M is then the direct sum of M_1 and M_2. Conversely, let M be an algebra over K and M_1, M_2 two two-sided ideals of M such that M is the direct sum of M_1 and M_2. Then $M_1M_2 \subset M_1 \cap M_2 = \{0\}$; then, if x_1, y_1 belong to M_1 and x_2, y_2 to M_2, then $(x_1 + x_2)(y_1 + y_2) = x_1y_1 + x_2y_2$, so that M is identified with the product algebra $M_1 \times M_2$. Every left (resp. right, two-sided) ideal of M_1 is a left (resp. right, two-sided) ideal of M. We leave to the reader the task of formulating the analogous results in the case of an arbitrary finite family of algebras.

Let M be an algebra over K and suppose that the K-module M admits a basis $(a_\lambda)_{\lambda \in L}$. There exists a unique system $(\gamma_{\lambda\mu\nu})_{(\lambda,\mu,\nu) \in L \times L \times L}$ of elements of K such that $a_\lambda a_\mu = \sum_\nu \gamma_{\lambda\mu\nu} a_\nu$ for all λ, μ in L. The $\gamma_{\lambda\mu\nu}$ are called the *constants of structure of* M *with respect to the basis* (a_λ).

Let M be an algebra over K, K_0 a commutative ring with unit element and ρ a homomorphism of K_0 into K mapping unit element to unit element. Then M can be considered as an algebra over K_0 by writing $\alpha . x = \rho(\alpha) . x$ for $\alpha \in K_0$, $x \in M$. This is the case in particular when K_0 is a subring of K containing the unit element and ρ is taken to be the inclusion mapping of K_0 into K.

Let M be an algebra over K, K_1 a commutative ring with unit element and σ a homomorphism of K into K_1 mapping unit element to unit element. Let $M_{(K_1, \sigma)} = M_{(K_1)}$ be the K_1-module derived from M by extending the ring of scalars to K_1 (*Algebra*, Chapter II, § 5). The product on M defines canonically a K_1-bilinear mapping of $M_{(K_1)} \times M_{(K_1)}$ into $M_{(K_1)}$ (*Algebra*, Chapter IX, § 1, no. 4) such that $M_{(K_1)}$ is given the structure of an algebra over K_1 (which is said to be *derived from* M *by extending the ring of scalars to* K_1). This is the case in particular when K is a subring of K_1 containing the unit element and σ the inclusion mapping of K into K_1.

2. LIE ALGEBRAS

DEFINITION 1. *An algebra* $\mathfrak{g}$ *over* K *is called a Lie algebra over* K *if its multiplication (denoted by* $(x, y) \mapsto [x, y]$*) satisfies the identities:*

$$[x, x] = 0 \tag{1}$$

$$[x, [y, z]] + [y, [z, x]] + [z, [x, y]] = 0 \tag{2}$$

for all x, y, z *in* $\mathfrak{g}$.

The product $[x, y]$ is called the *bracket* of x and y. Identity (2) is called the *Jacobi identity.*

The bracket $[x, y]$ is an alternating bilinear function of x and y. We have the identity:

$$[x, y] = -[y, x] \tag{3}$$

so that the Jacobi identity can be written:

$$[x, [y, z]] = [[x, y], z] + [y, [x, z]]. \tag{4}$$

Every subalgebra and every quotient algebra of a Lie algebra is a Lie algebra. Every product of Lie algebras is a Lie algebra. If $\mathfrak{g}$ is a Lie algebra, the opposite algebra $\mathfrak{g}^0$ is a Lie algebra and the mapping $x \mapsto -x$ an isomorphism of $\mathfrak{g}$ onto $\mathfrak{g}^0$, by virtue of identity (3).

Example 1. Let L be an associative algebra over K. The bracket $[x, y] = xy - yx$ is a bilinear function of x and y. It is easily verified that the law of composition $(x, y) \mapsto [x, y]$ on the K-module L makes L into a Lie algebra over K.

Example 2. In Example 1 choose L to be the associative algebra of endomorphisms of a K-module E. We obtain the *Lie algebra of endomorphisms of* E, denoted by $\mathfrak{gl}(E)$. (If $E = K^n$, the Lie algebra $\mathfrak{gl}(E)$ is denoted by $\mathfrak{gl}(n, K)$.)

Every Lie subalgebra of $\mathfrak{gl}(E)$ is a Lie algebra over K. In particular:

(1) If E is given a (not necessarily associative) algebra structure, the derivations of E form a Lie algebra over K.

(2) If E admits a finite basis, the endomorphisms of E of zero trace form a Lie algebra over K denoted by $\mathfrak{sl}(E)$ (or $\mathfrak{sl}(n, K)$ if $E = K^n$).

(3) The set $\mathbf{M}_n(K)$ of square matrices of order n can be considered as a Lie

algebra over K canonically isomorphic to $\mathfrak{gl}(n, K)$. Let (E_{ij}) be the canonical basis of $\mathbf{M}_n(K)$ (*Algebra*, Chapter II, § 10, no. 3). It follows easily that:

$$(5)\quad \begin{cases} [E_{ij}, E_{kl}] = 0 & \text{if } j \neq k \text{ and } i \neq l \\ [E_{ij}, E_{jl}] = E_{il} & \text{if } i \neq l \\ [E_{ij}, E_{ki}] = -E_{kj} & \text{if } j \neq k \\ [E_{ij}, E_{ji}] = E_{ii} - E_{jj} & \end{cases}$$

The Lie subalgebra of $\mathbf{M}_n(K)$ consisting of the triangular matrices (resp. triangular matrices of zero trace, resp. triangular matrices of zero diagonal) is denoted by $\mathfrak{t}(n, K)$ (resp. $\mathfrak{st}(n, K)$, resp. $\mathfrak{n}(n, K)$) (*Algebra*, Chapter II, § 10, no. 7).

**Example* 3. Let V be an infinitely differentiable real manifold. The differential operators with infinitely differentiable real coefficients constitute an associative algebra over $\mathbf{R}$ and hence, by Example 1, a Lie algebra Δ over $\mathbf{R}$. The bracket of two infinitely differentiable vector fields on V is an infinitely differentiable vector field and hence the infinitely differentiable vector fields on V constitute a Lie subalgebra $\mathfrak{f}$ of Δ. If V is a real *Lie group*, the left invariant vector fields constitute a Lie subalgebra $\mathfrak{g}$ of $\mathfrak{f}$ called the *Lie algebra* of V. The vector space $\mathfrak{g}$ is identified with the tangent space to V at e (the identity element of V). Let V′ be another real Lie group, e' its identity element and $\mathfrak{g}'$ its Lie algebra. Every analytic homomorphism of V into V′ defines a linear mapping of the tangent space to V at e into the tangent space to V′ at e'; this mapping is a homomorphism of the Lie algebra $\mathfrak{g}$ into the Lie algebra $\mathfrak{g}'$. If V is the linear group of a finite-dimensional real vector space E there exists a canonical isomorphism of $\mathfrak{gl}(E)$ onto the Lie algebra $\mathfrak{g}$ of V, under which $\mathfrak{g}$ is identified with $\mathfrak{gl}(E)$.*

DEFINITION 2. *Let $\mathfrak{g}$ be a Lie algebra and x an element of $\mathfrak{g}$. The linear mapping $y \mapsto [x, y]$ of $\mathfrak{g}$ into $\mathfrak{g}$ is called the adjoint linear mapping of x and is denoted by* $\mathrm{ad}_{\mathfrak{g}}\, x$ *or* $\mathrm{ad}\, x$.

PROPOSITION 1. *Let $\mathfrak{g}$ be a Lie algebra. For all $x \in \mathfrak{g}$,* $\mathrm{ad}\, x$ *is a derivation. The mapping $x \mapsto \mathrm{ad}\, x$ is a homomorphism of the Lie algebra $\mathfrak{g}$ into the Lie algebra $\mathfrak{d}$ of derivations of $\mathfrak{g}$. If $D \in \mathfrak{d}$ and $x \in \mathfrak{g}$,* $[D, \mathrm{ad}\, x] = \mathrm{ad}(Dx)$.

Identity (4) can be written:

$$(\mathrm{ad}\, x).[y, z] = [(\mathrm{ad}\, x).y, z] + [y, (\mathrm{ad}\, x).z]$$

or:

$$(\mathrm{ad}[x, y]).z = (\mathrm{ad}\, x).((\mathrm{ad}\, y).z) - (\mathrm{ad}\, y).((\mathrm{ad}\, x).z)$$

whence the first two assertions. On the other hand, if $D \in \mathfrak{d}$, $x \in \mathfrak{g}$, $y \in \mathfrak{g}$, then $[D, \mathrm{ad}\, x].y = D([x, y]) - [x, Dy] = [Dx, y] = (\mathrm{ad}\, Dx).y$, whence the last assertion.

The mapping $\mathrm{ad}\, x$ is also called the *inner derivation* defined by x.

3. COMMUTATIVE LIE ALGEBRAS

DEFINITION 3. *Two elements x, y of a Lie algebra are said to be permutable if* $[x, y] = 0$. $\mathfrak{g}$ *is said to be commutative if any two of its elements are permutable.*

Example 1. Let L be an associative algebra and $\mathfrak{g}$ the Lie algebra defined by it (no. 2, *Example* 1). Two elements x, y are permutable in $\mathfrak{g}$ if and only if $xy = yx$ in L.

**Example* 2. If a real Lie group G is commutative, its Lie algebra is commutative.*

Every K-module can obviously be given a unique commutative Lie algebra structure over K.

If $\mathfrak{g}$ is a Lie algebra, every monogenous submodule of $\mathfrak{g}$ is a commutative Lie subalgebra of $\mathfrak{g}$.

4. IDEALS

It follows from identity (3) that in a Lie algebra $\mathfrak{g}$ there is no distinction between left ideals and right ideals, every ideal being two-sided. We therefore speak simply of ideals.

**Example*. Let G be a Lie group, $\mathfrak{g}$ its Lie algebra and H a Lie subgroup of G. Every left invariant vector field on H defines canonically a left invariant vector field on G, whence there is a canonical injection of the Lie algebra $\mathfrak{h}$ of H into $\mathfrak{g}$; $\mathfrak{h}$ is identified with a Lie subalgebra of $\mathfrak{g}$ under this injection. If H is normal in G, the canonical image of $\mathfrak{h}$ in $\mathfrak{g}$ is an ideal of $\mathfrak{g}$.*

An ideal of $\mathfrak{g}$ is a submodule of $\mathfrak{g}$ which is stable under the inner derivations of $\mathfrak{g}$.

DEFINITION 4. *A submodule of* $\mathfrak{g}$ *which is stable under* every *derivation of* $\mathfrak{g}$ *is called a characteristic ideal of* $\mathfrak{g}$.

PROPOSITION 2. *Let* $\mathfrak{g}$ *be a Lie algebra,* $\mathfrak{a}$ *an ideal* (resp. *a characteristic ideal*) *of* $\mathfrak{g}$ *and* $\mathfrak{b}$ *a characteristic ideal of* $\mathfrak{a}$. *Then* $\mathfrak{b}$ *is an ideal* (resp. *a characteristic ideal*) *of* $\mathfrak{g}$.

Every inner derivation (resp. every derivation) of $\mathfrak{g}$ leaves $\mathfrak{a}$ stable and induces on $\mathfrak{a}$ a derivation and hence leaves $\mathfrak{b}$ stable.

Let $\mathfrak{g}$ be a Lie algebra. If $\mathfrak{a}$ and $\mathfrak{b}$ are ideals of $\mathfrak{g}$, $\mathfrak{a} + \mathfrak{b}$ and $\mathfrak{a} \cap \mathfrak{b}$ are ideals of $\mathfrak{g}$.

Let $\mathfrak{a}$ and $\mathfrak{b}$ be two submodules of $\mathfrak{g}$. By an abuse of notation, the submodule of $\mathfrak{g}$ generated by the elements of the form $[x, y] (x \in \mathfrak{a}, y \in \mathfrak{b})$ is denoted by $[\mathfrak{a}, \mathfrak{b}]$. We have $[\mathfrak{a}, \mathfrak{b}] = [\mathfrak{b}, \mathfrak{a}]$ by identity (3). If $z \in \mathfrak{g}$, $[z, \mathfrak{a}]$, or $[\mathfrak{a}, z]$, denotes the submodule $[Kz, \mathfrak{a}] = (\operatorname{ad} z)(\mathfrak{a})$.

PROPOSITION 3. *If* $\mathfrak{a}$ and $\mathfrak{b}$ *are ideals* (resp. *characteristic ideals*) *of* $\mathfrak{g}$, $[\mathfrak{a}, \mathfrak{b}]$ *is an ideal* (resp. *a characteristic ideal*) *of* $\mathfrak{g}$.

Let D be an inner derivation (resp. a derivation) of $\mathfrak{g}$. If $x \in \mathfrak{a}$ and $y \in \mathfrak{b}$, then

$$D([x, y]) = [Dx, y] + [x, Dy] \in [\mathfrak{a}, \mathfrak{b}].$$

Hence the proposition.

If $\mathfrak{a}$ is a submodule of $\mathfrak{g}$, the set of $x \in \mathfrak{g}$ such that $(\mathrm{ad}\, x) . \mathfrak{a} \subset \mathfrak{a}$ is a subalgebra $\mathfrak{n}$ of $\mathfrak{g}$ called the *normalizer* of $\mathfrak{a}$ in $\mathfrak{g}$. If further $\mathfrak{a}$ is a subalgebra of $\mathfrak{g}$, then $\mathfrak{a} \subset \mathfrak{n}$ and $\mathfrak{a}$ is an ideal of $\mathfrak{n}$.

5. DERIVED SERIES, LOWER CENTRAL SERIES

The characteristic ideal $[\mathfrak{g}, \mathfrak{g}]$ is called the *derived ideal* of a Lie algebra $\mathfrak{g}$ and denoted by $\mathscr{D}\mathfrak{g}$.

Every submodule of $\mathfrak{g}$ containing $\mathscr{D}\mathfrak{g}$ is an ideal of $\mathfrak{g}$.

The *derived series* of $\mathfrak{g}$ is the decreasing sequence $\mathscr{D}^0\mathfrak{g}, \mathscr{D}^1\mathfrak{g}, \ldots$ of characteristic ideals of $\mathfrak{g}$ defined inductively as follows: (1) $\mathscr{D}^0\mathfrak{g} = \mathfrak{g}$; (2) $\mathscr{D}^{p+1}\mathfrak{g} = [\mathscr{D}^p\mathfrak{g}, \mathscr{D}^p\mathfrak{g}]$.

The *lower central series* of $\mathfrak{g}$ is the decreasing sequence $\mathscr{C}^1\mathfrak{g}, \mathscr{C}^2\mathfrak{g}, \ldots$ of characteristic ideals of $\mathfrak{g}$ defined inductively as follows: (1) $\mathscr{C}^1\mathfrak{g} = \mathfrak{g}$; (2) $\mathscr{C}^{p+1}\mathfrak{g} = [\mathfrak{g}, \mathscr{C}^p\mathfrak{g}]$. Then $\mathscr{C}^2\mathfrak{g} = \mathscr{D}\mathfrak{g}$ and $\mathscr{C}^{p+1}\mathfrak{g} \supset \mathscr{D}^p\mathfrak{g}$ for all p, as is immediately seen by induction on p.

PROPOSITION 4. *Let $\mathfrak{g}$ and $\mathfrak{h}$ be two Lie algebras over* K *and f a homomorphism of $\mathfrak{g}$ onto $\mathfrak{h}$. Then $f(\mathscr{D}^p\mathfrak{g}) = \mathscr{D}^p\mathfrak{h}$, $f(\mathscr{C}^p\mathfrak{g}) = \mathscr{C}^p\mathfrak{h}$.*

If $\mathfrak{a}$ and $\mathfrak{b}$ are submodules of $\mathfrak{g}$, it follows immediately that

$$f([\mathfrak{a}, \mathfrak{b}]) = [f(\mathfrak{a}), f(\mathfrak{b})].$$

The proposition is then immediate by induction on p.

COROLLARY. *Let $\mathfrak{g}$ be a Lie algebra and $\mathfrak{a}$ an ideal of $\mathfrak{g}$. For the Lie algebra $\mathfrak{g}/\mathfrak{a}$ to be commutative, it is necessary and sufficient that $\mathfrak{a} \supset \mathscr{D}\mathfrak{g}$.*

To say that $\mathfrak{g}/\mathfrak{a}$ is commutative amounts to saying that $\mathscr{D}(\mathfrak{g}/\mathfrak{a}) = \{0\}$. But $\mathscr{D}(\mathfrak{g}/\mathfrak{a})$ is, by Proposition 4, the canonical image of $\mathscr{D}\mathfrak{g}$ in $\mathfrak{g}/\mathfrak{a}$.

6. UPPER CENTRAL SERIES

Let $\mathfrak{g}$ be a Lie algebra and P a subset of $\mathfrak{g}$. The *centralizer* of P in $\mathfrak{g}$ is the set of elements of $\mathfrak{g}$ which are permutable with those of P. This centralizer is the intersection of the kernels of the ad y, where y runs through P; it is therefore a subalgebra of $\mathfrak{g}$.

PROPOSITION 5. *Let $\mathfrak{g}$ be a Lie algebra and $\mathfrak{a}$ an ideal* (resp. *a characteristic ideal*) *of $\mathfrak{g}$. The centralizer $\mathfrak{a}'$ of $\mathfrak{a}$ in $\mathfrak{g}$ is an ideal* (resp. *a characteristic ideal*) *of $\mathfrak{g}$.*

Let D be an inner derivation (resp. a derivation) of $\mathfrak{g}$. If $x \in \mathfrak{a}'$ and $y \in \mathfrak{a}$, then

$$[Dx, y] = D([x, y]) - [x, Dy] = 0;$$

hence $Dx \in \mathfrak{a}'$. Hence the proposition.

Let $\mathfrak{g}$ be a Lie algebra. The centralizer of $\mathfrak{g}$ in $\mathfrak{g}$ is called the *centre* of $\mathfrak{g}$, that is the characteristic ideal of $x \in \mathfrak{g}$ such that $[x, y] = 0$ for all $y \in \mathfrak{g}$. The centre of $\mathfrak{g}$ is the kernel of the homomorphism $x \mapsto \operatorname{ad} x$.

The *upper central series* of $\mathfrak{g}$ is the increasing sequence $\mathscr{C}_0\mathfrak{g}, \mathscr{C}_1\mathfrak{g}, \ldots$ of characteristic ideals of $\mathfrak{g}$ defined inductively as follows: (1) $\mathscr{C}_0\mathfrak{g} = \{0\}$; (2) $\mathscr{C}_{p+1}\mathfrak{g}$ is the inverse image under the canonical mapping of $\mathfrak{g}$ onto $\mathfrak{g}/\mathscr{C}_p\mathfrak{g}$ of the centre of $\mathfrak{g}/\mathscr{C}_p\mathfrak{g}$ The ideal $\mathscr{C}_1\mathfrak{g}$ is the centre of $\mathfrak{g}$.

7. EXTENSIONS

DEFINITION 5. *Let $\mathfrak{a}$ and $\mathfrak{b}$ be two Lie algebras over* K. *An extension of $\mathfrak{b}$ by $\mathfrak{a}$ is a sequence:*

$$\mathfrak{a} \xrightarrow{\lambda} \mathfrak{g} \xrightarrow{\mu} \mathfrak{b}$$

where $\mathfrak{g}$ is a Lie algebra over K, *μ a surjective homomorphism of $\mathfrak{g}$ onto $\mathfrak{b}$ and λ an injective homomorphism of $\mathfrak{a}$ onto the kernel of μ.*

The kernel $\mathfrak{n}$ of μ is called the *kernel* of the extension. The homomorphism λ is an isomorphism of $\mathfrak{a}$ onto $\mathfrak{n}$ and the homomorphism μ defines an isomorphism of $\mathfrak{g}/\mathfrak{n}$ onto $\mathfrak{b}$ when passing to the quotient.
By an abuse of language, $\mathfrak{g}$ is also called an *extension of* $\mathfrak{b}$ *by* $\mathfrak{a}$.

Two extensions:

$$\mathfrak{a} \xrightarrow{\lambda} \mathfrak{g} \xrightarrow{\mu} \mathfrak{b}, \quad \mathfrak{a} \xrightarrow{\lambda'} \mathfrak{g}' \xrightarrow{\mu'} \mathfrak{b}$$

are said to be *equivalent* if there exists a homomorphism f of $\mathfrak{g}$ into $\mathfrak{g}'$ such that the following diagram:

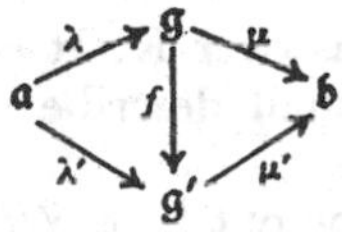

is commutative (that is such that $f \circ \lambda = \lambda'$, $\mu' \circ f = \mu$). We show that such a homomorphism is necessarily *bijective*. First f is injective. For if $x \in \mathfrak{g}$ is such that $f(x) = 0$, then $\mu(x) = \mu'(f(x)) = 0$ and hence $x = \lambda(y)$ for some $y \in \mathfrak{a}$; then $\lambda'(y) = f(\lambda(y)) = f(x) = 0$, hence $y = 0$ and hence $x = 0$. On the other hand, f is surjective. For $\mu' \circ f = \mu$ is surjective and hence $f(\mathfrak{g}) + \lambda'(\mathfrak{a}) = \mathfrak{g}'$; on the other hand $f(\mathfrak{g}) \supset f(\lambda(\mathfrak{a})) = \lambda'(\mathfrak{a})$.
It follows from this that the relation just defined between two extensions of $\mathfrak{b}$ by $\mathfrak{a}$ is an *equivalence relation.*

PROPOSITION 6. *Let*

$$\mathfrak{a} \xrightarrow{\lambda} \mathfrak{g} \xrightarrow{\mu} \mathfrak{b}$$

be an extension of $\mathfrak{b}$ by $\mathfrak{a}$ and $\mathfrak{n}$ its kernel.

(a) *If there exists a subalgebra* $\mathfrak{m}$ *of* $\mathfrak{g}$ *supplementary to* $\mathfrak{n}$ *in* $\mathfrak{g}$, *the restriction of* μ *to* $\mathfrak{m}$ *is an isomorphism of* $\mathfrak{m}$ *onto* $\mathfrak{b}$. *If* ν *denotes the inverse isomorphism of this restriction,* ν *is a homomorphism of* $\mathfrak{b}$ *into* $\mathfrak{g}$ *and* $\mu \circ \nu$ *is the identity automorphism of* $\mathfrak{b}$.

(b) *Conversely, if there exists a homomorphism* ν *of* $\mathfrak{b}$ *into* $\mathfrak{g}$ *such that* $\mu \circ \nu$ *is the identity automorphism of* $\mathfrak{b}$, *then* $\nu(\mathfrak{b})$ *is a supplementary subalgebra of* $\mathfrak{n}$ *in* $\mathfrak{g}$.

The assertions of (a) are immediate. On the other hand, let ν be a homomorphism of $\mathfrak{b}$ into $\mathfrak{g}$ such that $\mu \circ \nu$ is the identity automorphism of $\mathfrak{b}$. Then $\nu(\mathfrak{b})$ is a subalgebra of $\mathfrak{g}$ and $\mathfrak{g}$ is the direct sum of $\nu(\mathfrak{b})$ and $\overset{-1}{\mu}(0) = \mathfrak{n}$ (*Algebra*, Chapter VIII, § 1, no. 1).

DEFINITION 6. *Let*

$$\mathfrak{a} \xrightarrow{\lambda} \mathfrak{g} \xrightarrow{\mu} \mathfrak{b}$$

be an extension of $\mathfrak{b}$ *by* $\mathfrak{a}$ *and* $\mathfrak{n}$ *its kernel. This extension is called inessential* (resp. *trivial*) *if there exists a subalgebra* (resp. *an ideal*) *of* $\mathfrak{g}$ *supplementary to* $\mathfrak{n}$ *in* $\mathfrak{g}$. *This extension is called central if* $\mathfrak{n}$ *is contained in the centre of* $\mathfrak{g}$.

If the extension is trivial, let $\mathfrak{m}$ be an ideal of $\mathfrak{g}$ supplementary to $\mathfrak{n}$ in $\mathfrak{g}$. Then (cf. no. 1) $\mathfrak{g}$ is canonically identified with the Lie algebra $\mathfrak{m} \times \mathfrak{n}$ and hence with the Lie algebra $\mathfrak{a} \times \mathfrak{b}$. Conversely, let $\mathfrak{a}$ and $\mathfrak{b}$ be two Lie algebras; then $\mathfrak{a} \times \mathfrak{b}$ is a trivial extension of $\mathfrak{a}$ by $\mathfrak{b}$.

An inessential central extension is trivial. For let $\mathfrak{g}$ be a Lie algebra, $\mathfrak{n}$ an ideal of $\mathfrak{g}$ contained in the centre of $\mathfrak{g}$ and $\mathfrak{m}$ a subalgebra of $\mathfrak{g}$ supplementary to $\mathfrak{n}$ in $\mathfrak{g}$. Then $[\mathfrak{m}, \mathfrak{g}] = [\mathfrak{m}, \mathfrak{m}] + [\mathfrak{m}, \mathfrak{n}] = [\mathfrak{m}, \mathfrak{m}] \subset \mathfrak{m}$ and hence $\mathfrak{m}$ is an ideal of $\mathfrak{g}$.

8. SEMI-DIRECT PRODUCTS

Let $\mathfrak{a}$ and $\mathfrak{b}$ be two Lie algebras over K. It is not easy to construct all the extensions of $\mathfrak{b}$ by $\mathfrak{a}$. But we shall describe quite simply all the *inessential* extensions of $\mathfrak{b}$ by $\mathfrak{a}$.

Let $\mathfrak{g}$ be an inessential extension of $\mathfrak{b}$ by $\mathfrak{a}$. We identify $\mathfrak{a}$ with an ideal of $\mathfrak{g}$, $\mathfrak{b}$ with a subalgebra of $\mathfrak{g}$ supplementary to $\mathfrak{a}$ and the module $\mathfrak{g}$ with the module $\mathfrak{a} \times \mathfrak{b}$. For all $b \in \mathfrak{b}$, let ϕ_b be the restriction to $\mathfrak{a}$ of $\mathrm{ad}_{\mathfrak{g}}\, b$; this is a derivation of $\mathfrak{a}$ and the mapping $b \mapsto \phi_b$ is a homomorphism of $\mathfrak{b}$ into the Lie algebra of derivations of $\mathfrak{a}$. On the other hand, for a, a' in $\mathfrak{a}$ and b, b' in $\mathfrak{b}$, we have:

$$(6) \qquad \begin{aligned}[(a, b), (a', b')] &= [a + b, a' + b'] \\ &= [a, a'] + [a, b'] + [b, a'] + [b, b'] \\ &= ([a, a'] + \phi_b a' - \phi_{b'} a, [b, b']).\end{aligned}$$

Conversely, let $\mathfrak{a}$ and $\mathfrak{b}$ be Lie algebras over K and $b \mapsto \phi_b$ a homomorphism of $\mathfrak{b}$ into the Lie algebra of derivations of $\mathfrak{a}$. On the product $\mathfrak{g}$ *of the* K-*modules* $\mathfrak{a}$ and $\mathfrak{b}$ we define the bracket of two elements by writing:

$$[(a, b), (a', b')] = ([a, a'] + \phi_b a' - \phi_{b'} a, [b, b'])$$

for all a, a' in $\mathfrak{a}$, b, b' in $\mathfrak{b}$. It is immediate that this bracket is an alternating bilinear function of (a, b), (a', b'); we show that, given 3 elements (a, b), (a', b'), (a'', b'') of $\mathfrak{a} \times \mathfrak{b}$:

$$(7) \qquad [(a, b), [(a', b'), (a'', b'')]] + [(a', b'), [(a'', b''), (a, b)]] + [(a'', b''), [(a, b), (a', b')]] = 0.$$

As the left hand side of (7) is an alternating trilinear function of (a, b), (a', b'), (a'', b''), it suffices to make the verification when this system of elements takes one of the following forms:

$$(8) \qquad (a, 0), (a', 0), (a'', 0)$$

$$(9) \qquad (a, 0), (a', 0), (0, b'')$$

$$(10) \qquad (a, 0), (0, b'), (0, b'')$$

$$(11) \qquad (0, b), (0, b'), (0, b'').$$

In cases (8) and (11), relation (7) is an immediate consequence of the Jacobi identity in $\mathfrak{a}$ and $\mathfrak{b}$. In case (9), we have

$$\begin{aligned}
[(a, 0), [(a', 0), (0, b'')]] &= [(a, 0), (-\phi_{b''}a', 0)] = (-[a, \phi_{b''}a'], 0)\\
[(a', 0), [(0, b''), (a, 0)]] &= [(a', 0), (\phi_{b''}a, 0)] = ([a', \phi_{b''}a], 0)\\
[(0, b''), [(a, 0), (a', 0)]] &= [(0, b''), ([a, a'], 0)] = (\phi_{b''}([a, a']), 0)
\end{aligned}$$

and relation (7) follows from the equation:

$$\phi_{b''}([a, a']) = [\phi_{b''}a, a'] + [a, \phi_{b''}a'].$$

In case (10), we have:

$$\begin{aligned}
[(a, 0), [(0, b'), (0, b'')]] &= [(a, 0), (0, [b', b''])] = (-\phi_{[b', b'']}a, 0)\\
[(0, b'), [(0, b''), (a, 0)]] &= [(0, b'), (\phi_{b''}a, 0)] = (\phi_{b'}\phi_{b''}a, 0)\\
[(0, b''), [(a, 0), (0, b')]] &= [(0, b''), (-\phi_{b'}a, 0)] = (-\phi_{b''}\phi_{b'}a, 0)
\end{aligned}$$

and relation (7) follows from the equation:

$$\phi_{[b', b'']} = \phi_{b'}\phi_{b''} - \phi_{b''}\phi_{b'}.$$

Hence a Lie algebra structure has been defined on $\mathfrak{g}$. The mapping $(a, b) \mapsto b$ of $\mathfrak{g}$ onto $\mathfrak{b}$ is a homomorphism μ whose kernel $\mathfrak{n}$ is the ideal of elements of $\mathfrak{g}$ of the form $(a, 0)$. The mapping $a \mapsto (a, 0)$ is an isomorphism λ of $\mathfrak{a}$ onto $\mathfrak{n}$. Hence:

$$(12) \qquad \mathfrak{a} \xrightarrow{\lambda} \mathfrak{g} \xrightarrow{\mu} \mathfrak{b}$$

is an extension of $\mathfrak{b}$ by $\mathfrak{a}$ of kernel $\mathfrak{n}$, which is said to be *canonically defined by* $\mathfrak{a}$, $\mathfrak{b}$, ϕ. The mapping $b \mapsto (0, b)$ is an isomorphism ν of $\mathfrak{b}$ onto a subalgebra of $\mathfrak{g}$ supplementary to $\mathfrak{n}$ in $\mathfrak{g}$; hence the extension is inessential.

If $\mathfrak{a}$ is identified with $\mathfrak{n}$ under λ and $\mathfrak{b}$ with $\nu(\mathfrak{b})$ under ν, then, for $a \in \mathfrak{a}$ and $b \in \mathfrak{b}$:

$$(\operatorname{ad} b).a = [(0, b), (a, 0)] = (\phi_b a, 0) = \phi_b a.$$

When $\phi = 0$, $\mathfrak{g}$ is the product Lie algebra of $\mathfrak{b}$ and $\mathfrak{a}$. In the general case, $\mathfrak{g}$ is called the ***semi-direct product of*** $\mathfrak{b}$ ***by*** $\mathfrak{a}$ (corresponding to the homomorphism $b \mapsto \phi_b$ of $\mathfrak{b}$ into the Lie algebra of derivations of $\mathfrak{a}$).

We have therefore established the following proposition:

PROPOSITION 7. *Let $\mathfrak{a}$ and $\mathfrak{b}$ be two Lie algebras over* K,

$$\mathfrak{a} \xrightarrow{\lambda} \mathfrak{g} \xrightarrow{\mu} \mathfrak{b}$$

an inessential extension of $\mathfrak{b}$ by $\mathfrak{a}$, ν an isomorphism of $\mathfrak{b}$ onto a subalgebra of $\mathfrak{g}$ such that $\mu \circ \nu$ is the identity automorphism of $\mathfrak{b}$ and ϕ the corresponding homomorphism of $\mathfrak{b}$ into the Lie algebra of derivations of $\mathfrak{a}$. Let

$$\mathfrak{a} \xrightarrow{\lambda_0} \mathfrak{g}_0 \xrightarrow{\mu_0} \mathfrak{b}$$

be the inessential extension of $\mathfrak{b}$ by $\mathfrak{a}$ canonically defined by ϕ. Then the mapping $(a, b) \mapsto \lambda(a) + \nu(b)$ is an isomorphism f of $\mathfrak{g}_0$ onto $\mathfrak{g}$ and the following diagram

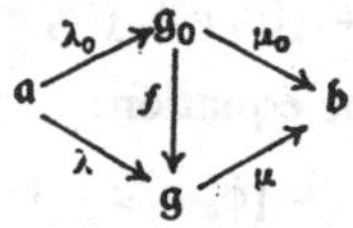

is commutative, so that the two extensions are equivalent.

Example 1. Let $\mathfrak{g}$ be a Lie algebra over K and D a derivation of $\mathfrak{g}$. Let $\mathfrak{h}$ be the *commutative* Lie algebra K. The mapping $\lambda \mapsto \lambda D (\lambda \in K)$ is a homomorphism of $\mathfrak{h}$ into the Lie algebra of derivations of $\mathfrak{g}$. We form the corresponding semi-direct product $\mathfrak{k}$ of $\mathfrak{h}$ by $\mathfrak{g}$. Let x_0 be the element $(0, 1)$ of $\mathfrak{k}$. For all $x \in \mathfrak{g}$, $Dx = [x_0, x]$.

Example 2. Let $\mathfrak{g}$ be a Lie algebra over K, M a K-module and ρ a homomorphism of $\mathfrak{g}$ into $\mathfrak{gl}(M)$. If M is considered as a commutative Lie algebra, the Lie algebra of derivations of M is $\mathfrak{gl}(M)$. We can therefore form the semi-direct product $\mathfrak{h}$ of $\mathfrak{g}$ by M corresponding to ρ.

In particular, let $\mathfrak{g} = \mathfrak{gl}(M)$ and ρ be the identity mapping of $\mathfrak{gl}(M)$. The semi-direct product of $\mathfrak{g}$ by M is then denoted by $\mathfrak{af}(M)$ (or $\mathfrak{af}(n, K)$ if $M = K^n$). An element of $\mathfrak{af}(M)$ is an ordered pair (m, u), where $m \in M$, $u \in \mathfrak{gl}(M)$; and the bracket is defined by

$$[(m, u), (m', u')] = (u(m') - u'(m), [u, u']).$$

*When M is a finite-dimensional vector space over **R**, $\mathfrak{af}(M)$ is canonically identified with the Lie algebra of the *affine group* of M.*

Let $\mathfrak{t}$ be a Lie algebra over K. A linear mapping θ of $\mathfrak{t}$ into $\mathfrak{af}(M)$ can be written $x \mapsto ((\zeta(x), \eta(x))$, where ζ is a linear mapping of $\mathfrak{t}$ into M and η a linear mapping of $\mathfrak{t}$ into $\mathfrak{gl}(M)$. We examine the conditions that ζ and η must satisfy for θ to be a homomorphism. For $x \in \mathfrak{t}$, $y \in \mathfrak{t}$, we must have

$$\theta([x, y]) = [\theta(x), \theta(y)]$$

that is

$$\begin{aligned}(\zeta([x, y]), \eta([x, y])) &= [(\zeta(x), \eta(x)), (\zeta(y), \eta(y))] \\ &= (\eta(x).\zeta(y) - \eta(y).\zeta(x), [\eta(x), \eta(y)]).\end{aligned}$$

Hence for θ to be a homomorphism of $\mathfrak{t}$ into $\mathfrak{af}(M)$, it is necessary and sufficient that η be a homomorphism of $\mathfrak{t}$ into $\mathfrak{gl}(M)$ and that ζ satisfy the relation:

$$\zeta([x, y]) = \eta(x).\zeta(y) - \eta(y).\zeta(x). \tag{13}$$

Let N be the K-module M $\times$ K. We take $\mathfrak{t}$ to be the subalgebra of $\mathfrak{gl}(N)$ consisting of the $w \in \mathfrak{gl}(N)$ such that $w(N) \subset M$. For all $w \in \mathfrak{t}$, let $\eta(w) \in \mathfrak{gl}(M)$ be the restriction of w to M and let $\zeta(w) = w(0, 1) \in M$. For $w_1 \in \mathfrak{t}$, $w_2 \in \mathfrak{t}$,

$$\zeta([w_1, w_2]) = w_1(\zeta(w_2)) - w_2(\zeta(w_1)) = \eta(w_1).\zeta(w_2) - \eta(w_2).\zeta(w_1).$$

Hence the mapping $w \mapsto (\zeta(w)), \eta(w))$ is a homomorphism θ of $\mathfrak{t}$ into $\mathfrak{af}(M)$. Clearly θ is *bijective*. Let $\phi = \theta^{-1}$. If $(m, u) \in \mathfrak{af}(M)$, $\phi(m, u)$ is the element w of $\mathfrak{t}$ defined by

$$w(m', \lambda) = (u(m') + \lambda m, 0).$$

$\mathfrak{af}(M)$ is often identified with the subalgebra $\mathfrak{t}$ of $\mathfrak{gl}(N)$ under the isomorphism ϕ.

*When M is a finite-dimensional vector space over **R**, the homomorphism ϕ of $\mathfrak{af}(M)$ into $\mathfrak{gl}(N)$ corresponds to a canonical homomorphism ψ of the affine group A of M into the group **GL**(N); if $a \in A$, $\psi(a)$ is the unique element g of **GL**(N) such that $g(m, 1) = (a(m), 1)$ for all $m \in M$. This homomorphism is injective and $\psi(A)$ is the set of automorphisms of N which leave invariant all the linear varieties of N parallel to M.*

9. CHANGE OF BASE RING

Let K_0 be a commutative ring with unit element and ρ a homomorphism of K_0 into K mapping unit element to unit element. Let $\mathfrak{g}$ be a Lie algebra over K. Let $\mathfrak{g}'$ be the algebra obtained by considering $\mathfrak{g}$ as an algebra over K_0 by means of ρ (cf. no. 1). Then $\mathfrak{g}'$ is a Lie algebra. The subalgebras (resp. ideals) of $\mathfrak{g}$ are subalgebras (resp. ideals) of $\mathfrak{g}'$. If $\mathfrak{a}$ and $\mathfrak{b}$ are submodules of $\mathfrak{g}$, the bracket $[\mathfrak{a}, \mathfrak{b}]$ is the same in $\mathfrak{g}$ and in $\mathfrak{g}'$; for $[\mathfrak{a}, \mathfrak{b}]$ is the set of elements of the form $\sum_{i=1}^{n} [x_i, y_i]$ where $x_i \in \mathfrak{a}$, $y_i \in \mathfrak{b}$. It follows that $\mathscr{D}^p\mathfrak{g} = \mathscr{D}^p\mathfrak{g}'$, $\mathscr{C}^p\mathfrak{g} = \mathscr{C}^p\mathfrak{g}'$ for all p.

The centralizer of a subset is the same in $\mathfrak{g}$ and $\mathfrak{g}'$. Hence $\mathscr{C}_p\mathfrak{g} = \mathscr{C}_p\mathfrak{g}'$ for all p.

Let K_1 be a commutative ring with unit element and σ a homomorphism of K into K_1 mapping unit element to unit element. Let $\mathfrak{g}$ be a Lie algebra over K. Let $\mathfrak{g}_{(K_1)}$ be the algebra over K_1 derived from $\mathfrak{g}$ by extending the base ring (cf. no. 1). Then $\mathfrak{g}_{(K_1)}$ is a Lie algebra. If $\mathfrak{a}$ is a subalgebra (resp. an ideal) of $\mathfrak{g}$, the canonical image of $\mathfrak{a}_{(K_1)}$ in $\mathfrak{g}_{(K_1)}$ is a subalgebra (resp. an ideal) of $\mathfrak{g}_{(K_1)}$. If $\mathfrak{a}$ and $\mathfrak{b}$ are submodules of $\mathfrak{g}$, the canonical image in $\mathfrak{g}_{(K_1)}$ of $[\mathfrak{a}, \mathfrak{b}]_{(K_1)}$ is equal to the bracket of the canonical images of $\mathfrak{a}_{(K_1)}$ and $\mathfrak{b}_{(K_1)}$. It follows that $\mathscr{D}^p(\mathfrak{g}_{(K_1)})$ is the canonical image of $(\mathscr{D}^p\mathfrak{g})_{(K_1)}$ and that $\mathscr{C}^p(\mathfrak{g}_{(K_1)})$ is the canonical image of $\mathscr{C}^p(\mathfrak{g}_{(K_1)})$.

If K is a field, K_1 an extension field of K and σ the canonical injection of K into K_1, then with the usual identifications we have

$$[\mathfrak{a}, \mathfrak{b}]_{(K_1)} = [\mathfrak{a}_{(K_1)}, \mathfrak{b}_{(K_1)}], \qquad \mathscr{D}^p(\mathfrak{g}_{(K_1)}) = (\mathscr{D}^p\mathfrak{g})_{(K_1)},$$
$$\mathscr{C}^p(\mathfrak{g}_{(K_1)}) = (\mathscr{C}^p\mathfrak{g})_{(K_1)}.$$

These results are completed in § 2, no. 9.

If M is a finite-dimensional vector space over the field K, $M_{(K_1)}$ is a finite-dimensional vector space over K_1 and the associative algebra $\mathscr{L}(M_{(K_1)})$ is canonically identified with the associative algebra $\mathscr{L}(M)_{(K_1)}$. Hence the Lie algebra $\mathfrak{gl}(M_{(K_1)})$ is canonically identified with the Lie algebra $\mathfrak{gl}(M)_{(K_1)}$.

§ 2. ENVELOPING ALGEBRA OF A LIE ALGEBRA

1. DEFINITION OF THE ENVELOPING ALGEBRA

Let $\mathfrak{g}$ be a Lie algebra over K. For any associative algebra with unit element L over K, an α-mapping of $\mathfrak{g}$ into L is a K-linear mapping σ of $\mathfrak{g}$ into L such that

$$\sigma([x, y]) = \sigma(x)\sigma(y) - \sigma(y)\sigma(x) \quad (x, y \text{ in } \mathfrak{g})$$

(in other words a homomorphism of $\mathfrak{g}$ into the Lie algebra associated with L).

If L′ is another associative algebra with unit element over K and τ a homomorphism of L into L′ mapping 1 to 1, then $\tau \circ \sigma$ is an α-mapping of $\mathfrak{g}$ into L′. We shall look for an associative algebra with unit element and an α-mapping of $\mathfrak{g}$ into this algebra which are *universal* (*Set Theory*, Chapter IV, § 3, no. 1).

DEFINITION 1. *Let $\mathfrak{g}$ be a Lie algebra over* K, T *the tensor algebra of the* K*-module* $\mathfrak{g}$ *and* J *the two-sided ideal of* T *generated by the tensors* $x \otimes y - y \otimes x - [x, y]$ *where* $x \in \mathfrak{g}, y \in \mathfrak{g}$. *The associative algebra* $U = T/J$ *is called the enveloping algebra of* $\mathfrak{g}$. *The restriction to* $\mathfrak{g}$ *of the canonical mapping of* T *onto* U *is called the canonical mapping of* $\mathfrak{g}$ *into* U.

Let T_+ be the two-sided ideal of T consisting of the tensors whose component of order 0 is zero. Let $T_0 = K.1$ be the set of elements of T of order 0. Let U_+ and U_0 be the canonical images of T_+ and T_0 in U. As $J \subset T_+$, the decomposition into a direct sum $T = T_0 + T_+$ implies a decomposition into a direct sum $U = U_0 + U_+$. The algebra U therefore has a unit element distinct from 0 and $U_0 = K.1$. For all $x \in U$, the component of x in U_0 is called the *constant term* of x. The elements with constant term zero form a two-sided ideal of U, namely the two-sided ideal U^+ generated by the canonical image of $\mathfrak{g}$ in U.

The associative algebra U is generated by 1 and the canonical image of $\mathfrak{g}$ in U.

If $x \in \mathfrak{g}$ and $y \in \mathfrak{g}$, $x \otimes y - y \otimes x$ and $[x, y]$ are congruent in T modulo J; hence, if σ_0 denotes the canonical mapping of $\mathfrak{g}$ into U,

$$\sigma_0(x)\sigma_0(y) - \sigma_0(y)\sigma_0(x) = \sigma_0([x, y])$$

in U. In other words, σ_0 is an α-mapping of $\mathfrak{g}$ into U.

PROPOSITION 1. *Let σ be an α-mapping of $\mathfrak{g}$ into the associative algebra* L *with unit element. There exists one and only one homomorphism τ of* U *into* L, *mapping* 1 *to* 1, *such that $\sigma = \tau \circ \sigma_0$, where σ_0 denotes the canonical mapping of $\mathfrak{g}$ into* U.

Let τ' be the unique homomorphism of T into L which extends σ and maps 1 to 1. Then, for x, y in $\mathfrak{g}$,

$$\tau'(x \otimes y - y \otimes x - [x, y]) = \sigma(x)\sigma(y) - \sigma(y)\sigma(x) - \sigma([x, x]) = 0;$$

hence τ' is zero on J and defines on passing to the quotient a homomorphism τ of U into L, mapping 1 to 1, such that $\sigma = \tau \circ \sigma_0$. The uniqueness of τ is immediate since $\sigma_0(\mathfrak{g})$ and 1 generate the algebra U.

Let $\mathfrak{g}'$ be another Lie algebra over K, U′ its enveloping algebra and σ_0' the canonical mapping of $\mathfrak{g}'$ into U′. Let ϕ be a homomorphism of $\mathfrak{g}$ into $\mathfrak{g}'$. Then $\sigma_0' \circ \phi$ is an α-mapping of $\mathfrak{g}$ into U′; hence there exists one and only one homomorphism $\tilde{\phi}$ of U into U′ mapping 1 to 1 and such that the diagram

$$\begin{array}{ccc} \mathfrak{g} & \xrightarrow{\phi} & \mathfrak{g}' \\ {\scriptstyle \sigma_0}\downarrow & & \downarrow{\scriptstyle \sigma_0'} \\ U & \xleftarrow{\tilde{\phi}} & U' \end{array}$$

is commutative. This homomorphism maps the elements of U whose constant term is zero to elements of U′ whose constant term is zero. If $\mathfrak{g}''$ is another Lie algebra over K and ϕ' is a homomorphism of $\mathfrak{g}'$ into $\mathfrak{g}''$, then $(\phi' \circ \phi)^{\sim} = \tilde{\phi}' \circ \tilde{\phi}$.

2. ENVELOPING ALGEBRA OF A PRODUCT OF LIE ALGEBRAS

Let $\mathfrak{g}_1, \mathfrak{g}_2$ be two Lie algebras over K, U_i the enveloping algebra of $\mathfrak{g}_i$ and σ_i the canonical mapping of $\mathfrak{g}_i$ into U_i $(i = 1, 2)$. Let $\mathfrak{g} = \mathfrak{g}_1 \times \mathfrak{g}_2$, U be its enveloping algebra and σ the canonical mapping of $\mathfrak{g}$ into U. The canonical injections of $\mathfrak{g}_1$ and $\mathfrak{g}_2$ into $\mathfrak{g}$ define canonical homomorphisms of U_1 and U_2 into U whose images commute and hence a homomorphism ϕ of the algebra $U_1 \otimes_K U_2$ into the algebra U, mapping 1 to 1.

PROPOSITION 2. *The homomorphism* ϕ *is an algebra isomorphism*

The mapping σ': $(x_1, x_2) \mapsto \sigma_1(x_1) \otimes 1 + 1 \otimes \sigma_2(x_2)$ $(x_1 \in \mathfrak{g}_1, x_2 \in \mathfrak{g}_2)$ is an α-mapping of $\mathfrak{g}$ into $U_1 \otimes_K U_2$ and hence there exists (no. 1, Proposition 1) a unique homomorphism τ of U into $U_1 \otimes_K U_2$ mapping 1 to 1, such that:

$$\sigma' = \tau \circ \sigma. \tag{1}$$

Now $\phi \circ \tau \circ \sigma = \phi \circ \sigma' = \sigma$ and $\tau \circ \phi \circ \sigma' = \tau \circ \sigma = \sigma'$, hence $\phi \circ \tau$ and $\tau \circ \phi$ are the identity mappings of U and $U_1 \otimes_K U_2$ respectively. Hence the proposition.

$U_1 \otimes_K U_2$ is identified with U under the isomorphism ϕ. Then the canonical mapping of $\mathfrak{g}$ into U is identified by (1) with the mapping:

$$(x_1, x_2) \mapsto \sigma_1(x_1) \otimes 1 + 1 \otimes \sigma_2(x_2).$$

Analogously, if $\mathfrak{g}_1, \ldots, \mathfrak{g}_n$ are Lie algebras over K with enveloping algebras $U_1, \ldots, U_n$, the enveloping algebra U of $\mathfrak{g}_1 \times \cdots \times \mathfrak{g}_n$ is canonically identified with $U_1 \otimes_K \cdots \otimes_K U_n$ and the canonical mapping of $\mathfrak{g}_1 \times \cdots \times \mathfrak{g}_n$ into U is identified with the mapping:

$$(x_1, \ldots, x_n) \mapsto \sigma_1(x_1) \otimes 1 \otimes \cdots \otimes 1 + \cdots + 1 \otimes \cdots \otimes 1 \otimes \sigma_n(x_n)$$

(σ_i denoting the canonical mapping of $\mathfrak{g}_i$ into U_i).

3. ENVELOPING ALGEBRA OF A LIE SUBALGEBRA

Let $\mathfrak{g}$ be a Lie algebra over K, $\mathfrak{h}$ a subalgebra of $\mathfrak{g}$ and σ, σ' the canonical mappings of $\mathfrak{g}$, $\mathfrak{h}$ into their enveloping algebras U, V. Then the canonical injection i of $\mathfrak{h}$ into $\mathfrak{g}$ defines a homomorphism $\tilde{i}$, called *canonical*, of V into U such that $\sigma \circ i = \tilde{i} \circ \sigma'$. The algebra $\tilde{i}(V)$ is generated by 1 and $\sigma(\mathfrak{h})$. We shall see (no. 7, Corollary to Theorem 5) that $\tilde{i}$ is injective in important cases.

If $\mathfrak{h}$ is an ideal of $\mathfrak{g}$, the left ideal of U generated by $\sigma(\mathfrak{h})$ coincides with the right ideal generated by $\sigma(\mathfrak{h})$, in other words it is a two-sided ideal R. This follows since, for $x \in \mathfrak{h}$ and $x' \in \mathfrak{g}$,

$$\sigma(x)\sigma(x') = \sigma(x')\sigma(x) + \sigma([x, x'])$$

and $[x, x'] \in \mathfrak{h}$.

PROPOSITION 3. *Let* $\mathfrak{h}$ *be an ideal of* $\mathfrak{g}$, p *the canonical homomorphism of* $\mathfrak{g}$ *onto* $\mathfrak{g}/\mathfrak{h}$ *and* W *the enveloping algebra of* $\mathfrak{g}/\mathfrak{h}$. *The homomorphism:*

$$\tilde{p}: U \rightarrow W$$

defined canonically by p *is surjective and its kernel is the ideal* R *of* U *generated by* $\sigma(\mathfrak{h})$.

Let σ'' be the canonical mapping of $\mathfrak{g}/\mathfrak{h}$ into W. The commutative diagram:

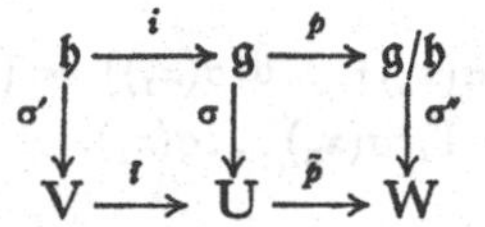

$$
\begin{array}{ccccc}
\mathfrak{h} & \xrightarrow{\ i\ } & \mathfrak{g} & \xrightarrow{\ p\ } & \mathfrak{g}/\mathfrak{h} \\
{\scriptstyle \sigma'}\downarrow & & {\scriptstyle \sigma}\downarrow & & \downarrow{\scriptstyle \sigma''} \\
V & \xrightarrow{\ t\ } & U & \xrightarrow{\ \tilde{p}\ } & W
\end{array}
$$

proves that $\tilde{p}$ is zero on $\sigma(\mathfrak{h})$ and hence on R. Let ψ be the canonical homomorphism of U onto U/R. There exists a homomorphism ϕ of U/R

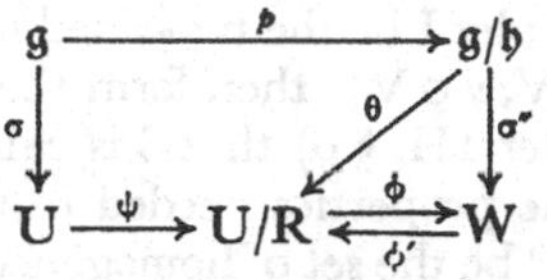

into W such that $\tilde{p} = \phi \circ \psi$. The mapping $\psi \circ \sigma$ of $\mathfrak{g}$ into U/R is an α-mapping and is zero on $\mathfrak{h}$ and hence defines an α-mapping θ of $\mathfrak{g}/\mathfrak{h}$ into U/R such that $\theta \circ p = \psi \circ \sigma$. Then $\phi \circ \theta \circ p = \phi \circ \psi \circ \sigma = \sigma'' \circ p$. Hence $\phi \circ \theta = \sigma''$. There exists (no. 1, Proposition 1) one and only one homomorphism ϕ' of W into U/R mapping 1 to 1 and such that $\theta = \phi' \circ \sigma''$. Then $\phi' \circ \phi \circ \theta = \phi' \circ \sigma'' = \theta$ and $\phi \circ \phi' \circ \sigma'' = \phi \circ \theta = \sigma''$ and hence $\phi' \circ \phi$ and $\phi \circ \phi'$ are the identity mappings of U/R and W respectively. This completes the proof.

U/R is identified with W under the isomorphism ϕ. Then the canonical mapping σ'' of $\mathfrak{g}/\mathfrak{h}$ into W is identified with θ, that is with the mapping of $\mathfrak{g}/\mathfrak{h}$ into U/R derived from σ by taking quotients.

4. ENVELOPING ALGEBRA OF THE OPPOSITE LIE ALGEBRA

Let $\mathfrak{g}$ be a Lie algebra over K, $\mathfrak{g}^0$ the opposite Lie algebra and σ and σ_0 the canonical mappings of $\mathfrak{g}$ and $\mathfrak{g}^0$ into their enveloping algebras U and V. Then σ is an α-mapping of $\mathfrak{g}^0$ into the associative algebra U^0 opposite to the associative algebra U. Hence there exists one and only one homomorphism ϕ of V into U^0 mapping 1 to 1 and such that $\sigma = \phi \circ \sigma_0$.

PROPOSITION 4. *The homomorphism ϕ is an isomorphism of V onto U^0.*

There exists a homomorphism ϕ' of U into V^0 mapping 1 to 1 and such that $\sigma_0 = \phi' \circ \sigma$. ϕ' can be considered as a homomorphism of U^0 into V. Then $\sigma_0 = \phi' \circ \phi \circ \sigma_0$ and $\sigma = \phi \circ \phi' \circ \sigma$ and hence $\phi' \circ \phi$ and $\phi \circ \phi'$ are the identity mappings of V and U. Hence the proposition.

V is identified with U^0 under the isomorphism ϕ. Then σ_0 is identified with σ.

With this identification, the isomorphism $\theta: x \mapsto -x$ of $\mathfrak{g}$ onto $\mathfrak{g}^0$ defines an isomorphism $\tilde{\theta}$ of U onto $V = U^0$. This isomorphism can be considered as an antiautomorphism of U. It is called the ***principal antiautomorphism*** of U. If $x_1, \ldots, x_n$ are in $\mathfrak{g}$, then:

$$(2) \qquad \tilde{\theta}(\sigma(x_1)\ldots\sigma(x_n)) = \tilde{\theta}(\sigma(x_n))\ldots\tilde{\theta}(\sigma(x_1)) = (-\sigma(x_n))\ldots(-\sigma(x_1)) = (-1)^n\sigma(x_n)\ldots\sigma(x_1).$$

5. SYMMETRIC ALGEBRA OF A MODULE

Let V be a K-module. V can be considered in a unique way as a commutative Lie algebra. The enveloping algebra of V can then be obtained as follows: let T be the tensor algebra of V; let I be the two-sided ideal of T generated by the tensors $x \otimes y - y \otimes x$ ($x \in V, y \in V$); then form the algebra $S = T/I$.

Recalling (*Algebra*, Chapter III, § 6) that S is called the *symmetric algebra* of V, we summarize briefly the properties needed in this chapter, the proofs of which are immediate. Let T^n be the set of homogeneous tensors of order n in T. Then $I = (I \cap T^2) + (I \cap T^3) + \cdots$ and hence S is the direct sum of the canonical images S^n of the T^n. The elements of S^n are called homogeneous of degree n. $S^0 = K.1$, S^1 is identified with V and $S^nS^p \subset S^{n+p}$. The algebra S is generated by 1 and $S^1 = V$. Clearly any two elements of S^1 are permutable and hence S is commutative. If V is a *free* K-module with basis $(x_\lambda)_{\lambda\in\Lambda}$, the canonical homomorphism f of the polynomial algebra $K[X_\lambda]_{\lambda\in\Lambda}$ onto S which maps 1 to 1 and X_λ to x_λ for all $\lambda \in \Lambda$ is an isomorphism: for by the universal property of S (no. 1, Proposition 1) there exists a homomorphism g of S into $K[X_\lambda]_{\lambda\in\Lambda}$ which maps 1 to 1 and x_λ to X_λ for all $\lambda \in \Lambda$ and f and g are inverse homomorphisms of one another.

Let $S'^n \subset T^n$ be the set of homogeneous symmetric tensors of order n (*Algebra*, Chapter III, § 5, no. 1, Definition 2). If K is a field of characteristic 0, S'^n and $I \cap T^n$ are supplementary in T^n. For let $(x_\lambda)_{\lambda\in\Lambda}$ be a basis of V. We give Λ a total ordering (*Set Theory*, Chapter III, § 2, no. 3, Theorem 1). Let Λ_n be the set of increasing sequences of n elements of Λ. For $M = (\lambda_1, \ldots, \lambda_n) \in \Lambda_n$, let

$$y_M = \frac{1}{n!}\sum_{\sigma\in\mathfrak{S}_n} x_{\lambda_{\sigma(n)}} \otimes \cdots \otimes x_{\lambda_{\sigma(n)}}.$$

The y_M for $M \in \Lambda_n$ form a system of generators of the vector K-space S'^n. Now their canonical images in S^n constitute, by the above paragraph, a basis of S^n. Hence $(y_M)_{M\in\Lambda_n}$ is a basis of a supplementary subspace of $I \cap T^n$ in T^n (*Algebra*, Chapter II, § 1, no. 6, Proposition 4), which establishes our assertion.

Thus, when K is a field of characteristic 0, the restriction to S'^n of the canonical mapping $T^n \to S^n$ is an isomorphism of the space S'^n onto the space S^n and therefore has an inverse isomorphism. The inverse isomorphisms thus obtained for each n define a canonical isomorphism of the space S onto the space $S' = \sum_{n\geqslant 0} S'^n$ of symmetric tensors.

6. FILTRATION OF THE ENVELOPING ALGEBRA

Let $\mathfrak{g}$ be a Lie algebra over K and T the tensor algebra of the K-module $\mathfrak{g}$. Let T^n be the submodule of T consisting of the homogeneous tensors of order n and $T_n = \sum_{i \leqslant n} T^i$. Then $T_n \subset T_{n+1}$, $T_0 = K.1$, $T_{-1} = \{0\}$ and $T_n T_p \subset T_{n+p}$. Let U_n be the canonical image of T_n in the enveloping algebra U of $\mathfrak{g}$. Then $U_n \subset U_{n+1}$, $U_0 = K.1$, $U_{-1} = \{0\}$ and $U_n U_p \subset U_{n+p}$; hence U can be described as an algebra *filtered by the* U_n (*Commutative Algebra*, Chapter III, § 2, no. 1); the elements of U_n will be said to be *of filtration* $\leqslant n$.

Let G^n be the K-module U_n/U_{n-1} and let G be the K-module the direct sum of the G^n. Multiplication on U defines, on taking quotients, a bilinear mapping of $G^n \times G^m$ into G^{n+m} and hence a bilinear mapping of $G \times G$ into G, which is associative. Thus G is given an associative K-algebra structure. Then $G^n G^m \subset G^{n+m}$. The elements of G^n are said to be of *degree n*. The graded algebra thus obtained is just the graded algebra associated with the filtered algebra U (*Commutative Algebra*, Chapter III, § 2, no. 3).

Let ϕ_n be the composition of the canonical K-linear mappings

$$T^n \to U_n \to G^n.$$

As T^n is supplementary to T_{n-1} in T_n, ϕ_n is surjective. The ϕ_n define a K-linear mapping ϕ of $\sum_n T^n = T$ onto $\sum_n G^n = G$.

PROPOSITION 5. *The mapping ϕ of T onto G is an algebra homomorphism mapping 1 to 1 and is zero on the two-sided ideal generated by the tensors $x \otimes y - y \otimes x$ ($x \in \mathfrak{g}, y \in \mathfrak{g}$).*

If $t \in T^n$ and $t' \in T^p$, then $\phi(t)\phi(t') = \phi(tt')$ by definition of the multiplication on G. Hence ϕ is an algebra homomorphism and clearly $\phi(1) = 1$. If x, y are in $\mathfrak{g}$, then $x \otimes y - y \otimes x \in T^2$ and the canonical image of this element in U_2 is equal to that of $[x, y]$ and therefore belongs to U_1. Hence $\phi(x \otimes y - y \otimes x) = 0$, which proves the proposition.

Let S be the symmetric algebra of the K-module $\mathfrak{g}$ and τ the canonical homomorphism of T onto S. Proposition 5 proves that there exists a unique homomorphism ω, called *canonical*, of the algebra S onto the algebra G, mapping 1 to 1, such that $\phi = \omega \circ \tau$. We have $\omega(S^n) = \phi(T^n) = G^n$. Let τ_n be the restriction of τ to T^n, ω_n the restriction of ω to S^n, ψ_n the canonical mapping of T^n into U_n and θ_n the canonical mapping of U_n onto G^n. The definition of ω_n proves that the following diagram is commutative:

$$\begin{array}{ccc}
 & \overset{\psi_n}{\longrightarrow} U_n \overset{\theta_n}{\searrow} & \\
T^n & & G^n \\
 & \underset{\tau_n}{\searrow} S^n \underset{\omega_n}{\longrightarrow} &
\end{array}$$

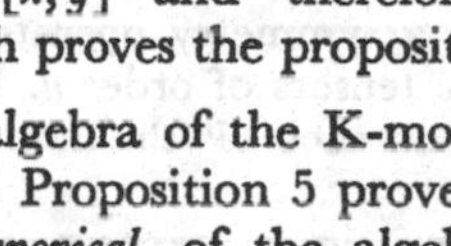

PROPOSITION 6. *If* K *is Noetherian and* $\mathfrak{g}$ *is a finitely generated module, the ring* U *is right and left Noetherian.*

S is a finitely generated algebra over K and hence a Noetherian ring (*Commutative Algebra*, Chapter III, § 2, no. 10, Corollary 3 to Theorem 2). Hence G, which is isomorphic to a quotient ring of S, is Noetherian. Hence U is right and left Noetherian (*Commutative Algebra*, Chapter III, § 2, no. 10, *Remark* 2).

COROLLARY. *Suppose that* K *is a field and that* $\mathfrak{g}$ *is finite-dimensional over* K. *Let* $I_1, \ldots, I_m$ *be right* (resp. *left*) *ideals of finite codimension in* U. *Then the product ideal* $I_1 I_2 \ldots I_m$ *is of finite codimension.*

By induction on m it suffices to consider the case of, for example, two right ideals. The right U-module I_1 is generated by a finite number of elements $u_1, \ldots, u_p$ (Proposition 6). Let $v_1, \ldots, v_q$ be elements of U whose classes modulo I_2 generate the vector space U/I_2. Then the canonical images in I_1/I_1I_2 of the $u_i v_j$ generate the vector space I_1/I_1I_2, which is therefore finite-dimensional. Hence $\dim_K(U/I_1I_2) = \dim_K(U/I_1) + \dim_K(I_1/I_1I_2) < +\infty$.

Remark. Let $\mathfrak{g}'$ be another Lie algebra over the ring K, U′ its enveloping algebra, U'_n the set of elements of U′ of filtration $\leqslant n$ and U^n (resp. U'^n) the set of canonical images in U (resp. U′) of the homogeneous symmetric tensors of $\mathfrak{g}$ (resp. $\mathfrak{g}'$) of order n. Let η be a homomorphism of $\mathfrak{g}$ into $\mathfrak{g}'$ and let $\tilde{\eta}$ be the corresponding homomorphism of U into U′. Then

$$\tilde{\eta}(U_n) \subset U'_n, \qquad \tilde{\eta}(U^n) \subset U'^n.$$

In particular, the principal antiautomorphism of U leaves U_n and U^n stable. The K-linear mapping of T^n onto itself which maps

$$x_1 \otimes x_2 \otimes \cdots \otimes x_n \quad \text{to} \quad x_n \otimes x_{n-1} \otimes \cdots \otimes x_1$$

for all $x_1, \ldots, x_n$ in $\mathfrak{g}$ is a symmetry operator and hence leaves fixed the homogeneous symmetric tensors of order n. Hence the principal antiautomorphism of U induces on each U^n the homothety of ratio $(-1)^n$.

7. THE POINCARÉ-BIRKHOFF-WITT THEOREM

THEOREM 1. *Let* $\mathfrak{g}$ *be a Lie* K*-algebra, U its enveloping algebra,* G *the graded algebra associated with the filtered algebra* U *and* S *the symmetric algebra of the* K*-module* $\mathfrak{g}$. *If* $\mathfrak{g}$ *is a free* K*-module, the canonical homomorphism* ω: S $\rightarrow$ G *is an isomorphism.*

Let $(x_\lambda)_{\lambda \in \Lambda}$ be a basis of the K-module $\mathfrak{g}$; we give Λ a total ordering (*Set Theory*, Chapter III, § 2, no. 3, Theorem 1). Let P be the polynomial algebra $K[z_\lambda]_{\lambda \in \Lambda}$ in indeterminates z_λ in one-to-one correspondence with the x_λ. For every sequence $M = (\lambda_1, \lambda_2, \ldots, \lambda_n)$ of elements of Λ, let z_M denote the monomial $z_{\lambda_1} z_{\lambda_2} \ldots z_{\lambda_n}$ and x_M the tensor $x_{\lambda_1} \otimes x_{\lambda_2} \otimes \cdots \otimes x_{\lambda_n}$.

The z_M, for M increasing, form a basis of the K-module P (we make the convention that $\varnothing$ is an increasing sequence and that $z_\varnothing = 1$). Let P_p be the submodule of polynomials of degree $\leqslant p$. We shall first prove several lemmas. (To abbreviate, we write $\lambda \leqslant M$ if $\lambda \leqslant \mu$ for every index μ of the sequence M.)

Lemma 1. *For every integer $p \geqslant 0$, there exists a unique homomorphism f_p of the* K-*module* $\mathfrak{g} \otimes_K P_p$ *into the* K-*module* P *satisfying the following conditions:*

(A_p) $\quad f_p(x_\lambda \otimes z_M) = z_\lambda z_M \quad$ *for* $\lambda \leqslant M$, $z_M \in P_p$;

(B_p) $\quad f_p(x_\lambda \otimes z_M) - z_\lambda z_M \in P_q \quad$ *for* $z_M \in P_q$, $q \leqslant p$;

(C_p) $\quad f_p(x_\lambda \otimes f_p(x_\mu \otimes z_N)) = f_p(x_\mu \otimes f_p(x_\lambda \otimes z_N)) + f_p([x_\lambda, x_\mu] \otimes z_N)$

for $z_N \in P_{p-1}$. (*The terms appearing in* (C_p) *are meaningful by* (B_p).)

Moreover, the restriction of f_p to $\mathfrak{g} \otimes P_{p-1}$ coincides with f_{p-1}.

The last assertion follows from the others since the restriction of f_p to $\mathfrak{g} \otimes_K P_{p-1}$ satisfies conditions (A_{p-1}), (B_{p-1}) and (C_{p-1}). We shall prove the existence and uniqueness of f_p by induction on p. For $p = 0$, condition (A_0) gives $f_0(x_\lambda \otimes 1) = z_\lambda$ and conditions (B_0) and (C_0) are then obviously satisfied. Suppose now that the existence and uniqueness of f_{p-1} are proved. We show that f_{p-1} admits a unique extension f_p to $\mathfrak{g} \otimes_K P_p$ satisfying conditions (A_p), (B_p) and (C_p).

We must define $f_p(x_\lambda \otimes z_M)$ for an increasing sequence M of p elements.

If $\lambda \leqslant M$, the value is given by condition (A_p). Otherwise, M can be written uniquely in the form (μ, N), where $\mu < \lambda$, $\mu \leqslant N$. Then

$$z_M = z_\mu z_N = f_{p-1}(x_\mu \otimes z_N)$$

by (A_{p-1}), so that the left hand side of (C_p) is $f_p(x_\lambda \otimes z_M)$. Now the right hand side of (C_p) is already defined: for (B_{p-1}) allows us to write:

$$f_p(x_\lambda \otimes z_N) = f_{p-1}(x_\lambda \otimes z_N) = z_\lambda z_N + w$$

where $w \in P_{p-1}$; hence the right hand side of (C_p) becomes:

$$z_\lambda z_\mu z_N + f_{p-1}(x_\mu \otimes w) + f_{p-1}([x_\lambda, x_\mu] \otimes z_N).$$

Thus f_p is defined uniquely and obviously satisfies conditions (A_p) and (B_p). Condition (C_p) is satisfied if $\mu < \lambda$, $\mu \leqslant N$. As $[x_\mu, x_\lambda] = -[x_\lambda, x_\mu]$, condition (C_p) is also satisfied for $\lambda < \mu$, $\lambda \leqslant N$. As (C_p) is trivially satisfied for $\lambda = \mu$, (C_p) is therefore satisfied if $\lambda \leqslant N$ or $\mu \leqslant N$. If none of these inequalities holds, $N = (\nu, Q)$, where $\nu \leqslant Q$, $\nu < \lambda$, $\nu < \mu$. Writing henceforth to abbreviate $f_p(x \otimes z) = xz$ for $x \in \mathfrak{g}$ and $z \in P_p$, we have by the induction hypothesis:

$$x_\mu z_N = x_\mu(x_\nu z_Q) = x_\nu(x_\mu z_Q) + [x_\mu, x_\nu] z_Q.$$

Now $z_\mu z_Q$ is of the form $z_\mu z_Q + w$, where $w \in P_{p-2}$. (C_p) can be applied to $x_\lambda(x_\nu(z_\mu z_Q))$, since $\nu \leqslant Q$ and $\nu < \mu$, and to $x_\lambda(x_\nu w)$ by the induction hypothesis, and hence to $x_\lambda(x_\nu(x_\mu z_Q))$. Hence:

$$x_\lambda(x_\mu z_N) = x_\nu(x_\lambda(x_\mu z_Q)) + [x_\lambda, x_\nu](x_\mu z_Q) + [x_\mu, x_\nu](x_\lambda z_Q) + [x_\lambda, [x_\mu, x_\nu]] z_Q.$$

Exchanging λ and μ and subtracting:

$$\begin{aligned} x_\lambda(x_\mu z_N) - x_\mu(x_\lambda z_N) &= x_\nu(x_\lambda(x_\mu z_Q) - x_\mu(x_\lambda z_Q)) \\ &\quad + [x_\lambda, [x_\mu, x_\nu]]z_Q - [x_\mu, [x_\lambda, x_\nu]]z_Q \\ &= x_\nu([x_\lambda, x_\mu]z_Q) + [x_\lambda, [x_\mu, x_\nu]]z_Q + [x_\mu, [x_\nu, x_\lambda]]z_Q \\ &= [x_\lambda, x_\mu](x_\nu z_Q) + ([x_\nu, [x_\lambda, x_\mu]] \\ &\quad + [x_\lambda, [x_\mu, x_\nu]] + [x_\mu, [x_\nu, x_\lambda]])z_Q \end{aligned}$$

and hence by the Jacobi identity

$$x_\lambda(x_\mu z_N) - x_\mu(x_\lambda z_N) = [x_\lambda, x_\mu]z_N$$

which completes the proof of Lemma 1.

Lemma 2. *There exists an α-mapping σ of $\mathfrak{g}$ into $\mathscr{L}_K(P)$ such that:*

(1) $\sigma(x_\lambda)z_M = z_\lambda z_M$ *for* $\lambda \leqslant M$;

(2) $\sigma(x_\lambda)z_M \equiv z_\lambda z_M \pmod{P_p}$ *if* M *has p elements.*

By Lemma 1 there exists a homomorphism f of the K-module $\mathfrak{g} \otimes_K P_p$ into P satisfying, for all p, conditions (A_p), (B_p), (C_p) (where f_p is replaced by f). This homomorphism defines a homomorphism σ of the K-module $\mathfrak{g}$ into the K-module $\mathscr{L}_K(P)$ and σ is an α-mapping because of condition (C_p). Finally, σ satisfies properties (1) and (2) of the lemma because of conditions (A_p) and (B_p).

Lemma 3. *Let t be a tensor in* $T_n \cap J$. *The homogeneous component t_n of t of order n is in the kernel* I *of the canonical homomorphism* $T \to S$.

We write t_n in the form $\sum_{i=1}^{r} x_{M_i}$, where the M_i are sequences of n elements of Λ. The mapping σ extends to a homomorphism of the algebra T into the algebra $\mathscr{L}_K(P)$ (which we shall also denote by σ), which is zero on J. By Lemma 2, $\sigma(t).1$ is a polynomial whose terms of highest degree are $\sum_{i=1}^{r} z_{M_i}$. As $t \in J$, $\sigma(t) = 0$ and hence $\sum_{i=1}^{r} z_{M_i} = 0$ in P. Now P is canonically identified with S, since $\mathfrak{g}$ has basis (x_λ). Hence the canonical image of t_n in S is zero, that is $t_n \in I$.

We can now prove Theorem 1. It is necessary to prove that the canonical homomorphism of S onto G is injective. In other words, if $t \in T^n$ and ψ denotes the canonical homomorphism of T onto U, it is necessary to show that the condition $\psi(t) \in U_{n-1}$ implies $t \in I$. Now $\psi(t) \in U_{n-1}$ means that there exists a tensor $t' \in T_{n-1}$ such that $t - t' \in J$. The tensor $t - t'$ admits t as homogeneous component of order n and hence $t \in I$ by Lemma 3.

Corollary 1. *Suppose that $\mathfrak{g}$ is a free* K*-module. Let* W *be a sub-*K*-module of* T^n.

If, in the notation of diagram (3), *the restriction of* τ_n *to* W *is an isomorphism of* W *onto* S^n, *then the restriction of* ψ_n *to* W *is an isomorphism of* W *onto a supplement of* U_{n-1} *in* U_n.

The restriction to W of $\omega_n \circ \tau_n$ is a bijection of W onto G^n; so is the restriction $\theta_n \circ \psi_n$ to W. Hence the corollary.

COROLLARY 2. *If* $\mathfrak{g}$ *is a free* K-*module, the canonical mapping of* $\mathfrak{g}$ *into its enveloping algebra is injective.*

This follows from Corollary 1 taking $W = T^1$.

When $\mathfrak{g}$ is a free K-module (in particular when K is a field), $\mathfrak{g}$ is identified with a submodule of U under the canonical mapping of $\mathfrak{g}$ into U. This convention is adopted from the following corollary onwards.

COROLLARY 3. *If* $\mathfrak{g}$ *admits a totally ordered basis* $(x_\lambda)_{\lambda \in \Lambda}$, *the elements* $x_{\lambda_1}x_{\lambda_2}\dots x_{\lambda_n}$ *of the enveloping algebra* U, *where* $(\lambda_1, \dots, \lambda_n)$ *is an arbitrary increasing finite sequence of elements of* Λ, *form a basis of the* K-*module* U.

Let Λ_n be the set of increasing sequences of n elements of Λ. For $M = (\lambda_1, \dots, \lambda_n) \in \Lambda_n$, let $y_M = x_{\lambda_1} \otimes x_{\lambda_2} \otimes \dots \otimes x_{\lambda_n}$. Let W be the submodule of T^n with basis $(y_M)_{M \in \Lambda_n}$. Corollary 1 shows that the restriction of ψ_n to W is an isomorphism of W onto a supplement of U_{n-1} in U_n. But

$$\psi_n(y_M) = x_{\lambda_1}x_{\lambda_2}\dots x_{\lambda_n},$$

whence the corollary.

COROLLARY 4. *Let* $S'^n \subset T^n$ *be the set of homogeneous symmetric tensors of order* n. *Suppose that* K *is a field of characteristic* 0. *Then the composite mapping of the canonical mappings*

$$S^n \to S'^n \to U_n$$

is an isomorphism of the vector space S^n *onto a supplement of* U_{n-1} *in* U_n.

This follows from Corollary 1 taking $W = S'^n$.

Suppose henceforth that K is a field of characteristic 0. Let η_n be the mapping of S^n into U_n just defined. Let $U^n = \eta_n(S^n)$. The vector space U is the direct sum of the U^n. The η_n define an isomorphism η of the vector space $S = \sum_n S^n$ onto the vector space $U = \sum_n U^n$, called the *canonical isomorphism of* S *onto* U; this is *not* an algebra isomorphism. We have the commutative diagram:

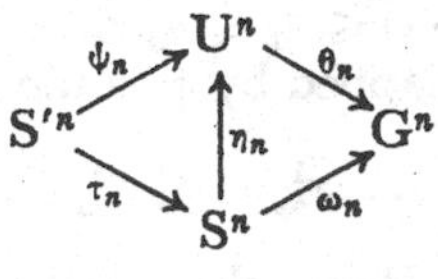

where each arrow represents a vector space isomorphism. If $x_1, x_2, \ldots, x_n$ are in $\mathfrak{g}$, η_n maps the product $x_1x_2\ldots x_n$, calculated in S, to the element

$$\frac{1}{n!}\sum_{\sigma\in\mathfrak{S}_n} x_{\sigma(1)}x_{\sigma(2)}\ldots x_{\sigma(n)}$$

calculated in U.

COROLLARY 5. *Let $\mathfrak{h}$ be a subalgebra of the Lie algebra $\mathfrak{g}$ and U′ its enveloping algebra. Suppose that the* K*-modules $\mathfrak{h}$ and $\mathfrak{g}/\mathfrak{h}$ are free (for example if* K *is a field). Let $(x_\alpha)_{\alpha\in L}$ be a basis of $\mathfrak{h}$ and $(y_\beta)_{\beta\in M}$ a family of elements of $\mathfrak{g}$ whose canonical images in $\mathfrak{g}/\mathfrak{h}$ form a basis of $\mathfrak{g}/\mathfrak{h}$.*

(a) *The canonical homomorphism of* U′ *into* U *is injective.*

(b) *If* M *is totally ordered, the elements $y_{\beta_1}\ldots y_{\beta_q}$, where $\beta_1 \leqslant \cdots \leqslant \beta_q$, form a basis of* U *considered as a left or right module over* U′.

We give L ∪ M a total ordering such that every element of L is less than every element of M. The elements $x_{\alpha_1}x_{\alpha_2}\ldots x_{\alpha_p}$ calculated in U′ (where $\alpha_1 \leqslant \cdots \leqslant \alpha_p$) form a basis of U′ (Corollary 3). The elements

$$x_{\alpha_1}\ldots x_{\alpha_p}y_{\beta_1}\ldots y_{\beta_q}$$

calculated in U (where $\alpha_1 \leqslant \cdots \leqslant \alpha_p \leqslant \beta_1 \leqslant \cdots \leqslant \beta_q$) similarly form a basis of U. Hence the canonical homomorphism of U′ into U maps the elements of a basis of U′ to linearly independent elements of U and is therefore injective. It is moreover seen that the $y_{\beta_1}\ldots y_{\beta_q}$ (where $\beta_1 \leqslant \cdots \leqslant \beta_q$) form a basis of U considered as a left U′-module. Ordering L ∪ M so that every element of M is less than every element of L, it is similarly seen that the $y_{\beta_1}\ldots y_{\beta_q}$ (where $\beta_1 \leqslant \cdots \leqslant \beta_q$) form a basis of U considered as a right U′-module.

Under the conditions of Corollary 5, U′ is identified with the subalgebra of U generated by $\mathfrak{h}$ by means of the canonical homomorphism of U′ into U.

COROLLARY 6. *Suppose that the* K*-module $\mathfrak{g}$ is the direct sum of subalgebras $\mathfrak{g}_1, \mathfrak{g}_2, \ldots, \mathfrak{g}_n$ and that each $\mathfrak{g}_i$ is a free* K*-module. Let U_i be the enveloping algebra of $\mathfrak{g}_i$ $(1 \leqslant i \leqslant n)$. Let ϕ be the* K*-linear mapping of the* K*-module $U_1 \otimes_K \cdots \otimes_K U_n$ into* U *defined by the multilinear mapping $(u_1, \ldots, u_n) \mapsto u_1\ldots u_n$ of $U_1 \times \cdots \times U_n$ into* U. *Then ϕ is a* K*-module isomorphism.*

Let $(x^i_\lambda)_{\lambda\in L_i}$ be a basis of $\mathfrak{g}_i$. We totally order $L_1 \cup \cdots \cup L_n$ so that every element of L_i exceeds every element of L_j for $i \geqslant j$. Then the elements:

$$(x^1_{\lambda_1}x^1_{\lambda_2}\ldots x^1_{\lambda_p}) \otimes \cdots \otimes (x^n_{\nu_1}x^n_{\nu_2}\ldots x^n_{\nu_q}),$$

where $\lambda_1 \leqslant \lambda_2 \leqslant \cdots \leqslant \lambda_p \leqslant \cdots \leqslant \nu_1 \leqslant \nu_2 \leqslant \cdots \leqslant \nu_q$, constitute a basis of $U_1 \otimes_K \cdots \otimes_K U_n$. They are mapped by ϕ to the elements:

$$x^1_{\lambda_1}x^1_{\lambda_2}\ldots x^1_{\lambda_p}\ldots x^n_{\nu_1}x^n_{\nu_2}\ldots x^n_{\nu_q}$$

which constitute a basis of U. Hence the corollary.

COROLLARY 7. *If K is an integral domain and $\mathfrak{g}$ is a free K-module, the algebra U has no divisors of zero.*

G is isomorphic to a polynomial algebra over K (Theorem 1) and is therefore an integral domain (*Algebra*, Chapter IV, § 1, no. 4, Theorem 1). Hence the corollary (*Commutative Algebra*, Chapter III, § 2, no. 3, Proposition 1).

8. EXTENSION OF DERIVATIONS

Lemma 4. *Let V be a K-module and T the tensor algebra of V. Let u be an endomorphism of V. There exists one and only one derivation of T which extends u. This derivation commutes with the symmetry operators on T.*

Let $F = V \times V \times \cdots \times V$ (n factors). The mapping

$$(x_1, \ldots, x_n) \mapsto ux_1 \otimes x_2 \otimes \cdots \otimes x_n + x_1 \otimes ux_2 \otimes \cdots \otimes x_n + \cdots + x_1 \otimes x_2 \otimes \cdots \otimes ux_n$$

of F into $\bigotimes^n V$ is multilinear. Hence there exists an endomorphism u_n of $\bigotimes^n V$ such that:

$$u_n(x_1 \otimes \cdots \otimes x_n) = ux_1 \otimes \cdots \otimes x_n + \cdots + x_1 \otimes \cdots \otimes ux_n$$

for all $x_1, \ldots, x_n$ in V. Then $u_1 = u$. Let v be the endomorphism of the K-module T which coincides with u_n on each $T^n = \bigotimes^n V$ and which is zero on $T^0 = K.1$. We show that v is a derivation of T. If $x_1, \ldots, x_n, y_1, \ldots, y_p$ are elements of V, then

$$\begin{aligned} v((x_1 \otimes \cdots \otimes x_n) \otimes (y_1 \otimes \cdots \otimes y_p)) &= \sum_{i=1}^{n} x_1 \otimes \cdots \otimes x_{i-1} \otimes ux_i \otimes x_{i+1} \otimes \cdots \otimes x_n \otimes y_1 \otimes \cdots \otimes y_p \\ &\quad + \sum_{j=1}^{p} x_1 \otimes \cdots \otimes x_n \otimes y_1 \otimes \cdots \otimes y_{j-1} \otimes uy_j \otimes y_{j+1} \otimes \cdots \otimes y_p \\ &= v(x_1 \otimes \cdots \otimes x_n) \otimes (y_1 \otimes \cdots \otimes y_p) + (x_1 \otimes \cdots \otimes x_n) \otimes v(y_1 \otimes \cdots \otimes y_p). \end{aligned}$$

By linearity it follows that v is a derivation. The uniqueness of v is obvious. Finally, clearly u_n commutes with all the symmetry operators on $\bigotimes^n V$, whence the last assertion.

PROPOSITION 7. *Let $\mathfrak{g}$ be a Lie algebra, U its enveloping algebra, σ the canonical mapping of $\mathfrak{g}$ into U and D a derivation of $\mathfrak{g}$.*

(a) *There exists one and only one derivation D_U of U such that $\sigma \circ D = D_U \circ \sigma$ (that is such that D_U extends D, when $\mathfrak{g}$ can be identified with a submodule of U under σ).*

(b) *D_U leaves stable U_n and the set U^n of images in U of the homogeneous symmetric tensors of order n over $\mathfrak{g}$.*

(c) D_U *commutes with the principal antiautomorphism of* U.

(d) *If* D *is the inner derivation of* $\mathfrak{g}$ *defined by an element* x *of* $\mathfrak{g}$, D_U *is the inner derivation of* U *defined by* $\sigma(x)$.

Let D_T be the derivation of the tensor algebra T of $\mathfrak{g}$ which extends D (Lemma 4). The two-sided ideal J of T generated by the

$$x \otimes y - y \otimes x - [x, y]$$

(x, y in $\mathfrak{g}$) is stable under D_T. For:

$$D_T(x \otimes y - y \otimes x - [x, y]) = Dx \otimes y - y \otimes Dx - [Dx, y] \\ + x \otimes Dy - Dy \otimes x - [x, Dy].$$

On passing to the quotient, D_T defines a derivation D_U of U such that $\sigma \circ D = D_U \circ \sigma$. The uniqueness of D_U is immediate since 1 and $\sigma(\mathfrak{g})$ generate the algebra U. Assertion (b) is obvious. Let A be the principal antiautomorphism of U. We now prove (c). If $x_1, \ldots, x_n$ are in $\mathfrak{g}$, then

$$\begin{aligned} D_U A(\sigma(x_1)\ldots\sigma(x_n)) &= D_U((-1)^n \sigma(x_n)\ldots\sigma(x_1)) \\ &= (-1)^n \sum_{i=1}^{n} \sigma(x_n)\ldots D_U(\sigma(x_i))\ldots\sigma(x_1) \\ &= (-1)^n \sum_{i=1}^{n} \sigma(x_n)\ldots\sigma(Dx_i)\ldots\sigma(x_1) \\ &= A\left(\sum_{i=1}^{n} \sigma(x_1)\ldots\sigma(Dx_i)\ldots\sigma(x_n)\right) \\ &= AD_U(\sigma(x_1)\ldots\sigma(x_n)). \end{aligned}$$

Finally, let $x \in \mathfrak{g}$. Let Δ be the inner derivation $y \mapsto \sigma(x)y - y\sigma(x)$ of U (*Algebra*, Chapter IV, § 4, no. 3, *Example* 2). Then, for $x' \in \mathfrak{g}$,

$$(\Delta \circ \sigma)(x') = \sigma(x)\sigma(x') - \sigma(x')\sigma(x) = \sigma([x, x']) = (\sigma \circ \operatorname{ad} x)(x'),$$

whence $\Delta \circ \sigma = \sigma \circ \operatorname{ad} x$. This completes the proof.

Applying Proposition 7 to the case of a commutative Lie algebra, it is seen that every endomorphism u of a K-module can be extended uniquely to a derivation of the symmetric algebra of this module; this derivation is derived on passing to the quotient from the derivation of the tensor algebra which extends u.

We again take a Lie algebra $\mathfrak{g}$ over K and let D be a derivation of $\mathfrak{g}$. We use the earlier notation T, S, U, G. Let D_T, D_S be the derivations of T, S which extend D and let D_U be the unique derivation of U such that $\sigma \circ D = D_U \circ \sigma$. Since D_U leaves the U_n stable, D_U defines on taking quotients a derivation D_G of G. Since D_U and D_S are derived from D_T when passing to quotients, the commutative diagram (3) proves that D_G can also be derived from D_S by the homomorphism ω defined in no. 6. If further K is a field of characteristic 0, the isomorphisms of diagram (4) map one into another the restrictions of D_T, D_S, D_U, D_G to S'^n, S^n, U^n, G^n. Hence the *canonical isomorphism of* S *onto* U *maps* D_S *to* D_U.

9. EXTENSION OF THE BASE RING

Let $\mathfrak{g}$ be a Lie algebra over K, T its tensor algebra, J the two-sided ideal of T generated by the $x \otimes y - y \otimes x - [x, y]$ (x, y in $\mathfrak{g}$) and $U = T/J$. Let K_1 be a commutative ring with unit element and σ a homomorphism of K into K_1 mapping 1 to 1. Then the tensor algebra of $\mathfrak{g}_{(K_1)}$ is canonically identified with $T_{(K_1)}$. Let J′ be the two-sided ideal of $T_{(K_1)}$ generated by the

$$x' \otimes y' - y' \otimes x' - [x', y']$$

(x', y' in $\mathfrak{g}_{(K_1)}$). Clearly the canonical image of $J_{(K_1)}$ in $T_{(K_1)}$ is contained in J′. To see that it is equal to J′, it suffices to show that, if x' and y' denote two elements of $\mathfrak{g}_{(K_1)}$, $x' \otimes y' - y' \otimes x' - [x', y']$ belongs to this image. Now

$$x' = \sum_i x_i \otimes \lambda_i, y' = \sum_j y_j \otimes \mu_j \quad (x_i, y_j \text{ in } \mathfrak{g}, \lambda_i, \mu_j \text{ in } K_1);$$

whence

$$x' \otimes y' - y' \otimes x' - [x', y'] = \sum_{i,j} (x_i \otimes y_j - y_j \otimes x_i - [x_i, y_j]) \otimes \lambda_i \mu_j$$

which proves our assertion. Then it can be seen that $U_{(K_1)} = (T/J)_{(K_1)}$ is canonically identified with $T_{(K_1)}/J'$: *the enveloping algebra of* $\mathfrak{g}_{(K_1)}$ *is canonically identified with* $U_{(K_1)}$ and the canonical mapping of $\mathfrak{g}_{(K_1)}$ into its enveloping algebra is identified with $\sigma \otimes 1$ (where σ denotes the canonical mapping of $\mathfrak{g}$ into U).

§ 3. REPRESENTATIONS

1. REPRESENTATIONS

Definition 1. *Let* $\mathfrak{g}$ *be a Lie algebra over* K *and* M *a* K*-module. A homomorphism of* $\mathfrak{g}$ *into the Lie Algebra* $\mathfrak{gl}(M)$ *is called a representation of* $\mathfrak{g}$ *on the module* M. *An injective representation is called faithful. If* K *is a field, the dimension (finite or infinite) of* M *over* K *is called the dimension of the representation. The representation* $x \mapsto \text{ad}\, x$ *of* $\mathfrak{g}$ *on the* K*-module* $\mathfrak{g}$ *is called the adjoint representation of* $\mathfrak{g}$.

A representation of $\mathfrak{g}$ on M is thus a K-linear mapping ρ of $\mathfrak{g}$ into the endomorphism module of M such that

$$\rho([x, y]).m = \rho(x)\rho(y).m - \rho(y)\rho(x).m$$

for all $x \in \mathfrak{g}$, $y \in \mathfrak{g}$, $m \in M$.

**Example.* Let G be a real Lie group, $\mathfrak{g}$ its Lie algebra and θ an analytic representation of G on a finite-dimensional real vector space E. Then the corresponding homomorphism of $\mathfrak{g}$ into $\mathfrak{gl}(E)$ is a representation of $\mathfrak{g}$ on E.$_*$

Let U be the enveloping algebra of $\mathfrak{g}$. Proposition 1 of § 2, no. 1 defines a one-to-one correspondence between the set of representations of $\mathfrak{g}$ on M and the set of representations of U on M. On the other hand we know (*Algebra*, Chapter VIII, § 13, no. 1) that there is an equivalence between the notion of representation of the associative algebra U and that of left U-module.

DEFINITION 2. *Let $\mathfrak{g}$ be a Lie algebra over* K *and* U *its enveloping algebra. A unitary left module over* U *is called a left $\mathfrak{g}$-module, or simply a $\mathfrak{g}$-module.*

If M is a $\mathfrak{g}$-module and $x \in U$, x_M will denote the homothety of M defined by x (cf. *Algebra*, Chapter VIII, § 1, no. 2).

A unitary right module over U is called a right $\mathfrak{g}$-module. Such a module is identified with a left U^0-module, that is (§ 2, no. 4) with a left $\mathfrak{g}^0$-module.

Let ϕ be the principal antiautomorphism of U. If M is a right $\mathfrak{g}$-module, a left $\mathfrak{g}$-module structure is defined on M by writing $a.m = m.\phi(a)$ for $m \in M$ and $a \in U$.

The notions and results of the theory of modules can be translated into the language of representations:

(1) Two representations ρ and ρ' of $\mathfrak{g}$ on M and M' are called *similar* or *isomorphic* if the $\mathfrak{g}$-modules M and M' are isomorphic. For this it is necessary and sufficient that there exist an isomorphism u of the K-module M onto the K-module M' such that

$$\rho'(x) = u \circ \rho(x) \circ u^{-1}$$

for all $x \in \mathfrak{g}$.

(2) For all $i \in I$, let ρ_i be a representation of $\mathfrak{g}$ on M_i. Let M be the $\mathfrak{g}$-module the direct sum of the $\mathfrak{g}$-modules M_i. There is a corresponding representation ρ of $\mathfrak{g}$ on M, called the *direct sum* of the ρ_i and denoted by $\sum_{i \in I} \rho_i$ (or $\rho_1 + \cdots + \rho_n$ in the case of n representations $\rho_1, \ldots, \rho_n$). If $m = (m_i)_{i \in I}$ is an element of M and $x \in \mathfrak{g}$, then $\rho(x).m = (\rho_i(x).m_i)_{i \in I}$.

(3) A representation ρ of $\mathfrak{g}$ on M is called *simple* or *irreducible* if the associated $\mathfrak{g}$-module is simple. It amounts to the same to say that there exists no sub-K-module of M (other than {0} and M) stable under all the $\rho(x)$, $x \in \mathfrak{g}$. A class of simple $\mathfrak{g}$-modules (*Algebra*, Chapter VIII, § 3, no. 2) defines a *class of simple representations of* $\mathfrak{g}$.

(4) A representation ρ of $\mathfrak{g}$ on M is called *semi-simple* or *completely reducible* if the associated $\mathfrak{g}$-module is semi-simple. It amounts to the same to say that ρ is similar to a direct sum of simple representations or that every sub-K-module of

M stable under the $\rho(x)$ $(x \in \mathfrak{g})$ has a supplement stable under the $\rho(x)$ $(x \in \mathfrak{g})$ (cf. *Algebra*, Chapter VIII, § 3, no. 3).

(5) Let $\mathcal{S}$ be a class of simple representations of $\mathfrak{g}$ corresponding to a class C of simple $\mathfrak{g}$-modules. On the other hand let ρ be a representation of $\mathfrak{g}$ on M. The isotypical component M_C of species C of the $\mathfrak{g}$-module M (*Algebra*, Chapter VIII, § 3, no. 4) is also called the *isotypical component of* M *of species* $\mathcal{S}$. This component is the sum of the sub-K-modules of M stable under the $\rho(x)$ and on which the $\rho(x)$ induce a representation of class $\mathcal{S}$; it is the direct sum of certain of these submodules; if M_C is of length n, ρ is said to *contain* $\mathcal{S}$ *n times*. The sum of the different M_C is direct; it is equal to M if and only if ρ is semi-simple.

(6) Let ρ, ρ' be two representations of $\mathfrak{g}$. ρ' is called a *subrepresentation* (resp. *quotient representation*) of ρ if the module of ρ' is a submodule (resp. quotient module) of the module of ρ.

Let M be a K-module. The zero representation of $\mathfrak{g}$ on M defines on M a $\mathfrak{g}$-module structure. With this structure M is called a *trivial* $\mathfrak{g}$-module.

Let M be a $\mathfrak{g}$-module. The quotient $\mathfrak{g}$-modules of the sub-$\mathfrak{g}$-modules of M are also the sub-$\mathfrak{g}$-modules of the quotient modules of M: they are obtained by considering two sub-$\mathfrak{g}$-modules U, U′ of M such that $U \supset U'$ and forming the $\mathfrak{g}$-module U/U'. Then if all the simple modules of the above type are isomorphic to a given simple $\mathfrak{g}$-module N, M is called a *pure* $\mathfrak{g}$-module *of species* N. If ρ and σ are the representations of $\mathfrak{g}$ corresponding to M and N, we also say that ρ is *pure of species* σ.

Let M′ be a sub-$\mathfrak{g}$-module of M. For M to be pure of species N, it is necessary and sufficient that M′ and M/M′ be pure of species N. For the condition is obviously necessary. Suppose that it holds and let U, U′ be sub-$\mathfrak{g}$-modules of M such that $U' \subset U$ and U/U' is simple; let ϕ be the canonical homomorphism of M onto M/M′; if $\phi(U) \neq \phi(U')$, U/U' is isomorphism to $\phi(U)/\phi(U')$ and hence isomorphic to N; if $\phi(U) = \phi(U')$, then $U \subset U' + M'$, hence U/U' is isomorphic to a simple submodule of $(U' + M')/U'$ and the latter module is itself isomorphic to $M'/(U' \cap M')$; hence U/U' is again isomorphic to N, so that M is pure of species N.

Henceforth let M be a $\mathfrak{g}$-module and suppose that the set of sub-$\mathfrak{g}$-modules of M which are pure of species N admits a maximal element M′. Then every submodule M″ of M which is pure of species N is contained in M′. For $M''/(M' \cap M'')$ and M′ are pure of species N, hence $M' + M''$ is pure of species N by the above and hence $M' + M'' \subset M'$.

Suppose that the $\mathfrak{g}$-module M admits a Jordan-Hölder series $(M_i)_{0 \leqslant i \leqslant n}$. For M to be pure of species N, it is necessary and sufficient that M_0/M_1, $M_1/M_2, \ldots, M_{n-1}/M_n$ be isomorphic to N; for the condition is obviously necessary and its sufficiency follows immediately by induction on n from what we have seen above.

PROPOSITION 1. *Let* $\mathfrak{g}$ *be a Lie algebra over* K *and* $\mathfrak{a}$ *an ideal of* $\mathfrak{g}$. *Let* M *be a* $\mathfrak{g}$*-module* *and* N *a simple* $\mathfrak{a}$*-module. Consider* M *as an* $\mathfrak{a}$*-module and suppose that the set of sub-*$\mathfrak{a}$*-modules of* M *which are pure of species* N *admits a maximal element* M′. *Then* M′ *is a sub-*$\mathfrak{g}$*-module of* M.

Let $y \in \mathfrak{g}$. Let ϕ be the canonical mapping of M onto M/M′ and f the mapping $m \mapsto \phi(y_M.m)$ of M′ into M/M′. It suffices to show that $f(M') = \{0\}$. Let $x \in \mathfrak{a}$. Then, for $m \in M$,

$$x_{M/M'}.f(m) = \phi(x_M y_M.m) = \phi(y_M x_M.m) + \phi([x, y]_M.m).$$

Now $[x, y] \in \mathfrak{a}$, whence $\phi([x, y]_M.m) = 0$; on the other hand,

$$\phi(y_M x_M.m) = f(x_M.m).$$

Hence $x_{M/M'}.f(m) = f(x_M.m)$. It follows that $f(M')$ is a sub-$\mathfrak{a}$-module of M/M′ isomorphic to a quotient of M′ and hence pure of species N; hence $f(M') = \{0\}$.

COROLLARY. *Let* $\mathfrak{g}$ *be a Lie algebra over* K *and* $\mathfrak{a}$ *an ideal of* $\mathfrak{g}$. *Let* M *be a simple* $\mathfrak{g}$*-module, of finite length as a* K*-module. There exists a simple* $\mathfrak{a}$*-module* N *such that* M *is a pure* $\mathfrak{a}$*-module of species* N.

Since the $\mathfrak{a}$-module M is of finite length, there exists a minimal element N in the set of sub-$\mathfrak{a}$-modules of M: it is a simple sub-$\mathfrak{a}$-module of M. The largest sub-$\mathfrak{a}$-module of M which is pure of species N is therefore $\neq \{0\}$ and is a sub-$\mathfrak{g}$-module of M (Proposition 1) and is therefore identical with M.

2. TENSOR PRODUCT OF REPRESENTATIONS

We have defined in no. 1 the direct sum of a family of representations of $\mathfrak{g}$. We shall now define other operations on representations.

Let $\mathfrak{g}_1$, $\mathfrak{g}_2$ be two Lie algebras over K and M_i a $\mathfrak{g}_i$-module ($i = 1, 2$). Let U_i be the enveloping algebra of $\mathfrak{g}_i$ and σ_i the canonical mapping of $\mathfrak{g}_i$ into U_i. Then M_i is a left U_i-module and hence $M_1 \otimes_K M_2$ has a canonical left $(U_1 \otimes_K U_2)$-module structure. Now $U_1 \otimes_K U_2$ is the enveloping algebra of $\mathfrak{g}_1 \times \mathfrak{g}_2$ and the mapping $(x_1, x_2) \mapsto \sigma_1(x_1) \otimes 1 + 1 \otimes \sigma_2(x_2)$ is the canonical mapping of $\mathfrak{g}_1 \times \mathfrak{g}_2$ into this enveloping algebra (§ 2, no. 2). Hence there exists a $(\mathfrak{g}_1 \times \mathfrak{g}_2)$-module structure on $M = M_1 \otimes_K M_2$ such that:

$$(1) \qquad \begin{aligned}(x_1, x_2)_M.(m_1 \otimes m_2) &= (\sigma_1(x_1) \otimes 1 + 1 \otimes \sigma_2(x_2)).(m_1 \otimes m_2)\\ &= ((x_1)_{M_1}.m_1) \otimes m_2 + m_1 \otimes ((x_2)_{M_2}.m_2).\end{aligned}$$

This structure defines a representation of $\mathfrak{g}_1 \times \mathfrak{g}_2$ on M.

If now $\mathfrak{g}_1 = \mathfrak{g}_2 = \mathfrak{g}$, the homomorphism $x \mapsto (x, x)$ of $\mathfrak{g}$ into $\mathfrak{g} \times \mathfrak{g}$, composed with the above representation, defines a representation of $\mathfrak{g}$ on M and hence a $\mathfrak{g}$-module structure on M such that:

$$(2) \qquad x_M.(m_1 \otimes m_2) = (x_{M_1}.m_1) \otimes m_2 + m_1 \otimes (x_{M_2}.m_2).$$

By an analogous argument we see that:

PROPOSITION 2. *Let* $\mathfrak{g}$ *be a Lie algebra over* K *and* M_i *a* $\mathfrak{g}$*-module* ($1 \leqslant i \leqslant n$). *On the tensor product* $M_1 \otimes_K M_2 \otimes \cdots \otimes M_n$, *there exists one and only one* $\mathfrak{g}$*-module structure such that*

$$(3) \qquad x_M.(m_1 \otimes \cdots \otimes m_m) = \sum_{i=1}^{n} m_1 \otimes \cdots \otimes (x_{M_i}.m_i) \otimes \cdots \otimes m_n$$

for all $x \in \mathfrak{g}$, $m_1 \in M_1, \ldots, m_n \in M_n$.

The corresponding representation is called the *tensor product* of the given representations of $\mathfrak{g}$ on the M_i.

In particular, if M is a $\mathfrak{g}$-module, Proposition 2 defines a $\mathfrak{g}$-module structure on each $M_p = \bigotimes^p M$ and hence on the tensor algebra T of M.

Formula (3) shows that, for all $x \in \mathfrak{g}$, x_T is the unique *derivation* of the algebra T which extends x_M. We know (§ 2, no. 8) that x_T defines on passing to the quotient a derivation of the symmetric algebra S of M. Hence S can be considered as a quotient $\mathfrak{g}$-module of T and the x_S are derivations of S.

Still more particularly, consider $\mathfrak{g}$ as a $\mathfrak{g}$-module by means of the adjoint representation of $\mathfrak{g}$. Let U be the enveloping algebra of $\mathfrak{g}$. By Proposition 7 of § 2, x_M defines on passing to the quotients a derivation of U which is just the inner derivation defined by $\sigma(x)$ (σ denoting the canonical mapping of $\mathfrak{g}$ into U). Then U can be considered as a quotient $\mathfrak{g}$-module of T. If K is a field of characteristic 0, the canonical isomorphism of S onto U is a $\mathfrak{g}$-module isomorphism (§ 2, no. 8).

3. REPRESENTATIONS ON HOMOMORPHISM MODULES

Again let $\mathfrak{g}_1$ and $\mathfrak{g}_2$ be two Lie algebras over K and M_i a $\mathfrak{g}_i$-module ($i = 1, 2$). Let U_i be the enveloping algebra of $\mathfrak{g}_i$ and σ_i the canonical mapping of $\mathfrak{g}_i$ into U_i. Then M_i is a left U_i-module and hence $\mathscr{L}_K(M_1, M_2)$ has a canonical left $(U_1^0 \otimes U_2)$-module structure. Now $U_1^0 \otimes_K U_2$ is the enveloping algebra of $\mathfrak{g}_1^0 \times \mathfrak{g}_2$ and the mapping

$$(x_1, x_2) \mapsto \sigma_1(x_1) \otimes 1 + 1 \otimes \sigma_2(x_2)$$

is the canonical mapping of $\mathfrak{g}_1^0 \times \mathfrak{g}_2$ into this enveloping algebra. Hence there exists a $(\mathfrak{g}_1^0 \times \mathfrak{g}_2)$-module structure on $M = \mathscr{L}_K(M_1, M_2)$ such that

$$(4) \qquad \begin{aligned} ((x_1, x_2)_M.u).m_1 &= ((\sigma_1(x_1) \otimes 1 + 1 \otimes \sigma_2(x_2)).u).m_1 \\ &= u((x_1)_{M_1}.m_1) + (x_2)_{M_2}.u(m_1) \end{aligned}$$

for all $u \in \mathscr{L}_K(M_1, M_2)$, $m_1 \in M_1$. This structure defines a representation of $\mathfrak{g}_1^0 \times \mathfrak{g}_2$ on M.

If now $\mathfrak{g}_1 = \mathfrak{g}_2 = \mathfrak{g}$, the homomorphism $x \mapsto (-x, x)$ of $\mathfrak{g}$ into $\mathfrak{g}^0 \times \mathfrak{g}$, composed with the above representation, defines a representation of $\mathfrak{g}$ on M and hence a $\mathfrak{g}$-module structure on M such that

$$(x_{M}.u).m_1 = x_{M_2}.u(m_1) - u(x_{M_1}.m_1) \tag{5}$$

or

$$x_{M}.u = x_{M_2}u - ux_{M_1}. \tag{6}$$

Combining these results with Proposition 2, we see that:

PROPOSITION 3. *Let $\mathfrak{g}$ be a Lie algebra over* K *and* M_i *a $\mathfrak{g}$-module* $(1 \leqslant i \leqslant n+1)$. *Let* N *be the* K*-module* $\mathscr{L}_K(M_1, \ldots, M_n; M_{n+1})$ *of multilinear mappings of* $\prod_{i=1}^{n} M_i$ *into* M_{n+1}. *There exists one and only one $\mathfrak{g}$-module structure on* N *such that*

$$(x_N.u)(m_1, \ldots, m_n) = -\sum_{i=1}^{n} u(m_1, \ldots, x_{M_i}.m_i, \ldots, m_n) + x_{M_{n+1}}.u(m_1, \ldots, m_n) \tag{7}$$

for all $x \in \mathfrak{g}$, $u \in N$ *and* $m_i \in M_i$ $(1 \leqslant i \leqslant n)$.

In particular, let $\mathfrak{g}$ be a Lie algebra over K and M a $\mathfrak{g}$-module and consider K as a trivial $\mathfrak{g}$-module. Proposition 3 defines a $\mathfrak{g}$-module structure on $\mathscr{L}_K(M, K) = M^*$. The corresponding representation is called the *dual* representation of the representation $x \mapsto x_M$. We have:

$$(x_{M^*}.f)(m) = -f(x_M.m) \tag{8}$$

for all $x \in \mathfrak{g}$, $f \in M^*$, $m \in M$. In other words:

$$x_{M^*} = -{}^t x_M. \tag{9}$$

When K is a field and M is finite-dimensional, the $\mathfrak{g}$-module M is simple (resp. semi-simple) if and only if the $\mathfrak{g}$-module M^* is simple (resp. semi-simple).

PROPOSITION 4. *Let* M_1, M_2 *be two $\mathfrak{g}$-modules. The canonical* K*-linear mappings* (*Algebra*, Chapter II, § 4, no. 2, Proposition 2 and no. 1, Proposition 1):

$$M_1^* \otimes_K M_2 \xrightarrow{\Phi} \mathscr{L}_K(M_1, M_2), \qquad \mathscr{L}_K(M_1, M_2^*) \xrightarrow{\Psi} (M_1 \otimes_K M_2)^*$$

(*where the second is bijective*) *are $\mathfrak{g}$-module homomorphisms.*

We write

$$N = M_1^* \otimes M_2, \quad P = \mathscr{L}(M_1, M_2), \quad Q = \mathscr{L}(M_1, M_2^*), \quad R = (M_1 \otimes M_2)^*.$$

Then, for $x \in \mathfrak{g}, f \in M_1^*, m_1 \in M_1, m_2 \in M_2$,

$$\begin{aligned}((\phi x_N)(f \otimes m_2)).m_1 &= (\phi(x_{M_1^*}f \otimes m_2 + f \otimes x_{M_2}m_2)).m_1 \\ &= \langle x_{M_1^*}f, m_1\rangle m_2 + \langle f, m_1\rangle x_{M_2}m_2 \\ ((x_P\phi)(f \otimes m_2)).m_1 &= x_{M_2}(\phi(f \otimes m_2).m_1) - \phi(f \otimes m_2)(x_{M_1}m_1) \\ &= \langle f, m_1\rangle x_{M_2}m_2 - \langle f, x_{M_1}m_1\rangle m_2\end{aligned}$$

and hence $\phi x_N = x_P\phi$. On the other hand, for $x \in \mathfrak{g}$, $u \in \mathscr{L}(M_1, M_2^*)$, $m_1 \in M_1$, $m_2 \in M_2$:

$$\begin{aligned}(\psi x_Q u)(m_1 \otimes m_2) &= \langle (x_Q u).m_1, m_2\rangle = \langle x_{M_2^*}um_1 - ux_{M_1}m_1, m_2\rangle \\ (x_R\psi u)(m_1 \otimes m_2) &= -\langle \psi u, x_{M_1}m_1 \otimes m_2 + m_1 \otimes x_{M_2}m_2\rangle \\ &= -\langle ux_{M_1}m_1, m_2\rangle - \langle um_1, x_{M_2}m_2\rangle\end{aligned}$$

and hence $\psi x_Q = x_R\psi$, which completes the proof.

The $\mathfrak{g}$-modules $\mathscr{L}(M_1, M_2^*)$ and $(M_1 \otimes M_2)^*$ are identified under the isomorphism ψ. If M_1 and M_2 have finite bases, ϕ is an isomorphism (*Algebra*, Chapter II, § 4, no. 2, Proposition 2), which allows us to identify the $\mathfrak{g}$-modules $M_1^* \otimes M_2$ and $\mathscr{L}(M_1, M_2)$; in that case, we can therefore identify the $\mathfrak{g}$-modules $M_1^* \otimes M_2^*$, $\mathscr{L}(M_1, M_2^*)$ and $(M_1 \otimes M_2)^*$.

4. EXAMPLES

Example 1. Let $\mathfrak{g}$ be a Lie algebra over K and M a $\mathfrak{g}$-module. The $\mathfrak{g}$-module structure on M and the trivial $\mathfrak{g}$-module structure on K define a $\mathfrak{g}$-module structure on the K-module $N = \mathscr{L}(M, M; K)$ of bilinear forms on M. Then

(10) $$(x_N.\beta)(m, m') = -\beta(x_M.m, m') + \beta(m, x_M.m')$$

for all $x \in \mathfrak{g}$, m, m' in M, $\beta \in N$. If β is a given element of N, the set of $x \in \mathfrak{g}$ such that $x_N.\beta = 0$ is a subalgebra of $\mathfrak{g}$.

Let M be a K-module and β a bilinear form on M. By the above, the set of $x \in \mathfrak{gl}(M)$ such that

$$\beta(x.m, m') + \beta(m, x.m') = 0$$

for all $m \in M$ and $m' \in M$ is a Lie subalgebra of $\mathfrak{gl}(M)$. Suppose that K is a field, M is finite-dimensional and β is non-degenerate. Then every $x \in \mathfrak{gl}(M)$ admits a left adjoint x^* (relative to β) which is everywhere defined and the subalgebra in question is the set of $x \in \mathfrak{gl}(M)$ such that $x^* = -x$. By this process we can construct two important examples of Lie algebras:

(a) Take $M = K^n$ and

$$\beta((\xi_1, \ldots, \xi_n), (\eta_1, \ldots, \eta_n)) = \xi_1\eta_1 + \cdots + \xi_n\eta_n.$$

We canonically identify $\mathfrak{gl}(K^n)$ with $\mathbf{M}_n(K)$. Then the Lie algebra obtained is the Lie algebra of skew-symmetric matrices. *(When $K = \mathbf{R}$, this algebra is the Lie algebra of the orthogonal group $\mathbf{O}(n, \mathbf{R})$).*

(b) Take $M = K^{2m}$ and

$$\beta((\xi_1, \ldots, \xi_{2m}), (\eta_1, \ldots, \eta_{2m})) = \xi_1\eta_{m+1} - \eta_1\xi_{m+1} + \cdots + \xi_m\eta_{2m} - \eta_m\xi_{2m}.$$

The matrix of β with respect to the canonical basis of K^{2m} is the matrix $\begin{pmatrix} 0 & I_m \\ -I_m & 0 \end{pmatrix}$. Let $U = \begin{pmatrix} A & B \\ C & D \end{pmatrix}$ be the matrix with respect to the canonical basis of K^{2m} of an element u of $\mathfrak{gl}(M)$ (A, B, C, D lying in $\mathbf{M}_m(K)$). By formula (50) of *Algebra*, Chapter IX, § 1, no. 10, u^* has with respect to the same basis the matrix

$$\begin{pmatrix} 0 & -I_m \\ I_m & 0 \end{pmatrix}\begin{pmatrix} {}^tA & {}^tC \\ {}^tB & {}^tD \end{pmatrix}\begin{pmatrix} 0 & I_m \\ -I_m & 0 \end{pmatrix} = \begin{pmatrix} {}^tD & -{}^tB \\ -{}^tC & {}^tA \end{pmatrix}.$$

The condition $u^* = -u$ is therefore equivalent to the conditions

$$D = -{}^tA \qquad B = {}^tB \qquad C = {}^tC.$$

When $K = \mathbf{R}$, the Lie algebra obtained is the Lie algebra of the symplectic group $\mathbf{Sp}(2m, \mathbf{R})$.

Example 2. We preserve the notation of *Example* 1.

The $\mathfrak{g}$-module structure on M defines on the K-module $P = \mathscr{L}_K(M, M)$ of endomorphisms of M a $\mathfrak{g}$-module structure. By (6), for all $x \in \mathfrak{g}$ and $u \in P$:

$$(11) \qquad x_P . u = [x_M, u] = (\operatorname{ad} x_M) . u$$

where $\operatorname{ad} x_M$ denotes the image of x_M under the adjoint representation of $\mathfrak{gl}(M)$. In other words:

$$(12) \qquad x_P = \operatorname{ad} x_M$$

in $\mathscr{L}(\mathscr{L}(M, M)) = \mathscr{L}(\mathfrak{gl}(M))$.

5. INVARIANT ELEMENTS

DEFINITION 3. *Let $\mathfrak{g}$ be a Lie algebra and M a $\mathfrak{g}$-module. An element $m \in M$ is called invariant (with respect to the $\mathfrak{g}$-module structure on M or with respect to the corresponding representation of $\mathfrak{g}$) if $x_M . m = 0$ for all $x \in \mathfrak{g}$.*

Let G be a connected real Lie group, $\mathfrak{g}$ its Lie algebra, θ an analytic representation of G on a finite-dimensional real vector space E and ρ the corresponding representation of $\mathfrak{g}$ on E. Let $m \in E$. The element m is invariant with respect to ρ if and only if $\theta(g) . m = m$ for all $g \in G$. This justifies the use of the word "invariant".

Example 1. Let M, N be two $\mathfrak{g}$-modules and $P = \mathscr{L}_K(M, N)$. For an element f of P to be invariant, it is necessary and sufficient, by (6), that f be a homomorphism of the $\mathfrak{g}$-module M into the $\mathfrak{g}$-module N. In particular, if $M = N$ and $x_M = x_N$ for all $x \in \mathfrak{g}$, f is invariant if and only if f is permutable with the x_M.

Example 2. Let M be a K-module with a finite basis. If M has a $\mathfrak{g}$-module structure, $\mathscr{L}(M, M)$ and $M^* \otimes M$ have $\mathfrak{g}$-module structures and the canonical mapping of $M^* \otimes M$ into $\mathscr{L}(M, M)$ is a $\mathfrak{g}$-module isomorphism (Proposition 4). As $1 \in \mathscr{L}(M, M)$ is obviously an invariant (cf. *Example* 1), the corresponding element u of $M^* \otimes M$ is an invariant. If $(e_i)_{1 \leqslant i \leqslant n}$ is a basis of M and $(e^i_*)_{1 \leqslant i \leqslant n}$ is the dual basis, we have $u = \sum_{i=1}^{n} e_i^* \otimes e_i$.

Example 3. Let M be a $\mathfrak{g}$-module. Let β be a bilinear form on M and f the corresponding element of $\mathscr{L}(M, M^*)$. For β to be invariant, it is necessary and sufficient that f be a $\mathfrak{g}$-module homomorphism (Proposition 4 and *Example* 1). Suppose that K is a field and that $\dim_K M < +\infty$. A *non-degenerate* invariant bilinear form β on M defines an *isomorphism* of the $\mathfrak{g}$-module M onto the $\mathfrak{g}$-module M^* and hence an isomorphism of the $\mathfrak{g}$-module $M \otimes M$ onto the $\mathfrak{g}$-module $M^* \otimes M$. Thus, by *Example* 2, giving β defines canonically an invariant element c in the $\mathfrak{g}$-module $M \otimes M$, which can be constructed as follows: let $(e_i)_{1 \leqslant i \leqslant n}$ be a basis of M and $(e'_i)_{1 \leqslant i \leqslant n}$ the basis of M such that $\beta(e_i, e'_j) = \delta_{ij}$; then $c = \sum_{i=1}^{n} e_i \otimes e'_i$.

PROPOSITION 5. *Let $\mathfrak{g}$ be a Lie K-algebra, $\mathfrak{h}$ an ideal of $\mathfrak{g}$, ρ a representation of $\mathfrak{g}$ on M and ρ' the restriction of ρ to $\mathfrak{h}$. Then the set N of elements of M invariant with respect to ρ' is stable under $\rho(\mathfrak{g})$.*

Let $n \in N$ and $y \in \mathfrak{g}$; for all $x \in \mathfrak{h}$, $[x, y] \in \mathfrak{h}$ and hence

$$\rho(x)\rho(y)n = \rho([x, y])n + \rho(y)\rho(x)n = 0;$$

hence $\rho(y)n \in N$.

PROPOSITION 6. *Let M be a semi-simple $\mathfrak{g}$-module. Then the submodule M_0 of invariant elements of M admits one and only one supplement stable under the x_M, namely the submodule M_1 generated by the $x_M . m$ ($x \in \mathfrak{g}$, $m \in M$).*

Let M' be a submodule of M which is stable under the x_M and a supplement of M_0 in M. For all $m \in M$, $m = m_0 + m'$ with $m_0 \in M_0$, $m' \in M'$, and hence $x_M m = x_M m' \in M'$. Hence $M_1 \subset M'$. Let M_2 be a submodule of M' stable under the x_M and supplementary to M_1 in M'. For all $m \in M_2$,

$$x_M m \in M_2 \cap M_1 = \{0\}$$

for all $x \in \mathfrak{g}$, hence $m \in M_0$ and hence $m = 0$. Hence $M_2 = \{0\}$, which proves that $M_1 = M'$.

6. INVARIANT BILINEAR FORMS

Let $\mathfrak{g}$ be a Lie algebra over K. The adjoint representation of $\mathfrak{g}$ on $\mathfrak{g}$ and the zero representation of $\mathfrak{g}$ on K define a $\mathfrak{g}$-module structure on the K-module

$N = \mathscr{L}(\mathfrak{g}, \mathfrak{g}; K)$ of bilinear forms on $\mathfrak{g}$. Briefly we say that a bilinear form β on $\mathfrak{g}$ is *invariant* if it is invariant under the representation $x \mapsto x_N$. By formula (10) the necessary and sufficient condition for this to be so is that:

$$\beta([x, y], z) = \beta(x, [y, z]) \tag{13}$$

for all x, y, z in $\mathfrak{g}$.

Now let $\mathfrak{d}$ be the Lie algebra of derivations of $\mathfrak{g}$. The identity representation of $\mathfrak{d}$ and the zero representation of $\mathfrak{d}$ on K define a representation $D \mapsto D_N$ of $\mathfrak{d}$ on N. Briefly we say that a bilinear form on $\mathfrak{g}$ is *completely invariant* if it is invariant under the representation $D \mapsto D_N$. A completely invariant bilinear form is invariant. For a bilinear form β on $\mathfrak{g}$ to be completely invariant, it is necessary and sufficient that:

$$\beta(Dx, y) + \beta(x, Dy) = 0 \tag{14}$$

for all x, y in $\mathfrak{g}$ and $D \in \mathfrak{d}$.

PROPOSITION 7. *Let $\mathfrak{g}$ be a Lie algebra, β an invariant symmetric bilinear form on $\mathfrak{g}$ and $\mathfrak{a}$ an ideal of $\mathfrak{g}$.*

(a) *The orthogonal $\mathfrak{a}'$ of $\mathfrak{a}$ with respect to β is an ideal of $\mathfrak{g}$.*

(b) *If $\mathfrak{a}$ is characteristic and β is completely invariant, $\mathfrak{a}'$ is characteristic.*

(c) *If β is non-degenerate, $\mathfrak{a} \cap \mathfrak{a}'$ is commutative.*

Let D be a derivation of $\mathfrak{g}$. Suppose that $\mathfrak{a}$ is stable under D and that $\beta(Dx, y) + \beta(x, Dy) = 0$ for x, y in $\mathfrak{g}$. Then $z \in \mathfrak{a}'$ implies $Dz \in \mathfrak{a}'$, since, for all $t \in \mathfrak{a}$, $Dt \in \mathfrak{a}$ and hence $\beta(Dz, t) = -\beta(z, Dt) = 0$. Thus $\mathfrak{a}'$ is stable under D. This establishes (a) and (b).

Now let $\mathfrak{b}$ be an ideal of $\mathfrak{g}$ and suppose that the restriction of β to $\mathfrak{b}$ is zero. For x, y in $\mathfrak{b}$ and $z \in \mathfrak{g}$, $\beta([x, y], z) = \beta(x, [y, z]) = 0$, for $[y, z] \in \mathfrak{b}$. Thus $[\mathfrak{b}, \mathfrak{b}]$ is orthogonal to $\mathfrak{g}$. If β is non-degenerate, $\mathfrak{b}$ is therefore commutative. This result applied to $\mathfrak{a} \cap \mathfrak{a}'$ proves (c).

DEFINITION 4. *Let $\mathfrak{g}$ be a Lie* K*-algebra and* M *a $\mathfrak{g}$-module. Suppose that* M, *considered as a* K*-module, admits a finite basis. The bilinear form associated with the $\mathfrak{g}$-module* M *(or with the corresponding representation) is the symmetric bilinear form $(x, y) \mapsto \mathrm{Tr}(x_M y_M)$ on $\mathfrak{g}$. If the representation in question is the adjoint representation, the associated bilinear form is called the Killing form of $\mathfrak{g}$.*

PROPOSITION 8. *Let $\mathfrak{g}$ be a Lie algebra and* M *a $\mathfrak{g}$-module. Supose that* M, *considered as a* K*-module, admits a finite basis. The bilinear form associated with* M *is invariant.*

For x, y, z in $\mathfrak{g}$, we have:

$$\begin{aligned}\mathrm{Tr}([x, y]_M z_M) &= \mathrm{Tr}(x_M y_M z_M) - \mathrm{Tr}(y_M x_M z_M) = \mathrm{Tr}(x_M y_M z_M) - \mathrm{Tr}(x_M z_M y_M) \\ &= \mathrm{Tr}(x_M [y, z]_M).\end{aligned}$$

PROPORTION 9. *Suppose that* K *is a field and that the Lie algebra $\mathfrak{g}$ is finite-dimensional* *over* K. *Let* $\mathfrak{a}$ *be an ideal of* $\mathfrak{g}$, β *the Killing form of* $\mathfrak{g}$ *and* β' *the Killing form of* $\mathfrak{a}$. *Then* β' *is the restriction of* β *to* $\mathfrak{a}$.

Let u be an endomorphism of the vector space $\mathfrak{g}$ which leaves $\mathfrak{a}$ stable. Let v be the restriction of u to $\mathfrak{a}$ and w the endomorphism of the vector space $\mathfrak{g}/\mathfrak{a}$ derived from u when passing to the quotient. Then $\operatorname{Tr} u = \operatorname{Tr} v + \operatorname{Tr} w$ as is seen by taking a basis $(x_1, \ldots, x_n)$ of $\mathfrak{g}$ of which the first p elements form a basis of $\mathfrak{a}$. Then let $x \in \mathfrak{a}$, $y \in \mathfrak{a}$ and apply the above formula to the case where $u = (\operatorname{ad}_{\mathfrak{g}} x)(\operatorname{ad}_{\mathfrak{g}} y)$. Then $v = (\operatorname{ad}_{\mathfrak{a}} x)(\operatorname{ad}_{\mathfrak{a}} y)$ and $w = 0$. Hence $\beta(x,y) = \beta'(x,y)$.

PROPOSITION 10. *Suppose that* K *is a field and that the Lie algebra* $\mathfrak{g}$ *is finite-dimensional over* K. *The Killing form* β *of* $\mathfrak{g}$ *is completely invariant.*

Let D be a derivation of $\mathfrak{g}$. There exists a Lie algebra $\mathfrak{g}'$ containing $\mathfrak{g}$ as an ideal of codimension 1 and an element x_0 of $\mathfrak{g}'$ such that $Dx = [x_0, x]$ for all $x \in \mathfrak{g}$ (§ 1, no. 8, Example 1). Let β' be the Killing form of $\mathfrak{g}'$. For x, y in $\mathfrak{g}$, $\beta'([x, x_0], y) = \beta'(x, [x_0, y])$, that is $\beta'(Dx, y) + \beta'(x, Dy) = 0$. Now the restriction of β' to $\mathfrak{g}$ is β (Proposition 9). Hence the proposition.

7. CASIMIR ELEMENT

PROPOSITION 11. *Let* $\mathfrak{g}$ *be a Lie algebra over a field* K, U *its enveloping algebra,* $\mathfrak{h}$ *a finite-dimensional ideal of* $\mathfrak{g}$ *and* β *an invariant bilinear form on* $\mathfrak{g}$, *whose restriction to* $\mathfrak{h}$ *is non-degenerate. Let* $(e_i)_{1 \leqslant i \leqslant n}$, $(e'_j)_{1 \leqslant j \leqslant n}$ *be two bases of* $\mathfrak{h}$ *such that* $\beta(e_i, e'_j) = \delta_{ij}$. *Then the element* $c = \sum_{i=1}^{n} e_i e'_i$ *of* U *belongs to the centre of* U *and is independent of the choice of basis* (e_i).

For $x \in \mathfrak{g}$ let $x_{\mathfrak{h}}$ be the restriction to $\mathfrak{h}$ of $\operatorname{ad}_{\mathfrak{g}} x$. Then $x \mapsto x_{\mathfrak{h}}$ is a representation of $\mathfrak{g}$ on the vector space $\mathfrak{h}$ and the restriction β' of β to $\mathfrak{h}$ is invariant under this representation. By no. 5, *Example* 3, the tensor $\sum_{i=1}^{n} e_i \otimes e'_i$ is independent of the choice of basis (e_i) and is an invariant element of the tensor algebra of $\mathfrak{h}$. It is also an element of the tensor algebra T of $\mathfrak{g}$, which is invariant for the representation derived from the adjoint representation of $\mathfrak{g}$. Its canonical image in U, that is c, is therefore independent of the choice of basis (e_i) and is an invariant for the representation of $\mathfrak{g}$ on U considered at the end of no. 2. This element is therefore permutable with every element of $\mathfrak{g}$ and therefore belongs to the centre of U.

When β is the bilinear form associated with a $\mathfrak{g}$-module M, the element c of Proposition 11 is called the *Casimir element* associated with M (or with the corresponding representation). This element exists if the restriction of β to $\mathfrak{h}$ is non-degenerate.

PROPOSITION 12. *Let* $\mathfrak{g}$ *be a Lie algebra over a field* K, $\mathfrak{h}$ *an ideal of* $\mathfrak{g}$ *of finite dimension n and* M *a* $\mathfrak{g}$*-module of finite dimension over* K. *Let c be the Casimir element (assumed to exist) associated with* M *and* $\mathfrak{h}$.

(a) $\mathrm{Tr}(c_{\mathrm{M}}) = n$.

(b) *If* M *is simple and n is not divisible by the characteristic of* K, c_{M} *is an automorphism of* M.

In the notation of Proposition 11,

$$\mathrm{Tr}(c_{\mathrm{M}}) = \sum_{i=1}^{n} \mathrm{Tr}((e_i)_{\mathrm{M}}(e'_i)_{\mathrm{M}}) = \sum_{i=1}^{n} \beta(e_i, e'_i) = n.$$

Hence, if n is not divisible by the characteristic of K, $c_{\mathrm{M}} \neq 0$. On the other hand, as c belongs to the centre of U, c_{M} is permutable with all the x_{M}, $x \in \mathfrak{g}$. If further M is simple, c_{M} is therefore invertible in $\mathscr{L}(\mathrm{M})$ (*Algebra*, Chapter VIII, § 4, no. 3, Proposition 2).

8. EXTENSION OF THE BASE RING

Let K_1 be a commutative ring with unit element and ϕ a homomorphism of K into K_1 mapping 1 to 1. Let $\mathfrak{g}$ be a Lie K-algebra, U its enveloping algebra and M a left $\mathfrak{g}$-module, that is a left U-module. Then $\mathrm{M}_{(\mathrm{K}_1)}$ has a canonical left $\mathrm{U}_{(\mathrm{K}_1)}$-module structure and hence a left $\mathfrak{g}_{(\mathrm{K}_1)}$-module structure. Let ρ and $\rho_{(\mathrm{K}_1)}$ be the representations of $\mathfrak{g}$ and $\mathfrak{g}_{(\mathrm{K}_1)}$ corresponding to M and $\mathrm{M}_{(\mathrm{K}_1)}$: $\rho_{(\mathrm{K}_1)}$ is said to be derived from ρ by *extending the base ring* and the results of *Algebra*, Chapter VIII, § 13, no. 4 can be applied. If $x \in \mathfrak{g}$, $\rho_{(\mathrm{K}_1)}(x)$ is just the endomorphism $\rho(x) \otimes 1$ of $\mathrm{M}_{(\mathrm{K}_1)} = \mathrm{M} \otimes_{\mathrm{K}} \mathrm{K}_1$.

Suppose that K is a field, that K_1 is an extension of K and that ϕ is the canonical injection of K into K_1. Let V and V′ be vector subspaces of M. Let $\mathfrak{a}$ be the vector subspace of $\mathfrak{g}$ consisting of the $x \in \mathfrak{g}$ such that $\rho(x)(\mathrm{V}) \subset \mathrm{V}'$. Let $\mathfrak{a}'$ be the vector subspace of $\mathfrak{g}_{(\mathrm{K}_1)}$ consisting of the $x' \in \mathfrak{g}_{(\mathrm{K}_1)}$ such that $\rho_{(\mathrm{K}_1)}(x')(\mathrm{V}_{(\mathrm{K}_1)}) \subset \mathrm{V}'_{(\mathrm{K}_1)}$. Then $\mathfrak{a}' = \mathfrak{a}_{(\mathrm{K}_1)}$. For clearly $\mathfrak{a}_{(\mathrm{K}_1)} \subset \mathfrak{a}'$. Now let $x' \in \mathfrak{a}'$. We may write $x' = \sum_{i=1}^{n} \lambda_i x_i$, where the x_i are in $\mathfrak{g}$ and the λ_i are elements of K_1 linearly independent over K. For all $u \in \mathrm{V}$, $\rho(x').u \in \mathrm{V}'_{(\mathrm{K}_1)}$, that is $\sum_{i=1}^{n} \lambda_i \rho(x_i).u \in \mathrm{V}'_{(\mathrm{K}_1)}$, whence $\rho(x_i).u \in \mathrm{V}'$, hence $x_i \in \mathfrak{a}$ and $x' \in \mathfrak{a}_{(\mathrm{K}_1)}$. This shows that $\mathfrak{a}' = \mathfrak{a}_{(\mathrm{K}_1)}$. In particular, the *centre* of $\mathfrak{g}_{(\mathrm{K}_1)}$ is derived from the centre of $\mathfrak{g}$ by extending K to K_1: it suffices to apply the above to the adjoint representation of $\mathfrak{g}$. It follows that $\mathscr{C}_p(\mathfrak{g}_{(\mathrm{K}_1)}) = (\mathscr{C}_p\mathfrak{g})_{(\mathrm{K}_1)}$ for all p. Similarly, let $\mathfrak{h}$ be a subalgebra of $\mathfrak{g}$ and $\mathfrak{n}$ the *normalizer* of $\mathfrak{h}$ in $\mathfrak{g}$. Then the normalizer of $\mathfrak{h}_{(\mathrm{K}_1)}$ in $\mathfrak{g}_{(\mathrm{K}_1)}$ is $\mathfrak{n}_{(\mathrm{K}_1)}$.

Let K, K_1, $\mathfrak{g}$, ρ, M be as in the last paragraph. Let $\mathfrak{b}$ be a vector subspace of $\mathfrak{g}$ and W a vector subspace of M. Let V be the vector subspace of M consisting of the $m \in \mathrm{M}$ such that $\rho(\mathfrak{b}).m \subset \mathrm{W}$. Let V′ be the vector subspace of $\mathrm{M}_{(\mathrm{K}_1)}$ consisting of the $m' \in \mathrm{M}_{(\mathrm{K}_1)}$ such that $\rho_{(\mathrm{K}_1)}(\mathfrak{b}_{(\mathrm{K}_1)}).m' \subset \mathrm{W}_{(\mathrm{K}_1)}$. As above it is seen that $V' = V_{(K_1)}$. In particular, the vector subspace of *invariants* of $M_{(K_1)}$ is derived from the vector subspace of invariants of M by extending the base field from K to K_1.

Let K, K_1 and ϕ be as at the beginning of this no. Let $\mathfrak{g}$ be a Lie K-algebra and M and N $\mathfrak{g}$-modules. If M and N are isomorphic $\mathfrak{g}$-modules, $M_{(K_1)}$ and $N_{(K_1)}$ are isomorphic $\mathfrak{g}_{(K_1)}$-modules. Conversely:

PROPOSITION 13. *Let* K *be a field,* K_1 *an extension of* K, $\mathfrak{g}$ *a Lie* K*-algebra and* M, N *two* $\mathfrak{g}$*-modules of finite dimension over* K. *If* $M_{(K_1)}$ *and* $N_{(K_1)}$ *are isomorphic* $\mathfrak{g}_{(K_1)}$*-modules,* M *and* N *are isomorphic* $\mathfrak{g}$*-modules.*

The proof is in two steps.

(1) Suppose first that K_1 is an extension of K of *finite degree n*. Let U be the enveloping algebra of $\mathfrak{g}$, so that the enveloping algebra of $\mathfrak{g}_{(K_1)}$ is $U_{(K_1)} = U \otimes_K K_1$ (§ 2, no. 9). As $M_{(K_1)}$ and $N_{(K_1)}$ are isomorphic as $U_{(K_1)}$-modules they are *a fortiori* isomorphic as U-modules; but as U-modules they are respectively isomorphic to M^n and N^n. Now M and N are U-modules of finite length; M (resp. N) is therefore the direct sum of a family $(P_i^{r_i})_{1 \leqslant i \leqslant p}$ (resp. $(Q_j^{s_j})_{1 \leqslant j \leqslant q}$) of submodules such that the P_i (resp. Q_j) are indecomposable and two P_i (resp. Q_j) of different indices are not isomorphic (*Algebra*, Chapter VIII, § 2, no. 2, Theorem 1). Then M^n (resp. N^n) is isomorphic to the direct sum of the $P_i^{nr_i}$ (resp. $Q_j^{ns_j}$); it follows (*loc. cit.*) that $p = q$ and that after permuting the Q_j if necessary $nr_i = ns_i$ and P_i is isomorphic to Q_i for $1 \leqslant i \leqslant p$, hence M is isomorphic to N.

(2) *General case.* Let P be the $\mathfrak{g}$-module $\mathscr{L}_K(M, N)$ and Q the subspace of invariants of P, that is the set of homomorphisms of the $\mathfrak{g}$-module M into the $\mathfrak{g}$-module N. In the $\mathfrak{g}_{(K_1)}$-module $\mathscr{L}_{K_1}(M_{(K_1)}, N_{(K_1)}) = (\mathscr{L}_K(M, N))_{(K_1)}$, the subspace of invariants is $Q_{(K_1)}$. The hypothesis that $M_{(K_1)}$ and $N_{(K_1)}$ are isomorphic implies that M and N have the same dimension over K and that there exists in $Q_{(K_1)}$ an element g which is an isomorphism of $M_{(K_1)}$ onto $N_{(K_1)}$. Let $(f_1, \ldots, f_d)$ be a basis of Q over K and choose bases of M and N over K. If $\lambda_k \in K_1$ for $1 \leqslant k \leqslant d$, the matrix of $f = \sum_{k=1}^{d} \lambda_k f_k$ with respect to these bases has determinant which is a polynomial $D(\lambda_1, \ldots, \lambda_d)$ with coefficients *in* K. When $f = g$, this determinant is non-zero and hence the coefficients of D are not all non-zero. Therefore, if Ω is the algebraic closure of K, there exists (since Ω is infinite) elements $\mu_k \in \Omega$ $(1 \leqslant k \leqslant d)$ such that $D(\mu_1, \ldots, \mu_d) \neq 0$ (*Algebra*, Chapter IV, § 2, no. 5, Proposition 8). If K_2 is the algebraic extension of K generated by the μ_k $(1 \leqslant k \leqslant d)$, it follows that $\sum_{k=1}^{d} \mu_k f_k$ is an isomorphism of $M_{(K_2)}$ onto $N_{(K_2)}$; but K_2 is of finite degree over K (*Algebra*, Chapter V, § 3, no. 2, Proposition 5) and hence M and N are isomorphic by the first part of the argument.

Again let K, K_1 and ϕ be as the beginning of this no. Let ρ be a representation of $\mathfrak{g}$ on a K-module M with a finite basis $(x_1, \ldots, x_n)$. Then the bilinear form on $\mathfrak{g}_{(K_1)}$ associated with $\rho_{(K_1)}$ is derived from the bilinear form associated with ρ by extending the base ring to K_1 (for, if $u \in \mathscr{L}_K(M)$, u has the same matrix with respect to $(x_1, \ldots, x_n)$ as $u \otimes 1$ with respect to $(x_1 \otimes 1, \ldots, x_n \otimes 1)$ and hence u and $u \otimes 1$ have the same trace). In particular, if the K-module $\mathfrak{g}$ has a finite basis, the *Killing form* of $\mathfrak{g}_{(K_1)}$ is derived from that of $\mathfrak{g}$ by extending the base ring to K_1.

§ 4. NILPOTENT LIE ALGEBRAS

Recall that henceforth K *denotes a commutative field. In the rest of the chapter the Lie algebras are assumed to be finite-dimensional over* K.

1. DEFINITION OF NILPOTENT LIE ALGEBRAS

DEFINITION 1. *A Lie algebra $\mathfrak{g}$ is called nilpotent if there exists a decreasing finite sequence of ideals $(\mathfrak{g}_i)_{1 \leqslant i \leqslant p}$ of $\mathfrak{g}$ with $\mathfrak{g}_0 = \mathfrak{g}$, $\mathfrak{g}_p = \{0\}$, such that $[\mathfrak{g}, \mathfrak{g}_i] \subset \mathfrak{g}_{i+1}$ for $0 \leqslant i < p$.*

A commutative Lie algebra is nilpotent.

PROPOSITION 1. *Let $\mathfrak{g}$ be a Lie algebra. The following conditions are equivalent:*

(a) *$\mathfrak{g}$ is nilpotent;*

(b) *$\mathscr{C}^k\mathfrak{g} = \{0\}$ for sufficiently large k;*

(c) *$\mathscr{C}_k\mathfrak{g} = \mathfrak{g}$ for sufficiently large k;*

(d) *there exists an integer k such that* $\operatorname{ad} x_1 \circ \operatorname{ad} x_2 \circ \cdots \circ \operatorname{ad} x_k = 0$ *for all elements $x_1, x_2, \ldots, x_k$ in $\mathfrak{g}$;*

(e) *there exists a decreasing sequence of ideals $(\mathfrak{g}_i)_{0 \leqslant i \leqslant n}$ of $\mathfrak{g}$ with $\mathfrak{g}_0 = \mathfrak{g}$, $\mathfrak{g}_n = \{0\}$, such that $[\mathfrak{g}, \mathfrak{g}_i] \subset \mathfrak{g}_{i+1}$ and $\dim \mathfrak{g}_i/\mathfrak{g}_{i+1} = 1$ for $0 \leqslant i < n$.*

If $\mathscr{C}^k\mathfrak{g} = \{0\}$ (resp. $\mathscr{C}_k\mathfrak{g} = \mathfrak{g}$), clearly the sequence $\mathscr{C}^1\mathfrak{g}, \ldots, \mathscr{C}^k\mathfrak{g}$ (resp. $\mathscr{C}_k\mathfrak{g}, \mathscr{C}_{k-1}\mathfrak{g}, \ldots, \mathscr{C}_0\mathfrak{g}$) has the properties of Definition 1 and hence $\mathfrak{g}$ is nilpotent. Conversely, suppose that there exists a sequence $(\mathfrak{g}_i)_{0 \leqslant i \leqslant p}$ with the properties of Definition 1. It is seen by induction on n that $\mathfrak{g}_i \supset \mathscr{C}^{i+1}\mathfrak{g}$ and $\mathfrak{g}_{p-i} \subset \mathscr{C}_i\mathfrak{g}$. Hence $\mathscr{C}^{p+1}\mathfrak{g} = \{0\}$ and $\mathscr{C}_p\mathfrak{g} = \mathfrak{g}$. We have thus proved that conditions (a), (b) and (c) are equivalent. On the other hand, $\mathscr{C}^i\mathfrak{g}$ is the set of linear combinations of elements of the form

$$[x_1, [x_2, \ldots, [x_{i-2}, [x_{i-1}, x_i]] \ldots]]$$

where $x_1, x_2, \ldots, x_i$ run through $\mathfrak{g}$. Hence conditions (b) and (d) are equivalent. Finally, if there exists a sequence $(\mathfrak{g}_i)_{0 \leqslant i \leqslant p}$ of ideals with the properties of

Definition 1, there exists a decreasing sequence $(\mathfrak{h}_i)_{0 \leqslant i \leqslant n}$ of vector subspaces of $\mathfrak{g}$ of dimensions $n, n-1, n-2, \ldots, 0$ and a sequence of indices

$$i_0 < i_1 < \cdots < i_p$$

with $\mathfrak{g}_0 = \mathfrak{h}_{i_0}, \mathfrak{g}_1 = \mathfrak{h}_{i_1} \ldots, \mathfrak{g}_p = \mathfrak{h}_{i_p}$; then as $[\mathfrak{g}, \mathfrak{h}_{i_k}] \subset \mathfrak{h}_{i_{k+1}}$ the $\mathfrak{h}_i$ are ideals and $[\mathfrak{g}, \mathfrak{h}_i] \subset \mathfrak{h}_{i+1}$ for all i. Hence conditions (a) and (e) are equivalent.

COROLLARY 1. *The centre of a non-zero nilpotent Lie algebra is non-zero.*

COROLLARY 2. *The Killing form of a nilpotent Lie algebra is zero.*

For all x and y in a nilpotent Lie algebra $\operatorname{ad} x \circ \operatorname{ad} y$ is nilpotent and hence of zero trace.

PROPOSITION 2. *Subalgebras, quotient algebras and* central *extensions of a nilpotent Lie algebra are nilpotent. A finite product of nilpotent Lie algebras is a nilpotent Lie algebra.*

Let $\mathfrak{g}$ be a Lie algebra, $\mathfrak{g}'$ a subalgebra of $\mathfrak{g}$, $\mathfrak{h}$ an ideal of $\mathfrak{g}$, $\mathfrak{k} = \mathfrak{g}/\mathfrak{h}$ and ϕ the canonical mapping of $\mathfrak{g}$ onto $\mathfrak{k}$. If $\mathfrak{g}$ is nilpotent, then $\mathscr{C}^k\mathfrak{g} = \{0\}$ for some integer k, hence $\mathscr{C}^k\mathfrak{g}' \subset \mathscr{C}^k\mathfrak{g} = \{0\}$ and $\mathscr{C}^k\mathfrak{k} = \phi(\mathscr{C}^k\mathfrak{g}) = \{0\}$ and hence $\mathfrak{g}'$ and $\mathfrak{k}$ are nilpotent. If $\mathfrak{k}$ is nilpotent and $\mathfrak{h}$ is contained in the centre of $\mathfrak{g}$, then $\mathscr{C}^k\mathfrak{k} = \{0\}$ for some integer k, hence $\mathscr{C}^k\mathfrak{g} \subset \mathfrak{h}$ and therefore $\mathscr{C}^{k+1}\mathfrak{g} \subset [\mathfrak{h}, \mathfrak{g}] = \{0\}$, so that $\mathfrak{g}$ is nilpotent. Finally, the assertion concerning products follows for example from the assertion (a) $\Leftrightarrow$ (d) of Proposition 1.

Definition 1 and Proposition 2 show that nilpotent Lie algebras are precisely the algebras obtained from commutative Lie algebras by a sequence of central extensions.

PROPOSITION 3. *Let $\mathfrak{g}$ be a nilpotent Lie algebra and $\mathfrak{h}$ a subalgebra of $\mathfrak{g}$ distinct from $\mathfrak{g}$. The normalizer of $\mathfrak{h}$ in $\mathfrak{g}$ is distinct from $\mathfrak{h}$.*

Let k be the greatest integer such that $\mathscr{C}^k\mathfrak{g} + \mathfrak{h} \neq \mathfrak{h}$. Then

$$[\mathscr{C}^k\mathfrak{g} + \mathfrak{h}, \mathfrak{h}] \subset \mathscr{C}^{k+1}\mathfrak{g} + \mathfrak{h} \subset \mathfrak{h}$$

and hence the normalizer of $\mathfrak{h}$ in $\mathfrak{g}$ contains $\mathscr{C}^k\mathfrak{g} + \mathfrak{h}$.

2. ENGEL'S THEOREM

Lemma 1. *Let* V *be a vector space over* K. *If x is a nilpotent endomorphism of* V, *the mapping $y \mapsto [x, y]$ of $\mathscr{L}(\mathrm{V})$ into $\mathscr{L}(\mathrm{V})$ is nilpotent.*

If f denotes this mapping, $f^m(y)$ is a sum of terms of the form $\pm x^i y x^j$ with $i + j = m$. If $x^k = 0$, then $f^{2k-1}(y) = 0$ for all y.

THEOREM 1 (Engel). *Let* V *be a vector space over* K *and $\mathfrak{g}$ a finite-dimensional subalgebra of $\mathfrak{gl}(\mathrm{V})$ whose elements are nilpotent endomorphisms of* V. *If* $\mathrm{V} \neq \{0\}$, *there exists $u \neq 0$ in* V *such that $x.u = 0$ for all $x \in \mathfrak{g}$.*

The proof proceeds by induction on the dimension n of $\mathfrak{g}$. The theorem is obvious if $n = 0$. Suppose that it is true for algebras of dimension $<n$.

Let $\mathfrak{h}$ be a Lie subalgebra of $\mathfrak{g}$ of dimension $m < n$. If $x \in \mathfrak{h}$, and $\mathrm{ad}_{\mathfrak{g}}\, x$ maps $\mathfrak{h}$ into itself and defines on passing to the quotient an endomorphism $\sigma(x)$ of the space $\mathfrak{g}/\mathfrak{h}$. By Lemma 1 $\mathrm{ad}_{\mathfrak{g}}\, x$ is nilpotent and hence $\sigma(x)$ is nilpotent. By the induction hypothesis there exists a non-zero element of $\mathfrak{g}/\mathfrak{h}$ which is annihilated by all the $\sigma(x)$, $x \in \mathfrak{h}$. It follows that $\mathfrak{h}$ is an ideal in a certain $(m + 1)$-dimensional subalgebra of $\mathfrak{g}$.

We conclude (by iteration starting with $\mathfrak{h} = \{0\}$) that $\mathfrak{g}$ has an ideal $\mathfrak{h}$ of dimension $n - 1$. Let $a \in \mathfrak{g}$, $a \notin \mathfrak{h}$. We again use the induction hypothesis: the $u \in V$ such that $x.u = 0$ for all $x \in \mathfrak{h}$ form a non-zero vector subspace U of V. This subspace is stable under a (§ 3, no. 5, Proposition 5). Since a is a nilpotent endomorphism of V, there exists a non-zero element of U which is annihilated by a and hence by every element of $\mathfrak{g}$.

Corollary 1. *For a Lie algebra $\mathfrak{g}$ to be nilpotent, it is necessary and sufficient that, for all $x \in \mathfrak{g}$, $\mathrm{ad}\, x$ be nilpotent.*

The condition is necessary (Proposition 1). Suppose that its sufficiency has been proved for Lie algebras of dimension $<n$ ($n \neq 0$). Let $\mathfrak{g}$ be an n-dimensional Lie algebra such that, for all $x \in \mathfrak{g}$, $\mathrm{ad}\, x$ is nilpotent. Theorem 1, applied to the set of $\mathrm{ad}\, x$ ($x \in \mathfrak{g}$), proves that the centre $\mathfrak{c}$ of $\mathfrak{g}$ is non-zero. Then $\mathfrak{g}$ is a central extension of the Lie algebra $\mathfrak{g}/\mathfrak{c}$, which is nilpotent by our induction hypothesis. The proof is completed by applying Proposition 2.

Corollary 2. *Let $\mathfrak{g}$ be a Lie algebra and $\mathfrak{h}$ an ideal of $\mathfrak{g}$. Suppose that $\mathfrak{g}/\mathfrak{h}$ is nilpotent and that, for all $x \in \mathfrak{g}$, the restriction of* $\mathrm{ad}\, x$ *to $\mathfrak{h}$ is nilpotent. Then $\mathfrak{g}$ is nilpotent.*

Let $x \in \mathfrak{g}$. As $\mathfrak{g}/\mathfrak{h}$ is nilpotent, there exists an integer k such that $(\mathrm{ad}\, x)^k(\mathfrak{g}) \subset \mathfrak{h}$. By hypothesis there exists an integer k' such that $(\mathrm{ad}\, x)^{k'}(\mathfrak{h}) = \{0\}$. Hence $(\mathrm{ad}\, x)^{k+k'}(\mathfrak{g}) = \{0\}$. Corollary 2 is hence a consequence of Corollary 1.

Corollary 3. *Let* V *be a vector space and $\mathfrak{g}$ a finite-dimensional subalgebra of* $\mathfrak{gl}(V)$ *whose elements are nilpotent endomorphisms of* V. *Then $\mathfrak{g}$ is a nilpotent Lie algebra.*

This follows immediately from Lemma 1 and Corollary 1.

Example. The algebra $\mathfrak{n}(n, K)$ (§ 1, no. 2, *Example* 2 (3)) is nilpotent.

3. THE LARGEST NILPOTENCY IDEAL OF A REPRESENTATION

Lemma 2. *Let $\mathfrak{g}$ be a Lie algebra, $\mathfrak{a}$ an ideal of $\mathfrak{g}$ and* M *a simple $\mathfrak{g}$-module. If, for all $x \in \mathfrak{a}$, x_M is nilpotent, then $x_M = 0$ for all $x \in \mathfrak{a}$.*

Let N be the subspace of M consisting of the $m \in M$ such that $x_M.m = 0$ for all $x \in \mathfrak{a}$. By Theorem 1, $N \neq \{0\}$. On the other hand, for all $y \in \mathfrak{g}$, N is stable under y_M (§ 3, no. 5, Proposition 5). Hence $N = M$, which proves the lemma.

Lemma 3. *Let $\mathfrak{g}$ be a Lie algebra, $\mathfrak{a}$ an ideal of $\mathfrak{g}$,* M *a $\mathfrak{g}$-module of finite dimension over*

K *and* $(M_i)_{0 \leqslant i \leqslant n}$ *a Jordan-Hölder series of the* $\mathfrak{g}$*-module* M. *The following conditions are equivalent:*

(a) *for all* $x \in \mathfrak{a}$, x_M *is nilpotent;*

(b) *for all* $x \in \mathfrak{a}$, x_M *is in the Jacobson radical of the associative algebra* A *generated by* 1 *and the* y_M *where* $y \in \mathfrak{g}$;

(c) *for all* $x \in \mathfrak{a}$,

$$x_M(M_0) \subset M_1, x_M(M_1) \subset M_2, \ldots, x_M(M_{n-1}) \subset M_n.$$

If these conditions are fulfilled, $\mathfrak{a}$ *is orthogonal to* $\mathfrak{g}$ *with respect to the bilinear form associated with the* $\mathfrak{g}$*-module* M.

(b) $\Rightarrow$ (a): as A is finite-dimensional over K, the Jacobson radical of A is a nilpotent ideal (*Algebra*, Chapter VIII, § 6, no. 4, Theorem 3) and hence every element of this radical is nilpotent.

(a) $\Rightarrow$ (c): each $Q_i = M_i/M_{i+1}$ $(0 \leqslant i < n)$ is a simple $\mathfrak{g}$-module. For all $x \in \mathfrak{a}$, the endomorphism x_{Q_i} (which is derived from x_M by restricting to M_i and passing to the quotient) is nilpotent if condition (a) holds and hence zero by Lemma 2; in other words, $x_M(M_i) \subset M_{i+1}$.

(c) $\Rightarrow$ (b): suppose condition (c) holds; let $x \in \mathfrak{a}$ and $z \in A$. Then $z(M_i) \subset M_i$ $(0 \leqslant i < n)$ and hence $(zx_M)^n(M) = \{0\}$; thus Ax_M is a left nilideal of A and hence is contained in the Jacobson radical of A (*Algebra*, Chapter VIII, § 6, no. 3, Corollary 3 to Theorem 1).

Finally, suppose conditions (a), (b) and (c) hold. Let $x \in \mathfrak{a}$ and $y \in \mathfrak{g}$. We have just seen that $y_M x_M$ is nilpotent and hence $\mathrm{Tr}(y_M x_M) = 0$, which proves the last assertion of the lemma.

PROPOSITION 4. *Let* $\mathfrak{g}$ *be a Lie algebra,* M *a* $\mathfrak{g}$*-module of finite dimension over* K *and* A *the associative algebra generated by* 1 *and the set of* x_M $(x \in \mathfrak{g})$.

(a) *The ideals* $\mathfrak{a}$ *of* $\mathfrak{g}$ *such that* x_M *is nilpotent for all* $x \in \mathfrak{a}$ *are all contained in one of them,* $\mathfrak{n}$.

(b) *The ideal* $\mathfrak{n}$ *is the set of* $x \in \mathfrak{g}$ *such that* x_M *belongs to the Jacobson radical of* A.

(c) *Let* $(M_i)_{0 \leqslant i \leqslant n}$ *be a Jordan-Hölder series of the* $\mathfrak{g}$*-module* M; *then* $\mathfrak{n}$ *is also the set of* $x \in \mathfrak{g}$ *such that* $(x)_{M_i/M_{i+1}} = 0$ *for all* i.

(d) $\mathfrak{n}$ *is orthogonal to* $\mathfrak{g}$ *with respect to the bilinear form associated with* ρ.

The set of $x \in \mathfrak{g}$ such that x_M belongs to the Jacobson radical of A is obviously an ideal of $\mathfrak{g}$. The proposition then follows immediately from Lemma 3.

DEFINITION 2. *The ideal* $\mathfrak{n}$ *of Proposition 4 is called the largest nilpotency ideal for the* $\mathfrak{g}$*-module* M *or the largest nilpotency ideal of the corresponding representation.*

Clearly $\mathfrak{n}$ contains the kernel of this representation. It equals it when M is semi-simple (Proposition 4 (c)), but not in general. It should be noted that an element of x of $\mathfrak{g}$ such that x_M is nilpotent does not necessarily belong to $\mathfrak{n}$.

We also note that a particular case of Lemma 3 immediately gives the following result:

PROPOSITION 5. *Let* V *be a vector space of finite dimension* n *over* K *and* $\mathfrak{g}$ *a Lie subalgebra of* $\mathfrak{gl}(V)$ *whose elements are nilpotent endomorphisms of* V. *Then there exists a decreasing sequence of vector subspaces* $V_0, V_1, \ldots, V_n$ *of* V, *of dimensions* $n, n-1, \ldots, 0$, *such that* $x(V_i) \subset V_{i+1}$ *for all* $x \in \mathfrak{g}$ *and* $i = 0, 1, \ldots, n-1$.

4. THE LARGEST NILPOTENT IDEAL OF A LIE ALGEBRA

Let $\mathfrak{g}$ be a Lie algebra and $\mathfrak{a}$ an ideal of $\mathfrak{g}$. For $\mathfrak{a}$ to be nilpotent, it is necessary and sufficient that, for all $x \in \mathfrak{a}$, $\operatorname{ad}_{\mathfrak{g}} x$ be nilpotent; the condition is obviously sufficient and is necessary, for, if $\mathfrak{a}$ is nilpotent and $x \in \mathfrak{a}$, $\operatorname{ad}_{\mathfrak{a}} x$ is nilpotent and $\operatorname{ad}_{\mathfrak{g}} x$ maps $\mathfrak{g}$ into $\mathfrak{a}$, hence $\operatorname{ad}_{\mathfrak{g}} x$ is nilpotent. Then Proposition 4 applied to the adjoint representation of $\mathfrak{g}$ gives the following result:

PROPOSITION 6. *Let* $\mathfrak{g}$ *be a Lie algebra and* E *the associative subalgebra of* $\mathscr{L}(\mathfrak{g})$ *generated by* 1 *and the* $\operatorname{ad}_{\mathfrak{g}} x$ $(x \in \mathfrak{g})$. *Let* R *be the Jacobson radical of* E.

(a) *The set* $\mathfrak{n}$ *of* $y \in \mathfrak{g}$ *such that* $\operatorname{ad}_{\mathfrak{g}} y \in R$ *is the largest nilpotent ideal of* $\mathfrak{g}$.

(b) *It is orthogonal to* $\mathfrak{g}$ *under the Killing form.*

It should be noted that $\mathfrak{g}/\mathfrak{n}$ can have non-zero nilpotent ideals.

5. EXTENSION OF THE BASE FIELD

Let $\mathfrak{g}$ be a Lie K-algebra, K_1 an extension of K and $\mathfrak{g}' = \mathfrak{g}_{(K_1)}$. As $\mathscr{C}^k\mathfrak{g}' = (\mathscr{C}^k\mathfrak{g})_{(K_1)}$, $\mathfrak{g}$ is nilpotent if and only if $\mathfrak{g}'$ is nilpotent.

Let M be a $\mathfrak{g}$-module of finite dimension over K, $\mathfrak{n}$ the largest nilpotency, ideal for M and $M' = M_{(K_1)}$. Let $(M_i)_{0 \leqslant i \leqslant n}$ be a Jordan-Hölder series of the $\mathfrak{g}$-module M. Then $x_M(M_i) \subset M_{i+1}$ for all i and all $x \in \mathfrak{n}$, hence

$$x'_{M'}((M_i)_{(K_1)}) \subset (M_{i+1})_{(K_1)}$$

for all i and all $x' \in \mathfrak{n}_{(K_1)}$; hence $x'_{M'}$ is nilpotent for $x' \in \mathfrak{n}_{(K_1)}$ so that $\mathfrak{n}_{(K_1)}$ is contained in the largest nilpotency ideal $\mathfrak{n}'$ for M'. We shall now see that, *if* K_1 *is separable over* K, *then* $\mathfrak{n}' = \mathfrak{n}_{(K_1)}$. Let E be the associative K-algebra generated by 1 and the x_M $(x \in \mathfrak{g})$, E' the associative K-algebra generated by 1 and the $x_{M'}$ $(x' \in \mathfrak{g}')$ and R and R' the Jacobson radicals of E and E'. The algebra E' is canonically identified with $E_{(K_1)}$. Then $R' = R_{(K_1)}$ (*Algebra*, Chapter VIII, § 7, no. 2, Corollary 2 (c) to Proposition 3). Then let $y' \in \mathfrak{n}'$ and write $y' = \sum_{i=1}^{n} \lambda_i y_i$, where the y_i are in $\mathfrak{g}$ and the $\lambda_i \in K_1$ are linearly independent over K. Then $y'_{M'} = \sum_{i=1}^{n} \lambda_i (y_i)_{M'}$ and $y'_{M'} \in R' = R_{(K_1)}$. Hence $(y_i)_M \in R$ and therefore $y_i \in \mathfrak{n}$ for all i. It follows that $y' \in \mathfrak{n}_{(K_1)}$, whence $\mathfrak{n}' \subset \mathfrak{n}_{(K_1)}$.

In particular, if K_1 is separable over K, the largest nilpotent ideal of $\mathfrak{g}_{(K_1)}$ is derived from that of $\mathfrak{g}$ by extending the base field from K to K_1.

§5. SOLVABLE LIE ALGEBRAS

Recall that K *henceforth denotes a field of characteristic* 0 *and that all Lie algebras are assumed to be finite-dimensional over* K.†

1. DEFINITION OF SOLVABLE LIE ALGEBRAS

DEFINITION 1. *A Lie algebra $\mathfrak{g}$ is called solvable if its kth derived algebra $\mathscr{D}^k\mathfrak{g}$ is zero for sufficiently large k.*

A nilpotent Lie algebra is solvable.

PROPOSITION 1. *Subalgebras and quotient algebras of a solvable Lie algebra are solvable. Every extension of a solvable algebra by a solvable algebra is solvable. Every finite product of solvable algebras is solvable.*

Let $\mathfrak{g}$ be a Lie algebra, $\mathfrak{g}'$ a subalgebra, $\mathfrak{h}$ an ideal of $\mathfrak{g}$, $\mathfrak{k} = \mathfrak{g}/\mathfrak{h}$ and ϕ the canonical mapping of $\mathfrak{g}$ onto $\mathfrak{k}$. If $\mathfrak{g}$ is solvable then $\mathscr{D}^k\mathfrak{g} = \{0\}$ for some integer k, hence $\mathscr{D}^k\mathfrak{g}' \subset \mathscr{D}^k\mathfrak{g} = \{0\}$, $\mathscr{D}^k\mathfrak{k} = \phi(\mathscr{D}^k\mathfrak{g}) = \{0\}$ and hence $\mathfrak{g}'$ and $\mathfrak{k}$ are solvable. If $\mathfrak{h}$ and $\mathfrak{k}$ are solvable there exist integers s, t such that

$$\mathscr{D}^s\mathfrak{h} = \mathscr{D}^t\mathfrak{k} = \{0\};$$

then $\mathscr{D}^t\mathfrak{g} \subset \mathfrak{h}$, hence $\mathscr{D}^{s+t}\mathfrak{g} = \mathscr{D}^s(\mathscr{D}^t\mathfrak{g}) \subset \mathscr{D}^s\mathfrak{h} = \{0\}$ and $\mathfrak{g}$ is solvable. The last assertion follows from the second by induction on the number of factors.

PROPOSITION 2. *Let $\mathfrak{g}$ be a Lie algebra. The following conditions are equivalent:*

(a) *$\mathfrak{g}$ is solvable;*

(b) *there exists a decreasing sequence $\mathfrak{g} = \mathfrak{g}_0 \supset \mathfrak{g}_1 \supset \cdots \supset \mathfrak{g}_n = \{0\}$ of ideals of $\mathfrak{g}$ such that the algebras $\mathfrak{g}_i/\mathfrak{g}_{i+1}$ are commutative ($i = 0, 1, \ldots, n-1$);*

(c) *there exists a decreasing sequence $\mathfrak{g} = \mathfrak{g}'_0 \supset \mathfrak{g}'_1 \supset \cdots \supset \mathfrak{g}'_p = \{0\}$ of subalgebras of $\mathfrak{g}$ such that $\mathfrak{g}'_{i+1}$ is an ideal of $\mathfrak{g}'_i$ and $\mathfrak{g}'_i/\mathfrak{g}'_{i+1}$ is commutative ($i = 0, 1, \ldots, p-1$);*

(d) *there exists a decreasing sequence $\mathfrak{g} = \mathfrak{g}''_0 \supset \mathfrak{g}''_1 \supset \cdots \supset \mathfrak{g}''_q = \{0\}$ of subalgebras of $\mathfrak{g}$ such that $\mathfrak{g}''_{i+1}$ is an ideal of $\mathfrak{g}''_i$ of codimension 1 ($i = 0, 1, \ldots, q-1$).*

(a) $\Rightarrow$ (b): it suffices to consider the sequence of derived ideals of $\mathfrak{g}$.

(b) $\Rightarrow$ (c): this is obvious.

(c) $\Rightarrow$ (d): suppose that condition (c) holds; every vector subspace of $\mathfrak{g}'_i$ containing $\mathfrak{g}'_{i+1}$ is an ideal of $\mathfrak{g}'_i$, whence immediately (d).

(d) $\Rightarrow$ (a): this follows immediately from the fact that an extension of a solvable algebra by a solvable algebra is solvable.

---

† The reader will note that the hypothesis on the characteristic of K is not used in nos. 1 and 2 of this paragraph.

Examples of solvable Lie algebras

I. Let $\mathfrak{g}$ be a 2-dimensional vector space over K and (e_1, e_2) a basis of $\mathfrak{g}$. There exists one and only one alternating bilinear multiplication $(x, y) \mapsto [x, y]$ on $\mathfrak{g}$ such that $[e_1, e_2] = e_2$. It is easily verified that $\mathfrak{g}$ is thus given a solvable Lie algebra structure. Now let $\mathfrak{h}$ be a non-commutative Lie algebra of dimension 2 over K. We show that $\mathfrak{h}$ is isomorphic to $\mathfrak{g}$. Let (f_1, f_2) be a basis of $\mathfrak{h}$. The element $[f_1, f_2]$ is non-zero (otherwise $\mathfrak{h}$ would be commutative) and hence it generates a 1-dimensional subspace $\mathfrak{k}$ of $\mathfrak{h}$. Then $[\mathfrak{h}, \mathfrak{h}] = \mathfrak{k}$. Let (e'_1, e'_2) be a basis of $\mathfrak{h}$ such that $e'_2 \in \mathfrak{k}$. Then $[e'_1, e'_2] = \lambda e'_2$ with $\lambda \neq 0$. Replacing e'_1 by $\lambda^{-1}e_1$, it can be assumed that $\lambda = 1$, whence our assertion.

II. Formulae (5) of § 1 prove that $\mathscr{D}\mathfrak{t}(n, \mathrm{K}) = \mathfrak{n}(n, \mathrm{K})$. As $\mathfrak{n}(n, \mathrm{K})$ is nilpotent and hence solvable, $\mathfrak{t}(n, \mathrm{K})$ is solvable. Therefore $\mathfrak{st}(n, \mathrm{K})$ is solvable. In particular, $\mathfrak{st}(2, \mathrm{K})$ is isomorphic to the algebra of Example I.

2. RADICAL OF A LIE ALGEBRA

Let $\mathfrak{a}$, $\mathfrak{b}$ be two solvable ideals of a Lie algebra $\mathfrak{g}$. The algebra $(\mathfrak{a} + \mathfrak{b})/\mathfrak{b}$ is isomorphic to $\mathfrak{a}/(\mathfrak{a} \cap \mathfrak{b})$ and hence is solvable and $\mathfrak{a} + \mathfrak{b}$, which is an extension of $(\mathfrak{a} + \mathfrak{b})/\mathfrak{b}$ by $\mathfrak{b}$, is also solvable (Proposition 1). It follows that a maximal solvable ideal of $\mathfrak{g}$ contains every solvable ideal of $\mathfrak{g}$ and hence $\mathfrak{g}$ has a largest solvable ideal. This enables us to make the following definition:

DEFINITION 2. *The radical of a Lie algebra is its largest solvable ideal.*

PROPOSITION 3. *The radical $\mathfrak{r}$ of a Lie algebra $\mathfrak{g}$ is the smallest ideal of $\mathfrak{g}$ such that $\mathfrak{g}/\mathfrak{r}$ has radical* $\{0\}$.

Let $\mathfrak{a}$ be an ideal of $\mathfrak{g}$ and ϕ the canonical mapping of $\mathfrak{g}$ onto $\mathfrak{g}/\mathfrak{a}$. If the radical of $\mathfrak{g}/\mathfrak{a}$ is zero, then $\phi(\mathfrak{r})$, which is a solvable ideal of $\mathfrak{g}/\mathfrak{a}$, is zero; hence $\mathfrak{r} \subset \mathfrak{a}$. On the other hand, the inverse image $\overset{-1}{\phi}(\mathfrak{r}')$ of the radical $\mathfrak{r}'$ of $\mathfrak{g}/\mathfrak{r}$ is an ideal of $\mathfrak{g}$ which is solvable by Proposition 1 and hence is equal to $\mathfrak{r}$; therefore $\mathfrak{r}' = \{0\}$.

PROPOSITION 4. *Let $\mathfrak{g}_1, \ldots, \mathfrak{g}_n$ be Lie algebras. The radical $\mathfrak{r}$ of the product of the $\mathfrak{g}_i$ is the product of the radicals $\mathfrak{r}_i$ of the $\mathfrak{g}_i$.*

The product $\mathfrak{r}'$ of the $\mathfrak{r}_i$ is a solvable ideal (Proposition 1) and hence $\mathfrak{r}' \subset \mathfrak{r}$. The canonical image of $\mathfrak{r}$ in $\mathfrak{g}_i$ is a solvable ideal of $\mathfrak{g}_i$ and hence is contained in $\mathfrak{r}_i$; hence $\mathfrak{r} \subset \mathfrak{r}'$.

3. NILPOTENT RADICAL OF A LIE ALGEBRA

DEFINITION 3. *Let $\mathfrak{g}$ be a Lie algebra. The nilpotent radical of $\mathfrak{g}$ is the intersection of the kernels of the finite-dimensional simple representations of $\mathfrak{g}$.*

Remarks. (1) Let $\mathfrak{s}$ be the nilpotent radical of $\mathfrak{g}$. As every decreasing sequence of vector subspaces of $\mathfrak{g}$ is stationary, there exists a finite number of finite-dimensional simple representations of $\mathfrak{g}$ whose kernels have intersection $\mathfrak{s}$.

The direct sum of these representations is semi-simple and has kernel $\mathfrak{s}$. It follows that the set of kernels of finite-dimensional semi-simple representations of $\mathfrak{g}$ has a least element, namely $\mathfrak{s}$.

(2) By Proposition 4 (c) of § 4, no. 3, $\mathfrak{s}$ is also the *intersection of the largest nilpotency ideals* of the finite-dimensional representations of $\mathfrak{g}$. In particular, $\mathfrak{s}$ is contained in the largest nilpotent ideal of $\mathfrak{g}$ and is therefore a *nilpotent ideal* of $\mathfrak{g}$.

(3) Every linear form λ on $\mathfrak{g}$ which is zero on $\mathscr{D}\mathfrak{g}$ is a simple representation (with space K) of $\mathfrak{g}$, whence $\lambda(\mathfrak{s}) = \{0\}$. It follows that $\mathfrak{s} \subset \mathscr{D}\mathfrak{g}$. On the other hand, $\mathfrak{s}$ is contained in the radical $\mathfrak{r}$ of $\mathfrak{g}$ by Remark 2. We shall prove that $\mathfrak{s} = \mathfrak{r} \cap \mathscr{D}\mathfrak{g}$.

Lemma 1. *Let* V *be a finite-dimensional vector space over* K, $\mathfrak{g}$ *a subalgebra of* $\mathfrak{gl}(V)$ *such that* V *is a simple* $\mathfrak{g}$*-module and* $\mathfrak{a}$ *a commutative ideal of* $\mathfrak{g}$. *Then* $\mathfrak{a} \cap \mathscr{D}\mathfrak{g} = \{0\}$.
Let S be the subalgebra of $\mathscr{L}(V)$ generated by 1 and $\mathfrak{a}$.

If $\mathfrak{b}$ is an ideal of $\mathfrak{g}$ contained in $\mathfrak{a}$ such that $\mathrm{Tr}\, bs = 0$ for all $b \in \mathfrak{b}$ and all $s \in S$, then in particular, by definition of S, $\mathrm{Tr}(b^n) = 0$ for every integer $n > 0$ and hence b is nilpotent (*Algebra*, Chapter VII, § 5, no. 5, Corollary 4 to Proposition 13); as the elements of $\mathfrak{b}$ are all nilpotent, $\mathfrak{b} = \{0\}$ (§ 4, no. 3, Lemma 2). We first apply this to the ideal $[\mathfrak{g}, \mathfrak{a}]$ of $\mathfrak{g}$. If $x \in \mathfrak{g}$, $a \in \mathfrak{a}$, $s \in S$, then $\mathrm{Tr}[x, a]s = \mathrm{Tr}(xas - axs) = \mathrm{Tr}\, x(as - sa) = 0$ since $as = sa$; hence $[\mathfrak{g}, \mathfrak{a}] = \{0\}$. Hence the elements of $\mathfrak{g}$ commute with those of $\mathfrak{a}$ and hence also with those of S. If x, y belong to $\mathfrak{g}$ and $s \in S$, then

$$\mathrm{Tr}[x, y]s = \mathrm{Tr}(xys - yxs) = \mathrm{Tr}\, x(ys - sy) = 0$$

since $ys = sy$; then taking $\mathfrak{b}$ to be the ideal $\mathscr{D}\mathfrak{g} \cap \mathfrak{a}$, it follows that $\mathscr{D}\mathfrak{g} \cap \mathfrak{a} = \{0\}$.

THEOREM 1. *Let* $\mathfrak{g}$ *be a Lie algebra,* $\mathfrak{r}$ *its radical and* $\mathfrak{s}$ *its nilpotent radical. Then* $\mathfrak{s} = \mathscr{D}\mathfrak{g} \cap \mathfrak{r}$.

We already know that $\mathfrak{s} \subset \mathscr{D}\mathfrak{g} \cap \mathfrak{r}$. Hence it will suffice to show that if ρ is a finite-dimensional simple representation of $\mathfrak{g}$ then $\rho(\mathscr{D}\mathfrak{g} \cap \mathfrak{r}) = \{0\}$. Let k be the least integer $\geqslant 0$ such that $\rho(\mathscr{D}^{k+1}\mathfrak{r}) = \{0\}$; we write $\mathfrak{g}' = \rho(\mathfrak{g})$, $\mathfrak{a}' = \rho(D^k\mathfrak{r})$; as $\mathscr{D}^k\mathfrak{r}$ is an ideal of $\mathfrak{g}$, $\mathfrak{a}'$ is an ideal of $\mathfrak{g}'$; this ideal is commutative since $\rho(\mathscr{D}^{k+1}\mathfrak{r}) = \{0\}$. If V is the space of ρ, $\mathfrak{g}' \subset \mathfrak{gl}(V)$ and V is a simple $\mathfrak{g}'$-module. Then $\rho(\mathscr{D}\mathfrak{g} \cap \mathscr{D}^k\mathfrak{r}) \subset \mathscr{D}\mathfrak{g}' \cap \mathfrak{a}' = \{0\}$. If $k > 0$, then $\mathscr{D}^k\mathfrak{r} \subset \mathscr{D}\mathfrak{g}$ and $\rho(\mathscr{D}^k\mathfrak{r}) = \{0\}$, contrary to the definition of k. Hence $k = 0$, that is

$$\rho(\mathscr{D}\mathfrak{g} \cap \mathfrak{r}) = \{0\}.$$

COROLLARY 1. *Let* $\mathfrak{g}$ *be a solvable Lie algebra. The nilpotent radical of* $\mathfrak{g}$ *is* $\mathscr{D}\mathfrak{g}$. *If* ρ *is a finite-dimensional simple representation of* $\mathfrak{g}$, $\rho(\mathfrak{g})$ *is commutative and the associative algebra* L *generated by* 1 *and* $\rho(\mathfrak{g})$ *is a field of finite degree over* K.

Here $\mathfrak{r} = \mathfrak{g}$, whence $\mathfrak{s} = \mathscr{D}\mathfrak{g}$. Hence $\rho(\mathscr{D}\mathfrak{g}) = \{0\}$, which shows that $\mathfrak{g}' = \rho(\mathfrak{g})$ is commutative. Every element $\neq 0$ of L is invertible by Schur's Lemma; hence L is a field.

Corollary 2 (Lie's Theorem). *Let* $\mathfrak{g}$ *be a solvable Lie algebra; suppose that* K *is algebraically closed. Let* M *be a* $\mathfrak{g}$*-module of finite dimension over* K *and let* $(M_i)_{0 \leqslant i \leqslant r}$ *be a Jordan-Hölder series of* M. *Then* M_{i-1}/M_i *is of dimension* 1 *over* K *for* $1 \leqslant i \leqslant r$ *and, for all* $x \in \mathfrak{g}$, $x_{M_{i-1}/M_i} = \lambda_i(x).1$, *where* λ_i *is a linear form on* $\mathfrak{g}$ *which is zero on* $\mathscr{D}\mathfrak{g}$. *In particular, every simple* $\mathfrak{g}$*-module of finite dimension over* K *is in fact of dimension* 1.

Let ρ_i be the representation of $\mathfrak{g}$ on M_{i-1}/M_i. The associative algebra L_i generated by 1 and $\rho_i(\mathfrak{g})$ is a field, a finite extension of K and therefore equal to K; and M_{i-1}/M_i is a simple L_i-module, whence $\dim M_{i-1}/M_i = 1$. The rest of the corollary is obvious.

Remarks. (1) If $(M_i)_{0 \leqslant i \leqslant r}$ is replaced by another Jordan-Hölder series of M, the sequence $(\lambda_1, \ldots, \lambda_r)$ is replaced by a sequence of the form $(\lambda_{\pi(1)}, \ldots, \lambda_{\pi(r)})$, where π is a permutation of $\{1, \ldots, r\}$, as follows from the Jordan-Hölder Theorem.

(2) Let $(e_1, \ldots, e_r)$ be a basis of M such that $e_i \in M_{i-1}$, $e_i \notin M_i$ $(1 \leqslant i \leqslant r)$. If $x \in \mathfrak{g}$ the endomorphism of M corresponding to x is represented with respect to this basis by a triangular matrix whose diagonal coefficients are

$$\lambda_1(x), \ldots, \lambda_r(x).$$

Corollary 3. *Suppose that* K *is algebraically closed. If* $\mathfrak{g}$ *is an* r*-dimensional solvable Lie algebra, every ideal of* $\mathfrak{g}$ *is a term of a decreasing sequence of ideals of dimensions* $r, r-1, \ldots, 0$.

Every ideal is part of a Jordan-Hölder series of $\mathfrak{g}$, considered as the space of the adjoint representation (*Algebra*, Chapter I, § 6, no. 14, Corollary to Theorem 8); then it suffices to apply Corollary 2.

Corollary 4. *Suppose that* $K = \mathbf{R}$. *Let* $\mathfrak{g}$ *be a solvable Lie algebra. Every simple representation of* $\mathfrak{g}$ *is of dimension* $\leqslant 2$. *Every ideal of* $\mathfrak{g}$ *is a term of a decreasing sequence* $(\mathfrak{g}_i)_{0 \leqslant i \leqslant m}$ *of ideals such that* $\mathfrak{g}_0 = \mathfrak{g}$, $\mathfrak{g}_m = \{0\}$, $\dim \mathfrak{g}_{i-1}/\mathfrak{g}_i \leqslant 2$ $(1 \leqslant i \leqslant m)$.

This is proved in a similar way to that for Corollaries 2 and 3, using the fact that every algebraic extension of $\mathbf{R}$ is of degree $\leqslant 2$.

Corollary 5. *For a Lie algebra* $\mathfrak{g}$ *to be solvable, it is necessary and sufficient that* $\mathscr{D}\mathfrak{g}$ *be nilpotent.*

The condition is necessary by Corollary 1. It is sufficient since $\mathfrak{g}/\mathscr{D}\mathfrak{g}$ is commutative.

Corollary 6. *Let* ρ *be a finite-dimensional representation of a Lie algebra* $\mathfrak{g}$. *Let*

$\mathfrak{r}$ be the radical of $\mathfrak{g}$. Every element $x \in \mathfrak{r}$ such that $\rho(x)$ is nilpotent belongs to the largest nilpotency ideal $\mathfrak{n}$ of ρ.

Let V be the space of ρ; let $(V_i)_{0 \leqslant i \leqslant r}$ be a Jordan-Hölder series for the $\mathfrak{r}$-module structure on V and let ρ_i be the representation of $\mathfrak{r}$ with space V_i/V_{i-1} $(1 \leqslant i \leqslant r)$. If $\rho(x)$ is nilpotent, so is $\rho_i(x)$; as for all i the algebra generated by $\rho_i(x)$ is a field, $\rho_i(x) = 0$. Conversely, if $\rho_i(x) = 0$ for all i, $\rho(x) = 0$. This shows that the set $\mathfrak{a}$ of $x \in \mathfrak{r}$ such that $\rho(x)$ is nilpotent is an ideal of $\mathfrak{r}$. On the other hand, $[\mathfrak{g}, \mathfrak{a}] \subset \mathscr{D}\mathfrak{g} \cap \mathfrak{r} \subset \mathfrak{n} \cap \mathfrak{r} \subset \mathfrak{a}$ and hence $\mathfrak{a}$ is an ideal of $\mathfrak{g}$. This proves that $\mathfrak{a} \subset \mathfrak{n}$.

COROLLARY 7. *Let $\mathfrak{g}$ be a Lie algebra and $\mathfrak{r}$ its radical. The following four sets are identical:* (a) *the largest nilpotent ideal of $\mathfrak{g}$;* (b) *the largest nilpotent ideal of $\mathfrak{r}$;* (c) *the set of $x \in \mathfrak{r}$ such that $\mathrm{ad}_{\mathfrak{g}}\, x$ is nilpotent;* (d) *the set of $x \in \mathfrak{r}$ such that $\mathrm{ad}_{\mathfrak{r}}\, x$ is nilpotent.*

Let these sets be denoted by $\mathfrak{a}$, $\mathfrak{b}$, $\mathfrak{c}$, $\mathfrak{d}$. The inclusions $\mathfrak{a} \subset \mathfrak{b} \subset \mathfrak{d} \subset \mathfrak{c}$ are clear. $\mathfrak{c} \subset \mathfrak{a}$ by Corollary 6 applied to the adjoint representation of $\mathfrak{g}$.

4. A CRITERION FOR SOLVABILITY

Lemma 2. *Let x be an endomorphism of a finite-dimensional vector space* V *and s* (resp. n) *its semi-simple* (resp. *nilpotent*) *component* (cf. *Algebra*, Chapter VIII, § 9, no. 4, Definition 4). *Let* ad x, ad s, ad n *be the respective images of x, s, n in the adjoint representation of* $\mathfrak{gl}(V)$. *Then* ad s (resp. ad n) *is the semi-simple* (resp. *nilpotent*) *component of* ad x *and is equal to a polynomial in* ad x *with coefficients in* K *and no constant term.*

We know that $\mathrm{ad}\, x = \mathrm{ad}\, s + \mathrm{ad}\, n$, $[\mathrm{ad}\, s, \mathrm{ad}\, n] = 0$ and ad n is nilpotent (§ 4, Lemma 1). We show that ad s is semi-simple. It suffices to do this when K is algebraically closed (cf. *Algebra*, Chapter VIII, § 9, no. 2, Proposition 3). Then let $(e_i)_{1 \leqslant i \leqslant n}$ be a basis of V such that $s(e_i) = \lambda_i e_i$ $(\lambda_i \in K)$. Let (E_{ij}) be the canonical basis of $\mathbf{M}_n(K) = \mathfrak{gl}(V)$. By formulae (5) of § 1,

$$(\mathrm{ad}\, s) . E_{ij} = (\lambda_i - \lambda_j) E_{ij}$$

and hence ad s is semi-simple. The last assertion of the lemma follows from *Algebra*, Chapter VIII, § 9, no. 4, Proposition 8.

Lemma 3. *Let* M *be a finite-dimensional vector space,* A *and* B *two vector subspaces of* $\mathfrak{gl}(M)$ *such that* $B \subset A$ *and* T *the set of* $t \in \mathfrak{gl}(M)$ *such that* $[t, A] \subset B$. *If* $z \in T$ *is such that* $\mathrm{Tr}(zu) = 0$ *for all* $u \in T$, *then z is nilpotent.*

It suffices to prove this when K is algebraically closed, which we shall assume henceforth. Let s and n be the semi-simple and nilpotent components of z and let (e_i) be a basis of M such that $s(e_i) = \lambda_i e_i$ $(\lambda_i \in K)$. Let $V \subset K$ be the vector space over $\mathbf{Q}$ generated by the λ_i. We need to prove that $V = \{0\}$. Let f be a $\mathbf{Q}$-linear form on V and let t be the endomorphism of M such that

$te_i = f(\lambda_i)e_i$. If (E_{ij}) is the canonical basis of $\mathfrak{gl}(M)$ defined by $E_{ij}e_k = \delta_{jk}e_i$, then

$$(\text{ad } s)E_{ij} = (\lambda_i - \lambda_j)E_{ij}$$
$$(\text{ad } t)E_{ij} = (f(\lambda_i) - f(\lambda_j))E_{ij}.$$

There exists a polynomial P with no constant term and with coefficients in K such that $P(\lambda_i - \lambda_j) = f(\lambda_i) - f(\lambda_j)$ for all i and j (for if $\lambda_i - \lambda_j = \lambda_h - \lambda_k$, then $f(\lambda_i) - f(\lambda_j) = f(\lambda_h) - f(\lambda_k)$ and, if $\lambda_i - \lambda_j = 0, f(\lambda_i) - f(\lambda_j) = 0$). Then ad $t = P$ (ad s). On the other hand, ad s is a polynomial with no constant term in ad z. Now $(\text{ad } z)(A) \subset B$, whence also $(\text{ad } t)(A) \subset B$. By the hypothesis $0 = \text{Tr}(zt) = \sum \lambda_i f(\lambda_i)$, whence $0 = f(\text{Tr}(zt)) = \sum f(\lambda_i)^2$. Since the $f(\lambda_i)$ are rational numbers, $f = 0$, which completes the proof.

THEOREM 2 (Cartan's criterion). *Let* $\mathfrak{g}$ *be a Lie algebra,* M *a finite-dimensional vector space,* ρ *a representation of* $\mathfrak{g}$ *on* M *and* β *the bilinear form on* $\mathfrak{g}$ *associated with* ρ. *Then* $\rho(\mathfrak{g})$ *is solvable if and only if* $\mathscr{D}\mathfrak{g}$ *is orthogonal to* $\mathfrak{g}$ *with respect to* β.

It can obviously be reduced to the case where $\mathfrak{g}$ is a Lie subalgebra of $\mathfrak{gl}(M)$ and ρ is the identity mapping. If $\mathfrak{g}$ is solvable, $\mathscr{D}\mathfrak{g}$ is contained in the largest nilpotency ideal of the identity representation of $\mathfrak{g}$ (Theorem 1) and hence is orthogonal to $\mathfrak{g}$ with respect to β (§ 4, Proposition 4 (d)). Suppose that $\mathscr{D}\mathfrak{g}$ is orthogonal to $\mathfrak{g}$ with respect to β. We prove that $\mathfrak{g}$ is solvable. Let T be the set of $t \in \mathfrak{gl}(M)$ such that $[t, \mathfrak{g}] \subset \mathscr{D}\mathfrak{g}$. If $t \in T$ and x, y belong to $\mathfrak{g}$, then $[t, x] \in \mathscr{D}\mathfrak{g}$ and hence

$$\text{Tr}(t[x, y]) = \beta([t, x], y) = 0$$

whence by linearity $\text{Tr}(tu) = 0$ for all $u \in \mathscr{D}\mathfrak{g}$. Also, clearly $\mathscr{D}\mathfrak{g} \subset T$. Hence (Lemma 3) every element of $\mathscr{D}\mathfrak{g}$ is nilpotent. It follows that $\mathscr{D}\mathfrak{g}$ is nilpotent (§ 4, Corollary 3 to Theorem 1) and hence that $\mathfrak{g}$ is solvable (no. 3, Corollary 5 to Theorem 1).

5. FURTHER PROPERTIES OF THE RADICAL

PROPOSITION 5. *Let* $\mathfrak{g}$ *be a Lie algebra and* $\mathfrak{r}$ *its radical.*

(a) *If* ρ *is a finite-dimensional representation of* $\mathfrak{g}$ *and* β *is the associated bilinear form,* $\mathfrak{r}$ *and* $\mathscr{D}\mathfrak{g}$ *are orthogonal with respect to* β.

(b) $\mathfrak{r}$ *is the orthogonal of* $\mathscr{D}\mathfrak{g}$ *with respect to the Killing form.*

Let x, y be in $\mathfrak{g}$, $z \in \mathfrak{r}$. Then $[y, z] \in \mathscr{D}\mathfrak{g} \cap \mathfrak{r}$ and hence

$$\beta([x, y], z) = \beta(x, [y, z]) = 0$$

(Theorem 1). Hence (a).

Let $\mathfrak{r}'$ be the orthogonal of $\mathscr{D}\mathfrak{g}$ with respect to the Killing form. It is an ideal of $\mathfrak{g}$ (§ 3, no. 6, Proposition 7 (a)) which contains $\mathfrak{r}$ by the above. On the other hand, the image $\mathfrak{s}$ of $\mathfrak{r}'$ under the adjoint representation of $\mathfrak{g}$ is solvable (Theorem 2) and hence $\mathfrak{r}'$ is solvable being a central extension of $\mathfrak{s}$. Hence $\mathfrak{r}' \subset \mathfrak{r}$.

COROLLARY 1. *Let $\mathfrak{g}$ be a Lie algebra. Then $\mathfrak{g}$ is solvable if and only if $\mathscr{D}\mathfrak{g}$ is orthogonal to $\mathfrak{g}$ with respect to the Killing form.*

This is an immediate consequence of Proposition 5 (b).

COROLLARY 2. *The radical $\mathfrak{r}$ of a Lie algebra $\mathfrak{g}$ is a characteristic ideal.*

$\mathscr{D}\mathfrak{g}$ is a characteristic ideal and the Killing form is completely invariant (§ 3, no. 6, Proposition 10). Hence the orthogonal of $\mathscr{D}\mathfrak{g}$ with respect to the Killing form is a characteristic ideal (§ 3, no. 6, Proposition 7 (b)).

COROLLARY 3. *Let $\mathfrak{g}$ be a Lie algebra, $\mathfrak{r}$ its radical and $\mathfrak{a}$ an ideal of $\mathfrak{g}$. Then the radical of $\mathfrak{a}$ is equal to $\mathfrak{r} \cap \mathfrak{a}$.*

$\mathfrak{r} \cap \mathfrak{a}$ is a solvable ideal of $\mathfrak{a}$ and hence is contained in the radical $\mathfrak{r}'$ of $\mathfrak{a}$. Conversely, $\mathfrak{r}'$ is an ideal of $\mathfrak{g}$ (Corollary 2 and § 1, no. 4, Proposition 2) and hence $\mathfrak{r}' \subset \mathfrak{r}$.

Corollary 2 can be made more precise as follows:

PROPOSITION 6. *Let $\mathfrak{g}$ be a Lie algebra, $\mathfrak{r}$ its radical and $\mathfrak{n}$ its largest nilpotent ideal. Every derivation of $\mathfrak{g}$ maps $\mathfrak{r}$ into $\mathfrak{n}$.*

Let D be a derivation of $\mathfrak{g}$. Let $\mathfrak{g}' = \mathfrak{g} + Kx_0$ be a Lie algebra in which $\mathfrak{g}$ is an ideal of codimension 1 such that $Dx = [x_0, x]$ for all $x \in \mathfrak{g}$ (§ 1, no. 8, *Example* 1). By Corollary 3 to Proposition 5, $\mathfrak{r}$ is contained in the radical $\mathfrak{r}'$ of $\mathfrak{g}'$. Then $D(\mathfrak{r}) = [x_0, \mathfrak{r}] \subset [\mathfrak{g}', \mathfrak{g}'] \cap \mathfrak{r}' = \mathfrak{s}'$. For all $x \in \mathfrak{s}'$, $\mathrm{ad}_{\mathfrak{g}'} x$ is nilpotent (Theorem 1). Hence, for all $x \in \mathfrak{s}' \cap \mathfrak{g}$, $\mathrm{ad}_{\mathfrak{g}} x$ is nilpotent. Hence $D(\mathfrak{r})$ is contained in the nilpotent ideal $\mathfrak{s}' \cap \mathfrak{g}$ of $\mathfrak{g}$.

COROLLARY. *The largest nilpotent ideal of a Lie algebra is a characteristic ideal.*

Remark. To summarize some of the above results, note that, if $\mathfrak{r}$, $\mathfrak{n}$, $\mathfrak{s}$, $\mathfrak{k}$ denote respectively the radical of $\mathfrak{g}$, the largest nilpotent ideal of $\mathfrak{g}$, the nilpotent radical of $\mathfrak{g}$ and the orthogonal of $\mathfrak{g}$ with respect to the Killing form, then

$$\mathfrak{r} \supset \mathfrak{k} \supset \mathfrak{n} \supset \mathfrak{s}.$$

The inclusion $\mathfrak{r} \supset \mathfrak{k}$ follows from Proposition 5 (b). The inclusion $\mathfrak{k} \supset \mathfrak{n}$ follows from § 4, no. 4, Proposition 6 (b). The inclusion $\mathfrak{n} \supset \mathfrak{s}$ has been pointed out in *Remark* 2 of no. 3.

6. EXTENSION OF THE BASE FIELD

Let $\mathfrak{g}$ be a Lie K-algebra and K_1 an extension of K. Clearly $\mathfrak{g}_{(K_1)}$ is solvable if and only if $\mathfrak{g}$ is solvable, since $\mathscr{D}^n(\mathfrak{g}_{(K_1)}) = (\mathscr{D}^n\mathfrak{g})_{(K_1)}$.

Let $\mathfrak{r}$ be the radical of $\mathfrak{g}$. Then $\mathfrak{r}_{(K_1)}$ *is the radical of* $\mathfrak{g}_{(K_1)}$. For let β be the Killing form of $\mathfrak{g}$. As $\mathfrak{r}$ is the orthogonal of $\mathscr{D}\mathfrak{g}$ with respect to β (Proposition 5 (b)), $\mathfrak{r}_{(K_1)}$ is the orthogonal of $(\mathscr{D}\mathfrak{g})_{(K_1)} = \mathscr{D}(\mathfrak{g}_{(K_1)})$ with respect to the form derived from β by extension from K to K_1, that is the Killing form of $\mathfrak{g}_{(K_1)}$ (§ 3, no. 8). Our assertion then follows from a further application of Proposition 5 (b).

§ 6. SEMI-SIMPLE LIE ALGEBRAS

Recall that K *denotes a field of characteristic* 0 *and that all Lie algebras are assumed to be finite-dimensional over* K.

1. DEFINITION OF SEMI-SIMPLE LIE ALGEBRAS

Definition 1. *Let* $\mathfrak{g}$ *be a Lie algebra.* $\mathfrak{g}$ *is called semi-simple if and only if the only commutative ideal of* $\mathfrak{g}$ *is* $\{0\}$.

Remarks. (1) The algebra $\{0\}$ is semi-simple. An algebra of dimension 1 or 2 is not semi-simple (cf. § 5, no. 1, *Example* 1). There exist semi-simple algebras of dimension 3 (cf. no. 7).

(2) A semi-simple algebra has zero centre and hence its adjoint representation is faithful.

(3) If $\mathfrak{g}_1, \ldots, \mathfrak{g}_n$ are semi-simple, $\mathfrak{g} = \mathfrak{g}_1 \times \cdots \times \mathfrak{g}_n$ is semi-simple: for if $\mathfrak{a}$ is a commutative ideal of $\mathfrak{g}$, the projections of $\mathfrak{a}$ onto $\mathfrak{g}_1, \ldots, \mathfrak{g}_n$ reduce to $\{0\}$.

Theorem 1. *Let* $\mathfrak{g}$ *be a Lie algebra. The following conditions are equivalent:*

(a) $\mathfrak{g}$ *is semi-simple.*

(b) *The radical* $\mathfrak{r}$ *of* $\mathfrak{g}$ *is zero.*

(c) *The Killing form* β *of* $\mathfrak{g}$ *is non-degenerate.*

Moreover, a semi-simple Lie algebra is equal to its derived ideal.

(a) $\Rightarrow$ (b): for if $\mathfrak{r} \neq \{0\}$, the last non-zero derived algebra of $\mathfrak{r}$ is a commutative ideal of $\mathfrak{g}$.

(b) $\Rightarrow$ (c): this follows from Proposition 5 (b) of § 5, no. 5 (which at the same time proves the last assertion of the theorem).

(c) $\Rightarrow$ (a): this follows from Proposition 6 (b) of § 4, no. 4.

Corollary. *Let* $\mathfrak{g}$ *be a semi-simple Lie algebra and* ρ *a representation of* $\mathfrak{g}$ *on a finite-dimensional vector space* V. *Then* $\rho(\mathfrak{g}) \subset \mathfrak{sl}(V)$.

The linear form $x \mapsto \operatorname{Tr} \rho(x)$ $(x \in \mathfrak{g})$ is zero when x is of the form $[y, z]$ $(y \in \mathfrak{g}, z \in \mathfrak{g})$ and hence on $\mathscr{D}\mathfrak{g} = \mathfrak{g}$.

PROPOSITION 1. *Let $\mathfrak{g}$ be a semi-simple Lie algebra and ρ a finite-dimensional faithful representation of $\mathfrak{g}$. Then the bilinear form on $\mathfrak{g}$ associated with ρ is non-degenerate.*

The orthogonal of $\mathfrak{g}$ with respect to this form is a solvable ideal (§ 5, no. 4, Theorem 2) and hence is zero.

COROLLARY 1. *Let $\mathfrak{g}$ be a Lie algebra, β its Killing form and $\mathfrak{a}$ a semi-simple subalgebra of $\mathfrak{g}$. The orthogonal $\mathfrak{h}$ of $\mathfrak{a}$ with respect to β is a supplementary subspace of $\mathfrak{a}$ in $\mathfrak{g}$ and $[\mathfrak{a}, \mathfrak{h}] \subset \mathfrak{h}$. If $\mathfrak{a}$ is an ideal of $\mathfrak{g}$, so is $\mathfrak{h}$, which is then the centralizer of $\mathfrak{a}$ in $\mathfrak{g}$.*

Let β' be the restriction of β to $\mathfrak{a}$: it is the bilinear form associated with the representation $x \mapsto \text{ad}_{\mathfrak{g}}\, x$ of $\mathfrak{a}$ on the space $\mathfrak{g}$. This representation is faithful and hence β' is non-degenerate (Proposition 1). Hence $\mathfrak{h}$ is supplementary to $\mathfrak{a}$ in $\mathfrak{g}$. On the other hand, if x, y are in $\mathfrak{a}$ and $z \in \mathfrak{h}$, then

$$\beta(x, [y, z]) = \beta([x, y], z) = 0,$$

since $[x, y] \in \mathfrak{a}$, and hence $[y, z] \in \mathfrak{h}$, which proves that $[\mathfrak{a}, \mathfrak{h}] \subset \mathfrak{h}$. If $\mathfrak{a}$ is an ideal of $\mathfrak{g}$, we know that $\mathfrak{h}$ is an ideal of $\mathfrak{g}$ (§ 3, Proposition 7) and $\mathfrak{g}$ is identified with $\mathfrak{a} \times \mathfrak{h}$. As the centre of $\mathfrak{a}$ is zero, the centralizer of $\mathfrak{a}$ in $\mathfrak{g}$ is $\mathfrak{h}$.

COROLLARY 2. *Every extension of a semi-simple Lie algebra by a semi-simple Lie algebra is semi-simple and trivial.*

This follows immediately from Corollary 1.

COROLLARY 3. *If $\mathfrak{g}$ is semi-simple, every derivation of $\mathfrak{g}$ is inner.*

ad $\mathfrak{g}$ is isomorphic to $\mathfrak{g}$ and hence semi-simple and is an ideal of the Lie algebra $\mathfrak{d}$ of derivations of $\mathfrak{g}$ (§ 1, Proposition 1). If $D \in \mathfrak{d}$ commutes with the elements of ad $\mathfrak{g}$, then, for all $x \in \mathfrak{g}$, $\text{ad}\, D(x) = [D, \text{ad}\, x] = 0$, whence $D(x) = 0$; hence $D = 0$. Corollary 3 then follows from Corollary 1.

2. SEMI-SIMPLICITY OF REPRESENTATIONS

Lemma 1. *Let $\mathfrak{g}$ be a semi-simple Lie algebra. The adjoint representation of $\mathfrak{g}$ is semi-simple. Every ideal and every quotient algebra of $\mathfrak{g}$ is semi-simple.*

Let $\mathfrak{a}$ be an ideal of $\mathfrak{g}$. The orthogonal $\mathfrak{b}$ of $\mathfrak{a}$ in $\mathfrak{g}$ with respect to the Killing form is an ideal of $\mathfrak{g}$ and $\mathfrak{a} \cap \mathfrak{b}$ is a commutative ideal (§ 3, no. 6, Proposition 7) and hence zero. Hence $\mathfrak{b}$ is supplementary to $\mathfrak{a}$ in $\mathfrak{g}$. Moreover, as the Killing form of $\mathfrak{g}$ is non-degenerate, so are its restrictions to $\mathfrak{a}$ and $\mathfrak{b}$ (*Algebra*, Chapter IX, § 4, no. 1, Corollary to Proposition 1) and hence $\mathfrak{a}$ and $\mathfrak{b}$ are semi-simple (no. 1, Theorem 1 and § 3, no. 6, Proposition 9).

Lemma 2. *Let $\mathfrak{g}$ be a Lie algebra. Then the following two conditions are equivalent:*

(a) *All finite-dimensional linear representations of $\mathfrak{g}$ are semi-simple.*

(b) *Given a linear representation ρ of $\mathfrak{g}$ on a finite-dimensional vector space* V *and* *a vector subspace* W *of codimension* 1 *such that* $\rho(x)(V) \subset W$ *for all* $x \in \mathfrak{g}$, *there exists a supplementary line of* W *which is stable under* $\rho(\mathfrak{g})$ (*and hence annihilated by* $\rho(\mathfrak{g})$).

Clearly (a) implies (b). Suppose that (b) holds. Let σ be a finite-dimensional representation of $\mathfrak{g}$ on a vector space M and N a vector subspace which is stable under $\sigma(\mathfrak{g})$. Let μ be the representation of $\mathfrak{g}$ on $\mathscr{L}(M)$ canonically derived from σ (§ 3, no. 3): recall that $\mu(x) = \mathrm{ad}_{\mathscr{L}(M)}\rho(x)$. Let V (resp. W) be the subspace of $\mathscr{L}(M)$ consisting of the linear mappings of M into N whose restriction to N is a homothety (resp. zero); then W is of codimension 1 in V and $\mu(x)(V) \subset W$ for all $x \in \mathfrak{g}$. By condition (b), there exists $u \in V$ which is annihilated by $\mu(x)$ for all $x \in \mathfrak{g}$ and whose restriction to N is a non-zero homothety. By multiplying u by a suitable scalar, it can be assumed that u is a projector of M onto N. To say that $\mu(x).u = 0$ means that u is permutable with $\sigma(x)$. Hence the kernel of u is a supplement of N in M which is stable under $\sigma(x)$ for all $x \in \mathfrak{g}$. Hence σ is semi-simple.

Lemma 3. *Let* $\mathfrak{g}$ *be a semi-simple Lie algebra*, ρ *a linear representation of* $\mathfrak{g}$ *on a finite-dimensional vector space* V *and* W *a subspace of* V *of codimension* 1 *such that* $\rho(x)(V) \subset W$ *for all* $x \in \mathfrak{g}$. *Then there exists a supplementary line of* W *which is stable under* $\rho(\mathfrak{g})$.

For all $x \in \mathfrak{g}$ let $\sigma(x)$ be the restriction of $\rho(x)$ to W. Suppose first that σ is simple. If $\sigma = 0$, then $\rho(x)\rho(y) = 0$ for all x, y in $\mathfrak{g}$, hence $\rho(\mathfrak{g}) = \rho(\mathscr{D}\mathfrak{g}) = \{0\}$ and our assertion is obvious. If $\sigma \neq 0$, let $\mathfrak{n}$ be the kernel of σ and let $\mathfrak{m}$ be a supplementary ideal of $\mathfrak{n}$ in $\mathfrak{g}$ (Lemma 1); then $\mathfrak{m} \neq \{0\}$ and the restriction of σ to $\mathfrak{m}$ is faithful; the restriction to $\mathfrak{m}$ of the bilinear form associated with σ is non-degenerate (Proposition 1) and hence the Casimir element c associated with $\mathfrak{m}$ and σ can be formed. By Proposition 12 of § 3, no. 7, $\sigma(c)$ is an automorphism of W. On the other hand, $\rho(c)(V) \subset W$. Hence the kernel Z of $\rho(c)$ is a supplementary line of W; since c belongs to the centre of the enveloping algebra of $\mathfrak{g}$, $\rho(c)$ is permutable with $\rho(x)$ for all $x \in \mathfrak{g}$ and hence Z is stable under $\rho(\mathfrak{g})$.

In the general case we argue by induction on the dimension of V. Let T be a minimal non-zero stable subspace of W. Let ρ' be the quotient representation on $V' = V/T$. Then, for all $x \in \mathfrak{g}$, $\rho'(x)(V') \subset W'$, where $W' = W/T$ is of codimension 1 in V'. By the induction hypothesis there exists a line Z' which is supplementary to W' and stable under $\rho'(\mathfrak{g})$. Its inverse image Z in V is stable under $\rho(\mathfrak{g})$, contains T as subspace of codimension 1, $Z \cap W = T$, and hence $\rho(x)(Z) \subset T$ for all $x \in \mathfrak{g}$. By what was proved above, there exists a supplementary line of T in Z which is stable under $\rho(\mathfrak{g})$; this line is supplementary to W in V, which completes the proof.

THEOREM 2 (H. Weyl). *Every finite-dimensional linear representation of a semi-simple Lie algebra is completely reducible.*

This follows from Lemmas 2 and 3.

Definition 2. *A Lie algebra $\mathfrak{g}$ is called simple if the only ideals of $\mathfrak{g}$ are $\{0\}$ and $\mathfrak{g}$ and if further $\mathfrak{g}$ is non-commutative.*

A simple Lie algebra is semi-simple. The algebra $\{0\}$ is not simple.

Proposition 2. *For a Lie algebra $\mathfrak{g}$ to be semi-simple, it is necessary and sufficient that it be a product of simple algebras.*

The condition is sufficient (no. 1, *Remark* 3). Conversely, suppose that $\mathfrak{g}$ is semi-simple. Since the adjoint representation of $\mathfrak{g}$ is semi-simple, $\mathfrak{g}$ is the direct sum of minimal non-zero ideals $\mathfrak{a}_1, \ldots, \mathfrak{a}_m$. Then $\mathfrak{g}$ is identified with the product algebra of the $\mathfrak{a}_i$ (§ 1, no. 1). Every ideal of $\mathfrak{a}_i$ is then an ideal of $\mathfrak{g}$ and hence zero or equal to $\mathfrak{a}_i$. On the other hand $\mathfrak{a}_i$ is non-commutative. Hence the $\mathfrak{a}_i$ are simple Lie algebras.

Corollary 1. *A semi-simple Lie algebra is the product of its simple ideals $\mathfrak{g}_i$. Every ideal of $\mathfrak{g}$ is the product of certain of the $\mathfrak{g}_i$.*

$\mathfrak{g} = \mathfrak{a}_1 \times \cdots \times \mathfrak{a}_m$, where the $\mathfrak{a}_i$ are simple. As the centre of $\mathfrak{a}_i$ is zero, the centralizer of $\mathfrak{a}_i$ in $\mathfrak{g}$ is the product of the $\mathfrak{a}_j$ for $j \neq i$. Then let $\mathfrak{a}$ be an ideal of $\mathfrak{g}$. If it does not contain $\mathfrak{a}_i$, then $\mathfrak{a} \cap \mathfrak{a}_i = \{0\}$, hence $[\mathfrak{a}, \mathfrak{a}_i] = \{0\}$ and $\mathfrak{a}$ is contained in the product of the $\mathfrak{a}_j$ for $j \neq i$. It follows that $\mathfrak{a}$ is the product of certain of the $\mathfrak{a}_i$. Hence the simple ideals of $\mathfrak{g}$ are precisely the $\mathfrak{a}_i$.

The simple ideals of a semi-simple Lie algebra are called the *simple components* of $\mathfrak{g}$.

Corollary 2. *Let $\mathfrak{g}$, $\mathfrak{g}'$ be two Lie algebras, $\mathfrak{r}$ and $\mathfrak{r}'$ their radicals and f a homomorphism of $\mathfrak{g}$ onto $\mathfrak{g}'$. Then $\mathfrak{r}' = f(\mathfrak{r})$.*

As $f(\mathfrak{r})$ is solvable, $f(\mathfrak{r}) \subset \mathfrak{r}'$. On the other hand, $\mathfrak{g}/\mathfrak{r}$ is semi-simple (§ 5, no. 2, Proposition 3), hence $\mathfrak{g}'/f(\mathfrak{r})$, which is isomorphic to a quotient of $\mathfrak{g}/\mathfrak{r}$, is semi-simple (Lemma 1) and hence $f(\mathfrak{r}) \supset \mathfrak{r}'$ (§ 5, no. 2, Proposition 3).

Remarks. (1) Theorem 2 admits a converse: if every finite-dimensional representation of $\mathfrak{g}$ is semi-simple, $\mathfrak{g}$ is semi-simple. For since the adjoint representation is semi-simple, every ideal of $\mathfrak{g}$ admits a supplementary ideal and hence can be considered as a quotient of $\mathfrak{g}$. If $\mathfrak{g}$ is not semi-simple then $\mathfrak{g}$ admits a non-zero commutative quotient and therefore a quotient of dimension 1. Now the Lie algebra K of dimension 1 admits non-semi-simple representations, for example

$$\lambda \mapsto \begin{pmatrix} 0 & 0 \\ \lambda & 0 \end{pmatrix}.$$

(2) Let $\mathfrak{g}$ be a Lie algebra over K and σ a representation of $\mathfrak{g}$ on a vector

space M. On the other hand let f be a K-linear mapping of $\mathfrak{g}$ into M such that:

$$(1) \qquad f([x, y]) = \sigma(x) . f(y) - \sigma(y) . f(x)$$

for all x, y in $\mathfrak{g}$. By § 1, no. 8, *Example* 2, being given σ and f is equivalent to being given a homomorphism $x \mapsto (f(x), \sigma(x))$ of $\mathfrak{g}$ into $\mathfrak{af}(M)$. On the other hand we have seen (*loc. cit.*) that the element $(f(x), \sigma(x))$ of $\mathfrak{af}(M)$ is canonically identified with the element $\rho(x)$ of $\mathfrak{gl}(N)$ (where $N = M \times K$) which induces $\sigma(x)$ on M and maps the element (0, 1) of N to $f(x)$. And ρ is then a representation of $\mathfrak{g}$ on N such that $\rho(x)(N) \subset M$ for all $x \in \mathfrak{g}$.

Then, if $\mathfrak{g}$ is *semi-simple*, there exists (Lemma 3) a line Z which is supplementary to M in N and annihilated by $\rho(\mathfrak{g})$. In other words, *there exists an element* $m_0 \in M$ such that $(-m_0, 1) \in N$ is annihilated by $\rho(x)$, that is *such that*

$$(2) \qquad f(x) = \sigma(x) . m_0$$

for all $x \in \mathfrak{g}$.

*Suppose that $K = \mathbf{R}$. Let G be a connected Lie group with Lie algebra $\mathfrak{g}$. Consider an analytic homomorphism ϕ of G into the affine group A of M corresponding to a homomorphism $x \mapsto (f(x), \sigma(x))$ of $\mathfrak{g}$ into $\mathfrak{af}(M)$. The above results can be interpreted by saying that if $\mathfrak{g}$ is semi-simple $\phi(G)$ leaves a point of M fixed. For let H be the set of elements of $\mathbf{GL}(N)$ which leave stable all the linear varieties of N parallel to M. There exists (§ 1, no. 8, *Example* 2) a canonical isomorphism ψ of A onto H. Let Z be a supplementary line of M in N. To say that $\rho(\mathfrak{g})$ annihilates Z amounts to saying that $(\psi \circ \phi)(G)$ leaves the points of Z fixed and hence (taking account of the definition of ψ) that $\phi(G)$ leaves fixed the projection onto M of the point of intersection of Z and $M \times \{1\}$.*

3. SEMI-SIMPLE ELEMENTS AND NILPOTENT ELEMENTS IN SEMI-SIMPLE LIE ALGEBRAS

PROPOSITION 3. *Let* M *be a finite-dimensional vector space over* K *and* $\mathfrak{g}$ *a semi-simple subalgebra of* $\mathfrak{gl}(M)$. *Then* $\mathfrak{g}$ *contains the semi-simple and nilpotent components of its elements.*

If K_1 is an extension of K, the Killing form of $\mathfrak{g}_{(K_1)}$ is the extension to $\mathfrak{g}_{(K_1)}$ of that of $\mathfrak{g}$ (§ 3, no. 8) and hence is non-degenerate; therefore $\mathfrak{g}_{(K_1)}$ is semi-simple. It therefore suffices to prove Proposition 3 when the base field is algebraically closed, which we shall henceforth assume to be the case.

For every subspace N of M, let $\mathfrak{g}_N$ be the subalgebra of $\mathfrak{gl}(M)$ consisting of the elements which leave N stable and whose restriction to N has trace zero. As $\mathfrak{g} = \mathscr{D}\mathfrak{g}$, $\mathfrak{g} \subset \mathfrak{g}_N$ if N is stable under $\mathfrak{g}$. Then let $\mathfrak{g}^*$ be the intersection of the normalizer of $\mathfrak{g}$ in $\mathfrak{gl}(M)$ and the algebras $\mathfrak{g}_N$ where N runs through the set of subspaces of M which are stable under $\mathfrak{g}$. As the semi-simple (resp. nilpotent) component s (resp. n) of $x \in \mathfrak{gl}(\mathrm{M})$ is a polynomial in x with no constant term and ad s (resp. ad n) is the semi-simple (resp. nilpotent) part of ad x (§ 5, no. 4, Lemma 2), clearly $x \in \mathfrak{g}^*$ implies $s \in \mathfrak{g}^*$ and $n \in \mathfrak{g}^*$; it therefore suffices to show that $\mathfrak{g}^* = \mathfrak{g}$. Since $\mathfrak{g}$ is a semi-simple ideal of $\mathfrak{g}^*$, $\mathfrak{g}^* = \mathfrak{a} \times \mathfrak{g}$ (no. 1, Corollary 1 to Proposition 1). Let $a \in \mathfrak{a}$ and let N be a subspace which is minimal among the non-zero subspaces of M which are stable under $\mathfrak{g}$. The restriction of a to N is a scalar multiple of the identity by Burnside's Theorem, has trace zero by construction and hence is zero since K is of characteristic 0. As M is the direct sum of subspaces such as N, it follows that $a = 0$ and hence $\mathfrak{g}^* = \mathfrak{g}$.

COROLLARY. *An element x of $\mathfrak{g}$ is a semi-simple* (resp. *nilpotent*) *endomorphism of* M *if and only if* $\mathrm{ad}_{\mathfrak{g}}\, x$ *is a semi-simple* (resp. *nilpotent*) *endomorphism of* $\mathfrak{g}$.

Let s (resp. n) be the semi-simple (resp. nilpotent) component of $x \in \mathfrak{g}$. Then $s \in \mathfrak{g}$ and $n \in \mathfrak{g}$ (Proposition 3). Then $\mathrm{ad}_{\mathfrak{g}}\, s$ (resp. $\mathrm{ad}_{\mathfrak{g}}\, n$) is the semi-simple (resp. nilpotent) component of $\mathrm{ad}_{\mathfrak{g}}\, x$, by Lemma 2 of § 5, no. 4. If x is semi-simple (resp. nilpotent) so then is $\mathrm{ad}_{\mathfrak{g}}\, x$. If now $\mathrm{ad}_{\mathfrak{g}}\, x$ is semi-simple (resp. nilpotent), it is equal to $\mathrm{ad}_{\mathfrak{g}}\, s$ (resp. $\mathrm{ad}_{\mathfrak{g}}\, n$) and hence $x = s$ (resp. $x = n$) since the adjoint representation of $\mathfrak{g}$ is faithful.

DEFINITION 3. *Let $\mathfrak{g}$ be a semi-simple Lie algebra. An element x of $\mathfrak{g}$ is called semi-simple* (resp. *nilpotent*) *if, for every $\mathfrak{g}$-module* M *of finite dimension over* K, x_{M} *is a semi-simple* (resp. *nilpotent*) *endomorphism of* M.

PROPOSITION 4. *Let $\mathfrak{g}$, $\mathfrak{g}'$ be semi-simple Lie algebras and f a homomorphism of $\mathfrak{g}$ into $\mathfrak{g}'$. If $x \in \mathfrak{g}$ is semi-simple* (resp. *nilpotent*), *so is $f(x)$. If f is surjective, every semi-simple* (resp. *nilpotent*) *element of $\mathfrak{g}'$ is the image under f of a semi-simple* (resp. *nilpotent*) *element of $\mathfrak{g}$.*

If ρ is a representation of $\mathfrak{g}'$, $\rho \circ f$ is a representation of $\mathfrak{g}$, whence the first assertion. If f is surjective, there exists a homomorphism g of $\mathfrak{g}'$ into $\mathfrak{g}$ such that $f \circ g$ is the identity homomorphism of $\mathfrak{g}'$ (no. 1, Corollary 2 to Proposition 1) and the second assertion then follows from the first.

THEOREM 3. *Let $\mathfrak{g}$ be a semi-simple Lie algebra.*

(a) *Let $x \in \mathfrak{g}$. If there exists a faithful representation ρ of $\mathfrak{g}$ such that $\rho(x)$ is a semi-simple* (resp. *nilpotent*) *endomorphism, then x is semi-simple* (resp. *nilpotent*).

(b) *Every element of $\mathfrak{g}$ can be written uniquely as the sum of a semi-simple element and a nilpotent element which commute with one another.*

Suppose that the hypothesis of (a) holds. Let σ be a representation of $\mathfrak{g}$, $\mathfrak{b}$ the supplementary ideal of the kernel of σ and α the projection of $\mathfrak{g}$ onto $\mathfrak{b}$. Then $\mathrm{ad}_{\mathfrak{g}}\, x$ is semi-simple (resp. nilpotent) by the Corollary to Proposition 3 and hence $\mathrm{ad}_{\mathfrak{b}}\, \alpha(x)$ is semi-simple (resp. nilpotent). As $\sigma(x) = \sigma(\alpha(x))$, the first assertion follows from the Corollary to Proposition 3. The second result then follows from Proposition 3 applied to a faithful representation.

4. REDUCTIVE LIE ALGEBRAS

DEFINITION 4. *A Lie algebra is called reductive if its adjoint representation is semi-simple.*

PROPOSITION 5. *Let* $\mathfrak{g}$ *be a Lie algebra and* $\mathfrak{r}$ *its radical. The following conditions are equivalent:*

(a) $\mathfrak{g}$ *is reductive.*

(b) $\mathscr{D}\mathfrak{g}$ *is semi-simple.*

(c) $\mathfrak{g}$ *is the product of a semi-simple algebra and a commutative algebra.*

(d) $\mathfrak{g}$ *has a finite-dimensional representation such that the associated bilinear form is non-degenerate.*

(e) $\mathfrak{g}$ *has a faithful semi-simple finite-dimensional representation.*

(f) *The nilpotent radical of* $\mathfrak{g}$ *is zero.*

(g) $\mathfrak{r}$ *is the centre of* $\mathfrak{g}$.

(a) $\Rightarrow$ (b): if the adjoint representation of $\mathfrak{g}$ is semi-simple, $\mathfrak{g}$ is a direct sum of minimal non-zero ideals $\mathfrak{a}_i$ and hence $\mathfrak{g}$ is isomorphic to the product of the $\mathfrak{a}_i$; and $\mathfrak{a}_i$ has no ideals other than $\{0\}$ and $\mathfrak{a}_i$ and hence is simple or commutative of dimension 1. Therefore $\mathscr{D}\mathfrak{g}$ is equal to the product of those $\mathfrak{a}_i$ which are simple and hence is semi-simple.

(b) $\Rightarrow$ (c): if $\mathscr{D}\mathfrak{g}$ is semi-simple, $\mathfrak{g}$ is isomorphic to the product of $\mathscr{D}\mathfrak{g}$ by a Lie algebra $\mathfrak{h}$ (no. 1, Corollary 1 to Proposition 1); $\mathfrak{h}$ is isomorphic to $\mathfrak{g}/\mathscr{D}\mathfrak{g}$ and hence is commutative.

(c) $\Rightarrow$ (d): let $\mathfrak{g}_1$ and $\mathfrak{g}_2$ be two Lie algebras, ρ_i a finite-dimensional representation of $\mathfrak{g}_i$ and β_i the bilinear form on $\mathfrak{g}_i$ associated with ρ_i $(i = 1, 2)$; ρ_1 and ρ_2 can be considered as representations of $\mathfrak{g} = \mathfrak{g}_1 \times \mathfrak{g}_2$; let ρ be their direct sum. Clearly the bilinear form on $\mathfrak{g}$ associated with ρ is the direct sum of β_1 and β_2 and hence is non-degenerate if β_1 and β_2 are non-degenerate. Then to prove the implication (c) $\Rightarrow$ (d) it suffices to consider the 2 following cases: (1) $\mathfrak{g}$ is semi-simple; then the adjoint representation admits as associated form the Killing form, which is non-degenerate; (2) $\mathfrak{g} = K$; then the identity representation of $\mathfrak{g}$ on K has an associated bilinear form which is non-degenerate.

(d) $\Rightarrow$ (e): let ρ be a finite-dimensional representation of $\mathfrak{g}$ and β the associated bilinear form; by Proposition 4 of § 4, no. 3, there exists a finite-dimensional semi-simple representation σ of $\mathfrak{g}$ such that the kernel $\mathfrak{n}$ of σ is orthogonal to $\mathfrak{g}$ with respect to β. If β is non-degenerate, then $\mathfrak{n} = \{0\}$ and hence σ is faithful.

(e) $\Rightarrow$ (f): this is obvious.

(f) $\Rightarrow$ (g): if the nilpotent radical of $\mathfrak{g}$ is zero, $\mathscr{D}\mathfrak{g} \cap \mathfrak{r}$ is zero (§ 5, no. 3, Theorem 1); as $[\mathfrak{g}, \mathfrak{r}] \subset \mathscr{D}\mathfrak{g} \cap \mathfrak{r}$, $\mathfrak{r}$ is the centre of $\mathfrak{g}$.

(g) ⇒ (a): if $\mathfrak{r}$ is the centre of $\mathfrak{g}$, the adjoint representation of $\mathfrak{g}$ is identified with a representation of $\mathfrak{g}/\mathfrak{r}$, which is a semi-simple Lie algebra (§ 5, no. 2, Proposition 3); this representation is therefore semi-simple (Theorem 2).

Remark. If a Lie algebra $\mathfrak{g}$ can be decomposed as a product $\mathfrak{a} \times \mathfrak{b}$ of a commutative Lie algebra $\mathfrak{a}$ and a semi-simple Lie algebra $\mathfrak{b}$, this decomposition is unique. More precisely, the centre of $\mathfrak{g}$ is equal to the product of the centres of $\mathfrak{a}$ and $\mathfrak{b}$ and is hence equal to $\mathfrak{a}$. And $\mathscr{D}\mathfrak{g} = \mathscr{D}\mathfrak{a} \times \mathscr{D}\mathfrak{b} = \mathfrak{b}$.

COROLLARY. (a) *Every finite product of reductive algebras is a reductive algebra.*

(b) *If $\mathfrak{g}$ is a reductive Lie algebra of centre $\mathfrak{c}$, every ideal of $\mathfrak{g}$ is a direct factor, the product of its intersections with $\mathfrak{c}$ and $\mathscr{D}\mathfrak{g}$, and is a reductive Lie algebra.*

(c) *Every quotient of a reductive Lie algebra is a reductive Lie algebra.*

Assertion (a) follows for example from condition (c) of Proposition 5.

Suppose that $\mathfrak{g}$ is reductive. Let $\mathfrak{a}$ be an ideal of $\mathfrak{g}$. Since the adjoint representation of $\mathfrak{g}$ is semi-simple, $\mathfrak{a}$ has a supplementary ideal $\mathfrak{b}$ and $\mathfrak{g}$ is identified with $\mathfrak{a} \times \mathfrak{b}$. For all $x \in \mathfrak{g}$, let $\rho(x)$ be the restriction of $\mathrm{ad}_{\mathfrak{g}}\, x$ to $\mathfrak{a}$. Then ρ is a semi-simple representation of $\mathfrak{g}$ which is zero on $\mathfrak{b}$ and defines on passing to the quotient the adjoint representation on $\mathfrak{a}$. Hence $\mathfrak{a}$ is reductive. Similarly, $\mathfrak{g}/\mathfrak{a}$ and $\mathfrak{b}$, which are isomorphic, are reductive. Finally, let $\mathfrak{d}, \mathfrak{d}'$ be the centres of $\mathfrak{a}$ and $\mathfrak{b}$; then $\mathfrak{a} = \mathfrak{d} \times \mathscr{D}\mathfrak{a}$, $\mathfrak{b} = \mathfrak{d}' \times \mathscr{D}\mathfrak{b}$, $\mathfrak{d} \times \mathfrak{d}' = \mathfrak{c}$, $\mathscr{D}\mathfrak{a} \times \mathscr{D}\mathfrak{b} = \mathscr{D}\mathfrak{g}$; hence $\mathfrak{a} = (\mathfrak{a} \cap \mathfrak{c}) + (\mathfrak{a} \cap \mathscr{D}\mathfrak{g})$.

PROPOSITION 6. *Let $\mathfrak{g}$ be a Lie algebra, $\mathfrak{r}$ its radical and $\mathfrak{s}$ its nilpotent radical.*

(a) $\mathfrak{s} = [\mathfrak{g}, \mathfrak{r}] = \mathscr{D}\mathfrak{g} \cap \mathfrak{r}$.

(b) *$\mathfrak{s}$ is the intersection of the orthogonals of $\mathfrak{g}$ with respect to the bilinear forms associated with the finite-dimensional representations of $\mathfrak{g}$.*

Clearly $[\mathfrak{g}, \mathfrak{r}] \subset \mathscr{D}\mathfrak{g} \cap \mathfrak{r}$. Now $\mathscr{D}\mathfrak{g} \cap \mathfrak{r} = \mathfrak{s}$ by Theorem 1 of § 5, no. 3. Let $\mathfrak{g}' = \mathfrak{g}/[\mathfrak{g}, \mathfrak{r}]$ and f be the canonical homomorphism of $\mathfrak{g}$ onto $\mathfrak{g}'$; then $f(\mathfrak{r})$ is the radical $\mathfrak{r}'$ of $\mathfrak{g}'$ (Corollary 3 to Proposition 2, no. 2), hence $[\mathfrak{g}', \mathfrak{r}'] = \{0\}$ and $\mathfrak{r}'$ is the centre of $\mathfrak{g}'$; therefore (Proposition 5) $\mathfrak{g}'$ has a finite-dimensional faithful semi-simple representation, whence $\mathfrak{s} \subset [\mathfrak{g}, \mathfrak{r}]$. This proves (a).

Let $\mathfrak{t}$ be the intersection of the orthogonals of $\mathfrak{g}$ with respect to the bilinear forms associated with the finite-dimensional representations of $\mathfrak{g}$. Then $\mathfrak{s} \subset \mathfrak{t}$ (§ 4, no. 3, Proposition 4 (d)). On the other hand, $\mathfrak{g}/\mathfrak{s}$ has a finite-dimensional faithful semi-simple representation and hence (Proposition 5) a finite-dimensional representation ρ such that the associated bilinear form is non-degenerate; considered as a representation of $\mathfrak{g}$, ρ has an associated bilinear form β on $\mathfrak{g}$ and the orthogonal of $\mathfrak{g}$ with respect to β is $\mathfrak{s}$, whence $\mathfrak{t} \subset \mathfrak{s}$. Hence $\mathfrak{t} = \mathfrak{s}$.

Even if $\mathfrak{s} \neq \{0\}$ there may exist non-degenerate symmetric bilinear forms on $\mathfrak{g}$ (Exercise 18 (c)). Such forms, of course, are not associated with any representation of $\mathfrak{g}$.

COROLLARY. *Let* $\mathfrak{g}$, $\mathfrak{g}'$ *be Lie algebras,* $\mathfrak{s}$ (resp. $\mathfrak{s}'$) *the nilpotent radical of* $\mathfrak{g}$ (resp. $\mathfrak{g}'$) *and f a homomorphism of* $\mathfrak{g}$ *onto* $\mathfrak{g}'$.

(a) *Then* $\mathfrak{s}' = f(\mathfrak{s})$.

(b) $\mathfrak{g}'$ *is reductive if and only if the kernel of f contains* $\mathfrak{s}$.

If $\mathfrak{r}, \mathfrak{r}'$ are the radicals of $\mathfrak{g}, \mathfrak{g}'$, then $\mathfrak{s}' = [\mathfrak{g}', \mathfrak{r}'] = [f(\mathfrak{g}), f(\mathfrak{r})] = f([\mathfrak{g}, \mathfrak{r}]) = f(\mathfrak{s})$. Assertion (b) is an immediate consequence of (a).

5. APPLICATION: A CRITERION FOR SEMI-SIMPLICITY OF REPRESENTATIONS

THEOREM 4. *Let* $\mathfrak{g}$ *be a Lie algebra,* $\mathfrak{r}$ *its radical,* ρ *a finite-dimensional representation of* $\mathfrak{g}$, $\mathfrak{g}' = \rho(\mathfrak{g})$ *and* $\mathfrak{r}' = \rho(\mathfrak{r})$. *Then the following conditions are equivalent:*

(a) ρ *is semi-simple;*

(b) $\mathfrak{g}'$ *is reductive and its centre consists of semi-simple endomorphisms;*

(c) $\mathfrak{r}'$ *consists of semi-simple endomorphisms;*

(d) *the restriction of* ρ *to* $\mathfrak{r}$ *is semi-simple.*

(a) ⇒ (b): if ρ is semi-simple, $\mathfrak{g}'$ is reductive (Proposition 5); the associative algebra generated by 1 and $\mathfrak{g}'$ is semi-simple (*Algebra*, Chapter VIII, § 5, no. 1, Proposition 3), hence its centre is semi-simple (*loc. cit.*, § 5, no. 4, Proposition 12) and hence the elements of this centre are semi-simple (*loc. cit.*, § 9, no. 1, Proposition 2).

(b) ⇒ (c): if $\mathfrak{g}'$ is reductive, its centre is equal to its radical, that is $\mathfrak{r}'$, whence the implication (b) ⇒ (c).

(c) ⇒ (d): suppose that $\mathfrak{r}'$ consists of semi-simple endomorphisms. As $[\mathfrak{g}', \mathfrak{r}']$ consists of nilpotent endomorphisms (no. 4, Proposition 6), $[\mathfrak{g}', \mathfrak{r}'] = \{0\}$. Then the implication (c) ⇒ (d) follows from *Algebra*, Chapter VIII, § 9, no. 2, Theorem 1.

(d) ⇒ (a): let $\mathfrak{s}$ be the nilpotent radical of $\mathfrak{g}$ and ρ' the restriction of ρ to $\mathfrak{r}$. The elements of $\rho(\mathfrak{s})$ are nilpotent and hence $\mathfrak{s}$ is contained in the largest nilpotency ideal of ρ'. As ρ' is semi-simple, $\rho'(\mathfrak{s}) = \{0\}$ and $\mathfrak{g}'$ is reductive (Corollary to Proposition 6), so that $\mathfrak{g}' = \mathfrak{a}' \times \mathfrak{r}'$ with $\mathfrak{a}'$ semi-simple (Proposition 5). Let A′ (resp. R′) be the associative algebra generated by 1 and $\mathfrak{a}'$ (resp. $\mathfrak{r}'$). It is semi-simple (*Algebra*, Chapter VIII, § 5, no. 1, Proposition 3), hence $A' \otimes_K R'$ is semi-simple (*loc. cit.*, § 7, no. 6, Corollary 4 to Theorem 3) and hence the associative algebra generated by 1 and $\mathfrak{g}'$, which is a quotient of $A' \otimes_K R'$, is semi-simple, which proves that ρ is semi-simple.

COROLLARY 1. *Let* $\mathfrak{g}$ *be a Lie algebra and* ρ *and* ρ' *two finite-dimensional semi-simple representations of* $\mathfrak{g}$. *Then the tensor product of* ρ *and* ρ' *is semi-simple.*

Let $\mathfrak{r}$ be the radical of $\mathfrak{g}$. For $x \in \mathfrak{r}$, $\rho(x)$ and $\rho'(x)$ are semi-simple (Theorem 4), hence $\rho(x) \otimes 1 + 1 \otimes \rho'(x)$ is semi-simple (*Algebra*, Chapter VIII, § 9, Corollary to Theorem 1) and hence the tensor product of ρ and ρ' is semi-simple (Theorem 4).

COROLLARY 2. *Let $\mathfrak{g}$ be a Lie algebra, ρ a semi-simple representation of $\mathfrak{g}$ on a finite-dimensional vector space* V, T *and* S *the tensor and symmetric algebras of* V *and* σ_T, σ_S *the representations of $\mathfrak{g}$ on* T *and* S *canonically derived from ρ. Then σ_T and σ_S are semi-simple and, more precisely, direct sums of finite-dimensional simple representations.*

Let T^n be the subspace of T consisting of the homogeneous tensors of order n. This subspace is stable under σ_T and the representation defined by σ_T on T^n is semi-simple (Corollary 1). Hence the corollary for σ_T and therefore for σ_S, which is a quotient representation of σ_T.

COROLLARY 3. *Let $\mathfrak{g}$ be a Lie algebra and ρ and ρ' two finite-dimensional semi-simple representations of $\mathfrak{g}$ on spaces* M *and* M'. *Then the representation of $\mathfrak{g}$ on $\mathscr{L}_K(M, M')$ canonically derived from ρ and ρ' is semi-simple.*

The $\mathfrak{g}$-module $\mathscr{L}_K(M, M')$ is canonically identified with the $\mathfrak{g}$-module $M^* \otimes_K M'$ (§ 3, no. 3, Proposition 4), so that Corollary 3 follows from Corollary 1.

COROLLARY 4. *Let $\mathfrak{g}$ be a Lie algebra, $\mathfrak{a}$ an ideal of $\mathfrak{g}$ and ρ a semi-simple representation of $\mathfrak{g}$.*

(a) *The restriction ρ' of ρ to $\mathfrak{a}$ is semi-simple.*

(b) *If ρ is simple, ρ' is a sum of simple representations isomorphic to one another.*

Passing to the quotient by the kernel of ρ, ρ can be assumed to be faithful. Then $\mathfrak{g}$ is reductive. Let $\mathfrak{g} = \mathfrak{g}_1 \times \mathfrak{g}_2$, where $\mathfrak{g}_1$ is the centre of $\mathfrak{g}$ and $\mathfrak{g}_2$ is semi-simple. Then $\mathfrak{a} = \mathfrak{a}_1 \times \mathfrak{a}_2$, where $\mathfrak{a}_1 \subset \mathfrak{g}_1$, $\mathfrak{a}_2 \subset \mathfrak{g}_2$ and $\mathfrak{a}_1$ is the centre of $\mathfrak{a}$. The elements of $\rho(\mathfrak{g}_1)$, and in particular those of $\rho(\mathfrak{a}_1)$, are semi-simple (Theorem 4) and hence ρ' is semi-simple (Theorem 4). Hence (a). Assertion (b) follows from (a), using § 3, no. 1, Corollary to Proposition 1.

6. SUBALGEBRAS REDUCTIVE IN A LIE ALGEBRA

DEFINITION 5. *Let $\mathfrak{g}$ be a Lie algebra and $\mathfrak{h}$ a Lie subalgebra of $\mathfrak{g}$. $\mathfrak{h}$ is called reductive in $\mathfrak{g}$ if the representation $x \mapsto \mathrm{ad}_{\mathfrak{g}}\, x$ of $\mathfrak{h}$ is semi-simple.*

This representation admits as subrepresentation the adjoint representation of $\mathfrak{h}$. Hence, if $\mathfrak{h}$ is reductive in $\mathfrak{g}$, $\mathfrak{h}$ is reductive. On the other hand, to say that a Lie algebra is reductive in itself is equivalent to saying that it is reductive.

PROPOSITION 7. *Let $\mathfrak{g}$ be a Lie algebra, $\mathfrak{h}$ a subalgebra reductive in $\mathfrak{g}$, ρ a representation of $\mathfrak{g}$ on a vector space* V *and* W *the sum of the finite-dimensional subspaces of* V *which are simple $\mathfrak{h}$-modules. Then* W *is stable under $\rho(\mathfrak{g})$.*

Let W_0 be a finite-dimensional simple sub-$\mathfrak{h}$-module of V. We need to prove that $\rho(x)(W_0) \subset W$ for all $x \in \mathfrak{g}$. Let M denote the vector space $\mathfrak{g}$ considered as an $\mathfrak{h}$-module by means of the representation $x \mapsto \mathrm{ad}_{\mathfrak{g}}\, x$ of $\mathfrak{h}$ on $\mathfrak{g}$. Then $M \otimes_K W_0$ is a semi-simple $\mathfrak{h}$-module (Corollary 1 to Theorem 4). Let θ be the K-linear mapping of $M \otimes_K W_0$ into V defined by $\theta(x \otimes w) = \rho(x)w$. This is an $\mathfrak{h}$-module homomorphism, for if $y \in \mathfrak{h}$ then:

$$\begin{aligned}\theta([y, x] \otimes w + x \otimes \rho(y)w) &= \rho([y, x])w + \rho(x)\rho(y)w \\ &= \rho(y)\rho(x)w = \rho(y)\theta(x \otimes w).\end{aligned}$$

Hence $\theta(M \otimes_K W_0)$ is a finite-dimensional semi-simple $\mathfrak{h}$-module. Hence $\theta(M \otimes_K W_0) \subset W$, that is $\rho(x)(W_0) \subset W$ for all $x \in \mathfrak{g}$.

COROLLARY 1. *Let $\mathfrak{g}$ be a Lie algebra, $\mathfrak{h}$ a subalgebra reductive in $\mathfrak{g}$ and ρ a finite-dimensional semi-simple representation of $\mathfrak{g}$. Then the restriction of ρ to $\mathfrak{h}$ is semi-simple.*

It suffices to consider the case where ρ is simple. We adopt the notation V, W of Proposition 4. Let W_1 be a subspace of V minimal among the non-zero subspaces stable under $\rho(\mathfrak{h})$. Then $W_1 \subset W$, hence $W \neq \{0\}$ and hence $W = V$.

COROLLARY 2. *Let $\mathfrak{g}$ be a Lie algebra, $\mathfrak{h}$ a subalgebra reductive in $\mathfrak{g}$ and $\mathfrak{k}$ a subalgebra of $\mathfrak{h}$ reductive in $\mathfrak{h}$. Then $\mathfrak{k}$ is reductive in $\mathfrak{g}$.*

The representation $x \mapsto \mathrm{ad}_{\mathfrak{g}}\, x$ of $\mathfrak{h}$ on $\mathfrak{g}$ is semi-simple and hence its restriction to $\mathfrak{k}$ is semi-simple (Corollary 1).

7. EXAMPLES OF SEMI-SIMPLE LIE ALGEBRAS

PROPOSITION 8. *Let V be a finite-dimensional vector space. Then $\mathfrak{gl}(V)$ is reductive. its centre is the set of homotheties of V, its derived algebra is $\mathfrak{sl}(V)$ and the latter is semi-simple.*

The identity representation of $\mathfrak{gl}(V)$ is simple, hence $\mathfrak{gl}(V)$ is reductive and therefore $\mathfrak{gl}(V)$ is the direct sum of its centre $\mathfrak{c}$ and its derived algebra $\mathscr{D}(\mathfrak{gl}(V))$. The centre $\mathfrak{c}$ is the set of homotheties (*Algebra*, Chapter II, § 2, no. 5, Corollary 1 to Proposition 5). Clearly $\mathscr{D}(\mathfrak{gl}(V)) \subset \mathfrak{sl}(V)$. As $\mathfrak{sl}(V) \cap \mathfrak{c} = \{0\}$, $\mathscr{D}(\mathfrak{gl}(V)) = \mathfrak{sl}(V)$. Hence $\mathfrak{sl}(V)$ is semi-simple.

Example. We identify $\mathfrak{sl}(K^2)$ with the Lie algebra of matrices of order 2 and trace zero. We write

$$X = \begin{pmatrix} 0 & 1 \\ 0 & 0 \end{pmatrix} \qquad Y = \begin{pmatrix} 0 & 0 \\ 1 & 0 \end{pmatrix} \qquad H = \begin{pmatrix} 1 & 0 \\ 0 & -1 \end{pmatrix}.$$

Then X, Y, H form a basis of $\mathfrak{sl}(K^2)$ and

$$[H, X] = 2X \qquad [H, Y] = -2Y \qquad [X, Y] = H.$$

As an algebra of dimension 1 or 2 is not semi-simple (no. 1, *Remark* 1), $\mathfrak{sl}(K^2)$ is simple. In fact, $\mathfrak{sl}(V)$ is simple for $\dim V \geqslant 2$, as we shall see later (cf. also Exercises 21 and 24).

PROPOSITION 9. *Let* V *be a vector space of finite dimension n over* K *and* β *a non-degenerate symmetric* (resp. *alternating*) *bilinear form on* V. *Let* $\mathfrak{g}$ *be the Lie algebra consisting of the* $x \in \mathfrak{gl}(V)$ *such that* $\beta(xm, m') + \beta(m, xm') = 0$ *for all m, m' in* V. *Then* $\mathfrak{g}$ *is reductive;* $\mathfrak{g}$ *is even semi-simple except in the case where* β *is symmetric and* $n = 2$.

For all $u \in \mathfrak{gl}(V)$ let u^* denote its adjoint relative to β; then $\mathrm{Tr}(u) = \mathrm{Tr}(u^*)$ by Proposition 7 of *Algebra*, Chapter IX, § 1, no. 8. The condition

$$\beta(um, m') + \beta(m, um') = 0$$

for all m, m' in V means that $u + u^* = 0$. In particular, if $v \in \mathfrak{gl}(V)$ then $(v - v^*)^* = v^* - v$ and hence $v - v^* \in \mathfrak{g}$. Then let u be an element of $\mathfrak{g}$ orthogonal to $\mathfrak{g}$ with respect to the bilinear form ϕ associated with the identity representation of $\mathfrak{g}$. For all $v \in \mathfrak{gl}(V)$, $\mathrm{Tr}\, u(v - v^*) = 0$, hence

$$\mathrm{Tr}(uv) = \mathrm{Tr}(uv^*) = \mathrm{Tr}(uv^*)^* = \mathrm{Tr}(vu^*) = -\mathrm{Tr}(vu) = -\mathrm{Tr}(uv)$$

and hence $\mathrm{Tr}(uv) = 0$. It follows that $u = 0$, so that ϕ is non-degenerate. Hence $\mathfrak{g}$ is reductive (Proposition 5). It remains to show that the centre of $\mathfrak{g}$ is zero (except when β is symmetric and $n = 2$). By extending the base field, we can assume that K is algebraically closed.

(a) When β is symmetric, it can be identified with the bilinear form on K^n with matrix I_n with respect to the canonical basis (*Algebra*, Chapter IX, § 6, Corollary 1 to Theorem 1). Under these conditions $\mathfrak{g}$ is identified with the Lie algebra of skew-symmetric matrices (§ 3, no. 4, *Example* 1). Let $U = (u_{ij}) \in \mathfrak{g}$; we use the fact that U commutes with the matrix $(v_{ij}) \in \mathfrak{g}$ all of whose elements are zero except $v_{i_0j_0}$ and $v_{j_0i_0}$ $(i_0 \neq j_0)$ which are equal respectively to 1 and -1. We find that $u_{i_0j} = u_{j_0j} = u_{ii_0} = u_{ij_0} = 0$ for $i \neq i_0, j_0$ and $j \neq i_0, j_0$. If $n > 2$, there exist, for all distinct indices i_0 and j, distinct indices i and j_0 such that $i \neq i_0, j_0 \neq j, j_0 \neq i_0$; hence $u_{i_0j} = 0$. This proves that an element of the centre of $\mathfrak{g}$ is zero.

(b) When β is alternating and $n = 2m$, β can be identified with the bilinear form on K^{2m} with matrix $\begin{pmatrix} 0 & I_m \\ -I_m & 0 \end{pmatrix}$ with respect to the canonical basis (*Algebra*, Chapter IX, § 5, Corollary 1 to Theorem 1). Under these conditions $\mathfrak{g}$ is identified with the Lie algebra of matrices of the form $U = \begin{pmatrix} A & B \\ C & D \end{pmatrix}$ where $D = -{}^tA$, B and C are symmetric (A, B, C, D in $\mathbf{M}_m(K)$) (§ 3, no. 4, *Example* 1). We use first the fact that U commutes with the matrix $\begin{pmatrix} X & 0 \\ 0 & -{}^tX \end{pmatrix}$, where $X \in \mathbf{M}_m(K)$. Then $AX = XA$, $CX = -{}^tXC$, $XB = -B.{}^tX$; as these equalities must hold for all X, it follows that A is a scalar matrix λI_m. We now use the fact that U commutes with the matrix $\begin{pmatrix} 0 & Y \\ 0 & 0 \end{pmatrix}$, where

Y is a symmetric matrix of $\mathbf{M}_m(\mathrm{K})$. Then $\lambda Y = YC = CY = 0$. This proves first that $\lambda = 0$. Moreover, for all $X \in \mathbf{M}_m(\mathrm{K})$, $X + {}^tX$ is symmetric and hence $XC = -{}^tXC$. Using the equation $CX = -{}^tXC$ obtained above, we see that C commutes with every element of $\mathbf{M}_m(\mathrm{K})$ and hence that C is a scalar matrix, necessarily zero since $YC = 0$. It is similarly shown that $B = 0$.

For β symmetric and $n = 2$, $\mathfrak{g}$ is of dimension 1 and hence commutative. For the other cases cf. Exercises 25 and 26.

8. THE LEVI–MALCEV THEOREM

Let E be a complete normed vector space over $\mathbf{R}$ and u a continuous endomorphism of E. We have seen (*Functions of a real variable*, Chapter IV, § 2, no. 6) that the sequence $\frac{u^n}{n!}$ is summable in $\mathscr{L}(\mathrm{E})$ and we wrote

$$e^u = \exp u = \sum_{n=0}^{\infty} \frac{u^n}{n!}.$$

Now let E be a vector space over the field K and u a *nilpotent* endomorphism of E. The series $\sum_{n=0}^{\infty} \frac{u^n}{n!}$ has only a finite number of non-zero terms and we can therefore write

$$e^u = \exp u = \sum_{n=0}^{\infty} \frac{u^n}{n!}.$$

This definition agrees with the above if $\mathrm{K} = \mathbf{R}$ and if E is complete and normed. If v is another nilpotent endomorphism of E which commutes with u, then:

$$e^u e^v = \left(\sum_{n=0}^{\infty} \frac{u^n}{n!}\right)\left(\sum_{p=0}^{\infty} \frac{v^p}{p!}\right) = \sum_{n,p=0}^{\infty} \frac{u^n v^p}{n!\,p!} \tag{3}$$

$$= \sum_{q=0}^{\infty} \frac{1}{q!}\left(\sum_{n+p=q} \binom{q}{n} u^n v^p\right) = \sum_{q=0}^{\infty} \frac{1}{q!}(u+v)^q = e^{u+v}.$$

In particular, $e^u e^{-u} = e^{-u} e^u = e^0 = 1$ and hence e^u is always an automorphism of E.

If further E is a (not necessarily associative) algebra and u is a (nilpotent) *derivation* of E, then e^u is an *automorphism of the algebra* E. For if $x, y \in \mathrm{E}$ then

$$u^p(xy) = \sum_{r+s=p} \binom{p}{r} u^r(x) u^s(y)$$

for every integer $p \geqslant 0$ (Leibniz's formula). It follows that

$$e^u(xy) = \sum_{p \geqslant 0} \frac{1}{p!} u^p(xy) = \sum_{p \geqslant 0} \sum_{r+s=p} \frac{u^r(x)}{r!} \frac{u^s(y)}{s!}$$

$$= \sum_{r,s=0}^{\infty} \frac{u^r(x)}{r!} \frac{u^s(y)}{s!} = e^u(x) e^u(y)$$

whence our assertion.

Now let $\mathfrak{g}$ be a Lie algebra. If x belongs to the nilpotent radical of $\mathfrak{g}$, the derivation $\mathrm{ad}_{\mathfrak{g}}\, x$ of $\mathfrak{g}$ is nilpotent. We can therefore make the following definition:

Definition 6. *A special automorphism of* $\mathfrak{g}$ *is an automorphism of* $\mathfrak{g}$ *of the form* $e^{\mathrm{ad}\, x}$, *where* x *is in the nilpotent radical of* $\mathfrak{g}$.

Clearly a special automorphism leaves every ideal of $\mathfrak{g}$ stable.

Definition 7. *Let* $\mathfrak{g}$ *be a Lie algebra and* $\mathfrak{r}$ *its radical. A Levi subalgebra of* $\mathfrak{g}$ *is any subalgebra of* $\mathfrak{g}$ *supplementary to* $\mathfrak{r}$.

A Levi subalgebra is isomorphic to $\mathfrak{g}/\mathfrak{r}$ and hence is semi-simple. As a semi-simple subalgebra has only 0 in common with $\mathfrak{r}$, every semi-simple subalgebra $\mathfrak{h}$ such that $\mathfrak{g} = \mathfrak{r} + \mathfrak{h}$ is a Levi subalgebra; consequently the image of a Levi subalgebra under a surjective homomorphism is a Levi subalgebra.

Theorem 5 (Levi–Malcev). *A Lie algebra* $\mathfrak{g}$ *always has a Levi subalgebra* $\mathfrak{s}$. *Every Levi subalgebra of* $\mathfrak{g}$ *is the image of* $\mathfrak{s}$ *under a special automorphism.*

Let $\mathfrak{r}$ denote the radical of $\mathfrak{g}$. We first treat two special cases.

(a) $[\mathfrak{g}, \mathfrak{r}] = \{0\}$.

By Proposition 5, $\mathfrak{g}$ is then the product of its centre $\mathfrak{r}$ by $\mathscr{D}\mathfrak{g}$ which is semi-simple. Hence $\mathscr{D}\mathfrak{g}$ is a Levi subalgebra. Moreover, if $\mathfrak{s}'$ is a semi-simple subalgebra, then $\mathfrak{s}' = \mathscr{D}\mathfrak{s}'$ (Theorem 1), hence $\mathfrak{s}' \subset \mathscr{D}\mathfrak{g}$ and $\mathscr{D}\mathfrak{g}$ is the unique Levi subalgebra of $\mathfrak{g}$.

(b) $[\mathfrak{g}, \mathfrak{r}] \neq \{0\}$ and the only ideals of $\mathfrak{g}$ contained in $\mathfrak{r}$ are $\{0\}$ and $\mathfrak{r}$.

Then $[\mathfrak{g}, \mathfrak{r}] = \mathfrak{r}$, $[\mathfrak{r}, \mathfrak{r}] = \{0\}$ and the centre of $\mathfrak{g}$ is zero. Let M (resp. N) be the subspace of $\mathscr{L}(\mathfrak{g})$ consisting of the linear mappings of $\mathfrak{g}$ into $\mathfrak{r}$ whose restriction to $\mathfrak{r}$ is a homothety (resp. zero); N is therefore of codimension 1 in M. For $m \in \mathrm{M}$, let $\lambda(m)$ denote the ratio of the homothety of $\mathfrak{r}$ defined by m. Let σ be the representation of $\mathfrak{g}$ on $\mathscr{L}(\mathfrak{g})$ canonically derived from the adjoint representation; recall that $\sigma(x).u = [\mathrm{ad}_{\mathfrak{g}}\, x, u]$ for all $x \in \mathfrak{g}$ and all $u \in \mathscr{L}(\mathfrak{g})$.

$$\begin{array}{c} \mathscr{L}(\mathfrak{g}) \\ \cup \\ \mathrm{M} \\ \cup \\ \mathrm{N} \\ \cup \\ \mathrm{P} \end{array}$$

Clearly $\sigma(x)(\mathrm{M}) \subset \mathrm{N}$ for all $x \in \mathfrak{g}$. Moreover, if $x \in \mathfrak{r}$, $y \in \mathfrak{g}$ and $u \in \mathrm{M}$, then

(4) $\qquad (\sigma(x).u)(y) = [x, u(y)] - u([x, y]) = -\lambda(u)[x, y]$

since $[\mathfrak{r}, \mathfrak{r}] = \{0\}$; and (4) can be written:

(5) $$(x).u = -\mathrm{ad}(\lambda(u).x).$$

As the centre of $\mathfrak{g}$ is zero, the mapping $x \mapsto \mathrm{ad}_{\mathfrak{g}}\, x$ defines a bijection ϕ of $\mathfrak{r}$ onto a subspace P of $\mathscr{L}(\mathfrak{g})$. This subspace is stable under $\sigma(\mathfrak{g})$ and contained in N since $\mathfrak{r}$ is a commutative ideal and (5) shows that $\sigma(x)(\mathrm{M}) \subset \mathrm{P}$ for $x \in \mathfrak{r}$. The representation of $\mathfrak{g}$ on $\mathrm{M}/\mathrm{P} = \mathrm{V}$ derived from σ is therefore zero on $\mathfrak{r}$ and defines a representation σ' of the semi-simple algebra $\mathfrak{g}/\mathfrak{r}$ on V. For all $y \in \mathfrak{g}/\mathfrak{r}$, the space $\sigma'(y)(\mathrm{V})$ is contained in N/P, which is of codimension 1 in V. Consequently (no. 2, Lemma 3) there exists $u_0 \in \mathrm{M}$ such that $\lambda(u_0) = -1$ and $\sigma(x).u_0 \in \mathrm{P}$ for all $x \in \mathfrak{g}$. The mapping $x \mapsto \overset{-1}{\phi}(\sigma(x).u_0)$ is a linear mapping of $\mathfrak{g}$ into $\mathfrak{r}$. By (5) its restriction to $\mathfrak{r}$ is the identity mapping of $\mathfrak{r}$. Hence its kernel is a subspace $\mathfrak{s}$ of $\mathfrak{g}$ supplementary to $\mathfrak{r}$ in $\mathfrak{g}$. As $\mathfrak{s}$ is the set of $x \in \mathfrak{g}$ such that $\sigma(x).u_0 = 0$, $\mathfrak{s}$ is a subalgebra of $\mathfrak{g}$ and therefore a Levi subalgebra of $\mathfrak{g}$.

Let $\mathfrak{s}'$ be another Levi subalgebra. For all $x \in \mathfrak{s}'$, let $h(x)$ be the unique element of $\mathfrak{r}$ such that $x + h(x) \in \mathfrak{s}$. Since $\mathfrak{s}$ is a subalgebra and $\mathfrak{r}$ is commutative, for x, y in $\mathfrak{s}'$:

$$[x + h(x), y + h(y)] = [x, y] + [x, h(y)] + [h(x), y] \in \mathfrak{s}$$

hence

$$h([x, y]) = (\mathrm{ad}\, x).h(y) - (\mathrm{ad}\, y).h(x).$$

By *Remark* 2 of no. 2, there exists $a \in \mathfrak{r}$ such that $h(x) = -[x, a]$ for all $x \in \mathfrak{s}'$. Then:

(6) $$x + h(x) = x + [a, x] = (1 + \mathrm{ad}\, a).x.$$

As $\mathfrak{r}$ is commutative, $(\mathrm{ad}\, a)^2 = 0$ and hence $1 + \mathrm{ad}\, a = e^{\mathrm{ad}\, a}$. As $\mathfrak{r} = [\mathfrak{g}, \mathfrak{r}]$, $e^{\mathrm{ad}\, a}$ is a special automorphism of $\mathfrak{g}$. By (6) this special automorphism maps $\mathfrak{s}'$ to $\mathfrak{s}$.

(c) General case:

We argue by induction on the dimension n of the radical. There is nothing to prove if $n = 0$ and hence it can be assumed that the theorem holds for Lie algebras whose radical is of dimension $< \dim \mathfrak{r}$. By (a) it suffices to consider the case where $[\mathfrak{g}, \mathfrak{r}] \neq \{0\}$. As $[\mathfrak{g}, \mathfrak{r}]$ is nilpotent (no. 4, Proposition 6), its centre $\mathfrak{c}$ is $\neq \{0\}$. Let $\mathfrak{m}$ be a minimal non-zero ideal of $\mathfrak{g}$ contained in $\mathfrak{c}$. If $\mathfrak{m} = \mathfrak{r}$, we have case (b). Therefore let $\mathfrak{m} \neq \mathfrak{r}$ and let f be the canonical mapping of $\mathfrak{g}$ onto $\mathfrak{g}' = \mathfrak{g}/\mathfrak{m}$. The radical of $\mathfrak{g}'$ is $\mathfrak{r}' = \mathfrak{r}/\mathfrak{m}$. By the induction hypothesis, $\mathfrak{g}'$ has a Levi subalgebra $\mathfrak{h}'$. Then $\mathfrak{h} = \overset{-1}{f}(\mathfrak{h}')$ is a subalgebra of

$\mathfrak{g}$ containing $\mathfrak{m}$ such that $\mathfrak{h}/\mathfrak{m} = \mathfrak{h}'$ is semi-simple and hence having $\mathfrak{m}$ as radical. By the induction hypothesis $\mathfrak{h} = \mathfrak{m} + \mathfrak{s}$ where $\mathfrak{s}$ is a semi-simple subalgebra. Then the equality $\mathfrak{g}' = \mathfrak{r}' + \mathfrak{h}'$ implies $\mathfrak{g} = \mathfrak{r} + \mathfrak{h} = \mathfrak{r} + \mathfrak{m} + \mathfrak{s} = \mathfrak{r} + \mathfrak{s}$ and hence $\mathfrak{s}$ is a Levi subalgebra of $\mathfrak{g}$.

Let $\mathfrak{s}'$ be another Levi subalgebra of $\mathfrak{g}$. Then $f(\mathfrak{s})$ and $f(\mathfrak{s}')$ are two Levi subalgebras of $\mathfrak{g}'$ and there exists by the induction hypothesis $a' \in [\mathfrak{g}', \mathfrak{r}']$ such that $e^{\mathrm{ad}\, a'}(f(\mathfrak{s}')) = f(\mathfrak{s})$. If $a \in [\mathfrak{g}, \mathfrak{r}]$ is such that $f(a) = a'$, it follows that:

$$\mathfrak{s}_1 = e^{\mathrm{ad}\, a}(\mathfrak{s}') \subset \mathfrak{m} + \mathfrak{s} = \mathfrak{h}.$$

Then $\mathfrak{s}_1$ and $\mathfrak{s}$ are two Levi subalgebras of $\mathfrak{h}$ and by the induction hypothesis there exists $b \in \mathfrak{m}$ such that $e^{\mathrm{ad}\, b}(\mathfrak{s}_1) = \mathfrak{s}$. Hence $\mathfrak{s} = e^{\mathrm{ad}\, b}.e^{\mathrm{ad}\, a}(\mathfrak{s}')$. Finally, as $\mathfrak{m}$ is in the centre of $[\mathfrak{g}, \mathfrak{r}]$, $e^{\mathrm{ad}\, b}.e^{\mathrm{ad}\, a} = e^{\mathrm{ad}(b+a)}$ and $b + a \in [\mathfrak{g}, \mathfrak{r}]$, which completes the proof.

COROLLARY 1. *Let $\mathfrak{s}$ be a Levi subalgebra of $\mathfrak{g}$ and $\mathfrak{h}$ a semi-simple subalgebra of $\mathfrak{g}$.*

(a) *There exists a special automorphism of $\mathfrak{g}$ mapping $\mathfrak{h}$ onto a subalgebra of $\mathfrak{s}$.*

(b) *$\mathfrak{h}$ is contained in a Levi subalgebra of $\mathfrak{g}$.*

Let $\mathfrak{r}$ be the radical of $\mathfrak{g}$ and $\mathfrak{a} = \mathfrak{h} + \mathfrak{r}$, which is a subalgebra of $\mathfrak{g}$. Then $\mathfrak{a}/\mathfrak{r}$ is semi-simple and $\mathfrak{r}$ is solvable, hence $\mathfrak{r}$ is the radical of $\mathfrak{a}$ and $\mathfrak{h}$ is a Levi subalgebra of $\mathfrak{a}$. On the other hand, $\mathfrak{a} \cap \mathfrak{s} = \mathfrak{h}'$ is a supplementary subalgebra to $\mathfrak{r}$ in $\mathfrak{a}$ and hence also a Levi subalgebra of $\mathfrak{a}$. Then there exists (Theorem 5) $a \in [\mathfrak{a}, \mathfrak{r}]$ such that $e^{\mathrm{ad}_\mathfrak{a} a}$ maps $\mathfrak{h}$ onto $\mathfrak{h}'$. Now $a \in [\mathfrak{g}, \mathfrak{r}]$; $e^{\mathrm{ad}_\mathfrak{g} a}$ maps $\mathfrak{h}$ onto a subalgebra of $\mathfrak{s}$ and $e^{-\mathrm{ad}_\mathfrak{g} a}(\mathfrak{s})$ is a Levi subalgebra of $\mathfrak{g}$ containing $\mathfrak{h}$.

COROLLARY 2. *For a subalgebra $\mathfrak{h}$ of $\mathfrak{g}$ to be a Levi subalgebra of $\mathfrak{g}$, it is necessary and sufficient that $\mathfrak{h}$ be a maximal semi-simple subalgebra of $\mathfrak{g}$.*

This follows immediately from Corollary 1.

COROLLARY 3. *Let $\mathfrak{g}$ be a Lie algebra and $\mathfrak{m}$ an ideal of $\mathfrak{g}$ such that $\mathfrak{g}/\mathfrak{m}$ is semi-simple. Then $\mathfrak{g}$ contains a subalgebra supplementary to $\mathfrak{m}$ in $\mathfrak{g}$. In other words, every extension of a semi-simple Lie algebra is inessential.*

Let $\mathfrak{s}$ be a Levi subalgebra of $\mathfrak{g}$ (Theorem 5). Its canonical image in $\mathfrak{g}/\mathfrak{m}$ is a Levi subalgebra and therefore equal to $\mathfrak{g}/\mathfrak{m}$, hence $\mathfrak{g} = \mathfrak{s} + \mathfrak{m}$. Then an ideal of $\mathfrak{s}$ supplementary in $\mathfrak{s}$ to the ideal $\mathfrak{m} \cap \mathfrak{s}$ is a subalgebra of $\mathfrak{g}$ supplementary to $\mathfrak{m}$ in $\mathfrak{g}$.

COROLLARY 4. *Let $\mathfrak{g}$ be a Lie algebra, $\mathfrak{r}$ its radical, $\mathfrak{s}$ a Levi subalgebra of $\mathfrak{g}$ and $\mathfrak{m}$ an ideal of $\mathfrak{g}$. Then $\mathfrak{m}$ is the direct sum of $\mathfrak{m} \cap \mathfrak{r}$ which is its radical and $\mathfrak{m} \cap \mathfrak{s}$ which is a Levi subalgebra of $\mathfrak{m}$.*

We know that $\mathfrak{m} \cap \mathfrak{s}$ is the radical of $\mathfrak{m}$ (§ 5, no. 5, Corollary 3 to Proposition 5). Let $\mathfrak{h}$ be a Levi subalgebra of $\mathfrak{m}$ and $\mathfrak{s}'$ a Levi subalgebra of $\mathfrak{g}$ containing $\mathfrak{h}$ (Corollary 1). The algebra $\mathfrak{m} \cap \mathfrak{s}'$ is an ideal of $\mathfrak{s}'$, is therefore semi-simple, and contains $\mathfrak{h}$ and is therefore equal to $\mathfrak{h}$. Hence $\mathfrak{m}$ is the direct sum of $\mathfrak{m} \cap \mathfrak{r}$ and $\mathfrak{m} \cap \mathfrak{s}'$. There exists a special automorphism mapping $\mathfrak{s}'$ onto $\mathfrak{s}$; this automorphism leaves $\mathfrak{r}$ and $\mathfrak{m}$ invariant; hence $\mathfrak{m}$ is the direct sum of $\mathfrak{m} \cap \mathfrak{r}$ and $\mathfrak{m} \cap \mathfrak{s}$ and $\mathfrak{m} \cap \mathfrak{s}$ is a Levi subalgebra of $\mathfrak{m}$.

9. THE INVARIANTS THEOREM

Let $\mathfrak{g}$ be a Lie algebra and ρ a representation of $\mathfrak{g}$ on a vector space M. For every class δ of simple representations of $\mathfrak{g}$ let M_δ be the isotypical component of M of species δ. The subspace M_0 of invariant elements of M is just M_{δ_0}, where δ_0 denotes the class of the zero representation of $\mathfrak{g}$ on a space of dimension 1.

Lemma 4. *Let* ρ, σ, τ *be representations of* $\mathfrak{g}$ *on vector spaces* M, N, P. *Suppose that we are given a* K*-bilinear mapping* $(m, n) \mapsto m.n$ *of* $M \times N$ *into* P *such that*

$$(\rho(x)m).n + m.(\sigma(x)n) = \tau(x)(m.n)$$

for all $m \in M$, $n \in N$, $x \in \mathfrak{g}$.

(a) *If* $m_0 \in M_0$, *the mapping* $n \mapsto m_0.n$ *is a* $\mathfrak{g}$*-module homomorphism.*

(b) *If* $n \in N_\delta$, *then* $m_0.n \in P_\delta$.

(c) *If* M *is a (not necessarily associative) algebra and the* $\rho(x)$ *are derivations of* M, M_0 *is a subalgebra of* M *and each* M_δ *is a right and left* M_0*-module.*

For $m_0 \in M_0$, $n \in N$ and $x \in \mathfrak{g}$,

$$\tau(x)(m_0.n) = m_0.(\sigma(x)n),$$

whence (a). Assertion (b) follows from (a) (*Algebra*, Chapter VIII, § 3, no. 4, Proposition 10). If $N = P = M$ and $\sigma = \tau = \rho$, assertion (b) gives assertion (c) as a special case.

Lemma 5. *Suppose further that* σ *and* τ *are semi-simple and hence* N (resp. P) *is the direct sum of the* N_δ (resp. P_δ). *For all* $n \in N$ (resp. $p \in P$), *let* $n^\natural$ (resp. $p^\natural$) *be its component in* N_0 (resp. P_0). *Let* $m_0 \in M_0$. *Then for all* $n \in N$, $(m_0.n)^\natural = m_0.n^\natural$.

By linearity it suffices to consider the case where $n \in N_\delta$. If $\delta \neq \delta_0$, $n^\natural = 0$ and $m_0.n \in P_\delta$ (Lemma 4), hence $(m_0.n)^\natural = 0 = m_0.n^\natural$. If $\delta = \delta_0$, $n^\natural = n$ and $m_0.n \in P_0$ (Lemma 4), hence $(m_0.n)^\natural = m_0.n = m_0.n^\natural$.

THEOREM 6. *Let* $\mathfrak{g}$ *be a Lie algebra,* V *a semi-simple* $\mathfrak{g}$*-module of finite dimension over* K, S *the symmetric algebra of* V *and* x_S *the derivation of* S *which extends* x_V (*so that* $x \mapsto x_S$ *is a representation of* $\mathfrak{g}$ *on* S).

(a) *The algebra* S_0 *of invariants of* S *is generated by a finite number of elements.*

(b) *For every class* δ *of simple representations of* $\mathfrak{g}$ *of finite dimension over* K, *let* S_δ *be the isotypical component of* S *of species* δ. *Then* S_δ *is a finitely generated* S_0*-module.*

Let $\bar{S} \subset S$ be the ideal of elements of S with no constant term. Let I be the ideal of S generated by $S_0 \cap \bar{S}$ and let $(s_1, s_2, \ldots, s_p)$ be a finite system of generators *of the ideal* I (*Commutative Algebra*, Chapter III, § 3). It can be assumed that the s_i belong to $S_0 \cap \overline{S}$ and are homogeneous (the x_S preserve degrees and hence each S_δ is a graded submodule). Let S_1 be the subalgebra of S generated by 1 and the s_i. Then $S_1 \subset S_0$. We show that $S_1 = S_0$. For this we prove that every homogeneous element s of S_0 is in S_1, arguing by induction on the degree n of s. If $n = 0$, our assertion is obvious. Suppose therefore $n > 0$ and that our assertion has been proved when the degree of s is $<n$. As $s \in I$, $s = \sum_{i=1}^{p} s_i s'_i$, where the s'_i are elements of S which can be assumed to be homogeneous, with $\deg(s'_i) = \deg(s) - \deg(s_i) < n$. Lemma 5 can be applied, as the $\mathfrak{g}$-module S is semi-simple (no. 5, Corollary 2 to Theorem 4); in the notation of this lemma,

$$s = s^\natural = \sum_{i=1}^{p} (s_i s'_i)^\natural = \sum_{i=1}^{p} s_i s'^\natural_i.$$

The $s'^\natural_i$ are elements of S_0 which are homogeneous and of degree $<n$ (since each S_δ is a graded submodule). Hence they are in S_1 by the induction hypothesis. Hence $s \in S_1$, which proves (a).

We now consider a simple representation of $\mathfrak{g}$ of class δ on a finite-dimensional space M. Let $L = \mathscr{L}_K(M, S)$. For all $s \in S$ and all $f \in L$, let sf be the element of L defined by $(sf)(m) = s.f(m)$ $(m \in M)$; an S-module structure is thus defined on L; as M is finite-dimensional over K, clearly L is a finitely generated S-module and hence a Noetherian S-module since the ring S is Noetherian. On the other hand, L has a canonical $\mathfrak{g}$-module structure. For every integer $n \geqslant 0$ let S^n be the set of homogeneous elements of S of degree n; then the $\mathfrak{g}$-module $\mathscr{L}_K(M, S^n)$ is semi-simple (no. 5, Corollary 3 to Theorem 4) and hence the $\mathfrak{g}$-module L is semi-simple. Moreover, for $s \in S$, $f \in L$, $x \in \mathfrak{g}$ and $m \in M$,

$$\begin{aligned}(x_L(sf))(m) &= x_S((sf)(m)) - (sf)(x_M m)\\ &= x_S(s.f(m)) - s.f(x_M m)\\ &= (x_S s).f(m) + s.(x_S f(m)) - s.f(x_M m)\\ &= ((x_S s)f)(m) + (s(x_L f))(m)\end{aligned}$$

and hence $x_L(sf) = (x_S s)f + s(x_L f)$. We can therefore apply Lemma 5.

The subset L_0 of invariant elements of L is just the set of homomorphisms of the $\mathfrak{g}$-module M into the $\mathfrak{g}$-module S. Hence, if ϕ denotes the canonical homomorphism of $M \otimes_K L$ onto S, $\phi(M \otimes_K L_0) = S_\delta$. As ϕ is obviously an S-module homomorphism, it suffices to show that L_0 is a finitely generated S_0-module. Let J be the sub-S-module of L generated by L_0. Since L is a Noetherian S-module, there exists a finite sequence $(f_1, \ldots, f_q)$ of elements of L_0 generating the S-module J. Let L_1 be the S_0-module generated by

$f_1, \ldots, f_q$. Then $L_1 \subset L_0$. On the other hand, if $f \in L_0$, then $f = \sum_{i=1}^{q} s_i f_i$ with $s_i \in S$ for all i and hence by Lemma 5 whose notation we adopt:

$$f = f^{\natural} = \left(\sum_{i=1}^{q} s_i f_i\right)^{\natural} = \sum_{i=1}^{q} s_i^{\natural} f_i \in L_1.$$

Hence $L_0 = L_1$, so that L_0 is a finite generated S_0-module.

10. CHANGE OF BASE FIELD

Let K_1 be a commutative extension of K. For a Lie algebra $\mathfrak{g}$ over K to be semi-simple, it is necessary and sufficient that $\mathfrak{g}_{(K_1)}$ be semi-simple; for the Killing form β_1 of $\mathfrak{g}_{(K_1)}$ is derived from the Killing form β of $\mathfrak{g}$ by extending the base field from K to K_1; hence β_1 is non-degenerate if and only if β is non-degenerate (*Algebra*, Chapter IX, § 1, no. 4, Corollary to Proposition 3).

If $\mathfrak{g}_{(K_1)}$ is simple, $\mathfrak{g}$ is semi-simple by the above and cannot be a product of two non-zero ideals, hence $\mathfrak{g}$ is simple. On the other hand if $\mathfrak{g}$ is simple $\mathfrak{g}_{(K_1)}$ (which is semi-simple) may be not simple (Exercises 17 and 26 (b)).

Let $\mathfrak{g}$ be a Lie algebra and $\mathfrak{r}$ its radical. Then $\mathfrak{r}_{(K_1)}$ is the radical of $\mathfrak{g}_{(K_1)}$ (§ 5, no. 6). Therefore, if $\mathfrak{s}$ denotes the nilpotent radical of $\mathfrak{g}$, the nilpotent radical of $\mathfrak{g}_{(K_1)}$ is $[\mathfrak{g}_{(K_1)}, \mathfrak{r}_{(K_1)}] = [\mathfrak{g}, \mathfrak{r}]_{(K_1)} = \mathfrak{s}_{(K_1)}$. It follows that $\mathfrak{g}$ is reductive if and only if $\mathfrak{g}_{(K_1)}$ is reductive.

Let $\mathfrak{g}$ be a Lie algebra and $\mathfrak{h}$ a subalgebra. Recall that a representation of $\mathfrak{h}$ is semi-simple if and only if the representation of $\mathfrak{h}_{(K_1)}$ derived by extending the base field to K_1 is semi-simple. Hence $\mathfrak{h}$ is reductive in $\mathfrak{g}$ if and only if $\mathfrak{h}_{(K_1)}$ is reductive in $\mathfrak{g}_{(K_1)}$.

Now let K_0 be a subfield of K such that $[K:K_0]$ is finite. Let $\mathfrak{g}$ be a Lie algebra and $\mathfrak{g}_0$ the (finite-dimensional) Lie algebra derived from $\mathfrak{g}$ by restricting the base field from K to K_0. Every commutative ideal of $\mathfrak{g}$ is a commutative ideal of $\mathfrak{g}_0$; conversely, if $\mathfrak{a}_0$ is a commutative ideal of $\mathfrak{g}_0$ the smallest vector subspace over K of $\mathfrak{g}$ containing $\mathfrak{a}_0$ is a commutative ideal of $\mathfrak{g}$; hence $\mathfrak{g}$ is semi-simple if and only if $\mathfrak{g}_0$ is semi-simple. If $\mathfrak{g}_0$ is simple, clearly $\mathfrak{g}$ is simple. Conversely, suppose that $\mathfrak{g}$ is simple. We show that $\mathfrak{g}_0$ is simple. Let $\mathfrak{a}_0$ be a simple component of $\mathfrak{g}_0$. For all $\lambda \in K^*$, $\lambda\mathfrak{a}_0$ is an ideal of $\mathfrak{g}_0$ and

$$[\mathfrak{a}_0, \lambda\mathfrak{a}_0] = \lambda[\mathfrak{a}_0, \mathfrak{a}_0] = \lambda\mathfrak{a}_0 \neq \{0\},$$

hence $\lambda\mathfrak{a}_0 \supset \mathfrak{a}_0$ and therefore $\lambda\mathfrak{a}_0 = \mathfrak{a}_0$ since $\dim_{K_0}(\lambda\mathfrak{a}_0) = \dim_{K_0}\mathfrak{a}_0$. Now the vector subspace of $\mathfrak{g}$ generated by $\mathfrak{a}_0$ is a non-zero ideal of $\mathfrak{g}$ and hence is the whole of $\mathfrak{g}$. Hence $\mathfrak{g} = \mathfrak{a}_0$, which proves our assertion.

§ 7. ADO'S THEOREM

Recall that K *denotes a field of characteristic* 0 *and that all Lie algebras are assumed to be finite-dimensional over* K.

1. COEFFICIENTS OF A REPRESENTATION

Let U be an associative algebra with unit element over K, U* the dual of the vector space U and ρ a representation of U on a vector space E. For $e \in E$ and $e' \in E^*$, let $\theta(e, e') \in U^*$ be the corresponding *coefficient* of ρ (*Algebra*, Chapter VIII, § 13, no. 3). Recall that $\theta(e, e')(x) = \langle \rho(x)e, e' \rangle$ and that the mapping $e \mapsto \theta(e, e')$ is for fixed e' a homomorphism of the U-module E into the U-module U* of the coregular representation of U (*loc. cit.*, Proposition 1); therefore the vector subspace of U* generated by the coefficients of ρ (a subspace which we shall denote by $C(\rho)$ in this paragraph) is a sub-U-module of U*. If $(e'_i)_{i \in I}$ is a family of elements generating E* over K, the mapping $e \mapsto (\theta(e, e'_i))$ is an *injective* U-homomorphism of E into $C(\rho)^I$, for $\theta(e, e'_i) = 0$ for all i implies $\langle e, e'_i \rangle = \langle \rho(1)e, e'_i \rangle = \theta(e, e'_i)(1) = 0$ for all i and hence $e = 0$.

In particular, if U is the enveloping algebra of a Lie algebra $\mathfrak{g}$ and ρ is a representation of $\mathfrak{g}$ (identified with a representation of U) on an n-dimensional vector space E, the $\mathfrak{g}$-module E is isomorphic to a sub-$\mathfrak{g}$-module of $(C(\rho))^n$.

2. THE EXTENSION THEOREM

Let $\mathfrak{g} = \mathfrak{h} + \mathfrak{g}'$ be a Lie algebra which is the direct sum of an ideal $\mathfrak{g}'$ and a subalgebra $\mathfrak{h}$, U the enveloping algebra of $\mathfrak{g}$ and $U' \subset U$ the enveloping algebra of $\mathfrak{g}'$. There exists one and only one $\mathfrak{g}$-module structure on U' such that: (α) for $x \in \mathfrak{g}'$ and $u \in U'$, $x_{U'}u = -ux$; (β) for $x \in \mathfrak{h}$ and $u \in U'$, $x_{U'}u = xu - ux$ (the latter element is certainly in U' since the inner derivation of U defined by x leaves $\mathfrak{g}'$ and hence U' stable). For conditions (α) and (β) define uniquely a linear mapping $x \mapsto x_{U'}$ of $\mathfrak{g}$ into $\mathscr{L}_K(U')$. It therefore suffices to verify that $[x, y]_{U'} = [x_{U'}, y_{U'}]$; it is only necessary to consider the following cases:

(1) $x \in \mathfrak{g}', y \in \mathfrak{g}'$: then

$$[x, y]_{U'}u = -u(xy - yx) = (x_{U'}y_{U'} - y_{U'}x_{U'})u;$$

(2) $x \in \mathfrak{h}, y \in \mathfrak{g}'$: then

$$\begin{aligned}[x, y]_{U'}u &= -u(xy - yx) = x(-uy) - (-uy)x + (xu - ux)y \\ &= (x_{U'}y_{U'} - y_{U'}x_{U'})u;\end{aligned}$$

(3) $x \in \mathfrak{h}, y \in \mathfrak{h}$: then $[x, y]_{U'}$ and $[x_{U'}, y_{U'}]$ are two derivations of U' whose restrictions to $\mathfrak{g}'$ coincide with those of $\mathrm{ad}_{\mathfrak{g}}[x, y]$ and $[\mathrm{ad}_{\mathfrak{g}}x, \mathrm{ad}_{\mathfrak{g}}y]$; hence these derivations are equal.

We shall also consider the dual representation $x \mapsto -{}^t x_{U'}$ of $\mathfrak{g}$ on U'*. For $x \in \mathfrak{g}'$, $-{}^t x_{U'}$ is the transpose of right multiplication by x in U'; the corresponding representation of U' is therefore the coregular representation of U'.

Definition 1. *Let* $\mathfrak{g}$ *be a Lie algebra,* $\mathfrak{g}'$ *a subalgebra of* $\mathfrak{g}$ *and* ρ' *a representation of* $\mathfrak{g}'$ *on* V'. *A representation* ρ *of* $\mathfrak{g}$ *on* V *is called an extension of* ρ' *to* $\mathfrak{g}$ *if there exists an injective homomorphism of the* $\mathfrak{g}'$*-module into the* $\mathfrak{g}'$*-module* V. *We also say that the* $\mathfrak{g}$*-module* V *is an extension of the* $\mathfrak{g}'$*-module* V'.

If ρ' is finite-dimensional and $\mathfrak{g}'$ is a solvable ideal of $\mathfrak{g}$, it is necessary for the existence of a *finite-dimensional* extension that $[\mathfrak{g}, \mathfrak{g}']$ be contained in the largest nilpotency ideal of ρ' (§ 5, no. 3, Theorem 1).

Theorem 1 (Zassenhaus). *Let* $\mathfrak{g} = \mathfrak{g}' + \mathfrak{h}$ *be a Lie algebra which is the direct sum of an ideal* $\mathfrak{g}'$ *and a subalgebra* $\mathfrak{h}$ *and* ρ' *a finite-dimensional representation of* $\mathfrak{g}'$ *whose largest nilpotency ideal contains* $[\mathfrak{h}, \mathfrak{g}']$.

(a) *There exists a finite-dimensional extension* ρ *of* ρ' *to* $\mathfrak{g}$ *whose largest nilpotency ideal contains that of* ρ'.

(b) *If for all* $x \in \mathfrak{h}$ *the restriction to* $\mathfrak{g}'$ *of* $\mathrm{ad}_{\mathfrak{g}}x$ *is nilpotent,* ρ *can be chosen such that moreover the largest nilpotency ideal of* ρ *contains* $\mathfrak{h}$.

Let U' be the enveloping algebra of $\mathfrak{g}'$. Suppose that U' and U'* have the $\mathfrak{g}$-module structures defined at the beginning of this no.

$$\begin{array}{c} U'^* \\ \cup \\ S \\ \cup \\ C(\rho') \end{array}$$

Let $I \subset U'$ be the kernel of ρ' (identified with a representation of U'). It is a two-sided ideal of U' of finite codimension. The subspace $C(\rho')$ of U'* (cf. no. 1) is orthogonal to I. Let S be the sub-$\mathfrak{g}$-module of U'* generated by $C(\rho')$.

We now show that S is *finite-dimensional* over K. Let V' be the space on which ρ' operates and $V' = V'_0 \supset V'_1 \supset \cdots \supset V'_d = \{0\}$ a Jordan-Hölder series of the $\mathfrak{g}'$-module V'. Let ρ'_i be the representation of $\mathfrak{g}'$ on V'_{i-1}/V'_i derived from ρ' $(1 \leqslant i \leqslant d)$. Let $I' \subset U'$ be the intersection of the kernels of the ρ'_i (identified with representations of U'). Then

$$I'^d \subset I \subset I'$$

and $I' \cap \mathfrak{g}'$ is the largest nilpotency ideal of ρ'. By § 2, no. 6, Corollary to Proposition 6, I'^d is of finite codimension in U'. If $x \in \mathfrak{h}$, the derivation $u \mapsto xu - ux$ of U' maps $\mathfrak{g}'$ into $[\mathfrak{h}, \mathfrak{g}'] \subset I'$, hence U' into I' and hence I'^d into I'^d. On the other hand, clearly I'^d is a sub-$\mathfrak{g}'$-module of U'. Hence I'^d is a sub-$\mathfrak{g}$-module of U'. The orthogonal of I'^d in U'* is a finite-dimensional sub-$\mathfrak{g}$-module which contains $C(\rho')$ and therefore S. This shows that S is finite-dimensional over K. For $x \in I' \cap \mathfrak{g}'$, x^d is obviously contained in the annihilator of the $\mathfrak{g}$-module U'/I'^d and hence also in the annihilator of the $\mathfrak{g}$-module S.

We saw in no. 1 that the $\mathfrak{g}'$-module V' is isomorphic to a sub-$\mathfrak{g}'$-module of a product $(C(\rho'))^n$. Hence the $\mathfrak{g}$-module S^n provides a finite-dimensional extension ρ of ρ' to $\mathfrak{g}$. Moreover, $\rho(x)$ is nilpotent for $x \in I' \cap \mathfrak{g}'$; as $I' \cap \mathfrak{g}'$

is an ideal of $\mathfrak{g}$ (for it contains $[\mathfrak{h}, \mathfrak{g}']$ by hypothesis), we see that $I' \cap \mathfrak{g}'$ is contained in the largest nilpotency ideal of ρ. Thus (a) is proved.

Suppose finally that for all $x \in \mathfrak{h}$ the restriction to $\mathfrak{g}'$ of $\mathrm{ad}_{\mathfrak{g}} x$ is nilpotent. As the elements of $\mathfrak{h}$ operate on U' by derivations, there exists for all $u \in U'$ and all $x \in \mathfrak{h}$ an integer e such that $(x_{U'})^e.u = 0$; the endomorphisms derived from $x_{U'}$ on U'/I'^d and on S (which are finite-dimensional spaces) are therefore nilpotent. Thus $\rho(x)$ is nilpotent for all $x \in \mathfrak{h}$. We have seen earlier that this is also true for $x \in I' \cap \mathfrak{g}'$. As $I' \cap \mathfrak{g}'$ is an ideal of $\mathfrak{g}'$ containing $[\mathfrak{h}, \mathfrak{g}']$, the sum $\mathfrak{h} + (I' \cap \mathfrak{g}')$ is also an ideal of $\mathfrak{g}$. Assertion (b) of Theorem 1 then results from the following lemma:

Lemma 1. *Let $\mathfrak{g} = \mathfrak{g}' + \mathfrak{h}$ be a Lie algebra which is the sum of an ideal $\mathfrak{g}'$ and a subalgebra $\mathfrak{h}$. Let σ be a finite-dimensional representation of $\mathfrak{g}$. Suppose that $\sigma(x)$ is nilpotent for all $x \in \mathfrak{g}'$ and all $x \in \mathfrak{h}$. Then $\sigma(x)$ is nilpotent for all $x \in \mathfrak{g}$.*

Passing to the quotient by the kernel of σ, σ may be assumed to be faithful. Then $\mathfrak{g}'$ and $\mathfrak{h}$ are nilpotent and hence $\mathfrak{g}$, which is an extension of a quotient of $\mathfrak{h}$ by $\mathfrak{g}'$, is solvable. Then $\mathfrak{h}$ and $\mathfrak{g}'$ are contained in the largest nilpotency ideal of σ (§ 5, no. 3, Corollary 6 to Theorem 1).

For an improvement of Theorem 1, cf. Exercise 4.

3. ADO'S THEOREM

PROPOSITION 1. *Let $\mathfrak{g}$ be a Lie algebra, $\mathfrak{n}$ its largest nilpotent ideal, $\mathfrak{a}$ a nilpotent ideal of $\mathfrak{g}$ and ρ a finite-dimensional representation of $\mathfrak{a}$ such that every element of $\rho(\mathfrak{a})$ is nilpotent. Then ρ admits a finite-dimensional extension σ to $\mathfrak{g}$ such that every element of $\sigma(\mathfrak{n})$ is nilpotent.*

Let $\mathfrak{a} = \mathfrak{n}_0 \subset \mathfrak{n}_1 \subset \cdots \subset \mathfrak{n}_p = \mathfrak{n}$ be a sequence of subalgebras of $\mathfrak{n}$ such that $\mathfrak{n}_{i-1}$ is an ideal of $\mathfrak{n}_i$ of codimension 1 for $1 \leqslant i \leqslant p$ (§ 4, no. 1, Proposition 1 (e)). The algebra $\mathfrak{n}_i$ is therefore the direct sum of $\mathfrak{n}_{i-1}$ and a 1-dimensional subalgebra. As $\mathrm{ad}_{\mathfrak{n}} x$ is nilpotent for all $x \in \mathfrak{n}$, it is possible (Theorem 1) to find one by one finite-dimensional extensions $\rho_1, \rho_2, \ldots, \rho_p = \rho'$ of ρ to $\mathfrak{n}_1, \mathfrak{n}_2, \ldots, \mathfrak{n}_p = \mathfrak{n}$ such that every element of $\rho'(\mathfrak{n})$ is nilpotent.

Let $\mathfrak{r}$ be the radical of $\mathfrak{g}$ and let $\mathfrak{n} = \mathfrak{r}_0 \subset \mathfrak{r}_1 \subset \cdots \subset \mathfrak{r}_q = \mathfrak{r}$ be a sequence of subalgebras of $\mathfrak{r}$ such that $\mathfrak{r}_{i-1}$ is an ideal of $\mathfrak{r}_i$ of codimension 1 for $1 \leqslant i \leqslant q$ (§ 5, no. 1, Proposition 2 (d)). The algebra $\mathfrak{r}_i$ is thus the direct sum of $\mathfrak{r}_{i-1}$ and a 1-dimensional subalgebra. As $[\mathfrak{r}, \mathfrak{r}] \subset \mathfrak{n}$, it is possible (Theorem 1) to find one by one finite-dimensional extensions $\rho'_1, \rho'_2, \ldots, \rho'_q = \rho''$ of ρ' to $\mathfrak{r}_1, \mathfrak{r}_2, \ldots, \mathfrak{r}_q = \mathfrak{r}$ such that every element of $\rho''(\mathfrak{n})$ is nilpotent.

Finally $\mathfrak{g}$ is the direct sum of $\mathfrak{r}$ and a subalgebra $\mathfrak{s}$ (§ 6, no. 8, Theorem 5). As $[\mathfrak{s}, \mathfrak{r}] \subset \mathfrak{n}$, it is possible (Theorem 1) to find a finite-dimensional extension σ of ρ'' to $\mathfrak{g}$ such that every element of $\sigma(\mathfrak{n})$ is nilpotent.

THEOREM 2. *Every Lie algebra has a finite-dimensional faithful linear representation.*

More precisely:

THEOREM 3 (Ado). *Let $\mathfrak{g}$ be a Lie algebra and $\mathfrak{n}$ its largest nilpotent ideal. There exists a finite-dimensional faithful representation ρ of $\mathfrak{g}$ such that every element of $\rho(\mathfrak{n})$ is nilpotent.*

The 1-dimensional Lie algebra K admits finite-dimensional faithful representations τ such that every element of $\tau(K)$ is nilpotent, for example the representation

$$\lambda \mapsto \begin{pmatrix} 0 & 0 \\ \lambda & 0 \end{pmatrix}.$$

It is easily deduced that the centre $\mathfrak{c}$ of $\mathfrak{g}$, which is a product of 1-dimensional algebras, admits a finite-dimensional faithful representation σ such that every element of $\sigma(\mathfrak{c})$ is nilpotent. Let σ_1 be a finite-dimensional extension of σ to $\mathfrak{g}$ such that every element of $\sigma_1(\mathfrak{n})$ is nilpotent (Proposition 1); if $\mathfrak{k}$ denotes the kernel of σ_1, then $\mathfrak{k} \cap \mathfrak{c} = \{0\}$. On the other hand, let σ_2 be the adjoint representation of $\mathfrak{g}$, whose kernel is $\mathfrak{c}$; every element of $\sigma_2(\mathfrak{n})$ is nilpotent. The direct sum ρ of σ_1 and σ_2 is finite-dimensional; every element of $\rho(\mathfrak{n})$ is nilpotent; and the kernel of ρ, contained in $\mathfrak{k}$ and in $\mathfrak{c}$, is zero, so that ρ is faithful.

EXERCISES

§ 1

1. Let $\mathfrak{g}$ be a Lie algebra and $\mathfrak{a}$ and $\mathfrak{b}$ ideals (resp. characteristic ideals) of $\mathfrak{g}$. Then the set of $x \in \mathfrak{g}$ such that $[x, \mathfrak{b}] \subset \mathfrak{a}$ is an ideal (resp. a characteristic ideal) of $\mathfrak{g}$ called the *transporter* of $\mathfrak{b}$ into $\mathfrak{a}$. Show that $\mathscr{C}_{p+1}\mathfrak{g}$ is the transporter of $\mathfrak{g}$ into $\mathscr{C}_p\mathfrak{g}$.

2. If $\mathfrak{g}$ is an n-dimensional Lie algebra over a field K and the centre of $\mathfrak{g}$ is of dimension $\geqslant n - 1$, $\mathfrak{g}$ is commutative.

3. Let M be a not necessarily associative algebra over a field K, $(e_i)_{i \in \mathrm{I}}$ a basis of the vector space M and c_{ijk} the constants of structure of M relative to the basis (e_i). For M to be a Lie algebra, it is necessary and sufficient that the c_{ijk} satisfy the following conditions: (a) $c_{iik} = 0$; (b) $c_{ijk} = -c_{jik}$; (c) $\sum_{r \in \mathrm{I}} (c_{ijr}c_{rkl} + c_{jkr}c_{ril} + c_{kir}c_{rjl}) = 0$ for all i, j, k, l in I.

4. (*a*) Suppose that 2 is invertible in K. Let $\mathfrak{a}$ be a Lie algebra over K. On the K-module $\mathfrak{a}' = \mathfrak{a} \times \mathfrak{a}$ we define a bracket by the formula:

$$[(x_1, x_2), (y_1, y_2)] = ([x_1, y_1] + [x_2, y_2], [x_1, y_2] + [x_2, y_1]).$$

Then $\mathfrak{a}'$ is a Lie K-algebra and the mappings $x \mapsto \frac{1}{2}(x, x)$, $x \mapsto \frac{1}{2}(x, -x)$ are isomorphisms of $\mathfrak{a}$ onto ideals $\mathfrak{b}$, $\mathfrak{c}$ of $\mathfrak{a}'$ of which $\mathfrak{a}'$ is the direct sum. (Consider the quadratic extension K' of K with basis 1 and k where $k^2 = 1$. Form $\mathfrak{a}_{(\mathrm{K}_1)}$ and then restrict the ring of scalars to K. Then observe that $\frac{1}{2}(1 + k)$, $\frac{1}{2}(1 - k)$ are idempotents of K' and that $(1 + k)(1 - k) = 0$.)

(*b*) Let $\mathfrak{a}$ be a real Lie algebra, $\mathfrak{g} = \mathfrak{a}_{(\mathbf{C})}$ and $\mathfrak{b}$ the real Lie algebra derived from $\mathfrak{g}$ by restricting the field of scalars to $\mathbf{R}$. Show that $\mathfrak{b}_{(\mathbf{C})}$ is isomorphic to $\mathfrak{a}'_{(\mathbf{C})}$, where $\mathfrak{a}'$ is the algebra introduced in (*a*). (Consider $\mathrm{V} = \mathbf{C} \otimes_{\mathbf{R}} \mathbf{C}$ as a vector space over $\mathbf{C}$ by $(z \otimes z')z'' = z \otimes z'z''$. Observe that V is the complexification of the real vector subspace of V generated by $1 \otimes 1$ and $i \otimes i$ and that this subspace is identified with the quadratic extension of $\mathbf{R}$ with basis 1 and k where $k^2 = 1$.)

5. Let V and W be vector spaces of dimensions $n+1$ and n over a field K. Show that $\mathfrak{gl}(W)$ is isomorphic to a subalgebra of $\mathfrak{sl}(V)$. (W can be assumed to be a hyperplane of V. Let $e \in V$, $e \notin W$. For $x \in \mathfrak{gl}(W)$, let $y(x)$ be the element of $\mathfrak{gl}(V)$ which extends x and is such that $y(e) = -\mathrm{Tr}(x).e$. Show that the mapping $x \mapsto y(x)$ is an isomorphism of $\mathfrak{gl}(W)$ onto a subalgebra of $\mathfrak{sl}(V)$.

6. For a Lie algebra $\mathfrak{g}$ to be associative, it is necessary and sufficient that $\mathscr{D}\mathfrak{g}$ be contained in the centre of $\mathfrak{g}$.

¶ 7. Let A be a ring (possibly without unit element) such that the relation $2a = 0$ in A implies $a = 0$. Let U be a subring of A and an ideal of the Lie **Z**-algebra associated with A.

(*a*) If x, y belong to U, then $(xy - yx)\mathrm{A} \subset \mathrm{U}$ (write:

$$(xy - yx)s = x(ys) - (ys)x - y(xs - sx))$$

and $\mathrm{A}(xy - yx)\mathrm{A} \subset \mathrm{U}$ (write:

$$r(xy - yx)s = (xy - yx)sr + r[(xy - yx)s] - [(xy - yx)s]r).$$

(*b*) Suppose that U is commutative. Let $x \in \mathrm{U}$, $s \in \mathrm{A}$ and $y = xs - sx$. Show that $(yt)^3 = 0$ for all $t \in \mathrm{A}$. (The element x commutes with $xs - sx$ and $xs^2 - s^2x$; deduce that $2(xs - sx)^2 = 0$, whence $y^2 = 0$. Similarly $(yt - ty)^2 = 0$ for all $t \in \mathrm{A}$.)

(*c*) Suppose henceforth that the only two-sided ideals of A are $\{0\}$ and A and that for every non-zero element y of A there exists $t \in \mathrm{A}$ such that yt is not nilpotent. Show that, if $\mathrm{U} \neq \mathrm{A}$, U is contained in the centre of A. (Show, using (*a*), that U is commutative and then use (*b*).)

(*d*) Let V be an ideal of the Lie **Z**-algebra associated with A. Then either V is contained in the centre of A or $\mathrm{V} \supset [\mathrm{A}, \mathrm{A}]$. (Apply (*c*) to the set U of $t \in \mathrm{A}$ such that $[t, \mathrm{A}] \subset \mathrm{V}$; use the identity $[xy, z] = [x, yz] + [y, zx]$.)

8. If x_1, x_2, x_3, x_4 are 4 elements of a Lie algebra, then

$$[[[x_1, x_2], x_3], x_4] + [[[x_2, x_1], x_4], x_3] + [[[x_3, x_4], x_1], x_2] + [[[x_4, x_3], x_2], x_1] = 0.$$

9. Let $\mathfrak{g}$ be a Lie algebra in which $[[x, y], y] = 0$ for all x and y.

(*a*) Show that $3[[x, y], z] = 0$ for all x, y, z in $\mathfrak{g}$. (Observe that $[[x, y], z]$ is an alternating trilinear function of (x, y, z) and apply the Jacobi identity.)

(*b*) Show similarly, using (*a*) and Exercise 8, that $[[[x, y], z], t] = 0$ for all x, y, z, t in $\mathfrak{g}$.

10. Let L be an associative algebra and $\mathfrak{g}$ the associated Lie algebra. Every derivation of L is a derivation of $\mathfrak{g}$, but the converse is not true. (Consider the identity mapping of a commutative associative algebra.)

11. Let $\mathfrak{g}$ be a Lie algebra and $\mathfrak{h}$ an ideal of $\mathfrak{g}$ such that $\mathscr{D}\mathfrak{h} = \mathfrak{h}$. Show that $\mathfrak{h}$ is a characteristic ideal of $\mathfrak{g}$.

12. Let $\mathfrak{g}$ be a Lie algebra.

(*a*) For a derivation D of $\mathfrak{g}$ to commute with all the inner derivations of $\mathfrak{g}$, it is necessary and sufficient that D map $\mathfrak{g}$ into its centre.

(*b*) Suppose henceforth that $\mathfrak{g}$ has centre zero, so that $\mathfrak{g}$ is identified with an ideal of the Lie algebra $\mathfrak{D}$ of all derivations of $\mathfrak{g}$. Show that the centralizer of $\mathfrak{g}$ in $\mathfrak{D}$ is zero (use (*a*)). In particular, the centre of $\mathfrak{D}$ is zero.

(*c*) Show that a derivation Δ of $\mathfrak{D}$ such that $\Delta(\mathfrak{g}) = \{0\}$ is zero. ($[\Delta(\mathfrak{D}), \mathfrak{g}] \subset \Delta([\mathfrak{D}, \mathfrak{g}]) + [\mathfrak{D}, \Delta(\mathfrak{g})] \subset \Delta(\mathfrak{g}) = \{0\}$ and hence $\Delta(\mathfrak{D}) = \{0\}$ by (*b*).)

(*d*) Show that, if further $\mathfrak{g} = \mathscr{D}\mathfrak{g}$, every derivation Δ of $\mathfrak{D}$ is inner (use (*c*) and Exercise 11).

13. Let $\mathfrak{g}$ be a Lie algebra and $\mathfrak{a}$ and $\mathfrak{b}$ two submodules of $\mathfrak{g}$. For every integer $i \geqslant 0$ we define the submodules $\mathfrak{m}_i = \mathfrak{m}_i(\mathfrak{a}, \mathfrak{b})$ and $\mathfrak{n}_i = \mathfrak{n}_i(\mathfrak{a}, \mathfrak{b})$ as follows: $\mathfrak{m}_1 = \mathfrak{b}$, $\mathfrak{m}_{i+1} = [\mathfrak{m}_i, \mathfrak{b}]$, $\mathfrak{n}_1 = \mathfrak{a}$, $\mathfrak{n}_{i+1} = [\mathfrak{n}_i, \mathfrak{b}]$. Show that

$$[\mathfrak{a}, \mathfrak{m}_i] \subset \mathfrak{n}_{i+1}$$

for $i = 1, 2, \ldots$ (argue by induction on i, observing that

$$\mathfrak{m}_i([\mathfrak{a}, \mathfrak{b}], \mathfrak{b}) = \mathfrak{m}_i(\mathfrak{a}, \mathfrak{b})$$

and $\mathfrak{n}_i([\mathfrak{a}, \mathfrak{b}], \mathfrak{b}) = \mathfrak{n}_{i+1}(\mathfrak{a}, \mathfrak{b})$).

¶ 14. Let $\mathfrak{a}$ be a Lie algebra and $\mathfrak{b}$ a subalgebra of $\mathfrak{a}$. A *composition series* joining $\mathfrak{a}$ to $\mathfrak{b}$ is a decreasing sequence $(\mathfrak{a}_i)_{0 \leqslant i \leqslant n}$ of subalgebras of $\mathfrak{a}$ such that $\mathfrak{a}_0 = \mathfrak{a}$, $\mathfrak{a}_n = \mathfrak{b}$, $\mathfrak{a}_{i+1}$ is an ideal of $\mathfrak{a}_i$ for $0 \leqslant i < n$. $\mathfrak{b}$ is called a *subinvariant* subalgebra of $\mathfrak{a}$ if there exists a composition series joining $\mathfrak{a}$ to $\mathfrak{b}$.

(*a*) Let $\mathfrak{b}$ be a subinvariant subalgebra of $\mathfrak{a}$ and $(\mathfrak{a}_i)_{0 \leqslant i \leqslant n}$ a composition series joining $\mathfrak{a}$ to $\mathfrak{b}$. Deduce from Exercise 13 that $[\mathscr{C}^{n+p}\mathfrak{b}, \mathfrak{a}] \subset \mathscr{C}^p\mathfrak{b}$ by observing that $[\mathfrak{a}_i, \mathfrak{b}] \subset \mathfrak{a}_{i+1}$ for $1 \leqslant i < n$.

(*b*) Deduce from (*a*) that the intersection $\mathscr{C}^\infty\mathfrak{b}$ of the $\mathscr{C}^p\mathfrak{b}$ $(p = 0, 1, 2, \ldots)$ is an ideal of $\mathfrak{a}$.

(*c*) Deduce from (*b*) that a subinvariant subalgebra $\mathfrak{c}$ of $\mathfrak{a}$ such that $\mathfrak{c} = \mathscr{D}\mathfrak{c}$ is an ideal of $\mathfrak{a}$.

(*d*) When K is a field and $\dim_K \mathfrak{a} < +\infty$, show that the intersection $\mathscr{D}^\infty\mathfrak{b}$ of the $\mathscr{D}^p\mathfrak{b}$ $(p = 0, 1, 2, \ldots)$ is an ideal of $\mathfrak{a}$. (Show that this intersection is a subinvariant subalgebra of $\mathfrak{a}$ and apply (*c*).)

¶ 15. (*a*) Let $\mathfrak{a}$ be a Lie algebra, $\mathfrak{b}$ a subalgebra of $\mathfrak{a}$, $\mathfrak{c}$ an ideal of $\mathfrak{b}$ and $\mathfrak{z}$ the centralizer of $\mathfrak{c}$ in $\mathfrak{a}$. Then $[\mathfrak{z}, \mathfrak{b}] \subset \mathfrak{z}$. Deduce that $\mathfrak{b} + \mathfrak{z}$ is a subalgebra of $\mathfrak{a}$ in which $\mathfrak{z}$ is an ideal.

(*b*) Let $\mathfrak{g}$ be a Lie algebra, $\mathfrak{a}$ a subinvariant subalgebra of $\mathfrak{g}$ and $\mathfrak{b}$ an ideal of $\mathfrak{a}$. If the centralizer of $\mathfrak{b}$ in $\mathfrak{a}$ and the centralizer of $\mathfrak{a}$ in $\mathfrak{g}$ are zero, the centralizer $\mathfrak{z}$ of $\mathfrak{b}$ in $\mathfrak{g}$ is zero. (Let $\mathfrak{h} = \mathfrak{a} + \mathfrak{z}$, which is a subalgebra by (*a*); $\mathfrak{a}$ is subinvariant in $\mathfrak{h}$; if $\mathfrak{z} \neq \{0\}$, then $\mathfrak{a} \neq \mathfrak{h}$, hence $\mathfrak{a}$ is an ideal in $\mathfrak{a}'$, with $\mathfrak{a} \neq \mathfrak{a}' \subset \mathfrak{h}$; let $a' \in \mathfrak{a}'$, $a' \notin \mathfrak{a}$ and $a' = a + z$, $a \in \mathfrak{a}$, $z \in \mathfrak{z}$, $z \neq 0$; show that $[z, \mathfrak{a}]$ is contained in $\mathfrak{a}$ and commutes with $\mathfrak{b}$, whence $[z, \mathfrak{a}] = \{0\}$ giving a contradiction.)

(*c*) Let $\mathfrak{g}$ be a Lie algebra with centre zero, $\mathfrak{D}_1$ the Lie algebra of derivations of $\mathfrak{g}$, $\mathfrak{D}_2$ the Lie algebra of derivations of $\mathfrak{D}_1, \ldots$ $\mathfrak{g}$ is identified with an ideal of $\mathfrak{D}_1$, $\mathfrak{D}_1$ with an ideal of $\mathfrak{D}_2, \ldots$ (cf. Exercise 12 (*b*)). Using (*b*) and Exercise 12 (*b*), show that the centralizer of $\mathfrak{g}$ in $\mathfrak{D}_i$ is zero for all i. (For a sequel to this exercise, cf. § 4, Exercise 15.)

16. Let $\mathfrak{g}$ be a Lie algebra and $\mathfrak{D}$ the Lie algebra of derivations of $\mathfrak{g}$. The identity mapping of $\mathfrak{D}$ defines a semi-direct product $\mathfrak{h}$ of $\mathfrak{D}$ by $\mathfrak{g}$ called the *holomorph* of $\mathfrak{g}$. $\mathfrak{g}$ is identified with an ideal and $\mathfrak{D}$ with a subalgebra of $\mathfrak{h}$.

(*a*) Show that the centralizer $\mathfrak{g}^*$ of $\mathfrak{g}$ in $\mathfrak{h}$ is the set of $\mathrm{ad}_{\mathfrak{g}}\, x - x (x \in \mathfrak{g})$.

(*b*) For every element $x + d$ of $\mathfrak{h}$ $(x \in \mathfrak{g}, d \in \mathfrak{D})$, we write

$$\theta(x + d) = \mathrm{ad}_{\mathfrak{g}}\, x - x + d.$$

Show that θ is an automorphism of $\mathfrak{h}$ of order 2 which maps $\mathfrak{g}$ onto $\mathfrak{g}^*$, so that $\mathfrak{h}$ can be considered as the holomorph of $\mathfrak{g}^*$.

(*c*) For every ideal $\mathfrak{k} \neq \{0\}$ of $\mathfrak{h}$, $\mathfrak{k} \cap (\mathfrak{g} + \mathfrak{g}^*) \neq \{0\}$. (If $\mathfrak{k} \cap \mathfrak{g} = \{0\}$, $\mathfrak{k} \subset \mathfrak{g}^*$.)

(*d*) If $\mathfrak{g}$ is commutative, every ideal $\mathfrak{k} \neq \{0\}$ of $\mathfrak{h}$ contains $\mathfrak{g}$. If K is a field and $\dim_K \mathfrak{g} < +\infty$, deduce that, if $\mathfrak{k} \neq \mathfrak{g}$, $\mathfrak{h}$ cannot be the holomorph of $\mathfrak{k}$ (use (*b*)).

17. Suppose that K is a field. Let T be an indeterminate and $L = K((T))$; L has canonically a valuation (order of the formal power series) which makes it a complete valued field.

(*a*) Let V be a finite-dimensional vector space over K. Then $V_{(L)}$ has canonically a complete metrizable topological vector space structure (*Topological Vector Spaces*, Chapter I, § 2, no. 3, Theorem 2). Let $(x_p)_{p \in \mathbf{Z}}$ be a family of elements of V such that $x_p = 0$ for p less than some rational integer. Then the series $\sum_{-\infty}^{+\infty} x_p T^p$ converges in $V_{(L)}$ and every element of $V_{(L)}$ can be expressed uniquely in this way. (Take a basis of V.)

(*b*) The vector space $\mathscr{L}_L(V_{(L)}) = \mathscr{L}_K(V))_{(L)}$ has a canonical complete metrizable topological vector space structure. Every endomorphism of $V_{(L)}$ can be expressed uniquely in the form $\sum_{-\infty}^{+\infty} u_p T^p$, where $u_p \in \mathscr{L}_K(V)$ and the u_p are zero for p less than some rational integer.

(*c*) If K is of characteristic 0 and $u \in \mathscr{L}_K(V)$, we write

$$e^{Tu} = 1 + \frac{1}{1!} uT + \frac{1}{2!} u^2T^2 + \cdots \in \mathscr{L}_L(V_{(L)}).$$

If $x \in V$ is such that $ux = 0$, then $e^{Tu}.x = x$.

(*d*) If W is another finite-dimensional vector space over K and $v \in \mathscr{L}_K(W)$, then $e^{T(u\otimes 1+1\otimes v)} = e^{Tu} \otimes e^{Tv}$.

(*e*) Deduce from (*c*) and (*d*) that, if V has a (not necessarily associative) algebra structure and u is a derivation of V, then e^{Tu} is an automorphism of the algebra $V_{(L)}$.

¶ 18. Let $\mathfrak{H}$ be a complex Hilbert space, Y and Z continuous Hermitian operators on $\mathfrak{H}$ and $B = [Z, Y]$. Suppose that Y and B are permutable.

(*a*) Show that $[Z, Y^n] = nBY^{n-1}$ (argue by induction on n).

(*b*) Let f be a continuously differentiable function of a real variable. Deduce from (*a*) that $[Z, f(Y)] = Bf'(Y)$.

(*c*) Deduce from (*b*) that $B = 0$. (Show that, if $B \neq 0$, f can be chosen such that $\|f(Y)\| \leqslant 1$ and $\|Bf'(Y)\|$ is arbitrarily large.)

(*d*) Let $\mathfrak{g}$ be the Lie algebra of continuous operators T on $\mathfrak{H}$ such that $T^* = -T$. Deduce from (*c*) that the conditions $Y \in \mathfrak{g}$, $Z \in \mathfrak{g}$, $[[Z, Y], Y] = 0$ imply $[Z, Y] = 0$.

19. (*a*) Let L be an associative algebra over K. Given k elements $x_1, \ldots, x_k$ of L, we write:

$$f_k(x_1, \ldots, x_k) = \sum_{\sigma \in \mathfrak{S}} x_{\sigma(1)} \ldots x_{\sigma(k)}.$$

In the algebra $L \otimes_K K[T_1, \ldots, T_k]$, where the T_i are indeterminates, $f_k(x_1, \ldots, x_k)$ is the coefficient of $T_1 \ldots T_k$ in $(x_1T_1 + \cdots + x_kT_k)^k$; it is also the coefficient of $T_1 \ldots T_k$ in:

$$\sum_{j=0}^{k-1} (x_1T_1 + \cdots + x_{k-1}T_{k-1})^j x_kT_k(x_1T_1 + \cdots + x_{k-1}T_{k-1})^{k-j-1}.$$

Given two elements x, y in L, we define $g_{i,k-i}(x, y)$ as the coefficient of $T_1^iT_2^{k-i}$ in $(xT_1 + yT_2)^k$. Show that

$$i!(k-i)!g_{i,k-i}(x, y) = f_k(t_1, \ldots, t_k)$$

where $t_h = x$ for $h \leqslant i$ and $t_h = y$ for $h \geqslant i + 1$.

(*b*) If L is considered as a Lie algebra over K, then in $\mathscr{L}_K(L)$:

$$(\operatorname{ad} x)^n = \sum_{i=0}^{n} (-1)^{n-i}\binom{n}{i} L_x^i R_x^{n-i},$$

where L_x (resp. R_x) is the multiplication $y \mapsto xy$ (resp. $y \mapsto yx$).

Deduce that, if K is of characteristic p (p prime), then:

$$(1) \qquad (\operatorname{ad} x)^p.y = (\operatorname{ad} x^p).y$$

$$(2) \qquad (\operatorname{ad} x)^{p-1}.y = \sum_{i=0}^{p-1} x^i y x^{p-1-i}.$$

Deduce from (2) that:

$$f_{p-1}(\mathrm{ad}\, x_1, \ldots, \mathrm{ad}\, x_{p-1}).y = f_p(x_1, \ldots, x_{p-1}, y)$$

and:

$$g_{i,p-1-i}(\mathrm{ad}\, x, \mathrm{ad}\, y).y = (p-i)g_{i,p-i}(x, y).$$

Conclude that for any two elements x, y of L:

$$(x+y)^p = x^p + y^p + \Lambda_p(x, y) \tag{3}$$

where:

$$\Lambda_p(x, y) = \sum_{i=1}^{p-1} (p-i)^{-1} g_{i,p-1-i}(\mathrm{ad}\, x, \mathrm{ad}\, y).y$$

belongs to the Lie subalgebra of L generated by x and y ("*Jacobson's formulae*").

20. Let $\mathfrak{g}$ be a Lie algebra over a ring K such that $p\mathrm{K} = \{0\}$ (p prime). A mapping $x \mapsto x^{[p]}$ of $\mathfrak{g}$ into itself is called a *p-mapping* if it satisfies the relations

$$\mathrm{ad}\, x^{[p]} = (\mathrm{ad}\, x)^p$$
$$(\lambda x)^{[p]} = \lambda^p x^{[p]}$$
$$(x+y)^{[p]} = x^{[p]} + y^{[p]} + \Lambda_p(x, y) \quad \text{(cf. Exercise 19 (b))}$$

for x, y in $\mathfrak{g}$ and $\lambda \in \mathrm{K}$. A *Lie p-algebra* over K is a set with the structure defined by giving it a Lie algebra structure over K and a p-mapping. Every associative algebra over K is a Lie p-algebra with $[x, y] = xy - yx$ and $x^{[p]} = x^p$ (Exercise 19). A mapping u of a Lie p-algebra $\mathfrak{g}$ into a Lie p-algebra $\mathfrak{g}'$ is called a *p-homomorphism* if u is a Lie algebra homomorphism and $u(x^{[p]}) = (u(x))^{[p]}$. Show that in a Lie p-algebra $\mathfrak{g}$ every p-mapping is of the form $x \mapsto x^{[p]} + f(x)$, where f is a mapping of $\mathfrak{g}$ into its centre, which is semi-linear for the endomorphism $\lambda \mapsto \lambda^p$ of K. More generally, if u is a homomorphism of $\mathfrak{g}$ onto a Lie p-algebra $\mathfrak{g}'$, $u(x^{[p]}) - (u(x))^{[p]}$ belongs to the centre of $\mathfrak{g}'$.

21. (*a*) Let K be a field of characteristic $p > 0$ and L a not necessarily associative algebra over K. Show that the derivations of L form a Lie p-algebra with the p-mapping $\mathrm{D} \mapsto \mathrm{D}^p$.

(*b*) Let L be the algebra $\mathrm{K}[\mathrm{X}]/(\mathrm{X}^p)$ and $\mathfrak{w}$ the p-algebra of derivations of L. Let D_i ($0 \leqslant i \leqslant p-1$) denote the derivation of L such that $\mathrm{D}_i(\mathrm{X}) = \mathrm{X}^i$. Show that the D_i form a basis of $\mathfrak{w}$ over K and that $[\mathrm{D}_i, \mathrm{D}_j] = (j-i)\mathrm{D}_{i+j-1}$ if $i + j \leqslant p$, $[\mathrm{D}_i, \mathrm{D}_j] = 0$ if $i + j > p$, and $\mathrm{D}_i^p = 0$, except for $i = 1$, in which case $\mathrm{D}_1^p = \mathrm{D}_1$.

(*c*) Show that, if $p > 3$, the Lie algebra $\mathfrak{w}$ only admits the ideals $\{0\}$ and $\mathfrak{w}$. (Using the multiplication table of the D_i, show that every non-zero ideal of $\mathfrak{w}$ contains a non-zero multiple of D_{p-1}.)

(*d*) Let V_k be the subspace of $\mathfrak{w}$ with basis the D_i of index $\geqslant k$. Show that, if $p \geqslant 5$, V_k is identical with the set of $Z \in \mathfrak{w}$ whose centralizer is of dimension $\geqslant k + 1$. Deduce that, if $p \geqslant 5$, the automorphism group of $\mathfrak{w}$ is solvable.

22. Let $\mathfrak{g}$ be a Lie p-group over a ring K of characteristic p (p prime). Let $x \mapsto x^p$ denote the p-mapping. For every subset E of $\mathfrak{g}$, let E^p denote the *submodule of* $\mathfrak{g}$ generated by the x^p, where $x \in E$. An ideal $\mathfrak{a} \subset \mathfrak{g}$ is called a *p-ideal* if $\mathfrak{a}^p \subset \mathfrak{a}$.

(*a*) For an ideal $\mathfrak{a} \subset \mathfrak{g}$ to be a p-ideal, it is necessary and sufficient that it be the kernel of a p-homomorphism (Exercise 20).

(*b*) The sum of two p-ideals is a p-ideal. The sum of a p-ideal and a p-subalgebra is a p-subalgebra.

(*c*) Let $\mathfrak{h}$ be a Lie subalgebra of $\mathfrak{g}$. The smallest Lie p-subalgebra containing $\mathfrak{h}$ is $\mathfrak{h} + \mathfrak{h}^p + \mathfrak{h}^{p^2} + \cdots = \overline{\mathfrak{h}}$; if we write $\mathfrak{h}_i = \mathfrak{h} + \mathfrak{h}^p + \cdots + \mathfrak{h}^{p^i}$, $\mathfrak{h}_i$ is an ideal of $\overline{\mathfrak{h}}$ and $\overline{\mathfrak{h}}/\mathfrak{h}$ is commutative. If $\mathfrak{h}$ is an ideal of $\mathfrak{g}$, so are the $\mathfrak{h}_i$ and $\overline{\mathfrak{h}}$ and the latter is the smallest p-ideal containing $\mathfrak{h}$.

(*d*) The normalizer and centralizer of an arbitrary submodule of $\mathfrak{g}$ are p-subalgebras of $\mathfrak{g}$.

23. Let $\mathfrak{g}$ be a finite-dimensional commutative Lie p-algebra over a perfect field K of characteristic $p > 0$.

(*a*) Show that $\mathfrak{g}$ is in a unique way the direct sum of two p-subalgebras $\mathfrak{h}$, $\mathfrak{k}$ such that on $\mathfrak{h}$ the p-mapping is bijective and on $\mathfrak{k}$ it is nilpotent (consider the iterations of the p-mapping on $\mathfrak{g}$; cf. *Algebra*, Chapter VIII, § 2, no. 2, Lemma 2 and Chapter VII, § 5, Exercise 20 (*a*)). $\mathfrak{h}$ is called the *p-core* of $\mathfrak{g}$.

(*b*) Suppose that K is algebraically closed. Show that there exists a basis (e_i) of $\mathfrak{h}$ such that $e_i^p = e_i$ for all i (consider a p-subalgebra of $\mathfrak{h}$ generated by a single element and show that it contains some x such that $x^p = x \neq 0$; then argue by induction on the dimension of $\mathfrak{h}$). Deduce that the p-subalgebras of $\mathfrak{h}$ are finite in number and are in one-to-one correspondence with the vector subspaces of the vector space over the prime field $\mathbf{F}_p$ generated by the e_i.

(*c*) Show that there exists a basis (f_{ij}) of $\mathfrak{k}$ ($1 \leqslant i \leqslant k$, $1 \leqslant j \leqslant s_i$ for all i) where the sequence of s_i is decreasing and which is such that $f_{1j}^p = 0$ for $1 \leqslant j \leqslant s_1$, $f_{ij}^p = f_{i-1,j}$ for $2 \leqslant i \leqslant k$, $1 \leqslant j \leqslant s_i$ (method of *Algebra*, Chapter VII, § 5, Exercise 20 (b)).

¶ 24. Let K be a (commutative) field of (arbitrary) characteristic p and X_i, Y_i, Z_i $3n$ distinct indeterminates; to abbreviate we shall denote the systems $(X_i)_{1 \leqslant i \leqslant n}$, $(Y_i)_{1 \leqslant i \leqslant n}$, $(Z_i)_{1 \leqslant i \leqslant n}$ by $\mathbf{x}$, $\mathbf{y}$, $\mathbf{z}$ respectively. Let $A_{\mathbf{x},\mathbf{y},\mathbf{z}}$ be the algebra of formal power series in the $3n$ indeterminates X_i, Y_i, Z_i with coefficients in K; we shall denote by $A_{\mathbf{x},\mathbf{y}}$ (resp. $A_{\mathbf{x}}$) the subalgebra of $A_{\mathbf{x},\mathbf{y},\mathbf{z}}$ consisting of the power series not containing $\mathbf{z}$ (resp. $\mathbf{y}$, $\mathbf{z}$). An element of $A_{\mathbf{x},\mathbf{y},\mathbf{z}}$ (resp. $A_{\mathbf{x},\mathbf{y}}$, $A_{\mathbf{x}}$) will be denoted by $u(\mathbf{x}, \mathbf{y}, \mathbf{z})$ (resp. $u(\mathbf{x}, \mathbf{y})$, $u(\mathbf{x})$);

a system $(u_i(\mathbf{x}, \mathbf{y}, \mathbf{z}))_{1 \leqslant i \leqslant n}$ of n elements of $A_{\mathbf{x},\mathbf{y},\mathbf{z}}$ will be denoted by $u(\mathbf{x}, \mathbf{y}, \mathbf{z})$; analogous notation will be used for formal power series containing only one or two of the three systems $\mathbf{x}$, $\mathbf{y}$, $\mathbf{z}$. If $u(\mathbf{x}, \mathbf{y}, \mathbf{z}) \in A_{\mathbf{x},\mathbf{y},\mathbf{z}}$ and $\mathbf{f} = (f_i)$, $\mathbf{g} = (g_i)$, $\mathbf{h} = (h_i)$ are three systems of n elements of $A_{\mathbf{x},\mathbf{y},\mathbf{z}}$ which are formal power series with no constant term, we denote by $u(\mathbf{f}(\mathbf{x}, \mathbf{y}, \mathbf{z}), \mathbf{g}(\mathbf{x}, \mathbf{y}, \mathbf{z}), \mathbf{h}(\mathbf{x}, \mathbf{y}, \mathbf{z}))$ the formal power series obtained by substituting f_i for X_i, g_i for Y_i, h_i for Z_i $(1 \leqslant i \leqslant n)$ in u. For every system $\alpha = (\alpha_1, \ldots, \alpha_n)$ of n integers $\geqslant 0$, let $\mathbf{x}^\alpha$ denote the monomial $X_1^{\alpha_1} \ldots X_n^{\alpha_n}$; $\mathbf{y}^\alpha$ and $\mathbf{z}^\alpha$ are defined similarly. Let ε_i denote the system α for which $\alpha_j = 0$ if $j \neq i$, $\alpha_i = 1$. We denote by $\mathbf{e}$ the system $(0, \ldots, 0)$ of n elements of $A_{\mathbf{x},\mathbf{y},\mathbf{z}}$.

(*a*) A *formal group law* over K (or, by an abuse of language, a *formal group* over K) of dimension n is a system $G = \mathbf{f}(\mathbf{x}, \mathbf{y})$ of n elements of $A_{\mathbf{x},\mathbf{y}}$ with the following properties: (1) $\mathbf{f}(\mathbf{x}, \mathbf{f}(\mathbf{y}, \mathbf{z})) = \mathbf{f}(\mathbf{f}(\mathbf{x}, \mathbf{y}), \mathbf{z})$; (2) $\mathbf{f}(\mathbf{e}, \mathbf{y}) = \mathbf{y}$, $\mathbf{f}(\mathbf{x}, \mathbf{e}) = \mathbf{x}$. Show that necessarily $f_i(\mathbf{x}, \mathbf{y}) = X_i + Y_i + g_i(\mathbf{x}, \mathbf{y})$, where g_i contains only monomials of degree $\geqslant 2$, each of which contains at least one X_j and at least one Y_j. Show that there exists one and only one system $\mathbf{h}(\mathbf{x})$ of n elements of $A_{\mathbf{x}}$ such that $\mathbf{f}(\mathbf{x}, \mathbf{h}(\mathbf{x})) = \mathbf{f}(\mathbf{h}(\mathbf{x}), \mathbf{x}) = \mathbf{e}$ (*Algebra*, Chapter IV, § 5, no. 9, Proposition 10). G is called *commutative* if $\mathbf{f}(\mathbf{y}, \mathbf{x}) = \mathbf{f}(\mathbf{x}, \mathbf{y})$.

(*b*) For all $u \in A_{\mathbf{x}}$, let $L_{\mathbf{y}}u$ denote the element $u(\mathbf{f}(\mathbf{y}, \mathbf{x}))$ of $A_{\mathbf{x},\mathbf{y}}$. A derivation D of $A_{\mathbf{x}}$ can be canonically extended to a derivation of $A_{\mathbf{x},\mathbf{y}}$ (also denoted D by an abuse of notation) by the condition that $D(Y_i) = 0$ for all i (*Algebra*, Chapter IV, § 5, no. 8, Proposition 6). D is said to be *left invariant* for the formal group G in question if $L_{\mathbf{y}}D = DL_{\mathbf{y}}$. We denote by $D^{(\mathbf{e})}$ the linear mapping of $A_{\mathbf{x},\mathbf{y}}$ into $A_{\mathbf{y}}$ defined by $D^{(\mathbf{e})}(u) = (Du)(\mathbf{e}, \mathbf{y})$; it maps $A_{\mathbf{x}}$ into K and is determined by its restriction $A_{\mathbf{x}}$. Show that, for D to be left invariant, it is necessary and sufficient that for all $v \in A_{\mathbf{x}}$, $D^{(\mathbf{e})}(L_{\mathbf{y}}v) = (Dv)(\mathbf{y})$.

Let D_i be the derivation $\partial/\partial X_i$ of $A_{\mathbf{x}}$ $(1 \leqslant i \leqslant n)$; there exists one and only one left invariant derivation T_i such that $T_i^{(\mathbf{e})} = D_i^{(\mathbf{e})}$. Show that the T_i are linearly independent over K and that every left invariant derivation is a linear combination of the T_i with coefficients in K. Deduce that, for the bracket $[D, D']$ and the p-mapping $D \mapsto D^p$ (when $p > 0$), the set $\mathfrak{g}$ of left invariant derivations is a *Lie algebra* (a *Lie p-algebra* if $p > 0$) called the *Lie algebra of the formal Lie group* G.

(*c*) Show that for every formal power series $u \in A_{\mathbf{x}}$:

$$u(\mathbf{f}(\mathbf{x}, \mathbf{y})) = u(0) + \sum_{i=1}^{n} Y_i T_i u + o_2(\mathbf{x}, \mathbf{y}) \tag{1}$$

where in the series o_2 all the monomials are of degree $\geqslant 2$ in the Y_j. Deduce that

$$u(\mathbf{f}(\mathbf{x}, \mathbf{f}(\mathbf{y}, \mathbf{z}))) = u(0) + \sum_{i=1}^{n} (Y_i + Z_i) T_i u + \sum_{i,j} Y_i Z_j ((T_i T_j)(u)) + o_3(\mathbf{x}, \mathbf{y}, \mathbf{z}) \tag{2}$$

where in the series o_3 all the monomials are of degree $\geqslant 3$ in the set of Y_i and Z_i. Deduce that:

$$(3)\quad u(\mathbf{f}(\mathbf{x},\mathbf{y})) - u(\mathbf{f}(\mathbf{y},\mathbf{x})) = \sum_{i<j} (X_iY_j - X_jY_i)([T_i, T_j]^{(e)}(u)) + o'_3(\mathbf{x},\mathbf{y})$$

where all the terms of o'_3 are of total degree $\geqslant 3$. Show that, if G is commutative, $\mathfrak{g}$ is commutative.

(*d*) Let G′ be another formal group of dimension m, whose group law is $\mathbf{f}'$. A *formal homomorphism* of G into G′ is a system $\mathbf{F} = (F_j(\mathbf{x}))_{1 \leqslant j \leqslant m}$ of elements of $A_{\mathbf{x}}$ such that $\mathbf{f}'(\mathbf{F}(\mathbf{x}), \mathbf{F}(\mathbf{y})) = \mathbf{F}(\mathbf{f}(\mathbf{x},\mathbf{y}))$. For every left invariant derivation $D \in \mathfrak{g}$, show that $v \mapsto D(v \circ \mathbf{F})$ is a left invariant derivation belonging to the Lie algebra $\mathfrak{g}'$ of G′. If it is denoted by $\mathbf{F}^*(D)$, $\mathbf{F}^*$ is a homomorphism (a p-homomorphism for $p > 0$) of $\mathfrak{g}$ into $\mathfrak{g}'$. If $(T_i)_{1 \leqslant i \leqslant n}$, $(T'_j)_{1 \leqslant j \leqslant m}$ are the bases of $\mathfrak{g}$ and $\mathfrak{g}'$ such that $T_i^{(e)} = D_i^{(e)}$ and $T_j'^{(e)} = D_j^{(e)}$, show that the matrix of $\mathbf{F}^*$ relative to these bases is the matrix $(\partial F_j/\partial X_i)_0$ (constant term of $\partial F_j/\partial X_i$, where i is the index of the rows and j that of the columns).

25. The *formal linear group* in n variables over the field K, denoted by $\mathbf{GL}(n, K)$ (when no confusion can arise), is the formal group of dimension n^2 whose group law is defined by $f_{ij}(\mathbf{u},\mathbf{v}) = -\delta_{ij} + \sum_{k=1}^{n} (\delta_{ik} + U_{ik})(\delta_{kj} + V_{kj})$ (δ_{ij} the Kronecker index; the $3n^2$ indeterminates of the general definition of Exercise 24 are here denoted by U_{ij}, V_{ij}, W_{ij} with 2 indices varying from 1 to n). Show that, if $D_{ij} = \partial/\partial U_{ij}$, the left invariant derivations X_{ij} such that $X_{ij}^{(e)} = D_{ij}^{(e)}$ are given by

$$X_{ij} = (1 + U_{ii})D_{ij} + \sum_{k \neq i} U_{ki}D_{kj}.$$

The Lie algebra of $\mathbf{GL}(n, K)$ is identified with $\mathfrak{gl}(n, K)$ by identifying X_{ij} with the element E_{ij} of the canonical basis (use formula (3) of Exercise 24). If K is of characteristic $p > 0$, $X_{ii}^p = X_{ii}$ and $X_{ij}^p = 0$ for $i \neq j$.

¶ 26. (*a*) Let K be a field of characteristic $\neq 2$, E an n-dimensional vector space over K and Φ a non-degenerate symmetric bilinear form on E. Suppose that E has an orthogonal basis with respect to Φ and let R denote the (diagonal) matrix of Φ with respect to this basis. We know (*Algebra*, Chapter IX, § 4, Exercise 11) that every matrix (relative to the basis in question) U of the orthogonal group $\mathbf{G}(\Phi)$ such that $\det(I + U) \neq 0$ can be written

$$U = (I - R^{-1}S)^{-1}(I + R^{-1}S),$$

where $S = R(U - I)(U + I)^{-1}$ is an alternating matrix such that $\det(I - R^{-1}S) \neq 0$; conversely, for every matrix S satisfying these conditions, $U = (I - R^{-1}S)^{-1}(I + R^{-1}S)$ is a matrix of $\mathbf{G}(\Phi)$ such that $\det(I + U) \neq 0$. Let S_{ij}, S'_{ij}, S''_{ij} be $3n(n-1)/2$ indeterminates ($1 \leqslant i < j \leqslant n$)

and let L be a field containing the ring of formal power series $\mathrm{K}[[\mathrm{S}_{ij}, \mathrm{S}'_{ij}, \mathrm{S}''_{ij}]]$. If we denote by S, S', S'' the matrices

$$\sum_{i<j} \mathrm{S}_{ij}(E_{ij} - E_{ji}), \qquad \sum_{i<j} \mathrm{S}'_{ij}(E_{ij} - E_{ji}), \qquad \sum_{i<j} \mathrm{S}''_{ij}(E_{ij} - E_{ji}),$$

then $\det(I - R^{-1}S) \neq 0$ and the analogues for S' and S'' hold; if we write

$$U = (I - R^{-1}S)^{-1}(I + R^{-1}S), \qquad U' = (I - R^{-1}S')^{-1}(I + R^{-1}S'),$$

then also $\det(I + UU') \neq 0$. We write

$$F(S, S') = R(UU' - I)(UU' + I)^{-1} = (f_{ij}(S, S'));$$

the $f_{ij}(S, S')$ belong to the ring of formal power series $\mathrm{K}[[\mathrm{S}_{ij}, \mathrm{S}'_{ij}]]$ and, by considering $F(S, S')$ as a formal power series with coefficients in the algebra of matrices of order n over K, we can write:

$$F(S, S') = S + S' + o_2(S, S')$$

where in the elements of the matrix $o_2(S, S')$ the terms are all of degree $\geqslant 1$ in the S_{ij} and of degree $\geqslant 1$ in the S'_{ij}; then similarly:

$$U = I + 2R^{-1}S + o'_2(S)$$

the elements of $o'_2(S)$ being formal power series of order $\geqslant 2$. Show that $F(S, S')$ is a formal group law over K of dimension $n(n-1)/2$; this formal group is called the *formal orthogonal group* (corresponding to Φ) and is denoted by $\mathbf{G}(\Phi)$ by an abuse of notation.

(*b*) If we write $M(S) = (m_{ij}(S)) = (I - R^{-1}S)^{-1}(I + R^{-1}S)$, M is a formal homomorphism of $\mathbf{G}(\Phi)$ into $\mathbf{GL}(n, \mathrm{K})$. Let $\mathfrak{o}(\Phi)$ be the Lie algebra of $\mathbf{G}(\Phi)$ and let $(\mathrm{T}_{ij})_{i<j}$ denote the basis of this algebra such that $\mathrm{T}_{ij}^{(e)} = \mathrm{D}_{ij}^{(e)}$ (cf. Exercise 24 (*d*)); if an element $\mathrm{D} = \sum_{i<j} c_{ij}\mathrm{T}_{ij}$ of $\mathfrak{o}(\Phi)$ is identified with the matrix $D = \sum_{i<j} c_{ij}(E_{ij} - E_{ji})$, show that (in the notation of Exercise 24 (*d*) with the identification made in Exercise 25 of the Lie algebra of $\mathbf{GL}(n, \mathrm{K})$ with $\mathfrak{gl}(n, \mathrm{K})$):

$$M^*(D) = 2R^{-1}D.$$

Deduce that M^* is an isomorphism of $\mathfrak{o}(\Phi)$ onto the subalgebra of $\mathfrak{gl}(n, \mathrm{K})$ consisting of the matrices X such that ${}^tXR + RX = 0$ (for every ordered pair (a, b) of elements of E, identified with matrices with one column, consider $\Phi(a + M(S).a, b + M(S).b) = {}^ta.(I + {}^tM(S))R(I + M(S)).b$ as a formal power series reduced to its constant term and use the fact that the subalgebra of $X \in \mathfrak{gl}(n, \mathrm{K})$ such that ${}^tXR + RX = 0$ is of dimension $n(n-1)/2$).

(*c*) Define similarly the *formal symplectic group* in $2n$ variables over an arbitrary (commutative) field and show that its Lie algebra is identified with the subalgebra of $\mathfrak{gl}(2n, K)$ consisting of the matrices X such that

$$^{t}XA + AX = 0, \quad \text{where} \quad A = \begin{pmatrix} 0 & I_n \\ -I_n & 0 \end{pmatrix}.$$

27. Let K be a field of characteristic $p > 0$ and G the 2-dimensional formal group defined by $f_1(\mathbf{x}, \mathbf{y}) = X_1 + Y_1 + X_1Y_1, f_2(\mathbf{x}, \mathbf{y}) = X_2 + Y_2(1 + X_1^p.)$ Show that G is not commutative but that its Lie algebra is commutative.

§ 2

1. We adopt the notation T, J, T_n of Definition 1 and no. 6. Let t be a homogeneous tensor of T of order p and σ a permutation of $\{1, 2, \ldots, p\}$. Then $t - \sigma t \in J + T_{p-1}$. (Reduce it to the case where σ is a transposition of two consecutive integers.)

2. Suppose that K is a field. Let $\mathfrak{g}$ be a Lie K-algebra and U its enveloping algebra.

(*a*) Let u, v be in U. If u is of filtration $>n$, and v of filtration $>p$, then uv is of filtration $>n + p$ (use Theorem 1).

(*b*) Deduce from (*a*) that the only invertible elements of U are the scalars.

(*c*) Deduce from (*b*) that the Jacobson radical of U is zero.

3. Suppose that K is a field. Let $\mathfrak{g}$ be a Lie K-algebra and U its enveloping algebra. The left regular representation of U corresponds to a representation ρ of $\mathfrak{g}$ on the space U. Show that the set U_+ of elements of U with no constant term is stable under ρ and that U_+ has no supplement in U stable under ρ. In particular, ρ is not semi-simple. (This will be considered again in Theorem 2 of § 6.)

4. Suppose that K is a field.

(*a*) Verify that there exists a 3-dimensional Lie algebra $\mathfrak{g}$ with basis (x, y, z) such that $[x, y] = z$, $[x, z] = [y, z] = 0$. Let U be the enveloping algebra of $\mathfrak{g}$. Show that the centre of U is the subalgebra generated by 1 and z. (Consider, for all $\xi \in \mathfrak{g}$, the derivation of the symmetric algebra S of $\mathfrak{g}$ which extends $\mathrm{ad}_{\mathfrak{g}}\, \xi$ and look for the elements of S annihilated by these derivations; then apply the final remark of no. 8.)

(*b*) Verify that there exists a 3-dimensional Lie algebra $\mathfrak{g}$ with basis (x, y, z) such that $[x, y] = y$, $[x, z] = z$, $[y, z] = 0$. Show that the centre of the enveloping algebra of $\mathfrak{g}$ reduces to the scalars. (Same method.)

5. Suppose that K is of characteristic $p > 0$ (p prime). Let $\mathfrak{g}$ be a Lie algebra over K with a basis, canonically identified with a submodule of its enveloping algebra U. For an endomorphism ϕ of the K-module $\mathfrak{g}$ to be a p-mapping, it is necessary and sufficient that $x \mapsto x^p - \phi(x)$ be a semi-linear mapping (for

$\lambda \mapsto \lambda^p$) of $\mathfrak{g}$ into the centre of U. Deduce that if (b_λ) is a basis of $\mathfrak{g}$, for there to exist a p-mapping on $\mathfrak{g}$, it is necessary and sufficient that, for all λ, there exist $c_\lambda \in \mathfrak{g}$ such that $(\operatorname{ad} b_\lambda)^p = \operatorname{ad} c_\lambda$; if this is so, there then exists one and only one p-mapping $x \mapsto x^{[p]}$ such that $b_\lambda^{[p]} = c_\lambda$ for all λ.

6. Let $\mathfrak{g}$ be a Lie p-algebra over a ring K such that $pK = \{0\}$ (p prime), U its enveloping algebra and σ the canonical mapping of $\mathfrak{g}$ into U. Let J be the two-sided ideal of U generated by the elements $(\sigma(x))^p - \sigma(x^{[p]})$ where x runs through $\mathfrak{g}$. The associative algebra $\tilde{U} = U/J$ is called the *restricted enveloping algebra* of $\mathfrak{g}$. The mapping σ defines when passing to the quotient a mapping $\tilde{\sigma}$ (called canonical) of $\mathfrak{g}$ into $\tilde{U}$, which is a p-homomorphism (when $\tilde{U}$ is considered as a Lie p-algebra).

(*a*) The algebra $\tilde{U}$ and the mapping $\tilde{\sigma}$ are solutions of a universal mapping problem: for every p-homomorphism f of $\mathfrak{g}$ into an associative algebra B over K (considered as a Lie p-algebra), there exists one and only one K-homomorphism f' of $\tilde{U}$ into B (with their associative algebra structures) such that $f = f' \circ \tilde{\sigma}$.

(*b*) Show that, if $(b_\lambda)_{\lambda \in \Lambda}$ is a basis of $\mathfrak{g}$ (where Λ is totally ordered), $\tilde{\sigma}$ is injective and that, if x and $\tilde{\sigma}(x)$ are identified for $x \in \mathfrak{g}$, the elements $\Pi_\lambda b_\lambda^{\nu_\lambda}$ (where the λ are in increasing order, the ν_λ are zero except for a finite number and $0 \leqslant \nu_\lambda < p$ for all λ) form a basis of $\tilde{U}$. (Canonically identifying $\mathfrak{g}$ with a submodule of U, we write $\phi(x) = x^p - x^{[p]}$ for all $x \in \mathfrak{g}$. For every composite index $\alpha = (\alpha_\lambda) \in \mathbf{N}^{(L)}$, let $\alpha_\lambda = \beta_\lambda + p\gamma_\lambda$ with $0 \leqslant \beta_\lambda < p$ and let

$$T_\alpha = (\Pi_\lambda b_\lambda^{\nu_\lambda})(\Pi_\lambda(\phi(b_\lambda))^{\gamma_\lambda}).$$

Show that the T_α form a basis of U and the T_α such that $\gamma = (\gamma_\lambda) \neq 0$ form a basis of J; observe that the $\phi(b_\lambda)$ belong to the centre of U.)

7. Let $\mathfrak{g}$ be a Lie p-algebra over a ring K such that $pK = \{0\}$ (p prime). A derivation D of $\mathfrak{g}$ is called a *p-derivation* if $D(x^p) = (\operatorname{ad} x)^{p-1}.Dx$ for all $x \in \mathfrak{g}$. Every inner derivation is a p-derivation.

(*a*) If L is an associative K-algebra, every derivation of L is a p-derivation when L is considered as a Lie p-algebra. (Use formula (2) of Exercise 19 of § 1.)

(*b*) Suppose that $\mathfrak{g}$ has a basis. For a derivation of $\mathfrak{g}$ to be a p-derivation, it is necessary and sufficient that it can be extended to a derivation of the restricted enveloping algebra of $\mathfrak{g}$. Deduce that the p-derivations of $\mathfrak{g}$ form a Lie p-subalgebra of the p-algebra of derivations of $\mathfrak{g}$.

(*c*) If D is a p-derivation of $\mathfrak{g}$, the kernel of D is a p-subalgebra of $\mathfrak{g}$.

(*d*) For every derivation D of $\mathfrak{g}$, $D(x^p) - (\operatorname{ad} x)^{p-1}.\ Dx$ belongs to the centre of $\mathfrak{g}$ for all $x \in \mathfrak{g}$ (use formula (2) of Exercise 19 of § 1, applied to $L = \mathscr{L}(\mathfrak{g})$).

8. Show that Theorem 1 remains valid if the module $\mathfrak{g}$ is a direct sum of

monogenous modules. (Replace the module P in the proof by the symmetric algebra of $\mathfrak{g}$.)

9. Let k be the field with two elements, V the vector space k^3, (x_1, x_2, x_3) its canonical basis and K the exterior algebra of V, which is an 8-dimensional commutative algebra over k. Let $\mathfrak{g}$ be the Lie K-algebra admitting the basis $(e_1, e_2, e_3, e_{12}, e_{13}, e_{23})$ such that

$$[e_1, e_2] = [e_2, e_1] = e_{12} \quad [e_1, e_3] = [e_3, e_1] = e_{13}, \quad [e_2, e_3] = [e_3, e_2] = e_{23}$$

and the other brackets are zero. Let $\mathfrak{h}$ be the ideal of $\mathfrak{g}$ generated by $u = x_1e_1 + x_2e_2 + x_3e_3$. As a module, $\mathfrak{h}$ is generated by u, $[u, e_1] = x_2e_{12} + x_3e_{13}$, $[u, e_2] = x_1e_{12} + x_3e_{23}$, $[u, e_3] = x_1e_{13} + x_2e_{23}$. Let

$$v = x_1x_2e_{12} + x_1x_3e_{13} + x_2x_3e_{23}.$$

(*a*) Show that $v \notin \mathfrak{h}$. (Consider the K-linear form ϕ on $\mathfrak{g}$ such that $\phi(e_1) = \phi(e_2) = \phi(e_3) = 0$, $\phi(e_{12}) = x_3$, $\phi(e_{13}) = x_2$, $\phi(e_{23}) = x_1$.)

(*b*) Let f be a linear mapping of $\mathfrak{g}$ into an associative K-algebra such that $f[(x, y]) = f(x)f(y) - f(y)f(x)$ for all x, y in $\mathfrak{g}$. Show that $f(v) = f(u)^2$.

(*c*) Deduce from (*a*) and (*b*) that the canonical mapping of the Lie algebra $\mathfrak{g}/\mathfrak{h}$ into its enveloping algebra is not injective.

10. Let $\mathfrak{g}$ be a finite-dimensional Lie algebra over a field, U its enveloping algebra and U_n the set of elements of U of filtration $\leqslant n$.

(*a*) Let $x \in U_n$, $y \in U_n$ be two non-zero elements. Show that there exists $u \in U$, $v \in V$ such that $ux = vy$. (Compare $\dim(U_px) = \dim U_p$, $\dim(U_py) = \dim U_p$ and $\dim U_{p+n}$. Deduce that $U_px \cap U_py \neq \{0\}$ for p sufficiently large.)

(*b*) Show that U admits a field of left quotients (*Algebra*, Chapter I, § 9, Exercise 8) which is at the same time a field of right quotients.

§ 3

1. Let $\mathfrak{g}$ be a Lie algebra, ρ a representation of $\mathfrak{g}$ on a K-module M and σ the associated representation of $\mathfrak{g}$ on the tensor algebra of M. Show that the submodule of symmetric tensors and the submodule of skew-symmetric tensors are stable under σ.

2. Let $\mathfrak{g}$ be a Lie algebra, ρ a representation of $\mathfrak{g}$ on a K-module M and σ the associated representation of $\mathfrak{g}$ on $Q = \mathscr{L}(M, M; M)$. For $f \in Q$ to be invariant under σ, it is necessary and sufficient that the $\rho(x)$ be derivations of M with the multiplication defined by f.

3. Let V be a finite-dimensional vector space over a perfect field K.

(*a*) The identity representation of $\mathfrak{gl}(V)$ defines canonically representations of $\mathfrak{gl}(V)$ on $\bigotimes^p V$ and $\bigotimes^q V^*$ and hence a representation $u \mapsto u_q^p$ of $\mathfrak{gl}(V)$ on

$V_q^p = \left(\bigotimes^p V\right) \otimes \left(\bigotimes^q V^*\right)$. Show that u_q^p is semi-simple if u is. (Reduce it, by extending the base field, to the case where K is algebraically closed.) Show that u_q^p is nilpotent if u is. Deduce that, if s and n are the semi-simple and nilpotent components of u, s_q^p and n_q^p are the semi-simple and nilpotent components of u_q^p.

(*b*) Deduce from (*a*) that, if an element of V_q^p is annihilated by u_q^p, it is annihilated by s_q^p and n_q^p.

(*c*) Deduce from (*b*) and Exercise 2 that, if V has a not necessarily associative algebra structure and u is a derivation of V, then s and n are derivations of V.

¶ 4. Suppose that K is a field. Let $\mathfrak{g}$ be a Lie K-algebra.

(*a*) Let M and N be $\mathfrak{g}$-modules, where N is finite-dimensional over K. Let $(f_j)_{1 \leqslant j \leqslant n}$ be a basis of N over K. If there are elements $e_1, \ldots, e_n$ of M, not all zero such that $\sum_{i=1}^{n} e_i \otimes f_i$ is invariant in $M \otimes N$, then the subspace M_1 of M generated by the e_i is stable under the x_M; if further N is simple the representation of $\mathfrak{g}$ on M_1 is dual to the representation of $\mathfrak{g}$ on N.

(*b*) Let M_1, M_2, N, N* be $\mathfrak{g}$-modules of finite dimension over K. Suppose that M_1 and M_2 are simple and that the representations of $\mathfrak{g}$ on N and N* are dual. For the $\mathfrak{g}$-module M_1 to be isomorphic to a sub-$\mathfrak{g}$-module of $M_2 \otimes N$, it is necessary and sufficient that M_2 be isomorphic to a sub-$\mathfrak{g}$-module of $M_1 \otimes N^*$. (Use Proposition 4; consider the representation of $\mathfrak{g}$ on $M_1^* \otimes M_2 \otimes N$ and apply (*a*) to the dual representation of the latter.)

¶ 5. Suppose that K is a field of characteristic 0. Let V be a finite-dimensional vector K-space, $\mathfrak{g} = \mathfrak{sl}(V)$, U the enveloping algebra of $\mathfrak{g}$, U_n the set of elements of filtration $\leqslant n$ in U, U^n the set of images in U of homogeneous symmetric tensors of order n over $\mathfrak{g}$ and $(\alpha_1, \ldots, \alpha_m)$ a basis of the vector space $\mathfrak{g}$. Let $W_s = \bigotimes^s V$. The identity representation of $\mathfrak{g}$ defines a representation of $\mathfrak{g}$ on W_s which extends to a homomorphism π_s of U into the algebra $M_s = \mathscr{L}(W_s)$.

(*a*) Let M'_s be the subspace of M_s generated by the elements of the form $\beta_1 \otimes \beta_2 \otimes \cdots \otimes \beta_s$, where $\beta_i \in \mathscr{L}(V)$ and $\beta_i = 1$ for at least one i. Let z be an element of U_s and $z_0 = \sum_{1 \leqslant i_1, \ldots, i_s \leqslant m} a_{i_1 \ldots i_s} \alpha_{i_1} \ldots \alpha_{i_s}$ its component in U^s, where the $a_{i_1 \ldots i_s}$ are symmetric with respect to permutation of indices. Show that

$$\pi_s(z) \equiv s! \sum_{1 \leqslant i_1, \ldots, i_s \leqslant m} a_{i_1 \ldots i_s} \alpha_{i_1} \otimes \cdots \otimes \alpha_{i_s} \quad (\text{mod. } M'_s).$$

(*b*) Deduce from (*a*) that for all $z \in U$ there exists a finite-dimensional representation π of U such that $\pi(z) \neq 0$. (For a sequel to this exercise, cf. § 7, Exercise 3.)

6. Suppose that K is a field. Let $\mathfrak{g}$ be a Lie K-algebra, $x \mapsto x_M$ a finite-dimensional representation of $\mathfrak{g}$, $(e_1, \ldots, e_n)$ a basis of M and $(e_1^*, \ldots, e_n^*)$ the dual basis.

(*a*) If f is a linear form on $\bigotimes^r$ M, then:

$$f = \sum_{1 \leqslant i_1, \ldots, i_r \leqslant n} f(e_{i_1} \otimes \cdots \otimes e_{i_r})(e_{i1}^* \otimes \cdots \otimes e_{i_r}^*).$$

(*b*) Let β be a non-degenerate bilinear form on M which is invariant under $\mathfrak{g}$. Let $(e'_1, \ldots, e'_n)$ be the basis of M such that $\beta(e_i, e'_j) = \delta_{ij}$. Let f be an invariant linear form on $\bigotimes^r$ M. Deduce from (*a*) that

$$\sum_{1 \leqslant i_1, \ldots, i_r \leqslant n} f(e_{i_1} \otimes \cdots \otimes e_{i_r})(e'_{i_1} \otimes \cdots \otimes e'_{i_r})$$

is an invariant element of $\bigotimes^r$ M independent of the choice of basis (e_i).

(*c*) Let U be the enveloping algebra of $\mathfrak{g}$ and U* the dual space of U. The adjoint representation of $\mathfrak{g}$ can be extended to a representation $x \mapsto x_U$ of $\mathfrak{g}$ on U defined by $x_U u = xu - ux$ for all $u \in$ U. Hence U* has a $\mathfrak{g}$-module structure. For an element f of U* to be invariant, it is necessary and sufficient that $f(uv) = f(vu)$ for all u, v in U.

(*d*) Let $\mathfrak{h}$ be a finite-dimensional ideal of $\mathfrak{g}$ and γ an invariant bilinear form on $\mathfrak{g}$ whose restriction to $\mathfrak{h}$ is non-degenerate. Let $(y_i)_{1 \leqslant i \leqslant n}$, $(y'_j)_{1 \leqslant j \leqslant n}$ be two bases of $\mathfrak{h}$ such that $\gamma(y_i, y'_j) = \delta_{ij}$. Let r be an integer $\geqslant 1$. Let f be a linear form on U such that $f(uv) = f(vu)$ for all u, v in U. Deduce from (b) and (*c*) that

$$\sum_{1 \leqslant i_1, \ldots, i_r \leqslant n} f(y_{i_1} y_{i_2} \cdots y_{i_r}) y'_{i_1} y'_{i_2} \cdots y'_{i_r}$$

is an element of the centre of U independent of the choice of basis (y_i). In particular recover the Casimir elements.

(*e*) Let θ be a permutation of $\{1, \ldots, r\}$. Arguing similarly, show that

$$\sum_{1 \leqslant i_1, \ldots, i_r \leqslant n} f(y_{i_{\theta(1)}} y_{i_{\theta(2)}} \cdots y_{i_{\theta(r)}}) y'_{i_1} y'_{i_2} \cdots y'_{i_r}$$

is an element of the centre of U independent of the choice of basis (y_i).

7. Let $\mathfrak{g}$ be a complex Lie algebra, β its Killing form and $\mathfrak{g}_0$ the Lie algebra derived from $\mathfrak{g}$ by restricting the base field to **R**. Show that the Killing form of $\mathfrak{g}_0$ is twice the real part of β.

8. Let $\mathfrak{g}$ be a real Lie algebra. The multilinear form

$$(x_1, \ldots, x_n) \mapsto \mathrm{Tr}(\mathrm{ad}\, x_1 \circ \mathrm{ad}\, x_2 \circ \cdots \circ \mathrm{ad}\, x_n)$$

on $\mathfrak{g}^n$ is invariant.

9. Let $\mathfrak{g}$ be a Lie K-algebra, ρ and ρ' semi-simple representations of $\mathfrak{g}$ on K-modules V, V' and ϕ a homomorphism of the $\mathfrak{g}$-module V onto the $\mathfrak{g}$-module V'. Show that the image under ϕ of the set of invariants of V is the set of invariants of V' (use Proposition 6).

10. Suppose that K is a field. Let $\mathfrak{g}$ be a Lie K-algebra and M_1, M_2 two non-isomorphic simple $\mathfrak{g}$-modules of finite dimension over K. If K_1 is a separable extension of K, show that there exists no simple $\mathfrak{g}_{(K_1)}$-module isomorphic both to a sub-$\mathfrak{g}_{(K_1)}$-module of $M_{1(K_1)}$ and to a sub-$\mathfrak{g}_{(K_1)}$-module of $M_{2(K_1)}$. (Use Proposition 4 and note that the existence of an invariant element $\neq 0$ in $M^*_{1(K_1)} \otimes_{K_1} M_{2(K_1)}$ implies the existence of an invariant element $\neq 0$ in $M_1^* \otimes_K M_2$ (*Algebra*, Chapter II, § 5, no. 3, Theorem 1.)

11. Let $\mathfrak{w}$ be the Lie p-algebra defined in Exercise 21 of § 1. Show that the Killing form of $\mathfrak{w}$ is zero.

¶ 12. Suppose that K is a field. Let $\mathfrak{g}$ be a Lie K-algebra and M a $\mathfrak{g}$-module. Let $C^p(\mathfrak{g}, M)$ denote the space of alternating linear mappings of $\mathfrak{g}^p$ into M. We write $C^0(\mathfrak{g}, M) = M$ and, for $p < 0$, $C^p(\mathfrak{g}, M) = \{0\}$. Let $C^*(\mathfrak{g}, M)$ be the direct sum of the $C^p(\mathfrak{g}, M)$. The elements of $C^*(\mathfrak{g}, M)$ are called the *cochains of* $\mathfrak{g}$ *with values in* M; those of $C^p(\mathfrak{g}, M)$ are said to be of degree p. For all $y \in \mathfrak{g}$, we denote by $i(y)$ the endomorphism of $C^*(\mathfrak{g}, M)$ which maps each subspace $C^p(\mathfrak{g}, M)$ into $C^{p-1}(\mathfrak{g}, M)$ and which, for $p > 0$, is given by the formula:

(1) $$(i(y)f)(x_1, \ldots, x_{p-1}) = f(y, x_1, \ldots, x_{p-1}).$$

We have $i(y)^2 = 0$.

(*a*) The adjoint representation of $\mathfrak{g}$ and the representation of $\mathfrak{g}$ on M define a representation of $\mathfrak{g}$ on the space of multilinear mappings of $\mathfrak{g}^p$ into M. Show that $C^p(\mathfrak{g}, M)$ is stable under this representation. Let θ be the representation of $\mathfrak{g}$ on $C^*(\mathfrak{g}, M)$ thus defined. Show that:

$$\theta(x)i(y) - i(y)\theta(x) = i([x, y])$$

for all x, y in $\mathfrak{g}$.

(*b*) Show that there exists one and only one endomorphism d of $C^*(\mathfrak{g}, M)$, mapping $C^p(\mathfrak{g}, M)$ into $C^{p+1}(\mathfrak{g}, M)$, such that

(3) $$di(y) + i(y)d = \theta(y)$$

for all $y \in \mathfrak{g}$. (Argue by induction on the degree of the cochains.) Show that, for $f \in C^p(\mathfrak{g}, M)$:

$$df(x_1, x_2, \ldots, x_{p+1}) = \sum_{i<j} (-1)^{i+j} f([x_i, x_j], x_1, \ldots, \hat{x}_i, \ldots, \hat{x}_j, \ldots, x_{p+1})$$
$$+ \sum_i (-1)^{i+1} (x_i)_M f(x_1, \ldots, \hat{x}_i, \ldots, x_{p+1})$$

(where the symbol ^ over a letter means that it is omitted).

(*c*) Show that, for all $y \in \mathfrak{g}$,

$$d\theta(y) = \theta(y)d. \tag{4}$$

(Show first, using (2) and (3), that $d\theta(y) - \theta(y)d$ commutes with every $i(x)$. Then argue by induction on the degree of the cochains.)

(*d*) Show that $d^2 = 0$. (Show first, using (3) and (4), that d^2 commutes with every $i(x)$. The argue by induction on the degree of the cochains.)

(*e*) The restriction of d to $C^p(\mathfrak{g}, M)$ has a kernel $Z^p(\mathfrak{g}, M)$ whose elements are called *cocycles* of degree p with values in M. The restriction of d to $C^{p-1}(\mathfrak{g}, M)$ has image $B^p(\mathfrak{g}, M)$, whose elements are called *coboundaries* of degree p with values in M. Then $B^p(\mathfrak{g}, M) \subset Z^p(\mathfrak{g}, M)$. The quotient space

$$Z^p(\mathfrak{g}, M)/B^p(\mathfrak{g}, M) = H^p(\mathfrak{g}, M)$$

is called the *cohomology space* of $\mathfrak{g}$ of degree p with values in M. The direct sum of the $H^p(\mathfrak{g}, M)$ is denoted by $H^*(\mathfrak{g}, M)$. Show that $H^0(\mathfrak{g}, M)$ is identified with the set of invariants of M. Let ϕ be a homomorphism of the $\mathfrak{g}$-module M into the G-module N. For all $f \in C^p(\mathfrak{g}, M)$, $\phi \circ f \in C^p(\mathfrak{g}, N)$, so that ϕ can be extended to a K-linear mapping $\phi' : C^*(\mathfrak{g}, M) \to C^*(\mathfrak{g}, N)$. Show that $\phi' \circ d = d \circ \phi'$. Deduce that ϕ' defines a homomorphism

$$\tilde{\phi} : H^*(\mathfrak{g}, M) \to H^*(\mathfrak{g}, N)$$

which is said to be associated with ϕ.

(*f*) Let L be a sub-$\mathfrak{g}$-module of M and N the quotient $\mathfrak{g}$-module M/L. The canonical homomorphisms $L \xrightarrow{i} M \xrightarrow{p} N$ define homomorphisms

$$C^*(\mathfrak{g}, L) \xrightarrow{i'} C^*(\mathfrak{g}, M) \xrightarrow{p'} C^*(\mathfrak{g}, N),$$

$$H^n(\mathfrak{g}, L) \xrightarrow{i^n} H^n(\mathfrak{g}, M) \xrightarrow{p^n} H^n(\mathfrak{g}, N).$$

Show that i' is injective and has image the kernel of p' and that p' is surjective. For all $c \in H^n(\mathfrak{g}, N)$, let $z \in Z^n(\mathfrak{g}, N)$ be a representative of c and $a \in C^n(\mathfrak{g}, M)$ be such that $p'(a) = z$; show that $da \in Z^{n+1}(\mathfrak{g}, L)$ and that its class in $H^{n+1}(\mathfrak{g}, L)$ depends only on c; denote this class by $\delta^n c$. Show that the sequence of homomorphisms:

$$0 \longrightarrow H^0(\mathfrak{g}, L) \xrightarrow{i^0} H^0(\mathfrak{g}, M) \xrightarrow{p^0} H^0(\mathfrak{g}, N) \xrightarrow{\delta^0}$$

$$\longrightarrow H^1(\mathfrak{g}, L) \xrightarrow{i^1} H^1(\mathfrak{g}, M) \xrightarrow{p^1} H^1(\mathfrak{g}, N) \xrightarrow{\delta^1} \cdots$$

is an exact sequence.

(*g*) With the notation of (*f*), the exact sequence:

$$0 \to \mathscr{L}(N, L) \to \mathscr{L}(N, M) \to \mathscr{L}(N, N) \to 0$$

defines an exact sequence

$$H^0(\mathfrak{g}, \mathscr{L}(N, M)) \to H^0(\mathfrak{g}, \mathscr{L}(N, N)) \to H^1(\mathfrak{g}, \mathscr{L}(N, L)).$$

The identity mapping of N is an invariant u of $\mathscr{L}(N, N)$ and hence an element of $H^0(\mathfrak{g}, \mathscr{L}(N, N))$; let c be its image in $H^1(\mathfrak{g}, \mathscr{L}(N, L))$. Then for the existence in M of a sub-$\mathfrak{g}$-module supplementary to L it is necessary and sufficient that $c = 0$. (This condition means that u is the image of an element of $H^0(\mathfrak{g}, \mathscr{L}(N, M))$, that is a homomorphism v of the $\mathfrak{g}$-module N into the $\mathfrak{g}$-module M; then $v(N)$ is the desired supplement.)

(*h*) Show that $Z^1(\mathfrak{g}, \mathfrak{g})$ (where $\mathfrak{g}$ is considered as a $\mathfrak{g}$-module by means of the adjoint representation) is identified with the vector space of derivations of $\mathfrak{g}$ and that $B^1(\mathfrak{g}, \mathfrak{g})$ is identified with the vector space of inner derivations of $\mathfrak{g}$.

(*i*) Let $\mathfrak{a}$ and $\mathfrak{b}$ be two Lie K-algebras with $\mathfrak{b}$ commutative and $\mathfrak{b} \xrightarrow{\lambda} \mathfrak{g} \xrightarrow{\mu} \mathfrak{a}$ an extension of $\mathfrak{a}$ by $\mathfrak{b}$. For all $x \in \mathfrak{g}$, the restriction of $\mathrm{ad}_{\mathfrak{g}}\, x$ to $\mathfrak{b}$ depends only on the class of x modulo $\mathfrak{b}$, whence there is an $\mathfrak{a}$-module structure on $\mathfrak{b}$. Let ν be a K-linear mapping of $\mathfrak{a}$ into $\mathfrak{g}$ such that $\mu \circ \nu$ is the identity mapping of $\mathfrak{a}$. For x, y in $\mathfrak{a}$, we write $f(x, y) = [\nu x, \nu y] - \nu([x, y])$. Show that $f \in Z^2(\mathfrak{a}, \mathfrak{b})$ and that the class c of f in $H^2(\mathfrak{a}, \mathfrak{b})$ does not depend on the choice of ν. For two extensions of $\mathfrak{a}$ by $\mathfrak{b}$, which define the same $\mathfrak{a}$-module structure on $\mathfrak{b}$, to be equivalent, it is necessary and sufficient that the corresponding class $c \in H^2(\mathfrak{a}, \mathfrak{b})$ be the same. For the extension to be inessential, it is necessary and sufficient that $c = 0$. If B is an $\mathfrak{a}$-module and $c \in H^2(\mathfrak{a}, B)$, there exists an extension of $\mathfrak{a}$ by B (considered as a commutative Lie algebra) defining the given $\mathfrak{a}$-module structure on B and the given element c of $H^2(\mathfrak{a}, B)$.

(*j*) Let $\mathfrak{g}$ be a finite-dimensional Lie algebra over K, U its enveloping algebra, $\mathfrak{h}$ an ideal of $\mathfrak{g}$, β an invariant bilinear form on $\mathfrak{g}$ whose restriction to $\mathfrak{h}$ is non-degenerate, $(y_i)_{1 \leqslant i \leqslant n}$ and $(y'_i)_{1 \leqslant i \leqslant n}$ two bases of $\mathfrak{h}$ such that $\beta(y_i, y'_j) = \delta_{ij}$, $t = \sum_{i=1}^{n} y_i y'_i \in U$, M a $\mathfrak{g}$-module and ρ the endomorphism of $C^*(\mathfrak{g}, M)$ which maps each $C^p(\mathfrak{g}, M)$ into $C^{p-1}(\mathfrak{g}, M)$ and which, for $p > 0$, is defined by:

$$(\rho f)(x_1, \ldots, x_{p-1}) = \sum_{k=1}^{n} (y_k)_M f(y'_k, x_1, \ldots, x_{p-1}).$$

Finally let Γ be the endomorphism of $C^*(\mathfrak{g}, M)$ extending t_M which maps each cochain f of degree >0 onto the cochain $t_M \circ f$. Show that $\rho d + d\rho = \Gamma$. (If for $x \in \mathfrak{g}$ we write:

$$[x, y_k] = \sum_{l=1}^{n} a_{kl} y_l, \qquad [x, y'_k] = \sum_{l=1}^{n} a'_{kl} y'_l,$$

show first that $a_{kl} = -a'_{lk}$.) Deduce that $\Gamma d = d\Gamma$. If M is simple and finite-dimensional over K, β is the associated bilinear form and dim $\mathfrak{h}$ is not divisible by the characteristic of K, show that $H^p(\mathfrak{g}, M) = \{0\}$ for all p. (Using Proposition 12, show that Γ is an automorphism of $C^*(\mathfrak{g}, M)$ and hence induces automorphisms of $Z^p(\mathfrak{g}, M)$ and $B^p(\mathfrak{g}, M)$. For

$$f \in Z^p(\mathfrak{g}, M), \qquad \Gamma f = (d\rho + \rho d)f = d\rho f \in B^p(\mathfrak{g}, M),$$

whence $Z^p(\mathfrak{g}, M) = B^p(\mathfrak{g}, M)$.)

§ 4

The conventions of § 4 remain valid unless otherwise mentioned.

1. Let $\mathfrak{g}$ be a nilpotent Lie algebra and p (resp. q) the smallest integer such that $\mathscr{C}^p\mathfrak{g} = \{0\}$ (resp. $\mathscr{C}_q\mathfrak{g} = \mathfrak{g}$). Show that $p = q + 1$ and that $\mathscr{C}_i\mathfrak{g} \supset \mathscr{C}^{p-i}\mathfrak{g}$. (Use the argument of Proposition 1.)

2. Let $\mathfrak{g}$ be a semi-direct product of an algebra $\mathfrak{h}$ of dimension 1 and a commutative ideal $\mathfrak{g}'$. Let $x \in \mathfrak{h}$, $x \neq 0$, and u be the restriction of $\mathrm{ad}_{\mathfrak{g}}\, x$ to $\mathfrak{g}'$.

(*a*) For $\mathfrak{g}$ to be nilpotent, it is necessary and sufficient that u be nilpotent.

(*b*) For the Killing form of $\mathfrak{g}$ to be zero, it is necessary and sufficient that $\mathrm{Tr}(u^2) = 0$.

(*c*) Deduce from (*a*) and (*b*) that there exist non-nilpotent Lie algebras whose Killing forms are zero.

(*d*) Deduce from (*a*) that in a nilpotent Lie algebra such that $\mathscr{C}^{p-1}\mathfrak{g} \neq \{0\}$, $\mathscr{C}^p\mathfrak{g} = \{0\}$ it is possible that $\mathscr{C}_i\mathfrak{g} \neq \mathscr{C}^{p-i}\mathfrak{g}$.

3. (*a*) Let $\mathfrak{g}$ be a nilpotent Lie algebra, $\mathfrak{z}$ its centre and $\mathfrak{b}$ a non-zero ideal of $\mathfrak{g}$. Show that $\mathfrak{z} \cap \mathfrak{b} \neq \{0\}$. (Consider $\mathfrak{b}$ as a $\mathfrak{g}$-module by means of the adjoint representation.)

(*b*) If in a Lie algebra $\mathfrak{g}$ an ideal $\mathfrak{b}$ is contained in $\mathscr{C}_{i+1}\mathfrak{g}$ but not in $\mathscr{C}_i\mathfrak{g}$, show that the ideals $\mathfrak{b} \cap \mathscr{C}_k\mathfrak{g}$ are all distinct for $0 \leqslant k \leqslant i + 1$. (Apply (*a*).)

4. Let $\mathfrak{g}$ be a nilpotent Lie algebra.

(*a*) Every subalgebra of $\mathfrak{g}$ is subinvariant. (Use Proposition 3.)

(*b*) Let $\mathfrak{b}$ be a vector subspace of $\mathfrak{g}$ such that $\mathfrak{b} + \mathscr{D}\mathfrak{g} = \mathfrak{g}$. Show that the subalgebra of $\mathfrak{g}$ generated by $\mathfrak{b}$ is $\mathfrak{g}$. (Apply (*a*) to this subalgebra. Reduce it thus to the case where $\mathfrak{b}$ is an ideal of $\mathfrak{g}$ and use Proposition 4 of § 1.) Deduce that the minimum number of generators of $\mathfrak{g}$ is $\dim \mathfrak{g}/\mathscr{D}\mathfrak{g}$.

5. (*a*) Show that a Lie algebra $\mathfrak{g}$ in which every subalgebra is subinvariant is nilpotent. (Show that if $\dim \mathfrak{g} > 1$ every $x \in \mathfrak{g}$ belongs to an ideal $\mathfrak{h} \neq \mathfrak{g}$ of $\mathfrak{g}$ which has the same property as $\mathfrak{g}$ and hence is nilpotent by induction on the dimension of $\mathfrak{g}$; hence $\mathrm{ad}_{\mathfrak{g}}\, x$ is nilpotent.)

(*b*) Show that if in a Lie algebra $\mathfrak{g}$ every subalgebra distinct from $\mathfrak{g}$ is distinct from its normalizer, $\mathfrak{g}$ is nilpotent. (Reduce it to (*a*).)

6. Let $\mathfrak{g}$ be a nilpotent Lie algebra and $\mathfrak{a}$ a commutative ideal of $\mathfrak{g}$. The following conditions are equivalent: (*a*) $\mathfrak{a}$ is a maximal commutative ideal of $\mathfrak{g}$; (*b*) $\mathfrak{a}$ is a maximal commutative subalgebra of $\mathfrak{g}$: (*c*) $\mathfrak{a}$ is equal to its centralizer $\mathfrak{a}'$ in $\mathfrak{g}$. (The implications not (*a*) $\Rightarrow$ not (*b*) $\Rightarrow$ not (*c*) are obvious. If $\mathfrak{a}' \neq \mathfrak{a}$, there exists by Proposition 1 an ideal $\mathfrak{a}''$ of $\mathfrak{g}$ such that $\mathfrak{a} \subset \mathfrak{a}'' \subset \mathfrak{a}'$ and $\dim \mathfrak{a}''/\mathfrak{a} = 1$. Then $\mathfrak{a}'' = \mathfrak{a} + Kx$, hence $[\mathfrak{a}'', \mathfrak{a}''] \subset [\mathfrak{a}, \mathfrak{a}] + [x, \mathfrak{a}] = \{0\}$ and hence (*a*) is false.)

7. (*a*) Let V be a finite-dimensional vector space over K, $\mathfrak{g}$ a Lie algebra of nilpotent endomorphisms of V and $(V_r)_{0 \leqslant r \leqslant n}$ a decreasing sequence of vector subspaces of V, with $V_0 = V$, $V_n = \{0\}$ and $\mathfrak{g}(V_r) \subset V_{r+1}$ for $0 \leqslant r < n$. Show by induction on i that $(\mathscr{D}^i\mathfrak{g})(V_r) \subset V_{r+2^i}$. If $\dim V \geqslant 2^i$, the subspace of elements of V annihilated by $\mathscr{D}^i\mathfrak{g}$ is of dimension $\geqslant 2^i$. If $\dim V \leqslant 2^i$, $\mathscr{D}^i\mathfrak{g} = \{0\}$.

(*b*) Let $\mathfrak{g}$ be a nilpotent Lie algebra and i an integer $\geqslant 0$. If $\dim \mathscr{D}^i\mathfrak{g} > 2^i + 1$ the centre of $\mathscr{D}^i\mathfrak{g}$ is of dimension $\geqslant 2^i$. If $\dim \mathscr{D}^i\mathfrak{g} \leqslant 2^i + 1$, $\mathscr{D}^i\mathfrak{g}$ is commutative. (Apply (*a*) to the restrictions to $\mathscr{D}^i\mathfrak{g}$ of the $\mathrm{ad}_{\mathfrak{g}}\, x$, $x \in \mathfrak{g}$.)

(*c*) Let $\mathfrak{g}$ be a nilpotent Lie algebra and i an integer $\geqslant 0$. If $\mathscr{D}^i\mathfrak{g}$ is not commutative, $\mathscr{D}^i\mathfrak{g}/\mathscr{D}^{i+1}\mathfrak{g}$ is of dimension $\geqslant 2^i + 1$. (Reduce it, by passing to the quotient, to the case where $\dim \mathscr{D}^{i+1}\mathfrak{g} = 1$ and use (*b*).)

8. Let $\mathfrak{g}$ be a Lie algebra, $\mathfrak{m}$ an ideal of codimension 1, x an element of $\mathfrak{g}$ not belonging to $\mathfrak{m}$ and $z \in \mathfrak{g}$.

(*a*) Show that the linear mapping D which reduce to 0 on $\mathfrak{m}$ and maps x to z is a derivation if z belongs to the centralizer $\mathfrak{a}$ of $\mathfrak{m}$ in $\mathfrak{g}$.

(*b*) Let q be the largest integer such that $\mathfrak{a} \subset \mathscr{C}^q\mathfrak{g}$. Show that if further $z \notin \mathscr{C}^{q+1}\mathfrak{g}$, D is not an inner derivation of $\mathfrak{g}$.

(*c*) Deduce from (*a*) and (*b*) and Exercise 7 (*c*) that if $\mathfrak{g}$ is a nilpotent Lie algebra of dimension >1, the vector space of inner derivations of $\mathfrak{g}$ is of codimension $\geqslant 2$ in the space of all derivations of $\mathfrak{g}$.

9. (*a*) Verify that the following multiplication tables define two nilpotent Lie algebras $\mathfrak{g}_3$, $\mathfrak{g}_4$ of dimensions 3 and 4:

$$\mathfrak{g}_3\colon [x_1, x_2] = x_3, \qquad [x_1, x_3] = [x_2, x_3] = 0$$

$$\mathfrak{g}_4\colon [x_1, x_2] = x_3, \quad [x_1, x_3] = x_4, \quad [x_1, x_4] = [x_2, x_3] = [x_2, x_4] = [x_3, x_4] = 0.$$

(*b*) Show that $\mathfrak{g}_3$ is isomorphic to $\mathfrak{n}(3, K)$ and also to $\mathfrak{sl}(2, K)$ if K is of characteristic 2.

(*c*) Let $\mathfrak{g}_1$ be the unique Lie algebra of dimension 1. Show that the nilpotent Lie algebras of dimension $\leqslant 4$ are given by the following table:

$$\text{dimension 1}: \mathfrak{g}_1;\ \text{dimension 2}: (\mathfrak{g}_1)^2;$$
$$\text{dimension 3}: (\mathfrak{g}_1)^3, \mathfrak{g}_3;\ \text{dimension 4}: (\mathfrak{g}_1)^4, \mathfrak{g}_3 \times \mathfrak{g}_1, \mathfrak{g}_4.$$

(Use Exercise 7 (*c*). If $\mathfrak{g}$ is nilpotent and 3-dimensional and $\dim \mathscr{D}\mathfrak{g} = 1$, observe that $\mathscr{D}\mathfrak{g}$ is contained in the centre of $\mathfrak{g}$ (Exercise 3 (*a*)), whence

$\mathfrak{g} = \mathfrak{g}_3$. If $\dim \mathfrak{g} = 4$, $\dim \mathscr{D}\mathfrak{g} = 1$, observe that the bracket on $\mathfrak{g}$ defines an alternating bilinear form on $\mathfrak{g}/\mathscr{D}\mathfrak{g}$, which is necessarily degenerate and hence that there exists a subalgebra $\mathfrak{h}$ of $\mathfrak{g}$ contained in the centre of $\mathfrak{g}$ with $\dim \mathfrak{h} = 1$, $\mathfrak{h} \cap \mathscr{D}\mathfrak{g} = \{0\}$, whence $\mathfrak{g} = \mathfrak{h} \times \mathfrak{g}_3$. If $\dim \mathfrak{g} = 4$, $\dim \mathscr{D}\mathfrak{g} = 2$, $\mathscr{D}\mathfrak{g}$ is commutative; applying Theorem 1 to the restrictions to $\mathscr{D}\mathfrak{g}$ of the $\mathrm{ad}_{\mathfrak{g}}\, x$ $(x \in \mathfrak{g})$, show that there exists a commutative ideal $\mathfrak{h}$ of $\mathfrak{g}$ with $\mathfrak{h} \supset \mathscr{D}\mathfrak{g}$, $\dim \mathfrak{h} = 3$; let $x \in \mathfrak{g}$, $x \notin \mathfrak{h}$; choose a basis of $\mathfrak{h}$ such that the restriction of $\mathrm{ad}_{\mathfrak{g}}\, x$ to $\mathfrak{h}$ has a Jordan matrix with respect to this basis; then $\mathfrak{g} = \mathfrak{g}_4$.)

10. Let $\mathscr{D}$ be the Lie algebra of derivations of the algebra $\mathfrak{g}_4$ (Exercise 9). Show that $\mathscr{D}$ is of dimension 7 and of zero centre and that the ideal $\mathscr{D}' \subset \mathscr{D}$ of inner derivations has no supplementary subalgebra in $\mathscr{D}$.

¶ 11. Let L be a commutative ring, A a left Artinian algebra over L and γ a mapping of A × A into L. Define on A the internal law of composition $(a, b) \mapsto a * b = ab + \gamma(a, b)ba$. For every subset E of A, let $\tilde{\mathrm{E}}$ denote the subring (*with no unit element*) of A generated by E. Show that if E consists of nilpotent elements and is stable under the law $*$, then $\tilde{\mathrm{E}}$ is nilpotent (that is, there exists $n > 0$ such that $\tilde{\mathrm{E}}^n = \{0\}$). Proceed as follows:

1. Suppose first that A is a simple ring and hence isomorphic to $\mathscr{L}_{\mathrm{D}}(\mathrm{T})$, where T is a left vector space of finite dimension m over a field D. Argue by induction on m. Let Φ be the set of subsets $\mathrm{F} \subset \mathrm{E}$ which are stable under $*$ and such that $\tilde{\mathrm{F}}$ is nilpotent. Show that Φ admits a maximal element M (note that $\tilde{\mathrm{F}}^m = \{0\}$ for all $\mathrm{F} \in \Phi$). Suppose $\tilde{\mathrm{M}} \neq \tilde{\mathrm{E}}$. Show that there exists $a \in \mathrm{E}$ such that $a \notin \tilde{\mathrm{M}}$ and $t * a \in \tilde{\mathrm{M}}$ for all $t \in \mathrm{M}$ (observe that there can be no infinite sequence (a_n) such that $a_n \in \mathrm{E}$, $a_n \notin \mathrm{M}$, $a_n = t_{n-1} * a_{n-1}$ with $t_{n-1} \in \mathrm{M}$). Let S be the subspace of T the direct sum of the $u(\mathrm{T})$ with $u \in \tilde{\mathrm{M}}$; show that $\mathrm{S} \neq \{0\}$ and $\mathrm{S} \neq \mathrm{T}$ and that $a(\mathrm{S}) \subset \mathrm{S}$. Let N be the set of $v \in \mathrm{E}$ such that $v(\mathrm{S}) \subset \mathrm{S}$. By using the induction hypothesis and considering the elements of N as operating on S and on T/S, show that $\mathrm{N} \in \Phi$, which implies a contradiction.

2. In the general case, use the fact that the Jacobson radical of A is nilpotent. Deduce from this result a new proof of Engel's Theorem.

12. Let $\mathfrak{g}$ be a Lie algebra and $\mathscr{C}^{\infty}\mathfrak{g}$ the intersection of the $\mathscr{C}^{p}\mathfrak{g}$.

(*a*) The Lie algebra $\mathfrak{g}/\mathscr{C}^{\infty}\mathfrak{g}$ is nilpotent.

(*b*) Show that there exists a nilpotent subalgebra $\mathfrak{h}$ of $\mathfrak{g}$ such that $\mathfrak{g} = \mathfrak{h} + \mathscr{C}^{\infty}\mathfrak{g}$. (Argue by induction on $\dim \mathfrak{g}$. Suppose that $\mathfrak{g}$ is not nilpotent. Let $x \in \mathfrak{g}$ be such that $\mathrm{I} = \bigcap_k (\mathrm{ad}\, x)^k(\mathfrak{g}) \neq \{0\}$. Let $\mathfrak{n}$ be the union of the kernels of the $(\mathrm{ad}\, x)^k$. Show that this is a subalgebra of $\mathfrak{g}$ and that $\mathfrak{g}$ is the direct sum of I and $\mathfrak{n}$. By the induction hypothesis $\mathfrak{n}$ is the sum of $\mathscr{C}^{\infty}\mathfrak{n}$ and a nilpotent subalgebra $\mathfrak{h}$. Finally, $\mathscr{C}^{\infty}\mathfrak{n} \subset \mathscr{C}^{\infty}\mathfrak{g}$ and $\mathrm{I} \subset \mathscr{C}^{\infty}\mathfrak{g}$.)

(*c*) Verify that the following multiplication table defines a (solvable) Lie algebra $\mathfrak{g}$ of dimension 5:

$$[x_1, x_2] = x_5, \quad [x_1, x_3] = x_3, \quad [x_1, x_4] = -x_4, \quad [x_3, x_4] = x_5$$

and the other $[x_i, x_j]$ are zero. Show that for this Lie algebra

$$\mathscr{C}^\infty\mathfrak{g} = \mathrm{K}x_3 + \mathrm{K}x_4 + \mathrm{K}x_5$$

but that there exists no supplementary subalgebra of $\mathscr{C}^\infty\mathfrak{g}$ in $\mathfrak{g}$.

¶ 13. Let $\mathfrak{g}$ be a Lie algebra and $\mathfrak{z}$ the centralizer of $\mathscr{C}^\infty\mathfrak{g}$ in $\mathfrak{g}$. If $\mathfrak{z} \not\subset \mathscr{C}^\infty\mathfrak{g}$, $\mathfrak{g}$ has non-zero centre. (Write $\mathfrak{g} = \mathscr{C}^\infty\mathfrak{g} + \mathfrak{k}$, where $\mathfrak{k}$ is a nilpotent subalgebra of $\mathfrak{g}$ (Exercise 12). Let $\mathfrak{g}_1 = \mathfrak{z} + \mathfrak{k}$. Let $\mathfrak{h}$ be a nilpotent subalgebra such that $\mathfrak{g}_1 = \mathscr{C}^\infty\mathfrak{g}_1 + \mathfrak{h}$. Show that $\mathfrak{g} = \mathscr{C}^\infty\mathfrak{g} + \mathfrak{h}$ and that $\mathscr{C}^\infty\mathfrak{g}_1 \subset \mathfrak{z} \cap \mathscr{C}^\infty\mathfrak{g}$. Let y be an element of $\mathfrak{z}$ not belonging to $\mathscr{C}^\infty\mathfrak{g}$. Write $y = x + x'$ where $x \in \mathfrak{h}$, $x' \in \mathscr{C}^\infty\mathfrak{g}_1$; then $x \in \mathfrak{z}$ and $x \neq 0$ and hence $\mathfrak{z} \cap \mathfrak{h}$ is a non-zero ideal of $\mathfrak{h}$; using Exercise 3 (*a*), deduce that there exists a non-zero element of $\mathfrak{g}$ permutable with $\mathscr{C}^\infty\mathfrak{g}$ and $\mathfrak{h}$.)

¶ 14. Let $\mathfrak{g}$ be a Lie algebra and $\mathfrak{b}$ a subinvariant subalgebra of $\mathfrak{g}$ such that the centralizer $\mathfrak{z}(\mathfrak{b})$ of $\mathfrak{b}$ in $\mathfrak{g}$ is zero.

(*a*) The centralizer $\mathfrak{z}(\mathscr{C}^\infty\mathfrak{b})$ of $\mathscr{C}^\infty\mathfrak{b}$ in $\mathfrak{g}$ is contained in $\mathscr{C}^\infty\mathfrak{b}$. (If $\mathfrak{z}(\mathscr{C}^\infty\mathfrak{b}) \not\subset \mathscr{C}^\infty\mathfrak{b}$, $\mathfrak{z}(\mathscr{C}^\infty\mathfrak{b}) \not\subset \mathfrak{b}$ by Exercise 13; $\mathscr{C}^\infty\mathfrak{b}$ is an ideal of $\mathfrak{g}$ (§ 1, Exercise 14); let $\mathfrak{c} = \mathfrak{b} + \mathfrak{z}(\mathscr{C}^\infty\mathfrak{b})$; $\mathfrak{c}$ is a subalgebra, $\mathfrak{b}$ is subinvariant in $\mathfrak{c}$; $\mathfrak{b}$ is an ideal of $\mathfrak{b}_1 \subset \mathfrak{c}$, where $\mathfrak{b} \neq \mathfrak{b}_1$; let $y \in \mathfrak{b}_1$, $y \notin \mathfrak{b}$; $y = x + x'$, where $x \in \mathfrak{z}(\mathscr{C}^\infty\mathfrak{b})$, $x' \in \mathfrak{b}$; then $x \in \mathfrak{z}(\mathscr{C}^\infty\mathfrak{b}) \cap \mathfrak{b}_1$, $x \notin \mathfrak{b}$; $\mathfrak{d} = \mathrm{K}x + \mathfrak{b}$ is a subalgebra; $\mathscr{C}^\infty\mathfrak{d} = \mathscr{C}^\infty\mathfrak{b}$ because $\mathfrak{c} \cap \mathfrak{z}(\mathscr{C}^\infty\mathfrak{b}) \subset \mathscr{C}^\infty\mathfrak{b}$, whence $\mathscr{C}^k\mathfrak{d} \subset \mathscr{C}^k\mathfrak{b}$ for all k; $\mathfrak{z}(\mathscr{C}^\infty\mathfrak{d}) \cap \mathfrak{d} \not\subset \mathscr{C}^\infty\mathfrak{d}$ and hence the centre of $\mathfrak{d}$ is non-zero by Exercise 13; a fortiori, $\mathfrak{z}(\mathfrak{b}) \neq \{0\}$, which is absurd.)

(*b*) Deduce from (*a*) that, if $\mathfrak{D}$ denotes the Lie algebra of derivations of $\mathscr{C}^\infty\mathfrak{b}$ and $\mathfrak{a}$ the centre of $\mathscr{C}^\infty\mathfrak{b}$, then $\dim \mathfrak{g} \leqslant \dim \mathfrak{D} + \dim \mathfrak{a}$ (let $\mathfrak{g}$ operate on $\mathscr{C}^\infty\mathfrak{b}$ by the adjoint representation).

15. Let $\mathfrak{l}$ be a Lie algebra with zero centre, $\mathfrak{D}_1$ the Lie algebra of derivations of $\mathfrak{l}$, $\mathfrak{D}_2$ the Lie algebra of derivations of $\mathfrak{D}_1, \ldots$. Then $\mathfrak{l}$ is an ideal of $\mathfrak{D}_1$, $\mathfrak{D}_1$ is an ideal of $\mathfrak{D}_2$, etc. (§ 1, Exercise 15 (*c*)). Let $\mathfrak{D}$ be the Lie algebra of derivations of $\mathscr{C}^\infty\mathfrak{l}$ and $\mathfrak{a}$ the centre of $\mathscr{C}^\infty\mathfrak{l}$.

(*a*) Then $\dim \mathfrak{D}_i \leqslant \dim \mathfrak{D} + \dim \mathfrak{a}$. (Use Exercise 14 above and Exercise 15 of § 1.)

(*b*) Deduce from (*a*) that for i sufficiently large all the derivations of $\mathfrak{D}_i$ are inner.

16. Let $\mathfrak{g}_1$ and $\mathfrak{g}_2$ be Lie K-algebras and $\mathfrak{n}_1$ and $\mathfrak{n}_2$ their largest nilpotent ideals. Show that the largest nilpotent ideal of $\mathfrak{g}_1 \times \mathfrak{g}_2$ is $\mathfrak{n}_1 \times \mathfrak{n}_2$.

17. We reconsider the Lie algebra $\mathfrak{g}_3$ of Exercise 9 (*a*) with its basis (x_1, x_2, x_3).

(*a*) Let V be the vector space K[X]. Let D be the differential operator with respect to X on V and M the operator of multiplication by X on V. Show that if K is of characteristic 0 the mapping

$$\alpha x_1 + \beta x_2 + \gamma x_3 \mapsto \alpha D + \rho M + \gamma$$

is an infinite-dimensional simple representation ρ of $\mathfrak{g}$ on V.

(*b*) If K is of characteristic $p > 0$, the ideal (X^p) of K[X] is stable under $\rho(\mathfrak{g})$. Under the quotient representation of $\mathfrak{g}$ on $K[X]/(X^p)$, no line is stable.

18. Let $\mathfrak{g}$ be a 7-dimensional vector space over K with a basis $(e_i)_{1\leqslant i\leqslant 7}$. An alternating bracket is defined on $\mathfrak{g}$ by the formulae:

(1) $$[e_i, e_j] = \alpha_{ij}e_{i+j} \quad (1 \leqslant i < j \leqslant 7, i + j \leqslant 7)$$

and all the other brackets $[e_i, e_j]$ are zero for $i < j$.

(*a*) For the Jacobi identity to hold, it is necessary and sufficient that:

(2) $$-\alpha_{23}\alpha_{15} + \alpha_{13}\alpha_{24} = 0$$

(3) $$\alpha_{12}\alpha_{34} - \alpha_{24}\alpha_{16} + \alpha_{14}\alpha_{25} = 0.$$

(*b*) Suppose henceforth that all the α_{ij} are $\neq 0$. Show that the ideals $\mathscr{C}^2\mathfrak{g}$, $\mathscr{C}^3\mathfrak{g}$, $\mathscr{C}^4\mathfrak{g}$, $\mathscr{C}^5\mathfrak{g}$, $\mathscr{C}^6\mathfrak{g}$ admit the following bases: $(e_3, e_4, e_5, e_6, e_7)$, (e_4, e_5, e_6, e_7), (e_5, e_6, e_7), (e_6, e_7), (e_7). Show that the centralizer $\mathfrak{h}$ of $\mathscr{C}^5\mathfrak{g}$ admits the basis $(e_2, e_3, e_4, e_5, e_6, e_7)$.

(*c*) Let $(e'_i)_{1\leqslant i\leqslant 7}$ be another basis of $\mathfrak{g}$ such that $\mathscr{C}^6\mathfrak{g}$ admits the basis (e'_7), $\mathscr{C}^5\mathfrak{g}$ admits the basis $(e'_6, e'_7), \ldots, \mathfrak{h}$ admits the basis $(e'_2, e'_3, e'_4, e'_5, e'_6, e'_7)$. Show that

$$[e'_i, e'_j] = \alpha'_{ij}e'_{i+j} + \beta e'_{i+j+1} + \gamma e'_{i+j+2} + \cdots + \lambda e'_7,$$

where

$$\alpha'_{14}\alpha'_{25}\alpha'^{-1}_{16}\alpha'^{-1}_{24} = \alpha_{14}\alpha_{25}\alpha^{-1}_{16}\alpha^{-1}_{24}.$$

(*d*) Deduce that there exists, if K is infinite, an infinity of non-isomorphic nilpotent Lie algebras of dimension 7 over K and that there exist complex nilpotent Lie algebras which cannot be derived from real nilpotent Lie algebras by extending the base field from **R** to **C**.

19. (*a*) Let $\mathfrak{g}$ be a Lie algebra. Let $\mathfrak{D}$ be the Lie algebra of derivations of $\mathfrak{g}$. Characteristic ideals $\mathfrak{g}^{[k]}$ of $\mathfrak{g}$ are defined inductively as follows: $\mathfrak{g}^{[0]} = \mathfrak{g}$ and $\mathfrak{g}^{[k+1]}$ is the subspace of $\mathfrak{g}$ generated by the Dx ($D \in \mathfrak{D}$, $x \in \mathfrak{g}^{[k]}$). The following conditions are equivalent: (1) every derivation of $\mathfrak{g}$ is nilpotent: (2) $\mathfrak{g}^{[k]} = \{0\}$ for k sufficiently large; (3) the holomorph of $\mathfrak{g}$ is nilpotent. If they are fulfilled, $\mathfrak{g}$ is said to be *characteristically nilpotent*. Such an algebra is nilpotent.

(*b*) Verify that an 8-dimensional Lie algebra is defined by the following multiplication table:

$$[x_1, x_2] = x_3, \quad [x_1, x_3] = x_4, \quad [x_1, x_4] = x_5, \quad [x_1, x_5] = x_6$$
$$[x_1, x_6] = x_8, \quad [x_1, x_7] = x_8, \quad [x_2, x_3] = x_5, \quad [x_2, x_4] = x_6$$
$$[x_2, x_5] = x_7, \quad [x_2, x_6] = 2x_8, \quad [x_3, x_4] = -x_7 + x_8, \quad [x_3, x_5] = -x_8$$

and $[x_i, x_j] = 0$ for $i + j > 8$. Show that

$$\mathscr{C}^2\mathfrak{g} = \sum_{i=3}^{8} \mathrm{K}x_i, \quad \mathscr{C}^3\mathfrak{g} = \sum_{i=4}^{8} \mathrm{K}x_i, \quad \mathscr{C}^4\mathfrak{g} = \sum_{i=5}^{8} \mathrm{K}x_i, \quad \mathscr{C}^5\mathfrak{g} = \sum_{i=6}^{8} \mathrm{K}x_i,$$
$$\mathscr{C}^6\mathfrak{g} = \mathrm{K}x_8, \quad \mathscr{C}^7\mathfrak{g} = \{0\}, \quad [\mathscr{C}^2\mathfrak{g}, \mathscr{C}^2\mathfrak{g}] = \mathrm{K}x_7 + \mathrm{K}x_8,$$

and that the transporter of $\mathscr{C}^2\mathfrak{g}$ into $\mathscr{C}^4\mathfrak{g}$ is $\sum_{i=2}^{8} \mathrm{K}x_i$. Deduce that every derivation D of $\mathfrak{g}$ is defined by formulae $\mathrm{D}x_i = \sum_{i=j}^{8} u_{ij}x_j$. Then show that $u_{ii} = 0$ for all i if the characteristic of K differs from 2, so that $\mathfrak{g}$ is characteristically nilpotent.

(*c*) If the characteristic of K is $\neq 2$, show that the Lie algebra $\mathfrak{g}$ of (*b*) is not the derived Lie algebra of any Lie algebra. (If $\mathfrak{g} = \mathscr{D}\mathfrak{h}$, show first that $\mathfrak{h}$ is nilpotent. Observe that $\dim \mathfrak{g}/\mathscr{D}\mathfrak{g} = 2$, which with Exercise 7 (*c*) leads to a contradiction.)

¶ 20. Let $\mathfrak{g}$ be a Lie algebra and U its enveloping algebra.

(*a*) Let $\mathfrak{g}'$ be an ideal of $\mathfrak{g}$, $\mathrm{U}' \subset \mathrm{U}$ its enveloping algebra, $x \in \mathfrak{g}$ and $a_1, \ldots, a_n$ in U. Suppose that $a_i = x$ for the indices $j_1, \ldots, j_p$ and $a_i \in \mathrm{U}'$ for the other indices (let $k_1, \ldots, k_q$ be these indices, with $k_1 < k_2 < \cdots < k_q$). Then $a_1a_2 \ldots a_n - x^p a_{k_1}a_{k_2} \ldots a_{k_q}$ is a sum of terms of the form $x^{p'}b$, where $b \in \mathrm{U}'$ and $p' < p$. (Argue by induction on p.)

(*b*) Suppose henceforth that $\mathfrak{g}$ is nilpotent and K is of characteristic 0. Let $\mathfrak{g}'$ be an ideal of codimension 1 in $\mathfrak{g}$, U′ its enveloping algebra, x an element of $\mathfrak{g}$ not belonging to $\mathfrak{g}'$ and U_0 (resp. U'_0) the subalgebra of U (resp. U′) annihilated by a set $\mathfrak{d}$ of derivations of $\mathfrak{g}$ (which extend to derivations of U) mapping $\mathfrak{g}$ into $\mathfrak{g}'$. Suppose that U_0 is contained in the centre of U and $\mathrm{U}_0 \not\subset \mathrm{U}'$. Show that there exist $a_1 \in \mathrm{U}'_0$, $a_2 \in \mathrm{U}'$ such that $a_1 \neq 0$ and $a = xa_1 + a_2 \in \mathrm{U}_0$. (Let $x^m b_m + x^{m-1}b_{m-1} + \cdots + b_0 \in \mathrm{U}_0$ with $b_m, \ldots, b_0$ in U′, $m > 0$, $b_m \neq 0$. Show, using (*a*), that, for every derivation $\mathrm{D} \in \mathfrak{d}$ of $\mathfrak{g}$, $\mathrm{D}b_m = 0$, $m(\mathrm{D}x)b_m + \mathrm{D}b_{m-1} = 0$ and hence $\mathrm{D}(mxb_m + b_{m-1}) = 0$.) Show that U_0 is contained in the algebra $\mathrm{K}[a, a_1^{-1}, \mathrm{U}'_0]$ generated by a, a_1^{-1}, U'_0 in the field of fractions of U_0 (which can be formed by Corollary 7 to Theorem 1 of § 2). Deduce that the field of fractions of U_0 is the field generated by a and U'_0. Show that a is transcendental over $\mathrm{K}(\mathrm{U}'_0)$.

(*c*) Let $\{0\} = \mathfrak{g}_0 \subset \mathfrak{g}_1 \subset \cdots \subset \mathfrak{g}_n = \mathfrak{g}$ be a sequence of ideals of $\mathfrak{g}$ of dimensions $0, 1, \ldots, n$. Let x_j be an element of $\mathfrak{g}_j$ not belonging to $\mathfrak{g}_{j-1}$. Let $j_1 < j_2 < \cdots < j_q$ be the indices j such that there exists in U^j (enveloping algebra of $\mathfrak{g}_j$) an element of the centre U_0 of U not belonging to U^{j-1}. By (*b*) there exists $a_{j1} \in \mathrm{U}_0 \cap \mathrm{U}^j$ and $a_{j2} \in \mathrm{U}^{j-1}$ such that $a_{j1} \neq 0$ and $a_j = x_j a_{j1} + a_{j2} \in \mathrm{U}_0 \cap \mathrm{U}^j$. Show by induction on n that

$$\mathrm{U}_0 \subset \mathrm{K}[a_{j_1}, \ldots, a_{j_q}, a_{j_{11}}^{-1}, \ldots, a_{j_{q1}}^{-1}]$$

and that the field of fractions of U_0 is generated by the algebraically independent elements $a_{j_1}, \ldots, a_{j_q}$. In particular, the field of fractions of the centre of U is a pure transcendental extension of K.

¶ 21. Let $\mathfrak{g}$ be a Lie algebra and σ an automorphism of $\mathfrak{g}$.

(*a*) Suppose first that K is algebraically closed; for all $\lambda \in \mathrm{K}$ let L_λ be the set of $x \in \mathfrak{g}$ which are annihilated by a power of $\sigma - \lambda\mathrm{I}$; $\mathfrak{g}$ is the direct sum of the L_λ. Show that $[\mathrm{L}_\lambda, \mathrm{L}_\mu] \subset \mathrm{L}_{\lambda\mu}$. (Note that

$$(\sigma - \lambda\mu\mathrm{I})([x, y]) = [(\sigma - \lambda\mathrm{I})x, \sigma y] + [\lambda x, (\sigma - \mu\mathrm{I})y].)$$

(*b*) With an arbitrary field K, suppose that none of the eigenvalues of σ (in an algebraically closed extension of K) is a root of unity. Show that $\mathfrak{g}$ is nilpotent. (Reduce it to the case where K is algebraically closed. If $\lambda_1, \ldots, \lambda_m$ are the distinct eigenvalues of σ, the $\lambda_i\lambda_j, \lambda_i\lambda_j^2, \lambda_i\lambda_j^3, \ldots$ are not all eigenvalues and hence $\mathrm{ad}_{\mathfrak{g}} x$ is nilpotent for $x \in \mathrm{L}_{\lambda_i}$. Conclude using Exercise 11 applied to the set of $\mathrm{ad}_{\mathfrak{g}} x$, where x runs through the union of the L_{λ_i}.)

(*c*) With an arbitrary field K, suppose that $\sigma^q = \mathrm{I}$, where q is a prime number and that no eigenvalue of σ is equal to 1. Show that $\mathfrak{g}$ is nilpotent. (Same method as in (*b*), observing that, if $\mathfrak{g} \neq \{0\}$, q is not equal to the characteristic of K and that for every ordered pair of eigenvalues λ_i, λ_j of σ there exists an integer k such that $\lambda_i\lambda_j^k = 1$.)

22. (*a*) Let u, v be two endomorphisms of a vector space E of finite dimension over an algebraically closed field K. For every eigenvalue λ of u let E_λ be the subspace of E consisting of the vectors annihilated by a power of $u - \lambda\mathrm{I}$. Show that, if $(\mathrm{ad}\, u)^n v = 0$, the subspaces E_λ are stable under v.

(*b*) Deduce from (*a*) that, if $\mathfrak{g}$ is a nilpotent Lie algebra of endomorphisms of E, E is the direct sum of subspaces F_j $(1 \leqslant j \leqslant m)$ which are stable under $\mathfrak{g}$ and such that on each F_j the restriction of every element $u \in \mathfrak{g}$ can be written $\lambda_j(u)\mathrm{I} + u_j$, where $\lambda_j(u) \in \mathrm{K}$ and u_j is nilpotent.

(*c*) If K is of characteristic 2, $\mathrm{E} = \mathrm{K}^2$ and $\mathfrak{g}$ is nilpotent algebra $\mathfrak{sl}(2, \mathrm{K})$, show that $m = 1$ and that possibly $\lambda(u + v) \neq \lambda(u) + \lambda(v)$ for two elements u, v of $\mathfrak{g}$. (For a sequel to this exercise, cf. § 5, Exercise 12.)

23. Let $\mathfrak{g}$ be a Lie p-algebra over a perfect field K of characteristic $p > 0$. $\mathfrak{g}$ is called *p-unipotent* if for all $x \in \mathfrak{g}$ there exists m such that $x^{p^m} = 0$.

(*a*) Show that every p-unipotent Lie p-algebra is nilpotent.

(*b*) Suppose that $\mathfrak{g}$ is nilpotent and let $\mathfrak{h}$ denote the p-core of the centre of $\mathfrak{g}$ (§ 1, Exercise 23 (*a*)). Show that $\mathfrak{g}/\mathfrak{h}$ is p-unipotent.

(*c*) Let $\mathfrak{g}$ be the nilpotent Lie p-algebra with a basis of three elements e_1, e_2, e_3 such that $[e_1, e_2] = [e_1, e_3] = 0$, $[e_2, e_3] = e_1$, $e_1^p = e_1$, $e_2^p = e_3^p = 0$. Then $\mathfrak{h} = Ke_1$, but $\mathfrak{g}$ is not the direct sum of $\mathfrak{h}$ and a p-unipotent p-subalgebra.

(*d*) If $\mathfrak{g}$ is p-unipotent, show that in the restricted enveloping algebra of $\mathfrak{g}$ the two-sided ideal generated by $\mathfrak{g}$ is nilpotent. (Argue by induction on the dimension of $\mathfrak{g}$.)

24. Suppose that the field K is of characteristic 2. Show that in the nilpotent Lie algebra $\mathfrak{g}_4$ of Exercise 9 there exists no 2-mapping.

25. Let $\mathfrak{g}$ be a Lie p-algebra. Show that the largest nilpotent ideal of $\mathfrak{g}$ is a p-ideal (cf. § 1, Exercise 22).

26. Let G be a group and let $(H_n)_{n\geqslant 1}$ be a decreasing sequence of subgroups of G; suppose that $H_1 = G$ and that if we write $(x, y) = xyx^{-1}y^{-1}$, the relations $x \in H_i$, $y \in H_j$ imply $(x, y) \in H_{i+j}$.

(*a*) Let $G_i = H_i/H_{i+1}$; show that G_i is commutative and that the mapping $x, y \mapsto (x, y)$ defines on passing to the quotient a **Z**-bilinear mapping of $G_i \times G_j$ into G_{i+j}.

(b) Let $\mathrm{gr}(G) = \sum_{i=1}^{\infty} G_i$ and extend by linearity the mappings $G_i \times G_j \to G_{i+j}$ defined in (*a*) to a **Z**-bilinear mapping of $\mathrm{gr}(G) \times \mathrm{gr}(G)$ into $\mathrm{gr}(G)$. Show that $\mathrm{gr}(G)$ is a Lie **Z**-algebra with this mapping (to verify the Jacobi identity, use the following formula:

$$((x, y), z^y).((y, z), x^z).((z, x), y^x) = e \qquad x, y, z \text{ in } G,$$

where x^y denotes yxy^{-1} and e denotes the identity element of G).

(c) Suppose that there exists n such that $H_n = \{e\}$. Show that $\mathrm{gr}(G)$ is a nilpotent Lie **Z**-algebra.

27. Let A be an associative algebra with unit element 1 and let $A_0 = A \supset A_1 \supset \cdots \supset A_n \supset \cdots$ be a decreasing sequence of two-sided ideals of A such that $A_i.A_j \subset A_{i+j}$. Let G be a group with identity element e and let $f: G \to A$ be a mapping such that $f(e) = 1$, $f(xy) = f(x).f(y)$ and $1 - f(x) \in A_1$ for all $x \in G$. Let H_n denote the set of $x \in G$ such that $1 - f(x) \in A_n$. Show that the H_n satisfy the conditions of Exercise 26. Show that the mapping $x \mapsto f(x) - 1$ defines on taking quotients an injective homomorphism of the Lie algebra $\mathrm{gr}(G)$ into the Lie algebra associated with the graded ring $\mathrm{gr}(A) = \sum A_n/A_{n+1}$ (cf. *Commutative Algebra*, Chapter III).

§ 5

The conventions of § 5 remain valid unless otherwise mentioned.

1. Let $\mathfrak{g}$ be the 2-dimensional non-commutative solvable Lie algebra. Show that the Killing form of $\mathfrak{g}$ is non-zero, that every invariant bilinear form on $\mathfrak{g}$ is degenerate and that every derivation of $\mathfrak{g}$ is inner.

2. (*a*) Show that in the 3-dimensional solvable Lie algebra over **R** defined by the multiplication table $[x, y] = z$, $[x, z] = -y$, $[y, z] = 0$, there exists no decreasing sequence of ideals of dimensions 3, 2, 1, 0.

(*b*) Show that in the non-commutative 2-dimensional solvable Lie algebra $\mathfrak{g}$ there exists a sequence of ideals of dimensions 2, 1, 0 but that $\mathfrak{g}$ is not nilpotent.

3. Let $\mathfrak{g}$ be a solvable Lie algebra such that the conditions $x \in \mathfrak{g}$, $y \in \mathfrak{g}$, $[[x, y], y] = 0$ imply $[x, y] = 0$. Show that $\mathfrak{g}$ is commutative. (Let k be the largest integer such that $\mathscr{D}^{k-1}\mathfrak{g} \neq \{0\}$, $\mathscr{D}^k\mathfrak{g} = \{0\}$. Assuming $k \geqslant 2$, show first that $[\mathscr{D}^{k-2}\mathfrak{g}, \mathscr{D}^{k-1}\mathfrak{g}] = \{0\}$ and then that $[\mathscr{D}^{k-2}\mathfrak{g}, \mathscr{D}^{k-2}\mathfrak{g}] = \{0\}$, whence a contradiction.)

4. Show that the centre of $\mathfrak{st}(n, \mathrm{K})$ is zero and that that of $\mathfrak{n}(n, \mathrm{K})$ is of dimension 1.

5. Let $\mathfrak{g}$ be a Lie algebra and $(\mathscr{D}^0\mathfrak{g}, \mathscr{D}^1\mathfrak{g}, \ldots, \mathscr{D}^n\mathfrak{g})$ the sequence of derived algebras of $\mathfrak{g}$ $(n \geqslant 0, \mathscr{D}^{n-1}\mathfrak{g} \neq \mathscr{D}^n\mathfrak{g})$. Then $\dim \mathscr{D}^i\mathfrak{g}/\mathscr{D}^{i+1}\mathfrak{g} \geqslant 2^{i-1} + 1$ for $1 \leqslant i \leqslant n - 2$. (Taking quotients by $\mathscr{D}^n\mathfrak{g}$, reduce it to the case where $\mathfrak{g}$ is solvable. Then use the fact that $\mathscr{D}\mathfrak{g}$ is nilpotent and Exercise 7 (c) of § 4.)

6. (*a*) Verify that the following multiplication table defines a 5-dimensional solvable Lie algebra $\mathfrak{g}$:

$$[x_1, x_2] = x_5, \qquad [x_1, x_3] = x_3, \qquad [x_2, x_4] = x_4$$
$$[x_1, x_4] = [x_2, x_3] = [x_3, x_4] = [x_5, \mathfrak{g}] = 0.$$

(*b*) Show that the orthogonal of $\mathfrak{g}$ with respect to the Killing form is $\mathscr{D}\mathfrak{g} = \mathrm{K}x_3 + \mathrm{K}x_4 + \mathrm{K}x_5$. Deduce that $\mathscr{D}\mathfrak{g}$ is the largest nilpotent ideal of $\mathfrak{g}$.

(*c*) Show that there exists no supplementary subalgebra of $\mathscr{D}\mathfrak{g}$ in $\mathfrak{g}$. Deduce that $\mathfrak{g}$ is not the semi-direct product of a commutative algebra and a nilpotent ideal. (Show that this nilpotent ideal would necessarily be $\mathscr{D}\mathfrak{g}$.)

7. Let $\mathfrak{g}$ be the 3-dimensional solvable Lie algebra with basis (x, y, z) such that $[x, y] = z$, $[x, z] = y$, $[y, z] = 0$. Show that the linear mapping which maps x to $-x$, y to $-z$, z to y is an automorphism of $\mathfrak{g}$ of order 4. Compare this result with Exercise 21 (*c*) of § 4.

8. (*a*) Let $\mathfrak{g}_0$ be a 3-dimensional solvable real Lie algebra such that $\mathscr{D}\mathfrak{g}_0$ is commutative and 2-dimensional. Let $\mathfrak{g}$ be the algebra derived from $\mathfrak{g}_0$ by

extending the base field from **R** to **C**. For $x \in \mathfrak{g}$, let u_x be the restriction of $\mathrm{ad}_{\mathfrak{g}} x$ to $\mathscr{D}\mathfrak{g}$. Show that the eigenvalues of u_x either have the same absolute value or are linearly dependent over **R**. (We have $x = \lambda y + z$ with $z \in \mathscr{D}\mathfrak{g}$, $y \in \mathfrak{g}_0$, $\lambda \in \mathbf{C}$ and hence $u_x = \lambda u_y$. Now u_y is the **C**-linear extension to $\mathscr{D}\mathfrak{g}$ of an **R**-linear endomorphism of $\mathscr{D}\mathfrak{g}_0$.)

(*b*) Show that there exists a 3-dimensional solvable complex Lie algebra $\mathfrak{g}$, with $\mathscr{D}\mathfrak{g}$ commutative and 2-dimensional, and an element x of $\mathfrak{g}$ such that the restriction of $\mathrm{ad}_{\mathfrak{g}} x$ to $\mathscr{D}\mathfrak{g}$ has eigenvalues which neither have the same absolute value nor are linearly dependent over **R**. (Construct $\mathfrak{g}$ as a semi-direct product of a 1-dimensional algebra and a 2-dimensional commutative algebra.)

(*c*) Show that the algebra constructed in (*b*) cannot be derived from a real Lie algebra by extending the base field from **R** to **C**.

9. Let $\mathfrak{g}$ be a Lie algebra, $\mathfrak{r}$ its radical, $\mathfrak{n}$ its largest nilpotent ideal and D a derivation of $\mathfrak{g}$. Show that $\mathscr{D}\mathfrak{g} \cap \mathfrak{r} \subset \mathfrak{n}$. (Let $\mathfrak{d}$ be the Lie algebra of derivations of $\mathfrak{g}$ and $\mathfrak{r}'$ its radical. If $x \in \mathfrak{g}$ is such that $\mathrm{D}x \in \mathfrak{r}$, then

$$\mathrm{ad}(\mathrm{D}x) = [\mathrm{D}, \mathrm{ad}\, x] \in \mathscr{D}\mathfrak{d} \cap \mathfrak{r}'$$

(Corollary 2 to Proposition 5), hence $\mathrm{ad}(\mathrm{D}x)$ is nilpotent (Theorem 1) and hence $\mathrm{D}x \in \mathfrak{n}$.)

10. Let $\mathfrak{g}$ be a Lie algebra, $\mathfrak{r}$ its radical and $\mathfrak{a}$ a subinvariant subalgebra of $\mathfrak{g}$. Show that the radical of $\mathfrak{a}$ is $\mathfrak{a} \cap \mathfrak{r}$. (Apply Corollary 3 to Proposition 5 several times.)

11. Let $\mathfrak{g}$ be a Lie algebra, ρ a finite-dimensional representation of $\mathfrak{g}$ and E the associative endomorphism algebra generated by 1 and $\rho(\mathfrak{g})$. Show that the largest nilpotency ideal $\mathfrak{n}$ of ρ is equal to the set $\mathfrak{n}'$ of $x \in \mathfrak{g}$ such that $\mathrm{Tr}(\rho(x)u) = 0$ for all $u \in \mathrm{E}$. (To show that $\mathfrak{n}' \subset \mathfrak{n}$, show that $\mathfrak{n}'$ is an ideal and that for all $x \in \mathfrak{n}'$ the semi-simple component of $\rho(x)$ is zero, by noting that $\mathrm{Tr}((\rho(x))^n) = 0$ for every integer $n > 0$.)

12. Suppose that K is algebraically closed. Let $\mathfrak{g}$ be a nilpotent Lie K-algebra. Let ρ be a finite-dimensional representation of $\mathfrak{g}$ on a vector space V. For every linear form λ on $\mathfrak{g}$, let V^λ be the vector subspace of V consisting of the $\xi \in \mathrm{V}$ such that, for all $x \in \mathfrak{g}$, $(\rho(x) - \lambda(x)\mathrm{I})^n\xi = 0$ for sufficiently large n.

(*a*) The V^λ are stable under $\rho(\mathfrak{g})$ and the sum of the V^λ is direct. (Use Exercise 22 of § 4.)

(*b*) We have $\mathrm{V} = \sum \mathrm{V}^\lambda$. (If each $\rho(x)$ has a single eigenvalue, $\mathrm{V} = \mathrm{V}^{\lambda_0}$ by Corollary 2 to Theorem 1. If $\rho(x_0)$ has at least two distinct eigenvalues, V is the direct sum of two non-trivial subspaces which are stable under $\rho(\mathfrak{g})$. Then argue by induction on the dimension of V.)

13. Let $\mathfrak{g}$ be a Lie Algebra and $\mathfrak{D}$ the Lie algebra of derivations of $\mathfrak{g}$. For $\mathfrak{g}$ to be characteristically nilpotent, it is necessary and sufficient that $\mathfrak{D}$ be nilpotent and that $\dim \mathfrak{g} > 1$. (To see that the condition is sufficient, write $\mathfrak{g}$ as a sum of subspaces $\mathfrak{g}^{\lambda}$, applying Exercise 12 to the identity mapping of $\mathfrak{D}$. Show that $[\mathfrak{g}^{\lambda}, \mathfrak{g}^{\mu}] \subset \mathfrak{g}^{\lambda+\mu}$ and that each $\mathfrak{g}^{\lambda}$ is an ideal of $\mathfrak{g}$. Deduce that $\mathfrak{g}^{\lambda}$ is commutative for $\lambda \neq 0$. Using again the fact that $\mathfrak{D}$ is nilpotent, show that $\mathfrak{g} = \mathfrak{g}^0$ if $\dim \mathfrak{g} > 1$. Otherwise $\mathfrak{g} = \mathfrak{g}^0 \times \mathfrak{h}$, where $\mathfrak{h}$ is commutative. Show first that $\dim \mathfrak{h} \leqslant 1$. If $\dim \mathfrak{h} = 1$, note that there would exist a derivation D of $\mathfrak{g}$ such that $D(\mathfrak{g}^0) = \{0\}$ and $D(\mathfrak{h})$ is contained in the centre of $\mathfrak{g}^0$ (§ 4, Exercise 8 (*a*)).)

14. Let V be a finite-dimensional vector space over K and z an endomorphism of V. We adopt the notation z_q^p of Exercise 3 of § 3. An endomorphism z' of V is called a *replica* of z if for all p and q every zero of z_q^p is a zero of $z_q'^p$. Show that if $\operatorname{Tr}(zz') = 0$ for every replica z' of z, then z is nilpotent. (Use the proof of Lemma 3. In the notation of this proof, prove in particular that t is a replica of z.)

15. Let K be a field of characteristic 2. The identity representation of the nilpotent Lie algebra $\mathfrak{sl}(2, K)$ on K^2 defines a semi-direct product $\mathfrak{h}$ of $\mathfrak{sl}(2, K)$ by K^2. Show that $\mathfrak{h}$ is solvable but that $\mathscr{D}\mathfrak{h}$ is not nilpotent. Deduce that $\mathfrak{h}$ admits no faithful linear representation by triangular matrices. Show also that the conclusions of Exercise 5 are false.

16. Let $\mathfrak{g}$ be a Lie algebra over an arbitrary field K, A a commutative associative algebra over K and $\mathfrak{g}' = \mathfrak{g} \otimes_K A$, which can be considered as a Lie algebra over K.

(*a*) If D is a derivation of A, show that there exists one and only one derivation D′ of $\mathfrak{g}'$ such that $D'(x \otimes a) = x \otimes Da$ for $x \in \mathfrak{g}$, $a \in A$.

(*b*) Let p be a prime number, G a cyclic group of order p and s a generator of G. Suppose that K is of characteristic p and henceforth let A be the algebra of G over K. Show that there exists one and only one derivation D of A such that $D(s^k) = ks^{k-1}$ for all $k \in \mathbf{Z}$. Show that the K-linear combinations of the $x \otimes (s-1)^k$ $(k = 1, 2, \ldots, p-1, x \in \mathfrak{g})$ form a solvable ideal $\mathfrak{r}$ of $\mathfrak{g}'$ and that $\mathfrak{g}'/\mathfrak{r}$ is isomorphic to $\mathfrak{g}$.

(*c*) Take $\mathfrak{g}$ to be simple (cf. § 6, no. 2, Definition 2). Then $\mathfrak{r}$ is the radical of $\mathfrak{g}'$ but is not a characteristic ideal. (Observe that $D(x \otimes (s-1)) = 1 \otimes x$.)†

17. Let $\mathfrak{g}$ be a Lie algebra. Suppose that for every simple $\mathfrak{g}$-module M of finite dimension over K the x_M are permutable with one another. Show that $\mathfrak{g}$ is solvable. (Observe that $\mathscr{D}\mathfrak{g}$ is contained in the nilpotent radical and hence is solvable.)

† This result, unedited, was communicated to us by N. Jacobson.

§ 6

The conventions of § 6 remain valid unless otherwise mentioned.

¶ 1. Let $\mathfrak{g}$ be a semi-simple Lie algebra and ρ a finite-dimensional representation of $\mathfrak{g}$ on M.

(*a*) If ρ is simple and non-zero, $H^p(\mathfrak{g}, M) = \{0\}$ for all p. (Use Proposition 1 and Exercise 12 (*j*) of § 3.)

(*b*) For all ρ, $H^1(\mathfrak{g}, M) = \{0\}$. (If ρ is simple and non-zero, apply (*a*). If ρ is zero, use the fact that $\mathfrak{g} = \mathscr{D}\mathfrak{g}$. In the general case, argue by induction on the dimension of ρ: if N is a sub-$\mathfrak{g}$-module of M distinct from $\{0\}$ and M, use the exact sequence:

$$H^1(\mathfrak{g}, N) \to H^1(\mathfrak{g}, M) \to H^1(\mathfrak{g}, M/N)$$

established in Exercise 12 (*f*) of § 3.) Hence recover Remark 2 of no. 2.

(*c*) Deduce from (*b*) a proof of Theorem 2. (Use Exercise 12 (*g*) of § 3.) Deduce also from (*b*) that every derivation of $\mathfrak{g}$ is inner. (Use Exercise 12 (*h*) of § 3.)

(*d*) For all ρ, $H^2(\mathfrak{g}, M) = \{0\}$. (Arguing as for (*b*), it suffices to consider the case where $\rho = 0$. Let $c \in H^2(\mathfrak{g}, M)$. Consider, following Exercise 12 (*i*) of § 3, the central extension $\mathfrak{h}$ of $\mathfrak{g}$ by M defined by c. The adjoint representation of $\mathfrak{h}$ defines a representation of $\mathfrak{g}$ on $\mathfrak{h}$. By Theorem 2, M admits in $\mathfrak{h}$ a supplement which is stable under $\mathfrak{g}$ and hence the extension is trivial. Hence $c = 0$ by Exercise 12 (*i*) of § 3.)

(*e*) Deduce from (*b*) and (*d*) a proof of Theorem 5. (As in the text, it can be reduced to the case where the radical is commutative.)

2. Let $\mathfrak{g}$ be a Lie algebra, $\mathfrak{r}$ its radical and $(\mathfrak{a}_0, \mathfrak{a}_1, \ldots)$ a sequence of ideals of $\mathfrak{g}$ defined as follows: (1) $\mathfrak{a}_0 = \{0\}$; (2) $\mathfrak{a}_{i+1}/\mathfrak{a}_i$ is a maximal commutative ideal of $\mathfrak{g}/\mathfrak{a}_i$. Let p be the smallest integer such that $\mathfrak{a}_p = \mathfrak{a}_{p+1} = \cdots$. Show that $\mathfrak{r} = \mathfrak{a}_p$. (Show that $\mathfrak{g}/\mathfrak{a}_p$ is semi-simple.)

3. For a Lie algebra $\mathfrak{g}$ to be semi-simple, it is necessary and sufficient that it be reductive in every Lie algebra containing $\mathfrak{g}$ as a subalgebra. (If $\mathfrak{g}$ satisfies this condition, let M be a $\mathfrak{g}$-module of finite dimension over K. Considering M as a commutative algebra, form the semi-direct product of $\mathfrak{g}$ and M, in which $\mathfrak{g}$ is reductive. Deduce that M is semi-simple.)

4. Let $\mathfrak{g}$ be a semi-simple Lie algebra and ρ a non-zero simple representation of $\mathfrak{g}$ on a finite-dimensional space M. Let $\mathfrak{h}$ be the corresponding semi-direct product. Show that $\mathfrak{h} = \mathscr{D}\mathfrak{h}$, that the centre of $\mathfrak{h}$ is zero and that $\mathfrak{h}$ is not the product of a semi-simple algebra and a solvable algebra.

5. Let $\mathfrak{a}$ be a Lie algebra and $\mathfrak{r}$ its radical. If $\mathfrak{r}$ has a decreasing sequence of characteristic ideals $\mathfrak{r} = \mathfrak{r}_0 \supset \mathfrak{r}_1 \supset \cdots \supset \mathfrak{r}_n = \{0\}$ such that $\dim \mathfrak{r}_i/\mathfrak{r}_{i+1} = 1$ for $0 \leqslant i < n$, $\mathfrak{g}$ is the product of a semi-simple algebra and a solvable algebra.

(Let $\mathfrak{s}$ be a Levi subalgebra of $\mathfrak{g}$. For all $x \in \mathfrak{s}$, let $\rho(x)$ be the restriction of $\mathrm{ad}_{\mathfrak{g}} x$ to $\mathfrak{r}$. Then ρ is a direct sum of 1-dimensional representations, which are zero because $\mathfrak{s} = \mathscr{D}\mathfrak{s}$.)

6. Let $\mathfrak{g}$ be a Lie algebra and $\mathscr{D}^{\infty}\mathfrak{g}$ the intersection of the $\mathscr{D}^{p}\mathfrak{g}$ ($p = 1, 2, \ldots$). Show that $\mathfrak{g}/\mathscr{D}^{\infty}\mathfrak{g}$ is solvable and that $\mathscr{D}(\mathscr{D}^{\infty}\mathfrak{g}) = \mathscr{D}^{\infty}\mathfrak{g}$. For $\mathfrak{g}$ to be isomorphic to the product of a semi-simple algebra and a solvable algebra, it is necessary and sufficient that $\mathscr{D}^{\infty}\mathfrak{g}$ be semi-simple.

7. (*a*) Let $\mathfrak{g}$ be a Lie algebra, $\mathfrak{h}$ a semi-simple ideal of $\mathfrak{g}$ and $\mathfrak{a}$ the centralizer of $\mathfrak{h}$ in $\mathfrak{g}$, so that $\mathfrak{g}$ is identified with $\mathfrak{h} \times \mathfrak{a}$. Show that, for every ideal $\mathfrak{k}$ of $\mathfrak{g}$, then $\mathfrak{k} = (\mathfrak{k} \cap \mathfrak{h}) \times (\mathfrak{k} \cap \mathfrak{a})$. (Let $\mathfrak{k}_1$ be the canonical projection of $\mathfrak{k}$ onto $\mathfrak{h}$; it is an ideal of $\mathfrak{h}$ and hence $\mathscr{D}\mathfrak{k}_1 = \mathfrak{k}_1$; deduce that $\mathfrak{k}_1 \subset \mathfrak{k}$.)

(*b*) Let β be an invariant bilinear form on $\mathfrak{g}$. Show that $\mathfrak{h}$ and $\mathfrak{a}$ are orthogonal with respect to β (use the fact that $\mathfrak{h} = \mathscr{D}\mathfrak{h}$ and that $\beta = \beta_1 + \beta_2$, where β_1 (resp. β_2) is an invariant bilinear form whose restriction to $\mathfrak{a}$ (resp. $\mathfrak{h}$) is zero.

(*c*) Deduce from (*a*) that there exists in $\mathfrak{g}$ a largest semi-simple ideal. (Consider a maximal semi-simple ideal of $\mathfrak{g}$.)

8. Let $\mathfrak{g}$ be a Lie algebra. An ideal $\mathfrak{h}$ of $\mathfrak{g}$ is called *minimal* if $\mathfrak{h} \neq \{0\}$ and every ideal of $\mathfrak{g}$ contained in $\mathfrak{h}$ is equal to $\{0\}$ or $\mathfrak{h}$.

(*a*) Every simple ideal of $\mathfrak{g}$ is minimal.

(*b*) Let $\mathfrak{h}$ be a minimal ideal of $\mathfrak{g}$ and $\mathfrak{r}$ the radical of $\mathfrak{g}$. Then either $\mathfrak{h} \subset \mathfrak{r}$, in which case $\mathfrak{h}$ is abelian, or $\mathfrak{h} \cap \mathfrak{r} = \{0\}$, in which case $\mathfrak{h}$ is simple. (Use the fact that the derived ideal of a Lie algebra and the simple components of a semi-simple Lie algebra are characteristic ideals.)

9. For a Lie algebra $\mathfrak{g}$ to be reductive, it is necessary and sufficient that its centre $\mathfrak{c}$ be equal to its largest nilpotent ideal. (If the condition holds, let $\mathfrak{r}$ be the radical of $\mathfrak{g}$; $\mathscr{D}\mathfrak{r}$ is contained in the centre of $\mathfrak{r}$, hence $\mathfrak{r}$ is nilpotent and hence $\mathfrak{r} = \mathfrak{c}$.)

10. Let $\mathfrak{g}$ be a Lie algebra such that the conditions $x \in \mathfrak{g}, y \in \mathfrak{g}, [[x, y], y] = 0$ imply $[x, y] = 0$. Show that $\mathfrak{g}$ is reductive. (Show that the radical $\mathfrak{r}$ of $\mathfrak{g}$ is commutative using Exercise 3 of § 5. Then show that $[\mathfrak{g}, \mathfrak{r}] = 0$.)

¶ 11. (*a*) Let $\mathfrak{g}$ be a Lie algebra, $\mathfrak{r}$ its radical, $\mathfrak{s}$ a Levi subalgebra of $\mathfrak{g}$ and $\mathfrak{m}$ an ideal of $\mathfrak{g}$ containing $\mathfrak{r}$. There exists an ideal $\mathfrak{t}$ of $\mathfrak{s}$, supplementary to $\mathfrak{m}$ in $\mathfrak{g}$, such that $[\mathfrak{m}, \mathfrak{t}] \subset \mathfrak{r}$.

(*b*) Let $\mathfrak{g}$ be a Lie algebra and $\mathfrak{a}$ a subinvariant subalgebra of $\mathfrak{g}$. Then there exists a composition series $\mathfrak{g} = \mathfrak{g}_0 \supset \mathfrak{g}_1 \supset \cdots \supset \mathfrak{g}_k = \mathfrak{a}$ such that $\mathfrak{g}_i$ is the direct sum of $\mathfrak{g}_{i+1}$ and a subalgebra $\mathfrak{h}_i$ which is either 1-dimensional and contained in the radical $\mathfrak{r}(\mathfrak{g}_i)$ of $\mathfrak{g}_i$ or simple and such that $[\mathfrak{h}_i, \mathfrak{g}_{i+1}] \subset \mathfrak{r}(\mathfrak{g}_{i+1})$. (Reduce it to the case where $\mathfrak{a}$ is an ideal of $\mathfrak{g}$. Let $\mathfrak{g}' = \mathfrak{g}/\mathfrak{a}$ and $\mathfrak{r}'$ be the radical of $\mathfrak{g}'$. The algebra $\mathfrak{g}'/\mathfrak{r}'$ is the product of its simple ideals $\mathfrak{a}_1, \mathfrak{a}_2, \ldots, \mathfrak{a}_c$.

Take a composition series of $\mathfrak{g}'$ consisting of a composition series of $\mathfrak{r}'$ whose successive quotients are 1-dimensional and the inverse images of the ideals $\mathfrak{a}_1, \mathfrak{a}_1 \times \mathfrak{a}_2, \ldots, \mathfrak{a}_1 \times \mathfrak{a}_2 \times \cdots \times \mathfrak{a}_c$. Then take the inverse image in $\mathfrak{g}$ of this composition series.)

12. Let $\mathfrak{g}$ be a Lie algebra, $\mathfrak{r}$ its radical and D a derivation of $\mathfrak{g}$.

(*a*) If $D(\mathfrak{r}) = \{0\}$, D is inner. (Let $\mathfrak{c}$ be the centralizer of $\mathfrak{r}$ in $\mathfrak{g}$. The adjoint representation of $\mathfrak{g}$ defines a representation $x^* \mapsto \rho(x^*)$ of $\mathfrak{g}^* = \mathfrak{g}/\mathfrak{r}$ on $\mathfrak{c}$. On the other hand, $[D(\mathfrak{g}), \mathfrak{r}] \subset D([\mathfrak{g}, \mathfrak{r}]) + [\mathfrak{g}, D(\mathfrak{r})] = \{0\}$, hence $D(\mathfrak{g}) \subset \mathfrak{c}$ and hence D defines a linear mapping $D^* : \mathfrak{g}^* \to \mathfrak{c}$. Show that

$$D^*([x^*, y^*]) = \rho(x^*)D^*y^* - \rho(y^*)D^*x^*$$

for all x^*, y^* in $\mathfrak{g}^*$. Deduce that there exists $a \in \mathfrak{c}$ such that $D^*x^* = \rho(x^*)a$ for all $x^* \in \mathfrak{g}^*$ (cf. no. 2, *Remark* 2), whence $Dx = [x, a]$ for all $x \in \mathfrak{g}$.)

(*b*) If D coincides on $\mathfrak{r}$ with an inner derivation of $\mathfrak{g}$, D is inner. (Apply (a).)

13. Let $\mathfrak{g}$ be a Lie algebra over an algebraically closed field, $\mathfrak{r}$ its radical and ρ a finite-dimensional simple representation of $\mathfrak{g}$. Then $\rho(z)$ is scalar for $z \in \mathfrak{r}$. (Reduce it to the case where $\mathfrak{g}$ is reductive. In this case, $\mathfrak{r}$ is the centre of $\mathfrak{g}$.)

14. Let X be a nilpotent matrix in $\mathfrak{gl}(n, K)$. Show that there exist in $\mathfrak{gl}(n, K)$ two other matrices Y, H such that $[H, X] = X$, $[H, Y] = -Y$, $[X, Y] = H$. (Argue by induction on n, using the Jordan form of X; thus reduce it to the case where $X = E_{12} + E_{23} + \cdots + E_{n-1,n}$; then take Y to be a combination of the $E_{k,k-1}$ and H a combination of the E_{jj}.)

¶ 15. (*a*) Let X, Y be two matrices of $\mathfrak{gl}(n, K)$. Show that, if $[X, Y] = X$, X is nilpotent. (For every polynomial $f \in K[T]$, observe that

$$[f(X), Y] = Xf'(X).)$$

(*b*) Let $\mathfrak{g}$ be a Lie algebra and h, x two elements of $\mathfrak{g}$ such that $[h, x] = x$ and there exists $z \in \mathfrak{g}$ with $[x, z] = h$. Show that there exists $y \in \mathfrak{g}$ such that $[x, y] = h$ and $[h, y] = -y$. (Observe first that $[z, h] + z$ belongs to the centralizer $\mathfrak{n}$ of x in $\mathfrak{g}$. Then show that $\mathfrak{n}$ is stable under ad h and that the restriction to $\mathfrak{n}$ of $I - \operatorname{ad} h$ is bijective; to do this, note that ad x is nilpotent, using (*a*), and prove that, if $\mathfrak{g}_i$ is the image of $\mathfrak{g}$ under $(\operatorname{ad} x)^i$, $I - \operatorname{ad} h$ gives on taking quotients a bijection of $(\mathfrak{n} \cap \mathfrak{g}_{i-1})/(\mathfrak{n} \cap \mathfrak{g}_i)$; for this use the relation

$$[\operatorname{ad} z, (\operatorname{ad} x)^k] = -k(\operatorname{ad} x)^{k-1}(\operatorname{ad} h + \frac{k-1}{2} I).)$$

(*c*) Let $\mathfrak{g}$ be a subalgebra of $\mathfrak{gl}(n, K)$; suppose that, for every nilpotent matrix $X \in \mathfrak{g}$, there exist two other matrices H, Y in $\mathfrak{g}$ such that $[H, X] = X$, $[H, Y] = -Y$, $[X, Y] = H$. Let $\mathfrak{h}$ be a subalgebra of $\mathfrak{g}$ such that there exists a subspace $\mathfrak{m}$ supplementary to $\mathfrak{h}$ in $\mathfrak{g}$ and such that $[\mathfrak{h}, \mathfrak{m}] \subset \mathfrak{m}$. Show that $\mathfrak{h}$ has the same property as $\mathfrak{g}$ for all its nilpotent matrices (use (*b*)).

16. Let $\mathfrak{g}$ be a subalgebra of $\mathfrak{gl}(n, \mathrm{K})$ such that the identity representation of $\mathfrak{g}$ is semi-simple. Show that every nilpotent matrix $X \in \mathfrak{g}$ is contained in a 3-dimensional simple Lie subalgebra of $\mathfrak{g}$. (Show that $\mathfrak{g}$ is reductive in $\mathfrak{gl}(n, \mathrm{K})$; then use Exercises 14 and 15 (*c*).)

17. Show that the algebra derived from a real simple Lie algebra by extending the base field from **R** to **C** is not always simple. (Let $\mathfrak{g}$ be a real simple Lie algebra. If $\mathfrak{g}' = \mathfrak{g}_{(\mathbf{C})}$ is simple, let $\mathfrak{h}$ be the real Lie algebra derived from $\mathfrak{g}'$ by restricting the field of scalars to **R**. We know that $\mathfrak{h}$ is simple. Then $\mathfrak{h}_{(\mathbf{C})}$ is, by Exercise 4 of § 1, the product of two algebras isomorphic to $\mathfrak{g}'$. Cf. also Exercise 26 (*b*).)

18. (*a*) Let $\mathfrak{g}$ be a simple Lie algebra. Every invariant bilinear form on $\mathfrak{g}$ is either zero or non-degenerate. If K is algebraically closed, every invariant bilinear form β on $\mathfrak{g}$ is proportional to the Killing form β_0. (Consider the endomorphism σ of the vector space $\mathfrak{g}$ defined by $\beta(x, y) = \beta_0(\sigma x, y)$ and show that σ commutes with $\mathrm{ad}_{\mathfrak{g}} x$ and is hence scalar.) Show that this result may be false if K is not algebraically closed. (Use Exercise 17 and the fact that the dimension of the space of invariant bilinear forms does not change when the base field is extended.)

(*b*) Let $\mathfrak{g}$ be a semi-simple Lie algebra over an algebraically closed field. Deduce from (*a*) and Exercise 7 (*b*) that the dimension of the space of invariant bilinear forms on $\mathfrak{g}$ is equal to the number of simple components of $\mathfrak{g}$ and that all these forms are symmetric.

(*c*) Let $\mathfrak{g}$ be a simple Lie algebra, M the dual space of the vector space $\mathfrak{g}$, with the dual representation of the adjoint representation, and $\mathfrak{h}$ the semi-direct product of $\mathfrak{g}$ and M defined by $x \mapsto x_{\mathrm{M}}$. For y, z in $\mathfrak{g}$ and y', z' in M, we write $\beta(y + y', z + z') = \langle y, z' \rangle + \langle z, y' \rangle$. Show that β is a non-degenerate invariant symmetric bilinear form on $\mathfrak{h}$, which is associated with no representation of $\mathfrak{h}$. (Observe that the radical and nilpotent radical of $\mathfrak{h}$ are equal to M.)

19. Let $\mathfrak{g}$ be a reductive Lie algebra, U its enveloping algebra, Z the centre of U and V the subspace of U generated by the elements of the form $uv - vu$ ($u \in \mathrm{U}, v \in \mathrm{U}$).

(*a*) U is the direct sum of Z and V. (Apply Proposition 6 of § 3 to the representation $x \mapsto \mathrm{ad}_{\mathrm{U}} x$ of $\mathfrak{g}$ on U, observing that, in the decomposition $\mathrm{U} = \sum_n \mathrm{U}^n$ of § 2, no. 7, Corollary 4 to Theorem 1, the U^n are stable under this representation.)

(*b*) Let $u \mapsto u^\natural$ be the projector of U onto Z parallel to V. Show that $(uv)^\natural = (vu)^\natural$, $(zu)^\natural = zu^\natural$ for all $u \in \mathrm{U}$, $v \in \mathrm{U}$, $z \in \mathrm{Z}$.

(*c*) Let R be a two-sided ideal of U. Show that $R = (R \cap Z) + (R \cap V)$. (To show that the component in V of an element r of R belongs to R, reduce it to the case where K is algebraically closed; observe that R is stable under the representation ρ of U on U which extends $x \mapsto \mathrm{ad}_U x$; decompose r into its components in the U^n; finally, apply Corollary 2 to Theorem 1 of *Algebra*, Chapter VIII, § 4, no. 2 to the restriction of ρ to U^n.)

(*d*) Let R′ be an ideal of Z. Let R_1 be the two-sided ideal of U generated by R′. Show that $R_1 \cap Z = R'$. (Use (*b*).)

20. Suppose that K is algebraically closed. Let $\mathfrak{g}$ be a Lie algebra and $\mathfrak{c}$ its centre.

(*a*) If $\mathfrak{g}$ admits a finite-dimensional faithful simple representation, $\mathfrak{g}$ is reductive and $\dim \mathfrak{c} \leqslant 1$.

(*b*) Conversely, if $\mathfrak{g}$ is reductive and $\dim \mathfrak{c} \leqslant 1$, $\mathfrak{g}$ admits a finite-dimensional faithful simple representation. (If $\mathfrak{g}$ is simple, use the adjoint representation. If $\mathfrak{g} = \mathfrak{g}_1 \times \mathfrak{g}_2$ and $\mathfrak{g}_1$, $\mathfrak{g}_2$ admit finite-dimensional faithful simple representations ρ_1, ρ_2 on spaces M_1, M_2, show that

$$(x_1, x_2) \mapsto \rho_1(x_1) \otimes 1 + 1 \otimes \rho_2(x_2)$$

is a semi-simple representation ρ of $\mathfrak{g}$, then that the centralizer of the $\mathfrak{g}$-module $M_1 \otimes M_2$ reduces to the scalars and hence that ρ is simple. When $\mathfrak{g}_1$ and $\mathfrak{g}_2$ are semi-simple or $\mathfrak{g}_1$ is semi-simple and $\mathfrak{g}_2$ is commutative, show that ρ is faithful by considering the kernel of ρ which is an ideal of $\mathfrak{g}_1 \times \mathfrak{g}_2$.)

21. Let V be a vector space over K of finite dimension $n > 1$. Show that $\mathfrak{sl}(V)$ is simple. (Reduce it to the case where K is algebraically closed. Suppose that $\mathfrak{sl}(V) = \mathfrak{a} \times \mathfrak{b}$, with $\dim \mathfrak{a} = a > 0$, $\dim \mathfrak{b} = b > 0$. Let A (resp. B) be the associative algebra generated by 1 and $\mathfrak{a}$ (resp. $\mathfrak{b}$). Then V can be considered as a simple $(A \otimes_K B)$-module. Hence there exists an A-module P of finite dimension p over K and a B-module Q of finite dimension q over K such that V is $(A \otimes_K B)$-isomorphic to $P \otimes_K Q$ (*Algebra*, Chapter VIII, § 7, no. 7, Proposition 8 and no. 4, Theorem 2); P and Q are faithful and hence $a \leqslant p^2$, $b \leqslant q^2$. On the other hand, $a + b = n^2 - 1$ and $pq = n$, whence $(p^2 - 1)(q^2 - 1) \leqslant 2$, which is contradictory.)

22. (*a*) Let $\mathfrak{g}$ be a nilpotent Lie algebra over an algebraically closed field K of arbitrary characteristic. Let $\mathfrak{z}$ be its centre, ρ a finite-dimensional representation of $\mathfrak{g}$ and β the associated bilinear form. Show that $z \cap \mathscr{D}\mathfrak{g}$ is orthogonal to $\mathfrak{g}$ with respect to β. (Reduce it, using a Jordan-Hölder series, to the case where ρ is simple. Then note that, for $z \in \mathfrak{z} \cap \mathscr{D}\mathfrak{g}$, $\rho(z)$ is a scalar matrix with trace zero. Then use Exercise 22 (*b*) of § 4.) Deduce that, if β is non-degenerate, $\mathfrak{g}$ is commutative. (Use Exercise 3 (*a*) of § 4.)

(*b*) Suppose that K is of characteristic 2. Show that for the solvable Lie algebra $\mathfrak{gl}(2, \mathrm{K}) = \mathfrak{g}$ the bilinear form associated with the identity representation is non-degenerate, but that the centre $\mathfrak{z}$ is contained in $\mathscr{D}\mathfrak{g}$ and is $\neq \{0\}$.

(*c*) Let $\mathfrak{g}$ be the 6-dimensional Lie algebra over K with basis (a, b, c, d, e, f) and multiplication table $[a, b] = -[b, a] = d$, $[a, c] = -[c, a] = e$, $[b, c] = -[c, b] = f$ and the other brackets are 0. Let β be the bilinear form on $\mathfrak{g}$ such that $\beta(c, d) = \beta(d, c) = 1, \beta(a, f) = \beta(f, a) = 1, \beta(b, e) = \beta(e, b) = -1$ and the other values of β at ordered pairs of the basis of $\mathfrak{g}$ in question are 0. Show that β is invariant, $\mathfrak{g}$ nilpotent, $\mathfrak{z} = \mathscr{D}\mathfrak{g} \neq \{0\}$ and β non-degenerate.

23. Let $\mathfrak{g}$ be a 3-dimensional non-commutative Lie algebra over a field K of arbitrary characteristic.

(*a*) If $\mathfrak{g}$ has centre $\mathfrak{z} \neq \{0\}$, then $\dim \mathfrak{z} = 1$ (§ 1, Exercise 2) and $\dim \mathscr{D}\mathfrak{g} = 1$. If $\mathfrak{z} \neq \mathscr{D}\mathfrak{g}$, $\mathfrak{g}$ is the product of $\mathfrak{z}$ and the 2-dimensional non-commutative algebra. If $\mathfrak{z} = \mathscr{D}\mathfrak{g}$, $\mathfrak{g}$ is the 3-dimensional non-commutative nilpotent algebra (§ 4, Exercise 9 (*a*)).

(*b*) If $\mathfrak{z} = \{0\}$ but there exists in $\mathfrak{g}$ a 2-dimensional commutative subalgebra, this subalgebra is unique and is equal to $\mathscr{D}\mathfrak{g}$; for all $x \notin \mathscr{D}\mathfrak{g}$, the restriction u of $\operatorname{ad} x$ to $\mathscr{D}\mathfrak{g}$ is a bijection of this vector space, determined to within a multiplicative constant. Conversely, every automorphism u of a 2-dimensional vector space $\mathrm{K}a + \mathrm{K}b$ determines a Lie algebra structure on $\mathfrak{g} = \mathrm{K}a + \mathrm{K}b + \mathrm{K}c$ by the conditions $[a, b] = 0$, $[c, a] = u(a)$, $[c, b] = u(b)$. For two Lie algebras thus defined by automorphisms u_1, u_2 to be isomorphic, it is necessary and sufficient that the matrices of u_1 and u_2 be similar to within a scalar factor.

(*c*) Suppose that there exists in $\mathfrak{g}$ no commutative subalgebra of dimension >1. Show that there then exists $a \in \mathfrak{g}$ such that $a \notin (\operatorname{ad} a)\mathfrak{g}$ (assuming that an element x does not have this property, show, using the Jacobi identity, that another element has the desired property). There then exists a basis (a, b, c) of $\mathfrak{g}$ such that $[a, b] = c$, $[a, c] = \beta b$, $[b, c] = \gamma a$, with $\beta \neq 0$ and $\gamma \neq 0$, and $\mathfrak{g}$ is simple. Let ϕ be a canonical isomorphism of $\mathfrak{g}$ onto the exterior power $\bigwedge^2 \mathfrak{g}^*$ of the dual space of $\mathfrak{g}$, determined to within an invertible constant factor (*Algebra*, Chapter III, § 8, no. 5); let u be the linear mapping of $\mathfrak{g}$ into $\mathfrak{g}^*$ defined by the condition $\langle [x, y], u(z) \rangle = \langle x \wedge y, \phi(z) \rangle$; the bilinear form $\Phi(x, y) = \langle x, u(y) \rangle$ on $\mathfrak{g} \times \mathfrak{g}$ is symmetric and non-degenerate and 2Φ is the Killing form of $\mathfrak{g}$ to within a factor $\neq 0$. For two 3-dimensional simple Lie algebras over K to be isomorphic, it is necessary and sufficient that the corresponding bilinear forms be equivalent to within a constant factor $\neq 0$.

(*d*) If K is not of characteristic 2, the simple algebra $\mathfrak{g}$ defined in (*c*) is isomorphic to the Lie algebra $\mathfrak{o}(\Phi)$ of the formal orthogonal group $\mathbf{G}(\Phi)$ (§ 1, Exercise 26 (*a*)). For the existence in $\mathfrak{g}$ of vectors x such that $\operatorname{ad} x$ admits eigenvectors not proportional to x, it is necessary and sufficient that Φ be of index >0; $\mathfrak{g}$ then admits a basis (a, b, c) such that $[a, b] = b$, $[a, c] = -c$, $[b, c] = a$.

(*e*) If K is of characteristic 2, show that there is no 2-mapping on $\mathfrak{g}$ and therefore that $\mathfrak{g}$ is not the Lie algebra of a formal group. Show that $\mathfrak{g}$ admits derivations which are not inner.

¶ 24. Let K be a field of arbitrary characteristic p. We again adopt Definition 1.

(*a*) Show that, unless $p = n = 2$, the only non-trivial ideals of $\mathfrak{gl}(n, K)$ are the 1-dimensional centre $\mathfrak{z}$ with basis $\sum_{i=1}^{n} E_{ii}$ and the Lie algebra $\mathfrak{sl}(n, K)$ consisting of the matrices of trace 0. (Except in the exceptional case indicated, note that if an ideal $\mathfrak{a}$ contains one of the E_{ij}, it necessarily contains $\mathfrak{sl}(n, K)$. If $\mathfrak{a}$ contains an element not belonging to $\mathfrak{z}$, by multiplying this element by at most four suitably chosen elements E_{ij}, a non-zero multiple of one of the E_{ij} is obtained.) If n is not a multiple of p, show that $\mathfrak{gl}(n, K)$ is the direct sum of $\mathfrak{z}$ and $\mathfrak{sl}(n, K)$ and $\mathfrak{sl}(n, K)$ is simple. If on the contrary n is a multiple of p and $n > 2$, show that $\mathfrak{sl}(n, K)/\mathfrak{z}$ is simple (same methods).

(*b*) Show that, for n a multiple of p and $n > 2$, $\mathfrak{gl}(n, K)/\mathfrak{z}$ has radical $\{0\}$, but admits the quotient $\mathfrak{gl}(n, K)/\mathfrak{sl}(n, K)$ which is Abelian.

¶ 25. Let K be a field of arbitrary characteristic p. Let $\mathfrak{sp}(2n, K)$ denote the Lie algebra of the formal symplectic group in $2n$ variables over K (§ 1, Exercise 26 (*c*)). Show that this Lie algebra, which is identified with a subalgebra of $\mathfrak{gl}(2n, K)$, has basis consisting of the elements

$$H_i = E_{ii} - E_{i+n,i+n} \quad (1 \leqslant i \leqslant n),$$
$$F_{ij} = E_{ij} - E_{j+n,i+n} \quad (1 \leqslant i, j \leqslant n, i \neq j),$$
$$G'_{ij} = E_{i,j+n} + E_{j,i+n}, \quad G''_{ij} = E_{i+n,j} + E_{j+n,i} \quad (1 \leqslant i < j \leqslant n),$$
$$E_{i,i+n} \text{ and } E_{i+n,i} \quad (1 \leqslant i \leqslant n).$$

(*a*) Show that, if $p \neq 2$ and $n \geqslant 1$, the algebra $\mathfrak{sp}(2n, K)$ is simple (for $n = 1$, $\mathfrak{sp}(2, K) = \mathfrak{sl}(2, K)$). (Same method as in Exercise 24.)

(*b*) If $p = 2$ and $n \geqslant 3$, show that the H_i, F_{ij}, G'_{ij} and G''_{ij} form an ideal $\mathfrak{a}$ of dimension $n(2n-1)$; the elements of $\mathfrak{a}$ of the form

$$\sum_{i=1}^{n} \lambda_i H_i + \sum_{i \neq j} \alpha_{ij} F_{ij} + \sum_{i<j} \beta_{ij} G'_{ij} + \sum_{i<j} \gamma_{ij} G''_{ij}$$

such that $\sum_{i=1}^{n} \lambda_i = 0$ form an ideal $\mathfrak{b} \subset \mathfrak{a}$ of dimension $n(2n-1) - 1$ and the multiples of $\sum_{i=1}^{n} H_i$ an ideal $\mathfrak{c}$, the centre of $\mathfrak{sp}(2n, K)$; $\mathfrak{b}$, $\mathfrak{c}$ are the only ideals of $\mathfrak{a}$, $\mathfrak{b} = \mathscr{D}(\mathfrak{sp}(2n, \mathrm{K}))$ and $\mathfrak{b}/\mathfrak{c}$ is simple. (Same method.) What are the ideals of $\mathfrak{sp}(4, \mathrm{K})$ when K is of characteristic 2?

¶ 26. Let K be a field of characteristic $\neq 2$ and Φ a non-degenerate symmetric bilinear form on an n-dimensional vector space over K.

(*a*) Show that, if $n \geqslant 5$, the Lie algebra $\mathfrak{o}(\Phi)$ is simple. (By extending the base field, it can be reduced to the case where the index of Φ is $m = [n/2]$. If for example $n = 2m$ is even, $\mathfrak{o}(\Phi)$ has basis consisting of the elements

$$H_i = E_{ij} - E_{i+m,i+m}, \quad F_{ij} = E_{ij} - E_{j+m,i+m} \quad (1 \leqslant i, j \leqslant m, i \neq j),$$
$$G'_{ij} = E_{i,j+m} - E_{j,i+m} \quad \text{and} \quad G''_{ij} = E_{i+m,j} - E_{j+m,i} \quad (1 \leqslant i < j \leqslant m);$$

then argue as in Exercises 24 and 25. Proceed similarly when $n = 2m + 1$ is odd.)

(*b*) Let Δ be the discriminant of Φ with respect to a basis. Suppose that $n = 4$. Show that, if Δ is not a square in K, $\mathfrak{o}(\Phi)$ is simple. If Δ is a square, $\mathfrak{o}(\Phi)$ is the product of two isomorphic 3-dimensional simple Lie algebras. (Use the structure of $\mathbf{G}(\Phi)$ described in *Algebra*, Chapter IX, § 9, Exercise 16.) Deduce an example of a simple Lie algebra which becomes non-simple when extending the base field.

27. (*a*) Show that, in a finite-dimensional Lie p-algebra, the radical is a p-ideal. (Use Exercise 22 (*c*) of § 1.)

(*b*) For a Lie p-algebra $\mathfrak{g}$ to have zero radical, it is necessary and sufficient that $\mathfrak{g}$ contain no commutative p-ideal $\neq \{0\}$. (Use Exercise 22 (*c*) of § 1.)

§ 7

The conventions of § 7 remain valid unless otherwise mentioned.

1. (*a*) For a Lie K-algebra to be nilpotent, it is necessary and sufficient that it be isomorphic to a subalgebra of an algebra $\mathfrak{n}(n, \mathrm{K})$.

(*b*) Suppose that K is algebraically closed. For a Lie K-algebra to be solvable, it is necessary and sufficient that it be isomorphic to a subalgebra of an algebra $\mathfrak{t}(n, \mathrm{K})$.

2. Let $\mathfrak{g}$ be a Lie algebra and $\mathfrak{n}$ its largest nilpotent ideal. There exist a finite-dimensional vector space V and an isomorphism of $\mathfrak{g}$ onto a subalgebra of $\mathfrak{sl}(\mathrm{V})$, which maps every element of $\mathfrak{n}$ to a nilpotent endomorphism of V. (Use Ado's Theorem and Exercise 5 of § 1.)

3. Let $\mathfrak{g}$ be a Lie algebra and U its enveloping algebra.

(*a*) Show that for all $u \in \mathrm{U}$ ($u \neq 0$) there exists a finite-dimensional representation π of U such that $\pi(u) \neq 0$. (Use Exercise 2 and Exercise 5 (*b*) of § 3.)

(*b*) Deduce from (*a*) that if $\mathfrak{g}$ is semi-simple there exists an infinity of inequivalent finite-dimensional simple representations of $\mathfrak{g}$. (Let $\rho_1, \ldots, \rho_n$ be finite-dimensional simple representations of $\mathfrak{g}$, $N_1, \ldots, N_n$ the annihilators of the corresponding U-modules and $N = \bigcap_{i=1}^{n} N_i$. Show that N is of finite co-dimension in U. Let $n \in N$, $n \neq 0$. Apply (*a*) to n.)

¶ 4. Let $\mathfrak{g}$ be a Lie algebra, $\mathfrak{a}$ a subinvariant subalgebra of $\mathfrak{g}$, $\mathfrak{r}(\mathfrak{a})$ the radical of $\mathfrak{a}$ and ρ a finite-dimensional representation of $\mathfrak{a}$. For ρ to admit a finite-dimensional extension to $\mathfrak{g}$, it is necessary and sufficient that $\mathscr{D}\mathfrak{g} \cap \mathfrak{r}(\mathfrak{a})$ be contained in the largest nilpotency ideal of ρ. (For the sufficiency proceed as follows; let $\mathfrak{g} = \mathfrak{g}_0 \supset \mathfrak{g}_1 \supset \cdots \supset \mathfrak{g}_k = \mathfrak{a}$ be a decomposition series of $\mathfrak{g}$ with the properties of Exercise 11 (*b*) of § 6. Show by induction the existence of an extension ρ_i of ρ to $\mathfrak{g}_i$, whose largest nilpotency ideal contains $\mathscr{D}\mathfrak{g} \cap \mathfrak{r}(\mathfrak{g}_i)$. In the notation of Exercise 11 (*b*) of § 6, the passage from ρ_{i+1} to ρ_i follows from Theorem 1 when $\mathfrak{h}_i$ is simple, or when $\mathfrak{h}_i$ is 1-dimensional and

$$\mathscr{D}\mathfrak{g} \cap \mathfrak{r}(\mathfrak{g}_{i+1}) = \mathscr{D}\mathfrak{g} \cap \mathfrak{r}(\mathfrak{g}_i).$$

When $\mathscr{D}\mathfrak{g} \cap \mathfrak{r}(\mathfrak{g}_{i+1}) \neq \mathscr{D}\mathfrak{g} \cap \mathfrak{r}(\mathfrak{g}_i)$, it can be assumed that $\mathfrak{h}_i \subset \mathscr{D}\mathfrak{g} \cap \mathfrak{r}(\mathfrak{g}_i)$. Then $\mathfrak{r}(\mathfrak{g}_i) = \mathfrak{r}(\mathfrak{g}) \cap \mathfrak{g}_i$ (§ 5, Exercise 10) and Theorem 1 can be applied.)

¶ 5. Let K be a field of characteristic $p > 0$, $\mathfrak{g}$ a Lie algebra of finite dimension n over K and U the enveloping algebra of $\mathfrak{g}$.

(*a*) A *p-polynomial* (in one indeterminate) over K is any polynomial in K[X] whose only terms $\neq 0$ are of degree a power of p. Show that for every polynomial $f(X) \neq 0$ in K[X] there exists $g(X) \in K[X]$ such that $f(X)g(X)$ is a p-polynomial (consider the remainders in Euclidean division of the X^{p^k} by $f(X)$).

(*b*) Show that for every element $z \in \mathfrak{g}$ there exists a non-zero polynomial $f(X) \in K[X]$ such that $f(z)$ belongs to the centre C of U. (Consider the minimal polynomial of the endomorphism ad z and apply (*a*) to this polynomial, and also formula (1) of Exercise 19 of § 1.)

(*c*) Let $(e_i)_{1 \leqslant i \leqslant n}$ be a basis of $\mathfrak{g}$. For each i let $f_i \neq 0$ be a polynomial in K[X] such that $f_i(e_i) \in C$; let d_i be its degree and L the two-sided ideal of U generated by the $y_i = f_i(e_i)$. Show that the classes mod. L of the elements $e_1^{\alpha_1} \ldots e_1^{\alpha_n}$, where $0 \leqslant \alpha_i < d_i$ form a basis of U/L.

(*d*) Show that $\mathfrak{g}$ admits a finite-dimensional faithful linear representation (choose the f_i so that $d_i > 1$ for all i).

(*e*) Show that if $\mathfrak{g} \neq \{0\}$ there exists a non-semi-simple finite-dimensional linear representation of $\mathfrak{g}$. (Assuming that the f_i are of degrees $d_i > 1$, replace the f_i by f_i^2 in the definition of L and note that there then exist nilpotent elements $\neq 0$ in the centre of U/L and hence that U/L is not semi-simple.)

CHAPTER II

Free Lie Algebras

In this chapter,† *the letter* K *denotes a non-zero commutative ring. The unit element of* K *is denoted by* 1. *Unless otherwise mentioned, all cogebras, algebras and bigebras, all modules and all tensors products are over* K.

From § 6, K *will be assumed to be a field of characteristic* 0.

§ 1. ENVELOPING BIGEBRA OF A LIE ALGEBRA

Throughout this paragraph, $\mathfrak{g}$ will denote a Lie algebra over K, $U(\mathfrak{g})$ or simply U its enveloping algebra (Chapter I, § 2, no. 1), σ the canonical mapping of $\mathfrak{g}$ into $U(\mathfrak{g})$ (*loc. cit.*) and $(U_n)_{n \geqslant 0}$ the canonical filtration of U (*loc. cit.*, no. 6).

1. PRIMITIVE ELEMENTS OF A COGEBRA

Throughout this no. we consider a *cogebra* E (*Algebra*, Chapter III, § 11, no. 1) with coproduct

$$c \colon E \to E \otimes E$$

and *counit* ε (*loc. cit.*, no. 2). Recall that ε is a linear form on the K-module E such that (with the canonical identification of $E \otimes K$ and $K \otimes E$ with E):

$$\mathrm{Id}_E = (\varepsilon \otimes \mathrm{Id}_E) \circ c = (\mathrm{Id}_E \otimes \varepsilon) \circ c.$$

Let E^+ denote the kernel of ε and let u be an element of E such that

$$c(u) = u \otimes u \quad \text{and} \quad \varepsilon(u) = 1.$$

† The results of Chapters II and III depend on the first six books, on *Lie groups and Lie Algebras*, Chapter I, on *Commutative Algebra* and on *Differentiable and Analytic Manifolds*, Summary of Results; no. 9 of § 6 of Chapter III depends also on *Spectral Theories*, Chapter I.

The K-module E is the direct sum of E^+ and the submodule $K.u$ which is free with basis u; let $\pi_u: E \to E^+$ and $\eta_u: E \to K.u$ denote the projectors associated with this decomposition. Then

$$\pi_u(x) = x - \varepsilon(x).u, \qquad \eta_u(x) = \varepsilon(x).u. \tag{1}$$

DEFINITION 1. *An element x of* E *is called u-primitive if*

$$c(x) = x \otimes u + u \otimes x. \tag{2}$$

The u-primitive elements of E form a submodule of E, denoted by $P_u(E)$.

PROPOSITION 1. *Every u-primitive element of* E *belongs to* E^+.

(2) implies $x = \varepsilon(x).u + \varepsilon(u).x = \varepsilon(x).u + x$, whence $\varepsilon(x) = 0$.

Remark. If $x \in E$ and $c(x) = x' \otimes u + u \otimes x''$, where x', x'' are in E^+, then $x = \varepsilon(x').u + \varepsilon(u).x'' = x''$; similarly $x = x'$ and x is u-primitive.

For all $x \in E^+$, we write

$$c_u^+(x) = c(x) - x \otimes u - u \otimes x. \tag{3}$$

PROPOSITION 2. *We have*

$$(\pi_u \otimes \pi_u) \circ c = c_u^+ \circ \pi_u. \tag{4}$$

Let x be in E; then

$$\begin{aligned}(\pi_u \circ \pi_u)(c(x)) &= ((1 - \eta_u) \otimes (1 - \eta_u))(c(x)) \\ &= c(x) - (1 \otimes \eta_u)(c(x)) - (\eta_u \otimes 1)(c(x)) + (\eta_u \otimes \eta_u)(c(x)).\end{aligned}$$

As ε is counit of E,

$$(1 \otimes \eta_u)(c(x)) = x \otimes u, \qquad (\eta_u \otimes 1)(c(x)) = u \otimes x$$

whence

$$(\eta_u \otimes \eta_u)(c(x)) = (\eta_u \otimes 1)((1 \otimes \eta_u)(c(x))) = \varepsilon(x).u \otimes u;$$

from this we conclude

$$(\pi_u \otimes \pi_u)(c(x)) = c(x) - x \otimes u - u \otimes x + \varepsilon(x).u \otimes u.$$

On the other hand,

$$c_u^+(\pi_u(x)) = c(x) - x \otimes u - u \otimes x + \varepsilon(x).u \otimes u,$$

whence formula (4).

As E^+ is a direct factor submodule of E, $E^+ \otimes E^+$ can be identified with a direct factor submodule of $E \otimes E$. With this identification, $\pi_u \otimes \pi_u$ is a projector of $E \otimes E$ onto $E^+ \otimes E^+$. By formula (4), c_u^+ maps E^+ into $E^+ \otimes E^+$ and π_u *is a morphism of the cogebra* (E, c) *into the cogebra* (E^+, c_u^+).

PROPOSITION 3. *If the cogebra* (E, c) *is coassociative* (resp. *cocommutative*) (*Algebra*, Chapter III, § 11, no. 2), *so is the cogebra* (E^+, c_u^+).

This follows from the following lemma:

Lemma 1. *Let* π: E $\rightarrow$ E′ *be a surjective cogebra morphism. If* E *is coassociative* (resp. *cocommutative*), *so is* E′.

Let B be an associative K-algebra; the mapping $f \mapsto f \circ \pi$ is an *injective* algebra homomorphism of $\mathrm{Hom}_K(E', B)$ into $\mathrm{Hom}_K(E, B)$. It then suffices to apply Proposition 1 (resp. Proposition 2) of *Algebra*, Chapter III, § 11, no. 2.

2. PRIMITIVE ELEMENTS OF A BIGEBRA

Let E be a *bigebra* (*Algebra*, Chapter III, § 11, no. 4), c its coproduct, ε its counit and 1 its unit element. As $\varepsilon(1) = 1$ and $c(1) = 1 \otimes 1$, the results of the previous no. can be applied with $u = 1$. The 1-primitive elements of E (no. 1, Definition 1) are simply called *primitive* (cf. *Algebra*, Chapter III; § 11, no. 8), that is the elements x of E such that

$$c(x) = x \otimes 1 + 1 \otimes x. \tag{5}$$

We write simply π, η, P(E), c^+ instead of π_1, η_1, $P_1(E)$, c_1^+.

PROPOSITION 4. *The set* P(E) *of primitive elements of* E *is a Lie subalgebra of* E.

If x, y are in P(E), then

$$\begin{aligned} c(xy) &= c(x)c(y) = (x \otimes 1 + 1 \otimes x)(y \otimes 1 + 1 \otimes y) \\ &= xy \otimes 1 + 1 \otimes xy + x \otimes y + y \otimes x, \end{aligned}$$

whence

$$c([x, y]) = [x, y] \otimes 1 + 1 \otimes [x, y].$$

PROPOSITION 5. *Let* f: E $\rightarrow$ E′ *be a bigebra morphism. If* x *is a primitive element of* E, *then* $f(x)$ *is a primitive element of* E′ *and the restriction of* f *to* P(E) *is a Lie algebra homomorphism* $P(f)$: P(E) $\rightarrow$ P(E′).

Let c (resp. c') be the coproduct of E (resp. E′). Since f is a cogebra morphism $c' \circ f = (f \otimes f) \circ c$, whence

$$\begin{aligned} c'(f(x)) &= (f \otimes f)(c(x)) = (f \otimes f)(x \otimes 1 + 1 \otimes x) \\ &= f(x) \otimes 1 + 1 \otimes f(x), \end{aligned}$$

for x primitive. Hence f maps P(E) into P(E′) and $f([x, y]) = [f(x), f(y)]$ since f is an algebra homomorphism.

Remarks. (1) Let p be a prime number such that $p.1 = 0$ in K. The binomial formula and the congruences $\binom{p}{i} \equiv 0 \pmod{p}$ for $1 \leqslant i \leqslant p - 1$ imply that

P(E) is stable under the mapping $x \mapsto x^p$.

(2) By definition, the diagram

$$0 \longrightarrow \mathrm{P(E)} \longrightarrow \mathrm{E}^+ \xrightarrow{c^+} \mathrm{E}^+ \otimes \mathrm{E}^+$$

is an exact sequence. If K′ is a commutative ring and $\rho: \mathrm{K} \to \mathrm{K}'$ a ring homomorphism, $\rho^*(\mathrm{E}) = \mathrm{E} \otimes_{\mathrm{K}} \mathrm{K}'$ is a K′-bigebra and the inclusion $\mathrm{P(E)} \to \mathrm{E}$ defines a homomorphism of Lie K′-algebras

$$\alpha: \mathrm{P(E)} \otimes_{\mathrm{K}} \mathrm{K}' \to \mathrm{P}(\mathrm{E} \otimes_{\mathrm{K}} \mathrm{K}').$$

If K′ is *flat* over K (*Commutative Algebra*, Chapter I, § 2, no. 3, Definition 2), it follows from *loc. cit.* that the diagram

$$0 \longrightarrow \mathrm{P(E)} \otimes_{\mathrm{K}} \mathrm{K}' \longrightarrow \mathrm{E}^+ \otimes_{\mathrm{K}} \mathrm{K}' \xrightarrow{c^+ \otimes_{\mathrm{K}} \mathrm{Id}_{\mathrm{K}'}} (\mathrm{E}^+ \otimes_{\mathrm{K}} \mathrm{K}') \otimes_{\mathrm{K}'} (\mathrm{E}^+ \otimes_{\mathrm{K}} \mathrm{K}')$$

is an exact sequence, which implies that α is an *isomorphism.*

3. FILTERED BIGEBRAS

DEFINITION 2. *Let* E *be a bigebra with coproduct c. A filtration compatible with the bigebra structure on* E *is an increasing sequence* $(\mathrm{E}_n)_{n \geqslant 0}$ *of submodules of* E *such that*

$$(6) \qquad \begin{gathered} \mathrm{E}_0 = \mathrm{K}.1, \qquad \mathrm{E} = \bigcup_{n \geqslant 0} \mathrm{E}_n \\ \mathrm{E}_m . \mathrm{E}_n \subset \mathrm{E}_{m+n} \quad \text{for} \quad m \geqslant 0, n \geqslant 0 \\ c(\mathrm{E}_n) \subset \sum_{i+j=n} \mathrm{Im}(\mathrm{E}_i \otimes \mathrm{E}_j) \quad \text{for } n \geqslant 0.\dagger \end{gathered}$$

A bigebra with a filtration compatible with its bigebra structure is called a filtered bigebra.

Example. Let E be a graded bigebra (*Algebra*, Chapter III, § 11, no. 4, Definition 3) and $(\mathrm{E}^n)_{n \geqslant 0}$ its graduation. We write $\mathrm{E}_n = \sum_{i=0}^{n} \mathrm{E}^i$. The sequence (E_n) is a filtration compatible with the bigebra structure on E.

PROPOSITION 6. *Let* E *be a filtered bigebra and* $(\mathrm{E}_n)_{n \geqslant 0}$ *its filtration. For every integer* $n \geqslant 0$, *let* $\mathrm{E}_n^+ = \mathrm{E}_n \cap \mathrm{E}^+$. *Then* $\mathrm{E}_0^+ = \{0\}$ *and*

$$(7) \qquad c^+(\mathrm{E}_n^+) \subset \sum_{i=1}^{n-1} \mathrm{Im}(\mathrm{E}_i^+ \otimes \mathrm{E}_{n-i}^+) \quad \text{for } n \geqslant 0.\dagger$$

As $\mathrm{E}_0 = \mathrm{K}.1$, $\mathrm{E}_0^+ = 0$. If $x \in \mathrm{E}_n$, $\pi(x) = x - \varepsilon(x).1$ (formula (1)), whence $\pi(x) \in \mathrm{E}_n^+$ and $\pi(\mathrm{E}_n) \subset \mathrm{E}_n^+$. It follows that $\pi \otimes \pi$ maps $\mathrm{Im}(\mathrm{E}_i \otimes \mathrm{E}_j)$ into

† If A and B are two submodules of E, we denote by $\mathrm{Im}(\mathrm{A} \otimes \mathrm{B})$ the image of the canonical mapping $\mathrm{A} \otimes \mathrm{B} \to \mathrm{E} \otimes \mathrm{E}$.

$\operatorname{Im}(E_i^+ \otimes E_j^+)$ for $i \geqslant 0, j \geqslant 0$. As $c^+ = (\pi \otimes \pi) \circ c$ in E^+ (no. 1, Proposition 2), by (6)

$$c^+(E_n^+) \subset \sum_{i=0}^{n} \operatorname{Im}(E_i^+ \otimes E_{n-i}^+) = \sum_{i=1}^{n-1} \operatorname{Im}(E_i^+ \otimes E_{n-i}^+).$$

COROLLARY. *The elements of* E_1^+ *are primitive.*

If $x \in E_1^+$, then $c^+(x) = 0$ by (7), whence (5).

4. ENVELOPING BIGEBRA OF A LIE ALGEBRA

Recall that $\mathfrak{g}$ denotes a Lie algebra and U its enveloping algebra, with its canonical filtration $(U_n)_{n \geqslant 0}$.

PROPOSITION 7. *There exists on the algebra* U *one and only one coproduct* c *which makes* U *into a bigebra such that the elements of* $\sigma(\mathfrak{g})$ *are primitive. The bigebra* (U, c) *is cocommutative; its counit is the linear form* ε *such that the constant term* (Chapter I, § 2, no. 1) *of every element* x *of* U *is* $\varepsilon(x).1$. *The canonical filtration* $(U_n)_{n \geqslant 0}$ *of* U *is compatible with this bigebra structure.*

(a) Let $x \in \mathfrak{g}$; we write $c_0(x) = \sigma(x) \otimes 1 + 1 \otimes \sigma(x) \in U \otimes U$. If x, y are in $\mathfrak{g}$, then
$c_0(x)c_0(y) = (\sigma(x)\sigma(y)) \otimes 1 + 1 \otimes (\sigma(x)\sigma(y)) + \sigma(x) \otimes \sigma(y) + \sigma(y) \otimes \sigma(x)$,
whence

$$[c_0(x), c_0(y)] = c_0([x, y]).$$

By the universal property of U (Chapter I, § 2, no. 1, Proposition 1), there exists one and only one unital algebra homomorphism

$$c \colon U \to U \otimes U$$

such that $c(\sigma(x)) = \sigma(x) \otimes 1 + 1 \otimes \sigma(x)$ for $x \in \mathfrak{g}$. This proves the uniqueness assertion of Proposition 7.

(b) *We show that* c *is coassociative.* The linear mappings c' and c'' of U into $U \otimes U \otimes U$ defined by

$$c' = (c \otimes \mathrm{Id}_U) \circ c \quad \text{and} \quad c'' = (\mathrm{Id}_U \otimes c) \circ c$$

are unital algebra homomorphisms which coincide on $\sigma(\mathfrak{g})$ since, for $a \in \sigma(\mathfrak{g})$,

$$c'(a) = a \otimes 1 \otimes 1 + 1 \otimes a \otimes 1 + 1 \otimes 1 \otimes a = c''(a),$$

whence the result.

(c) *We show that* c *is cocommutative.* Let τ be the automorphism of $U \otimes U$ such that $\tau(a \otimes b) = b \otimes a$ for a, b in U. The mappings $\tau \circ c$ and c of U into $U \otimes U$ are unital algebra homomorphisms which coincide on $\sigma(\mathfrak{g})$, whence the result.

(d) *We show that* ε *is a counit for* c. The mappings $(\mathrm{Id}_U \otimes \varepsilon) \circ c$ and

$(\varepsilon \otimes \mathrm{Id}_{\mathrm{U}}) \circ c$, of U into U are unital algebra homomorphisms which coincide with Id_{U} on $\sigma(\mathfrak{g})$.

(e) We know that $\mathrm{U}_0 = \mathrm{K}.1$, $\mathrm{U}_n \subset \mathrm{U}_{n+1}$, $\mathrm{U} = \bigcup_{n \geq 0} \mathrm{U}_n$ and $\mathrm{U}_n . \mathrm{U}_m \subset \mathrm{U}_{n+m}$ (Chapter I, § 2, no. 6). Let $a_1, \ldots, a_n$ be in $\sigma(\mathfrak{g})$. Then

$$c(a_1 \ldots a_n) = \prod_{i=1}^{n} c(a_i) = \prod_{i=1}^{n} (a_i \otimes 1 + 1 \otimes a_i) \tag{8}$$

$$= \sum_{i=0}^{n} \sum_{\alpha \in \mathrm{I}(i)} (a_{\alpha(1)} \ldots a_{\alpha(i)}) \otimes (a_{\alpha(i+1)} \ldots a_{\alpha(n)}),$$

where $\mathrm{I}(i)$ denotes the set of permutations of $[1, n]$ which are increasing in each of the intervals $[1, i]$ and $[i + 1, n]$. As U_n is the K-module generated by the products of at most n elements of $\sigma(\mathfrak{g})$, formula (8) implies that the filtration (U_n) is compatible with the bigebra structure of (U, c).

Definition 3. *The bigebra* (U, c) *is called the* enveloping bigebra *of the Lie algebra* $\mathfrak{g}$.

Proposition 8. *Let* E *be a bigebra with coproduct denoted by* c_{E} *and let* h *be a Lie algebra homomorphism of* $\mathfrak{g}$ *into* P(E) (no. 2, Proposition 4). *The unital algebra homomorphism* $f: \mathrm{U} \to \mathrm{E}$ *such that* $f(\sigma(x)) = h(x)$ *for all* $x \in \mathfrak{g}$ *is a bigebra morphism.*

We show that $(f \otimes f) \circ c = c_{\mathrm{E}} \circ f$. These are two unital algebra homomorphisms of U into $\mathrm{E} \otimes \mathrm{E}$ and, for $a \in \sigma(\mathfrak{g})$,

$$(f \otimes f)(c(a)) = f(a) \otimes 1 + 1 \otimes f(a) = c_{\mathrm{E}}(f(a))$$

since $f(a) \in \mathrm{P}(\mathrm{E})$. Similarly if ε_{E} is the counit of E, $\varepsilon_{\mathrm{E}} \circ f$ is a unital algebra homomorphism $\mathrm{U} \to \mathrm{K}$ which is zero on $\sigma(\mathfrak{g})$ (no. 1, Proposition 1) and therefore coincides with ε.

It follows from Propositions 5 and 8 that the mapping $f \mapsto f \circ \sigma$ defines a one-to-one correspondence between bigebra homomorphisms $\mathrm{U}(\mathfrak{g}) \to \mathrm{E}$ and Lie algebra homomorphisms $\mathfrak{g} \to \mathrm{P}(\mathrm{E})$.

Corollary. *Let* $\mathfrak{g}_i$ $(i = 1, 2)$ *be a Lie algebra,* $\mathrm{U}(\mathfrak{g}_i)$ *its enveloping bigebra and* $\sigma_i: \mathfrak{g}_i \to \mathrm{U}(\mathfrak{g}_i)$ *the canonical mapping. For every Lie algebra homomorphism* $h: \mathfrak{g}_1 \to \mathfrak{g}_2$, *the unital algebra homomorphism* $\mathrm{U}(h): \mathrm{U}(\mathfrak{g}_1) \to \mathrm{U}(\mathfrak{g}_2)$ *such that* $\mathrm{U}(h) \circ \sigma_1 = \sigma_2 \circ h$ (Chapter I, § 2, no. 1) *is a bigebra morphism.*

5. STRUCTURE OF THE COGEBRA $\mathrm{U}(\mathfrak{g})$ IN CHARACTERISTIC 0

In this no., K will be assumed to be a *field of characteristic* 0.

Let $\mathrm{S}(\mathfrak{g})$ be the symmetric algebra of the vector space $\mathfrak{g}$, c_{S} its coproduct (*Algebra*, Chapter III, § 11, no. 1, *Example* 6) and η the canonical isomorphism of the vector space $S(\mathfrak{g})$ onto the vector space U (Chapter I, § 2, no. 7). Recall that if $x_1, \ldots, x_n$ are in $\mathfrak{g}$, then

$$(9) \qquad \eta(x_1 \ldots x_n) = \frac{1}{n!} \sum_{\tau \in \mathfrak{S}_n} \sigma(x_{\tau(1)}) \ldots \sigma(x_{\tau(n)}).$$

In particular, for $x \in \mathfrak{g}$ and $n \geqslant 0$,

$$(10) \qquad \eta(x^n) = \sigma(x)^n.$$

Note that by *Algebra*, Chapter III, § 6, no. 1, *Remark* 3, η is the unique linear mapping of $S(\mathfrak{g})$ into U satisfying condition (10).

PROPOSITION 9. *For every integer $n \geqslant 0$, let U^n be the vector subspace of* U *generated by the $\sigma(x)^n$ for $x \in \mathfrak{g}$.*

(a) *The sequence $(U^n)_{n \geqslant 0}$ is a graduation of the vector space* U *compatible with its cogebra structure.*

Let U *be given the graduation (U^n).*

(b) *The canonical mapping $\eta\colon S(\mathfrak{g}) \to U$ is an isomorphism of graded cogebras.*

Let $x \in \mathfrak{g}$ and $n \in \mathbf{N}$. Then

$$(11) \qquad c_S(x^n) = c_S(x)^n = (x \otimes 1 + 1 \otimes x)^n = \sum_{i=0}^{n} \binom{n}{i} x^i \otimes x^{n-i}$$

since c_S is an algebra homomorphism. Similarly, by (10),

$$(12) \qquad c(\eta(x^n)) = c(\sigma(x)^n) = c(\sigma(x))^n = (\sigma(x) \otimes 1 + 1 \otimes \sigma(x))^n$$
$$= \sum_{i=0}^{n} \binom{n}{i} \sigma(x)^i \otimes \sigma(x)^{n-i} = \sum_{i=0}^{n} \binom{n}{i} \eta(x^i) \otimes \eta(x^{n-i}),$$

whence

$$(\eta \otimes \eta)(c_S(x^n)) = c(\eta(x^n)).$$

As the x^n, for $x \in \mathfrak{g}$ and $n \in \mathbf{N}$, generate the vector space $S(\mathfrak{g})$, $(\eta \otimes \eta) \circ c_S = c \circ \eta$ and η is a cogebra isomorphism.

On the other hand, formula (10) shows that $\eta(S^n(\mathfrak{g})) = U^n$, which completes the proof of (a) and (b) taking account of the fact that the graduation of $S(\mathfrak{g})$ is compatible with its cogebra structure.

The graduation $(U^n)_{n \geqslant 0}$ of U is called the *canonical graduation*.

COROLLARY. *The canonical mapping σ defines an isomorphism of $\mathfrak{g}$ onto the Lie algebra* P(U) *of primitive elements of* U.

As c^+ is a graded homomorphism of degree 0,

$$P(U) = \sum_{n \geqslant 1} (P(U) \cap U^n).$$

It suffices to prove that if $n > 1$ and $a \in U^n$ is primitive, then $a = 0$. Now a can be written as $\sum_i \lambda_i a_i^n$, where $\lambda_i \in K$, $a_i \in \sigma(\mathfrak{g})$. By (12), the term of bidegree $(1, n-1)$ in $c^+(a)$ is $n \sum_i \lambda_i a_i \otimes a_i^{n-1}$. Hence $\sum_i \lambda_i a_i \otimes a_i^{n-1} = 0$. If $\mu : U \otimes U \to U$ is the linear mapping defined by multiplication on U, then

$$a = \sum_i \lambda_i a_i^n = \mu\Big(\sum_i \lambda_i a_i \otimes a_i^{n-1}\Big) = 0.$$

Remarks. (1) $U_n = \sum_{i=0}^{n} U^i$ (Chapter I, § 2, no. 7, Corollary 4 to Theorem 1).

(2) The mapping η is the unique morphism of graded cogebras of $S(\mathfrak{g})$ into U such that $\eta(1) = 1$ and $\eta(x) = \sigma(x)$ for $x \in \mathfrak{g}$. For if η' is a morphism satisfying these conditions, we prove by induction on n that $\eta'(x^n) = \eta(x^n)$ for $x \in \mathfrak{g}$ and $n > 1$. As $c_S^+(x^n) = \sum_{i=1}^{n-1} \binom{n}{i} x^i \otimes x^{n-i}$ by (3) and (11),

$$(\eta \otimes \eta)(c_S^+(x^n)) = (\eta' \otimes \eta')(c_S^+(x^n))$$

by the induction hypothesis. It follows that $c^+(\eta(x^n)) = c^+(\eta'(x^n))$; it follows that $\eta(x^n) - \eta'(x^n)$ is a primitive element of degree n and hence is zero (Corollary to Proposition 9).

(3) Let ψ be the canonical isomorphism of the bigebra $TS(\mathfrak{g})$ onto the bigebra $S(\mathfrak{g})$ (*Algebra*, Chapter IV, § 5, Corollary 1 to Proposition 12). The mapping

$$\eta \circ \psi : TS(\mathfrak{g}) \to U$$

is called *canonical*. It is the unique morphism η' of graded cogebras of $TS(\mathfrak{g})$ into U such that $\eta'(1) = 1$ and $\eta'(x) = \sigma(x)$ for all $x \in \mathfrak{g}$.

(4) Let V be a vector space. The primitive elements of the bigebra $S(V)$ are the elements of degree 1. This follows from the Corollary to Proposition 9 applied to the commutative Lie algebra V.

Let $(e_i)_{i \in I}$ be a basis of the vector K-space $\mathfrak{g}$, where the indexing set I is totally ordered. For all $\alpha \in \mathbf{N}^{(I)}$, we write

$$e_\alpha = \prod_{i \in I} \frac{\sigma(e_i)^{\alpha(i)}}{\alpha(i)!}. \tag{13}$$

The e_α, for $|\alpha| \leqslant n$, form a basis of the vector K-space U_n (Chapter I, § 2, no. 7, Corollary 3 to Theorem 1). Then

$$e_0 = 1, \qquad e_{\varepsilon_i} = \sigma(e_i) \quad \text{for } i \in I.$$

As the graded algebra associated with the filtered algebra U is commutative (*loc. cit.*, Theorem 1), for α, β in $\mathbf{N}^{(\mathrm{I})}$,

$$e_\alpha . e_\beta \equiv ((\alpha, \beta)) . e_{\alpha+\beta} \mod. \mathrm{U}_{|\alpha|+|\beta|-1}. \tag{14}$$

where

$$((\alpha, \beta)) = \prod_{i \in \mathrm{I}} \frac{(\alpha(i) + \beta(i))!}{\alpha(i)!\beta(i)!}.$$

On the other hand, we have immediately

$$\varepsilon(e_0) = 1, \qquad \varepsilon(e_\alpha) = 0 \quad \text{for } |\alpha| \geqslant 1. \tag{15}$$

Finally, formula (12) implies that, for $\alpha \in \mathbf{N}^{(\mathrm{I})}$,

$$c(e_\alpha) = \sum_{\beta+\gamma=\alpha} e_\beta \otimes e_\gamma. \tag{16}$$

This formula allows us to determine the algebra $\mathrm{U}' = \mathrm{Hom}(\mathrm{U}, \mathrm{K})$ *dual* to the cogebra U (*Algebra*, Chapter III, § 11, no. 2). For let $\mathrm{K}[[\mathrm{X}_i]]_{i \in \mathrm{I}}$ be the algebra of formal power series in indeterminates $(\mathrm{X}_i)_{i \in \mathrm{I}}$ (cf. *Algebra*, Chapter III, § 2, no. 11); if $\lambda \in \mathrm{U}'$, let f_λ denote the formal power series

$$f_\lambda = \sum_\alpha \langle \lambda, e_\alpha \rangle \mathrm{X}^\alpha, \quad \text{where } \mathrm{X}^\alpha = \prod_{i \in \mathrm{I}} \mathrm{X}_i^{\alpha(i)}$$

and the summation index α runs through $\mathbf{N}^{(\mathrm{I})}$.

PROPOSITION 10. *The mapping $\lambda \mapsto f_\lambda$ is an isomorphism of the algebra* U′ *onto the algebra of formal power series* $\mathrm{K}[[\mathrm{X}_i]]_{i \in \mathrm{I}}$.

Because (e_α) is a basis of U, the mapping $\lambda \mapsto f_\lambda$ is K-linear and bijective. On the other hand, for λ, μ in U′,

$$\begin{aligned} f_{\lambda\mu} &= \sum_\alpha \langle \lambda\mu, e_\alpha \rangle \mathrm{X}^\alpha = \sum_\alpha \langle \lambda \otimes \mu, c(e_\alpha) \rangle \mathrm{X}^\alpha \\ &= \sum_\alpha \langle \lambda \otimes \mu, \sum_{\beta+\gamma=\alpha} e_\beta \otimes e_\gamma \rangle \mathrm{X}^\alpha \qquad \text{(by (16))} \\ &= \sum_{\beta, \gamma} \langle \lambda, e_\beta \rangle \langle \mu, e_\gamma \rangle \mathrm{X}^{\beta+\gamma} = f_\lambda f_\mu, \end{aligned}$$

which shows that $\lambda \mapsto f_\lambda$ is an *algebra* isomorphism and completes the proof.

6. STRUCTURE OF FILTERED BIGEBRAS IN CHARACTERISTIC 0

In this no. we continue to assume that K is a *field of characteristic* 0.

If E is a bigebra, the canonical injection $\mathrm{P(E)} \to \mathrm{E}$ can be extended to a bigebra morphism $f_\mathrm{E}\colon \mathrm{U(P(E))} \to \mathrm{E}$ (no. 4, Proposition 8).

THEOREM 1. *Let* E *be a cocommutative bigebra.*

(a) *The bigebra morphism* f_E: U(P(E)) $\rightarrow$ E *is injective.*

(b) *If there exists on* E *a filtration compatible with its bigebra structure* (no. 3, Definition 2), *the morphism* f_E *is an isomorphism.*

(In case (b), the bigebra E is therefore identified with the enveloping bigebra of the Lie algebra of its primitive elements.)

Let c_E (resp. ε_E) be the coproduct (resp. counit) of E. We write $\mathfrak{g} = P(E)$; let $(e_i)_{i \in I}$ be a basis of the vector K-space $\mathfrak{g}$, where the indexing set I is totally ordered, and let $(e_\alpha)_{\alpha \in \mathbf{N}^{(I)}}$ be the basis introduced in the preceding no. We write $X_\alpha = f_E(e_\alpha)$ for $\alpha \in \mathbf{N}^{(I)}$. By (15) and (16), we have:

$$\varepsilon_E(X_0) = 1, \qquad \varepsilon_E(X_\alpha) = 0 \quad \text{for } |\alpha| \geqslant 1, \tag{17}$$

$$c_E(X_\alpha) = \sum_{\beta+\gamma=\alpha} X_\beta \otimes X_\gamma \quad \text{for } \alpha \in \mathbf{N}^{(I)}, \tag{18}$$

since f_E is a cogebra morphism.

We show that f_E *is injective.* This results from the following lemma:

Lemma 2. *Let* V *be a vector space,* E *a cogebra and* f: S(V) $\rightarrow$ E *a cogebra morphism. If the restriction of* f *to* $S^0(V) + S^1(V)$ *is injective, then* f *is injective.*

Let $n \geqslant 0$; we write $S_n = \sum_{i \geqslant n} S^i(V)$ and c_S for the coproduct of S(V) and show by induction on n that $f \mid S_n$ is injective. Since the assertion is trivial for $n = 0$ and $n = 1$, we assume that $n \geqslant 2$ and let $u \in S_n$ be such that $f(u) = 0$. Then

$$\begin{aligned} 0 = c_E(f(u)) &= (f \otimes f)(c_S(u)) \\ &= f(u) \otimes 1 + 1 \otimes f(u) + (f \otimes f)(c_S^+(u)) \\ &= (f \otimes f)(c_S^+(u)). \end{aligned}$$

As $c_S^+(u) \in S_{n-1} \otimes S_{n-1}$, by (11) the induction hypothesis shows that u is a primitive element of S(V), hence is of degree 1 (no. 5, *Remark* 4) and hence is zero, since $f \mid S^1(V)$ is injective.

It follows in particular that the family (X_α) is *free*.

We show that f_E *is surjective* if E has a filtration compatible with its bigebra structure. Let $(E_n)_{n \geqslant 0}$ be such a filtration and write $E_n^+ = E_n \cap \operatorname{Ker}(\varepsilon_E)$. We show by induction on n that E_n^+ is contained in the image of f_E. As $E = K.1 + \bigcup_{n \geqslant 0} E_n^+$, this will imply the surjectivity of f_E. The assertion is trivial for $n = 0$ and follows from the Corollary to Proposition 6 of no. 3 for $n = 1$; suppose henceforth that $n \geqslant 2$ and let $x \in E_n^+$. By Proposition 6 of no. 3,

$$c_E^+(x) \in \sum_{i=1}^{n-1} E_i^+ \otimes E_{n-i}^+$$

and by the induction hypothesis there exist scalars $\lambda_{\alpha,\beta}$, where α, β are in $\mathbf{N}^{(I)}$, which are zero except for a finite number, such that

$$c_E^+(x) = \sum_{\alpha,\beta\neq 0} \lambda_{\alpha,\beta} X_\alpha \otimes X_\beta. \tag{19}$$

Hence by formula (18)

$$(c_E^+ \otimes \mathrm{Id}_E)(c_E^+(x)) = \sum_{\alpha,\beta,\gamma\neq 0} \lambda_{\alpha+\beta,\gamma} X_\alpha \otimes X_\beta \otimes X_\gamma$$

$$(\mathrm{Id}_E \otimes c_E^+)(c_E^+(x)) = \sum_{\alpha,\beta,\gamma\neq 0} \lambda_{\alpha,\beta+\gamma} X_\alpha \otimes X_\beta \otimes X_\gamma.$$

By Proposition 3 of no. 1 and the linear independence of the X_α it follows that

$$\lambda_{\alpha+\beta,\gamma} = \lambda_{\alpha,\beta+\gamma} \quad \text{for } \alpha, \beta, \gamma \text{ in } \mathbf{N}^{(I)} - \{0\}. \tag{20}$$

On the other hand, the coproduct c_E is cocommutative; the same argument as above implies

$$\lambda_{\alpha,\beta} = \lambda_{\beta,\alpha} \quad \text{for } \alpha, \beta \text{ in } \mathbf{N}^{(I)} - \{0\}. \tag{21}$$

Suppose that there exists a family of scalars (μ_α) *with* $|\alpha| \geqslant 2$, *such that*

$$\mu_{\alpha+\beta} = \lambda_{\alpha,\beta} \quad \text{for } \alpha, \beta \text{ in } \mathbf{N}^{(I)} - \{0\}. \tag{22}$$

Then

$$c_E^+(x) = \sum_{\alpha,\beta\neq 0} \mu_{\alpha+\beta} X_\alpha \otimes X_\beta = \sum_{|\gamma|\geqslant 2} \mu_\gamma c_E^+(X_\gamma)$$

by formula (18), hence $y = x - \sum_{|\gamma|\geqslant 2} \mu_\gamma X_\gamma$ is primitive and hence belongs to $P(E) \subset \mathrm{Im}(f_E)$. Hence

$$x = y + \sum_{|\gamma|\geqslant 2} \mu_\gamma f_E(e_\gamma) \in \mathrm{Im}(f_E).$$

The proof will therefore be complete when we have proved the following lemma:

Lemma 3. *If a family of scalars* $(\lambda_{\alpha,\beta})$ *of finite support* (with α, β in $\mathbf{N}^{(I)} - \{0\}$) *satisfies relations* (20) *and* (21), *there exists a family* $(\mu_\alpha)_{|\alpha|\geqslant 2}$ *of finite support such that* $\mu_{\alpha+\beta} = \lambda_{\alpha,\beta}$ *for* α, β *non-zero.*

It suffices to prove that

$$\alpha + \beta = \gamma + \delta \tag{23}$$

implies $\lambda_{\alpha,\beta} = \lambda_{\gamma,\delta}$ for α, β, γ, δ non-zero. By Riesz's Decomposition Lemma (*Algebra*, Chapter VI, § 1, no. 10, Theorem 1) there exist π, ρ, σ, τ in $\mathbf{N}^{(I)}$ such that

$$\alpha = \pi + \sigma, \qquad \beta = \rho + \tau, \qquad \gamma = \pi + \rho, \qquad \delta = \sigma + \tau.$$

Suppose $\pi \neq 0$; as $\sigma + \beta = \rho + \delta$, relation (20) implies

$$\lambda_{\alpha,\beta} = \lambda_{\pi+\sigma,\beta} = \lambda_{\pi,\sigma+\beta} = \lambda_{\pi,\rho+\delta} = \lambda_{\pi+\rho,\delta} = \lambda_{\gamma,\delta}.$$

If on the other hand $\pi = 0$, then $\beta = \gamma + \tau$ and $\delta = \alpha + \tau$, whence

$$\lambda_{\alpha,\beta} = \lambda_{\alpha,\gamma+\tau} = \lambda_{\alpha+\tau,\gamma} = \lambda_{\delta,\gamma}$$

by (20), but also $\lambda_{\delta,\gamma} = \lambda_{\gamma,\delta}$ by (21), whence $\lambda_{\alpha,\beta} = \lambda_{\gamma,\delta}$.

§ 2. FREE LIE ALGEBRAS

1. REVISION OF FREE ALGEBRAS

Let X be a set. Recall the construction of the free magma M(X) constructed on X (*Algebra*, Chapter I, § 7, no. 1). By induction on the integer $n \geqslant 1$, we define the sets X_n by writing $X_1 = X$ and taking X_n to be the sum set of the sets $X_p \times X_{n-p}$ where $p = 1, 2, \ldots, n-1$; if X is finite, so is each X_n. The sum set of the family $(X_n)_{n \geqslant 1}$ is denoted by M(X); each of the sets X_n (and in particular X) is identified with a subset of M(X). Let w and w' be in M(X); let p and q denote the integers such that $w \in X_p$ and $w' \in X_q$ and let $n = p + q$; the image of the ordered pair (w, w') under the canonical injection of $X_p \times X_{n-p}$ into X_n is denoted by $w.w'$ and called the product of w and w'. Every mapping of X into a magma M can be extended in a unique way to a magma homomorphism of M(X) into M.

Let w be in M(X); the unique integer n such that $w \in X_n$ is called the *length* of w and denoted by $l(w)$. Then $l(w.w') = l(w) + l(w')$ for w, w' in M(X). The set X is the subset of M(X) consisting of the elements of length 1. Every element w of length $\geqslant 2$ can be written uniquely in the form $w = w'.w''$.

The algebra of the magma M(X) with coefficients in the ring K is denoted by Lib(X), or $\mathrm{Lib}_K(X)$ when it is necessary to indicate the ring K. The set M(X) is a basis of the K-module Lib(X) and X will therefore be identified with a subset of Lib(X). If A is an algebra, every mapping of X into A can be extended uniquely to a homomorphism of Lib(X) into A (*Algebra*, Chapter III, § 2, no. 7, Proposition 7).

2. CONSTRUCTION OF THE FREE LIE ALGEBRA

DEFINITION 1. *The free Lie algebra over the set* X *is the quotient algebra*

$$L(X) = \mathrm{Lib}(X)/\mathfrak{a},$$

where $\mathfrak{a}$ *is the two-sided ideal of* Lib(X) *generated by the elements of one of the forms*

(1) $$Q(a) = a.a \quad \text{for } a \text{ in } \mathrm{Lib}(X),$$

(2) $$J(a, b, c) = a.(b.c) + b.(c.a) + c.(a.b)$$

for a, b, c *in* Lib(X).

Clearly L(X) is a Lie K-lagebra; the composition of two elements u, v of L(X) will be denoted by $[u, v]$. When it is necessary to indicate the ring K, we write $L_K(X)$ for L(X).

The following proposition justifies the name *free* Lie algebra given to L(X).

PROPOSITION 1. *Let ψ be the canonical mapping of* Lib(X) *onto* L(X) *and ϕ the restriction of ψ to* X. *For every mapping f of* X *into a Lie algebra $\mathfrak{g}$, there exists one and only one homomorphism* $F: L(X) \to \mathfrak{g}$ *such that $f = F \circ \phi$.*

(a) *Existence of* F: let h be the homomorphism of Lib(X) into $\mathfrak{g}$ extending f (no. 1). For all a in Lib(X), $h(Q(a)) = h(a.a) = [h(a), h(a)] = 0$; similarly, the Jacobi identity satisfied by $\mathfrak{g}$ implies that $h(J(a, b, c)) = 0$ for a, b, c in Lib(X). It follows that $h(\mathfrak{a}) = 0$, whence there is a homomorphism F of L(X) into $\mathfrak{g}$ such that $h = F \circ \psi$. By restricting to X, we obtain $f = F \circ \phi$.

(b) *Uniqueness of* F: let $F': L(X) \to \mathfrak{g}$ be a homomorphism such that $f = F' \circ \phi$. The homomorphisms $F \circ \psi$ and $F' \circ \psi$ of Lib(X) into $\mathfrak{g}$ coincide on X and hence are equal; as ψ is surjective, $F = F'$.

COROLLARY 1. *The family $(\phi(x))_{x \in X}$ is free over* K *in* L(X).

Let $x_1, x_2, \ldots, x_n$ be distinct elements of X and $\lambda_1, \ldots, \lambda_n$ be elements of K such that

$$\lambda_1 . \phi(x_1) + \cdots + \lambda_n . \phi(x_n) = 0. \tag{3}$$

Let $\mathfrak{g}$ be the commutative Lie algebra with K as underlying module. For $i = 1, 2, \ldots, n$, there exists a homomorphism F_i of L(X) into $\mathfrak{g}$ such that $F_i(\phi(x_i)) = 1$ and $F_i(\phi(x)) = 0$ for $x \neq x_i$ (Proposition 1); applying F_i to relation (3), we obtain $\lambda_i = 0$.

COROLLARY 2. *Let $\mathfrak{a}$ be a Lie algebra. Every extension of* L(X) *by $\mathfrak{a}$ is inessential.*

Let $\mathfrak{a} \xrightarrow{\lambda} \mathfrak{g} \xrightarrow{\mu} L(X)$ be such an extension (Chapter I, § 1, no. 7). As μ is surjective, there exists a mapping f of X into $\mathfrak{g}$ such that $\phi = \mu \circ f$. Let F be the homomorphism of L(X) into $\mathfrak{g}$ such that $f = F \circ \phi$ (Proposition 1). Then $(\mu \circ F) \circ \phi = \mu \circ f = \phi$ and Proposition 1 shows that $\mu \circ F$ is the identity automorphism of L(X). The given extension is therefore inessential (Chapter I, § 1, no. 7, Proposition 6 and Definition 6).

As the ring K is non-zero, Corollary 1 to Proposition 1 shows that ϕ is *injective. Hence the set* X *can be identified by means of ϕ with its image in* L(X); with this convention, X generates L(X) and every mapping of X into a Lie algebra $\mathfrak{g}$ *can be extended* to a Lie algebra homomorphism of L(X) into $\mathfrak{g}$.

Remark. When X is empty, M(X) is empty and hence $L(X) = \{0\}$. If X consists of a single element x, the submodule $K.x$ of L(X) is a Lie subalgebra of

L(X); as X generates L(X), Corollary 1 to Proposition 1 shows that L(X) is a free module with basis $\{x\}$.

3. PRESENTATIONS OF A LIE ALGEBRA

Let $\mathfrak{g}$ be a Lie algebra and $\boldsymbol{a} = (a_i)_{i \in \mathrm{I}}$ a family of elements of $\mathfrak{g}$. Let $f_{\boldsymbol{a}}$ be the homomorphism of L(I) into $\mathfrak{g}$ mapping each $i \in \mathrm{I}$ to a_i. The image of $f_{\boldsymbol{a}}$ is the subalgebra of $\mathfrak{g}$ generated by $\boldsymbol{a}$; the elements of the kernel of $f_{\boldsymbol{a}}$ are called the *relators* of the family $\boldsymbol{a}$. The family $\boldsymbol{a}$ is called generating (resp. free, basic) if $f_{\boldsymbol{a}}$ is surjective (resp. injective, bijective).

Let $\mathfrak{g}$ be a Lie algebra. A *presentation* of $\mathfrak{g}$ is an ordered pair $(\boldsymbol{a}, \boldsymbol{r})$ consisting of a generating family $\boldsymbol{a} = (a_i)_{i \in \mathrm{I}}$ and a family $\boldsymbol{r} = (r_j)_{j \in \mathrm{J}}$ of relators of $\boldsymbol{a}$ generating the ideal of L(I) the kernel of $f_{\boldsymbol{a}}$. We also say that $\mathfrak{g}$ is presented by the family $\boldsymbol{a}$ related by the relators r_j ($j \in \mathrm{J}$).

Let I be a set and $\boldsymbol{r} = (r_j)_{j \in \mathrm{J}}$ be a family of elements of the free Lie algebra L(I); let $\mathfrak{a}_{\boldsymbol{r}}$ be the ideal of L(I) generated by $\boldsymbol{r}$. The quotient algebra $\mathrm{L}(\mathrm{I}, \boldsymbol{r}) = \mathrm{L}(\mathrm{I})/\mathfrak{a}_{\boldsymbol{r}}$ is called the Lie algebra defined by I and the family of relators $(r_j)_{j \in \mathrm{J}}$; we also say that $\mathrm{L}(\mathrm{I}, \boldsymbol{r})$ is defined by the presentation $(\mathrm{I}, \boldsymbol{r})$, or also by $(\mathrm{I}; (r_j = 0)_{j \in \mathrm{J}})$. When the family $\boldsymbol{r}$ is empty, $\mathrm{L}(\mathrm{I}, \boldsymbol{r}) = \mathrm{L}(\mathrm{I})$.

Let I and $\boldsymbol{r}$ be as above; let ξ_i denote the image of i in $\mathrm{L}(\mathrm{I}, \boldsymbol{r})$. The generating family $\boldsymbol{\xi} = (\xi_i)_{i \in \mathrm{I}}$ and the family of relators $\boldsymbol{r}$ constitute a presentation of $\mathrm{L}(\mathrm{I}, \boldsymbol{r})$. Conversely, if $\mathfrak{g}$ is a Lie algebra and $(\boldsymbol{a}, \boldsymbol{r})$, where $\boldsymbol{a} = (a_i)_{i \in \mathrm{I}}$, is a presentation of $\mathfrak{g}$, there exists a unique isomorphism $u\colon \mathrm{L}(\mathrm{I}, \boldsymbol{r}) \to \mathfrak{g}$ such that $u(\xi_i) = a_i$ for all $i \in \mathrm{I}$.

4. LIE POLYNOMIALS AND SUBSTITUTIONS

Let I be a set. Let T_i denote the canonical image of the element i of I in L(I) (which is also sometimes denoted by $\mathrm{L}((\mathrm{T}_i)_{i \in \mathrm{I}})$); the elements of L(I) are called *Lie polynomials* in the indeterminates $(\mathrm{T}_i)_{i \in \mathrm{I}}$.

Let $\mathfrak{g}$ be a Lie algebra. If $\boldsymbol{t} = (t_i)_{i \in \mathrm{I}}$ is a family of elements of $\mathfrak{g}$, let $f_{\boldsymbol{t}}$ denote the homomorphism of L(I) into $\mathfrak{g}$ such that $f_{\boldsymbol{t}}(\mathrm{T}_i) = t_i$ for $i \in \mathrm{I}$ (no. 2, Proposition 1). The image under $f_{\boldsymbol{t}}$ of the element P of L(I) is denoted by $\mathrm{P}((t_i)_{i \in \mathrm{I}})$. In particular, $\mathrm{P}((\mathrm{T}_i)_{i \in \mathrm{I}}) = \mathrm{P}$; the above element $\mathrm{P}((t_i)_{i \in \mathrm{I}})$ is sometimes called the element of $\mathfrak{g}$ obtained by substituting the t_i for the T_i in the Lie polynomial $\mathrm{P}((\mathrm{T}_i)_{i \in \mathrm{I}})$.

Let $\sigma\colon \mathfrak{g} \to \mathfrak{g}'$ be a Lie algebra homomorphism. For every family $\boldsymbol{t} = (t_i)_{i \in \mathrm{I}}$ of elements of $\mathfrak{g}$ and all $\mathrm{P} \in \mathrm{L}(\mathrm{I})$,

$$(4) \qquad \sigma(\mathrm{P}((t_i)_{i \in \mathrm{I}})) = \mathrm{P}((\sigma(t_i))_{i \in \mathrm{I}}),$$

for $\sigma \circ f_{\boldsymbol{t}}$ maps T_i to $\sigma(t_i)$ for $i \in \mathrm{I}$.

Let $(\mathrm{Q}_j)_{j \in \mathrm{J}}$ be a family of elements of L(I) and let $\mathrm{P} \in \mathrm{L}(\mathrm{J})$. By substituting the Q_j for the T_j in P, we obtain a Lie polynomial $R = P((Q_j)_{j \in J}) \in L(I)$. Then

$$(5) \qquad R((t_i)_{i \in I}) = P((Q_j((t_i)_{i \in I}))_{j \in J})$$

for every family $t = (t_i)_{i \in I}$ of elements of a Lie algebra $\mathfrak{g}$, as is seen by operating by the homomorphism f_t on the equation $R = P((Q_j)_{j \in J})$ and using (4).

Let $\mathfrak{g}$ be a Lie algebra, I a finite set and $P \in L(I)$. Suppose that $\mathfrak{g}$ is a free K-module. The mapping

$$\tilde{P} : \mathfrak{g}^I \to \mathfrak{g}$$

defined by $\tilde{P}((t_i)_{i \in I}) = P((t_i)_{i \in I})$ is then *polynomial*.† For the set F of mappings of $\mathfrak{g}^I$ into $\mathfrak{g}$ is a Lie algebra with the bracket defined by

$$(6) \qquad [\phi, \psi](t) = [\phi(t), \psi(t)];$$

the set F′ of polynomial mappings of $\mathfrak{g}^I$ into $\mathfrak{g}$ is a Lie subalgebra of F by the bilinearity of the bracket. Our assertion then follows from the fact that the mapping $P \mapsto \tilde{P}$ is a Lie algebra homomorphism and $\tilde{T}_i = pr_i \in F'$ for all i.

5. FUNCTORIAL PROPERTIES

PROPOSITION 2. *Let* X *and* Y *be two sets. Every mapping* $u: X \to Y$ *can be extended uniquely to a Lie algebra homomorphism* $L(u): L(X) \to L(Y)$. *For every mapping* $v: Y \to Z$, $L(v \circ u) = L(v) \circ L(u)$.

The existence and uniqueness of $L(u)$ follow from Proposition 1 of no. 2. The homomorphisms $L(v \circ u)$ and $L(v) \circ L(u)$ have the same restriction to X and hence are equal (Proposition 1).

COROLLARY. *If u is injective* (resp. *surjective, bijective*), *so is* $L(u)$.

Since the assertion is trivial for $X = \varnothing$, we assume $X \neq \varnothing$. If u is injective there exists a mapping v of Y into X such that $v \circ u$ is the identity mapping of X; by Proposition 2, $L(v) \circ L(u)$ is the identity of automorphism of L(X) and hence $L(u)$ is injective. When u is surjective, there exists a mapping w of Y into X such that $u \circ w$ is the identity mapping of Y; then $L(u) \circ L(w)$ is the identity mapping of L(Y), which proves that $L(u)$ is surjective.

† Recall (*Algebra*, Chapter IV, § 5, no. 10) the definition of polynomial mappings of a free module M into a module N: if q is an integer $\geqslant 0$, a mapping $f: M \to N$ is called *homogeneous polynomial of degree* q if there exists a multilinear mapping u of M^q into N such that

$$f(x) = u(x, \ldots, x) \quad \text{for all } x \in M.$$

A mapping of M into N is called *polynomial* if it is a finite sum of homogeneous polynomial mappings of suitable degrees.

Let X be a set and S a subset of X. The above corollary shows that the canonical injection of S into X can be extended to an isomorphism α of L(S) onto the Lie subalgebra L′(S) of L(X) generated by S; *we shall identify* L(S) *and* L′(S) by means of α.

Let $(S_\alpha)_{\alpha \in I}$ be a *right directed family* of subsets of X with union S. The relation $S_\alpha \subset S_\beta$ implies $L(S_\alpha) \subset L(S_\beta)$ and hence the family of Lie subalgebras $L(S_\alpha)$ of L(X) is right directed. Therefore $\mathfrak{g} = \bigcup_{\alpha \in I} L(S_\alpha)$ is a Lie subalgebra of L(X); then $S \subset \mathfrak{g}$, whence $L(S) \subset \mathfrak{g}$, and, as $L(S_\alpha) \subset L(S)$ for all $\alpha \in I$, $\mathfrak{g} \subset L(S)$. Hence

$$L\Big(\bigcup_{\alpha \in I} S_\alpha\Big) = \bigcup_{\alpha \in I} L(S_\alpha) \tag{7}$$

for every *right directed* family $(S_\alpha)_{\alpha \in I}$ of subsets of X.

Applying the above to the family of finite subsets of X, we see that every element of L(X) is of the form $P(x_1, \ldots, x_n)$ where P is a Lie polynomial in n indeterminates and $x_1, \ldots, x_n$ are elements of X.

PROPOSITION 3. *Let* K′ *be a non-zero commutative ring and* $u: K \to K'$ *a ring homomorphism. For every set* X *there exists one and only one Lie* K′*-algebra homomorphism*

$$v: L_K(X) \otimes K' \to L_{K'}(X)$$

such that $v(x \otimes 1) = x$ for $x \in X$. *Further, v is an isomorphism.*

Applying Proposition 1 to $\mathfrak{g} = L_{K'}(X)$ considered as a Lie K-algebra and the mapping $x \mapsto x$ of X into $\mathfrak{g}$, we obtain a K-homomorphism $L_K(X) \to L_{K'}(X)$, whence there is a K′-homomorphism $v: L_K(X) \otimes K' \to L_{K'}(X)$. The fact that v is unique and is an isomorphism follows from the fact that the ordered pair $(L_K(X) \otimes K', x \mapsto x \otimes 1)$ is a solution of the same universal mapping problem as the ordered pair $(L_{K'}(X), x \mapsto x)$.

Remark. Let $\mathfrak{h}'$ be a Lie K′-algebra and $\mathfrak{h}$ the Lie K-algebra derived from $\mathfrak{h}'$ by restricting the ring of scalars. If $P \in L_K(X)$, we can define $\tilde{P}: \mathfrak{h}^X \to \mathfrak{h}$ (no. 4). We see immediately that

$$\tilde{P} = (v(P \otimes 1))^\sim.$$

6. GRADUATIONS

Let Δ be a commutative monoid, written additively. Let ϕ_0 denote a mapping of X into Δ and ϕ the homomorphism of the free magma M(X) into Δ which extends ϕ_0. For all $\delta \in \Delta$, let $\mathrm{Lib}^\delta(X)$ be the submodule of Lib(X) with basis the subset $\phi^{-1}(\delta)$ of M(X). The family $(\mathrm{Lib}^\delta(X))_{\delta \in \Delta}$ is a graduation of the algebra Lib(X), that is

$$\mathrm{Lib}(X) = \bigoplus_{\delta \in \Delta} \mathrm{Lib}^\delta(X) \tag{8}$$

$$\mathrm{Lib}^\delta(X) . \mathrm{Lib}^{\delta'}(X) \subset \mathrm{Lib}^{\delta+\delta'}(X) \quad \text{for } \delta, \delta' \text{ in } \Delta \tag{9}$$

(*Algebra*, Chapter III, § 3, no. 1, *Example* 3).

Lemma 1. *The ideal* $\mathfrak{a}$ *of Definition* 1 *is graded.*

For a, b in Lib(X), let $B(a, b) = a.b + b.a$. The formulae

$$B(a, b) = Q(a + b) - Q(a) - Q(b) \tag{10}$$

$$Q(\lambda_1.w_1 + \cdots + \lambda_n.w_n) = \sum_i \lambda_i^2 Q(w_i) + \sum_{i<j} \lambda_i\lambda_j B(w_i, w_j) \tag{11}$$

for $w_1, \ldots, w_n$ in M(X) and $\lambda_1, \ldots, \lambda_n$ in K, show that the families $(Q(a))_{a \in \mathrm{Lib(X)}}$ and $(Q(w), B(w, w'))_{w, w' \in M(X)}$ generate the same submodule of Lib(X). As J is trilinear the ideal $\mathfrak{a}$ is generated by the homogeneous elements $Q(w)$, $B(w, w')$ and $J(w, w', w'')$, where w, w', w'' are in M(X), and hence is graded (*Algebra*, Chapter III, § 3, no. 3, Proposition 1).

Let the Lie algebra $L(X) = \mathrm{Lib}(X)/\mathfrak{a}$ be given the quotient graduation. The homogeneous component of L(X) of degree δ is denoted by $L^\delta(X)$; it is the submodule of L(X) generated by the images of the elements $w \in M(X)$ such that $\phi(w) = \delta$.

We shall make special use of the following two cases:

(a) *Total graduation:* we take $\Delta = \mathbf{N}$ and $\phi_0(x) = 1$ for all $x \in X$, whence $\phi(w) = l(w)$ for w in M(X). The K-module $L^n(X)$ is generated by the images of the elements of length n in M(X), which we shall call *alternants of degree n*. We shall see later that the module $L^n(X)$ is free and admits a basis consisting of alternants of degree n (no. 11, Theorem 1). Then $L(X) = \bigoplus_{n \geq 1} L^n(X)$ and $L^1(X)$ admits X as basis (no. 2, Corollary 1 to Proposition 1). By the construction of M(X),

$$L^n(X) = \sum_{p=1}^{n-1} [L^p(X), L^{n-p}(X)] \tag{12}$$

and in particular

$$[L^m(X), L^n(X)] \subset L^{m+n}(X). \tag{13}$$

(b) *Multigraduation:* we take Δ to be the free commutative monoid $\mathbf{N}^{(X)}$ constructed on X. The mapping ϕ_0 of X into Δ is defined by $(\phi_0(x))(x') = \delta_{xx'}$, where $\delta_{xx'}$ is the Kronecker symbol. For $w \in M(X)$ and $x \in X$, the integer $(\phi(w))(x)$ is "the number of occurrences of the letter x in w". For α in $\mathbf{N}^{(X)}$, we write $|\alpha| = \sum_{x \in X} \alpha(x)$, whence $|\phi(w)| = l(w)$ for all w in M(X). It follows that

$$L^n(X) = \bigoplus_{|\alpha| = n} L^\alpha(X); \tag{14}$$

obviously

$$[L^\alpha(X), L^\beta(X)] \subset L^{\alpha+\beta}(X) \quad \text{for } \alpha, \beta \text{ in } \mathbf{N}^{(X)}. \tag{15}$$

PROPOSITION 4. *Let* S *be a subset of* X. *If* $\mathbf{N}^{(S)}$ *is identified with its canonical image in* $\mathbf{N}^{(X)}$ (*Algebra*, Chapter I, § 7, no. 7), *then* $L(S) = \sum_{\alpha \in \mathbf{N}^{(S)}} L^{\alpha}(X)$. *Further, for all* $\alpha \in \mathbf{N}^{(S)}$, *the homogeneous component of degree* α *under the multigraduation on* $L(S)$ *is equal to* $L^{\alpha}(X)$.

Let $\alpha \in \mathbf{N}^{(S)}$. The module $L^{\alpha}(S)$ is generated by the images in $L(X)$ of the elements w in $M(S)$ such that $\phi(w) = \alpha$, that is (*Algebra*, § 7, no. 9, formulae (23) and (24)) the set *of* w in $M(X)$ such that $\phi(w) = \alpha$. Hence $L^{\alpha}(S) = L^{\alpha}(X)$.

The proposition follows from this and the relation $L(S) = \sum_{\alpha \in \mathbf{N}^{(S)}} L^{\alpha}(S)$.

COROLLARY. *For every family* $(S_i)_{i \in I}$ *of subsets of* X,

(16)
$$L\left(\bigcap_{i \in I} S_i\right) = \bigcap_{i \in I} L(S_i).$$

This follows from Proposition 4 and the obvious formula

(17)
$$\mathbf{N}^{(S)} = \bigcap_{i \in I} \mathbf{N}^{(S_i)}$$

where we have written $S = \bigcap_{i \in I} S_i$.

7. LOWER CENTRAL SERIES

PROPOSITION 5. *Let* $\mathfrak{g}$ *be a Lie algebra and* P *a submodule of* $\mathfrak{g}$. *We define the submodules* P_n *of* $\mathfrak{g}$ *by the formulae* $P_1 = P$ *and* $P_{n+1} = [P, P_n]$ *for* $n \geqslant 1$. *Then*

(18)
$$[P_m, P_n] \subset P_{m+n},$$

(19)
$$P_n = \sum_{p=1}^{n-1} [P_p, P_{n-p}] \quad \text{for } n \geqslant 2.$$

We prove (18) by induction on m. The case $m = 1$ is obvious. By the Jacobi identity,

$$[[P, P_m], P_n] \subset [P_m, [P, P_n]] + [P, [P_m, P_n]],$$

that is

$$[P_{m+1}, P_n] \subset [P_m, P_{n+1}] + [P, [P_m, P_n]].$$

The induction hypothesis implies that $[P_m, P_{n+1}] \subset P_{m+n+1}$ and $[P_m, P_n] \subset P_{m+n}$, whence

$$[P_{m+1}, P_n] \subset P_{m+n+1} + [P, P_{m+n}] = P_{m+n+1}.$$

By formula (18), $P_n \supset \sum_{p=1}^{n-1} [P_p, P_{n-p}] \supset [P_1, P_{n-1}] = P_n$, whence (19).

When we take $P = \mathfrak{g}$, the sequence (P_n) is the *lower central series* $(\mathscr{C}^n \mathfrak{g})$ of $\mathfrak{g}$ (Chapter I, § 1, no. 5). Hence:

PROPOSITION 6. *Let* $\mathfrak{g}$ *be a Lie algebra and* $(\mathscr{C}^n\mathfrak{g})_{n\geqslant 1}$ *the lower central series of* $\mathfrak{g}$. *Then*

$$[\mathscr{C}^m\mathfrak{g}, \mathscr{C}^n\mathfrak{g}] \subset \mathscr{C}^{m+n}\mathfrak{g} \quad \textit{for } m \geqslant 1 \textit{ and } n \geqslant 1.$$

Generalizing Definition 1 of Chapter I, § 4, no. 1, we shall say that a Lie algebra $\mathfrak{g}$ is *nilpotent* if $\mathscr{C}^n\mathfrak{g} = \{0\}$ for n sufficiently large. The *nilpotency class* of a nilpotent Lie algebra $\mathfrak{g}$ is the smallest integer n such that $\mathscr{C}^{n+1}\mathfrak{g} = \{0\}$.

PROPOSITION 7. *Let* X *be a set and* n *an integer* $\geqslant 1$.

(a) $L^{n+1}(X) = [L^1(X), L^n(X)]$.

(b) *The module* $L^n(X)$ *is generated by the elements* $[x_1, [x_2, \ldots, [x_{n-1}, x_n]\ldots]]$ *where* $(x_1, \ldots, x_n)$ *runs through the set of sequences of* n *elements of* X.

(c) *The lower central series of* L(X) *is given by* $\mathscr{C}^n(L(X)) = \sum_{p\geqslant n} L^p(X)$.

(a) We apply Proposition 5 with $\mathfrak{g} = L(X)$ and $P = L^1(X)$. By induction on n, we deduce from (12) (no. 6) and (19) the equality $P_n = L^n(X)$. The desired relation is then equivalent to the definition $[P, P_n] = P_{n+1}$.

(b) This follows from (a) by induction on n.

(c) Let $\mathfrak{g} = L(X)$ and $\mathfrak{g}_n = \sum_{p\geqslant n} L_p(X)$. Then $\mathfrak{g} = \mathfrak{g}_1$ and formula (13) of no. 6 implies $[\mathfrak{g}_n, \mathfrak{g}_m] \subset \mathfrak{g}_{n+m}$ and in particular $[\mathfrak{g}, \mathfrak{g}_n] \subset \mathfrak{g}_{n+1}$. By induction on n, $\mathscr{C}^n\mathfrak{g} \subset \mathfrak{g}_n$. On the other hand, from (a) we deduce that $L^n(X) \subset \mathscr{C}^n\mathfrak{g}$ by induction on n. As $\mathscr{C}^n\mathfrak{g}$ is an ideal of $\mathfrak{g}$, the relation $L^p(X) \subset \mathscr{C}^n\mathfrak{g}$ implies that

$$L^{p+1}(X) = [L^1(X), L^p(X)] \subset \mathscr{C}^n\mathfrak{g}$$

by (a). Hence $L^p(X) \subset \mathscr{C}^n\mathfrak{g}$ for $p \geqslant n$, whence $\mathfrak{g}_n \subset \mathscr{C}^n\mathfrak{g}$.

COROLLARY. *Let* $\mathfrak{g}$ *be a Lie algebra and* $(x_i)_{i\in I}$ *a generating family of* $\mathfrak{g}$. *The n-th term* $\mathscr{C}^n\mathfrak{g}$ *of the lower central series of* $\mathfrak{g}$ *is the module generated by the iterated brackets* $[x_{i_1}, [x_{i_2}, \ldots, [x_{i_{p-1}}, x_{i_p}]\ldots]]$ *for* $p \geqslant n$, *and* $i_1, \ldots, i_p$ *in* I.

Let f be the homomorphism of L(I) into $\mathfrak{g}$ such that $f(i) = x_i$ for all $i \in I$. As $(x_i)_{i\in I}$ generates $\mathfrak{g}$, $\mathfrak{g} = f(L(I))$, whence $\mathscr{C}^n\mathfrak{g} = f(\mathscr{C}^n(L(I)))$ by Proposition 4 of Chapter I, § 1, no. 5. The corollary then follows from assertions (b) and (c) of Proposition 7.

8. DERIVATIONS OF FREE LIE ALGEBRAS

PROPOSITION 8. *Let* X *be a set, let* M *be an* L(X)*-module and let* d *be a mapping of* X *into* M. *There exists one and only one linear mapping* D *of* L(X) *into* M *extending* d *and satisfying the relation:*

(20) $$D([a, a']) = a.D(a') - a'.D(a) \quad \textit{for } a, a' \textit{ in } L(X).$$

We define a Lie algebra $\mathfrak{g}$ with underlying module $M \times L(X)$ by means of the bracket

(21) $$[(m, a), (m', a')] = (a.m' - a'.m, [a, a']),$$

for a, a' in L(X) and m, m' in M (Chapter I, § 1, no. 8). Let f be the homomorphism of L(X) into $\mathfrak{g}$ such that $f(x) = (d(x), x)$ for all x in X; let $f(a) = (\mathrm{D}(a), u(a))$ for all a in L(X). By formula (21), u is a homomorphism of L(X) into itself; as $u(x) = x$ for x in X, $u(a) = a$ for all a in L(X), whence

$$f(a) = (\mathrm{D}(a), a). \tag{22}$$

By (21) and (22), relation (20) then follows from $f([a, a']) = [f(a), f(a')]$.

Conversely, let D′ be a mapping of L(X) into M satisfying relation (20′) analogous to (20) and extending d. Let $f'(a) = (\mathrm{D}'(a), a)$ for $a \in \mathrm{L(X)}$; by (20′) and (21), f' is a homomorphism of L(X) into $\mathfrak{g}$, coinciding with f on X, whence $f' = f$ and D′ = D.

Corollary. *Every mapping of* X *into* L(X) *can be extended uniquely to a derivation of* L(X).

When M is equal to L(X) with the adjoint representation, relation (20) means that D is a derivation.

9. ELIMINATION THEOREM

Proposition 9. *Let* S_1 *and* S_2 *be two disjoint sets and d a mapping of* $\mathrm{S}_1 \times \mathrm{S}_2$ *into* $\mathrm{L(S_2)}$. *Let* $\mathfrak{g}$ *be the quotient Lie algebra of* $\mathrm{L(S_1 \cup S_2)}$ *by the ideal generated by the elements* $[s_1, s_2] - d(s_1, s_2)$ *with* $s_1 \in \mathrm{S}_1$, $s_2 \in \mathrm{S}_2$; *let* ψ *be the canonical mapping of* $\mathrm{L(S_1 \cup S_2)}$ *onto* $\mathfrak{g}$.

(a) *For* $i = 1, 2$, *the restriction* ϕ_i *of* ψ *to* S_i *can be extended to an isomorphism of* $\mathrm{L(S}_i)$ *onto a subalgebra* $\mathfrak{a}_i$ *of* $\mathfrak{g}$.

(b) $\mathfrak{g} = \mathfrak{a}_1 + \mathfrak{a}_2$, $\mathfrak{a}_1 \cap \mathfrak{a}_2 = \{0\}$ *and* $\mathfrak{a}_2$ *is an ideal of* $\mathfrak{g}$.

For $i = 1, 2$, let ψ_i denote the homomorphism of $\mathrm{L(S}_i)$ into $\mathfrak{g}$ which extends ϕ_i and $\mathfrak{a}_i$ its image. Clearly $\phi_i(\mathrm{S}_i)$ generates $\mathfrak{a}_i$.

Let $s_1 \in \mathrm{S}_1$; we write $\mathrm{D} = \mathrm{ad}\, \phi_1(s_1)$. The derivation D of $\mathfrak{g}$ maps $\phi_2(\mathrm{S}_2)$ into $\mathfrak{a}_2$ by the relation

$$[\phi_1(s_1), \phi_2(s_2)] = \psi_2(d(s_1, s_2)) \quad \text{for } s_2 \in \mathrm{S}_2;$$

as the subalgebra $\mathfrak{a}_2$ of $\mathfrak{g}$ is generated by $\phi_2(\mathrm{S}_2)$, therefore $\mathrm{D}(\mathfrak{a}_2) \subset \mathfrak{a}_2$. The set of $x \in \mathfrak{g}$ such that ad x leaves $\mathfrak{a}_2$ invariant is a Lie subalgebra of $\mathfrak{g}$ which contains $\phi_1(\mathrm{S}_1)$ by the above and hence also $\mathfrak{a}_1$. Hence

$$[\mathfrak{a}_1, \mathfrak{a}_2] \subset \mathfrak{a}_2. \tag{23}$$

Therefore $\mathfrak{a}_1 + \mathfrak{a}_2$ is a Lie subalgebra of $\mathfrak{g}$ and, as it contains the generating set $\phi_1(\mathrm{S}_1) \cup \phi_2(\mathrm{S}_2)$,

$$\mathfrak{a}_1 + \mathfrak{a}_2 = \mathfrak{g}. \tag{24}$$

For all $s_1 \in \mathrm{S}_1$ there exists a derivation D_{s_1} of $\mathrm{L(S_2)}$ such that

$$\mathrm{D}_{s_1}(s_2) = d(s_1, s_2)$$

for all s_2 in S_2 (no. 8, Corollary to Proposition 8). The mapping $s_1 \mapsto D_{s_1}$ can be extended to a homomorphism D of $L(S_1)$ into the Lie algebra of derivations of $L(S_2)$. Let $\mathfrak{h}$ be the semi-direct product of $L(S_1)$ by $L(S_2)$ corresponding to D (Chapter I, § 1, no. 8). As a module $\mathfrak{h}$ is equal to $L(S_1) \times L(S_2)$ and in particular

$$[(s_1, 0), (0, s_2)] = (0, d(s_1, s_2)) \tag{25}$$

for $s_1 \in S_1$ and $s_2 \in S_2$.

From (25) we deduce the existence of a homomorphism f of $\mathfrak{g}$ into $\mathfrak{h}$ such that $f(\phi_1(s_1)) = (s_1, 0)$ and $f(\phi_2(s_2)) = (0, s_2)$ for $s_1 \in S_1$ and $s_2 \in S_2$. We deduce immediately the relation

$$f(\psi_1(a_1) + \psi_2(a_2)) = (a_1, a_2) \tag{26}$$

for $a_1 \in L(S_1)$ and $a_2 \in L(S_2)$.

Relation (26) shows that ψ_1 and ψ_2 are injective and that $\mathfrak{a}_1 \cap \mathfrak{a}_2 = \{0\}$. Formulae (23) and (24) then imply the proposition.

Proposition 10 (elimination theorem). *Let* X *be a set,* S *a subset of* X *and* T *the set of sequences* $(s_1, \ldots, s_n, x)$ *with* $n \geqslant 0$, $s_1, \ldots, s_n$ *in* S *and* x *in* X − S.†

(a) *The module* L(X) *is the direct sum of the subalgebra* L(S) *of* L(X) *and the ideal* $\mathfrak{a}$ *of* L(X) *generated by* X − S.

(b) *There exists a Lie algebra isomorphism* ϕ *of* L(T) *onto* $\mathfrak{a}$ *which maps* $(s_1, \ldots, s_n, x)$ *to* $(\operatorname{ad} s_1 \circ \cdots \circ \operatorname{ad} s_n)(x)$.

Let $\mathfrak{g}$ be the Lie algebra constructed as in Proposition 9 given

$$S_1 = S, \qquad S_2 = T, \qquad d(s, t) = (s, s_1, \ldots, s_n, x) \in T \subset L(T)$$

for $t = (s_1, \ldots, s_n, x)$ in T and $s \in S_1$. We identify L(S) and L(T) with their canonical images in $\mathfrak{g}$ (Proposition 9 (a)).

Let ψ be the mapping $(s_1, \ldots, s_n, x) \mapsto (\operatorname{ad} s_1 \circ \cdots \circ \operatorname{ad} s_n)(x)$ of T into L(X). Obviously $\psi(d(s, t)) = [s, \psi(t)]$ for $s \in S$ and $t \in T$ and hence there exists a homomorphism $\alpha: \mathfrak{g} \to L(X)$ whose restriction to S is the identity and whose restriction to T is ψ. Now $X - S \subset T$, whence there is a homomorphism $\beta: L(X) \to \mathfrak{g}$ whose restriction to $X = S \cup (X - S)$ is the identity.

We show that α is an isomorphism and β the inverse isomorphism. As $\psi(x) = x$ for x in X − S, we see that $\alpha \circ \beta$ coincides with the identity on X, whence $\alpha \circ \beta = \mathrm{Id}_{L(X)}$. On the other hand, $[s, t] = d(s, t)$ in $\mathfrak{g}$ for $s \in S$, $t \in T$ by construction; it follows that $t = (s_1, \ldots, s_n, x)$ is equal in $\mathfrak{g}$ to $(\operatorname{ad} s_1 \circ \cdots \circ \operatorname{ad} s_2)(x)$ whence $t = \beta(\alpha(t))$. As $\beta(\alpha(s)) = s$ for $s \in S$ and $S \cup T$ generates $\mathfrak{g}$, $\beta \circ \alpha = \mathrm{Id}_{\mathfrak{g}}$.

As α is an isomorphism of $\mathfrak{g}$ onto L(X), Proposition 9 shows that the restric-

† For $n = 0$, we obtain the elements of X − S, whence $X - S \subset T$.

tion of α to L(T) is an isomorphism ϕ of L(T) onto an ideal $\mathfrak{b}$ of L(X) such that the module L(X) is the direct sum of L(S) and $\mathfrak{b}$. Obviously

$$\phi(s_1, \ldots, s_n, x) = (\operatorname{ad} s_1 \circ \cdots \circ \operatorname{ad} s_n)(x)$$

for $(s_1, \ldots, s_n, x)$ in T.

Hence $\phi(\mathrm{T}) \subset \mathfrak{a}$, whereas $\mathfrak{b} \subset \mathfrak{a}$ since $\phi(\mathrm{T})$ generates the subalgebra $\mathfrak{b}$ of L(X). But $\mathfrak{b}$ is an ideal and $\mathrm{X} - \mathrm{S} \subset \phi(\mathrm{T}) \subset \mathfrak{b}$, whence $\mathfrak{a} \subset \mathfrak{b}$.

Corollary. *Let* $y \in \mathrm{X}$. *The free Lie algebra* L(X) *is the direct sum of the free submodule* K.y *and the Lie subalgebra admitting as basic family the family of* $((\operatorname{ad} y)^n . z)$ *where* $n \geqslant 0$ *and* $z \in \mathrm{X} - \{y\}$.

It suffices to put $\mathrm{S} = \{y\}$ in Proposition 10.

10. HALL SETS IN A FREE MAGMA

Let X be a set, M(X) the free magma constructed on X and $\mathrm{M}^n(\mathrm{X})$, where $n \in \mathbf{N}^*$, the set of elements of M(X) of length n (no. 1). If $w \in \mathrm{M}(\mathrm{X})$ and $l(w) \geqslant 2$, let $\alpha(w)$ and $\beta(w)$ denote the elements of M(X) determined by the relation $w = \alpha(w)\beta(w)$; then $l(\alpha(w)) < l(w)$, $l(\beta(w)) < l(w)$. Finally, for u, v in M(X), let $u^m v$ denote the element defined by induction on the integer $m \geqslant 0$ by $u^0 v = v$ and $u^{m+1} v = u(u^m v)$.

Definition 2. *A* Hall set relative to X *is any totally ordered subset* H *of* M(X) *satisfying the following conditions:*

(A) *If* $u \in \mathrm{H}$, $v \in \mathrm{H}$ *and* $l(u) < l(v)$, *then* $u < v$.

(B) $\mathrm{X} \subset \mathrm{H}$ *and* $\mathrm{H} \cap \mathrm{M}^2(\mathrm{X})$ *consists of the products* xy *with* x, y *in* X *and* $x < y$.

(C) *An element* w *of* M(X) *of length* $\geqslant 3$ *belongs to* H *if and only if it is of the form* $a(bc)$ *with* a, b, c *in* H, $bc \in \mathrm{H}$, $b \leqslant a < bc$ *and* $b < c$.

Proposition 11. *There exists a Hall set relative to* X.

We shall construct by induction on the integer n sets $\mathrm{H}_n \subset \mathrm{M}^n(\mathrm{X})$ and a total ordering on these sets:

(a) We write $\mathrm{H}_1 = \mathrm{X}$ and give it a total ordering.

(b) The set H_2 consists of the products xy with x, y in X and $x < y$. We give it a total ordering.

(c) Let $n \geqslant 3$ be such that the totally ordered sets $\mathrm{H}_1, \ldots, \mathrm{H}_{n-1}$ are already defined. The set $\mathrm{H}'_{n-1} = \mathrm{H}_1 \cup \cdots \cup \mathrm{H}_{n-1}$ has a total ordering which induces the given relations on $\mathrm{H}_1, \ldots, \mathrm{H}_{n-1}$ and is such that $w < w'$ if $l(w) < l(w')$. We define H_n to be the set of products $a(bc) \in \mathrm{M}^n(\mathrm{X})$ with a, b, c in H'_{n-1} satisfying the relations $bc \in \mathrm{H}'_{n-1}$, $b \leqslant a < bc$, $b < c$ and give H_n a total ordering.

We write $\mathrm{H} = \bigcup_{n \geqslant 1} \mathrm{H}_n$; we give H the total ordering defined thus: $w \leqslant w'$ if and only if $l(w) < l(w')$ or $l(w) = l(w') = n$ and $w \leqslant w'$ in the set H_n. It is immediate that H is a Hall set relative to X.

For every subset S of X, we identify the free magma M(S) with its canonical image in M(X).

PROPOSITION 12. *Let* H *be a Hall set relative to* X *and let* x, y *be in* X.

(a) $H \cap M(\{x\}) = \{x\}$.

(b) *Suppose that* $x < y$ *and let* d_y *be the homomorphism of* M(X) *into* **N** *such that* $d_y(y) = 1$ *and* $d_y(z) = 0$ *for* $z \in X$, $z \neq y$. *The set of elements* $w \in H \cap M(\{x, y\})$ *such that* $d_y(w) = 1$ *consists of the elements* $x^n y$ *with* n *an integer* $\geqslant 0$.

By Definition 2 (B), $x \in H$ and $H \cap M^2(\{x\}) = \varnothing$. If $w \in H \cap M(\{x\})$, where $n = l(w) \geqslant 3$, the elements $\alpha(w)$ and $\beta(w)$ also belong to $H \cap M(\{x\})$ by Definition 2 (C). It immediately follows by induction on n that $H \cap M^n(\{x\}) = \varnothing$ for $n \geqslant 2$, whence (a).

We now prove (b). By Definition 2 (B), $y \in H$ and $xy \in H$. We show by induction on n that $x^n y \in H$ for n an integer $\geqslant 2$. Now $x^n y = x(x(x^{n-2}y))$ and the induction hypothesis implies that $x^{n-2}y \in H$. Now $l(x) < l(x^{n-2}y)$ for $n > 2$ and $x < y$, whence $x < x^{n-2}y$ in any case; condition (C) of Definition 2 shows that $x^n y \in H$. On the other hand, certainly $d_y(x^n y) = 1$. Conversely, let $w \in H \cap M(\{x, y\})$, with $d_y(w) = 1$. If $l(w) = 1$, then $w = y$; if $l(w) = 2$, then $w = xy$ by Definition 2 (B). If $l(w) \geqslant 3$, then $w = a(bc)$, with a, b, c, bc in $H \cap M(\{x, y\})$ (Definition 2 (C)). $d_y(bc) = 0$ is impossible, since this would imply $bc \in M(\{x\})$, which is impossible by (a). Hence $d_y(bc) = 1$ and $d_y(a) = 0$, whence $a = x$ by (a). It follows immediately by induction on $n = l(w)$ that $w = x^{n-1}y$, which completes the proof of (b).

COROLLARY. *If* Card X $\geqslant 2$, *then* $H \cap M^n(X) \neq \varnothing$ *for every integer* $n \geqslant 1$.

PROPOSITION 13. *Let* X *be a finite set with at least two elements. Let* H *denote a Hall set relative to* X. *Then there exist a strictly increasing bijection* $p \mapsto w_p$ *of* **N** *onto* H *and a sequence* $(P_p)_{p \in \mathbf{N}}$ *of subsets of* H *with the following properties:*

(a) $P_0 = X$.

(b) *For every integer* $p \geqslant 0$, $w_p \in P_p$.

(c) *For every integer* $n \geqslant 1$, *there exists an integer* $p(n)$ *such that every element of* P_p *is of length* $> n$ *for all* $p \geqslant p(n)$.

(d) *For every integer* $p \geqslant 0$, *the set* P_{p+1} *consists of the elements of the form* $w_p^i w$, *where* $i \geqslant 0$, $w \in P_p$ *and* $w \neq w_p$.

As X is finite, each of the sets $M^n(X)$ is finite. Let $H_n = H \cap M^n(X)$ for all $n \geqslant 1$. The Corollary to Proposition 12 shows that the finite set H_n is non-empty. Let u_n be the cardinal of H_n; let $v_0 = 0$ and $v_n = u_1 + \cdots + u_n$ for $n \geqslant 1$. As H_n is a totally ordered finite set, there exists a strictly increasing bijection $p \mapsto w_p$ of the interval $[v_{n-1}, v_n - 1]$ of **N** onto H_n. It is immediate that $p \mapsto w_p$ is a strictly increasing bijection of **N** onto H.

Let $P_0 = X$ and for every integer $p \geqslant 1$ let P_p be the set of elements w of H such that $w \geqslant w_p$, and, either $w \in X$, or $\alpha(w) < w_p$ (note that if w is of length $\geqslant 2$ the relation $w \in H$ implies $\alpha(w) \in H$ by condition (C) of Definition 2). Then $w_p \in P_p$; this is clear if $w_p \in X$ and follows from the inequality $l(\alpha(w_p)) < l(w_p)$ and condition (A) of Definition 2 when $w_p \notin X$.

Hence conditions (a) and (b) are satisfied.

Let n be an integer $\geqslant 1$ and let $p \geqslant v_n$. For all $w \in P_p$, $l(w) \geqslant l(w_p) > n$ by the very definition of the mapping $p \mapsto w_p$. This establishes (c).

We now show that every element of the form $u = w_p^i w$ with $i \geqslant 0$, $w \in P_p$ and $w \neq w_p$ belongs to P_{p+1}. If $i \neq 0$, then $l(u) > l(w_p)$, whence $u > w_p$ and $u \geqslant w_{p+1}$; then $u \notin X$ and $\alpha(u) = w_p < w_{p+1}$, whence $u \in P_{p+1}$. If $i = 0$, then $u \in P_p$ and $u \neq w_p$; then $u > w_p$, whence $u \geqslant w_{p+1}$; if u does not belong to X, then $\alpha(w) < w_p$, whence $\alpha(w) < w_{p+1}$; then again $u \in P_{p+1}$.

Conversely, let $u \in P_{p+1}$. We distinguish two cases:

(α) There exists no element v of M(X) such that $u = w_p v$. By definition of P_{p+1}, $u > w_p$. Further, if $u \notin X$, then $\alpha(u) \neq w_p$ by the given hypothesis and $\alpha(u) < w_{p+1}$ since $u \in P_{p+1}$; hence $\alpha(u) < w_p$. Hence $u \in P_p$ and $u \neq w_p$.

(β) There exists v in M(X) such that $u = w_p v$. By Definition 2, of necessity, either $w_p \in X$, $v \in X$ and $w_p < v$, or $v \notin X$ and $\alpha(v) \leqslant w_p < v$. In either case, $v \in P_{p+1}$.

Then there exist an integer $i \geqslant 0$ and an element w of M(X) such that $u = w_p^i w$, and either $w \in X$ or $w \notin X$ and $\alpha(w) \neq w_p$. If $i = 0$, we have case (α) above, whence $w \in P_p$ and $w \neq w_p$. If $i > 0$, the proof of (β) above establishes, by induction on i, the relations $w \in P_{p+1}$ and $w \neq w_p$. Suppose $w \notin X$; from $w \in P_{p+1}$ it follows that $\alpha(w) \leqslant w_p$ and as $\alpha(w) \neq w_p$, we conclude that $w \in P_p$. This completes the proof of (d).

Example. Suppose that X has two elements x, y; let X be ordered such that $x < y$. The construction given in the proof of Proposition 11 gives a set H with 14 elements of length $\leqslant 5$ given in the following table:

H_1	$w_1 = x$	$w_2 = y$	
H_2	$w_3 = (xy)$		
H_3	$w_4 = (x(xy))$	$w_5 = (y(xy))$	
H_4	$w_6 = (x(x(xy)))$	$w_7 = (y(x(xy)))$	$w_8 = (y(y(xy)))$
H_5	$w_9 = (x(x(x(xy))))$	$w_{10} = (y(x(x(xy))))$	$w_{11} = (y(y(x(xy))))$
	$w_{12} = (y(y(y(xy))))$	$w_{13} = ((xy)(x(xy)))$	$w_{14} = ((xy)(y(xy)))$.

(The elements of H have been numbered according to the total ordering chosen on each H_n.

11. HALL BASES OF A FREE LIE ALGEBRA

We preserve the notation of the preceding no.

THEOREM 1. *Let* H *be a Hall set relative to* X *and* Ψ *the canonical mapping of* M(X) *into the free Lie algebra* L(X). *The restriction of* Ψ *to* H *is a basis of the module* L(X).

For every element w of H we write $\bar{w} = \Psi(w)$.

(A) *The case where* X *is finite.*

If X is empty, so are M(X) and therefore H and L(X) is zero. If X has a single element x, $H \cap M^n(X)$ is empty for $n \geqslant 2$ (Proposition 12 (a)). Therefore, $H = \{x\}$; we know also (no. 2, *Remark*) that the module L(X) is free and has basis $\{\bar{x}\}$. The theorem is therefore true when X has at most one element.

Suppose henceforth that X has at least two elements; choose sequences (w_p) and (P_p) with the properties stated in Proposition 13. For every integer $p \geqslant 0$, let L_p denote the submodule of L(X) generated by the elements $\bar{w}_i$ with $0 \leqslant i < p$ and $\mathfrak{g}_p$ the Lie subalgebra of L(X) generated by the family $(\bar{u})_{u \in P_p}$.

Lemma 2. *For every integer $p \geqslant 0$, the module L_p admits the family $(\bar{w}_i)_{0 \leqslant i < p}$ as basis, the Lie algebra $\mathfrak{g}_p$ admits $(\bar{u})_{u \in P_p}$ as basic family and the module* L(X) *is the direct sum of L_p and $\mathfrak{g}_p$.*

$L_0 = \{0\}$ and $\mathfrak{g}_0 = L(X)$ and the lemma is true for $p = 0$. We argue by induction on p. Suppose then that the lemma is true for some integer $p \geqslant 0$. Let $u_{i,w} = (\operatorname{ad} \bar{w}_p)^i . \bar{w} = \Psi(w_p^i w)$ for $i \geqslant 0$, $w \in P_p$, $w \neq w_p$. By the Corollary to Proposition 10 of no. 9, the free Lie algebra $\mathfrak{g}_p$ is the direct sum of the module T_p of basis $\{\bar{w}_p\}$ and a Lie subalgebra $\mathfrak{h}_p$ admitting

$$\mathscr{F} = (u_{i,w})_{i \geqslant 0,\, w \in P_p,\, w \neq w_p}$$

as basic family. By Proposition 13 (d), the family $(\bar{u})_{u \in P_{p+1}}$ is equal to $\mathscr{F}$ and hence is a basic family of $\mathfrak{h}_p = \mathfrak{g}_{p+1}$. Hence $L(X) = L_p \oplus T_p \oplus \mathfrak{g}_{p+1}$ and, as $L_{p+1} = L_p + T_p$, $L(X) = L_{p+1} \oplus \mathfrak{g}_{p+1}$ and $(\bar{w}_0, \bar{w}_1, \ldots, \bar{w}_{p-1}, \bar{w}_p)$ is a basis of the module L_{p+1}.

Let n be a positive integer. By Proposition 13 (c) there exists an integer $p(n)$ such that P_p has only elements of length $> n$ for $p \geqslant p(n)$. For $p \geqslant p(n)$, the Lie subalgebra $\mathfrak{g}_p$ of L(X) is generated by elements of degree $> n$ and hence $L^n(X) \cap \mathfrak{g}_p = \{0\}$. On the other hand, the elements $\bar{w}_i$ of L(X) are homogeneous and the family $(w_i)_{0 \leqslant i < p}$ is a basis of a supplementary module of $\mathfrak{g}_p$. It follows immediately that the family of elements $\bar{w}_i$ of degree n is a basis of the module $L^n(X)$ and that the sequence $(\bar{w}_i)_{i \geqslant 0}$ is a basis of the module L(X).

(B) *General case.*

If S is a subset of X, recall that M(S) is identified with the submagna of M(X) generated by S and L(S) is identified with the Lie subalgebra of L(X) generated by S; we have seen that if $w \in M(S)$ is of length $\geqslant 2$ then $\alpha(w) \in M(S)$ and $\beta(w) \in M(S)$. It follows immediately that $H \cap M(S)$ is a Hall set relative to S.

For every finite subset Φ of H there exists a finite subset S of X such that

$\Phi \subset M(S)$. Case (A) then shows that the elements $\bar{w}$ with $w \in \Phi$ are linearly independent in L(S) and hence in L(X). Therefore the family $(\bar{w})_{w \in H}$ is free.

For every element a of L(X) there exists a finite subset S of X such that $a \in L(S)$. By case (A), the subset $\Psi(H \cap M(S))$ of $\Psi(H)$ generates the module L(S) and hence a is a linear combination of elements of $\Psi(H)$. Hence $\Psi(H)$ generates the module L(X), which completes the proof.

COROLLARY. *The module* L(X) *is free and so is each of the submodules* $L^\alpha(X)$ *where* $\alpha \in \mathbf{N}^{(X)}$ *and* $L^n(X)$ *where* $n \in \mathbf{N}$. *The modules* $L^\alpha(X)$ *are of finite rank and so are the modules* $L^n(X)$ *if* X *is finite.*

There exists a Hall set H relative to X (Proposition 11). For all $w \in H$, the element $\Psi(w)$ of L(X) belongs to one of the modules $L^\alpha(X)$ (with $\alpha \in \mathbf{N}^{(X)}$) and the module L(X) is the direct sum of the submodules $L^\alpha(X)$. Further, for all $\alpha \in \mathbf{N}^{(X)}$, the set of elements of M(X) whose canonical image in $\mathbf{N}^{(X)}$ is equal to α is finite; this shows that each of the modules $L^\alpha(X)$ is free and of finite rank and that L(X) is free. Now $L^n(X) = \sum_{|\alpha|=n} L^\alpha(X)$ and hence $L^n(X)$ is free; when X is finite, the set of $\alpha \in \mathbf{N}^{(X)}$ such that $|\alpha| = n$ is finite and hence $L^n(X)$ is then of finite rank.

DEFINITION 3. *A Hall basis of a free Lie algebra* L(X) *is any basis of* L(X) *which is the canonical image of a Hall set relative to* X.

Remark. Suppose that X consists of two distinct elements x and y and let $L^{(\cdot,1)}$ be the submodule of L(X) the sum of the $L^\alpha(X)$ where $\alpha \in \mathbf{N}^X$ and $\alpha(y) = 1$. It follows immediately from Theorem 1 and Proposition 12 of no. 10 that the elements of $(\operatorname{ad} x)^n.y$ where n is an integer $\geqslant 0$ form a *basis* of the submodule $L^{(\cdot,1)}$. It follows that *the restriction to* $L^{(\cdot,1)}$ *of the mapping* ad x *is injective.*

§ 3. ENVELOPING ALGEBRA OF THE FREE LIE ALGEBRA

In this paragraph, $A(X) = A_K(X)$ *denotes the free associative algebra* Libas(X) *of the set* X *over the ring* K (*Algebra*, Chapter III, § 2, no. 7, Definition 2). X *is identified with its canonical image in* A(X); *recall that the* K-*module* A(X) *admits as basis the free monoid* Mo(X) *derived from* X; $A^+(X)$ *denotes the submodule of* A(X) *generated by the non-empty words.*

1. ENVELOPING ALGEBRA OF L(X)

THEOREM 1. *Let* $\alpha: L(X) \to A(X)$ *be the unique Lie algebra homomorphism extending the canonical injection of* X *into* A(X) (§ 2, no. 2, Proposition 1). *Let*

$\sigma\colon \mathrm{L(X)} \to \mathrm{U(L(X))}$ *be the canonical mapping of* L(X) *into its enveloping algebra and let* $\beta\colon \mathrm{U(L(X))} \to \mathrm{A(X)}$ *be the unique unital algebra homomorphism such that* $\beta \circ \sigma = \alpha$ (Chapter I, § 2, no. 1, Proposition 1). *Then:*

(a) α *is injective and* $\alpha(\mathrm{L(X)})$ *is a direct factor submodule of* A(X).

(b) β *is bijective.*

Let B be a unital K-algebra and ϕ a mapping of X into B; by Proposition 1 of § 2, no. 2, there exists a Lie algebra homomorphism $\psi\colon \mathrm{L(X)} \to \mathrm{B}$ such that $\psi \mid \mathrm{X} = \phi$; by Proposition 1 of Chapter 1, § 2, no. 1, there exists a unital algebra homomorphism $\theta\colon \mathrm{U(L(X))} \to \mathrm{B}$ such that $\theta \circ \sigma = \psi$ and hence such that $(\theta \circ \sigma) \mid \mathrm{X} = \phi$. As $\sigma(\mathrm{X})$ generates the unital algebra U(L(X)), the homomorphism θ is the unique unital algebra homomorphism satisfying the latter condition. This shows that the ordered pair $(\mathrm{U(L(X))}, \sigma \mid \mathrm{X})$ is a solution of the same universal mapping problem as A(X); taking ϕ to be the canonical injection of X into A(X), we deduce that β is an isomorphism, which proves (b).

Finally, as L(X) is a free K-module (§ 2, no. 11, Corollary to Theorem 1), σ is injective and $\sigma(\mathrm{L(X)})$ is a direct factor submodule of U(L(X)) (Chapter I, § 2, no. 7, Corollary 3 to Theorem 1). By (b), this proves (a).

COROLLARY 1. *There exists on the algebra* A(X) *a unique coproduct making* A(X) *into a bigebra such that the elements of* X *are primitive. Further,* β *is an isomorphism of the bigebra* U(L(X)) *onto* A(X) *with this bigebra structure.*

This follows from assertion (b) of the theorem and the fact that X generates the unital algebra A(X).

Henceforth A(X) is given this bigebra structure and L(X) *is identified* with its image under α, that is *with the Lie subalgebra of* A(X) *generated by* X.

COROLLARY 2. *If* K *is a field of characteristic* 0, L(X) *is the Lie algebra of primitive elements of* A(X).

This follows from Corollary 1 and the Corollary to Proposition 9 of § 1, no. 5.

Remarks. (1) Let K′ be a commutative ring containing K. If A(X), L(X) and $\mathrm{L_{K'}(X)}$ are identified with subsets of $\mathrm{A_{K'}(X)}$, we deduce from part (a) of Theorem 1 the relation

$$\mathrm{L(X)} = \mathrm{L_{K'}(X)} \cap \mathrm{A(X)}. \tag{1}$$

(2) Corollary 2 to Theorem 1 remains valid if it only assumes that the additive group of the ring K is torsion-free. For suppose first that $\mathrm{K} = \mathbf{Z}$; every primitive element of A(X) is a primitive element of $\mathrm{A_{\mathbf{Q}}(X)}$ and hence is in $\mathrm{L_{\mathbf{Q}}(X)} \cap \mathrm{A(X)} = \mathrm{L(X)}$ (Corollary 2 and formula (1)). In the general case, K is flat over **Z** and we apply *Remark* 2 of § 1, no. 2 and Proposition 3 of § 2, no. 5.

(3) Let Δ be a commutative monoid, ϕ_0 a mapping of X into Δ and $\phi\colon \mathrm{Mo}(X) \to \Delta$ the homomorphism of the associated monoid; if A(X) is given the graduation $(A^\delta(X))_{\delta \in \Delta}$ defined in *Algebra*, Chapter III, § 3, no. 1, *Example* 3 and L(X) the graduation $(L^\delta(X))_{\delta \in \Delta}$ defined in § 2, no. 6, we have immediately, for $\delta \in \Delta$, $L^\delta(X) \subset L(X) \cap A^\delta(X)$. As L is the sum of the $L^\delta(X)$ for $\delta \in \Delta$, and the sum of the $L(X) \cap A^\delta(X)$ for $\delta \in \Delta$ is direct, this implies

$$L^\delta(X) = L(X) \cap A^\delta(X). \tag{2}$$

(4) Let A be a unital associative algebra and $\boldsymbol{t} = (t_i)_{i \in I}$ a family of elements of A. We have a diagram

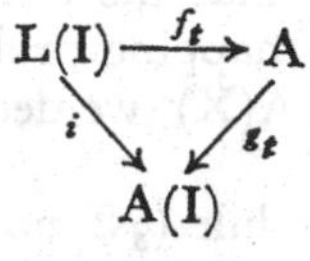

where i is the canonical injection, f_t is the Lie algebra homomorphism defined by $\boldsymbol{t}$ and g_t is the unital algebra homomorphism such that $g_t(i) = t_i$ for $i \in I$. The diagram is commutative for $g_t \circ i$ and f_t coincide on I. It follows that if $P \in L(I)$, the element $P((t_i)_{i \in I})$ defined in § 2, no. 4, coincides with the element $P((t_i)_{i \in I})$ defined in *Algebra*, Chapter III, § 2, no. 8, *Example* 2.

2. PROJECTOR OF $A^+(X)$ ONTO $L(X)$

Let π be the linear mapping of $A^+(X)$ into $L(X)$ defined by

$$\pi(x_1 \ldots x_n) = (\mathrm{ad}(x_1) \circ \cdots \circ \mathrm{ad}(x_{n-1}))(x_n) \tag{3}$$

for $n > 0$, $x_1, \ldots, x_n$ in X.

PROPOSITION 1. (a) *The restriction* π_0 *of* π *to* L(X) *is a derivation of* L(X).

(b) *For every integer* $n \geqslant 1$ *and all* u *in* $L^n(X)$, $\pi(u) = n.u$.

(a) Let E be the endomorphism algebra of the module L(X) and θ the homomorphism of A(X) into E such that $\theta(x) = \mathrm{ad}\, x$ for all $x \in X$. The restriction of θ to L(X) is a Lie algebra homomorphism of L(X) into E, which coincides on X with the adjoint representation of L(X), whence

$$\theta(u).v = [u, v] \quad \text{for } u, v \text{ in } L(X). \tag{4}$$

Let a be in A(X) and b in $A^+(X)$; then

$$\pi(a.b) = \theta(a).\pi(b). \tag{5}$$

It suffices to consider the case $a = x_1 \ldots x_p$, $b = x_{p+1} \ldots x_{p+q}$ with $p \geqslant 0$, $q \geqslant 1$ and $x_1, \ldots, x_{p+q}$ in X; but then (5) follows immediately from (3) since $\theta(x) = \mathrm{ad}\, x$ for $x \in X$.

Let u and v be in L(X); by (4) and (5),

$$\pi_0([u, v]) = \pi(uv - vu) = \theta(u).\pi(v) - \theta(v).\pi(u)$$
$$= [u, \pi(v)] - [v, \pi(u)] = [u, \pi_0(v)] + [\pi_0(u), v],$$

hence π_0 is a derivation of L(X).

(b) Let π_1 be the endomorphism of the module L(X) which coincides on $L^n(X)$ with multiplication by the integer $n \geqslant 1$. The formula $[L^n(X), L^m(X)] \subset L^{n+m}(X)$ shows that π_1 is a derivation (*Algebra*, Chapter III, § 10, no. 3, *Example* 6). The derivation $\pi_1 - \pi_0$ of L(X) is zero on X and, as X generates L(X), $\pi_0 = \pi_1$, whence (b).

COROLLARY. *Suppose that* K *is a* **Q**-*algebra. Let* P *be the linear mapping of* $A^+(X)$ *into itself such that*

$$\text{(6)} \qquad P(x_1 \ldots x_n) = \frac{1}{n}(\operatorname{ad} x_1 \circ \cdots \circ \operatorname{ad} x_{n-1})(x_n)$$

for $n \geqslant 1$ *and* $x_1, \ldots, x_n$ *in* X. *Then* P *is a projector of* $A^+(X)$ *onto* L(X).

The image of P is contained in L(X). Further, for all $n \geqslant 1$ and all u in $L^n(X)$, $P(u) = \frac{1}{n}\pi(u)$, whence $P(u) = u$ by Proposition 1. As

$$L(X) = \sum_{n \geqslant 1} L^n(X),$$

we see that the restriction of P to L(X) is the identity.

Remark. Suppose that K is a field of characteristic zero and let Q be the projector of A(X) = U(L(X)) onto L(X) associated with the canonical graduation of U(L(X)), cf. § 1, no. 5. *For* $\alpha \in \mathbf{N}^{(X)}$, $Q(A^\alpha(X)) \subset L^\alpha(X)$. For it suffices to verify that the image and the kernel of Q are graded submodules of A(X) with the graduation of type $\mathbf{N}^{(X)}$. This is obvious for the image, which is equal to L(X). On the other hand, let n be an integer $\geqslant 1$. The vector subspace of A(X) generated by the y^n, where $y \in L(X)$, is equal to the vector subspace of A(X) generated by the $\sum_{\sigma \in \mathfrak{S}_n} y_{\sigma(1)} y_{\sigma(2)} \ldots y_{\sigma(n)}$, where $y_1, y_2, \ldots, y_n$ are homogeneous elements of L(X); then this subspace is a graded submodule of A(X).

(Note that, if Card(X) $\geqslant 2$, the projectors P and Q *do not coincide* on $A^+(X)$. For let x, y be in X with $x \neq y$ and write

$$z = x[x, y] + [x, y]x = x^2y - yx^2.$$

Then $Q(z) = 0$ and $P(z) = \frac{1}{3}[x, [x, y]] \neq 0$, cf. § 2, no. 10, *Example* and no. 11, Theorem 1.)

3. DIMENSION OF THE HOMOGENEOUS COMPONENTS OF L(X)

Let X be a set, α an element of $\mathbf{N}^{(X)}$ and d an integer >0. We write $d|\alpha$ if there exists $\beta \in \mathbf{N}^{(X)}$ such that $\alpha = d\beta$. The element β, which is unique, is then denoted by α/d.

Lemma 1. *Let n be an integer >0, $T_1, \ldots, T_n$ indeterminates and $u_1, \ldots, u_n$ elements of* $\mathbf{Z}$. *Let $(c(\alpha))_{\alpha \in \mathbf{N}^n - \{0\}}$ be a family of elements of* $\mathbf{Z}$ *such that*

$$1 - \sum_{i=1}^{n} u_i T_i = \prod_{\alpha \neq 0} (1 - T^\alpha)^{c(\alpha)}. \tag{7}$$

For all $\alpha \in \mathbf{N}^n - \{0\}$,

$$c(\alpha) = \frac{1}{|\alpha|} \sum_{d|\alpha} \mu(d) \frac{(|\alpha|/d)!}{(\alpha/d)!} u^{\alpha/d} \tag{8}$$

where μ is the Möbius function (Appendix).

Formula (7) is equivalent, on taking logarithms on both sides (*Algebra*, Chapter IV, § 6, no. 9) to:

$$\log\left(1 - \sum_{i=1}^{n} u_i T_i\right) = \sum_{\alpha \neq 0} c(\alpha) \log(1 - T^\alpha). \tag{9}$$

Now

$$\begin{aligned} -\log\left(1 - \sum_{i=1}^{n} u_i T_i\right) &= \sum_{j \geqslant 1} \frac{1}{j} \left(\sum_{i=1}^{n} u_i T_i\right)^j \\ &= \sum_{j \geqslant 1} \frac{1}{j} \sum_{|\beta| = j} \frac{|\beta|!}{\beta!} u^\beta T^\beta \\ &= \sum_{|\beta| > 0} \frac{1}{|\beta|} \frac{|\beta|!}{\beta!} u^\beta T^\beta. \end{aligned} \tag{10}$$

On the other hand

$$\begin{aligned} -\sum_{\alpha \neq 0} c(\alpha) \log(1 - T^\alpha) &= \sum_{|\alpha| > 0,\, k \geqslant 1} \frac{1}{k} c(\alpha) T^{k\alpha} \\ &= \sum_{|\beta| > 0,\, k|\beta} \frac{1}{k} c\left(\frac{\beta}{k}\right) T^\beta. \end{aligned} \tag{11}$$

Hence (7) is equivalent to

$$\sum_{k|\beta} \left|\frac{\beta}{k}\right| c\left(\frac{\beta}{k}\right) = \frac{|\beta|!}{\beta!} u^\beta \quad \text{for all } \beta \in \mathbf{N}^n - \{0\}. \tag{12}$$

Let Λ be the set of $(\lambda_1, \lambda_2, \ldots, \lambda_n) \in \mathbf{N}^n - \{0\}$ such that the g.c.d. of $\lambda_1, \lambda_2, \ldots, \lambda_n$ is equal to 1. Every element of $\mathbf{N}^n - \{0\}$ can be written uniquely in the form $m\lambda$, where m is an integer $\geqslant 1$ and $\lambda \in \Lambda$. Condition (12) is is equivalent to

$$(13) \qquad \sum_{k|m} \left|\frac{m\lambda}{k}\right| c\left(\frac{m\lambda}{k}\right) = \frac{(m|\lambda|)!}{(m\lambda)!} u^{m\lambda} \text{ for all } \lambda \in \Lambda \text{ and all } m \geqslant 1.$$

By the Möbius inversion formula (Appendix), condition (13) is equivalent to

$$(14) \qquad |m\lambda| c(m\lambda) = \sum_{d|m} \mu(d) \frac{\left|\frac{m\lambda}{d}\right|!}{\left(\frac{m\lambda}{d}\right)!} u^{\frac{m\lambda}{d}}$$

for all $\lambda \in \Lambda$ and all $m \geqslant 1$.

THEOREM 2. *Let* X *be a finite set and* $n = \text{Card}(X)$.

(a) *For every integer* $r \geqslant 1$, *the* K-*module* $L^r(X)$ *is free of rank*

$$(15) \qquad c(r) = \frac{1}{r} \sum_{d|r} \mu(d) n^{r/d},$$

where μ *is the Möbius function.*

(b) *For all* $\alpha \in \mathbf{N}^X - \{0\}$, *the* K-*module* $L^\alpha(X)$ (§ 2, no. 6) *is free of rank*

$$(16) \qquad c(\alpha) = \frac{1}{|\alpha|} \sum_{d|\alpha} \mu(d) \frac{(|\alpha|/d)!}{(\alpha/d)!}.$$

We already know that the modules $L^r(X)$, where $r \in \mathbf{N}$, and $L^\alpha(X)$, where $\alpha \in \mathbf{N}^X$, are free (§ 2, no. 11, Corollary to Theorem 1). Consider the multi-graduation $(A^\alpha(X))_{\alpha \in \mathbf{N}^X}$ of A(X) defined by the canonical homomorphism ϕ of Mo(X) into $\mathbf{N}^X$ (*Algebra*, Chapter III, § 3, no. 1, *Example* 3); then $A^\alpha(X) \cap L(X) = L^\alpha(X)$ by *Remark* 3 of no. 1. For $\alpha \in \mathbf{N}^X$, the K-module $A^\alpha(X)$ admits as basis the set of words in which each letter x of X appears $\alpha(x)$ times. Let $d(\alpha)$ be the number of these words, that is the rank of $A^\alpha(X)$; we shall calculate in two different ways the formal power series

$$P((T_x)_{x \in X}) \in \mathbf{Z}[[(T_x)_{x \in X}]]$$

defined by

$$(17) \qquad P(T) = \sum_{\alpha \in \mathbf{N}^X} d(\alpha) T^\alpha.$$

(1) We have

$$P(T) = \sum_{m \in Mo(X)} T^{\phi(m)} = \sum_{r=0}^{\infty} \sum_{x_1, \ldots, x_r} T_{x_1} \ldots T_{x_r} = \sum_{r=0}^{\infty} \left(\sum_{x \in X} T_x\right)^r$$

whence

$$(18) \qquad P(T) = \left(1 - \sum_{x \in X} T_x\right)^{-1}.$$

(2) For all $\alpha \in \mathbf{N}^{\mathrm{X}} - \{0\}$, let $(e_{\alpha,j})_{1 \leqslant j \leqslant c(\alpha)}$ be a basis of $\mathrm{L}^{\alpha}(\mathrm{X})$ and give the set I of ordered pairs (α, j) such that $\alpha \in \mathrm{N}^{\mathrm{X}} - \{0\}$ and $1 \leqslant j \leqslant c(\alpha)$ a total ordering. By Theorem 1 of no. 1 and the Poincaré–Birkhoff–Witt Theorem (Chapter I, § 2, no. 7, Corollary 3 to Theorem 1), the elements

$$y_m = \prod_{(\alpha, j) \in \mathrm{I}} (e_{\alpha,j})^{m(\alpha,j)},$$

where the index m runs through $\mathbf{N}^{(\mathrm{I})}$, form a basis of A(X). Each y_m is of multidegree $\sum_{(\alpha,j)\in \mathrm{I}} m(\alpha, j)\alpha$. Let $u(m)$ denote this multidegree. It follows that

$$\mathrm{P}(\mathrm{T}) = \sum_{m \in \mathbf{N}^{(\mathrm{I})}} \mathrm{T}^{u(m)} = \sum_{m \in \mathbf{N}^{(\mathrm{I})}} \prod_{(\alpha,j)\in \mathrm{I}} \mathrm{T}^{m(\alpha,j)\alpha}$$

$$= \prod_{(\alpha,j)\in \mathrm{I}} \sum_{r=0}^{\infty} \mathrm{T}^{r\alpha} = \prod_{(\alpha,j)\in \mathrm{I}} (1 - \mathrm{T}^{\alpha})^{-1},$$

whence finally

$$\text{(19)} \qquad \mathrm{P}(\mathrm{T}) \prod_{\alpha \in \mathbf{N}^{\mathrm{X}} - \{0\}} (1 - \mathrm{T}^{\alpha})^{-c(\alpha)}.$$

Comparing (18) and (19), we obtain

$$\text{(20)} \qquad 1 - \sum_{x \in \mathrm{X}} \mathrm{T}_x = \prod_{\alpha \in \mathbf{N}^{\mathrm{X}} - \{0\}} (1 - \mathrm{T}^{\alpha})^{c(\alpha)}.$$

Lemma 1 then gives (b).

If we now substitute the same indeterminate U for the T_x for $x \in \mathrm{X}$ in formula (20), we obtain

$$1 - n\mathrm{U} = \prod_{\alpha \in \mathbf{N}^{\mathrm{X}} - \{0\}} (1 - \mathrm{U}^{|\alpha|})^{c(\alpha)} = \prod_{r > 0} (1 - \mathrm{U}^{r})^{c(r)}.$$

Applying Lemma 1 again, we deduce (a).

Examples. We have

$c(1) = n, \qquad c(2) = \frac{1}{2}(n^2 - n), \qquad c(3) = \frac{1}{3}(n^3 - n),$

$c(4) = \frac{1}{4}(n^4 - n^2), \qquad c(5) = \frac{1}{5}(n^5 - n), \qquad c(6) = \frac{1}{6}(n^6 - n^3 - n^2 + n).$

Remark. Let X be a set and let $\alpha \in \mathbf{N}^{(\mathrm{X})}$; the rank of the free K-module $\mathrm{L}^{\alpha}(\mathrm{X})$ is also given by formula (16). This follows immediately from Theorem 2 (b) and Proposition 4 of § 2, no. 6.

§ 4. CENTRAL FILTRATIONS

1. REAL FILTRATIONS

DEFINITION 1. *Let* G *be a group. A real filtration on* G *is a family* $(\mathrm{G}_{\alpha})_{\alpha \in \mathbf{R}}$ *of subgroups of* G *such that*

$$\text{(1)} \qquad \mathrm{G}_{\alpha} = \bigcap_{\beta < \alpha} \mathrm{G}_{\beta} \quad \textit{for all } \alpha \in \mathbf{R}.$$

Formula (1) implies $G_\alpha \subset G_\beta$ for $\beta < \alpha$ and hence the family (G_α) is *decreasing*. The filtration (G_α) is called *separated* if $\bigcap_\alpha G_\alpha$ reduces to the identity element and is called *exhaustive* if $G = \bigcup_\alpha G_\alpha$.

Remark. Let $(G_n)_{n \in \mathbf{Z}}$ be a decreasing sequence of subgroups of G. It is a decreasing filtration in the sense of *Commutative Algebra*, Chapter III, § 2, no. 1, Definition 1. For every integer n and all α in the interval $]n-1, n]$ of $\mathbf{R}$, we write $H_\alpha = G_n$, in particular $H_n = G_n$. It is immediate that we thus obtain a real filtration $(H_\alpha)_{\alpha \in \mathbf{R}}$ on G; such a filtration will be called an *integral filtration*. Hence decreasing filtrations in the sense of *Commutative Algebra*, Chapter III, § 2, can be identified with integral filtrations.

Let A be an algebra; a real filtration (A_α) on the additive group of A is called compatible with the algebra structure if $A_\alpha . A_\beta \subset A_{\alpha+\beta}$ for α, β in $\mathbf{R}$ and $K.A_\alpha \subset A_\alpha$ for $\alpha \in \mathbf{R}$. If the filtration is exhaustive, (A_α) is a fundamental system of neighbourhoods of 0 under the topology on A which is compatible with the algebra structure. Let B be a unital algebra; a real filtration (B_α) on the additive group of B is called compatible with the unital algebra structure if it is compatible with the algebra structure and $1 \in B_0$.

2. ORDER FUNCTION

Let G be a group with identity element e. Let (G_α) be a real filtration on G. For all x in G let I_x denote the set of real numbers α such that $x \in G_\alpha$. If $\alpha \in I_x$ and $\beta < \alpha$, then $\beta \in I_x$ and hence I_x is an interval (*General Topology*, Chapter IV, § 2, no. 4, Proposition 1). Using relation (1), we see that I_x contains its least upper bound when this is finite. Therefore I_x is of the form $]-\infty, v(x)] \cap \mathbf{R}$ with $v(x) \in \overline{\mathbf{R}}$; we have $v(x) = \sup\{\alpha \mid x \in G_\alpha\}$.

The mapping v of G into $\overline{\mathbf{R}}$ is called the *order function* associated with the real filtration (G_α) and $v(x)$ is called the *order* of x. This mapping has the following properties:

(a) *For $x \in G$ and $\alpha \in \mathbf{R}$, the relations $x \in G_\alpha$ and $v(x) \geqslant \alpha$ are equivalent.*

(b) *For x, y in G,*

$$v(x^{-1}) = v(x), \qquad v(e) = +\infty. \tag{2}$$

$$v(xy) \geqslant \inf(v(x), v(y)). \tag{3}$$

Further, we have equality in (3) *if $v(x) > v(y)$.*

(c) *For all $\alpha \in \mathbf{R}$, let G_α^+ denote the set of $x \in G$ such that $v(x) > \alpha$. Then $G_\alpha^+ = \bigcup_{\beta > \alpha} G_\beta$ and in particular G_α^+ is a subgroup of* G.

Conversely, let v be a mapping of G into $\overline{\mathbf{R}}$ satisfying relations (2) and (3). For all $\alpha \in \mathbf{R}$, let G_α be the set of $x \in G$ such that $v(x) \geqslant \alpha$. Then $(G_\alpha)_{\alpha \in \mathbf{R}}$ is a real filtration of G and v is the order function associated with this filtration.

For the filtration (G_α) to be integral, it is necessary and sufficient that v map G into $\mathbf{Z} \cup \{+\infty, -\infty\}$. For it to be exhaustive (resp. separated), it is necessary and sufficient that $\overset{-1}{v}(-\infty) = \varnothing$ (resp. $\overset{-1}{v}(+\infty) = \{e\}$).

Let A be a K-algebra (resp. unital K-algebra). By the above, the relation

$$\text{“}x \in A_\alpha \Leftrightarrow v(x) \geqslant \alpha \quad \text{for } x \in A \text{ and } \alpha \in \mathbf{R}\text{”}$$

defines a bijection of the set of exhaustive real filtrations $(A_\alpha)_{\alpha\in\mathbf{R}}$ compatible with the algebra (resp. unital algebra) structure on A onto the set of mappings $v: A \to \overline{\mathbf{R}}$ not taking the value $-\infty$ and satisfying axioms (4) to (7) (resp. (4) to (8)) below:

$$v(x+y) \geqslant \inf(v(x), v(y)) \quad (x, y \text{ in } A) \tag{4}$$

$$v(-x) = v(x) \quad (x \in A) \tag{5}$$

$$v(\lambda x) \geqslant v(x) \quad (\lambda \in K, x \in A) \tag{6}$$

$$v(xy) \geqslant v(x) + v(y) \quad (x, y \text{ in } A) \tag{7}$$

$$v(1) \geqslant 0. \tag{8}$$

Remark. If $v(x)$ is not everywhere equal to $+\infty$, conditions (7) and (8) imply $v(1) = 0$.

3. GRADED ALGEBRA ASSOCIATED WITH A FILTERED ALGEBRA

Let G be a *commutative* group with a real filtration $(G_\alpha)_{\alpha\in\mathbf{R}}$. As before we write

$$G_\alpha^+ = \bigcup_{\beta>\alpha} G_\alpha; \tag{9}$$

clearly G_α^+ is a subgroup of G_α. We write $\mathrm{gr}_\alpha(G) = G_\alpha/G_\alpha^+$ and

$$\mathrm{gr}(G) = \bigoplus_{\alpha\in\mathbf{R}} \mathrm{gr}_\alpha(G).$$

The *graded group associated with the filtered group* G is the group gr(G) with its natural graduation of type **R**.

Remark. When the filtration (G_α) is integral, $\mathrm{gr}_\alpha(G) = \{0\}$ for non-integral α and $\mathrm{gr}_n(G) = G_n/G_{n-1}$ for every integer n. The definition of the associated graded group therefore coincides essentially with that of *Commutative Algebra*, Chapter III, § 2, no. 3.

Let A be an algebra (resp. unital algebra) and $(A_\alpha)_{\alpha\in\mathbf{R}}$ a real filtration compatible with the algebra (resp. unital algebra) structure (no. 1). Then

$$A_\alpha . A_\beta \subset A_{\alpha+\beta}, \qquad A_\alpha^+ . A_\beta + A_\alpha . A_\beta^+ \subset A_{\alpha+\beta}^+,$$

and the bilinear mapping of $A_\alpha \times A_\beta$ into $A_{\alpha+\beta}$ the restriction of multiplication on A defines on taking quotients a bilinear mapping

$$\mathrm{gr}_\alpha(A) \times \mathrm{gr}_\beta(A) \to \mathrm{gr}_{\alpha+\beta}(A).$$

We derive a bilinear mapping of gr(A) × gr(A) into gr(A) which makes it a graded algebra (resp. unital graded algebra) of type **R**. If A is an associative (resp. commutative, resp. Lie) algebra, so is gr(A).

4. CENTRAL FILTRATIONS ON A GROUP

DEFINITION 2. *Let* G *be a group. A real filtration* (G_α) *on* G *is called central if* $G = \bigcup_{\alpha>0} G_\alpha$ *and the commutator* $(x, y) = x^{-1}y^{-1}xy$ *of an element* x *of* G_α *and an element* y *of* G_β *belongs to* $G_{\alpha+\beta}$.

In terms of the order function v, the above definition translates into the relations

$$(10) \qquad v(x) > 0, \qquad v((x,y)) \geqslant v(x) + v(y) \quad \text{for all } x, y \text{ in } G.$$

We deduce that $v((x,y)) > v(x)$ if $v(x) \neq +\infty$; if we write $x^y = y^{-1}xy$ (cf. *Algebra*, Chapter I, § 6, no. 2), then $x^y = x.(x, y)$, whence

$$(11) \qquad v(x^y) = v(x).$$

This relation expresses the fact that each of the subgroups G_α of G is *normal*. The G_α form a fundamental system of neighbourhoods of e for a *topology* compatible with the group structure on G (*General Topology*, Chapter III, § 1, no. 2, *Example*) said to be defined by the filtration (G_α).

In the rest of this no., G denotes a group with a *central* filtration (G_α). For all $\alpha \in \mathbf{R}$, we define the subgroup G_α^+ of G by

$$(12) \qquad G_\alpha^+ = \bigcup_{\beta>\alpha} G_\alpha.$$

In particular $G_\alpha^+ = G_\alpha = G$ for $\alpha \leqslant 0$. Recall that if A and B are two subgroups of G, (A, B) denotes the subgroup of G generated by the commutators (a, b) with $a \in A$ and $b \in B$. With this notation we have the formulae

$$(13) \qquad (G_\alpha, G_\beta) \subset G_{\alpha+\beta}$$

$$(13') \qquad (G_\alpha^+, G_\beta) \subset G_{\alpha+\beta}^+$$

$$(14) \qquad (G, G_\alpha) \subset G_\alpha^+.$$

By (14), G_α^+ is a normal subgroup of G_α for all $\alpha \in \mathbf{R}$ and the quotient group $\mathrm{gr}_\alpha(G) = G_\alpha/G_\alpha^+$ is commutative. We write $\mathrm{gr}(G) = \bigoplus_{\alpha \in \mathbf{R}} \mathrm{gr}_\alpha(G)$ and give this group the graduation of type **R** in which $\mathrm{gr}_\alpha(G)$ consists of elements of degree α. Then $\mathrm{gr}_\alpha(G) = \{0\}$ for $\alpha \leqslant 0$.

PROPOSITION 1. (i) *Let* α, β *be in* **R**. *There exists a biadditive mapping*

$$\phi_{\alpha\beta}: \mathrm{gr}_\alpha(G) \times \mathrm{gr}_\beta(G) \rightarrow \mathrm{gr}_{\alpha+\beta}(G)$$

which maps $(xG_\alpha^+, yG_\beta^+)$ *onto* $(x, y)G_{\alpha+\beta}^+$.

(ii) *Let* ϕ *be the biadditive mapping of* $\mathrm{gr}(G) \times \mathrm{gr}(G)$ *into* $\mathrm{gr}(G)$ *whose restriction to* $\mathrm{gr}_\alpha(G) \times \mathrm{gr}_\beta(G)$ *is* $\phi_{\alpha\beta}$ *for every ordered pair* (α, β). *The mapping* ϕ *gives* $\mathrm{gr}(G)$ *a Lie* **Z**-*algebra structure.*

(i) Recall the identity

$$(xx', y) = (x, y)^{x'}(x', y) \tag{15}$$

for x, x', y in G (*Algebra*, Chapter I, § 6, no. 2, formula (4 bis)).

For $x \in G_\alpha$ and $y \in G_\beta$, the class modulo $G_{\alpha+\beta}^+$ of the element (x, y) of $G_{\alpha+\beta}$ will be denoted by $f(x, y)$. For a in $G_{\alpha+\beta}$ and x' in G, $a^{-1}.a^{x'} = (a, x') \in G_{\alpha+\beta}^+$; in particular $f(x, y)$ is equal to the class modulo $G_{\alpha+\beta}^+$ of $(x, y)^{x'}$. Formula (15) therefore implies

$$f(xx', y) = f(x, y)f(x', y). \tag{16}$$

Now $(y, x) = (x, y)^{-1}$, whence

$$f(y, x) = f(x, y)^{-1}. \tag{17}$$

From (16) and (17) we deduce

$$f(x, yy') = f(x, y)f(x, y'). \tag{18}$$

We have to prove that the mapping $f\colon G_\alpha \times G_\beta \to \mathrm{gr}_{\alpha+\beta}(G)$ defines on taking quotients a mapping $\phi_{\alpha\beta}\colon \mathrm{gr}_\alpha(G) \times \mathrm{gr}_\beta(G) \to \mathrm{gr}_{\alpha+\beta}(G)$. By (16) and (18) it suffices to prove that $f(x, y) = 0$ if $x \in G_\alpha^+$ *or* $y \in G_\beta^+$, which follows from (13′).

(ii) As $(x, x) = e$, it follows from (17) that ϕ is an alternating **Z**-bilinear mapping. Hence it remains to prove that, for $u \in \mathrm{gr}_\alpha(G)$, $v \in \mathrm{gr}_\beta(G)$ and $w \in \mathrm{gr}_\gamma(G)$,

$$\phi(u, \phi(v, w)) + \phi(v, \phi(w, u)) + \phi(w, \phi(u, v)) = 0. \tag{19}$$

Let $x \in G_\alpha$, $y \in G_\beta$ and $z \in G_\gamma$ be elements representing respectively u, v and w. We know that x^y and x are two elements of G_α which are congruent modulo G_α^+ and hence x^y is a representative of u in G_α; as (y, z) is a representative of $\phi(v, w)$ in $G_{\beta+\gamma}$, we see that $(x^y, (y, z))$ is a representative of $\phi(u, \phi(v, w))$ in $G_{\alpha+\beta+\gamma}$. By cyclic permutation, we see that $(y^z, (z, x))$ and $(z^x, (x, y))$ represent respectively $\phi(v, \phi(w, u))$ and $\phi(w, \phi(u, v))$ in $G_{\alpha+\beta+\gamma}$. Relation (19) is then a consequence of the following identity (*Algebra*, Chapter I, § 6, no. 2, formula (15)):

$$(x^y, (y, z)).(y^z, (z, x)).(z^x, (x, y)) = e. \tag{20}$$

The Lie algebra gr(G) over **Z** defined in Proposition 1 is called the *graded Lie algebra associated with the filtered group* G.

5. AN EXAMPLE OF A CENTRAL FILTRATION

Let A be a unital associative algebra with a unital algebra filtration (A_α) such that $A_0 = A$; then A_α is a two-sided ideal of A for all $\alpha \in \mathbf{R}$. Let A^* denote the multiplicative group of invertible elements of A. For all $\alpha > 0$, let Γ_α denote the set of $x \in A^*$ such that $x - 1 \in A_\alpha$; we write $\Gamma = \bigcup_{\alpha>0} \Gamma_\alpha$ and $\Gamma_\beta = \Gamma$ for $\beta \leqslant 0$.

PROPOSITION 2. *The set* Γ *is a subgroup of* A^* *and* (Γ_α) *is a central filtration on* Γ.

$\Gamma = \bigcup_{\alpha>0} \Gamma_\alpha$ by construction and the relation $\Gamma_\alpha = \bigcap_{\beta<\alpha} \Gamma_\beta$ follows from $A_\alpha = \bigcap_{\beta<\alpha} A_\beta$.

We show that Γ_α is a subgroup of A^*. Now $1 \in \Gamma_\alpha$; let x, y be in Γ_α, whence $x - 1 \in A_\alpha$, $y - 1 \in A_\alpha$. As A_α is a two-sided ideal of A, the formulae

$$xy - 1 = (x-1)(y-1) + (x-1) + (y-1), \tag{21}$$

$$x^{-1} - 1 = -x^{-1}(x-1), \tag{22}$$

imply $xy - 1 \in A_\alpha$ and $x^{-1} - 1 \in A_\alpha$, whence $xy \in \Gamma_\alpha$ and $x^{-1} \in \Gamma_\alpha$.

As $\Gamma = \bigcup_{\alpha>0} \Gamma_\alpha$, this is a subgroup of A^*.

Finally let $\alpha > 0$, $\beta > 0$, $x \in \Gamma_\alpha$ and $y \in \Gamma_\beta$. Let $x - 1 = \xi$ and $y - 1 = \eta$. Then

$$(x, y) - 1 = x^{-1}y^{-1}(\xi\eta - \eta\xi); \tag{23}$$

by hypothesis, $\xi \in A_\alpha$ and $\eta \in A_\beta$, whence $\xi\eta - \eta\xi \in A_{\alpha+\beta}$. As $A_{\alpha+\beta}$ is a two-sided ideal of A, $(x, y) - 1 \in A_{\alpha+\beta}$, whence $(x, y) \in \Gamma_{\alpha+\beta}$.

Remark. Let $\alpha \geqslant 0$, $\beta \geqslant 0$ and $x \in \Gamma_\alpha$, $y \in \Gamma_\beta$. By formulae (21), (22) and (23),

$$x^{-1} - 1 \equiv -(x-1) \quad \text{mod. } A_{2\alpha} \tag{24}$$

$$xy - 1 \equiv (x-1) + (y-1) \quad \text{mod. } A_{\alpha+\beta} \tag{25}$$

$$(x, y) - 1 \equiv [(x-1), (y-1)] \quad \text{mod. } A_{\alpha+\beta+\inf(\alpha,\beta)}. \tag{26}$$

We prove for example (26). If $x - 1 = \xi$ and $y - 1 = \eta$, (23) gives:

$$(x, y) - 1 - [\xi, \eta] = ((x^{-1} - 1) + (y^{-1} - 1) + (x^{-1} - 1)(y^{-1} - 1))[\xi, \eta].$$

Now $[\xi, \eta] \in A_{\alpha+\beta}$, $(x^{-1} - 1) \in A_\alpha$, $(y^{-1} - 1) \in A_\beta$, whence we obtain (26).

Let G be a group and $\rho: G \to \Gamma$ a homomorphism. For all real α, we write $G_\alpha = \rho^{-1}(\Gamma_\alpha)$. As (Γ_α) is a central filtration on Γ, it is immediate that (G_α) is a central filtration on G.

PROPOSITION 3. (i) *For all* $\alpha \in \mathbf{R}$, *there exists a unique group homomorphism* $g_\alpha: \mathrm{gr}_\alpha(G) \to \mathrm{gr}_\alpha(A)$ *which maps the class modulo* G_α^+ *of an element* $a \in G_\alpha$ *to the class modulo* A_α^+ *of* $\rho(a) - 1$.

(ii) *Let g be the group homomorphism of* $\mathrm{gr}(G)$ *into* $\mathrm{gr}(A)$ *whose restriction to* $\mathrm{gr}_\alpha(G)$ *is* g_α *for all* α. *The mapping g is an injective homomorphism of Lie* $\mathbf{Z}$*-algebras.*

(i) Let $\alpha > 0$. By hypothesis, for all a in G_α, $\rho(a) - 1 \in A_\alpha$; let $p_\alpha(a)$ denote the class of $\rho(a) - 1$ modulo A_α^+. As $A_{2\alpha} \subset A_\alpha^+$, relation (25) implies $p_\alpha(ab) = p_\alpha(a) + p_\alpha(b)$. Then $a \in G_\alpha^+$ if and only if $\rho(a) - 1 \in A_\alpha^+$; therefore G_α^+ is the kernel of the homomorphism p_α of G_α into $\mathrm{gr}_\alpha(A)$. On passing to the quotient, p_α then defines an injective homomorphism g_α of $\mathrm{gr}_\alpha(G)$ into $\mathrm{gr}_\alpha(A)$.

For $\alpha \leqslant 0$, $\mathrm{gr}_\alpha(G) = \{0\}$ and the only choice is $g_\alpha = 0$.

(ii) As g_α is injective for all real α, g is injective. We show that g is a Lie algebra homomorphism. As $\mathrm{gr}_\alpha(G) = \{0\}$ for $\alpha \leqslant 0$, it suffices to establish the formula

$$p_{\alpha+\beta}((a, b)) = [p_\alpha(a), p_\beta(b)] \tag{27}$$

for $\alpha > 0$, $\beta > 0$, $\rightarrow \in G_\alpha$ and $b \in G_\beta$, which follows from (26).

6. INTEGRAL CENTRAL FILTRATIONS

Recall (no. 1, *Remark*) that a filtration (G_α) on the group G is called integral if $G_\alpha = G_n$ for every integer n and all $\alpha \in]n-1, n]$. To be given an integral central filtration on a group G is equivalent to being given a sequence $(G_n)_{n\geqslant 1}$ of subgroups of G satisfying the conditions

(i) $G_1 = G$

(ii) $G_n \supset G_{n+1}$ for all $n \geqslant 1$

(iii) $(G_m, G_n) \subset G_{m+n}$ for $m \geqslant 1$ and $n \geqslant 1$.

For every integer $n \geqslant 1$, G_n is a normal subgroup of G and the quotient $\mathrm{gr}_n(G) = G_n/G_{n+1}$ is commutative. On taking quotients, the mapping $(x, y) \mapsto (x, y) = x^{-1}y^{-1}xy$ of $G_m \times G_n$ into G_{m+n} allows us to define on $\mathrm{gr}(G) = \bigoplus_{n\geqslant 1} \mathrm{gr}_n(G)$ a graded Lie algebra structure of type $\mathbf{N}$ over the ring $\mathbf{Z}$.

Recall (*Algebra*, Chapter I, § 6, no. 3, Definition 5) that the *lower central series* of the group G is defined by

$$C^1G = G, \qquad C^{n+1} = (G, C^nG) \quad \text{for } n \geqslant 1. \tag{28}$$

The corresponding filtration is called the *lower central filtration* of G.

PROPOSITION 4. (i) *The lower central series of* G *is an integral central filtration on* G.

(ii) *If* $(G_n)_{n\in\mathbf{N}^*}$ *is an integral central filtration on* G, *then* $C^nG \subset G_n$ *for all* $n \in \mathbf{N}^*$.

Assertion (i) has been proved in *Algebra*, Chapter I, § 6, no. 3, formula (7).

We prove (ii) by induction on n; $C^1G = G = G_1$; for $n > 1$,

$$C^nG = (G, C^{n-1}G) \subset (G, G_{n-1}) \subset G_n.$$

PROPOSITION 5. *Let* G *be a group and* gr(G) *the graded Lie* **Z**-*algebra associated with the lower central filtration on* G. *Then* gr(G) *is generated by* $gr_1(G) = G/(G, G)$.

Let L be the Lie subalgebra of gr(G) generated by $gr_1(G)$; we show that $L \supset gr_n(G)$ by induction on n, the assertion being trivial for $n = 1$. Suppose that $n > 1$ and $L \supset gr_{n-1}(G)$. As $C^nG = (G, C^{n-1}G)$, the construction of the Lie algebra law on gr(G) shows immediately that

$$gr_n(G) = [gr_1(G), gr_{n-1}(G)] \subset L.$$

The above proof shows that the lower central series of the Lie algebra gr(G) (§ 2, no. 7) is given by

$$\mathscr{C}^n(gr(G)) = \sum_{m \geqslant n} gr_m(G). \tag{29}$$

Remark. Let k be a ring, n an integer >0 and A the set of lower triangular matrices with n rows and n columns and elements in k. For $p \geqslant 0$, let A_p be the set of $(x_{ij}) \in A$ such that $x_{ij} = 0$ for $i - j < p$. Then $A_0 = A$ and $A_pA_q \subset A_{p+q}$. Let $\Gamma_p = 1 + A_p$. Then Γ_1 is a subgroup of $\mathbf{GL}(n, k)$ called the *strict lower triangular group* of order n over k. By Proposition 2 of no. 5, (Γ_p) is an integral filtration on Γ_1. As $\Gamma_n = \{1\}$, we see that *the group* Γ_1 *is nilpotent* (*Algebra*, Chapter I, § 6, no. 3, Definition 6).

§ 5. MAGNUS ALGEBRAS

In this paragraph, X *denotes a set,* F(X) *the free group constructed on* X (*Algebra,* Chapter I, § 7, no. 5) *and* A(X) *the free associative algebra constructed on* X *with its total graduation* $(A^n(X))_{n \geqslant 0}$ (cf. *Algebra*, Chapter III, § 3, no. 1, *Example* 3). X *is identified with its images in* F(X) *and* A(X).

1. MAGNUS ALGEBRAS

Let $\hat{A}(X)$ be the product module $\prod_{n \geqslant 0} A^n(X)$. We define on $\hat{A}(X)$ a multiplication by the rule

$$(a.b)_n = \sum_{i=0}^{n} a_i . b_{n-i} \tag{1}$$

where $a = (a_n)$ and $b = (b_n)$ are in $\hat{A}(X)$. We know (*Commutative Algebra*, Chapter III, § 2, no. 12, *Example* 1) that $\hat{A}(X)$ is an associative algebra and that A(X) is identified with the subalgebra of $\hat{A}(X)$ consisting of the sequence all of whose terms are zero except for a finite number.

$\hat{A}(X)$ is given the product topology of the discrete topologies on the factors

$A^n(X)$; this topology makes $\hat{A}(X)$ into a complete Hausdorff topological algebra, when the ring K has the discrete topology, and $A(X)$ is dense in $\hat{A}(X)$ Let $a = (a_n) \in \hat{A}(X)$; the family $(a_n)_{n \geqslant 0}$ is summable and $a = \sum_{n \geqslant 0} a_n$.

For every integer $m \geqslant 0$, let $\hat{A}_m(X)$ denote the ideal consisting of the series $a = \sum_{n \geqslant m} a_n$ such that $a_n \in A^n(X)$ for all $n \geqslant m$. This sequence of ideals is a fundamental system of neighbourhoods of 0 in $\hat{A}(X)$ and an integral filtration on $\hat{A}(X)$. The order function associated with the above filtration is denoted by ω; then $\omega(0) = +\infty$ and $\omega(a) = m$ if $a = \sum_{n \geqslant m} a_n$ with $a_n \in A^n(X)$ for all $n \geqslant m$ and $a_m \neq 0$ (§ 4, nos. 1 and 2).

$\hat{A}(X)$ is called the *Magnus algebra* of the set X with coefficients in K. If there is any ambiguity over K we write $\hat{A}_K(X)$.

PROPOSITION 1. *Let* B *be a unital associative algebra with a real filtration* $(B_\alpha)_{\alpha \in \mathbf{R}}$ *such that* B *is Hausdorff and complete* (§ 4, nos. 1 and 2). *Let* f *be a mapping of* X *into* B *such that there exists* $\lambda > 0$ *for which* $f(X) \subset B_\lambda$. *Then* f *can be extended in one and only one way to a continuous unital homomorphism* $\hat{f}$ *of* $\hat{A}(X)$ *into* B.

Let f' be the unique unital algebra homomorphism of $A(X)$ into B extending f (*Algebra*, Chapter III, § 2, no. 7, Proposition 7). We show that f' is *continuous*: $f'(A^n(X)) \subset B_{n\lambda}$ whence $f'(\hat{A}_n(X) \cap A(X)) \subset B_{n\lambda}$. Therefore f' can be extended in one and only one way by continuity to a homomorphism $\hat{f} : \hat{A}(X) \to B$.

We preserve the hypotheses and notation of Proposition 1 and let $u \in \hat{A}(X)$. The element $\hat{f}(u)$ is denoted by $u((f(x))_{x \in X})$ and called the *result of substituting the* $f(x)$ *for the* x *in* u. In particular, $u((x)_{x \in X}) = u$. Now let $\boldsymbol{u} = (u_y)_{y \in Y}$ be a family of elements of $\hat{A}_1(X)$ and let $v \in \hat{A}(Y)$. The above allows us to define the element $v((u_y)_{y \in Y}) \in \hat{A}(X)$. It is denoted by $v \circ \boldsymbol{u}$. As $u_y((f(x))) \in B_\lambda$, the elements $u_y((f(x)))$ can be substituted for the y in v. The mappings $v \mapsto (v \circ \boldsymbol{u})((f(x)))$ and $v \mapsto v((u_y((f(x)))))$ are then two continuous homomorphisms of unital algebras of $\hat{A}(X)$ into B taking the same value $u_y((f(x)))$ at the element $y \in Y$. Therefore (Proposition 1)

$$(v \circ \boldsymbol{u})((f(x))) = v((u_y((f(x))))) \tag{2}$$

for all $v \in \hat{A}(Y)$.

2. MAGNUS GROUP

For all $a = (a_n)_{n \geqslant 0}$ in $\hat{A}(X)$, the element a_0 of K will be called the *constant term* of a and denoted by $\varepsilon(a)$. Formula (1) shows that ε is an algebra homomorphism of $\hat{A}(X)$ into K.

Lemma 1. *For an element* a *of* $\hat{A}(X)$ *to be invertible, it is necessary and sufficient that its constant term be invertible in* K.

If a is invertible in $\hat{A}(X)$, $\varepsilon(a)$ is invertible in K. Conversely, if $\varepsilon(a)$ is invertible in K, there exists $u \in \hat{A}_1(X)$ such that $a = \varepsilon(a)(1 - u)$; we write $b = \left(\sum_{n \geqslant 0} u^n\right)\varepsilon(a)^{-1}$. Then $ab = ba = 1$ and a is invertible.

The set of elements of $\hat{A}(X)$ of constant term 1 is therefore a subgroup of the multiplicative monoid $\hat{A}(X)$, called the *Magnus group* constructed on X (relative to K). In this chapter it will be denoted by $\Gamma(X)$ or simply Γ. For every integer $n \geqslant 1$, we denote by Γ_n the set of $a \in \Gamma$ such that $\omega(a - 1) \geqslant n$. By Proposition 2 of § 4, no. 5, the sequence $(\Gamma_n)_{n \geqslant 1}$ is an *integral central filtration on* Γ.

3. MAGNUS GROUP AND FREE GROUP

THEOREM 1. *Let r be a mapping of X into $\hat{A}(X)$ such that $\omega(r(x)) \geqslant 2$ for all $x \in X$. The unique homomorphism g of the free group F(X) into the Magnus group $\Gamma(X)$ such that $g(x) = 1 + x + r(x)$ for all $x \in X$ is injective.*

We first prove three lemmas.

Lemma 2. *Let n be a non-zero rational integer. In the ring of formal power series* $K[[t]]$ *we write* $(1 + t)^n = \sum_{j \geqslant 0} c_{j,n} t^j$. *There exists an integer $j \geqslant 1$ such that* $c_{j,n} \neq 0$.

If $n > 0$, then $c_{n,n} = 1$ by the binomial formula.

Suppose that $n < 0$ and let $m = -n$. If $c_{j,n} = 0$ for all $j \geqslant 1$, then $(1 + t)^n = 1$, whence, taking the inverse, $(1 + t)^m = 1$, which contradicts the formula $c_{m,m} = 1$.

Lemma 3. *Let $x_1, \ldots, x_s$ be elements of X such that $s \geqslant 1$ and $x_i \neq x_{i+1}$ for $1 \leqslant i \leqslant s - 1$; let $n_1, \ldots, n_s$ be non-zero rational integers. Then the element* $\prod_{i=1}^{s} (1 + x_i)^{n_i}$ *of $\hat{A}(X)$ is $\neq 1$.*

Let $\mathfrak{m}$ be a maximal ideal of K and k the field $K/\mathfrak{m}$; let $p\colon \hat{A}_K(X) \to \hat{A}_k(X)$ be the unique continuous homomorphism of unital K-algebras such that $p(x) = x$ for $x \in X$ (no. 1, Proposition 1). It suffices to prove that $p\left(\prod (1 + x_i)^{n_i}\right) \neq 1$ and the problem is reduced to the case where K is a field.

In the notation of Lemma 2:

$$\prod_{i=1}^{s} (1 + x_i)^{n_i} = \sum_{b_i \geqslant 0} c_{b_1, n_1} \ldots c_{b_s, n_s} x_1^{b_1} \ldots x_s^{b_s}.$$

By Lemma 2, there exist integers $a_i > 0$ such that $c_{a_i, n_i} \neq 0$ $(1 \leqslant i \leqslant s)$. By *Algebra*, Chapter I, § 7, no. 4, Proposition 6, no monomial $x_1^{b_1} \ldots x_s^{b_s}$ such that

$b_i \geq 0$ and $(b_1, \ldots, b_s) \neq (a_1, \ldots, a_s)$ can be equal to $x_1^{a_1} \ldots x_s^{a_s}$. It follows that the coefficient of $x_1^{a_1} \ldots x_s^{a_s}$ in $\prod_{i=1}^{s} (1 + x_i)^{n_i}$ is $c_{a_1, n_1} \ldots c_{a_s, n_s} \neq 0$, which implies the result.

Lemma 4. *Let* σ *be the continuous endomorphism of* $\hat{A}(X)$ *such that* $\sigma(x) = x + r(x)$ *for* $x \in X$ (no. 1, Proposition 1). *Then* σ *is an automorphism and* $\sigma(\hat{A}_m(X)) = \hat{A}_m(X)$ *for all* $m \in \mathbf{N}$.

$\sigma(x) \equiv x$ mod. $\hat{A}_2(X)$ for $x \in X$, whence, for $n \geq 1$ and $x_1, \ldots, x_n$ in X,

$$\sigma(x_1) \ldots \sigma(x_n) \equiv x_1 \ldots x_n \quad \text{mod. } \hat{A}_{n+1}(X);$$

it follows by linearity that $\sigma(a) \equiv a$ modulo $\hat{A}_{n+1}(X)$ for all $a \in A^n(X)$ and in particular $\sigma(A^n(X)) \subset \hat{A}_n(X)$. It follows that $\sigma(A^m(X)) \subset \hat{A}_n(X)$ for $m \geq n$, whence $\sigma(\hat{A}_n(X)) \subset \hat{A}_n(X)$. In other words, σ is compatible with the filtration $(\hat{A}_m(X))$ on A(X) and its restriction to the associated graded ring is the identity. Hence σ is bijective (*Commutative Algebra*, Chapter III, § 2, no. 8, Corollary 3 to Theorem 1).

Finally we prove Theorem 1. Let $w \neq 1$ be an element of F(X). By *Algebra*, Chapter I, § 7, no. 5, Proposition 7, there exist $x_1, \ldots, x_s$ in X and non-zero rational integers $n_1, \ldots, n_s$ such that $s \geq 1$, $x_i \neq x_{i+1}$ $(1 \leq i \leq s - 1)$ and

$$w = x_1^{n_1} \ldots x_s^{n_s}.$$

In the notation of Lemma 4,

$$g(w) = \prod (1 + \sigma(x_i))^{n_i} = \sigma\Big(\prod (1 + x_i)^{n_i}\Big),$$

hence $g(w) \neq 1$ by Lemmas 3 and 4.

4. LOWER CENTRAL SERIES OF A FREE GROUP

We shall prove the two following theorems:

THEOREM 2. *Suppose that in the ring* K *the relation* $n.1 = 0$ *implies* $n = 0$ *for every integer n. Let r be a mapping of* X *into* $\hat{A}(X)$ *such that* $\omega(r(x)) \geq 2$ *for* $x \in X$ *and let g be the homomorphism of* F(X) *into the Magnus group* $\Gamma(X)$ *such that* $g(x) = 1 + x + r(x)$ *for* $x \in X$. *For all* $n \geq 1$, $C^nF(X)$ *is the inverse image under g of the subgroup* $1 + \hat{A}_n(X)$ *of* $\Gamma(X)$.

THEOREM 3. *For all* $x \in X$, *let* $c(x)$ *be the canonical image of x in* $F(X)/(F(X), F(X))$. *Let* $\mathfrak{g}$ *be the graded Lie* **Z**-*algebra associated with the filtration* $(C^nF(X))_{n \geq 1}$ *of* F(X) (§ 4, no. 6). *The unique homomorphism of the free Lie* **Z**-*algebra* $L_{\mathbf{Z}}(X)$ *into* $\mathfrak{g}$ *which extends c is an isomorphism.*

Loosely speaking, the graded Lie **Z**-algebra associated with the free group F(X) (with the lower central series) is the free Lie **Z**-algebra $L_{\mathbf{Z}}(X)$.

We write $F(X) = F$, $\Gamma(X) = \Gamma$, $\hat{A}(X) = \hat{A}$, $\hat{A}_{\mathbf{Z}}(X) = \hat{A}_{\mathbf{Z}}$, $C^nF(X) = C^n$, $\Gamma_n = 1 + \hat{A}_n(X)$ and let $\alpha: L_{\mathbf{Z}}(X) \to \mathfrak{g}$ be the homomorphism introduced in the statement of Theorem 3.

(A) *Preliminary reductions.*

Let γ denote the homomorphism of F into Γ defined by $\gamma(x) = 1 + x$ for $x \in X$. By Lemma 4 there exists an automorphism σ of the algebra $\hat{A}$ compatible with the filtration on $\hat{A}$ and such that $\sigma(1 + x) = g(x)$ for all $x \in X$; then $\sigma(\Gamma_n) = \Gamma_n$ for all n. As the homomorphisms g and $\sigma \circ \gamma$ of F into Γ coincide on X, $g = \sigma \circ \gamma$ and hence $\overset{-1}{g}(\Gamma_n) = \overset{-1}{\gamma}(\Gamma_n)$. Under the hypotheses of Theorem 2, $\mathbf{Z}$ can be identified with a subring of K; the Magnus algebra $\hat{A}_{\mathbf{Z}}$ is therefore identified with a subring of $\hat{A}$ and the filtration on $\hat{A}_{\mathbf{Z}}$ is induced by that on $\hat{A}$. As γ maps F into $\hat{A}_{\mathbf{Z}}$, we see that it suffices to prove Theorems 2 and 3 under the supplementary hypotheses $K = \mathbf{Z}$, $r = 0$ and hence $g = \gamma$, *hypotheses which we shall henceforth make.*

(B) *Surjectivity of* α.

As X generates the group $F = C^1$, the set $c(X)$ generates the $\mathbf{Z}$-module $\mathfrak{g}^1 = C^1/C^2$. But $\mathfrak{g}^1$ generates the Lie $\mathbf{Z}$-algebra $\mathfrak{g}$ (§ 4, no. 6, Proposition 5) and hence $c(X)$ generates $\mathfrak{g}$, which proves that α is surjective.

(C) We identify the graded algebra $\mathrm{gr}(\hat{A})$ with $A(X)$ under the canonical isomorphisms $A^n(X) \to \hat{A}_n/\hat{A}_{n+1}$. For every integer $n \geqslant 1$, we write $F^n = \overset{-1}{\gamma}(\Gamma_n)$; we know (§ 4, no. 5) that $(F^n)_{n \geqslant 1}$ is an integral central filtration on F. Let $\mathfrak{g}'$ denote the associated graded Lie $\mathbf{Z}$-algebra (§ 4, no. 4). Let f be the Lie algebra homomorphism of $\mathfrak{g}'$ into $A(X)$ associated with γ (§ 4, no. 5, Proposition 3). Now $C^n \subset F^n$ for every integer $n \geqslant 1$ (§ 4, no. 6, Proposition 4) and hence there is a canonical homomorphism ε of $\mathfrak{g} = \bigoplus_{n \geqslant 1} C^n/C^{n+1}$ into $\mathfrak{g}' = \bigoplus_{n \geqslant 1} F^n/F^{n+1}$

$$L_{\mathbf{Z}}(X) \xrightarrow{\alpha} \mathfrak{g} \xrightarrow{\varepsilon} \mathfrak{g}' \xrightarrow{f} A(X).$$

We write $\beta = f \circ \varepsilon$; we give β explicitly as follows: if u is the class modulo C^{n+1} of an element w of C^n, then $\gamma(w) - 1$ is of order $\geqslant n$ in $\hat{A}$ and $\beta(u)$ is the homogeneous component of $\gamma(w) - 1$ of degree n. In particular,

$$\beta(c(x)) = x \quad \text{for all } x \in X. \tag{3}$$

(D) *Proof of Theorems* 2 *and* 3.

The Lie algebra homomorphism $\beta \circ \alpha: L_{\mathbf{Z}}(X) \to A(X)$ restricted to X is the identity by (3) and hence is the *canonical injection* (§ 3, no. 1). Therefore α is injective and hence bijective by (B); this proves Theorem 3. As $\beta \circ \alpha = f \circ \varepsilon \circ \alpha$ is injective and α is bijective, ε is injective. For all $n \geqslant 1$,

$$\varepsilon_n: C^n/C^{n+1} \to F^n/F^{n+1}$$

is injective and hence

$$C^n \cap F^{n+1} = C^{n+1}.$$

$C^1 = F = F^1$; if $C^n = F^n$, then $C^n \cap F^{n+1} = F^{n+1}$ whence $C^{n+1} = F^{n+1}$ which proves Theorem 2 by induction on $n \geqslant 1$.

COROLLARY. $\bigcap_{n \geqslant 1} C^n F(X) = \{e\}.$

Applying Theorem 2 with $K = \mathbf{Z}$ and $r = 0$,

$$\bigcap_{n \geqslant 1} C^n F(X) = \bigcap_{n \geqslant 1} \overset{-1}{g}(1 + \hat{A}_n(X)) = \overset{-1}{g}\Big(\bigcap_{n \geqslant 1} (1 + \hat{A}_n(X))\Big) = \overset{-1}{g}(1) = \{e\}.$$

Remark. Let H be a Hall set relative to X (§ 2, no. 10). Let M be the magma defined by the law of composition $(x, y) \mapsto (x, y) = x^{-1}y^{-1}xy$ on F(X) and let ϕ be the homomorphism of M(X) into M whose restriction to X is the identity. The elements of $\phi(H)$ are called the *basic commutators* of F(X) associated with the Hall set H. For every integer $n \geqslant 1$, let H_n be the subset of H consisting of the elements of length n; we know (§ 2, no. 11, Theorem 1) that the canonical mapping of H_n into $L_{\mathbf{Z}}(X)$ is a basis of the Abelian group $L^n_{\mathbf{Z}}(X)$. Moreover, $\phi(H_n) \subset C^n$; for all $m \in H_n$, let $\phi_n(m)$ denote the class mod. C^{n+1} of $\phi(m) \in C^n$. Theorem 3 then shows that ϕ_n *is a bijection of* H_n *onto a basis of the Abelian group* C^n/C^{n+1}. It follows immediately that, for all $w \in F(X)$ and all $i \geqslant 1$, there exists a unique element α_i of $\mathbf{Z}^{(H_i)}$ such that, for $n \geqslant 1$,

$$(4) \qquad w = \prod_{i=1}^{n} \prod_{m \in H_i} \phi(m)^{\alpha_i(m)} \quad \text{mod. } C^{n+1},$$

where the product is calculated according to the total ordering given on H.

Example. Suppose that X is a set with two elements x, y and let $H_1 = \{x, y\}$, $H_2 = \{xy\}$. Every element w of F(X) can therefore be written

$$w \equiv x^a y^b (x, y)^c \text{ mod. } C^3 \quad \text{with } a, b, c \text{ in } \mathbf{Z}.$$

For $w = (xy)^n$, $a = b = n$ and $c = n(1 - n)/2$ (cf. Exercise 9), whence

$$(xy)^n \equiv x^n y^n (x, y)^{n(1-n)/2} \quad \text{mod. } C^3.$$

5. *p*-FILTRATION OF FREE GROUPS

In this no., p denotes a prime number and we assume that $K = \mathbf{F}_p$. Let γ be the homomorphism of F(X) into $\Gamma(X)$ defined by $\gamma(x) = 1 + x$ for x in X; we write $F^{(p)}_n(X) = \overset{-1}{\gamma}(1 + \hat{A}_n(X))$. The sequence $(F^{(p)}_n(X))_{n \geqslant 1}$ is an integral central filtration on F(X), which is *separated* since γ is injective (no. 3, Theorem 1). It is called the *p-filtration* on F(X).

PROPOSITION 2. *Suppose that* X *is finite. For every integer* $n \geqslant 1$, *the group* $F(X)/F^{(p)}_n(X)$ *is a finite p-group of nilpotency class* $\leqslant n$.

Arguing by induction on n, it suffices to prove that $F_n^{(p)}(X)/F_{n+1}^{(p)}(X)$ is a finite commutative p-group for all $n \geqslant 1$. For all $w \in F_n^{(p)}(X)$, the element $\gamma(w) - 1$ of $\hat{A}(X)$ is of order $\geqslant n$; we denote by $\delta_n(w)$ the homogeneous component of $\gamma(w) - 1$ of degree n. The mapping $\delta_n: F_n^{(p)}(X) \to A^n(X)$ is a homomorphism with kernel $F_{n+1}^{(p)}(X)$ (§ 4, no. 5, Proposition 3) and hence $F_n^{(p)}(X)/F_{n+1}^{(p)}(X)$ is isomorphic to a subgroup of $A^n(X)$. Since X is finite, $A^n(X)$ is a finite-dimensional vector space over $\mathbf{F}_p$ and hence a finite commutative p-group and so is $F_n^{(p)}(X)/F_{n+1}^{(p)}(X)$.

PROPOSITION 3. *For all $w \neq 1$ in* F(X), *there exist a finite p-group* G *and a homomorphism f of* F(X) *into* G *such that* $f(w) \neq 1$.

There exist elements $x_1, \ldots, x_r$ of X and integers $n_1, \ldots, n_r$ such that $w = x_1^{n_1} \ldots x_r^{n_r}$. Let $Y = \{x_1, \ldots, x_r\}$. The canonical injection of Y into X extends to a homomorphism $\alpha: F(Y) \to F(X)$; on the other hand, let β be the homomorphism of F(X) into F(Y) whose restriction to Y is the identity and which maps $X - Y$ to $\{1\}$. Then $\beta(\alpha(y)) = y$ for $y \in Y$ and hence $\beta \circ \alpha$ is the identity automorphism of F(Y). There obviously exists w' in F(Y) such that $w = \alpha(w')$; then $\beta(w) = w' \neq 1$; now $\bigcap_{n \geqslant 1} F_n^{(p)}(Y) = \{1\}$ and there therefore exists an integer $n \geqslant 1$ such that $\beta(w) \notin F_n^{(p)}(Y)$. By Proposition 2, the group $G = F(Y)/F_n^{(p)}(Y)$ is a finite p-group. If f is the composition of β and the canonical homomorphism of F(Y) onto G, then $f(w) \neq 1$.

COROLLARY. *The intersection of the normal subgroups of finite index in* F(X) *is* $\{1\}$.

§ 6. THE HAUSDORFF SERIES

In this paragraph we assume that K *is a field of characteristic* 0.

1. EXPONENTIAL AND LOGARITHM IN FILTERED ALGEBRAS

Let A be a unital associative algebra which is Hausdorff and complete under a real filtration (A_α). We write $\mathfrak{m} = A_0^+ = \bigcup_{\alpha > 0} A_\alpha$.

For $x \in \mathfrak{m}$, the family $(x^n/n!)_{n \in \mathbf{N}}$ is summable. We write

$$e^x = \exp x = \sum_{n \geqslant 0} x^n/n!. \tag{1}$$

Then $\exp(x) \in 1 + \mathfrak{m}$ and the mapping $\exp: \mathfrak{m} \to 1 + \mathfrak{m}$ is called the *exponential mapping* of A.

For all $y \in 1 + \mathfrak{m}$, the family $((-1)^{n-1}(y-1)^n/n)_{n \geqslant 1}$ is summable. We write

$$\log y = \sum_{n \geqslant 1} (-1)^{n-1}(y-1)^n/n. \tag{2}$$

Then $\log y \in \mathfrak{m}$ and the mapping $\log: 1 + \mathfrak{m} \to \mathfrak{m}$ is called the *logarithmic mapping* of A.

PROPOSITION 1. *The exponential mapping is a homeomorphism of* $\mathfrak{m}$ *onto* $1 + \mathfrak{m}$ *and the logarithmic mapping is the inverse homeomorphism.*

For $x \in A_\alpha$, $\frac{x^n}{n!} \in A_{n\alpha}$. It follows that the series defining the exponential converges uniformly on each of the sets A_α for $\alpha > 0$; as A_α is open in $\mathfrak{m}$ and $\mathfrak{m} = \bigcup_{\alpha>0} A_\alpha$, the exponential mapping is continuous. It can be similarly shown that the logarithmic mapping is continuous.

Let e and l be the formal power series with no constant term

$$e(X) = \sum_{n\geqslant 1} \frac{X^n}{n!}, \qquad l(X) = \sum_{n\geqslant 1} (-1)^{n-1}X^n/n.$$

We know (*Algebra*, Chapter IV, § 6, no. 9) that $e(l(X)) = l(e(X)) = X$ on $\hat{A}(\{X\}) = K[[X]]$. By substitution (§ 5, no. 1), we deduce that

$$e(l(x)) = l(e(x)) = x$$

for $x \in \mathfrak{m}$; as

$$\exp x = e(x) + 1, \qquad \log(1 + x) = l(x)$$

it follows immediately that

$$\log \exp x = x, \qquad \exp \log(1 + x) = 1 + x$$

for x in $\mathfrak{m}$, whence the proposition.

Remarks. (1) If $x \in \mathfrak{m}$, $y \in \mathfrak{m}$ and x and y commute, then

$$\exp(x + y) = \exp(x) \exp(y),$$

since the family $\left(\frac{x^i}{i!}\cdot\frac{y^j}{j!}\right)_{i,j\in\mathbf{N}}$ is summable (cf. *Algebra*, Chapter IV, § 6, no. 9, Proposition 11).

(2) As the series e and l are without constant term and A_α is a closed ideal of A, $\exp A_\alpha \subset 1 + A_\alpha$ and $\log(1 + A_\alpha) \subset A_\alpha$ whence $\exp A_\alpha = 1 + A_\alpha$ and $\log(1 + A_\alpha) = A_\alpha$ for $\alpha > 0$.

(3) Let B be a complete Hausdorff filtered unital associative algebra and $\mathfrak{n} = \bigcup_{\alpha>0} B_\alpha$. Let f be a continuous unital homomorphism of A into B such that $f(\mathfrak{m}) \subset \mathfrak{n}$. Then $f(\exp x) = \exp f(x)$ for $x \in \mathfrak{m}$ and $f(\log y) = \log f(y)$ for $y \in 1 + \mathfrak{m}$; we show for example the first of these formulae:

$$f(\exp x) = \sum_{n\geqslant 0} f(x^n)/n! = \sum_{n\geqslant 0} f(x)^n/n! = \exp f(x).$$

(4) Let E be a unital associative algebra. If a is a nilpotent element of E, the family $\left(\frac{a^n}{n!}\right)_{n\in\mathbf{N}}$ has finite support and we write $\exp a = \sum_{n\geq 0} a^n/n!$. An element b is unipotent if $b - 1$ is nilpotent; then we write

$$\log b = \sum_{n\geq 1} (-1)^{n-1}(b-1)^n/n.$$

We deduce from the relations $e(l(\mathrm{X})) = l(e(\mathrm{X})) = \mathrm{X}$ that the mapping $a \mapsto \exp a$ is a bijection of the set of nilpotent elements of E onto the set of unipotent elements of E and that $b \mapsto \log b$ is the inverse mapping.

2. HAUSDORFF GROUP

Let X be a set. We use the notation of § 5, nos. 1 and 2. The free Lie algebra L(X) is identified with its canonical image in A(X) (§ 3, no. 1, Theorem 1). We denote by $\hat{\mathrm{L}}(\mathrm{X})$ the closure of L(X) in $\hat{\mathrm{A}}(\mathrm{X})$, that is the set of elements of $\hat{\mathrm{A}}(\mathrm{X})$ of the form $a = \sum_{n\geq 1} a_n$ such that $a_n \in \mathrm{L}^n(\mathrm{X})$ for all $n \geq 0$; this is filtered Lie subalgebra of $\hat{\mathrm{A}}(\mathrm{X})$.

THEOREM 1. *The restriction of the exponential mapping of* $\hat{\mathrm{A}}(\mathrm{X})$ *to* $\hat{\mathrm{L}}(\mathrm{X})$ *is a bijection of* $\hat{\mathrm{L}}(\mathrm{X})$ *onto a closed subgroup of the Magnus group* $\Gamma(\mathrm{X})$.

We write $\mathrm{A}(\mathrm{X}) = \mathrm{A}$, $\mathrm{A}^n(\mathrm{X}) = \mathrm{A}^n$, $\hat{\mathrm{A}}(\mathrm{X}) = \hat{\mathrm{A}}$, $\mathrm{L}^n(\mathrm{X}) = \mathrm{L}^n$, $\hat{\mathrm{L}}(\mathrm{X}) = \hat{\mathrm{L}}$, $\Gamma(\mathrm{X}) = \Gamma$. Let B be the algebra $\mathrm{A} \otimes \mathrm{A}$ with the graduation of type **N** defined by $\mathrm{B}^n = \sum_{i+j=n} \mathrm{A}^i \otimes \mathrm{A}^j$. Let $\hat{\mathrm{B}} = \prod_{n\geq 0} \mathrm{B}^n$ be the associated complete filtered algebra (*Commutative Algebra*, Chapter III, § 2, no. 12, *Example* 1). The coproduct $c\colon \mathrm{A} \to \mathrm{A} \otimes \mathrm{A}$ defined in § 3, no. 1, Corollary 1 to Theorem 1 is graded of degree 0 and hence extends by continuity to a homomorphism $\hat{c}\colon \hat{\mathrm{A}} \to \hat{\mathrm{B}}$ given by

$$\hat{c}\Big(\sum_{n\geq 0} a_n\Big) = \sum_{n\geq 0} c(a_n) \quad \text{for } a_n \in \mathrm{A}^n.$$

We also define continuous homomorphisms δ' and δ'' of $\hat{\mathrm{A}}$ into $\hat{\mathrm{B}}$ by

$$\delta'\Big(\sum_{n\geq 0} a_n\Big) = \sum_{n\geq 0} (a_n \otimes 1), \qquad \delta''\Big(\sum_{n\geq 0} a_n\Big) = \sum_{n\geq 0} (1 \otimes a_n) \quad \text{for } a_n \in \mathrm{A}^n.$$

By Corollary 2 to Theorem 1 of § 3, no. 1, L^n is the set of $a_n \in \mathrm{A}^n$ such that $c(a_n) = a_n \otimes 1 + 1 \otimes a_n$. It follows that $\hat{\mathrm{L}}$ is the set of $a \in \hat{\mathrm{A}}$ such that

$$\hat{c}(a) = \delta'(a) + \delta''(a). \tag{3}$$

Let Δ be the set of $b \in \hat{\mathrm{A}}$ of constant term equal to 1 and satisfying the relation

$$\hat{c}(b) = \delta'(b) . \delta''(b), \tag{4}$$

in other words, the set of $b = \sum_{n \geqslant 0} b_n$ such that $b_n \in A^n$ for all $n \geqslant 0$, $b_0 = 1$ and $c(b_n) = \sum_{i+j=n} b_i \otimes b_j$ for $n \geqslant 0$. The latter characterization shows that Δ is a closed subset of Γ; as $\hat{c}$, δ' and δ'' are ring homomorphisms and every element of $\delta'(\hat{A})$ commutes with every element of $\delta''(\hat{A})$, the restrictions to Γ of the mappings c and $\delta'\delta''$ are group homomorphisms and Δ is a subgroup of Γ.

By Proposition 1 of no. 1, the exponential mapping of $\hat{A}$ is a bijection of the set $\hat{A}^+$ of elements of $\hat{A}$ with no constant term onto Γ. Let $a \in \hat{A}^+$ and $b = \exp a$. As $\hat{c}$ is a continuous ring homomorphism,

$$\hat{c}(b) = \hat{c}\Big(\sum_{n \geqslant 0} a^n/n!\Big) = \sum_{n \geqslant 0} \hat{c}(a)^n/n! = \exp \hat{c}(a).$$

The relations

$$\delta'(b) = \exp \delta'(a), \qquad \delta''(b) = \exp \delta''(a)$$

are proved similarly and, as $\delta'(a)$ commutes with $\delta''(a)$ (no. 1, *Remark* 1),

$$\delta'(b)\,\delta''(b) = \exp\,(\delta'(a) + \delta''(a)).$$

Therefore a satisfies (3) if and only if b satisfies (4), which proves the theorem.

Remark. The above proof shows that $\exp(\hat{L})$ is the subgroup Δ of Γ consisting of the b satisfying (4).

Hence the group law of Δ can be transported by the exponential mapping to $\hat{L}$. In other words, $\hat{L}$ is a complete topological group with the law of composition $(a, b) \mapsto a \,\mathrm{H}\, b$ given by

$$a \,\mathrm{H}\, b = \log(\exp a . \exp b).$$

The topological group thus obtained is called the *Hausdorff group* (derived from X relative to K).

Let g be the homomorphism of the free group $F = F(X)$ into Γ such that $g(x) = \exp x$ for $x \in X$. As $\exp x - 1 - x = \sum_{n \geqslant 2} x^n/n!$ is of order $\geqslant 2$, g is injective by Theorem 1 of § 5, no. 3. Therefore *the mapping* $\log \circ g$ *is an injective homomorphism of* F *into the Hausdorff group which extends the canonical injection* $X \to \hat{L}$.

For every integer $m \geqslant 1$ we denote by $\hat{L}_m$ the set of elements of order $\geqslant m$ in $\hat{L}$ and by Γ_m the set of $u \in \Gamma$ such that $u - 1$ is of order $\geqslant m$. Then $\hat{L}_m = \overset{-1}{\exp}(\Gamma_m)$ by *Remark* 2 of no. 1; as $(\Gamma_m)_{m \geqslant 1}$ is an integral central filtration on Γ (§ 4, no. 5 Proposition 2), $(\hat{L}_m)_{m \geqslant 1}$ *is an integral central filtration on the group* $\hat{L}$.

3. LIE FORMAL POWER SERIES

Lemma 1. *Let* $\mathfrak{g}$ *be a filtered Lie algebra* (§ 4, no. 1), $(\mathfrak{g}_\alpha)_{\alpha \in \mathbf{R}}$ *its filtration and let* $\alpha \in \mathbf{R}$. *Let* P *be a homogeneous Lie polynomial of degree* n *in the indeterminates* $(T_i)_{i \in I}$ (§ 2, no. 4). *Then* $P((a_i)) \in \mathfrak{g}_{n\alpha}$ *for every family* $(a_i)_{i \in I}$ *of elements of* $\mathfrak{g}_\alpha$.

Every Lie polynomial of degree $n \geqslant 2$ is a finite sum of terms of the form $[Q, R]$ where Q and R are of degree $<n$ and the sum their degrees is equal to n (§ 2, no. 7, Proposition 7). The lemma follows by induction on n.

A *Lie formal power series*† (with coefficients in K) *in the indeterminates* $(T_i)_{i\in I}$ is any element of the Lie algebra $\hat{L}((T_i)_{i\in I}) = \hat{L}(I)$. Such an element u can be written uniquely as the sum of a summable family $(u_\nu)_{\nu\in \mathbf{N}^{(I)}}$ where $u_\nu \in L^\nu(I)$.

Suppose that I is *finite*. Let $\mathfrak{g}$ be a complete Hausdorff filtered Lie algebra such that $\mathfrak{g} = \bigcup_{\alpha>0} \mathfrak{g}_\alpha$; let $\boldsymbol{t} = (t_i)_{i\in I}$ be a family of elements of $\mathfrak{g}$.

PROPOSITION 2. *The homomorphism $f_{\boldsymbol{t}}: L(I) \to \mathfrak{g}$ such that $f_{\boldsymbol{t}}(T_i) = t_i$ (§ 2, no. 4) can be extended by continuity to one and only one continuous homomorphism $\hat{f}_{\boldsymbol{t}}$ of $\hat{L}(I)$ into $\mathfrak{g}$.*

There exists $\alpha > 0$ such that $t_i \in \mathfrak{g}_\alpha$ for all $i \in I$; hence $f_{\boldsymbol{t}}(L^\nu(I)) \subset \mathfrak{g}_{|\nu|\alpha}$ for all ν (Lemma 1), which implies the continuity of $f_{\boldsymbol{t}}$.

If $u \in \hat{L}(I)$, we write $u((t_i)) = \hat{f}_{\boldsymbol{t}}(u)$. In particular, taking $\mathfrak{g} = \hat{L}(I)$, $u = u((T_i))$; in the general case, $u((t_i))$ is called the result of substituting the t_i for the T_i in the Lie formal power series $u((T_i))$. If $u = \sum_{\nu\in\mathbf{N}^{(I)}} u_\nu$, where $u_\nu \in L^\nu(X)$, the family $(u_\nu((t_i)))_{\nu\in\mathbf{N}^{(I)}}$ is summable and

(5) $$u((t_i)) = \sum_{\nu\in\mathbf{N}^I} u_\nu((t_i)).$$

Let σ be a continuous homomorphism of $\mathfrak{g}$ into a complete Hausdorff filtered Lie algebra $\mathfrak{g}'$ such that $\mathfrak{g}' = \bigcup_{\alpha>0} \mathfrak{g}'_\alpha$. For every *finite* family $\boldsymbol{t} = (t_i)_{i\in I}$ of elements of $\mathfrak{g}$ and all $u \in \hat{L}(I)$,

(6) $$\sigma(u((t_i))) = u((\sigma(t_i))),$$

for the homomorphism $\sigma \circ \hat{f}_{\boldsymbol{t}}$ is continuous and maps T_i to $\sigma(t_i)$ for $i \in I$.

Let $\boldsymbol{u} = (u_j)_{j\in J}$ be a *finite* family of elements of $\hat{L}(I)$ and let $v \in \hat{L}(J)$; by substituting the u_j for the T_j in v, we obtain an element $w = v((u_j)_{j\in J})$ of $\hat{L}(I)$ denoted by $v \circ \boldsymbol{u}$. Then

(7) $$w((t_i)_{i\in I}) = v((u_j((t_i)_{i\in I}))_{j\in J})$$

for every *finite* family $\boldsymbol{t} = (t_i)_{i\in I}$ of elements of $\mathfrak{g}$, as is seen by operating with the continuous homomorphism $\hat{f}_{\boldsymbol{t}}$ on the equation $w = v((u_j)_{j\in J})$.

Let $u = \sum_{\nu\in\mathbf{N}^I} u_\nu \in \hat{L}(I)$, where $u_\nu \in L^\nu(I)$. The mapping $\tilde{u}\ (t_i) \mapsto u((t_i))$ of $\mathfrak{g}^I$ into $\mathfrak{g}$ is *continuous*: for in each of the open sets $\mathfrak{g}_\alpha$ with $\alpha > 0$ the family of $\tilde{u}_\alpha$ is uniformly summable and it suffices to prove that each $\tilde{u}_\nu$ is continuous, which is immediate by induction on $|\nu|$.

---

† A Lie formal power series is not in general a formal power series in the sense of *Algebra*, Chapter IV, § 6.

4. THE HAUSDORFF SERIES

Let $\{U, V\}$ be a set with two elements.

DEFINITION 1. *The element* $H = U \mathbin{\mathrm{H}} V = \log(\exp U . \exp V)$ (no. 2) *of the Lie algebra* $\hat{L}_{\mathbf{Q}}(\{U, V\})$ *is called the Hausdorff series in the indeterminates* U *and* V.

We denote by H_n (resp. $H_{r,s}$) the homogeneous component of H of total degree n (resp. multidegree (r, s)). Then

$$(8) \qquad H = \sum_{n \geqslant 0} H_n = \sum_{r,s \geqslant 0} H_{r,s}, \qquad H_n = \sum_{\substack{r+s=n\\ r,s\geqslant 0}} H_{r,s}.$$

THEOREM 2. *If* r *and* s *are two positive integers such that* $r + s \geqslant 1$, *then* $H_{r,s} = H'_{r,s} + H''_{r,s}$, *where*

$$(9) \quad (r+s)H'_{r,s} = \sum_{m\geqslant 1} \frac{(-1)^{m-1}}{m} \sum_{\substack{r_1+\cdots+r_m=r\\ s_1+\cdots+s_{m-1}=s-1\\ r_1+s_1\geqslant 1,\ldots,r_{m-1}+s_{m-1}\geqslant 1}} \left(\left(\prod_{i=1}^{m-1} \frac{(\operatorname{ad} U)^{r_i}}{r_i!} \frac{(\operatorname{ad} V)^{s_i}}{s_i!}\right) \frac{(\operatorname{ad} U)^{r_m}}{r_m!}\right)(V)$$

$$(10) \quad (r+s)H''_{r,s} = \sum_{m\geqslant 1} \frac{(-1)^{m-1}}{m} \sum_{\substack{r_1+\cdots+r_{m-1}=r-1\\ s_1+\cdots+s_{m-1}=s\\ r_1+s_1\geqslant 1,\ldots,r_{m-1}+s_{m-1}\geqslant 1}} \left(\prod_{i=1}^{m-1} \frac{(\operatorname{ad} U)^{r_i}}{r_i!} \frac{(\operatorname{ad} V)^{s_i}}{s_i!}\right)(U).$$

In $\hat{A}_{\mathbf{Q}}(\{U, V\})$, $\exp U . \exp V = 1 + W$, where $W = \sum_{r+s\geqslant 1} \frac{U^r}{r!} \frac{V^s}{s!}$, whence $H = \sum_{m\geqslant 1} (-1)^{m-1} W^m/m$ (no. 2), that is:

$$(11) \qquad H_{r,s} = \sum_{m\geqslant 1} \frac{(-1)^{m-1}}{m} \sum_{\substack{r_1+\cdots+r_m=r\\ s_1+\cdots+s_m=s\\ r_1+s_1\geqslant 1,\ldots,r_m+s_m\geqslant 1}} \prod_{i=1}^{m} \frac{U^{r_i}}{r_i!} \frac{V^{s_i}}{s_i!}.$$

The linear mapping P_n, defined by $P_n(x_1, \ldots, x_n) = \frac{1}{n}\left(\prod_{i=1}^{n-1} (\operatorname{ad} x_i)\right)(x_n)$ for $n \geqslant 1$ and $x_1, \ldots, x_n$ in $\{U, V\}$, is a projector of $A^n_{\mathbf{Q}}(\{U, V\})$ onto $L^n_{\mathbf{Q}}(\{U, V\})$ (§ 3, no. 2, Corollary to Proposition 1); as $H_{r,s}$ belongs to $L^{r+s}_{\mathbf{Q}}(\{U, V\})$, $H_{r,s} = P_{r+s}(H_{r,s})$. Now

$$(12) \quad P_{r+s}\left(\prod_{i=1}^{m} \frac{U^{r_i}}{r_i!} \frac{V^{s_i}}{r_i!}\right) = \frac{1}{r+s}\left(\left(\prod_{i=1}^{m-1} \frac{(\operatorname{ad} U)^{r_i}}{r_i!} \frac{(\operatorname{ad} V)^{s_i}}{s_i!}\right) \frac{(\operatorname{ad} U)^{r_m}}{r_m!} \frac{(\operatorname{ad} V)^{s_m-1}}{s_m!}\right)(V)$$

when $s_m \geqslant 1$ and

$$(13)\quad P_{r+s}\left(\prod_{i=1}^{m} \frac{U^{r_i}}{r_i!}\frac{V^{s_i}}{s_i!}\right) = \frac{1}{r+s}\left(\left(\prod_{i=1}^{m-1} \frac{(\operatorname{ad} U)^{r_i}}{r_i!}\frac{(\operatorname{ad} V)^{s_i}}{s_i!}\right)\frac{(\operatorname{ad} U)^{r_m-1}}{r_m!}\right)(U)$$

when $r_m \geqslant 1$ and $s_m = 0$. Moreover, obviously $(\operatorname{ad} t)^{p-1}.t = 0$ if $p \geqslant 2$ and $(\operatorname{ad} t)^0.t = t$. It follows that the two sides of (12) are zero when $s_m \geqslant 2$ and those of (13) are zero when $r_m \geqslant 2$. The theorem then follows since $H'_{r,s}$ is the sum of the terms of type (12) and $H''_{r,s}$ is the sum of the terms of type (13).

Remarks. (1) We have defined (§ 3, no. 2, *Remark*) a projector Q of A(X) onto L(X) such that $Q(a^m) = 0$ for $a \in L(X)$ and $m \geqslant 2$ and $Q(1) = 0$. Then $H = Q(\exp H) = Q(\exp U.\exp V)$, whence immediately

$$(14)\qquad H_{r,s} = Q\left(\frac{U^r}{r!}\frac{V^s}{s!}\right) \quad \text{for } r+s \geqslant 1.$$

(2) We have

$$(15)\quad H(U,V) \equiv U + V + \tfrac{1}{2}[U,V] + \tfrac{1}{12}[U,[U,V]] + \tfrac{1}{12}[V,[V,U]] - \tfrac{1}{24}[U,[V,[U,V]]]$$

modulo $\sum_{n\geqslant 5} L^n(\{U, V\})$.

(3) $H_{0,n} = H_{n,0} = 0$ for every integer $n \neq 1$, whence

$$(16)\qquad H(U,0) = H(0,U) = U.$$

On the other hand, as $[U, -U] = 0$,

$$(17)\qquad H(U,-U) = 0.$$

5. SUBSTITUTIONS IN THE HAUSDORFF SERIES

As K is a field containing $\mathbf{Q}$, the Hausdorff series can be considered as a Lie formal power series with coefficients in K. Therefore, if $\mathfrak{g}$ is a complete Hausdorff filtered Lie algebra with $\mathfrak{g} = \bigcup_{\alpha>0} \mathfrak{g}_\alpha$, then, for a, b in $\mathfrak{g}$, a and b can be substituted for U and V in H (cf. no. 3 and § 2, no. 5, *Remark*).

In particular, let A be a complete Hausdorff filtered unital associative algebra. We write $\mathfrak{m} = \bigcup_{\alpha>0} A_\alpha$ and $\mathfrak{m}_\alpha = A_\alpha \cap \mathfrak{m}$ for $\alpha \in \mathbf{R}$; hence $\mathfrak{m}_\alpha = A_\alpha$ for $\alpha > 0$ and $\mathfrak{m}_\alpha = \mathfrak{m}$ for $\alpha \leqslant 0$. With the bracket $[a, b] = ab - ba$, $\mathfrak{m}$ is a complete Hausdorff filtered Lie algebra, to which the above can be applied. With this notation, we have the following result which completes Proposition 1 of no. 1.

PROPOSITION 3. *If $a \in \mathfrak{m}$, $b \in \mathfrak{m}$, then* $\exp H(a, b) = \exp a.\exp b$.

Let a, b be in $\mathfrak{m}$; there exists $\alpha > 0$ such that $a \in A_\alpha$ and $b \in A_\alpha$. Then there exists a continuous homomorphism θ of the Magnus algebra $\hat{A}(\{U, V\})$ into A mapping U to a and V to b (§ 5, no. 1, Proposition 1).

The restriction of θ to $\hat{L}(\{U, V\})$ is a continuous homomorphism of Lie algebras of $L(\{U, V\})$ into $\mathfrak{m}$ which maps U (resp. V) to a (resp. b). By formula (6) of no. 3, therefore $\theta(H) = H(a, b)$. It then suffices to apply the continuous homomorphism θ to the two sides of the relation

$$\exp H(U, V) = \exp U . \exp V$$

taking account of *Remark* 3 of no. 1.

Remark 1. If a and b commute, then $H_{r,s}(a, b) = 0$ for $r + s \geqslant 2$, for every homogeneous Lie polynomial of degree $\geqslant 2$ is zero at (a, b). Then $H(a, b) = a + b$ and Proposition 3 recovers the formula

$$\exp(a + b) = \exp a . \exp b.$$

PROPOSITION 4. *Let $\mathfrak{g}$ be a complete Hausdorff filtered Lie algebra such that $\mathfrak{g} = \bigcup_{\alpha > 0} \mathfrak{g}_\alpha$. The mapping $(a, b) \mapsto H(a, b)$ is a group law on $\mathfrak{g}$ compatible with the topology on $\mathfrak{g}$ under which 0 is the identity element and $-a$ is the inverse of a for all $a \in \mathfrak{g}$.*

The mapping $(a, b) \mapsto H(a, b)$ of $\mathfrak{g} \times \mathfrak{g}$ into $\mathfrak{g}$ is continuous (no. 3); as the mapping $a \mapsto -a$ is obviously continuous, it suffices to prove the relations

$$H(H(a, b), c) = H(a, H(b, c)) \tag{18}$$

$$H(a, -a) = 0 \tag{19}$$

$$H(a, 0) = H(0, a) = a \tag{20}$$

for a, b, c in $\mathfrak{g}$. By formula (7) of no. 3, it suffices to prove these formulae when a, b, c are three indeterminates and $\mathfrak{g} = \hat{L}(\{a, b, c\})$. Now the restriction of the exponential mapping to $\hat{L}(\{a, b, c\})$ is an injection into the Magnus algebra $\hat{A}(\{a, b, c\})$ and by Proposition 3:

$$\exp H(H(a, b), c) = \exp H(a, b) . \exp c = \exp a . \exp b . \exp c$$
$$\exp H(a, H(b, c)) = \exp a . \exp H(b, c) = \exp a . \exp b . \exp c$$
$$\exp H(a, -a) = \exp a . \exp(-a) = \exp(a - a) = \exp 0$$
$$\exp H(a, 0) = \exp a . \exp 0 = \exp a$$
$$\exp H(0, a) = \exp 0 . \exp a = \exp a.$$

This establishes relations (18) to (20).

Remarks. (2) Take $\mathfrak{g}$ to be the Lie algebra $\hat{L}(X)$. The group law introduced in the above proposition coincides with the law defined in no. 2. In other words,

$$a \,\mathrm{H}\, b = H(a, b) \quad \text{for } a, b \text{ in } \hat{L}(X); \tag{21}$$

thus the Hausdorff group law is given by the Hausdorff series.

(3) Let $\mathfrak{g}$ be a Lie algebra with the integral filtration $(\mathscr{C}^n\mathfrak{g})$ defined by the lower central series. Suppose that there exists $m \geqslant 1$ such that $\mathscr{C}^m\mathfrak{g} = \{0\}$. With the topology derived from the filtration $(\mathscr{C}^n\mathfrak{g})_{n\geqslant 1}$, the Lie algebra $\mathfrak{g}$ is Hausdorff, complete and even discrete. Then $P(a_1, \ldots, a_r) = 0$ for $a_1, \ldots, a_r$ in $\mathfrak{g}$ and for every homogeneous Lie polynomial P of degree $\geqslant m$; in particular, $H_{r,s}(a, b) = 0$ for $r + s \geqslant m$ and the series $H(a, b) = \sum_{r,s} H_{r,s}(a, b)$ has only a finite number of non-zero terms. The group law $(a, b) \mapsto H(a, b)$ on $\mathfrak{g}$ is then a polynomial mapping (§ 2, no. 4).

PROPOSITION 5. *Let* $K_{r,s}$ *be the component of* $H(U + V, -U)$ *of multidegree* (r, s). *Then*

$$K_{n,1}(U, V) = \frac{1}{(n+1)!}(\operatorname{ad} U)^n(V) \quad \textit{for } n \geqslant 0.$$

We write $K(U, V) = H(U + V, -U)$, $K_1(U, V) = \sum_{n\geqslant 0} K_{n,1}(U, V)$. We denote by L (resp. R) left (resp. right) multiplication by U on $\hat{A}(\{U, V\})$.

We can write

$$\begin{aligned} e^U V e^{-U} &= \sum_{p,q} \frac{U^p}{p!} V \frac{(-U)^q}{q!} \\ &= \sum_{n\geqslant 0} \frac{1}{n!}\left(\sum_{p+q=n} \frac{n!}{p!q!}(L^p(-R)^q).V\right) \\ &= \sum_{n\geqslant 0} \frac{1}{n!}(L - R)^n.V \end{aligned}$$

and therefore

$$e^U V e^{-U} = \sum_{n\geqslant 0} \frac{1}{n!}(\operatorname{ad} U)^n V. \tag{22}$$

We now calculate modulo the ideal $\sum_{m\geqslant 0} \sum_{n\geqslant 2} A^{m,n}(\{U, V\})$ of $A(\{U, V\})$. For all $n \geqslant 1$,

$$(U + V)^n \equiv U^n + \sum_{i=1}^{n-1} U^i V U^{n-1-i}$$

whence

$$\begin{aligned} (\operatorname{ad} U)(U + V)^n &\equiv \left((L - R)\sum_{i=1}^{n-1} L^i R^{n-i}\right).V \\ &\equiv (L^n - R^n).V \\ &\equiv U^n V - V U^n. \end{aligned}$$

Therefore

$$(\operatorname{ad} U).e^{U+V} \equiv e^U V - V e^U \tag{23}$$

summing over n.

On the other hand, $K_1(U, V) \equiv K(U, V)$ and $e^{K_1(U,V)} \equiv 1 + K_1(U, V)$ and hence

$$K_1 \equiv e^K - 1 \equiv e^{U+V}e^{-U} - 1$$

by Proposition 3. We deduce that

$$\begin{aligned}(\mathrm{ad}\,U)K_1 &\equiv Ue^{U+V}e^{-U} - e^{U+V}e^{-U}U \equiv (Ue^{U+V} - e^{U+V}U)e^{-U} \\ &\equiv (e^UV - Ve^U)e^{-U} && \text{by (23)} \\ &\equiv e^UVe^{-U} - V \\ &\equiv \sum_{n\geqslant 1} \frac{1}{n!}(\mathrm{ad}\,U)^nV && \text{by (22)} \\ &\equiv (\mathrm{ad}\,U)\Big(\sum_{n\geqslant 0} \frac{(\mathrm{ad}\,U)^n}{(n+1)!}V\Big).\end{aligned}$$

It then suffices to apply the *Remark* of § 2, no. 11.

§ 7. CONVERGENCE OF THE HAUSDORFF SERIES (REAL OR COMPLEX CASE)

In this paragraph we assume that K is one of the fields **R** or **C** with its usual absolute value. Recall that a normable algebra over K is a (not necessarily associative) algebra over K with a topology $\mathscr{T}$ with the following properties:

(1) $\mathscr{T}$ can be defined by a norm:

(2) the mapping $(x, y) \mapsto xy$ of $A \times A$ into A is continuous.

A normed algebra over K is an algebra A over K with a norm such that $\|xy\| \leqslant \|x\|\,\|y\|$ for all x, y in A.

We denote by $\mathfrak{g}$ a complete normable Lie algebra over K. We choose a norm on $\mathfrak{g}$ and a number $M > 0$ such that

$$\|[x, y]\| \leqslant M\|x\|\,\|y\| \quad \text{for } x, y \text{ in } \mathfrak{g}. \tag{1}$$

1. CONTINUOUS-POLYNOMIALS WITH VALUES IN $\mathfrak{g}$

Let I be a *finite* set and let $P(\mathfrak{g}^I; \mathfrak{g})$ (resp. $\hat{P}(\mathfrak{g}^I; \mathfrak{g})$) be the vector space of *continuous-polynomials* (resp. *formal power series with continuous components*) on $\mathfrak{g}^I$ with values in $\mathfrak{g}$. Recall (*Differentiable and Analytic Manifolds*, R, Appendix) that $P(\mathfrak{g}^I; \mathfrak{g})$ has a graduation of type $\mathbf{N}^I$ and that $\hat{P}(\mathfrak{g}^I; \mathfrak{g})$ is identified with the completion of the vector space $P(\mathfrak{g}^I; \mathfrak{g})$ with the topology defined by the filtration associated with the graduation of $P(\mathfrak{g}^I; \mathfrak{g})$. Moreover, $P(\mathfrak{g}^I; \mathfrak{g})$ is a graded Lie algebra with the bracket defined by $[f, g](\boldsymbol{x}) = [f(\boldsymbol{x}), g(\boldsymbol{x})]$ for

f, g in $P(\mathfrak{g}^I; \mathfrak{g})$, $x \in \mathfrak{g}^I$; this Lie algebra structure can be extended by continuity to $\hat{P}(\mathfrak{g}^I; \mathfrak{g})$ and makes it into a complete Hausdorff filtered Lie algebra.

By proposition 2 of § 6, no. 3, there exists one and only one continuous Lie algebra homomorphism $\phi_I : u \mapsto \tilde{u}$ of $\hat{L}(I)$ into $\hat{P}(\mathfrak{g}^I; \mathfrak{g})$ mapping the indeterminate of index i to pr_i for all $i \in I$, since $\mathrm{pr}_i \in P(\mathfrak{g}^I; \mathfrak{g})$. It follows that $\tilde{u} \in P(\mathfrak{g}^I; \mathfrak{g})$ for $u \in L(I)$; more precisely, when $u \in L(I)$, $\tilde{u}$ is just the polynomial mapping $(t_i) \mapsto u((t_i))$ of § 2, no. 4. On the other hand, clearly ϕ_I is compatible with the multigraduations of $L(I)$ and $P(\mathfrak{g}^I; \mathfrak{g})$. If $u = \sum_{\nu \in \mathbf{N}^I} u_\nu$, where $u_\nu \in L^\nu(I)$ for $\nu \in \mathbf{N}^I$, then

$$\tilde{u} = \sum_{\nu \in \mathbf{N}^I}^{\nu \in \mathbf{N}} \tilde{u}_\nu, \quad \text{where } \tilde{u}_\nu \in P_\nu(\mathfrak{g}^I; \mathfrak{g}).$$

Let $\boldsymbol{u} = (u_j)_{j \in J}$ be a *finite* family of elements of $\hat{L}(I)$, let $v \in \hat{L}(J)$ and let $w = v \circ \boldsymbol{u}$ (§ 6, no. 3). We write $\tilde{\boldsymbol{u}} = (\tilde{u}_j)_{j \in J} \in \mathfrak{Z}$. Then

$$\tilde{v} \circ \tilde{\boldsymbol{u}} = (v \circ \boldsymbol{u})^\sim. \tag{2}$$

This follows by extending by continuity formula (7) of § 6, no. 3 and from *Differentiable and Analytic Manifolds*, R, Appendix, no. 6.

2. GROUP GERM DEFINED BY A COMPLETE NORMED LIE ALGEBRA

Let $H = \sum_{r,s \geqslant 0} H_{r,s} \in \hat{L}(U, V)$ be the Hausdorff series (§ 6, no. 4, Definition 1). We shall show that the corresponding formal power series

$$\tilde{H} = \sum_{r,s \geqslant 0} \tilde{H}_{r,s} \in \hat{P}(\mathfrak{g} \times \mathfrak{g}, \mathfrak{g}) \tag{3}$$

is *convergent* (*Differentiable and Analytic Manifolds*, R, 3.1.1).

We introduce the following formal power series $\eta \in \mathbf{Q}[[U, V]]$

$$\eta(U, V) = -\log(2 - \exp(U + V)) \tag{4}$$

$$= \sum_{m \geqslant 1} \frac{1}{m} (\exp(U + V) - 1)^m \tag{5}$$

$$= \sum_{m \geqslant 1} \frac{1}{m} \sum_{\substack{r_1, \ldots, r_m \\ s_1, \ldots, s_m \\ r_i + s_i \geqslant 1}} \frac{U^{r_1}}{r_1!} \frac{V^{s_1}}{s_1!} \frac{U^{r_2}}{r_2!} \cdots \frac{V^{s_m}}{s_m!}. \tag{6}$$

Hence

$$\eta(U, V) = \sum_{r,s \geqslant 0} \eta_{r,s} U^r V^s, \tag{7}$$

where

$$(8)\qquad \eta_{r,s} = \sum_{m\geqslant 1}\frac{1}{m}\sum_{\substack{r_1+\cdots+r_m=r\\ s_1+\cdots+s_m=s\\ r_i+s_i\geqslant 1}}\frac{1}{r_1!\ldots r_m!s_1!\ldots s_m!}.$$

Now let u and v be two positive real numbers such that $u + v < \log 2$; then $0 \leqslant \exp(u + v) - 1 < 1$; the series derived from (5) and (6) by substituting u for U and v for V are convergent and the above calculations imply that

$$(9)\qquad \sum_{r,s\geqslant 0}\eta_{r,s}u^r v^s = -\log(2 - \exp(u + v)) < +\infty.$$

Let $r, s \geqslant 0$ and let $\|\tilde{\mathrm{H}}_{r,s}\|$ be the norm of the continuous-polynomial $\tilde{\mathrm{H}}_{r,s}$ (*Differentiable and Analytic Manifolds*, R, Appendix, no. 2).

Lemma 1. $\qquad \|\tilde{\mathrm{H}}_{r,s}\| \leqslant \mathrm{M}^{r+s-1}\eta_{r,s}.$

Let r_i, s_i be in **N** for $1 \leqslant i \leqslant m$, with $s_m = 1$; we write $r = \sum_i r_i$, $s = \sum_i s_i$ and consider the following element of $\mathrm{L}(\{\mathrm{U}, \mathrm{V}\})$:

$$\mathrm{Z} = \left(\left(\sum_{i=1}^{m-1}(\mathrm{ad\,U})^{r_i}(\mathrm{ad\,V})^{s_i}\right)(\mathrm{ad\,U})^{r_m}\right)(\mathrm{V}).$$

Then $\tilde{\mathrm{Z}} = f \circ p$, where f is the following $(r + s)$-linear mapping of $\mathfrak{g}^{r+s}$ into $\mathfrak{g}$:

$(x_1, \ldots, x_r, y_1, \ldots, y_s) \mapsto$

$$(\mathrm{ad}(x_1)\circ\cdots\circ\mathrm{ad}(x_{r_1})\circ\mathrm{ad}(y_1)\circ\cdots\circ\mathrm{ad}(y_{s_1})\circ\mathrm{ad}(x_{r_1+1})\circ\cdots\circ\mathrm{ad}(x_r))(y_s)$$

and where p is the following mapping of $\mathfrak{g}^2$ into $\mathfrak{g}^{r+s}$:

$$(x, y) \mapsto (\underbrace{x, \ldots, x}_{r}, \underbrace{y, \ldots, y}_{s});$$

hence $\|\tilde{\mathrm{Z}}\| \leqslant \|f\| \leqslant \mathrm{M}^{r+s-1}$ (*Differentiable and Analytic Manifolds*, R, Appendix). Applying these inequalities to the various terms on the right hand side of formula (9) of § 6, no. 4, we obtain:

$$(10)\qquad \|(\mathrm{H}'_{r,s})^{\sim}\| \leqslant \frac{\mathrm{M}^{r+s-1}}{r+s}\sum_{m\geqslant 1}\frac{1}{m}\sum_{\substack{r_1+\cdots+r_m=r\\ s_1+\cdots+s_{m-1}=s-1\\ r_1+s_1\geqslant 1,\ldots,r_{m-1}+s_{m-1}\geqslant 1}}\frac{1}{r_1!\ldots r_m!s_1!\ldots s_{m-1}!}.$$

A similar argument gives

$$(11)\qquad \|(\mathrm{H}''_{r,s})^{\sim}\| \leqslant \frac{\mathrm{M}^{r+s-1}}{r+s}\sum_{m\geqslant 1}\frac{1}{m}\sum_{\substack{r_1+\cdots+r_{m-1}=r-1\\ s_1+\cdots+s_{m-1}=s\\ r_1+s_1\geqslant 1,\ldots,r_{m-1}+s_{m-1}\geqslant 1}}\frac{1}{r_1!\ldots r_{m-1}!s_1!\ldots s_{m-1}!}$$

whence, by (8)

$$\|\tilde{H}_{r,s}\| < \eta_{r,s}\frac{M^{r+s-1}}{r+s} \leqslant \eta_{r,s}M^{r+s-1},$$

which proves the lemma.

PROPOSITION 1. *The formal power series* $\tilde{H}$ *is a convergent series* (*Differentiable and Analytic Manifolds*, R, 3.1.1)*; its domain of absolute convergence* (*Differentiable and Analytic Manifolds*, R, 3.1.4) *contains the open set*

$$\Omega = \left\{(x, y) \in \mathfrak{g} \times \mathfrak{g} \mid \|x\| + \|y\| < \frac{1}{M}\log 2\right\}.$$

Let u, v be two real numbers >0 such that $u + v < \frac{1}{M}\log 2$; then (Lemma 1)

$$(12)\quad M\sum_{r,s\geqslant 0}\|\tilde{H}_{r,s}\|u^r v^s$$
$$\leqslant \sum_{r,s\geqslant 0}\eta_{r,s}M^{r+s}u^r v^s = -\log(2 - \exp M(u+v)) < +\infty$$

by (9).

Let $h\colon \Omega \to \mathfrak{g}$ denote the *analytic function* (*Differentiable and Analytic Manifolds*, R, 3.2.9) defined by $\tilde{H}$, that is by the formula

$$(13)\qquad h(x, y) = \sum_{r,s\geqslant 0}\tilde{H}_{r,s}(x, y) = \sum_{r,s\geqslant 0}H_{r,s}(x, y) \quad \text{for } (x, y) \in \Omega.$$

This function is called the *Hausdorff function* of $\mathfrak{g}$ relative to M (or simply the Hausdorff function of $\mathfrak{g}$ if no confusion can arise). Note that $H_{r,s}(U, -U) = 0$ if $r + s \geqslant 2$ and hence

$$(14)\qquad h(x, -x) = 0 \quad \text{for } \|x\| < \frac{1}{2M}\log 2.$$

Similarly

$$(15)\qquad h(0, x) = h(x, 0) = x \quad \text{for } \|x\| < \frac{1}{M}\log 2.$$

PROPOSITION 2. *Let*

$$\Omega' = \left\{(x, y, z) \in \mathfrak{g} \times \mathfrak{g} \times \mathfrak{g} \mid \|x\| + \|y\| + \|z\| < \frac{1}{M}\log\frac{3}{2}\right\}.$$

If $(x, y, z) \in \Omega'$, *then*

$$(16)\quad (x, y) \in \Omega, \qquad (h(x, y), z) \in \Omega, \qquad (y, z) \in \Omega, \qquad (x, h(y, z)) \in \Omega$$

and

$$(17)\qquad h(h(x, y), z) = h(x, h(y, z)).$$

Let $(x, y, z) \in \Omega'$; clearly $(x, y) \in \Omega$ and $(y, z) \in \Omega$. Moreover:

$$\|h(x, y)\| \leqslant \sum_{r,s} \|\tilde{\mathrm{H}}_{r,s}\| \, \|x\|^r \|y\|^s,$$

and hence by (13)

$$\|h(x, y)\| \leqslant -\frac{1}{\mathrm{M}} \log(2 - \exp \mathrm{M}(\|x\| + \|y\|)).$$

Now $\mathrm{M}(\|x\| + \|y\|) < \log \frac{3}{2} - \mathrm{M}\|z\|$; we write $u = \exp(\mathrm{M}\|z\|)$; then $1 \leqslant u \leqslant \frac{3}{2}$ and

$$\begin{aligned} \mathrm{M}(\|h(x, y)\| + \|z\|) &< -\log(2 - \exp(\log \tfrac{3}{2} - \mathrm{M}\|z\|)) + \mathrm{M}\|z\| \\ &= -\log\left(2 - \frac{3}{2u}\right) + \log u = \log \frac{2u^2}{4u - 3} \\ &= \log\left(2 + \frac{2(u-1)(u-3)}{4u-3}\right) \leqslant \log 2. \end{aligned}$$

We see similarly that $(x, h(y, z)) \in \Omega$.

We now prove (17). In the Lie algebra $\hat{\mathrm{L}}(\{\mathrm{U}, \mathrm{V}, \mathrm{W}\})$,

$$\mathrm{H}(\mathrm{H}(\mathrm{U}, \mathrm{V}), \mathrm{W}) = \mathrm{H}(\mathrm{U}, \mathrm{H}(\mathrm{V}, \mathrm{W}))$$

by Proposition 4 of § 6, no. 5. By no. 1, formula (2), we therefore have in $\hat{\mathrm{P}}(\mathfrak{g} \times \mathfrak{g} \times \mathfrak{g}, \mathfrak{g})$ the relation

$$\tilde{\mathrm{H}} \circ (\tilde{\mathrm{H}} \times \mathrm{Id}_{\mathfrak{g}}) = \tilde{\mathrm{H}} \circ (\mathrm{Id}_{\mathfrak{g}} \times \tilde{\mathrm{H}}).$$

By *Differentiable and Analytic Manifolds*, R, 3.1.9, there exists a number $\varepsilon > 0$ such that formula (17) is true when $\|x\|$, $\|y\|$ and $\|z\|$ are $\leqslant \varepsilon$. But the functions $(x, y, z) \mapsto h(h(x, y), z)$ and $(x, y, z) \mapsto h(x, h(y, z))$ are analytic functions on Ω' with values in $\mathfrak{g}$ (*Differentiable and Analytic Manifolds*, R, 3.2.7). As Ω' is connected and they coincide in a neighbourhood of 0, they are equal (*Differentiable and Analytic Manifolds*, R, 3.2.5).

The above results imply:

Let α be a real number such that $0 < \alpha \leqslant \frac{1}{3\mathrm{M}} \log \frac{3}{2}$. Let

$$\mathrm{G} = \{x \in \mathfrak{g} \mid \|x\| < \alpha\},$$

$\Theta = \{(x, y) \in \mathrm{G} \times \mathrm{G} \mid h(x, y) \in \mathrm{G}\}$ and $m : \Theta \to \mathrm{G}$ be the restriction of h to Θ. Then:

(1) Θ is open in $\mathrm{G} \times \mathrm{G}$ and m is analytic.

(2) $x \in \mathrm{G}$ implies $(0, x) \in \Theta$, $(x, 0) \in \Theta$ and $m(0, x) = m(x, 0) = x$.

(3) $x \in \mathrm{G}$ implies $-x \in \mathrm{G}$, $(x, -x) \in \Theta$, $(-x, x) \in \Theta$ and

$$m(x, -x) = m(-x, x) = 0.$$

(4) Let x, y, z be elements of G such that $(x, y) \in \Theta$, $(m(x, y), z) \in \Theta$, $(y, z) \in \Theta$ and $(x, m(y, z)) \in \Theta$. Then $m(m(x, y), z) = m(x, m(y, z))$.

**In other words* (Chapter III, § 1), *if we write* $-x = \sigma(x)$, *the quadruple* $(G, 0, \sigma, m)$ *is a Lie group germ over* K.*

3. EXPONENTIAL IN COMPLETE NORMED ASSOCIATIVE ALGEBRAS

In this no. we denote by A a *complete normed unital associative algebra* (*General Topology*, Chapter IX, § 3, no. 7). Then $\|x.y\| \leqslant \|x\| . \|y\|$ for x, y in A.

Let I be a *finite* set and let $\hat{P}(A^I; A)$ be the vector space of *formal power series with continuous components* on A^I with values in A (*Differentiable and Analytic Manifolds*, R, Appendix, no. 5) with the algebra structure obtained by writing

$$f.g = m \circ (f, g) \quad \text{for } f, g \text{ in } \hat{P}(A^I; A),$$

where $m: A \times A \to A$ denotes multiplication on A. Arguing as in no. 1 and using Proposition 1 of § 5, no. 1, we define a continuous homomorphism of unital algebras $u \mapsto \tilde{u}$ of $\hat{A}(I)$ into $\hat{P}(A^I; A)$ mapping the indeterminate of index i to pr_i; this homomorphism extends the Lie algebra homomorphism of $\hat{L}(I)$ into $\hat{P}(A^I; A)$ defined in no. 1. If $u = \sum_{\nu} u_\nu$ with $u_\nu \in A^\nu(I)$ for $\nu \in \mathbf{N}^I$, then $\tilde{u} = \sum_{\nu} \tilde{u}_\nu$, where $\tilde{u}_\nu$ is the polynomial mapping $(t_i)_{i \in I} \mapsto u_\nu((t_i))$.

Let $\boldsymbol{u} = (u_j)_{j \in I}$ be a finite family of elements of $\hat{A}(I)$, let $v \in \hat{A}(J)$ and write $w = v \circ \boldsymbol{u}$ (§ 5, no. 1). Then

$$(v \circ \boldsymbol{u})^\sim = \tilde{v} \circ \tilde{\boldsymbol{u}}. \tag{18}$$

This follows by extending by continuity formula (2) of § 5, no. 1 and from *Differentiable and Analytic Manifolds*, R, Appendix, no. 6.

In particular we take $I = \{U\}$, identify A and $A^{(U)}$ and consider the images $\tilde{e}$ and $\tilde{l}$ of the series $e(U) = \sum_{n \geqslant 1} U^n/n!$ and $l(U) = \sum_{n \geqslant 1} (-1)^{n-1} U^n/n$ in $\hat{P}(A; A)$. Then $\|\tilde{U}^n\| \leqslant 1$ for $\|x_1 \ldots x_n\| \leqslant \|x_1\| \ldots \|x_n\|$ for $x_1, \ldots, x_n$ in A. Therefore the *radius of absolute convergence of* $\tilde{e}$ (resp. $\tilde{l}$) *is infinite* (resp. $\geqslant 1$).

We shall denote by e_A (resp. l_A) the analytic mapping of A into A (resp. of B into A, where B is the open unit ball of A) defined by the convergent series $\tilde{e}$ (resp. $\tilde{l}$) and we shall write $\exp_A(x) = 1 + e_A(x)$ (for $x \in A$) and

$$\log_A(x) = l_A(x - 1)$$

(for $x \in A$, $\|x - 1\| < 1$). Then

$$\exp_A x = \sum_{n \geqslant 0} \frac{x^n}{n!} \quad (x \in A) \tag{19}$$

$$(20)\qquad \log_A x = \sum_{n\geqslant 1} (-1)^{n-1}\frac{(x-1)^n}{n} \quad (x \in A, \|x-1\| < 1).$$

As $(e \circ l)(U) = (l \circ e)(U) = U$ (cf. § 6, no. 1), by (18) $\tilde{e} \circ \tilde{l} = \tilde{l} \circ \tilde{e} = \mathrm{Id}_A$. Therefore (*Differentiable and Analytic Manifolds*, R, 3.1.9)

$$(21)\qquad \exp_A(\log_A(x)) = x \quad (x \in A, \|x-1\| \leqslant 1)$$

$$(22)\qquad \log_A(\exp_A(x)) = x \quad (x \in A, \|x\| < \log 2)$$

for $\|x\| < \log 2$ implies $\|\exp_A(x) - 1\| \leqslant \exp \|x\| - 1 < 1$.

Finally we consider A as a complete normed Lie algebra. Then

$$\|[x, y]\| = \|xy - yx\| \leqslant 2\|x\|.\|y\|.$$

Proposition 1 of no. 2 implies that the domain of absolute convergence of the formal power series $\tilde{H}$ contains the set

$$\Omega = \{x, y) \in A \times A \mid \|x\| + \|y\| < \tfrac{1}{2}\log 2\}.$$

Hence $\tilde{H}$ defines an analytic function $h\colon \Omega \to A$. Then $h(x, y) = \sum_{r,s} H_{r,s}(x, y)$ (cf. § 3, no. 1, *Remark* 4).

PROPOSITION 3. *For* $\|x\| + \|y\| < \frac{1}{2}\log 2$,

$$(23)\qquad \exp_A x.\exp_A y = \exp_A h(x, y).$$

It follows from (18) and the relation $e^U e^V = e^{H(U,V)}$ that

$$m \circ (1 + \tilde{e}, 1 + \tilde{e}) = (1 + \tilde{e}) \circ \tilde{H}$$

in $\hat{P}(A \times A; A)$. We therefore deduce from *Differentiable and Analytic Manifolds*, R, 3.1.9 that (23) is true for (x, y) sufficiently close to $(0, 0)$, whence the proposition follows by analytic continuation (*Differentiable and Analytic Manifolds*, R, 3.2.5).

§ 8. CONVERGENCE OF THE HAUSDORFF SERIES (ULTRAMETRIC CASE)

In this paragraph we assume that K is a non-discrete *complete valued field of characteristic zero*, with an *ultrametric* absolute value. We denote by p the characteristic of the residue field of K (*Commutative Algebra*, Chapter VI, § 3, no. 2).

If $p \neq 0$, we write $a = |p|$; we know (*Commutative Algebra*, Chapter VI, § 6, nos. 2 and 3) that $0 < a < 1$ and that there exists one and only one valuation v on K with values in $\mathbf{R}$ whose restriction to $\mathbf{Q}$ is the p-adic valuation v_p and which is such that $|x| = a^{v(x)}$ for all $x \in K$. Also we write:

$$(1)\qquad \theta = \frac{1}{p-1}.$$

If $p = 0$, we denote by a a real number such that $0 < a < 1$ and by v a valuation on K with values in **R** such that $|x| = a^{v(x)}$ for all $x \in$ K (*loc. cit.*). Then $v(x) = 0$ for $x \in \mathbf{Q}^*$. Also we write:

(2) $$\theta = 0.$$

1. p-ADIC UPPER BOUNDS OF THE SERIES exp, log AND H

In this no. we assume that $p \neq 0$.

Lemma 1. *Let n be an integer $\geqslant 0$ and let $n = n_0 + n_1 p + \cdots + n_k p^k$, with $0 \leqslant n_i \leqslant p - 1$, be the p-adic expansion of n. Let $\mathrm{S}(n) = n_0 + n_1 + \cdots + n_k$. Then*

(3) $$v_p(n!) = \frac{n - \mathrm{S}(n)}{p - 1}.$$

$v_p(n!) = \sum_{i=1}^{n} v_p(i)$ and the number of integers i between 1 and n for which $v_p(i) \geqslant j$ is equal to the integral part $[n/p^j]$ of n/p^j. Then

$$v_p(n!) = \sum_{j \geqslant 0} j([n/p^j] - [n/p^{j+1}]) = \sum_{j \geqslant 1} [n/p^j].$$

As $[n/p^j] = \sum_{i \geqslant j} n_i p^{i-j}$, the lemma follows.

Lemma 2. $v(n) \leqslant v(n!) \leqslant (n-1)\theta$ *and* $v(n) \leqslant (\log n)/(\log p)$ *for every integer* $n \geqslant 1$.

$v(n!) = v_p(n!) = (n - \mathrm{S}(n))\theta \leqslant (n-1)\theta$ by Lemma 1.

On the other hand, $n \geqslant p^{v(n)}$, whence $v(n) \leqslant (\log n)/(\log p)$.

Let $\mathrm{I} = \{\mathrm{U}, \mathrm{V}\}$ be a set of two elements and let

$$\mathrm{H} = \sum_{r,s \geqslant 0} \mathrm{H}_{r,s}(\mathrm{U}, \mathrm{V}) \in \hat{\mathrm{L}}_{\mathbf{Q}}(\mathrm{I})$$

be the Hausdorff series (§ 6, no. 4, Definition 1). Let $\mathbf{Z}_{(p)}$ be the local ring of **Z** relative to the prime ideal (p) and $(e_b)_{b \in \mathrm{B}}$ a basis of $\mathrm{L}_{\mathbf{Z}_{(p)}}(\mathrm{I})$ over **Z** (§ 2, no. 11, Theorem 1). It is also a basis of $\mathrm{L}_{\mathbf{Q}}(\mathrm{I})$ over **Q**.

PROPOSITION 1. *Let r and s be two integers $\geqslant 0$. If $\mathrm{H}_{r,s} = \sum_{b \in \mathrm{B}} \lambda_b e_b$, where $\lambda_b \in \mathbf{Q}$, is the decomposition of* H *with respect to the basis $(e_b)_{b \in \mathrm{B}}$, then*

(4) $$v_p(\lambda_b) \geqslant -(r + s - 1)\theta \quad \textit{for all } b \in \mathrm{B}.$$

The ring $\mathrm{A}_{\mathbf{Z}_{(p)}}(\mathrm{I})$ is identified with the sub-$\mathbf{Z}_{(p)}$-module of $\mathrm{A}_{\mathbf{Q}}(\mathrm{I})$ generated by the words $w \in \mathrm{Mo}(\mathrm{I})$. As $\mathrm{L}_{\mathbf{Z}_{(p)}}(\mathrm{I})$ is a direct factor of $\mathrm{A}_{\mathbf{Z}_{(p)}}(\mathrm{I})$,

(5) $$\mathrm{L}_{\mathbf{Z}_{(p)}}(\mathrm{I}) = \mathrm{A}_{\mathbf{Z}_{(p)}}(\mathrm{I}) \cap \mathrm{L}_{\mathbf{Q}}(\mathrm{I}).$$

Let f be the integer such that $f \leqslant (r+s-1)\theta < f+1$. Relation (4) is equivalent to $v_p(\lambda_b) \geqslant -f$ for all $b \in B$, that is $H_{r,s} \in p^{-f}L_{Z_{(p)}}$. But this is equivalent also by (5) to $H_{r,s} \in p^{-f}A_{Z_{(p)}}(I)$.

By formula (11) of § 6, no. 4, it suffices to show that, for every integer $m \geqslant 1$ and all integers $r_1, \ldots, r_m, s_1, \ldots, s_m$ such that

$$(6) \qquad r_1 + \cdots + r_m = r, \qquad s_1 + \cdots + s_m = s, \\ r_i + s_i \geqslant 1 \quad \text{for } 1 \leqslant i \leqslant m,$$

we have

$$(7) \qquad v_p(m.r! \ldots r_m! s_1! \ldots s_m!) \leqslant f.$$

But by Lemma 2, $v_p(r_i! s_i!) \leqslant (r_i + s_i - 1)\theta$ and $v_p(m) \leqslant v_p(m!) \leqslant (m-1)\theta$; the left hand side of (7) is therefore bounded above by

$$\theta(m - 1 + \sum_{i=1}^{m} (r_i + s_i - 1)) = \theta(r + s - 1);$$

as it is an integer, it is $\leqslant f$, which completes the proof.

2. NORMED LIE ALGEBRAS

DEFINITION 1. *A normed Lie algebra over* K *is a Lie algebra with a norm such that*

$$(8) \qquad \|x + y\| \leqslant \sup(\|x\|, \|y\|)$$

$$(9) \qquad \|[x, y]\| \leqslant \|x\| \cdot \|y\|$$

for all x, y in $\mathfrak{g}$.

Throughout the rest of this paragraph, $\mathfrak{g}$ denotes a complete normed Lie algebra.

For every finite set I we define as in § 7, no. 1 a continuous Lie algebra homomorphism $u \mapsto \tilde{u}$ of $\hat{L}(I)$ into $\hat{P}(\mathfrak{g}^I; \mathfrak{g})$. We see as in § 7 that if $u = \sum_\nu u_\nu$, with $u_\nu \in L^\nu(I)$ for $\nu \in \mathbf{N}^I$, then $\tilde{u} = \sum_\nu \tilde{u}_\nu$, where $\tilde{u}_\nu$ is the polynomial mapping $(t_i)_{i \in I} \mapsto u_\nu((t_i))$ defined in § 2, no. 4. The composition formula (2) of § 7, no. 1, remains valid.

3. GROUP DEFINED BY A COMPLETE NORMED LIE ALGEBRA

Let $H = \sum_{r \geqslant 0} H_{r,s} \in \hat{L}(\{U, V\})$ be the Hausdorff series (§ 6, no. 4, Definition 1). We shall show that the corresponding formal power series with continuous components

$$(10) \qquad \tilde{H} = \sum_{r,s \geqslant 0} \tilde{H}_{r,s} \in \hat{P}(\mathfrak{g} \times \mathfrak{g}, \mathfrak{g})$$

is *convergent* (*Differentiable and Analytic Manifolds*, R, 4.1.1).

Let $r \geqslant 0$, $s \geqslant 0$ be such that $r + s \neq 0$ and let $\|\tilde{H}_{r,s}\|$ be the norm of the continuous polynomial $\tilde{H}_{r,s}$ (*Differentiable and Analytic Manifolds*, R, Appendix, no. 2).

Lemma 3. $$\|\tilde{H}_{r,s}\| \leqslant a^{-(r+s-1)\theta}.$$

Let B be a Hall set relative to I and let $H_{r,s} = \sum_{b \in B} \lambda_b e_b$ be the decomposition of $H_{r,s}$ with respect to the corresponding basis of $L(\{U, V\})$. Then

(11) $$|\lambda_b| \leqslant a^{-(r+s-1)\theta}.$$

This is trivial for $p = 0$, for $\lambda_b \in \mathbf{Q}$; and it follows from Proposition 1 of no. 1 for $p \neq 0$.

Moreover,

(12) $$\|\tilde{e}_b\| \leqslant 1 \quad \text{for } b \in B.$$

We show more generally by induction on n that, for every alternant b of degree n in the two indeterminates U and V (§ 2, no. 6), $\|\tilde{b}\| \leqslant 1$. If $n = 1$, $\tilde{b}$ is one of the projections of $\mathfrak{g} \times \mathfrak{g}$ onto $\mathfrak{g}$ and hence is of norm $\leqslant 1$; if $n > 1$, there exist two alternants b_1 and b_2 of degrees $< n$ such that $b = [b_1, b_2]$. As the mapping $\gamma: (x, y) \mapsto [x, y]$ of $\mathfrak{g} \times \mathfrak{g}$ into $\mathfrak{g}$ is bilinear of norm $\leqslant 1$, we have (*Differentiable and Analytic Manifolds*, R, Appendix, no. 4)

(13) $$\|\tilde{b}\| = \|\gamma \circ (\tilde{b}_1, \tilde{b}_2)\| \leqslant \|\tilde{b}_1\| . \|\tilde{b}_2\| \leqslant 1.$$

Relations (11) and (12) imply the lemma.

PROPOSITION 2. *The formal power series* $\tilde{H}$ *is a convergent series* (*Differentiable and Analytic Manifolds*, R, 4.1.1). *If* G *is the ball* $\{x \in \mathfrak{g} \mid \|x\| < a^\theta\}$, *the domain of absolute convergence of* $\tilde{H}$ (*Differentiable and Analytic Manifolds*, R, 4.1.3) *contains* G × G.

If u and v are two real numbers >0 such that $u < a^\theta$ and $v < a^\theta$, then (Lemma 3)

(14) $$\|\tilde{H}_{r,s}\| u^r v^s \leqslant a^\theta (ua^{-\theta})^r (va^{-\theta})^s$$

and $\|H_{r,s}\| u^r v^s$ tends to 0 when $r + s$ tends to infinity.

We denote by $h: G \times G \to \mathfrak{g}$ the *analytic function* (*Differentiable and Analytic Manifolds*, R, 4.2.4) defined by $\tilde{H}$, that is by the formula

(15) $$h(x, y) = \sum_{r, s \geqslant 0} \tilde{H}_{r,s}(x, y) = \sum_{r, s \geqslant 0} H_{r,s}(x, y) \quad \text{for } (x, y) \in G \times G.$$

This function is called the *Hausdorff function* of $\mathfrak{g}$.

Let $(x, y) \in G \times G$. Then

(16) $$\|\tilde{H}_{r,s}(x, y)\| \leqslant \sup(\|x\|, \|y\|)$$

(17) $$\|h(x, y)\| \leqslant \sup(\|x\|, \|y\|).$$

(17) follows immediately from (16) and (16) is trivial for $r = s = 0$; if $r \geqslant 1$, then

$$\begin{aligned}\|\tilde{H}_{r,s}(x, y)\| &\leqslant \|\tilde{H}_{r,s}\| \, \|x\|^r \|y\|^s \\ &\leqslant \|x\| \left(\frac{\|x\|}{a^\theta}\right)^{r-1} \left(\frac{\|y\|}{a^\theta}\right)^s \\ &\leqslant \|x\|;\end{aligned}$$

we argue similarly if $s \geqslant 1$.

In particular, $\|h(x, y)\| < a^\theta$ for $(x, y) \in G \times G$.

PROPOSITION 3. *Let* G *be the ball* $\{x \in \mathfrak{g} \mid \|x\| < a^\theta\}$. *The analytic mapping*

$$h\colon G \times G \to G$$

makes G *into a group in which* 0 *is the identity element and* $-x$ *is the inverse of* x *for all* $x \in G$. *Moreover, if* R *is a real number such that* $0 < R < a^\theta$, *the ball*

$$\{x \in \mathfrak{g} \mid \|x\| < R\}$$

(resp. $\{x \in \mathfrak{g} \mid \|x\| \leqslant R\}$) *is an open subgroup of* G.

As $H(U, -U) = 0$ and $H(0, U) = H(U, 0) = U$, $h(x, -x) = 0$ and

$$h(0, x) = h(x, 0) = x$$

for all $x \in G$. It therefore remains to prove the associativity formula

$$(18) \qquad h(h(x, y), z) = h(x, h(y, z)) \quad \text{for } x, y, z \text{ in } G.$$

As

$$H(H(U, V), W) = H(U, H(V, W))$$

in $\hat{L}(\{U, V, W\})$ (§ 6, no. 5, Proposition 4), we have

$$(19) \qquad \tilde{H} \circ (\tilde{H} \times \mathrm{Id}_{\mathfrak{g}}) = \tilde{H} \circ (\mathrm{Id}_{\mathfrak{g}} \times \tilde{H})$$

in $\hat{P}(\mathfrak{g} \times \mathfrak{g} \times \mathfrak{g}; \mathfrak{g})$ (no. 2) and (19) implies (18) by (16) and *Differentiable and Analytic Manifolds*, R, 4.1.5.

In other words (Chapter III, § 1), G with the Hausdorff function is a Lie group.

4. EXPONENTIAL IN COMPLETE NORMED ASSOCIATIVE ALGEBRAS

In this no. A will denote a unital associative algebra with a norm $x \mapsto \|x\|$ satisfying the conditions:

$$\begin{aligned}\|x + y\| &\leqslant \sup(\|x\|, \|y\|) \\ \|xy\| &\leqslant \|x\| \cdot \|y\| \\ \|1\| &= 1\end{aligned}$$

for x, y in A, and *complete* with this norm. The results of the second and third paragraphs of § 7, no. 3, remain valid.

We take $I = \{U\}$ and consider the images $\tilde{e}$ and $\tilde{l}$ of the series $e(U) = \sum_{n \geqslant 1} \frac{U^n}{n!}$ and $l(U) = \sum_{n \geqslant 1} (-1)^{n-1} \frac{U^n}{n}$ in $\hat{P}(A; A)$. Then:

$$\left\| \left(\frac{U^n}{n!}\right)^{\sim} \right\| \leqslant a^{-(n-1)\theta} \tag{20}$$

$$\left\| \left(\frac{U^n}{n}\right)^{\sim} \right\| \leqslant a^{-\frac{\log n}{\log p}} \tag{21}$$

by Lemma 2 of no. 1. Hence the *radius of absolute convergence* of the series $\tilde{e}$ (resp. $\tilde{l}$) is $\geqslant a^\theta$ (resp. $\geqslant 1$) (*Differentiable and Analytic Manifolds*, R, 4.1.3). For $R > 0$, let $G_R = \{x \in A \mid \|x\| < R\}$; we write $G = G_\theta$. The series $\tilde{e}$ (resp. $\tilde{l}$) defines an analytic mapping e_A (resp. l_A) of G (resp. G_1) into A. We write:

$$\exp_A(x) = 1 + e_A(x) = \sum_{n \geqslant 0} \frac{x^n}{n!} \qquad \text{for } x \in G \tag{22}$$

$$\log_A(x) = l_A(x-1) = \sum_{n \geqslant 1} (-1)^{n-1} \frac{(x-1)^n}{n} \qquad \text{for } x - 1 \in G_1 \tag{23}$$

(we omit the index A when no confusion can arise). For $x \in G_R$ and $n \geqslant 1$,

$$\left\| \frac{x^n}{n} \right\| \leqslant \left\| \frac{x^n}{n!} \right\| < R^n a^{-(n-1)\theta} = R\left(\frac{R}{a^\theta}\right)^{n-1} \tag{24}$$

and hence $e_A(G_R) \subset G_R$, $l_A(G_R) \subset G_R$ for $R \leqslant a^\theta$.

Proposition 4. *Let R be a real number such that $0 < R \leqslant a^\theta$. The mapping $\exp_A$ defines an analytic isomorphism of G_R onto $1 + G_R$ and the inverse isomorphism is the restriction of $\log_A$ to $1 + G_R$.*

$e(l(X)) = l(e(X)) = X$. By (20), (21) and *Differentiable and Analytic Manifolds*, R, 4.1.5, we deduce that $e_A(l_A(x)) = l_A(e_A(x))$ for $x \in G_R$. Then

$$\exp_A(\log_A x) = x \quad \text{for } x \in 1 + G_R$$
$$\log_A(\exp_A x) = x \quad \text{for } x \in G_R$$

which completes the proof.

If A is given the bracket $[x, y] = xy - yx$, A becomes a complete normed Lie algebra, for $\|xy - yx\| \leqslant \sup(\|xy\|, \|yx\|) \leqslant \|x\| . \|y\|$. Proposition 2 of no. 3 implies that the domain of absolute convergence of $\tilde{H}$ contains $G \times G$ and $\tilde{H}$ therefore defines an analytic function $h: G \times G \to A$; then

$$h(x, y) = \sum_{r, s \geqslant 0} H_{r,s}(x, y). \tag{25}$$

PROPOSITION 5. *For x, y in* G,

(26) $$\exp_A . \exp_A y = \exp_A h(x, y).$$

$e^U e^V = e^{H(U, V)}$ and hence

$$m \circ (1 + \tilde{e}, 1 + \tilde{e}) = (1 + \tilde{e}) \circ \tilde{H}$$

in $\tilde{H}(A \times A; A)$ (where m denotes multiplication on A). The proposition then follows from Proposition 2, Lemma 3 and *Differentiable and Analytic Manifolds*, R, 4.1.5.

APPENDIX

MÖBIUS FUNCTION

Let n be an integer $\geqslant 1$. If n is divisible by the square of a prime number, we write $\mu(n) = 0$. If n is not divisible by the square of a prime number, we write $\mu(n) = (-1)^k$, where k is the number of prime divisors of n. The function $\mu: \mathbf{N}^* \to \{-1, 0, 1\}$ thus defined is called the *Möbius function*.

Recall that given two integers $n_1 \geqslant 1$, $n_2 \geqslant 1$, we write $n_1 | n_2$ if n_1 divides n_2.

PROPOSITION. (i) *The function μ is the unique mapping of $\mathbf{N}^*$ into $\mathbf{Z}$ such that $\mu(1) = 1$ and*

(1) $$\sum_{d|n} \mu(d) = 0$$

for every integer $n > 1$.

(ii) *Let s and t be two mappings of $\mathbf{N}^*$ into a commutative group written additively. In order that*

(2) $$s(n) = \sum_{d|n} t(d) \quad \textit{for every integer } n \geqslant 1,$$

it is necessary and sufficient that

(3) $$t(n) = \sum_{d|n} \mu(d) s\left(\frac{n}{d}\right) \quad \textit{for every integer } n \geqslant 1.$$

The uniqueness assertion in (i) is obvious, for (1) allows us to determine $\mu(n)$ by induction on n. We show that the function μ satisfies (1). Let n be an integer >1. Let P be the set of prime divisors of n and let $n = \prod_{p \in P} p^{v_p(n)}$ be the decomposition of n into prime factors. If d is a divisor of n, then $\mu(d) = 0$ unless d is of the form $\prod_{p \in H} p$, where H is a subset of P. Then

$$\sum_{d|n} \mu(d) = \sum_{H \subset P} (-1)^{\text{Card } H}$$
$$= \sum_{k=0}^{\text{Card } P} \binom{n}{k} (-1)^k = (1 - 1)^{\text{Card } P} = 0.$$

Let s and t be two mapping of $\mathbf{N}^*$ into a commutative group written additively. Let $n \in \mathbf{N}^*$. If (2) holds, then

$$\sum_{d|n} \mu(d) s\left(\frac{n}{d}\right) = \sum_{d|n} \mu(d) \sum_{\delta|(n/d)} t(\delta) = \sum_{d\delta|n} \mu(d) t(\delta)$$

$$= \sum_{\delta|n} t(\delta) \sum_{d|(n/\delta)} \mu(d) = t(n).$$

Conversely, if (3) holds, then

$$\sum_{d|n} t(d) = \sum_{d|n} \sum_{\delta|d} \mu(\delta) s\left(\frac{d}{\delta}\right) = \sum_{d|n} s(d) \sum_{\delta|(n/d)} \mu(\delta) = s(n),$$

which completes the proof.

Formula (3) is called the *Möbius inversion formula.*

EXERCISES

§ 1

1. Suppose that K is a field of characteristic 0. Let E be a cocommutative bigebra of finite rank over K. Show that $P(E) = \{0\}$ (apply Theorem 1 of no. 6).

2. Let E be a bigebra such that $P(E) = \{0\}$ and let $(E_n)_{n \geqslant 0}$ be a filtration of E compatible with its bigebra structure. Show by induction on n that $E_n^+ = \{0\}$ for all $n \geqslant 0$ and deduce that $E^+ = \{0\}$, i.e. that E reduces to K.

3. Let G be a monoid and $E = K[G]$ its algebra (*Algebra*, Chapter III, § 2, no. 6).

(*a*) Show that there exists on E one and only one cogebra structure such that $c(g) = g \otimes g$ for all $g \in G$; this structure is compatible with the algebra structure on E and makes E into a cocommutative bigebra whose counit ε is such that $\varepsilon(g) = 1$ for all $g \in G$.

(*b*) Show that every primitive element of E is zero. Deduce that, if $G \neq \{e\}$, the bigebra E admits no filtration compatible with its bigebra structure (apply Exercise 2).

(*c*) Suppose that K is an integral domain. Show that the elements of G are the only non-zero elements $x \in E$ such that $c(x) = x \otimes x$. Show that E has an inversion i (cf. *Algebra*, Chapter III, § 11, Exercise 4) if and only if G is a group and that in that case $i(g) = g^{-1}$ for all $g \in G$.

4. Let E be a cocommutative bigebra and let G be the set of $g \in E$ such that $\varepsilon(g) = 1$ and $c(g) = g \otimes g$. Show that G is stable under multiplication and that it is a group if E has an inversion. Show that, if K is a field, the elements of G are linearly independent over K.

5. Let E be a bigebra and let $E' = \mathrm{Hom}(E, K)$ be its dual with the algebra structure derived by duality from the cogebra structure on E (*Algebra*, Chapter III, § 11, no. 1). Let $\mathfrak{m}$ be the kernel of the homomorphism $u \mapsto u(1)$ of

E′ onto K; it is an ideal of E′. Show that, if $u \in \mathfrak{m}^2$ and $x \in P(E)$, then $u(x) = 0$. When K is a field show that P(E) is the orthogonal of $\mathfrak{m}^2$ in E^+; deduce that $\dim P(E) \leqslant \dim \mathfrak{m}/\mathfrak{m}^2$ and that we have equality if E is of finite rank over K.

6. Let E be a bigebra and let $u \in E' = \mathrm{Hom}(E, K)$ (cf. Exercise 5). Let $g \in E$ be such that $\varepsilon(g) = 1$ and $c(g) = g \otimes g$. Show that the mapping $u \mapsto u(g)$ is an algebra homomorphism of E′ into K and that the mapping $y \mapsto gy$ is an endomorphism of the cogebra E; show that this endomorphism is an automorphism if E has an inversion.

7. Suppose that K is a field. Let E be a bigebra satisfying the following conditions:
(i) E is cocommutative;
(ii) E has an inversion (*Algebra*, Chapter III, § 11, Exercise 4);
(iii) $P(E) = \{0\}$;
(iv) E is of finite rank over K.
Let E′ denote the dual K-algebra of the cogebra E. Let G denote the set of elements $g \in E$ such that $\varepsilon(g) = 1$ and $c(g) = g \otimes g$; it is a group (cf. Exercise 4).

(*a*) Let $g \in G$ and let $\mathfrak{m}_g$ be the kernel of the homomorphism $u \mapsto u(g)$ of E′ into K. Show that $\mathfrak{m}_g = \mathfrak{m}_g^2$ (reduce it to the case $g = 1$ using Exercise 6; then apply Exercise 5).

(*b*) Suppose that K is algebraically closed. Show that the ideals $\mathfrak{m}_g$ are the only maximal ideals of E′; deduce that their intersection is {0}, that E′ is identified with the product K^G and that E is identified with the bigebra K[G] of the finite group G (cf. Exercise 3).

8. Show that the enveloping bigebra of a Lie algebra $\mathfrak{g}$ has an inversion i and that $i(x) = -x$ for all $x \in \sigma(\mathfrak{g})$.

9. Suppose that K is a $\mathbf{Q}$-algebra. Let $\mathfrak{g}$ be a Lie algebra admitting a K-basis and let U be its enveloping bigebra with the canonical filtration $(U_n)_{n \geqslant 0}$. Let $x \in U^+$ and let n be an integer $\geqslant 1$. Show that x belongs to U_n^+ if and only if $c^+(x)$ belongs to $\sum_{\substack{i+j=n \\ i,j \geqslant 1}} \mathrm{Im}(U_i^+ \otimes U_j^+)$.

10. Suppose that K is a $\mathbf{Q}$-algebra. Let E be a cocommutative K-bigebra with a filtration compatible with its bigebra structure. Show that the morphism

$$f_E \colon U(P(E)) \to E$$

defined in no. 4, Proposition 8 is an isomorphism, if P(E) is assumed to be a free K-module (same proof as for Theorem 1).

¶ 11. Let E be a bigebra and let I be a finite set. We are given a basis (e_α)

of E indexed by the elements α of $\mathbf{N}^{\mathrm{I}}$ and such that:

(i) $e_0 = 1$;

(ii) $c(e_\gamma) = \sum_{\alpha+\beta=\gamma} e_\alpha \otimes e_\beta$ for all $\gamma \in \mathbf{N}^{\mathrm{I}}$.

Condition (ii) implies that E is isomorphic as a cogebra to the cogebra $\mathrm{TS}(\mathrm{K}^{\mathrm{I}})$ of symmetric tensors of K^{I} (cf. *Algebra,* Chapter IV, § 5, no. 7).

(*a*) Let $\mathrm{E}' = \mathrm{Hom}(\mathrm{E}, \mathrm{K})$ be the dual algebra of E and let $\Lambda = \mathrm{K}[[(\mathrm{X}_\iota)_{\iota\in\mathrm{I}}]]$ be the algebra of formal power series in indeterminates X_ι indexed by I. For all $u \in \mathrm{E}'$ let ϕ_u be the formal power series

$$\sum_\alpha u(e_\alpha)\boldsymbol{x}^\alpha,$$

where $\boldsymbol{x}_\alpha = \prod_{\iota\in\mathrm{I}} \mathrm{X}_\iota^{\alpha(\iota)}$.

Show that $u \mapsto \phi_u$ is an isomorphism of E′ onto Λ and that this isomorphism maps the topology of simple convergence on E′ to the product topology on Λ.

(*b*) Let $c_{\alpha\beta\gamma}$ be the constants of structure of the algebra E relative to the basis (e_α). Then

$$e_\alpha e_\beta = \sum_\gamma c_{\alpha\beta\gamma} e_\gamma.$$

We write $\boldsymbol{x}$ instead of $(\mathrm{X}_\iota)_{\iota\in\mathrm{I}}$ and $\boldsymbol{y}$ instead of $(\mathrm{Y}_\iota)_{\iota\in\mathrm{I}}$, where the Y_ι are new indeterminates. Show that there exists a family $\boldsymbol{f}(\boldsymbol{x},\boldsymbol{y}) = (f_\iota(\boldsymbol{x},\boldsymbol{y}))_{\iota\in\mathrm{I}}$ of formal power series in the variables $\boldsymbol{x}, \boldsymbol{y}$ such that

$$\boldsymbol{f}(\boldsymbol{x},\boldsymbol{y})^\gamma = \sum_{\alpha,\beta} c_{\alpha\beta\gamma}\boldsymbol{x}^\alpha\boldsymbol{y}^\beta \quad \text{for all } \gamma \in \mathbf{N}^{\mathrm{I}},$$

where we write, as above, $\boldsymbol{x}^\alpha = \prod_\iota \mathrm{X}_\iota^{\alpha(\iota)}$ and similarly for $\boldsymbol{y}^\beta$ and $\boldsymbol{f}(\boldsymbol{x},\boldsymbol{y})^\gamma$.

Show that $\boldsymbol{f}(\boldsymbol{x},\boldsymbol{y})$ is a *formal group law* over K of dimension $n = \mathrm{Card}(\mathrm{I})$, in the sense of Chapter I, § 1, Exercise 24†; define a Lie algebra isomorphism of this formal group law onto the Lie algebra P(E), which has basis the e_α with $|\alpha| = 1$.

(*c*) Conversely, show that every formal group law over K in the $\boldsymbol{x}$ and $\boldsymbol{y}$ can be obtained by the above procedure, which is unique up to isomorphism.

(*d*) When K is a $\mathbf{Q}$-algebra, apply the above to the enveloping bigebra of a Lie algebra $\mathfrak{g}$ with a finite basis over K. Deduce the existence (and uniqueness, up to isomorphism) of a formal group law with $\mathfrak{g}$ as Lie algebra.

(*e*) Give explicitly the bigebra corresponding to the formal group law in one parameter $f(x, y) = x + y$ (resp. $f(x, y) = x + y + xy$).

¶ 12. Suppose that K is a field of characteristic $p > 0$.

(*a*) Let $\mathfrak{g}$ be a Lie p-algebra (Chapter I, § 1, Exercise 20) and let $\tilde{\mathrm{U}}$ be its

† The exercise in question assumes that K is a field, but nothing is lost if this hypothesis is suppressed.

restricted enveloping bigebra (Chapter I, § 2, Exercise 6). Show that there exists on $\tilde{U}$ one and only one bigebra structure which is compatible with its algebra structure and for which the elements of $\mathfrak{g}$ are primitive. Show that $\tilde{U}$ is cocommutative and that $P(\tilde{U}) = \mathfrak{g}$.

(*b*) If n is an integer $\geqslant 0$, let $\tilde{U}_n$ denote the vector subspace of $\tilde{U}$ generated by the products $x_1 \ldots x_n$ where $x_i \in \mathfrak{g}$ for all i. Show that $(\tilde{U}_n)_{n \geqslant 0}$ is a filtration of $\tilde{U}$ compatible with its bigebra structure.

(*c*) Let E be a bigebra and let $\mathfrak{g} = P(E)$ be the Lie algebra of its primitive elements. The mapping $x \mapsto x^p$ leaves $\mathfrak{g}$ stable and gives $\mathfrak{g}$ a Lie p-algebra structure. Show that the canonical injection $\mathfrak{g} \to E$ can be extended uniquely to a bigebra morphism $\tilde{U} \to E$ and that this morphism is injective (use the fact that, if $(x_i)_{i \in I}$ is a basis of $\mathfrak{g}$ with a total ordering, the monomials $\prod x_i^{n_i}$ $(0 \leqslant n_i < p)$ form a basis of U (*loc. cit.*) and argue as in the proof of Lemma 2).

§ 2

1. Let $\mathfrak{g}$ be a free Lie algebra with basic family $(x_1, \ldots, x_n, \ldots)$. Let $\mathfrak{g}_n$ denote the subalgebra of $\mathfrak{g}$ generated by $x_1, \ldots, x_n$ and let $\mathfrak{b}_n$ denote the smallest ideal of $\mathfrak{g}_n$ containing x_n.

(*a*) Show that $\mathfrak{b}_n$ is the submodule of $\mathfrak{g}_n$ generated by the

$$(\mathrm{ad}(x_{i_1}) \circ \cdots \circ \mathrm{ad}(x_{i_k}))(x_n),$$

where $k \geqslant 0$ and $i_h \leqslant n$ for all h.

(*b*) Show that $\mathfrak{g}_n = \mathfrak{g}_{n-1} \oplus \mathfrak{b}_n$; deduce that $\mathfrak{g}$ is the direct sum of the $\mathfrak{b}_n$ for $n \geqslant 1$.

(*c*) Show, using (*a*) and (*b*), that the K-module $\mathfrak{g}$ is generated by the elements $[x_{i_1}, [x_{i_2}, \ldots, [x_{i_{k-1}}, x_{i_k}] \ldots]]$, where $k \geqslant 0$ and $i_h \leqslant i_k$ for $h < k$.

¶ 2. Let X be a countable set with at least two elements and let $\mathfrak{H}$ be the set of subsets of M(X) which are Hall sets (cf. Definition 2). Show that $\mathrm{Card}(\mathfrak{H}) = 2^{\aleph_0}$.

3. Let X be a set with a total ordering. Show that there exists a Hall set relative to X such that $H \cap M^3(X)$ consists of the products $z(yx)$ with $y < x$, $y \leqslant z$ and that $H \cap M^4(X)$ consists of the products $w(z(yx))$ with $w \geqslant z \geqslant y$, $y < x$ and the products $(ab)(cd)$ with $a < b$, $c < d$ and either $a < c$ or $a = c$ and $b < d$.

4. (*a*) Show that the Lie algebra defined by the presentation

$$\{x, y; [x, [x, y]] = [y, [x, y]] = 0\}$$

admits the basis $(x, y, [x, y])$. Define an embedding of this algebra in a matrix algebra.

(*b*) Do the same for the Lie algebra defined by the presentation

$$\{x, y; [x, [x, y]] - 2x = [y, [x, y]] + 2y = 0\}.$$

(*c*) Show that the Lie algebra with presentation

$$\{x, y, z; [x, y] - x = [y, z] - y = [z, x] - z = 0\}$$

reduces to 0.

5. Let $\mathfrak{g}$ be a free Lie algebra and let $\mathfrak{r}$ be an ideal of $\mathfrak{g}$. Suppose that $\mathfrak{r} = [\mathfrak{g}, \mathfrak{r}]$. Show that $\mathfrak{r} = \{0\}$ (use the fact that $\bigcap_{n \geqslant 0} \mathscr{C}^n\mathfrak{g} = \{0\}$).

6. Let E be a module. Let ME be the free algebra of E (*Algebra*, Chapter III, § 2, Exercise 13). Recall that $\mathrm{ME} = \bigoplus_{n \geqslant 1} \mathrm{E}_n$, where

$$\mathrm{E}_1 = \mathrm{E}, \qquad \mathrm{E}_n = \bigoplus_{p+q=n} \mathrm{E}_p \otimes \mathrm{E}_q \quad \text{if} \quad n \geqslant 2,$$

and the algebra structure on ME is defined by means of the canonical mappings $\mathrm{E}_n \otimes \mathrm{E}_m \rightarrow \mathrm{E}_{n+m}$.

(*a*) Let J be the smallest two-sided ideal of ME containing the elements of the form xx and $(xy)z + (yz)x + (zx)y$ with x, y, z in E. Let $\mathrm{LE} = \mathrm{ME}/\mathrm{J}$. Show that J is a graded ideal of ME; deduce a graduation $(\mathrm{L}^n\mathrm{E})_{n \geqslant 1}$ of LE.

(*b*) Show that LE is a Lie algebra; it is called the *free Lie algebra of the module* E. Show that for every Lie algebra $\mathfrak{g}$ and every linear mapping $f\colon \mathrm{E} \rightarrow \mathfrak{g}$ there exists one and only one Lie algebra homomorphism $\mathrm{F}\colon \mathrm{LE} \rightarrow \mathfrak{g}$ which extends f.

(*c*) Define isomorphisms $\mathrm{E} \rightarrow \mathrm{L}^1\mathrm{E}$ and $\wedge^2\mathrm{E} \rightarrow \mathrm{L}^2\mathrm{E}$. Show that $\mathrm{L}^3\mathrm{E}$ is identified with the quotient of $\mathrm{E} \otimes \wedge^2\mathrm{E}$ by the submodule generated by the elements

$$x \otimes (y \wedge z) + y \otimes (z \wedge x) + z \otimes (x \wedge y) \quad \text{for } x, y, z \text{ in E}.$$

(*d*) Show that, when E is a free module of basis X, LE is identified with the free Lie algebra L(X) defined in no. 2.

(*e*) Show that the canonical mapping of E into the enveloping algebra U(LE) of LE can be extended to an isomorphism of the tensor algebra TE onto U(LE).

(*f*) Let σ be the linear mapping of $\wedge^2\mathrm{E}$ into $\mathrm{T}^2\mathrm{E}$ such that

$$\sigma(x \wedge y) = x \otimes y - y \otimes x.$$

Construct a module E for which σ is not injective. Deduce that for this module the canonical mapping of LE into U(LE) is not injective (compare with Exercise 9 of Chapter I, § 2).

7. Suppose that K is a field. Let $\mathfrak{l}$ be a free Lie algebra with basic family $(x_i)_{i \in I}$ and let M be an $\mathfrak{l}$-module.

(*a*) Show that $H^2(\mathfrak{l}, M) = \{0\}$, cf. Chapter I, § 3, Exercise 12. (Use Corollary 2 to Proposition 1 and part (i) of the Exercise in question.)

(*b*) Show that, for every family $(m_i)_{i \in I}$ of elements of M, there exists a cocycle $\phi : \mathfrak{l} \to M$ of degree 1 such that $\phi(x_i) = m_i$ and that this cocycle is unique. Deduce an exact sequence:

$$0 \to H^0(\mathfrak{l}, M) \to M \to M^I \to H^1(\mathfrak{l}, M) \to 0.$$

If I is finite of cardinal n and M is of finite rank over K, then

$$\operatorname{rg} H^1(\mathfrak{l}, M) - \operatorname{rg} H^0(\mathfrak{l}, M) = (n-1) \operatorname{rg} M.$$

8. Suppose that K is a field. Let $\mathfrak{l}$ be a free Lie algebra with basic family $(\bar{x}_i)_{i \in I}$, let $\mathfrak{r}$ be an ideal of $\mathfrak{l}$ contained in $[\mathfrak{l}, \mathfrak{l}]$ and let $\mathfrak{g} = \mathfrak{l}/\mathfrak{r}$; let x_i denote the image of $\bar{x}_i$ in $\mathfrak{g}$.

Show the equivalence of the following properties:

(i) $\mathfrak{g}$ is a free Lie algebra.

(ii) $\mathfrak{g}$ admits (x_i) as basic family (i.e. $\mathfrak{r} = \{0\}$).

(iii) For every $\mathfrak{g}$-module M, $H^2(\mathfrak{g}, M) = \{0\}$.

(iv) If $\mathfrak{g}$ operates trivially on K, then $H^2(\mathfrak{g}, K) = \{0\}$.

(The implications (ii) $\Rightarrow$ (i) and (iii) $\Rightarrow$ (iv) are obvious and (i) $\Rightarrow$ (iii) follows from Exercise 7. To prove that (iv) $\Rightarrow$ (ii) it suffices to show that $\mathfrak{r} = [\mathfrak{l}, \mathfrak{r}]$, cf. Exercise 5; otherwise take a hyperplane $\mathfrak{h}$ of $\mathfrak{r}$ containing $[\mathfrak{l}, \mathfrak{r}]$ and note that the extension $\mathfrak{r}/\mathfrak{h} \to \mathfrak{l}/\mathfrak{h} \to \mathfrak{l}/\mathfrak{r} = \mathfrak{g}$ is essential.)

¶ 9. Let $\mathfrak{g} = \bigoplus_{n \geqslant 1} \mathfrak{g}_n$ be a graded Lie algebra. If $\mathfrak{h}$ is a graded subalgebra of $\mathfrak{g}$ such that $\mathfrak{g} = \mathfrak{h} + [\mathfrak{g}, \mathfrak{g}]$, show that $\mathfrak{h} = \mathfrak{g}$. If (x_i) is a family of homogeneous elements of $\mathfrak{g}$, deduce that (x_i) is a generating family if and only if the images of the x_i in $\mathfrak{g}/[\mathfrak{g}, \mathfrak{g}]$ generate $\mathfrak{g}/[\mathfrak{g}, \mathfrak{g}]$.

If K is a field on which $\mathfrak{g}$ operates trivially and $H^2(\mathfrak{g}, K) = \{0\}$, show that the family (x_i) is basic if and only if the images of the x_i in $\mathfrak{g}/[\mathfrak{g}, \mathfrak{g}]$ form a basis of this vector space. (Use Exercise 8.)†

10. Let $\mathfrak{l}$ be a free Lie algebra admitting a finite basic family with n elements and let u be a surjective endomorphism of $\mathfrak{l}$. Show that u is an isomorphism. (Write $\operatorname{gr}^n(\mathfrak{l}) = \mathscr{C}^n\mathfrak{l}/\mathscr{C}^{n+1}\mathfrak{l}$ and denote by $\operatorname{gr}^n(u)$ the endomorphism of $\operatorname{gr}^n(\mathfrak{l})$ derived from u; note that $\operatorname{gr}^n(u)$ is surjective; as $\operatorname{gr}^n(\mathfrak{l})$ is a free K-module of finite rank, deduce that $\operatorname{gr}^n(u)$ is bijective since the kernel of u is contained in $\bigcap_n \mathscr{C}^n\mathfrak{l}$, which is $\{0\}$.)

---

† When $\mathfrak{g}$ is not graded, we do not know whether the condition

"$H^2(\mathfrak{g}, M) = \{0\}$ for every $\mathfrak{g}$-module M"

implies that $\mathfrak{g}$ is a free Lie algebra.

If $(y_1, \ldots, y_m)$ is a generating family of $\mathfrak{l}$, show that $m \geqslant n$ and that we have equality if and only if $(y_1, \ldots, y_m)$ is a basic family.

¶ 11. Let Y and Z be two disjoint sets (with Y totally ordered), let $X = Y \cup Z$ and $\mathfrak{l} = L(X)$. Let $\mathfrak{a}_Y$ be the submodule of $\mathfrak{l}$ generated by Z and $[\mathfrak{l}, \mathfrak{l}]$; it is an ideal of the Lie algebra $\mathfrak{l}$. We propose to find explicitly a basic family of $\mathfrak{a}_Y$.

Let M be the set of positive integral functions on Y of finite support. If $m \in M$, we denote by θ^m the endomorphism of $\mathfrak{l}$ given by

$$\theta^m = \prod_{y \in Y} (\operatorname{ad} y)^{m(y)},$$

where the product is relative to the order relation given on Y. Let $\mathfrak{S}$ denote the subset of $Y \times M$ consisting of the ordered pairs (u, m) such that there exists $y \in Y$ for which $y > u$ and $m(y) \geqslant 1$. Show that the elements

$$\theta^m(u) \quad \text{for} \quad (u, m) \in \mathfrak{S} \quad \text{and} \quad \theta^m(z) \quad \text{for} \quad (z, m) \in Z \times M$$

form a basic family of the Lie algebra $\mathfrak{a}_Y$.

(Reduce the problem to the case where Y is finite and argue by induction on Card(Y). Use the Corollary to Proposition 10 applied to the greatest element y of Y and apply the induction hypothesis to $Y - \{y\}$.)

12. Let $X = \{x, y\}$ be a set of two elements. Show that the derived algebra of L(X) is a free Lie algebra admitting as basic family the elements $((\operatorname{ad} y)^q \circ (\operatorname{ad} x)^p)(y)$, for $p \geqslant 1$, $q \geqslant 0$. (Use Exercise 11.)

¶ 13. (In this exercise we assume that every projective module over K is *free*; this is so for example if K is a principal ideal domain.)

Let $\mathfrak{l} = \sum_{n=1}^{\infty} \mathfrak{l}_n$ be a graded Lie algebra admitting a basic family B consisting of homogeneous elements and let $\mathfrak{h}$ be a graded Lie subalgebra of $\mathfrak{l}$, which is a direct factor of $\mathfrak{l}$ considered as a module.

(*a*) For all $i \geqslant 0$, let $\mathfrak{l}^{(i)}$ be the graded subalgebra of $\mathfrak{l}$ such that

$$\mathfrak{l}_j^{(i)} = \mathfrak{h}_j \quad \text{if } j \leqslant i.$$
$$\mathfrak{l}_j^{(i)} = \mathfrak{l}_j \quad \text{if } j > i.$$

Then $\mathfrak{l} = \mathfrak{l}^{(0)} \supset \mathfrak{l}^{(1)} \supset \cdots \supset \mathfrak{h}$. Show, arguing by induction on i, the existence of a basic family $B^{(i)}$ of $\mathfrak{l}^{(i)}$, consisting of homogeneous elements, such that the elements of $B^{(i-1)}$ and of $B^{(i)}$ of degree $< i - 1$ are the same. (Suppose that $B^{(i-1)}$ has been constructed. Let $\mathfrak{m}_i$ be the intersection of $\mathfrak{l}_i = \mathfrak{l}_i^{(i-1)}$ with the subalgebra generated by the $\mathfrak{h}_j$ for $j < i$ and let $\mathfrak{b}_i$ be the submodule of $\mathfrak{l}_i$ generated by the elements of $B^{(i-1)}$ of degree i. The induction hypothesis implies that $\mathfrak{l}_i = \mathfrak{m}_i \oplus \mathfrak{b}_i$. As $\mathfrak{m}_i \subset \mathfrak{h}_i$, we can decompose $\mathfrak{b}_i$ as $\mathfrak{b}_i = \mathfrak{y}_i \oplus \mathfrak{z}_i$, so that $\mathfrak{h}_i = \mathfrak{m}_i \oplus \mathfrak{z}_i$; by the hypothesis on K the modules $\mathfrak{y}_i$ and $\mathfrak{z}_i$ are free;

changing the elements of $B^{(i-1)}$ of degree i if necessary, we can assume that $\mathfrak{z}_i$ is generated by a subset of $B^{(i-1)}$. Applying Exercise 11 to $\mathfrak{l}^{(i-1)}$ and its ideal $\mathfrak{l}^{(i)}$, we obtain a basic family $B^{(i)}$ of $\mathfrak{l}^{(i)}$ with the desired properties.)

(*b*) Deduce from (*a*) the fact that $\mathfrak{h}$ has a basic family consisting of homogeneous elements.

¶ 14. Suppose that K is a field. Let X be a set and x, y two elements of L(X) which are linearly independent over K. Show that the family (x, y) is free in the Lie algebra L(X). (Let x_p (resp. y_q) be the non-zero homogeneous component of x (resp. y) of highest degree. Adding if necessary a multiple of y to x, we can assume that x_p and y_q are linearly independent. The subalgebra of L(X) generated by x_p and y_q is graded and hence free, cf. Exercise 13. Deduce that (x_p, y_q) is a free family and pass from this to (x, y).)

¶ 15. Let $X = \{x, y\}$ be a set with two elements and let σ be an automorphism of the Lie algebra L(X). Show that, if K is a field, σ preserves the graduation of L(X) (apply Exercise 14 to the non-zero homogeneous components of $\sigma(x)$ and $\sigma(y)$ of highest degree); deduce an isomorphism of Aut(L(X)) onto $\mathbf{GL}(2, K)$. Extend these results to the case where the ring K has no nilpotent element $\neq 0$. When K contains an element $\varepsilon \neq 0$ of zero square, show that $x \mapsto x$, $y \mapsto y + \varepsilon[x, y]$ can be extended to an automorphism of L(X) which does not preserve the graduation of L(X).

¶ 16. Suppose that K is a $\mathbf{Q}$-algebra. If $\mathfrak{g}$ is a Lie algebra, we denote by $U(\mathfrak{g})$ its enveloping algebra, by σ the canonical mapping of $\mathfrak{g}$ into $U(\mathfrak{g})$ and by $S(\mathfrak{g})$ the symmetric algebra of the K-module $\mathfrak{g}$. There exists one and only one linear mapping

$$\eta(\mathfrak{g}) : S(\mathfrak{g}) \to U(\mathfrak{g})$$

such that $\eta(\mathfrak{g})(x^n) = \sigma(x)^n$ for all $x \in \mathfrak{g}$ and all $n \in \mathbf{N}$. Then

$$\eta(\mathfrak{g})(x_1 \ldots x_n) = \frac{1}{n!} \sum_{s \in \mathfrak{S}_n} \sigma(x_{s(1)} \ldots x_{s(n)}), \qquad x_i \in \mathfrak{g}.$$

We propose to show that $\eta(\mathfrak{g})$ is *bijective*.

(*a*) Prove that $\eta(\mathfrak{g})$ is surjective and that it is bijective when the K-module $\mathfrak{g}$ is free (use the Poincaré–Birkhoff–Witt Theorem).

(*b*) Let $\mathfrak{h}$ be an ideal of $\mathfrak{g}$. Let $\mathfrak{s}_\mathfrak{h}$ (resp. $\mathfrak{u}_\mathfrak{h}$) denote the ideal of $S(\mathfrak{g})$ (resp. the two-sided ideal of $U(\mathfrak{g})$) generated by $\mathfrak{h}$ (resp. $\sigma(\mathfrak{h})$). The diagram

$$\begin{array}{ccccccccc} 0 & \to & \mathfrak{s}_\mathfrak{h} & \to & S(\mathfrak{g}) & \to & S(\mathfrak{g}/\mathfrak{h}) & \to & 0 \\ & & & & \downarrow{\scriptstyle \eta(\mathfrak{g})} & & \downarrow{\scriptstyle \eta(\mathfrak{g}/\mathfrak{h})} & & \\ 0 & \to & \mathfrak{u}_\mathfrak{h} & \to & U(\mathfrak{g}) & \to & U(\mathfrak{g}/\mathfrak{h}) & \to & 0 \end{array}$$

is commutative and its rows are exact. Deduce that the image $\tilde{\mathfrak{u}}_\mathfrak{h}$ of $\mathfrak{s}_\mathfrak{h}$ under $\eta(\mathfrak{g})$ is contained in $\mathfrak{u}_\mathfrak{h}$ and that $\tilde{\mathfrak{u}}_\mathfrak{h} = \mathfrak{u}_\mathfrak{h}$ when $\eta(\mathfrak{g})$ and $\eta(\mathfrak{g}/\mathfrak{h})$ are injective (in particular, when the module $\mathfrak{g}$ and $\mathfrak{g}/\mathfrak{h}$ are free).

(*c*) Take $\mathfrak{g}$ to be a free Lie algebra with basic family $(X_1, \ldots, X_n, H)$, where $n \geqslant 0$, and take $\mathfrak{h}$ to be the ideal of $\mathfrak{g}$ generated by H. Show that $\mathfrak{g}$ and $\mathfrak{g}/\mathfrak{h}$ are free modules. Deduce that, in that case, $\tilde{\mathfrak{u}}_{\mathfrak{h}} = \mathfrak{u}_{\mathfrak{h}}$. In particular, there exists $x \in \mathfrak{s}_{\mathfrak{h}}$ such that $(\eta(\mathfrak{g}))(x) = \sigma(X_1) \ldots \sigma(X_n)\sigma(H)$.

(*d*) We return to the general case. Show that $\mathfrak{u}_{\mathfrak{h}}$ is generated over K by the elements of the form $\sigma(x_1) \ldots \sigma(x_n)\sigma(h)$, with $n \geqslant 0$, $x_i \in \mathfrak{g}$ and $h \in \mathfrak{h}$ (note that $\mathfrak{u}_{\mathfrak{h}}$ coincides with the left ideal generated by $\mathfrak{h}$). Show, using (*c*), that such an element belongs to $\tilde{\mathfrak{u}}_{\mathfrak{h}}$ (use a suitable homomorphism of a free Lie algebra into $\mathfrak{g}$). Deduce that $\tilde{\mathfrak{u}}_{\mathfrak{h}} = \mathfrak{u}_{\mathfrak{h}}$.

(*e*) Show that, if $\eta(\mathfrak{g})$ is bijective, so is $\eta(\mathfrak{g}/\mathfrak{h})$. Deduce finally that, for every Lie algebra $\mathfrak{g}$, $\eta(\mathfrak{g})$ *is bijective* (write $\mathfrak{g}$ as the quotient of a free Lie algebra) and that the canonical homomorphism

$$\omega\colon S(\mathfrak{g}) \to \operatorname{gr} U(\mathfrak{g}) \quad \text{(cf. Chapter I, § 2, no. 6)}$$

is an *isomorphism* ("Poincaré–Birkhoff–Witt Theorem" for Lie algebras over $\mathbf{Q}$-algebras).

§ 3

The letter X denotes a set.

1. (*a*) Show that every element $u \in A^+(X)$ can be written uniquely in the form $u = \sum_{x \in X} u_x x$, where $u_x \in A(X)$.

(*b*) Show that $\{0\}$ is the only submodule of $A^+(X)$ which is stable under all the mappings $u \mapsto u_x$ $(x \in X)$. (If $\mathfrak{a}$ is such a submodule and $\mathfrak{a} \neq \{0\}$, consider a non-zero element of $\mathfrak{a}$ of minimal degree.)

2. Let $\mathfrak{g}$ be a Lie algebra which is a free module; it is identified by means of $\sigma\colon \mathfrak{g} \to U\mathfrak{g}$ with its image in the enveloping algebra $U\mathfrak{g}$. Let $U^+\mathfrak{g}$ be the kernel of the canonical homomorphism $U\mathfrak{g} \to K$. Let i be a mapping of X into $\mathfrak{g}$ such that $i(X)$ generates $\mathfrak{g}$ as a Lie algebra.

(*a*) Show that $U^+\mathfrak{g}$ is generated by $i(X)$ as a left $U\mathfrak{g}$-module.

(*b*) Show the equivalence of the following properties:

(i) $\mathfrak{g}$ is free with basic family $i\colon X \to \mathfrak{g}$.

(ii) The family i is a basis of the left $U\mathfrak{g}$-module $U^+\mathfrak{g}$.

(The implication (i) $\Rightarrow$ (ii) follows from Exercise 1 (*a*) using the isomorphism $U\mathfrak{g} \to A(X)$. To show (ii) $\Rightarrow$ (i), apply Exercise 1 (*b*) to the kernel of the homomorphism $A(X) \to U\mathfrak{g}$ defined by i.)

3. Let $x \in X$. Show that the centralizer of x in $A(X)$ is the subalgebra generated by x. Deduce that the only elements of $L(X)$ which commute with x are the multiples of x. In particular, the centre of $L(X)$ is $\{0\}$ if $\operatorname{Card}(X) \geqslant 2$ and the centre of $L(X)/\left(\sum_{n>p} L^n(X)\right)$ reduces to the canonical image of $L^p(X)$.

¶ 4. Suppose that K is a field of characteristic $p > 0$.

(*a*) Let H be a basis of the Lie subalgebra L(X) of A(X). Show (using the isomorphism $U(L(X)) \to A(X)$ and the Poincaré–Birkhoff–Witt Theorem) that the elements h^{p^n} ($h \in H$, $n \in \mathbf{N}$) are linearly independent over K.

(*b*) If m is an integer $\geqslant 0$, let L_m denote the submodule of A(X) with basis the h^{p^n}, where $h \in H$, $n \leqslant m$. Show that L_m is a Lie subalgebra of A(X) and that, if $a \in L_{m-1}$, then $a^p \in L_m$ (argue by induction on m and use the *Jacobson formulae*, cf. Chapter I, § 1, Exercise 19). Deduce that L_m does not depend on the choice of H and that it is the Lie subalgebra of A(X) generated by the x^{p^n}, where $x \in X$, $n \leqslant m$.

(*c*) Let $L(X, p)$ be the union of the L_m for $m \geqslant 0$. Show that $L(X, p)$ is the smallest Lie p-subalgebra of A(X) containing X (cf. Chapter I, § 1, Exercise 20); it admits as basis the family of h^{p^n}, where $h \in H$, $n \in \mathbf{N}$.

(*d*) Let $\tilde{U}(L(X, p))$ be the restricted enveloping algebra of $L(X, p)$, cf. Chapter I, § 2, Exercise 6. Show that the injection of $L(X, p)$ into A(X) can be extended to an *isomorphism* of $\tilde{U}(L(X, p))$ onto A(X). (Use the Poincaré–Birkhoff–Witt Theorem and the exercise quoted above.) Deduce that $L(X, p)$ is the set of primitive elements of A(X), cf. § 1, Exercise 12.

(*e*) Let f be a mapping of X into a Lie p-algebra $\mathfrak{g}$. Show that f can be extended uniquely to a p-homomorphism $F: L(X, p) \to \mathfrak{g}$. (Begin by extending f to an algebra homomorphism of A(X) into $\tilde{U}\mathfrak{g}$.)

(The Lie p-algebra $L(X, p)$ is called the *free Lie p-algebra* on the set X.)

§ 4

In the following exercises the letter G *denotes a group.*

1. Let (G_n) be an integral central filtration on G. Show that the Lie algebra gr(G) is generated by $gr_1(G)$ if and only if $G_n = G_{n+1} . C^nG$ for all $n \geqslant 1$. In that case, show that $G_n = G_m . C^nG$ for $m > n$ and deduce that $(G_n) = (C^nG)$ if there exists an integer m such that $G_m = \{e\}$.

2. Let (G_α) be a real filtration on a group G and let v be the corresponding order function. Let H be a subgroup of G.

(*a*) Let $H_\alpha = H \cap G_\alpha$. Show that (H_α) is a real filtration of H and that the corresponding order function is the restriction of v to H. $H_\alpha^+ = H \cap G_\alpha^+$ and gr(H) is identified with a graded subgroup of gr(G).

(*b*) Suppose that H is normal and that $v(G) \cap \mathbf{R}$ is a *discrete* subset of $\mathbf{R}$. We write $(G/H)_\alpha = (G_\alpha H)/H$. Show that $((G/H)_\alpha)$ is a real filtration of G/H and that the corresponding order function $v_{G/H}$ is given by the formula

$$v_{G/H}(x) = \operatorname*{Sup}_{y \in x} v(y).$$

Show that, for all $\alpha \in \mathbf{R}$, there is an exact sequence

$$0 \to gr_\alpha(H) \to gr_\alpha(G) \to gr_\alpha(G/H) \to 0.$$

(*c*) With the hypotheses of (*b*), suppose that (G_α) is a *central* filtration. Show that so are the filtrations on H and G/H induced by (G_α) and that the Lie algebra gr(G/H) is identified with the quotient of gr(G) by the ideal gr(H).

¶ 3. Let (H_α) be a real filtration on a group H and let v_H be the corresponding order function. Suppose that G operates on H on the left and that, for all $g \in G$, the mapping $h \mapsto g(h)$ is an automorphism of the filtered group H.

(*a*) If $\alpha \in \mathbf{R}$, we denote by G_α the set of $g \in G$ such that

$$v_H(h^{-1}g(h)) \geqslant v_H(h) + \alpha \quad \text{for all } h \in H.$$

Show that (G_α) is a real filtration of G and that $G_\alpha = G$ if $\alpha \leqslant 0$. Let v denote the corresponding order function.

(*b*) Suppose that the filtration (H_α) is central and that $G_0^+ = G$. Show that (G_α) is a central filtration of G. If $\xi \in \mathrm{gr}_\alpha(G)$ and $\eta \in \mathrm{gr}_\beta(H)$, let g (resp. h) be a representative of ξ (resp. η) in G_α (resp. H_β); show that the image of $h^{-1}g(h)$ in $\mathrm{gr}_{\alpha+\beta}(H)$ is independent of the choice of g and h; if it is denoted by $D_\xi(\eta)$, show that D_ξ can be extended to a derivation of the Lie algebra gr(H) of degree α and that $\xi \mapsto D_\xi$ is a homomorphism of the Lie algebra gr(G) into the Lie algebra of derivations of gr(H). If $v_H(H) \cap \mathbf{R}$ is contained in a discrete subgroup Γ of $\mathbf{R}$, so is $v(G) \cap \mathbf{R}$ and, for all $\alpha \in \mathbf{R}$, the mapping $\xi \mapsto D_\xi$ defined above is injective.

4. Let H be a nilpotent group of class c and let G be the group of automorphisms of H which act trivially on H/(H, H). Show that G is nilpotent of class $\leqslant c - 1$. (Apply Exercise 3 to the lower central series of H and note that G operates trivially on gr(H).) Show that, if H is a finite p-group (p prime), so is G (same method).

5. Let K be a commutative field and let L be a finite Galois extension of K with Galois group G. Let v_L be a valuation of L with values in Z (*Commutative Algebra*, Chapter VI, § 3, no. 2) which is invariant under G. If $g \in G$, we write

$$v(g) = \sup_{x \in L^*} v_L\left(\frac{g(x) - x}{x}\right).$$

Show that v is the order function of a separated integral filtration (G_n) on G such that $G_0 = G$ and that the restriction of this filtration to G_1 is central (apply Exercise 3, taking H to be the ideal of v_L). Show that G_1 is a p-group (resp. reduces to the identity element) if the residue field of L is of characteristic $p > 0$ (resp. of characteristic zero); when the fields L and K have the same residue field, G_1 is the kernel of the homomorphism ϕ defined in *Commutative Algebra*, Chapter VI, § 8, Exercise 11 (*b*).

¶ 6. Let K[G] be the algebra of G over K and let I be the kernel of the canonical homomorphism $K[G] \to K$ ("augmentation ideal"). Then

$K[G] = K \oplus I$ and I admits as basis the family of $g - 1$ where $g \in G - \{e\}$.

(*a*) If n is an integer $\geqslant 0$, we denote by I^n the ideal of K[G] the n-th power of I. Let G_n be the set of $g \in G$ such that $g - 1 \in I^n$. Show that (G_n) is an integral central filtration on G. In particular, $G_n \supset C^nG$ for all n.

(*b*) Show that if $K = \mathbf{Z}$ the mapping $g \mapsto g - 1$ defines on taking quotients an isomorphism of G/(G, G) onto I/I^2. Deduce that $G_2 = C^2G$.†

(*c*) Suppose that K is a field of characteristic zero and that G is finite. Show that $I^n = I$ for all $n \geqslant 1$.

(*d*) Suppose that K is a field of characteristic $p > 0$ and that G is a p-group. Show that $I^n\{0\}$ for n sufficiently large. (Show first, using *Algebra*, Chapter I, § 6, no. 5, Proposition 11, that every simple K[G]-module is isomorphic to K; deduce that I is the radical of K[G] and hence is nilpotent since K[G] is of finite rank over K.)

(*e*) Suppose that $K = \mathbf{Z}$ and that G has the following property: for all $g \in G$ such that $g \neq 1$, there exists a prime number p, a p-group P and a homomorphism f: G → P such that $f(g) \neq e$. Show that in that case $\bigcap_n I^n = \{0\}$. (Reduce it immediately to the case where G = P. By applying (*d*) to the field $\mathbf{F}_p$, we see that there exists m such that $I^m \subset p.\mathbf{Z}[G]$ and, as I is a direct factor in $\mathbf{Z}[G]$, this implies that $I^m \subset pI$, whence $\bigcap_n I^{mn} \subset \bigcap_n p^nI$, which reduces to 0 since I is a finitely generated Abelian group.)

7. Let G be given the filtration (C^nG) and suppose that $gr_1(G) = G/(G, G)$ is cyclic. Show that $gr_n(G) = \{0\}$ for $n \geqslant 2$ (use Proposition 5) and deduce that $C^nG = (G, G)$ for $n \geqslant 2$.

8. Let $G = \mathbf{SL}_2(\mathbf{Z})$ and let $x = \begin{pmatrix} 1 & 1 \\ 0 & 1 \end{pmatrix}$, $y = \begin{pmatrix} 1 & 0 \\ 1 & 1 \end{pmatrix}$, $w = \begin{pmatrix} 0 & 1 \\ -1 & 0 \end{pmatrix}$.

(*a*) Verify the formulae $w^4 = 1$, $w = xy^{-1}x$, $wxw^{-1} = y^{-1}$.

(*b*) If $g = \begin{pmatrix} a & b \\ c & d \end{pmatrix}$ is an element of G, we write $l(g) = |a| + |b|$. Show that $l(g) = 1$ if and only if g is of the form y^nw^a, where $n \in \mathbf{Z}$ and $0 \leqslant a \leqslant 3$. If $l(g) \geqslant 2$, show that there exists a power h of x or y such that $l(gh) < l(g)$. Deduce that G is *generated by* $\{x, y\}$.

(*c*) Using (*a*) and (*b*), show that G/(G, G) is generated by the image ξ of x and that $\xi^{12} = e$.‡ Deduce that $C^nG = (G, G)$ for $n \geqslant 2$. (apply Exercise 7).

9. Let G be given the filtration (C^nG).

(*a*) Show that the ideal of gr(G) generated by $gr_2(G)$ is $\sum_{q \geqslant 2} gr_q(G)$.

---

† We do not know whether $G_n = C^nG$ for all n. It is true in any case when G is a *free group*, cf. § 5, Exercise 1.

‡ It can be shown that ξ is *of order* 12.

(*b*) Let $(x_i)_{i \in I}$ be a generating family of G and let $m \geqslant 1$. Suppose that, for all $(i, j) \in I^2$, $(x_i, x_j)^m \in C^3G$. Show that, for all $q \geqslant 2$ and all $u \in C^qG$, $u^m \in C^{q+1}G$ (use (*a*)).

10. Let x, y be in G and let r, s be two integers $\geqslant 1$. Suppose that $(x^r, y^s) = e$.

(*a*) Show that, for all $n \geqslant 1$, $(x, y)^{(rs)^n} \in C^{n+2}G$. (It can be assumed that G is generated by $\{x, y\}$; then apply Exercise 9 (*b*), noting that $(x^r, y^s) \equiv (x, y)^{rs}$ mod. C^3G.)

(*b*) Suppose that $x^r = y^s = e$; let t denote the g.c.d. of r and s. Show that $(x, y)^t \in C^3G$ and deduce that $(x, y)^{t^n} \in C^{n+2}G$ for all $n \geqslant 1$ (same method).

11. Let H be a subgroup of G and let m be an integer $\geqslant 1$. Suppose that G is generated by a family $(x_i)_{i \in I}$ such that $x_i^m \in H$ for all i.

(*a*) Let H be given the filtration induced by the filtration (C^nG) on G and let gr(H) be identified with a graded Lie subalgebra of gr(G), cf. Exercise 2. Show that, for all $n \geqslant 0$,

$$m^n . \mathrm{gr}_n(G) \subset \mathrm{gr}_n(H).$$

Deduce that, for all $z \in H.C^nG$, $z^m \in H.C^{n+1}G$.

(*b*) If G is nilpotent, show that there exists an integer $N \geqslant 0$ (depending only on the nilpotency class of G) such that $z^{m^N} \in H$ for all $z \in G$.

12. Suppose that G is nilpotent. Let H be a subgroup of G and let m be an integer $\geqslant 1$ and let x, y be elements of G such that $x^m \in H$ and $y^m \in H$. Show that there exists an integer $N \geqslant 0$ such that $(xy)^{m^N} \in H$. (Apply Exercise 11 to the group generated by $\{x, y\}$ and its intersection with H.)

13. (*a*) Let F be a free group with basic family $\{\bar{x}, \bar{y}\}$ of two elements, let c be an integer $\geqslant 2$ and let m be an integer $\geqslant 1$. We write $F^c = F/C^cF$ and denote by x, y the images of $\bar{x}, \bar{y}$ in F^c. Let F^c_m be the subgroup of F^c generated by $\{x^m, y\}$. Show that there exists an integer $N \geqslant 0$ such that $z^{m^N} \in F^c_m$ for all $z \in F^c$ (use Exercise 11).

(*b*) Let I^c (resp. I^c_m) be the normal subgroup of F^c (resp. F^c_m) generated by y. Show that, if N is chosen as above and $z \in I^c$, then $z^{m^N} \in I^c_m$ (note that F^c/I^c is an infinite cyclic group with generator the image of x and deduce that $F^c_m/I^c_m \rightarrow F^c/I^c$ is injective and hence that $I^c_m = F^c_m \cap I^c$). In particular $xy^{m^N}x^{-1} \in I^c_m$.

(*c*) Suppose that G is nilpotent. Let H be a subgroup of G and let L be a normal subgroup of H. Let $g \in G$ and let $m \geqslant 1$ be such that $g^m \in H$. Show that, if N is sufficiently large, $gl^{m^N} g^{-1} \in L$ for all $l \in L$. (If G is of class $< c$, choose N as in (*a*) and use the homomorphism $f\colon F^c \rightarrow G$ such that $f(x) = g, f(y) = l$; note that $f(F^c_m) \subset H, f(I^c_m) \subset L$ and apply (*b*) above.)

¶ 14. Let P be a set of prime numbers. An integer n is called a P-*integer* if it is $\neq 0$ and all its prime factors belong to P. An element $x \in G$ is called a P-*torsion* element if there exists a P-integer n such that $x^n = e$; G is called a P-

torsion (resp. P-*torsion free*) group if every element of G is a P-torsion element (resp. if no element of G other than e is a P-torsion element). G is called P-*divisible* if, for all $x \in G$ and every P-integer n, there exists $y \in G$ such that $x = y^n$.

Suppose that G is *nilpotent*.

(*a*) Let H be a subgroup of G and let H_P be the set of $x \in G$ such that there exists a P-integer n for which $x^n \in H$. Show that H_P is a subgroup of G (use Exercise 12) and that $(H_P)_P = H_P$. The group H_P is called the P-*saturation* of H in G. If G is P-divisible, so is H_P.

(*b*) Let L be a normal subgroup of H. Show that L_P is normal in H_P. (Use Exercise 13 (*c*) to prove that, if $g \in H_P$ and $l \in L$, then $glg^{-1} \in L_P$.)

In particular, if L is normal in G, so is L_P and G/L_P is P-torsion-free. Deduce that the set of P-torsion elements of G is the smallest normal subgroup N of G such that G/N is P-torsion-free.

(*c*) Suppose that G is P-torsion-free. Let n be a P-integer and let $x, y \in G$ be such that $x^n = y^n$. Show that $x = y$. (Apply Exercise 10 (*a*) with $r = s = n$. Deduce that there exists a P-integer N such that $(x, y)^N = e$, whence $(x, y) = e$ since G is P-torsion-free; then $(x^{-1}y)^n = e$, whence finally $x = y$.)

(*d*) Let H be a subgroup of G, let L be a nilpotent group and let f: H $\rightarrow$ L be a homomorphism. Let Γ be the graph of f in H $\times$ L and let Γ_P be its P-saturation in G $\times$ L. Show that Γ_P is contained in $H_P \times L$ and that $pr_1: \Gamma_P \rightarrow H_P$ is surjective if L is P-divisible and injective if L is P-torsion-free.

Suppose that L is P-divisible and P-torsion-free. Show that f can be extended uniquely to a homomorphism $f_P: H_P \rightarrow L$ and that the graph of f_P is Γ_P.

¶ 15. Preserving the notation of the above exercise, suppose that G is nilpotent. Let i: G $\rightarrow \overline{G}$ be a homomorphism of G into a nilpotent group $\overline{G}$. $(i, \overline{G})$ is called a P-*envelope* of G if the following conditions hold:

(i) $\overline{G}$ is P-divisible and P-torsion-free.

(ii) The kernel of i is the set of P-torsion elements of G.

(iii) The P-saturation of $i(G)$ in $\overline{G}$ is equal to $\overline{G}$.

(*a*) Let $(i, \overline{G})$ be a P-envelope of G and let L be a P-divisible P-torsion-free nilpotent group. Show that, for every homomorphism f: G $\rightarrow$ L, there exists one and only one homomorphism $\bar{f}: \overline{G} \rightarrow L$ such that $\bar{f} \circ i = f$ (reduce it to the case where G is P-torsion-free and use Exercise 14 (*d*)). Deduce that, if G has a P-envelope,† this is unique up to isomorphism.

(*b*) Let $(i, \overline{G})$ be a P-envelope of G, let H be a subgroup of G, let $\overline{H}$ be the P-envelope of $i(H)$ in $\overline{G}$ and let $i_H: H \rightarrow \overline{H}$ the homomorphism induced by i. Show that $(i_H, \overline{H})$ is a P-envelope of H.

† In fact, every nilpotent group has a P-envelope, cf. § 5, Exercise 6; see also M. Lazard, *Annales E. N. S.*, **71** (1954), pp. 101–190, Chapter II, § 3.

Suppose that H is normal in G. Then $\overline{H}$ is normal in $\overline{G}$, cf. Exercise 14 (*b*); if $i_{G/H}: G/H \to \overline{G}/\overline{H}$ is the homomorphism induced by i, show that $(i_{G/H}, \overline{G}/\overline{H})$ is a P-envelope of G/H.

16. We preserve the notation of the two previous exercises.

(*a*) Let S_P be the set of P-integers and let $\mathbf{Z}_P = S_P^{-1}\mathbf{Z}$ be the ring of fractions of $\mathbf{Z}$ defined by S_P (*Commutative Algebra*, Chapter II, § 2, no. 1). Suppose that G is nilpotent, P-torsion-free and P-divisible and let $t \in \mathbf{Z}_P$, $g \in G$. Show that there exists one and only one element h of G such that $h^s = g^{st}$ for all $s \in S_P$ such that $st \in \mathbf{Z}$. The element h is called the *t-th power* of g and denoted by g^t. The mapping $t \mapsto g^t$ is a homomorphism of $\mathbf{Z}_P$ into G. If t is invertible in $\mathbf{Z}_P$, $g \mapsto g^t$ is bijective. (Use Exercise 14 (*c*)).

(*b*) Let A be a commutative group and let i be the canonical mapping of A into $A_P = \mathbf{Z}_P \times A$. Show that (i, A_P) is a P-envelope of A.

(*c*) Let G (resp. $\overline{G}$) be the lower *strict triangular* group of order n over a ring k (resp. over the ring $k_P = k \otimes \mathbf{Z}_P$) and let i be the homomorphism of G into $\overline{G}$ defined by the canonical homomorphism of k into k_P (*Algebra*, Chapter II, § 10, no. 4). Show that $(i, \overline{G})$ is a P-envelope of G.

(*d*) Let G be a *finite* nilpotent group and hence the direct product of its Sylow p-groups G_p (*Algebra*, Chapter I, § 6, no. 7, Theorem 4). Let $\overline{G}$ be the product of the G_p for $p \notin P$ and let i be the canonical projection of G onto $\overline{G}$. Show that $(i, \overline{G})$ is a P-envelope of G.

¶ 17. Suppose that G is nilpotent. Let P be a set of prime numbers and let $\mathfrak{V}_p(G)$ be the set of normal subgroups of G of finite index whose index is a P-integer (cf. Exercise 14). Let $\mathscr{T}_p(G)$ be the topology defined on G by the filter base $\mathfrak{V}_p(G)$ (*General Topology*, Chapter III, § 1, no. 2, *Example*).

(*a*) Let N be a subgroup of G of finite index such that (G: N) is a P-integer and let N′ be the intersection of the conjugates of N. Show that (G: N′) is a P-integer (use the fact that the group G/N′ is the product of its Sylow groups); deduce that N is open with respect to $\mathscr{T}_P(G)$.

(*b*) Let H be a subgroup of G of finite index. Show that $\mathscr{T}_P(H)$ coincides with the topology induced on H by $\mathscr{T}_P(G)$. (Same method.)

(*c*) Let H be a normal subgroup of G such that G/H is isomorphic to $\mathbf{Z}$ and let x be a representative in G of a generator of G/H. Let $N \in \mathfrak{V}_P(H)$ and let $N' = \bigcap_{n \in \mathbf{Z}} x^n N x^{-1}$. Show that $N' \in \mathfrak{V}_P(H)$ and that $xN'x^{-1} = N'$. If N″ is the subgroup generated by N′ and x, show that (G: N″) = (H: N′). Deduce that $\mathscr{T}_P(H)$ is induced by $\mathscr{T}_P(G)$.

(*d*) Suppose that G is finite. Show that, if H is a subgroup of G, there exists a sequence of subgroups

$$H = H_0 \subset H_1 \subset \cdots \subset H_k = G$$

such that H_i is normal in H_{i+1} for $0 \leqslant i < k$ and that H_{i+1}/H_i is finite or isomorphic to $\mathbf{Z}$. Deduce, using (*b*) and (*c*), that $\mathscr{T}_P(H)$ is induced by $\mathscr{T}_P(G)$.

(*e*) With the hypotheses of (*d*), let P′ denote the complement of P in the set of prime numbers. Show that the closure of H under $\mathscr{T}_P(G)$ coincides with the P′-saturation of H in G (Exercise 14); in particular, G is Hausdorff if and only if it is P′-torsion-free.

18. The *upper central series* (Z_iG) of the group G is defined inductively as follows:

(i) $Z_iG = \{e\}$ if $i \leqslant 0$;

(ii) $Z_iG/Z_{i-1}G$ is the centre of $G/Z_{i-1}G$.

Then $\{e\} = Z_0G \subset Z_1G \subset \cdots$ and Z_1G is the centre of G; the Z_iG are characteristic subgroups of G.

(*a*) Show that G is nilpotent of class $\leqslant c$ if and only if $G = Z_cG$.

(*b*) Show that $(C^nG, Z_mG) \subset Z_{m-n}G$.

(*c*) Suppose that G is nilpotent and P-torsion-free (where P is a set of prime numbers, cf. Exercise 14). Show that the Z_iG are P-saturated. (It suffices to verify this for Z_1G; if n is a P-integer and $g \in G$ is such that $g^n \in Z_1G$, then $xg^nx^{-1} = g^n$ for all $x \in G$, whence $xgx^{-1} = g$ by Exercise 14 (*c*) and it follows that g belongs to Z_1G.)

§ 5

In the following exercises we assume the hypotheses and notation of § 5. F denotes the free group F(X) and g the unique homomorphism of F into the Magnus group $\Gamma(X)$ such that $g(x) = 1 + x$ for all $x \in X$ (cf. Theorem 1).

1. Let the algebra K[F] of F be given the filtration (I^n) consisting of the powers of the augmentation ideal I (§ 4, Exercise 6).

(*a*) Let $\tilde{g}\colon K[F] \to \hat{A}(X)$ be the unique algebra homomorphism extending $g\colon F \to \hat{A}(X)^*$. Show that $\tilde{g}$ maps I^n to the ideal $\hat{A}_n(X)$ (cf. no. 1) and defines when taking quotients an *isomorphism* $\tilde{g}_n$ of $K[F]/I^n$ onto $\hat{A}(X)/\hat{A}_n(X)$. (Define an inverse homomorphism of $\tilde{g}_n$ by means of the homomorphism of A(X) into K[F] which maps x to $x - 1$ for all $x \in X$.) Deduce that $\hat{A}(X)$ *is isomorphic to the Hausdorff completion of* K[F] under the topology defined by (I^n).

(*b*) Suppose that $K = \mathbf{Z}$. Show that the filtration (I^n) is separated (use Proposition 3 and Exercise 6 (*e*) of § 4) and that the filtration of F defined by it (Exercise 6 (*a*) of § 4) coincides with the filtration (C^nF). Deduce that $\tilde{g}\colon \mathbf{Z}[F] \to \hat{A}_{\mathbf{Z}}(X)$ is injective.

2. Let G be a group, let K[G] be its algebra over K and let I be its augmentation ideal (§ 4, Exercise 6). Let ε denote the canonical homomorphism of K[G] onto K; then $\operatorname{Ker}(\varepsilon) = I$.

(*a*) Let M be a left K[G]-module and let Z(G, M) be the group of *crossed homomorphisms* of G into M (*Algebra*, Chapter I, § 6, Exercise 7). If

$f \in \mathrm{Hom}_{K[G]}(I, M)$, let f_1 be the mapping $g \mapsto f(g - 1)$ of G into M. Show that f_1 is a crossed homomorphism (use the identity

$$gg' - 1 = g(g' - 1) + g - 1$$

to show that $f_1(gg') = g.f_1(g') + f_1(g)$) and that $f \mapsto f_1$ is an *isomorphism* of $\mathrm{Hom}_{K[G]}(I, M)$ onto Z(G, M).

(*b*) Take G to be the free group F = F(X). Show that, for every mapping $\theta: X \to M$, there exists one and only one element f_θ of Z(M, F) which extends θ. (Use the interpretation of Z(F, M) in terms of the semi-direct product of F by M, cf. *Algebra*, Chapter I, *loc. cit.*) Deduce, using (*a*), that the family $(x - 1)_{x \in X}$ is a *basis* of I as a left K[F]-module.

(*c*) By (*b*) every element $u \in K[F]$ can be written uniquely in the form

$$u = \varepsilon(u) + \sum_{x \in X} D_x(u)(x - 1),$$

where $D_x(u) \in K[F]$ and the $D_x(u)$ are zero except for a finite number. The mapping $u \mapsto D_x(u)$ is called the *partial derivative with respect to x*. It is characterized by the following properties:

(i) D_x is a K-linear mapping of K[F] into K[F].

(ii) $D_x(uv) = u.D_x(v) + \varepsilon(v).D_x(u)$ for u, v in K[F].

(iii) $D_x(x) = 1$ and $D_x(y) = 0$ if $y \in X - \{x\}$.

If $n \geqslant 1$, then

$$D_x(x^n) = 1 + x + \cdots + x^{n-1}$$
$$D_x(x^{-n}) = -x^{-n} - x^{-n+1} - \cdots - x^{-1}.$$

(*d*) Let ε_A be the canonical homomorphism of A(X) onto K. Show that every element u of A(X) can be written uniquely in the form

$$u = \varepsilon_A(u) + \sum_{x \in X} d_x(u).x,$$

where $d_x(u) \in A(X)$ and the $d_x(u)$ are zero except for a finite number. Show that $d_x: u \mapsto d_x(u)$ is an endomorphism of A(X) which can be extended by continuity to $\hat{A}(X)$ and enjoys properties analogous to (i), (ii), (iii) above. Show $d_x \circ \tilde{g} = \tilde{g} \circ D_x$, where $\tilde{g}$ is the homomorphism of K[F] into $\hat{A}(X)$ which extends g (cf. Exercise 1).†

¶ 3. We preserve the notation of Exercise 2.

(*a*) Let J be a right ideal of K[F] contained in I and let u be an element of I. Show that

$$u \in J.I \Leftrightarrow D_x(u) \in J \quad \text{for all } x \in X.$$

In particular, an element of I belongs to I^n ($n \geqslant 1$) if and only if all its partial derivatives belong to I^{n-1}.

† For more details on the D_x, see R. Fox, *Ann of Math.*, **57** (1953), pp. 547–560.

(*b*) Let R be a normal subgroup of F, let G = F/R and let J be the kernel of the canonical homomorphism $\gamma\colon K[F] \to K[G]$. Show that J is generated as a left (resp. right) ideal by the elements $r - 1$, where $r \in R$. Prove the exactness of the sequence

$$(*) \qquad 0 \longrightarrow J/(J.I) \xrightarrow{\alpha} I/(J.I) \xrightarrow{\beta} K[G] \xrightarrow{\delta} K \longrightarrow 0,$$

where α is derived by taking quotients from the inclusion of J in I and β is derived by taking quotients from the restriction of γ to I.

(*c*) For $u \in I$, $x \in X$, let $\overline{D}_x(u)$ denote the image of $D_x(u)$ in K[G] under γ. Show, using (*a*), that the family $(\overline{D}_x)_{x \in X}$ defines on taking quotients an *isomorphism* of I/(J.I) onto $K[G]^{(X)}$.

(*d*) For $r \in R$, let $\theta(r)$ be the image of $r - 1$ in J/(J.I). Show that

$$\theta(rr') = \theta(r) + \theta(r') \quad \text{and} \quad \theta(yry^{-1}) = y.\theta(r) \quad \text{for } r, r' \text{ in } R, y \in F.$$

(Use the identities

$$rr' - 1 = (r - 1)(r' - 1) + (r - 1) + (r' - 1)$$
$$yry^{-1} - 1 = y(r - 1)(y^{-1} - 1) + y(r - 1).)$$

Deduce that θ defines a homomorphism of R/(R, R) into J/(J.I) compatible with the action of G and show that the image of this homomorphism generates the K-module J/(J.I); when $K = \mathbf{Z}$, show that an *isomorphism* is thus obtained of R/(R, R) onto J/(J.I) (define the inverse homomorphism directly).

(*e*) Let $(r_\alpha)_{\alpha \in A}$ be a family of elements of R generating R as a normal subgroup of F. Show that the $\theta(r_\alpha)$ generate the K[G]-module J/(J.I).

(*f*) The matrix $(\overline{D}_x(r_\alpha))_{x \in X, \alpha \in A}$ defines a homomorphism

$$\rho\colon K[G]^{(A)} \to K[G]^{(X)}$$

of left K[G]-modules. Show that the sequence

$$(**) \qquad K[G]^{(A)} \xrightarrow{\rho} K[G]^{(X)} \xrightarrow{\delta} K[G] \xrightarrow{\varepsilon} K \longrightarrow 0$$

is exact, where δ is the homomorphism $(u_x) \mapsto \sum_{x \in X} u_x(\gamma(x) - 1)$.

(Transform the exact sequence (*) using (*c*), (*d*), (*e*) above.)†

4. For all $n \in \mathbf{N}$, let $\binom{T}{n}$ denote the polynomial

$$T(T - 1) \ldots (T - n + 1)/n!.$$

---

† For more details on this exercise, see K. W. Gruenberg, *Lecture Notes in Math.*, no. 143, Chapter 3, and R. Swan, *J. of Algebra*, **12** (1969), pp. 585–601.

If $f(T)$ is a polynomial, let Δf denote the polynomial $f(T+1) - f(T)$. Then $\Delta\binom{T}{0} = \Delta 1 = 0$ and $\Delta\binom{T}{n} = \binom{T}{n-1}$ if $n \geqslant 1$.

(*a*) Let $f \in \mathbf{Q}[T]$. Show the equivalence of the following properties:

(i) f maps $\mathbf{Z}$ into itself.

(ii) f is a linear combination with coefficients in $\mathbf{Z}$ of the $\binom{T}{n}$.

(iii) $f(0) \in \mathbf{Z}$ and Δf maps $\mathbf{Z}$ into itself.

(Argue by induction on $\deg(f)$, noting that $\deg(\Delta f) = \deg(f) - 1$ if $\deg(f) \neq 0$.)

A polynomial f satisfying the above properties is called *binomial*. The sum, product and composition of two binomial polynomials is a binomial polynomial.

(*b*) Let p be a prime number and let f be a binomial polynomial. Show that f maps the ring $\mathbf{Z}_p$ of p-adic integers into itself (use the continuity of f and the fact that $\mathbf{Z}$ is dense in $\mathbf{Z}_p$).

(*c*) Let P be a set of prime numbers, let S_p be the set of P-integers (§ 4, Exercise 14) and let $\mathbf{Z}_p = S_P^{-1}\mathbf{Z}$. Show that if f is a binomial polynomial, f maps $\mathbf{Z}_P$ into itself (apply (*b*) to the prime numbers p not belonging to P).

5. Let $\Gamma = \Gamma(X)$ be the Magnus group on X and let (Γ_n) be its natural filtration (no. 2). Show that the graded Lie algebra $\operatorname{gr}(\Gamma)$ corresponding to this filtration is isomorphic to the Lie subalgebra $\bigoplus_{n \geqslant 1} A_n(X)$ of $A(X)$. If $X \neq \emptyset$, this Lie algebra is not generated by its elements of degree 1; deduce that (Γ_n) is not the lower central series of Γ.

¶ 6. Let P be a set of prime numbers, let S_P be the set of P-integers (§ 4, Exercise 14) and let $\mathbf{Z}_P = S_P^{-1}\mathbf{Z}$.

(*a*) Suppose that, for all $s \in S_P$, the mapping $k \mapsto sk$ is a bijection of K onto itself; this is equivalent to saying that K can be given a $\mathbf{Z}_P$-algebra structure.

With the notation of Exercise 5, show that, for all $s \in S_P$, the mapping $a \mapsto a^s$ of Γ into itself is bijective and that the same is true for each quotient Γ/Γ_n for $n \geqslant 1$. If $t \in \mathbf{Z}_P$ and $a \in \Gamma$ (resp. $a \in \Gamma/\Gamma_n$), a^t is defined as in Exercise 16 of § 4; writing a in the form $1 + \alpha$, where $\alpha \in \hat{A}_1(X)$, show that

$$a^t = (1+\alpha)^t = \sum_{n=0}^{\infty} \binom{t}{n} \alpha^n.$$

(Note that the coefficients $\binom{t}{n}$ belong to $\mathbf{Z}_P$, cf. Exercise 4.)

(*b*) Suppose now that $K = \mathbf{Z}_P$. Let c be an integer $\geqslant 1$. The group $F^c = F/C^cF$ is identified with a subgroup of Γ/Γ_c by means of the homomorphism derived from g by taking quotients (notation of Theorem 2). The group Γ/Γ_c is a P-torsion-free P-divisible group of class $<c$. Let F^c_P be the P-saturation of F^c in Γ/Γ_c (§ 4, Exercise 14) and let i be the injection of F^c into F^c_P. The ordered pair (i, F^c_P) is a P-*envelope* of F^c (§ 4, Exercise 15). Deduce that *every nilpotent group has a* P-*envelope* (note that every nilpotent group of class $<c$ is the quotient of a group F^c for suitable X and use part (*b*) of Exercise 15 of § 4).

(*c*) If $n \leqslant c$, let $F^c_{P,n}$ be the intersection of F^c_P with Γ_n/Γ_c; if $n \geqslant c$, we write $F^c_{P,n} = \{e\}$. Show that $F^c_{P,n}$ is the P-saturation of C^nF^c in Γ/Γ_c. The filtration $(F^c_{P,n})$ is an integral central filtration of F^c_P; let $\mathrm{gr}(F^c_P)$ be the graded module associated with this filtration. Show that, if $n < c$, the image of $\mathrm{gr}_n(F^c_P)$ in $\mathrm{gr}_n(\Gamma) = A^n_{\mathbf{Z}_P}(X)$ is $S_P^{-1}.L^n_{\mathbf{Z}}(X) = L^n_{\mathbf{Z}_P}(X)$. Deduce that the Lie algebra $\mathrm{gr}(F^c_P)$ is generated by its elements of degree 1 and hence that $F^c_{P,n} = C^n(F^c_P)$ for all n (§ 4, Exercise 1). Show that the group F^c_P is generated by the $x^{1/s}$ for $x \in X$, $s \in S_P$ (note that the images of these elements in $\mathrm{gr}_1(F^c_P) \simeq \mathbf{Z}_P^{(X)}$ generate the group $\mathrm{gr}_1(F^c_P)$ and apply Corollary 3 to Proposition 8 of *Algebra*, Chapter I, § 6, no. 3).

(*d*) Let H be a Hall set relative to X (§ 2, no. 10) and let H(c) be the subset of H consisting of the elements of length $<c$. For $m \in \mathrm{H}(c)$, let $\phi_c(m)$ denote the image in F^c of the basic commutator $\phi(m)$ defined by m (cf. no. 4, *Remark*). Show that, for all $w \in F^c_P$, there exists a unique element $\alpha \in \mathbf{Z}_P^{(\mathrm{H}(c))}$ such that

$$w = \prod_{m \in \mathrm{H}(c)} \phi_c(m)^{\alpha(m)}.$$

(Use the determination of $\mathrm{gr}(F^c_P)$ obtained above.)

7. We preserve the notation of Exercise 6.

(*a*) Let G be a nilpotent group of class $<c$ and let $(i, \overline{G})$ be a P-envelope of G. Let $\rho\colon F^c \to G$ be a surjective homomorphism (such a homomorphism exists if X is suitably chosen) and let $\bar\rho$ be the corresponding homomorphism of F^c_P into $\overline{G}$ (§ 4, Exercise 15). The homomorphism $\bar\rho$ is surjective and maps $C^nF^c_P = F^c_{P,n}$ onto $C^n\overline{G}$. Show that $C^n\overline{G}$ *is the* P-*envelope of* $i(C^nG)$ *in* $\overline{G}$ and that *the Lie algebra* $\mathrm{gr}(\overline{G})$ *is identified with* $\mathbf{Z}_P \otimes \mathrm{gr}(G)$.

(*b*) Deduce that G is P-divisible and P-torsion-free if and only if the $C^nG/C^{n+1}G$ are.

¶ 8. Suppose that $K = \mathbf{Z}$ and that X is finite. For every integer $k \geqslant 0$, let $(e_{k,\alpha})$ denote the basis of $A_k(X)$.

(*a*) Let $(w_n)_{n \in \mathbf{N}}$ be a sequence of elements of F. Let $w_{k,\alpha}(n)$ be the coefficient of $e_{k,\alpha}$ in the term of $g(w_n) \in \hat{A}(X)$ of degree k. Then:

$$g(w_n) = \sum_{k,\alpha} w_{k,\alpha}(n) e_{k,\alpha} \quad \text{for all } n \in \mathbf{Z}.$$

The sequence (w_n) is called *typical* if, for every ordered pair (k, α) the function $w_{k,\alpha}: n \mapsto w_{k,\alpha}(n)$ is a *binomial polynomial of degree* $\leqslant k$, cf. Exercise 4. This condition is independent of the choice of the bases $(e_{k,\alpha})$. Show that it is equivalent to the existence of a sequence $(a_k)_{k\in\mathbf{N}}$ of elements of $\hat{A}_{\mathbf{Q}}(X)$ such that $\omega(a_k) \geqslant k$ for all k and

$$g(w_n) = \sum_{k=0}^{\infty} n^k a_k \quad \text{for all } n \in \mathbf{Z}.$$

(*b*) Show that, if (w_n) and (w'_n) are two typical sequences, so are (w_n^{-1}) and $(w_n w'_n)$.

(*c*) Let w be an element of C^kF for $k \geqslant 1$ and let f be a binomial polynomial of degree $\leqslant k$. Show that $(w^{f(n)})$ is a typical sequence. In particular if $z \in F$, the sequence (z^n) of powers of z is typical.

(*d*) Let (w_n) be a sequence of elements of F. Show that (w_n) is typical if and only if there exist $y_0, y_1, \ldots, y_n, \ldots$ in F, where $y_i \in C^iF$ for all $i \geqslant 1$ and

$$w_n \equiv y_0 y_1^n \ldots y_k^{\binom{n}{k}} \mod C^{k+1}F$$

for all $n \in \mathbf{Z}$ and all $k \geqslant 1$. (Determine the y_i from (w_n) by setting successively $n = 0, 1, \ldots$. For example $w_0 = y_0$, $w_1 = y_0 y_1, \ldots$)

(*e*) Let H be a Hall set relative to X and let (w_n) be a sequence of elements of F. Let $(\alpha(m, n))_{m\in H, n\in\mathbf{Z}}$ be the family of integers such that

$$w_n \equiv \prod_{m\in H} \phi(m)^{\alpha(m,n)} \mod C^cF$$

for all $n \in \mathbf{Z}$ and all $c \geqslant 1$ (cf. no. 4, *Remark*). Show that (w_n) is typical if and only if, for all $m \in H$, the function $n \mapsto \alpha(m, n)$ is a binomial polynomial of degree $\leqslant l(m)$, where $l(m)$ is the length of m.†

(*f*) A sequence (w_n) is called 1-*typical* if the corresponding functions $w_{k,\alpha}$ are binomial polynomials of degree $\leqslant k - 1$. Show that, if (w_n) is 1-typical, there exist $y_i \in C^{i+1}F$ for $i = 0, 1, \ldots$, such that

$$w_n \equiv y_0 y_1^n \ldots y_k^{\binom{n}{k}} \mod. C^{k+1}F$$

for all $n \in \mathbf{Z}$ and all $k \geqslant 1$.

¶ 9. Let X be the set with two elements $\{x, y\}$.

(*a*) Show that there exists a sequence $w_2, w_3, \ldots$ of elements of F such that $w_i \in C^iF$ for all i and

$$(xy)^n \equiv x^n y^n w_2^{\binom{n}{2}} \ldots w_i^{\binom{n}{i}} \mod C^{i+1}F$$

† For more details on typical sequences, see M. Lazard, *Annales E. N. S.*, **71** (1954), pp. 101–190, Chapter II, §§ 1, 2.

for all $n \in \mathbf{Z}$ and all $i \geqslant 1$. (Apply Exercise 8 (*d*) to the sequence of $x^{-n}(xy)^n$.)
Show that $w_2 = y^{-1}(y^{-1}, x^{-1})y \equiv (x, y)^{-1} \bmod C^3F$.

(*b*) Let G be a nilpotent group of class $\leqslant c$ and let x, y be in G. Deduce from the above the following formula ("*Hall's formula*");

$$(xy)^n = x^n y^n \prod_{i=2}^{c} w_i(x, y)^{\binom{n}{i}}.$$

(*c*) Let p be a prime number. Show that there exists $u_i \in C^iF$ $(2 \leqslant i \leqslant p - 1)$ and $w \in C^pF$ such that

$$(xy)^p = x^p y^p u_2^p \ldots u_{p-1}^p w.$$

(Use (*a*), noting that $\binom{p}{i}$ is divisible by p if $1 < i < p$.) Let $\bar{w}$ be the image of w in $\mathrm{gr}_p(F) = L_{\mathbf{Z}}^p(X)$ and let $\bar{w}_p$ be the image of $\bar{w}$ in $L_{\mathbf{F}_p}^p(X)$. Show that $\bar{w}_p = \Lambda_p(x, y)$, cf. Chapter I, § 1, Exercise 19. (Use the extension g of F to $\hat{A}(X)$ and compare the terms of degree p in $g((xy)^p)$ and $g(x^p y^p u_2^p \ldots u_{p-1}^p \omega)$. The first is equal to $(x + y)^p$ and the second is congruent mod. p to $x^p + y^p + \bar{w}_p$. Hence the result.)

Show that there exist $a \in C^2F$ and $w' \in C^pF$ such that

$$(xya)^p = x^p y^p w'.$$

(Use the above formula for $(xy)^p$.) Deduce that in a nilpotent group of class p the p-th powers form a subgroup.

(*d*) Show that there exists a sequence $v_3, v_4, \ldots$ of elements of F, where $v_i \in C^iF$ for all i and

$$(x^n, y) \equiv (x, y)^n v_3^{\binom{n}{2}} \ldots v_{i+1}^{\binom{n}{i}} \quad \bmod C^{i+2}F$$

for all $n \in \mathbf{N}$ and all $i \geqslant 1$. (Apply Exercise 8 (*f*) to the sequence of (x^n, y).)

Deduce that, if p is a prime number, there exists $t_i \in C^iF$ for $3 \leqslant i \leqslant p$ and $z \in C^{p+1}F$ such that

$$(x^p, y) = (x, y)^p t_3^p \ldots t_p^p z.$$

Show that the image $\bar{z}_p$ of z in $L_{\mathbf{F}_p}^{p+1}(X)$ is $(\mathrm{ad}\, x)^p(y)$. (Same method as for (*c*).)

¶ 10. Let p be a prime number and let G be a group with a real central filtration (G_α). Suppose that the relation $x \in G_\alpha$ implies $x^p \in G_{p\alpha}$, in which case the filtration (G_α) is called *restricted*.

(*a*) Show that the Lie algebra gr(G) associated with (G_α) is such that $p.\mathrm{gr}(G) = 0$ and hence can be given an *algebra* structure *over* $\mathbf{F}_p$.

(*b*) Let $\xi \in \mathrm{gr}_\alpha(G)$ and let x be a representative of ξ in G_α. Show that the image of x^p in $\mathrm{gr}_{p\alpha}(G)$ does not depend on the choice of x (use Exercise 9 (*c*)). If this image is denoted by $\xi^{[p]}$, prove that $\xi \mapsto \xi^{[p]}$ is a linear mapping of $\mathrm{gr}_\alpha(G)$ into $\mathrm{gr}_{p\alpha}(G)$ and that $(\xi + \xi')^{[p]} = \xi^{[p]} + \xi'^{[p]} + \Lambda_p(\xi, \xi')$. (Same method.)

Show that, if $\xi \in \mathrm{gr}_\alpha(G)$ and $\eta \in \mathrm{gr}_\beta(G)$, then

$$[\xi^{[p]}, \eta] = (\mathrm{ad}\ \xi)^p(\eta).$$

(Use Exercise 9 (*d*).)

(*c*) Show that there exists one and only one *p-mapping* of gr(G) into itself (Chapter I, § 1, Exercise 20) which extends the mappings $\xi \mapsto \xi^{[p]}$ defined on the $\mathrm{gr}_\alpha(G)$. With this structure, gr(G) is a *Lie p-algebra*, said to be associated with the restricted filtration (G_α).

¶ 11. (*a*) Let A be a filtered algebra satisfying the conditions of § 4, no. 5 and let $\Gamma = A^* \cap (1 + A_0^+)$. Γ is given the filtration induced by that on A (*loc. cit.*, Proposition 2). Show that, if K is a field of characteristic $p > 0$, (Γ_α) is a *restricted filtration* (Exercise 10) and the embedding of $\mathrm{gr}(\Gamma)$ in gr(A) defined in *loc. cit.*, Proposition 3 is compatible with the Lie p-algebra structures on $\mathrm{gr}(\Gamma)$ and gr(A).

(*b*) Suppose that $K = \mathbf{F}_p$. We apply the above to the filtered algebra $\hat{A}(K)$, in which $F = F(X)$ is embedded by means of the homomorphism g. The filtration (Γ_n) of Γ induces a filtration (F_n) on F, which is a restricted filtration. The Lie p-algebra gr(F) is identified with a Lie p-subalgebra of A(X) containing X. Deduce that gr(F) contains the free Lie p-algebra $L(X, p)$, cf. § 3, Exercise 4 (*e*).

(*c*) Let H be a Hall set relative to X; for every integer i, let H_i be the set of elements of H of length i. Let $w \in F$ and let $\alpha_i \in \mathbf{Z}^{(H_i)}$ be such that

$$w \equiv \prod_{i=1}^{n} \prod_{m \in H} \phi(m)^{\alpha_i(m)} \mod C^{n+1}F$$

for all n, cf. no. 4. For all $m \in H$, let $l(m)$ denote the length of m and $q(m, w)$ the largest power of p which divides $\alpha_i(m)$, where $i = l(w)$ (if $\alpha_i(m) = 0$, we make the convention $q(m, w) = +\infty$). Let $N = \operatorname{Inf}_{m \in H} (l(m) . q(m, w))$. Show that the $\phi(m)^{\alpha_i(m)}$ belong to F_N and even to F_{N+1} if $l(w) . q(m, w) > N$. Show that, if $l(w) . q(m, w) = N$, the image of $\phi(m)^{\alpha_i(m)}$ in $\mathrm{gr}_N(\hat{A}(X)) = A^N(X)$ is a non-zero multiple of $\overline{m}^{[q(m, w)]}$, where $\overline{m}$ is the element of L(X) defined by m (§ 2, no. 11). Now these elements are linearly independent (§ 3, Exercise 4); deduce that the image of w in $\mathrm{gr}_N(F)$ is $\neq 0$ and belongs to $L(X, p)$, whence the fact that $\mathrm{gr}(F) = L(X, p)$.

12. Let $\mathfrak{g}$ be a Lie algebra over a field k of characteristic $p > 0$.

(*a*) Let h be an integer $\leqslant p - 1$. $\mathfrak{g}$ is said to satisfy *Engel's h-th condition* if $(\mathrm{ad}\ x)^h(y) = 0$ for every ordered pair (x, y) of elements of $\mathfrak{g}$. Show that this is equivalent to

$$\sum_{\sigma \in \mathfrak{S}_h} \mathrm{ad}(x_{\sigma(1)}) \circ \cdots \circ \mathrm{ad}(x_{\sigma(h)}) = 0$$

for $x_1, \ldots, x_h$ in $\mathfrak{g}$ (apply the formula $(\mathrm{ad}\ x)^h = 0$ to the linear combinations of the x_i).

(*b*) Let x, y be elements of $\mathfrak{g}$ such that $\Lambda_p(ax, by) = 0$ for all a, b in k. Show that, if $\Lambda_p^{r,s}$ is the bihomogeneous component of Λ_p of bidegree (r, s), then $\Lambda_p^{r,s}(x, y) = 0$ for every ordered pair (r, s). Deduce (taking $r = p - 1$, $s = 1$) that $(\operatorname{ad} x)^{p-1}(y) = 0$. In particular, if $\Lambda_p(x, y) = 0$ for x, y in $\mathfrak{g}$, the Lie algebra $\mathfrak{g}$ satisfies Engel's $(p - 1)$-th condition.

(*c*) Let $\mathfrak{c}$ be the centre of $\mathfrak{g}$. Suppose that $(\operatorname{ad} x)^p = 0$ for all $x \in \mathfrak{g}$. Show that $\mathfrak{g}/\mathfrak{c}$ satisfies Engel's $(p - 1)$-th condition. (Show that $\Lambda_p(\operatorname{ad} x, \operatorname{ad} y) = 0$ for x, y in $\mathfrak{g}$ and apply (*b*) to the Lie algebra $\operatorname{ad} \mathfrak{g} = \mathfrak{g}/\mathfrak{c}$.)

(*d*) Let G be a group such that $x^p = 1$ for all $x \in G$ and let (G_α) be a real central filtration of G. The filtration (G_α) is *restricted* (Exercise 10) and $\operatorname{gr}(G)$ is a Lie p-algebra whose p-mapping is zero. Deduce that $\operatorname{gr}(G)$ satisfies Engel's $(p - 1)$-th condition.†

§ 6

1. Show that $\exp(U).\exp(V).\exp(-U) = \exp\left(\sum_{n=0}^{\infty} \frac{1}{n!} (\operatorname{ad} U)^n(V)\right)$. Deduce that

$$H(U, H(V, -U)) = \sum_{n=0}^{\infty} \frac{1}{n!} (\operatorname{ad} U)^n(V) = (\exp(\operatorname{ad} U))(V).$$

2. Show that $H(U, V) = \sum_{n=0}^{\infty} \frac{1}{n!} (\operatorname{ad} U)^n(H(V, U))$. (Apply Exercise 1, noting that $H(U, H(H(V, U), -U)) = H(U, V)$.)

3. Let $M(U, V) = H(-U, U + V)$. Then

$$\exp(M(U, V)) = \exp(-U).\exp(U + V).$$

Let $M_1(U, V)$ (resp. $H_1(U, V)$) denote the sum of the bihomogeneous terms of the series M (resp. H) whose degree in V is 1.

(*a*) Show that

$$M_1(U, V) = \sum_{n=0}^{\infty} \frac{(-1)^n}{(n + 1)!} (\operatorname{ad} U)^n(V) = (f(\operatorname{ad} U))(V)$$

where $f(T) = (1 - e^{-T})/T$.

(Same argument as for the proof of Proposition 5. It can also be reduced to Proposition 5 by using the identity

$$\exp(U).\exp(M(U, V)).\exp(-U) = \exp(H(U + V, -U)),$$

together with Exercise 1.)

† For properties of Lie algebras satisfying an Engel condition and their applications to the "restricted Burnside problem", see A. Kostrikin, *Izv. Akad. Nauk SSSR*, **23** (1959), pp. 3–34.

(*b*) Show that H(U, M(U, V)) = U + V. Deduce that

$$H_1(U, M_1(U, V)) = V.$$

(*c*) Let $g(T) = 1/f(T) = 1 + T/2 + \sum_{n=1}^{\infty} \frac{1}{(2n)!} b_{2n} T^{2n}$, where the b_{2n} are the *Bernouilli numbers* (*Functions of a Real Variable*, Chapter VI, § 1, no. 4). Show that

$$H_1(U, V) = (g(\mathrm{ad}\, U))(V) = V + \tfrac{1}{2}[U, V] + \sum_{n=1}^{\infty} \frac{1}{(2n)!} b_{2n} (\mathrm{ad}\, U)^{2n}(V).$$

Derive the first few terms of the expansion of $H_1(U, V)$:

$$H_1(U, V) = V + \tfrac{1}{2}(\mathrm{ad}\, U)(V) + \frac{1}{12}(\mathrm{ad}\, U)^2(V) - \frac{1}{720}(\mathrm{ad}\, U)^4(V) + \frac{1}{30240}(\mathrm{ad}\, U)^6(V) - \cdots$$

(*d*) Show that the sum of the bihomogeneous terms of H(U, V) whose degree in U is 1 is the series:

$$U + \tfrac{1}{2}[U, V] + \sum_{n=1}^{\infty} \frac{1}{(2n)!} b_{2n} (\mathrm{ad}\, V)^{2n}(U).$$

(Apply (*c*) to H(V, U) and use Exercise 2 to pass from H(V, U) to H(U, V).)

(*e*) Let W be another indeterminate and let X(U, V, W) be the sum of the trihomogeneous terms of H(U, V + W) whose degree in W is 1. Show that

$$X(U, H_1(V, W)) = H_1(H(U, V), W).$$

(Use the identity H(U, H(V, W)) = H(H(U, V), W).) Deduce that

$$X(U, V, W) = (g(\mathrm{ad}\, H(U, V)) \circ f(\mathrm{ad}\, V))(W).$$

¶ 4. Let X be a finite set and let $\hat{L} = \hat{L}(X)$. We give $\hat{L}$ the Hausdorff law of composition $(a, b) \mapsto a \,\mathrm{H}\, b$, which we denote simply by $a.b$. If $t \in K$, $a \in \hat{L}$, we write $a^t = ta$.

(*a*) Let $\mathscr{H}$ be a Hall set relative to X. If $m \in \mathscr{H}$, let m_L denote the corresponding element of the Lie algebra L(X) (§ 2, no. 11) and $m_{\mathscr{H}}$ the basic commutator of the group $\hat{L}$ defined by m (§ 5, no. 4, *Remark*). If m is of length l, show that $m_{\mathscr{H}} = m_L + \mu$, where μ is of filtration $\geqslant l + 1$ (argue by induction on l). Deduce that every element w of the group $\hat{L}$ can be written uniquely in the form

$$w = \prod_{m \in \mathscr{H}} m_{\mathscr{H}}^{\alpha(m)},$$

where $\alpha(m) \in \mathrm{K}$ and the product is convergent in the topological group $\hat{\mathrm{L}}$.

(*b*) Let **P** be the set of prime numbers and let c be an integer $\geqslant 1$. Let $\mathrm{K} = \mathbf{Q}$. Show that the group $\hat{\mathrm{L}}/\hat{\mathrm{L}}_c$ is identified with the **P**-envelope $\mathrm{F}^c_{\mathbf{P}}$ of the group $\mathrm{F}^c = \mathrm{F}(\mathrm{X})/\mathrm{C}^c\mathrm{F}(\mathrm{X})$, cf. § 5, Exercise 6.

(*c*) Take X to be a set of two elements $\{\mathrm{U}, \mathrm{V}\}$. Show the existence of two families $\alpha(m)$, $\beta(m)$ of rational numbers such that

$$\mathrm{U} + \mathrm{V} = \prod_{m \in \mathscr{H}} m_{\mathscr{H}}^{\alpha(m)}$$

$$[\mathrm{U}, \mathrm{V}] = \prod_{m \in \mathscr{H}} m_{\mathscr{H}}^{\beta(m)}.$$

For example

$$\mathrm{U} + \mathrm{V} = \mathrm{U}.\mathrm{V}.(\mathrm{U}, \mathrm{V})^{-\frac{1}{2}} \ldots$$

$$[\mathrm{U}, \mathrm{V}] = (\mathrm{U}, \mathrm{V}) \ldots$$

(*d*) Let $\mathfrak{g}$ be a nilpotent Lie **Q**-algebra with the Hausdorff group law. Let u, v be in $\mathfrak{g}$. Show that

$$u + v = \prod_{m \in \mathscr{H}} m_{\mathscr{H}}(u, v)^{\alpha(m)}$$

$$[u, v] = \prod_{m \in \mathscr{H}} m_{\mathscr{H}}(u, v)^{\beta(m)},$$

where α and β are the families of rational numbers defined above ("*inversion of the Hausdorff formula*"). (Use the continuous homomorphism $\phi: \hat{\mathrm{L}} \rightarrow \mathfrak{g}$ such that $\phi(\mathrm{U}) = u$, $\phi(\mathrm{V}) = v$ and note that it is also a homomorphism for the Hausdorff group law.)

(*e*) Let G be a torsion-free divisible nilpotent group (i.e. **P**-torsion-free and **P**-divisible). If $u \in \mathrm{G}$, $t \in \mathbf{Q}$, u^t is defined (§ 4, Exercise 16). If $u \in \mathrm{G}$, $v \in \mathrm{G}$, we define $u + v$ and $[u, v]$ by the formulae of (*d*) above. Show that G thus has a nilpotent *Lie* **Q**-*algebra* structure and that the corresponding Hausdorff law is the given group law on G (verify these assertions when G is a group $\mathrm{F}^c_{\mathbf{P}}$, cf. (*b*), and pass from this to the general case using homomorphisms of $\mathrm{F}^c_{\mathbf{P}}$ into G).

Let f be a mapping of G into a torsion-free divisible nilpotent group G′. Show that f is a (group) homomorphism if and only if it is a Lie algebra homomorphism. (The Hausdorff law therefore defines an *isomorphism* of the "category" of nilpotent Lie **Q**-algebras onto that of torsion-free divisible nilpotent groups.)

5. Let $\mathfrak{g}$ be a nilpotent Lie **Q**-algebra which we give the Hausdorff group law, denoted by $(x, y) \mapsto x.y$.

(*a*) Show that, if $x \in \mathfrak{g}$, $y \in \mathfrak{g}$, then

$$x.y.x^{-1} = \sum_{n=0}^{\infty} \frac{1}{n!} (\mathrm{ad}\, x)^n(y).$$

(Use Exercise 2.)

(*b*) Let $\mathfrak{h}$ be a subset of $\mathfrak{g}$. Show that $\mathfrak{h}$ is a Lie subalgebra (resp. an ideal) of $\mathfrak{g}$ if and only if it is a saturated subgroup (resp. a saturated normal subgroup) of the group $\mathfrak{g}$. (Use the formulae of Exercise 4 (*d*) to pass from the group law to the Lie algebra law.)

(*c*) Let $\mathfrak{h}$ be a subgroup of the group $\mathfrak{g}$. Show that the **P**-saturation of $\mathfrak{h}$ is $\mathbf{Q}.\mathfrak{h}$.

(*d*) Let $\mathfrak{h}$ be a Lie subalgebra of $\mathfrak{g}$. Show that the centralizer (resp. normalizer) of $\mathfrak{h}$ in the group $\mathfrak{g}$ is the set of $x \in \mathfrak{g}$ such that $(\mathrm{ad}\, \mathrm{x})(\mathfrak{h}) = 0$ (resp. $(\mathrm{ad}\, x)(\mathfrak{h}) \subset \mathfrak{h}$).

(*e*) Show that the lower central series of the Lie algebra $\mathfrak{g}$ coincides with that of the group $\mathfrak{g}$ and that the associated graded Lie algebra $\mathrm{gr}(\mathfrak{g})$ is the same from the "group" point of view as from the "Lie algebra" point of view.

¶ 6. Let G be a finitely generated torsion-free nilpotent group. Show that, if n is sufficiently large, G can be embedded in the lower strict triangular group of order n over **Z**. (Let $(i, \overline{\mathrm{G}})$ be a **P**-envelope of G, cf. § 5, Exercise 6; with its canonical Lie algebra structure (Exercise 4), $\overline{\mathrm{G}}$ is a finite-dimensional nilpotent Lie **Q**-algebra. Apply Ado's Theorem (Chapter I, § 7) to $\overline{\mathrm{G}}$ and deduce an embedding of $\overline{\mathrm{G}}$ in a strict triangular group over **Q**. Pass from **Q** to **Z** by conjugating by a suitable integral diagonal matrix.)

§ 7

1. Let $\mathfrak{g}$ be a complete normed Lie algebra over K such that

$$\|[x, y]\| \leqslant \|x\|\,\|y\|$$

for x, y in $\mathfrak{g}$. Let Θ denote the set of $x \in \mathfrak{g}$ such that $\|x\| < \frac{1}{3}\log\frac{3}{2}$. For x, y in Θ, $h(x, y) \in \Theta$, cf. no. 2.

(*a*) We write $f(\mathrm{T}) = (1 - e^{-\mathrm{T}})/\mathrm{T}$ and $g(\mathrm{T}) = 1/f(\mathrm{T})$. The series f converges throughout the complex plane and the series g on the open disc of radius 2π (cf. *Functions of a Real Variable*, Chapter VI, § 2, no. 3). Deduce that, if $z \in \Theta$, $f(\mathrm{ad}\, z)$ and $g(\mathrm{ad}\, z)$ are defined and that they are elements of $\mathscr{L}(\mathfrak{g}; \mathfrak{g})$ and inverses of one another.

(*b*) Let $\mathrm{D}_2 h(x, y)$ be the second partial derivative of h at a point (x, y) of $\Theta \times \Theta$ (*Differentiable and Analytic Manifolds*, R, 1.6.2). It is an element of $\mathscr{L}(\mathfrak{g}; \mathfrak{g})$. Show that

$$\mathrm{D}_2 h(x, y) = g(\mathrm{ad}\, h(x, y)) \circ f(\mathrm{ad}\, y).$$

(Use Exercise 3 (*e*) of § 6 to show this formula when x and y are sufficiently close to zero and pass from this to the general case by analytic continuation.)

Show that the formula

$$f(\mathrm{ad}\, h(x, y)) \circ \mathrm{D}_2 h(x, y) = f(\mathrm{ad}\, y)$$

is valid at every point of the domain of absolute convergence of the formal power series $\tilde{\mathrm{H}}$ (cf. Proposition 1).

§ 8

We suppose that the residual characteristic p of the field K is >0. We denote by $\mathfrak{o}_K$ the ring of the valuation v of K.

1. (*a*) An element π of K is called admissible if $v(\pi) = \theta$ and $\pi^{p-1}/p \equiv -1 \pmod{\pi\mathfrak{o}_K}$.
(Note that $\pi^{p-1}/p \in \mathfrak{o}_K$ since $v(\pi) = \theta$.)
Show that there exist fields K satisfying the conditions of the paragraph and containing an admissible element (adjoin to $\mathbf{Q}_p$ a $(p-1)$-th root of $-p$).

(*b*) Let π be an admissible element of K. Consider the following formal power series with coefficients in K:

$$e_\pi(U) = \frac{1}{\pi}e(\pi U) = \sum_{n=1}^{\infty} \pi^{n-1}U^n/n!,$$

$$l_\pi(U) = \frac{1}{\pi}l(\pi U) = \sum_{n=1}^{\infty} (-\pi)^{n-1}U^n/n.$$

Show that these series are inverses of one another and that their coefficients belong to $\mathfrak{o}_K$ (use Lemma 2).

(*c*) Let $\{U, V\}$ be a set with two elements and let $H(U, V)$ be the Hausdorff series. We write

$$H_\pi(U, V) = \frac{1}{\pi}H(\pi U, \pi V).$$

Then $H_\pi(U, V) \in \hat{L}_K(\{U, V\})$.
Show that

$$H_\pi(U, V) = l_\pi(e_\pi(U) + e_\pi(V) + \pi e_\pi(U).e_\pi(V)).$$

Deduce that $H_\pi \in \hat{L}_{\mathfrak{o}_K}(\{U, V\})$, i.e. that the coefficients of H_π belong to $\mathfrak{o}_K$; hence another proof of Proposition 1.

(*d*) Let $\tilde{e}_\pi$, $\tilde{l}_\pi$ and $\tilde{H}_\pi$ denote the series obtained by reducing e_π, l_π and H_π modulo $\pi\mathfrak{o}_K$; their coefficients belong to the ring $\mathfrak{o}_K/\pi\mathfrak{o}_K$.

Show that $\tilde{l}_\pi(U) = U - U^p$ (note that $v_p(n) < (n-1)\theta$ if $n \neq 1, p$). Deduce that

$$\tilde{e}_\pi(U) = U + U^p + \cdots + U^{p^n} + \cdots$$

and that

$$\tilde{H}_\pi(U, V) = \tilde{e}_\pi(U) + \tilde{e}_\pi(V) - (\tilde{e}_\pi(U) + \tilde{e}_\pi(V))^p,$$

or also

$$\tilde{H}_\pi(U, V) = U + V - \Lambda_p\left(\sum_{n=0}^{\infty} U^{p^n}, \sum_{n=0}^{\infty} V^{p^n}\right),$$

where Λ_p is defined by the formula

$$\Lambda_p(U, V) = (U + V)^p - U^p - V^p,$$ cf. Chapter I, § 1, Exercise 19.

In particular, $\tilde{H}_\pi(U, V)$ belongs to $\hat{L}_{F_p}(\{U, V\})$ and its homogeneous component of degree p is $-\Lambda_p(U, V)$. Then

$$\tilde{H}_\pi(U, V) = \tilde{H}_\pi(V, U)$$
$$\tilde{H}_\pi(U, \tilde{H}_\pi(V, W)) = \tilde{H}_\pi(\tilde{H}_\pi(U, V), W) \quad \text{in } \hat{L}_{F_p}(\{U, V, W\}),$$

where W is a third indeterminate.

(*e*) Let $C(U, V) = H(-U, H(-V, H(U, V)))$ ("Hausdorff commutator"). Then $\exp(C(U, V)) = \exp(-U)\exp(-V)\exp(U)\exp(V)$. We write

$$C_\pi(U, V) = \frac{1}{\pi} C(\pi U, \pi V).$$

Show that the coefficients of $C_\pi(U, V)$ belong to $\pi \mathfrak{o}_K$ (use the fact that $\tilde{H}_\pi(U, V) = \tilde{H}_\pi(V, U)$).†

2. We know (§ 6, Exercise 3) that if $n \geqslant 2$ the component of $H(U, V)$ of bidegree $(n, 1)$ is $\frac{1}{n!} b_n (\operatorname{ad} U)^n(V)$, where b_n is the n-th Bernoulli number. Deduce from this and Proposition 1 the inequality

$$v_p(b_n/n!) \leqslant n/(p-1).$$

Recover this result by means of the Clausen-von Staudt Theorem (*Functions of a Real Variable*, Chapter VI, § 2, Exercise 6) and show that there is equality if and only if $S(n) = p - 1$.

3. Suppose that K contains a primitive p-th root of unity w. We write $\pi = w - 1$. Using the formula

$$w^i - 1 = \pi(1 + w + \cdots + w^{i-1})$$

show that $v(w^i - 1) = v(\pi)$ for $1 \leqslant i \leqslant p - 1$ and that

$$\frac{w^i - 1}{\pi} \equiv i \pmod{\pi}.$$

Deduce, by means of the formula $p = \prod_{i=1}^{p-1} (w^i - 1)$, that π is *admissible* (Exercise 1).

¶ 4. Let c be an integer $\geqslant 1$ and let P be a set of prime numbers containing all prime numbers $\leqslant c$. Let $\mathbf{Z}_P = S_P^{-1}\mathbf{Z}$ (§ 4, Exercise 16). Show that the terms of degree $\leqslant c$ in the Hausdorff series $H(U, V)$ belong to $L_{\mathbf{Z}_P}(\{U, V\})$. Deduce

† For more details on this exercise, see M. Lazard, *Bull. Soc. Math. France*, **91** (1963), pp. 435–451.

that, if $\mathfrak{g}$ is a nilpotent Lie $\mathbf{Z}_P$-algebra of class $\leqslant c$, the law of composition $(u, v) \mapsto H(u, v)$ makes $\mathfrak{g}$ into a P-torsion-free P-divisible nilpotent group of class $\leqslant c$. Show that conversely every group with these properties can be obtained by this process. (Use "the inversion of the Hausdorff formula", cf. § 6, Exercise 4, and show that the exponents $\alpha(m)$, $\beta(m)$ which appear there belong to $\mathbf{Z}_P$ when $l(m) \leqslant c$.)

In particular, every p-group of order p^n and class $<p$ can be obtained by means of the Hausdorff law from a nilpotent Lie $\mathbf{Z}$-algebra of class $<p$ with p^n elements. (Take P to be the set of prime numbers distinct from p.)

APPENDIX

1. Let Φ_n be the cyclotomic polynomial of index n (*Algebra*, Chapter V, § 11, no. 2). Using the formula

$$X^n - 1 = \prod_{d|n} \Phi_d(X),$$

show that

$$\Phi_n(X) = \prod_{d|n} (X^{n/d} - 1)^{\mu(d)}.$$

(Apply the Möbius inversion formula to the multiplicative group of the field $\mathbf{Q}(X)$.)

2. Let D be the total algebra of the monoid $\mathbf{N}^*$ (*Algebra*, Chapter III, § 2, no. 10). If $n \in \mathbf{N}^*$, let n^ω denote its image in D, so that $1^\omega = 1$ and $(nm)^\omega = n^\omega m^\omega$ for n, m in $\mathbf{N}^*$.† Every element f of D can be written uniquely as a series $\sum_{n=1}^{\infty} a_n n^\omega$, where $a_n \in K$.

(*a*) Let $f = \sum_{n=1}^{\infty} a_n n^\omega$ be an element of D. Show that f is invertible in D if and only if a_1 is invertible in K. In particular, if K is a local ring, so is D.

(*b*) We write $\zeta = \sum_{n=1}^{\infty} n^\omega$ and $\mu = \sum_{n=1}^{\infty} \mu(n) n^\omega$. Show that ζ and μ are inverses of one another. Deduce that, if $s = \sum_n s(n) n^\omega$ and $t = \sum_n t(n) n^\omega$ are two elements of D, the relations $s = \zeta . t$ and $t = \mu . s$ are equivalent (variant of the Möbius inversion formula).

(*c*) Let P be the set of prime numbers. Show that the family of $(1 - p^\omega)$ for $p \in P$ is multipliable in D and that

$$\mu = \sum_{p \in P}^{\infty} (1 - p^\omega) \quad \text{and} \quad \zeta = \prod_{p \in P} \frac{1}{1 - p^\omega}.$$

† We often write $-s$ instead of ω; we then say that the elements of D are the *Dirichlet formal power series* with coefficients in K.

CHAPTER III

Lie Groups

Throughout the chapter, K *denotes either the valued field* **R** *of real numbers, or the valued field* **C** *of complex numbers, or a non-discrete complete ultrametric commutative field. We assume that* K *is of characteristic* 0 *from* § 4 *onwards, that* K = **R** *or* **C** *in* § 6, *that* K *is ultrametric in* § 7. *Unless otherwise mentioned, all the manifolds, all the algebras and all the vector spaces considered are over* K. *Recall that, when we speak of a manifold of class* C^r, $r \in N_K$, *that is* $r = \omega$ *if* $K \neq \mathbf{R}$ *and* $1 \leqslant r \leqslant \omega$ *if* $K = \mathbf{R}$.

The conventions on norms, normable spaces and normed spaces are those of *Differentiable and Analytic Manifolds*, R.

Recall that a *normable algebra* over K is a (not necessarily associative) algebra A over K, with a topology $\mathscr{T}$ which possesses the following properties:

(1) $\mathscr{T}$ can be defined by a norm;

(2) the mapping $(x, y) \mapsto xy$ of A × A into A is continuous.

The group of bicontinuous automorphisms of A is denoted by Aut(A). Every finite-dimensional algebra over K is a normable algebra with the canonical topology. A *normed algebra* over K is an algebra A over K with a norm such that $\|xy\| \leqslant \|x\| \|y\|$ for all x, y in A; the algebra A with the topology defined by this norm is a normable algebra. If A is a normable algebra, there exists a norm on A defining its topology and making A into a normed algebra.

If G is a group, e_G, or simply e, denotes the identity element of G. For $g \in G$, $\gamma(g)$, $\delta(g)$ and Int(g) denote the mappings $g' \mapsto gg'$, $g' \mapsto g'g^{-1}$ and $g' \mapsto gg'g^{-1}$ of G into G. If f is a mapping of G into a set E, $\check{f}$ denotes the mapping $g \mapsto f(g^{-1})$ of G into E.

§ 1. LIE GROUPS

1. DEFINITION OF A LIE GROUP

Let G be a set. A group structure and an analytic K-manifold structure on G are called *compatible* if the following condition holds:

(GL) The mapping $(g, h) \mapsto gh^{-1}$ of G × G into G is analytic.

DEFINITION 1. *A Lie group over* K *is a set* G *with a group structure and an analytic* K-*manifold structure such that these two structures are compatible.*

A Lie group over **R** (resp. **C**, $\mathbf{Q}_p$) is called a real (resp. complex, p-adic) Lie group.

Let G be a group with an analytic manifold structure. For g, h, g_0, h_0 in G,

$$gh^{-1} = (g_0h_0^{-1})h_0((g_0^{-1}g)(h_0^{-1}h)^{-1})h_0^{-1}. \tag{1}$$

It follows that G is a Lie group if and only if the following three conditions hold:

(GL_1) for all $g_0 \in G$, the mapping $g \mapsto g_0g$ of G into G is analytic;
(GL_2) for all $g_0 \in G$, the mapping $g \mapsto g_0gg_0^{-1}$ of G into G is analytic in an open neighbourhood of e;
(GL_3) the mapping $(g, h) \mapsto gh^{-1}$ of $G \times G$ into G is analytic in an open neighbourhood of (e, e).

Let G be a Lie group. For all $g \in G$, $\gamma(g)$ and $\delta(g)$ are automorphisms of the underlying manifold of G. It follows that this manifold is pure (*Differentiable and Analytic Manifolds*, R, 5.1.7). In particular, the dimension of G at g is equal to dim G for all $g \in G$ (recall that dim G is an integer $\geqslant 0$ or $+\infty$).

Since an analytic mapping is continuous, a Lie group is a topological group with the underlying topology of its manifold structure. Let G be a set. A topological group structure and an analytic K-manifold structure on G are called *compatible* if the group structure and the manifold structure are compatible and the topology on G is the topology underlying the manifold structure.

Lemma 1. *Let* G *be a Lie group,* U *an open neighbourhood of* e, E *a complete normed space and* $\phi: U \to E$ *a chart of the manifold* G. *There exists a neighbourhood* W *of* e *contained in* U *such that* $\phi \mid W$ *is an isomorphism of* W (*with the right uniform structure*) *onto* $\phi(W)$ (*with the uniform structure induced by that on* E).

It can be assumed that $\phi(e) = 0$. Let $U' = \phi(U)$. Let $\psi: U' \to U$ be the inverse mapping of ϕ. Let V be a symmetric open neighbourhood of e such that $V^2 \subset U$ and let $V' = \phi(V)$. We define mappings θ_1, θ_2 of $V' \times V'$ into $V' \times U'$ as follows:

$$\theta_1(x, y) = (x, \phi(\psi(x)\psi(y)^{-1}))$$
$$\theta_2(x, y) = (x, \phi(\psi(y)^{-1}\psi(x))).$$

It is immediately verified that $\theta_2(\theta_1(x, y)) = \theta_1(\theta_2(x, y)) = (x, y)$ for x, y sufficiently close to 0. On the other hand, θ_1 and θ_2 are analytic and hence strictly differentiable at (0, 0). Therefore (*Differentiable and Analytic Manifolds*, R, 1.2.2) there exist a neighbourhood W′ of 0 in V′ and constants $a > 0$, $b > 0$ such that

$$\begin{aligned} a(\|x_1 - x_2\| + \|\phi(\psi(x_1)\psi(y_1)^{-1}) - \phi(\psi(x_2)\psi(y_2)^{-1})\|) \\ \leqslant \|x_1 - x_2\| + \|y_1 - y_2\| \\ \leqslant b(\|x_1 - x_2\| + \|\phi(\psi(x_1)\psi(y_1)^{-1}) - \phi(\psi(x_2)\psi(y_2)^{-1})\|) \end{aligned}$$

for all x_1, x_2, y_1, y_2 in W'. Writing $x_1 = x_2 = y_2$, we obtain

$$(2) \qquad a\|\phi(\psi(x_1)\psi(y_1)^{-1})\| \leqslant \|x_1 - y_1\| \leqslant b\|\phi(\psi(x_1)\psi(y_1)^{-1})\|.$$

For $\delta > 0$, let N_δ be the set of ordered pairs $(x, y) \in W' \times W'$ such that $\|x - y\| \leqslant \delta$. The N_δ form a fundamental system of entourages in W'. We write $W = \psi(W')$. Let M_δ be the set of ordered pairs $(u, v) \in W \times W$ such that $\|\phi(uv^{-1})\| \leqslant \delta$. The M_δ form a fundamental system of entourages in W with the right uniform structure. But relation (2) proves that

$$N_\delta \subset (\phi \times \phi)(M_{a^{-1}\delta}), \qquad (\phi \times \phi)(M_\delta) \subset N_{b\delta}$$

and hence W has the property of the lemma.

PROPOSITION 1. *A Lie group is a complete metrizable topological group.*

Since e admits an open neighbourhood homeomorphic to an open ball of a normed space, e admits a countable fundamental system of neighbourhoods whose intersection is $\{e\}$. Hence G is metrizable (*General Topology*, Chapter III, § 1, Corollary to Proposition 2 and Chapter IX, § 3, Proposition 1). By Lemma 1, there exists a neighbourhood of e which is complete under the right uniform structure and hence G is complete (*General Topology*, Chapter III, § 3, Proposition 4.

PROPOSITION 2. *Let* G *be a Lie group.*

(i) *If* K = **R** *or* **C**, G *is locally connected.*

(ii) *If* K *is distinct from* **R** *and* **C**, G *is zero-dimensional* (*General Topology*, Chapter IX, § 6, Definition 5).

(iii) *Suppose that* K *is locally compact. For* G *to be locally compact, it is necessary and sufficient that* G *be finite-dimensional.*

(iv) *If* G *is generated by a subspace whose topology admits a countable base, then the topology on* G *admits a countable base.*

Let U be a neighbourhood of e. There exists an open neighbourhood U_1 of e contained in U and homeomorphic to an open ball of a normed space E over K. If K = **R** or **C**, U_1 is connected, which proves (i). Suppose that K is ultrametric. There exists a neighbourhood U_2 of e which is closed in G and such that $U_2 \subset U_1$. Then there exists a neighbourhood U_3 of e such that $U_3 \subset U_2$ and U_3 is open and closed relative to U_1. Then U_3 is closed relative to U_2 and hence G and open relative to U_1 and hence G. This proves (ii). For G to be locally compact, it is necessary and sufficient that E be locally compact; if K is locally compact, this amounts to saying that E is finite-dimensional (*Topological Vector Spaces*, Chapter I, § 2, Theorem 3), whence (iii). Suppose that G is generated by a subset V and let $W = V \cup V^{-1}$; then

$$G = W \cup W^2 \cup W^3 \cup \ldots;$$

if there exists a sequence dense in V, we see that there exists a sequence dense in G and, as G is metrizable (Proposition 1), the topology on G admits a countable base.

COROLLARY. *If* K = $\mathbf{R}$ *or* $\mathbf{C}$ *and* G *is connected and finite-dimensional, then* G *is locally connected and locally compact and its topology admits a countable base.*

Lemma 2. *Let* X *be a manifold of class* C^r, *e a point of* X, U *and* V *open neighbourhoods of e and m a mapping of class* C^r *of* U × U *into* X *satisfying the following conditions:*

(a) $m(e, x) = m(x, e) = x$ *for all* $x \in U$;

(b) $V \subset U$, $m(V \times V) \subset U$ *and* $m(m(x, y), z) = m(x, m(y, z))$ *for all* x, y, z *in* V.

Then there exists an open neighbourhood W *of e in* V *and an automorphism* θ *of the manifold* W *such that* $\theta(e) = e$, $\theta(\theta(x)) = x$ *and* $m(x, \theta(x)) = m(\theta(x), x) = e$ *for all* $x \in W$.

$m(e, y) = y$ for all $y \in U$ and hence, by the implicit function theorem, there exists an open neighbourhood W_1 of e in V and a mapping θ_1 of class C^r of W_1 into V such that $\theta_1(e) = e$, $m(x, \theta_1(x)) = e$ for all $x \in W_1$. Similarly, there exists an open neighbourhood W_2 of e in V and a mapping θ_2 of class C^r of W_2 into V such that $\theta_2(e) = e$, $m(\theta_2(x), x) = e$ for all $x \in W_2$. For $x \in W_1 \cap W_2$.

$$\begin{aligned}\theta_2(x) &= m(\theta_2(x), e) = m(\theta_2(x), m(x, \theta_1(x))) \\ &= m(m(\theta_2(x), x), \theta_1(x)) = m(e, \theta_1(x)) = \theta_1(x).\end{aligned}$$

Let $\theta(x)$ be the common value of $\theta_1(x)$ and $\theta_2(x)$ for $x \in W_1 \cap W_2$. Let W be the set of $x \in W_1 \cap W_2$ such that $\theta(x) \in W_1 \cap W_2$. The set W is open. For $x \in W$,

$$\theta(\theta(x)) = m(m(x, \theta(x)), \theta(\theta(x))) = m(x, m(\theta(x), \theta(\theta(x)))) = m(x, e) = x$$

and hence $\theta(x) \in W$. We see that $\theta \mid W$ defines an automorphism of the manifold W.

PROPOSITION 3. *Let* X *be an analytic manifold and m an analytic associative law of composition on* X *admitting an identity element. The set* G *of invertible elements of* X *is open in* X *and* G *is a Lie group with* $m \mid (G \times G)$ *and the manifold structure induced by that on* X.

By Lemma 2, G is a neighbourhood of the identity element. For all $g \in G$, the mapping $x \mapsto m(g, x)$ is an automorphism of the manifold X. Hence the image of G under this mapping is a neighbourhood of g, obviously contained in G. Therefore G is open in X. Clearly conditions (GL_1) and (GL_2) hold. Condition (GL_3) holds by Lemma 2.

Examples of Lie groups

(1) Let E be a complete normable space over K. The mapping $(x, y) \mapsto x - y$ of E × E into E is continuous and linear and hence analytic.

Hence E, with its additive group and analytic manifold structures, is a Lie group.

In particular, K is a Lie group.

(2) Let A be a complete normable unital associative algebra over K. The multiplication $(x, y) \mapsto xy$ of A × A into A is bilinear and continuous and hence analytic. Proposition 3 shows that the group A* of invertible elements of A is open in A (which also follows from *General Topology*, Chapter IX, § 3, Proposition 13) and that A* is a Lie group.

For example, let E be a complete normable space over K and let $A = \mathscr{L}(E)$ (*General Topology*, Chapter IX, § 3, Proposition 5). Then A* is the automorphism group **GL**(E) of E. *This group therefore has canonically a Lie group structure over* K. More particularly, **GL**(*n*, K), with the manifold structure induced by that on $\mathbf{M}_n(K)$, is a Lie group. For $n = 1$, we see that the multiplicative group K* is a Lie group with the manifold structure induced by that on K.

(3) Let G be a Lie group over K. Let K′ = **R** or **C** or a non-discrete complete ultrametric field and σ an isomorphism of the valued field K′ onto a valued subfield of K. Then the group G, with the K′-manifold structure obtained by restriction of scalars, is a Lie group over K′, which is said to be *derived from the Lie group* G *by restriction of scalars* (from K to K′ by means of σ). For example, every complex Lie group has canonically a real Lie group structure. Again, with every complex Lie group G is associated a complex Lie group called the *conjugate* of G, derived from G by means of the automorphism $z \mapsto \bar{z}$ of **C**.

2. MORPHISMS OF LIE GROUPS

DEFINITION 2. *Let* G *and* H *be Lie groups. A Lie group morphism of* G *into* H (*or simply a morphism of* G *into* H *if no confusion can arise*) *is a mapping of* G *into* H *which is a group homomorphism and is analytic. The automorphism group of* G *is denoted by* Aut(G).

The identity mapping of G is a morphism. The composition of two morphisms is a morphism. If f: G → H and f': H → G are two inverse morphisms, f and f' are Lie group isomorphisms.

Examples. (1) Let G be a Lie group. For all $x \in G$, Int(x) is an automorphism of the Lie group G.

(2) Let G be a Lie group. Let $G^{\vee}$ denote the opposite group to G, with the same manifold structure as G. It is immediate that $G^{\vee}$ is a Lie group (called the *opposite* Lie group to G) and that the mapping $g \mapsto g^{-1}$ is an isomorphism of the Lie group G onto the Lie group $G^{\vee}$.

(3) Let G be a Lie group and E a complete normable space. An *analytic linear representation* of G on E (or simply a linear representation of G on E when

no confusion can arise) is a morphism of the Lie group G into the Lie group **GL**(E), in other words an analytic mapping π of G into **GL**(E) such that $\pi(gg') = \pi(g)\pi(g')$ for g, g' in G. Suppose that E admits a finite basis $(e_1, e_2, \ldots, e_n)$ over K; let $(e_1^*, e_2^*, \ldots, e_n^*)$ be the dual basis; let ρ be a homomorphism of the group G into the group **GL**(E); then the following conditions are equivalent:

(i) ρ is an analytic linear representation;

(ii) for all $x \in E$ and $x' \in E'$, the function $g \mapsto \langle\rho(g)x, x'\rangle$ on G is analytic;

(iii) for all i and j, the function $g \mapsto \langle\rho(g)e_i, e_j^*\rangle$ on G is analytic.

For the implications (i) $\Rightarrow$ (ii) $\Rightarrow$ (iii) are clear. On the other hand, the functions $u \mapsto \langle ue_i, e_j^*\rangle$ form a coordinate system on $\mathscr{L}$(E); hence their restrictions to **GL**(E) form a coordinate system on **GL**(E), whence the implication (iii) $\Rightarrow$ (i).

Let G be a real Lie group, E a real complete normable space and ρ a homomorphism of the group G into the group **GL**(E). We shall see in § 8, Theorem 1 that, if ρ is continuous (when **GL**(E) has the topology derived from the norm on L(E)), then ρ is analytic. But note that this notion of continuity is different from that considered in *Integration*, Chapter VIII, § 2, Definition 1 (ii) (Exercise 1).

(4) Let G be a *real* Lie group and E a *complex* complete normable space. An *analytic linear representation of* G *on* E is a morphism of G into the underlying real Lie group of **GL**(E).

PROPOSITION 4. *Let* G *and* H *be Lie groups and* f *a homomorphism of the group* G *into the group* H. *For* f *to be analytic, it is necessary and sufficient that there exist a non-empty open subset* U *of* G *such that* $f \mid U$ *is analytic.*

The condition is obviously necessary. Suppose that it holds. For all $x_0 \in G$, $f(x_0x) = f(x_0)f(x)$ for all $x \in U$ and hence $f \mid x_0U$ is analytic. But the sets x_0U, where $x_0 \in G$, form an open covering of G.

Remark. If f is an immersion at e (resp. a submersion at e), clearly f is an immersion (resp. a submersion).

3. LIE SUBGROUPS

Let G be a Lie group and H a subgroup of G which is at the same time a submanifold of G. Then the mapping $(x, y) \mapsto xy^{-1}$ of H $\times$ H into G is analytic and hence the mapping $(x, y) \mapsto xy^{-1}$ of H $\times$ H into H is analytic. (*Differentiable and Analytic Manifolds*, R, 5.8.5). Thus H, with the group and manifold structures induced by those on G, is a Lie group.

DEFINITION 3. *Let* G *be a Lie group. A subset* H *of* G *is called a Lie subgroup if* H *is a subgroup and a submanifold of* G.

An open subgroup of G is a Lie subgroup of G. In particular, if G is a real or complex Lie group, its identity component is a Lie subgroup of G.

PROPOSITION 5. *Let* G *be a Lie group and* H *a Lie subgroup of* G.

(i) H *is closed in* G.

(ii) *The canonical injection of* H *into* G *is a Lie group morphism.*

(iii) *Let* L *be a Lie group and* f *a mapping of* L *into* G *such that* $f(L) \subset H$. *For* f *to be a morphism of* L *into* H, *it is necessary and sufficient that* f *be a morphism of* L *into* G.

By *Differentiable and Analytic Manifolds*, R, 5.8.3, H is locally closed. Hence H is closed (*General Topology*, Chapter III, § 2, Proposition 4). Assertion (ii) is obvious. Assertion (iii) follows from *Differentiable and Analytic Manifolds*, R, 5.8.5.

PROPOSITION 6. *Let* G *be a Lie group and* H *a subgroup of* G. *For* H *to be a Lie subgroup of* G, *it is necessary and sufficient that there exist a point* $h \in H$ *and an open neighbourhood* U *of* h *in* G *such that* $H \cap U$ *is a submanifold of* G.

The condition is obviously necessary. Suppose that it holds. For all $h' \in H$, the translation $\gamma(h'h^{-1})$ is an automorphism of the manifold G and maps the submanifold $H \cap U$ of U into the submanifold $(h'h^{-1}H) \cap (h'h^{-1}U)$ of $h'h^{-1}U$. As $h'h^{-1}H = H$ and $h'h^{-1}U$ is an open neighbourhood of h' in G, we see that every point of H has an open neighbourhood V such that $V \cap H$ is a submanifold of G. Hence H is a submanifold of G.

Let G be a Lie group and H a Lie subgroup of G. If L is a Lie subgroup of H, L is a Lie subgroup of G by *Differentiable and Analytic Manifolds*, R, 5.8.6. Let M be a Lie subgroup of G such that $M \subset H$. Then M is a Lie subgroup of H, for the canonical injection of M into H is obviously an immersion.

Let k be a non-discrete closed subfield of K. A Lie k-subgroup of G is a Lie subgroup of the underlying Lie k-group of G.

Remark. If "submanifold" is replaced by "quasi-submanifold" in Definition 3, we obtain the definition of *Lie quasi-subgroups* of G. (For finite-dimensional G, the Lie quasi-subgroups are just the Lie subgroups.) Suppose that K is of characteristic 0. Proposition 5 remains valid with the same proof for Lie quasi-subgroups. Proposition 6 remains valid with the same proof, replacing "Lie subgroup" by "Lie quasi-subgroup" and "submanifold" by "quasi-submanifold".

4. SEMI-DIRECT PRODUCTS OF LIE GROUPS

Let I be a *finite* set and $(L_i)_{i \in I}$ a family of Lie groups. The group and manifold structures on $L = \prod_{i \in I} L_i$ are compatible and L thus has a Lie group structure. L is called the *product* Lie group of the family of Lie groups $(L_i)_{i \in I}$.

Let L and M be Lie groups and σ a homomorphism of L into the automorphism group of the group M. Let S be the external semi-direct product of L by M relative to σ (*Algebra*, Chapter I, § 6, no. 1, Definition 2).

PROPOSITION 7. *If the mapping* $(m, l) \mapsto \sigma(l)m$ *of* M × L *into* M *is analytic, the group* S, *with the product manifold structure of* M *and* L, *is a Lie group.*

For l, l' in L and m, m' in M,

$$(m, l)(m', l')^{-1} = mll^{-1}m'^{-1} = m(\sigma(ll'^{-1})m'^{-1})ll'^{-1}$$
$$= (m(\sigma(ll'^{-1})m'^{-1}), ll'^{-1})$$

whence the proposition.

If the conditions of Proposition 7 hold, the Lie group S is called the *(external) semi-direct product Lie group of* L *by* M *relative to* σ.

Clearly the canonical injection of L (resp. M) into S is an isomorphism of L (resp. M) onto a Lie subgroup of S which we identify with L (resp. M). The canonical mapping of S onto L is a Lie group morphism.

Conversely, let G be a Lie group and L, M two Lie subgroups such that the group G is (algebraically) the semi-direct product of L by M(*Algebra*, Chapter I, § 6, no. 1). We write $\sigma(l)m = lml^{-1}$ for $l \in$ L and $m \in$ M. Then σ satisfies the conditions of Proposition 7. We can therefore form the semi-direct product Lie group S of L by M relative to σ. The mapping $j\colon (m, l) \mapsto ml$ of S onto G is a group isomorphism and is analytic. If j is a Lie group isomorphism, *the Lie group* G is called *the (internal) semi-direct product of* L *by* M and S and G are identified. For all $g \in$ G, we write $g = p(g)q(g)$, where $p(g) \in$ M and $q(g) \in$ L. For the Lie group G to be the semi-direct product of L by M, it is necessary and sufficient that one of the mappings $p\colon$ G → M and $q\colon$ G → L be analytic, in which case both are analytic; or alternatively, it is necessary and sufficient that $T_e(G)$ be the topological direct sum of $T_e(M)$ and $T_e(L)$ (for, if this condition holds, j is étale at e_S).

Example. Let E be a normable space, G = **GL**(E), T the translation group of E and A the permutation group of E generated by G and T. The group A is algebraically the semi-direct product of G by T. (If E is finite-dimensional, A is the affine group of E, cf. *Algebra*, Chapter II, § 9, no. 4). Let σ be the identity linear representation of G on E and S the external semi-direct product of G by E relative to σ. For all $x \in$ E, let t_x be the translation of E defined by x. The mapping $(x, u) \mapsto t_x \circ u$ is isomorphism Φ of the group S onto the group A. The mapping $(x, u) \mapsto \sigma(u)x = u(x)$ of E × $\mathscr{L}$(E) into E is continuous and bilinear and hence analytic; its restriction to E × G is therefore analytic. Thus the group S, with the product manifold structure of E and G, is a Lie group. We transport this structure to A by means of Φ. Then A becomes a Lie group, the internal semi-direct product of G by T as a Lie group.

PROPOSITION 8. *Let* G *and* H *be Lie groups,* $p\colon$ G → H *and* $s\colon$ H → G *Lie group morphisms such that* $p \circ s = \mathrm{id}_H$ *and* N = Ker p. *Then* N *is a Lie subgroup of* G, s *is an isomorphism of* H *onto a Lie subgroup of* G *and the Lie group* G *is the internal semi-direct product of* s(H) *by* N.

$T_e(p) \circ T_e(s) = \mathrm{id}_{T_e(H)}$ and hence p (resp. s) is a submersion (resp. an immersion). By *Differentiable and Analytic Manifolds*, R, 5.10.5, N is a Lie subgroup of G. On the other hand, s is a homeomorphism of H onto $s(H)$ and hence s is an isomorphism of H onto a Lie subgroup of G (*Differentiable and Analytic Manifolds*, R, 5.8.3). Finally, for all $g \in G$, $g = (s \circ p)(g).n$ for some $n \in N$; as $s \circ p$ is analytic, the Lie group G is the semi-direct product of $s(H)$ by N.

5. QUOTIENT OF A MANIFOLD BY A LIE GROUP

Let G be a Lie group, X a manifold of class C^r and $(g, x) \mapsto gx$ a law of left operation (*Algebra*, Chapter I, § 5, no. 1) of class C^r of G on X. For all $g \in G$, let $\tau(g)$ denote the automorphism $x \mapsto gx$ of X defined by g. For all $x \in X$, let $\rho(x)$ denote the orbital mapping $g \mapsto gx$ of G into X defined by x. Then

$$\rho(x) = \rho(gx) \circ \delta(g), \qquad \rho(x) = \tau(g) \circ \rho(x) \circ \gamma(g^{-1}) \tag{3}$$

for all $g \in G$ and $x \in X$. Hence

$$T_g(\rho(x)) = T_e(\rho(gx)) \circ T_g(\delta(g)) \tag{4}$$

$$T_g(\rho(x)) = T_x(\tau(g)) \circ T_e(\rho(x)) \circ T_g(\gamma(g^{-1})) \tag{5}$$

Proposition 9. *Let $x \in X$ and $g_0 \in G$.*

(i) *If $\rho(x)$ is an immersion* (resp. *a submersion, a subimmersion*) *at g_0, then, for all $g \in G$, $\rho(gx)$ is an immersion* (resp. *a submersion, a subimmersion*).

(ii) *If $\rho(x)$ is of rank k at g_0, then for all $g \in G$, $\rho(gx)$ is of constant rank equal to k.*

This follows immediately from formulae (4) and (5) since $T_g(\delta(g))$, $T_x(\tau(g))$ and $T_g(\gamma(g^{-1}))$ are isomorphisms.

Corollary. *Let $x \in X$. If K is of characteristic 0 and X is finite-dimensional, $\rho(x)$ is a subimmersion. If further $\rho(x)$ is injective, $\rho(x)$ is an immersion.*

This follows from Proposition 9 and *Differentiable and Analytic Manifolds*, R, 5.10.6.

Observe that, if η denotes the mapping $(g, x) \mapsto gx$ of $G \times X$ into X, then, for $g \in G$, $x \in X$, $u \in T_g(G)$, $v \in T_x(X)$,

$$T_{(g, x)}(\eta)(u, v) = T_{(g, x)}(\eta)(u, 0) + T_{(g, x)}(\eta)(0, x)$$

that is

$$T_{(g, x)}(\eta)(u, v) = T_g(\rho(x))u + T_x(\tau(g))v. \tag{6}$$

Proposition 10. *Let G be a Lie group and X a manifold of class C^r with a law of left operation of class C^r of G on X. Suppose that:*

(a) *the group G operates properly and freely on X;*

(b) *for all $x \in X$, $\rho(x)$ is an immersion* (*which is a consequence of* (a) *if K is of characteristic 0 and X is finite-dimensional*).

Then the equivalence relation defined by G *on* X *is regular* (*Differentiable and Analytic Manifolds*, R, 5.9.5). *There exists on the quotient set* X/G *one and only one manifold structure such that the canonical mapping* $\pi: X \to X/G$ *is a submersion. The underlying topology of this manifold structure is the quotient topology of that on* X; *it is Hausdorff. Finally,* (X, G, X/G, π) *is a principal left fibre bundle.*†

Let θ be the mapping $(g, x) \mapsto (x, gx)$ of $G \times X$ into $X \times X$. This mapping is of class C^r. We show that it is an immersion. For $u \in T_g(G)$ and $v \in T_x(X)$, by (6),

$$T_{(g,x)}(\theta)(u, v) = (v, T_g(\rho(x))u + T_x(\tau(g))v). \tag{7}$$

But $T_g(\rho(x))$ is injective by hypothesis (b) and hence $T_{(g,x)}(\theta)$ is injective. Its image is the topological direct sum of the subspace $H_{g,x}$ consisting of the vectors $(v, T_x(\tau(g))v)$ for $v \in T_x(X)$ and the subspace

$$I_{g,x} = \{0\} \times T_g(\rho(x))(T_g(G)).$$

By hypothesis (b), $T_g(\rho(x))(T_g(G))$ admits a topological supplement $J_{g,x}$ in $T_{gx}(X)$. Hence the image of $T_{(g,x)}(\theta)$ admits the topological supplement $\{0\} \times J_{g,x}$. Hence we have proved that θ is an immersion of $G \times X$ in $X \times X$.

As G operates freely on X, θ is injective. Let C be the graph of the equivalence relation R defined by G on X. As G operates properly, θ is a homeomorphism of $G \times X$ onto C (*General Topology*, Chapter I, § 10, Proposition 2). By *Differentiable and Analytic Manifolds*, R, 5.8.3, it is a submanifold of $X \times X$ and θ is an isomorphism of the manifold $G \times X$ onto the manifold C. The tangent space $T_{(x,gx)}(C)$ is identified with

$$T_{(g,x)}(\theta)(T_{(g,x)}(G \times X)) = H_{g,x} \oplus I_{g,x} \subset T_{(x,gx)}(X \times X).$$

Let pr_1 and pr_2 be the canonical projections of $X \times X$ onto the two factors. It is immediate that $T_{(x,gx)}(pr_1)$ maps $H_{g,x}$ onto $T_x(X)$ and that the kernel of $T_{(x,gx)}(pr_1) \mid T_{(x,gx)}(C)$ is $I_{g,x}$. Thus $pr_1 \mid C$ is a submersion of C onto X. By *Differentiable and Analytic Manifolds*, R, 5.9.5, R is regular. By definition, there therefore exists on the quotient set X/G one and only one manifold structure such that π is a submersion. The underlying topology of X/G is the quotient topology of that of X (*Differentiable and Analytic Manifolds*, R, 5.9.4). This topology is Hausdorff (*General Topology*, Chapter III, § 4, Proposition 3).

For all $b \in X/G$, there exist an open neighbourhood W of b and a morphism $\sigma: W \to X$ such that $\pi \circ \sigma = \mathrm{id}_W$ (*Differentiable and Analytic Manifolds*, R, 5.9.1). Let ϕ be the bijection $(g, w) \mapsto g\sigma(w)$ of $G \times W$ onto $\overset{-1}{\pi}(W)$. It is of class C^r. Then $\pi(g\sigma(w)) = w$ and

$$\theta^{-1}(\sigma(w), g\sigma(w)) = (g, \sigma(w))$$

† The principal fibre bundles defined in *Differentiable and Analytic Manifolds*, R, 6.2.1 are principal *right* fibre bundles. The definition of principal left fibre bundles can be deduced from this in an obvious way.

and hence the inverse bijection of ϕ is of class C^r. Clearly $\phi(gg', w) = g\phi(g', w)$ for $w \in W$, $g \in G$, $g' \in G$. Hence $(X, G, X/G, \pi)$ is a principal left fibre bundle.

Remark. With the above hypotheses, further let H be a manifold of class C^r and $(x, h) \mapsto m(x, h)$ a mapping of class C^r of $X \times H$ into X such that $m(gx, h) = gm(x, h)$ for $x \in X, g \in G, h \in H$. Let n be the mapping of $(X/G) \times H$ into X/G derived from m by taking quotients. We show that *n is of class* C^r. Consider the diagram

$$\begin{array}{ccc} X \times H & \xrightarrow{m} & X \\ \downarrow{\scriptstyle \pi\times 1} & & \downarrow{\scriptstyle \pi} \\ (X/G) \times H & \xrightarrow{n} & X/G \end{array}$$

It is commutative, $\pi \circ m$ is of class C^r and $\pi \times 1$ is a surjective submersion; it then suffices to apply *Differentiable and Analytic Manifolds*, R, 5.9.5.

Let G be a Lie group, X a manifold of class C^r and $(g, x) \mapsto xg$ a law of right operation of class C^r of G on X. Let $\tau(g)x = \rho(x)g = xg$ for $g \in G$, $x \in X$. Then this time

$$\rho(x) = \rho(xg) \circ \gamma(g^{-1}), \qquad \rho(x) = \tau(g) \circ \rho(x) \circ \delta(g) \tag{3'}$$

and hence

$$T_g(\rho(x)) = T_e(\rho(xg)) \circ T_g(\gamma(g^{-1})) \tag{4'}$$

$$T_g(\rho(x)) = T_x(\tau(g)) \circ T_e(\rho(x)) \circ T_g(\delta(g)). \tag{5'}$$

On the other hand, if η denotes the mapping $(g, x) \mapsto xg$ of $G \times X$ into X, formula (6) remains valid. Proposition 9, its Corollary and Proposition 10 remain equally true (with "principal left fibre bundle" replaced by "principal right fibre bundle" in the last).

6. HOMOGENEOUS SPACES AND QUOTIENT GROUPS

PROPOSITION 11. *Let X be a Lie group and G a Lie subgroup of X.*

(i) *There exists on the homogeneous set X/G one and only one analytic manifold structure such that the canonical projection π of X onto X/G is a submersion. The law of operation of X on X/G is analytic. For all $x \in X$, the kernel of $T_x(\pi)$ is obtained from $T_e(G)$ by $T_e(\gamma(x))$.*

(ii) *If G is normal in X, X/G is a Lie group with its group structure and the manifold structure defined in* (i). *The mapping π is a Lie group morphism.*

By *General Topology*, Chapter III, § 4, no. 1, *Example* 1, G operates properly and freely on X by right translation. Hence the first assertion of (i) follows from

Proposition 10 of no. 5. The second follows from the *Remark* of no. 5. Since π is a submersion, the kernel of $T_x(\pi)$ is the tangent space at x to

$$\overset{-1}{\pi}(\pi(x)) = xG = \gamma(x)(G)$$

and hence is obtained from $T_e(G)$ by $T_e(\gamma(x))$.

Suppose that G is normal. Let m be the mapping $(x, y) \mapsto xy^{-1}$ of $(X/G) \times (X/G)$ into X/G. Then $(m \circ (\pi \times \pi))(x, y) = \pi(xy^{-1})$ for all x, y in X. Hence $m \circ (\pi \times \pi)$ is analytic. As $\pi \times \pi$ is a surjective submersion, m is analytic (*Differentiable and Analytic Manifolds*, R. 5.9.5), whence (ii).

The homogeneous set X/G with the manifold structure defined in (i) is called the *quotient* (*left*) *Lie homogeneous space of* X *by* G. The (right) Lie homogeneous space G\X is defined analogously. When G is normal, the Lie group X/G defined in (ii) is called the *quotient Lie group of* X *by* G.

PROPOSITION 12. *Let* X *be a Lie group and* Y *a non-empty analytic manifold with a law of analytic left operation of* X *on* Y. *For all* $y \in Y$, *let* $\rho(y)$ *be the orbital mapping by* y *and* X_y *the stabilizer of* y *in* X. *The following conditions are equivalent:*

(i) *there exists* $y \in Y$ *such that* $\rho(y)$ *is a surjective submersion;*

(i′) *for all* $y \in Y$, $\rho(y)$ *is a surjective submersion;*

(ii) *there exists* $y \in Y$ *such that* X_y *is a Lie subgroup of* X *and the canonical mapping of* X/X_y *into* Y *is a manifold isomorphism;*

(ii′) *for all* $y \in Y$, X_y *is a Lie subgroup of* X *and the canonical mapping of* X/X_y *into* Y *is a manifold isomorphism;*

(iii) *the mapping* $(x, y) \mapsto (y, xy)$ *of* $X \times Y$ *into* $Y \times Y$ *is a surjective submersion.*

As the canonical mapping of X onto X/X_y is a submersion, the equivalences (i) $\Leftrightarrow$ (ii), (i′) $\Leftrightarrow$ (ii′) are immediate. (i) $\Leftrightarrow$ (i′) by Proposition 9 of no. 5. The equivalence (i′) $\Leftrightarrow$ (iii) follows from formula (7) of no. 5.

Under the conditions of Proposition 12, Y is called a (*left*) *Lie homogeneous space of* X. Right Lie homogeneous spaces of X are defined analogously.

Example. Let G be a Lie group. We make $G \times G$ operate on G on the left by $(g_1, g_2)x = g_1xg_2^{-1}$. Let ρ be the orbital mapping of e. Then the restrictions of $T_{(e,e)}(\rho)$ to $T_{(e,e)}(G \times \{e\}) = T_e(G) \times \{0\}$ and to

$$T_{(e,e)}(\{e\} \times G) = \{0\} \times T_e(G)$$

are isomorphisms of these spaces onto $T_e(G)$. Hence $T_{(e,e)}(\rho)$ is surjective and $\operatorname{Ker} T_{(e,e)}(\rho)$ admits for example the topological supplement $T_e(G) \times \{0\}$ in $T_{(e,e)}(G \times G)$. Thus ρ is a submersion at (e, e). Hence G is a left Lie homogeneous space of $G \times G$.

PROPOSITION 13. *Let* G *be a Lie group,* H *a normal Lie subgroup of* G, X *a manifold of class* C^r *and* $(g, x) \mapsto gx$ *a law of left operation of class* C^r *of* G *on* X. *Suppose that conditions* (a) *and* (b) *of Proposition* 10 *hold.*

(i) *The law of left operation* $(h, x) \mapsto hx$ *of* H *on* X *satisfies conditions* (a) *and* (b) *of Proposition* 10 (*so that we can consider the quotient manifolds* X/G *and* X/H).

(ii) *The law of left operation of* G *on* X *defines on taking quotients a law of left operation of class* C^r *of* G/H *on* X/H; *this law satisfies conditions* (a) *and* (b) *of Proposition* 10 (*so that we can consider the quotient manifold* (X/H)/(G/H)).

(iii) *The canonical mapping of* X *onto* X/H *defines on taking quotients a bijection of* X/G *onto* (X/H)/(G/H). *This bijection is an isomorphism of manifolds of class* C^r.

Clearly H operates freely on X; it operates properly by *General Topology*, Chapter III, § 4, no. 1, *Example* 1. The orbital mappings of H on X are immersions since the canonical injection of H into G is an immersion. This proves (i).

The law of left operation of G on X obviously defines, on taking quotients, a law of left operation of G/H on X/H. This law is of class C^r by *Differentiable and Analytic Manifolds*, R, 5.9.6. Let $g \in G$ and $x \in X$ be such that $(Hg)(Hx) = Hx$; then $H(gx) = Hx$ and hence $gx \in Hx$ and $g \in H$; this proves that G/H operates freely on X/H. The mapping $\theta: (g, x) \mapsto (x, gx)$ of $G \times X$ into $X \times X$ is closed; on the other hand, $\theta(Hg \times Hx) = Hx \times H(gx)$; it follows immediately that the mapping

$$(Hg, Hx) \mapsto (Hx, H(gx))$$

of $(G/H) \times (X/H)$ into $(X/H) \times (X/H)$ is closed; as moreover G/H operates freely on X/H, Theorem 1 (*c*) of *General Topology*, Chapter I, § 10, no. 2 proves that G/H operates properly on X/H.

Let π be the canonical mapping of X onto X/H, σ the canonical mapping of G onto G/H, x an element of X and $y = \pi(x)$.

$$\begin{array}{ccc} G & \xrightarrow{\rho(x)} & X \\ \sigma\downarrow & & \downarrow\pi \\ G/H & \xrightarrow{\rho(y)} & X/H \end{array}$$

Then $\pi \circ \rho(x) = \rho(y) \circ \sigma$ and hence

$$T_x(\pi) \circ T_e(\rho(x)) = T_e(\rho(y)) \circ T_e(\sigma).$$

Let $u \in T_e(G/H)$ be such that $T_e(\rho(y))u = 0$. There exists $v \in T_e(G)$ such that $u = T_e(\sigma)v$. Then $T_x(\pi)(T_e(\rho(x))v) = 0$, hence $T_e(\rho(x))v$ is tangent to Hx (*Differentiable and Analytic Manifolds*, R, 5.10.5) and therefore is of the form $T_e(\rho(x) \mid H)v'$ for some $v' \in T_e(H)$. As $T_e(\rho(x))$ is injective, it follows that $v = v'$, whence $v \in T_e(H)$ and therefore $u = 0$. Thus $T_e(\rho(y))$ is injective. The image of $T_e(\rho(y))$ is equal to that of $T_x(\pi) \circ T_e(\rho(x))$; now the image of $T_e(\rho(x))$ admits a topological supplement in $T_x(X)$ and contains the kernel of $T_x(\pi)$. It is therefore seen that $\rho(y)$ is an immersion, which completes the proof of (ii).

Assertion (iii) follows from the above and *Differentiable and Analytic Manifolds*, R, 5.9.7.

COROLLARY. *Let* G *be a Lie group and* H *and* L *normal Lie subgroups of* G *with* $L \subset H$. *Then* H/L *is a normal Lie subgroup of* G/L *and the canonical bijection of* G/H *onto* (G/L)/(H/L) *is a Lie group isomorphism.*

7. ORBITS

PROPOSITION 14. *Let* G *be a Lie group,* X *an analytic manifold and* $(g, x) \mapsto gx$ *a law of analytic left operation of* G *on* X. *Let* $x \in X$. *Suppose that the corresponding orbital mapping* $\rho(x)$ *is a subimmersion (which is always the case if* K *is of characteristic* 0 *and* X *is finite-dimensional* (Corollary to Proposition 9)). *Let* G_x *be the stabilizer of* x *in* G.

(i) G_x *is a Lie subgroup and* $T_e(G_x) = \operatorname{Ker} T_e(\rho(x))$.

(ii) *The canonical mapping* i_x *of the homogeneous space* G/G_x *into* X *is an immersion with image* Gx.

(iii) *If further the orbit* Gx *is locally closed and the topology on* G *admits a countable base, then* Gx *is a submanifold of* X, i_x *is an isomorphism of the manifold* G/G_x *onto the manifold* Gx *and* $T_x(Gx) = \operatorname{Im} T_e(\rho(x))$.

The inverse image of x under $\rho(x)$ is G_x. As $\rho(x)$ is a subimmersion, G_x is a submanifold and, for all $g \in G$, the tangent space J to $gG_x = \overset{-1}{\rho}_{(x)}(gx)$ at g is $\operatorname{Ker} T_g(\rho(x))$ (*Differentiable and Analytic Manifolds*, R, 5.10.5), whence (i). Let $\pi\colon G \to G/G_x$ be the canonical mapping. Then $i_x \circ \pi = \rho(x)$. As G/G_x is a quotient manifold of G, this equality proves that i_x is analytic. Further, the kernels of $T_g(\rho(x))$ and $T_g(\pi)$ are both equal to J. Hence $T_{\pi(g)}(i_x)$ is injective. The image of $T_{\pi(g)}(i_x)$ is equal to the image of $T_g(\rho(x))$ and hence admits a topological supplement. This proves (ii).

Suppose that Gx is locally closed. Every point of Gx then has a neighbourhood in Gx which is homeomorphic to a closed subspace of a complete metric space and hence is a Baire space. Hence Gx is a Baire space (*General Topology*, Chapter IX, § 5, Proposition 4). If G has a countable base, i_x is therefore a homeomorphism of G/G_x onto Gx (*General Topology*, Chapter IX, § 5). Then by (ii) and *Differentiable and Analytic Manifolds*, R, 5.8.3, i_x is an isomorphism of the manifolds G/G_x onto the manifold Gx and

$$T_x(Gx) = \operatorname{Im} T_{\pi(e)}(i_x) = \operatorname{Im} T_e(\rho(x)).$$

Remark. Let G be a finite-dimensional Lie group, X a manifold of class C^r and $(g, x) \mapsto gx$ a law of left operation of class C^r of G on X. Then Proposition 14 remains valid. The only point which needs a different proof is the fact that G_x is a Lie subgroup. But if $r \neq \omega$, $K = \mathbf{R}$; as clearly G_x is closed, G_x is a Lie subgroup by § 8, Theorem 2.

COROLLARY. *Let* G *be a Lie group whose topology admits a countable base and* X *a non-empty finite-dimensional analytic manifold with a law of analytic left operation of* G *on* X. *Suppose that* G *operates transitivity on* X *and that* K *is of characteristic* 0. *Then* X *is a Lie homogeneous space for* G.

Let $x \in X$. The orbit of x, equal to X, is closed and we can therefore apply Proposition 14 (iii).

8. VECTOR BUNDLES WITH OPERATORS

Let G be a Lie group, X a manifold of class C^r and $(g, x) \mapsto gx$ a law of left operation of class C^r of G on X. Let E be a vector bundle of class C^r, with base space X and $\pi: E \to X$ the projection of E onto X. For all $x \in X$, let E_x be the fibre of E at x. Let $(g, u) \mapsto gu$ be a law of left operation of G on E such that π is compatible with the operations of G on X and on E. For all $g \in G$ and all $x \in X$, the restriction to E_x of the mapping $u \mapsto gu$ is a bijection $\psi_{g,x}$ of E_x onto E_{gx}. We shall assume that, for all $g \in G$ and all $x \in X$, $\psi_{g,x}$ is continuous and linear and hence is an isomorphism of the normable space E_x onto the normable space E_{gx}.

Let ϕ be the automorphism $(g, x) \mapsto (g, gx)$ of the manifold $G \times X$. Let p be the canonical projection of $G \times X$ onto X and E′ the inverse image of E under p. Let $\psi: E' \to E'$ be the mapping the sum of the $\psi_{g,x}: E'_{(g,x)} \to E'_{(g,gx)}$.

DEFINITION 4. *If* ψ *is a* ϕ*-morphism of vector bundles of class* C^r, E *is called a vector* G*-bundle of class* C^r.

In other words, E is a vector G-bundle of class C^r if for all $(g_0, x_0) \in G \times X$ the following condition holds: there exists an open neighbourhood U of (g_0, x_0) in $G \times X$ such that, if $E' \mid U$ (resp. $E' \mid \phi(U)$) is identified with a trivial vector bundle of fibre M (resp. N) by means of a vector chart, the mapping $(g, x) \mapsto \psi_{g,x}$ of U into $\mathscr{L}(M, N)$ is of class C^r.

The mapping ψ is obviously bijective and it follows from the above local criterion that ψ^{-1} is a ϕ^{-1}-morphism of vector bundles so that ψ is a ϕ*-isomorphism* of vector bundles.

A *trivial vector* G*-bundle of base* X is a vector bundle $X \times F$ (where F is a complete normable space) with the law of operation $(g, (x, f)) \mapsto (gx, f)$ of G on $X \times F$.

We again assume the hypotheses and notation preceding Definition 4 and further take τ to be a vector functor of class C^r for isomorphisms (*Differentiable and Analytic Manifolds*, R, 7.6.6). Then τE is a vector bundle with base space X. For all $x \in X$, its fibre $(\tau E)_x$ is equal to $\tau(E_x)$. For all normable spaces N_1, N_2, let $\mathrm{Isom}(N_1, N_2)$ denote the set of isomorphisms of N_1 onto N_2. If $g \in G$, then

$$\tau(\psi_{g,x}) \in \mathrm{Isom}((\tau E)_x, (\tau E)_{gx}).$$

The $\tau(\psi_{g,x})$ define a law of left operation $(g, u) \mapsto gu$ of G on τE and the canonical projection of τE onto X is compatible with the operations of G on X and τE.

PROPOSITION 15. *If* E *is a vector* G*-bundle of class* C^r, τE *is a vector* G*-bundle of class* C^r.

Let g_0, x_0, U, M, N be as in the paragraph following Definition 4. Then the mapping $(g, x) \mapsto \tau(\psi_{g,x})$ of U into $\mathscr{L}(\tau M, \tau N)$ is the composition of the mapping $(g, x) \mapsto \psi_{g,x}$ of U into $\mathscr{L}(M, N)$ and the mapping $f \mapsto \tau(f)$ of Isom(M, N) into Isom(τM, τN); these two mappings are of class C^r and hence so is their composition, whence the proposition.

PROPOSITION 16. *Let* G *be a Lie group,* X *a manifold of class* C^r $(r \geqslant 2)$ *and* $(g, x) \mapsto gx$ *a law of left operation of class* C^r *of* G *on* X, *whence, by transporting the structure, there is a law of left operation of* G *on* TX. *Under this law,* TX *is a vector* G*-bundle of class* C^{r-1}.

Let pr_1 (resp. pr_2) be the canonical projection of G × X onto G (resp. X) and let E_1 (resp. E_2) be the inverse image of TG (resp. TX) relative to pr_1 (resp. pr_2). Then the vector bundle T(G × X) is the direct sum of E_1 and E_2. Let $i: E_2 \to T(G \times X)$ and $q: T(G \times X) \to E_2$ be the canonical vector bundle morphisms defined by this decomposition into a direct sum. Let ϕ be the mapping $(g, x) \mapsto (g, gx)$ of G × X into G × X. Then the mapping denoted by ψ in Definition 4 (where we put E = TX) is just $q \circ T(\phi) \circ i$. But $T(\phi)$ is a ϕ-morphism of vector bundles of class C^{r-1} (*Differentiable and Analytic Manifolds*, R, 8.1.2).

COROLLARY. *If* τ *is a vector functor of class* C^r *for isomorphisms,* $\tau(TX)$ *is a vector* G*-bundle of class* C^{r-1}.

This follows from Propositions 15 and 16.

Remark 1. If τ is a vector functor of class C^r for isomorphisms *in finite dimension* and E is of finite rank, τE is defined similarly and Proposition 15 remains valid; the Corollary to Proposition 16 remains valid provided X is finite-dimensional.

Examples. With the hypotheses and notation of Proposition 16, let F be a complete normable space. Then $\mathscr{L}((TX)^p; F)$ is a vector G-bundle of class C^{r-1}; so is $\mathrm{Alt}^p(TX; F)$ if K is of characteristic zero or X is finite-dimensional (cf. *Differentiable and Analytic Manifolds*, R, 7.7, 7.8). If X is finite-dimensional, $\bigotimes^p (TX) \otimes \bigotimes^q (TX)^*$ is a vector G-bundle of class C^{r-1}.

PROPOSITION 17. *Let* G *be a Lie group,* X *a left Lie homogeneous space of* G, x_0 *a point of* X, G_0 *the stabilizer of* x_0 *in* G, E *and* E′ *left vector* G*-bundles of class* C^r *and base space* X, E_0 (resp. E'_0) *the fibre of* E (resp. E′) *at* x_0 *and* f *an element of* $\mathscr{L}(E_0, E'_0)$ *such that* $f(gu) = gf(u)$ *for all* $u \in E_0$ *and* $g \in G_0$. *Then there exists one and only one morphism of* E *into* E′ *compatible with the operations of* G *and extending* f.

The uniqueness of this morphism is obvious. We prove its existence. Let g,

g' elements of G and $u \in E_0$ be such that $gu = g'u$. Then $g'^{-1}g \in G_0$ and $g'^{-1}gu = u$ and hence $g'^{-1}gf(u) = f(u)$, that is $gf(u) = g'f(u)$. Hence a mapping ϕ is defined of E into E′ by writing $\phi(gu) = gf(u)$. Clearly this mapping extends f and it is compatible with the operations of G. We show that ϕ is a vector bundle morphism of class C^r. Let $x_1 \in X$. There exists an open neighbourhood V of x_1 in X and a submanifold W of G such that the mapping $g \mapsto gx_0$ is an isomorphism θ of class C^r of W onto V. By shrinking V and W it can be assumed that:

(1) $E \mid V$ (resp. $E' \mid V$) is identified with a trivial vector bundle of fibre M (resp. M′);

(2) if ψ_g (resp. ψ'_g) denotes the mapping $u \mapsto gu$ of E_0 (resp. E'_0) into E_{gx_0} (resp. E'_{gx_0}), then the mappings $g \mapsto \psi_g$ and $g \mapsto \psi_g^{-1}$ (resp. $g \mapsto \psi'_g$ and $g \mapsto \psi_g'^{-1}$) of W into $\mathscr{L}(E_0, M)$ and $\mathscr{L}(M, E_0)$ (resp. $\mathscr{L}(E'_0, M')$ and $\mathscr{L}(M', E'_0)$) are of class C^r.

For $x \in V$, let $\phi_x: M \to N$ be the restriction of ϕ to $E_x = M$. Then ϕ_x is obtained by composing the following mappings:

(1) the mapping $(\psi_{\theta^{-1}_x})^{-1}$ of M into E_0;

(2) the mapping f of E_0 into E_0;

(3) the mapping $\psi'_{\theta^{-1}_x}$ of E_0 into M′.

Hence we see that the mapping $x \mapsto \phi_x$ of V into $\mathscr{L}(M, M')$ is of class C^r.

Corollary 1. *Let $E_0^{G_0}$ be the set of elements of E_0 which are invariant under G_0. For all $u \in E_0^{G_0}$, let σ_u be the mapping of X into E defined by $\sigma_u(gx_0) = gu$ for all $g \in G$.*

(i) *The G-invariant sections† of E are of class C^r.*

(ii) *$u \mapsto \sigma_u$ is a bijection of $E_0^{G_0}$ onto the set of G-invariant sections of E.*

Assertion (ii) is obvious. To prove (i) it is sufficient to prove that each section σ_u is of class C^r. Let E′ be the trivial G-bundle of base X and fibre $E_0^{G_0}$. Let f be the canonical injection of $E_0^{G_0}$ into E_0. By Proposition 17 there exists a morphism ϕ of E′ into E compatible with the operations of G and extending f. If $u \in E_0^{G_0}$ and $g \in G$, then

$$\sigma_u(gx_0) = gu = gf(u) = \phi(gu) = \phi((u, gx_0))$$

and hence $\sigma_u(x) = \phi((u, x))$ for all $x \in X$, which proves our assertion.

*For example, let G be a finite-dimensional real Lie group, G_0 a compact Lie subgroup of G and X the homogeneous space G/G_0. Let x_0 denote the canonical image of e in X. There exists a positive definite symmetric bilinear form on $T_{x_0}(X)$ invariant under G_0 (*Integration*, Chapter VII, § 3, Proposition 1). Applying the above to $(TX)^* \otimes (TX)^*$, we see that there exists on X an analytic Riemannian metric invariant under G_*.

---

† By a *section* of E we here mean a mapping σ (not necessarily of class C^r) of X into E such that $p \circ \sigma = \mathrm{Id}_X$, where p denotes the projection of E onto X.

Corollary 2. *Suppose that* G_0 *operates trivially on* E_0. *Let* E′ *be the trivial* G-*bundle with base space* X *and fibre* E_0. *There exists one and only one isomorphism of* E *onto* E′ *compatible with the operations of* G *and extending* Id_{E_0}.

This follows immediately from Proposition 17.

Remark 2. In this no., the laws of left operation can be replaced throughout by laws of right operation.

9. LOCAL DEFINITION OF A LIE GROUP

Proposition 18. *Let* G *be a group and* U *and* V *two subsets of* G *containing* e. *Suppose that* U *has an analytic manifold structure satisfying the following conditions:*

(i) $V = V^{-1}$, $V^2 \subset U$, V *is open in* U;

(ii) *the mapping* $(x, y) \mapsto xy^{-1}$ *of* $V \times V$ *into* U *is analytic;*

(iii) *for all* $g \in G$, *there exists an open neighbourhood* V′ *of* e *in* V *such that* $gV'g^{-1} \subset U$ *and such that the mapping* $x \mapsto gxg^{-1}$ *of* V′ *into* U *is analytic.*

Then there exists one and only one analytic manifold structure on G *with the following properties:*

(α) G *with this structure is a Lie group;*

(β) V *is open in* G;

(γ) *the manifold structures on* G *and* U *induce the same structure on* V.

(a) Let A be an open subset of V and v_0 an element of V such that $v_0A \subset V$. Then v_0A is the set of $v \in V$ such that $v_0^{-1}v \in A$ and hence is an open subset of V (taking account of (ii)). Moreover, (ii) implies that the mappings $v \mapsto v_0v$ of A onto v_0A and $v \mapsto v_0^{-1}v$ of v_0A onto A are inverse analytic bijections and hence analytic isomorphisms.

(b) We choose an open neighbourhood W of e in V such that $W = W^{-1}$, $W^3 \subset V$ and there exists a chart (W, ϕ, E) of the manifold U with domain W. For all $g \in G$, let ϕ_g be the mapping $h \mapsto \phi(g^{-1}h)$ of gW into E. We show that the charts ϕ_g of G are analytically compatible. Let g_1, g_2 be elements of G such that $g_1W \cap g_2W \neq \emptyset$, so that $g_2^{-1}g_1$ and $g_1^{-1}g_2$ belong to W^2. By (a), $W \cap g_1^{-1}g_2W$ is an open subset of W and hence

$$\phi_{g_1}(g_1W \cap g_2W) = \phi(W \cap g_1^{-1}g_2W)$$

is an open subset D of E. For $d \in D$,

$$(\phi_{g_2} \circ \phi_{g_1}^{-1})(d) = \phi(g_2^{-1}g_1\phi^{-1}(d));$$

by (a) we see that $\phi_{g_2} \circ \phi_{g_1}^{-1}$ is analytic.

(c) By (b) there exists on G an analytic manifold structure such that $(\phi_g)_{g\in G}$ is an atlas on G. For all $g_0 \in G$, the mapping $g \mapsto g_0g$ $(g \in G)$ leaves this atlas invariant and hence is an automorphism of G with this manifold structure. In particular condition (GL_1) is satisfied.

(d) Let $v_0 \in V$. By (ii) there exists an open neighbourhood A of e in W such that $v_0A \subset V$. This proves first that V is open in G. By (a) the mapping $v \mapsto v_0v$ of A onto v_0A is an analytic isomorphism with the structures induced by U. By (c) we see that the manifold structures on G and U induce the same structure on v_0A and hence finally on V.

(e) by (d), (ii) and (iii), we see that conditions (GL_2) and (GL_3) hold. Hence G is a Lie group.

(f) If a manifold structure on G is compatible with the group structure of G and is such that V is an open submanifold of G, then $(\phi_g)_{g \in G}$ is an atlas on G. Hence the uniqueness assertion of the proposition.

PROPOSITION 19. *Let* G *be a topological group,* H *a Lie group and* f *a homomorphism of the group* G *into the group* H. *Suppose that there exist an open neighbourhood of* e_G *in* G, *a chart* (V, ϕ, E) *of the manifold* H *at* e_H *and a closed vector subspace* F *of* E *admitting a topological supplement, such that* $f(U) \subset V$ *and* $\phi \circ f) \mid U$ *is a homeomorphism of* U *onto* $\phi(V) \cap F$. *Then there exists a unique manifold structure on* G *such that* f *is an immersion; this structure is the inverse image under* f *of the manifold structure on* H. *With this structure* G *is a Lie group.*

As translations of G (resp. H) are homeomorphisms (resp. analytic isomorphisms), f satisfies condition (R) of *Differentiable and Analytic Manifolds,* R, 5.8.1. The first two assertions of the proposition then follow from *Differentiable and Analytic Manifolds,* R, *loc. cit.* Consider the commutative diagram

$$\begin{array}{ccc} G \times G & \xrightarrow{m} & G \\ {\scriptstyle f \times f}\downarrow & & \downarrow{\scriptstyle f} \\ H \times H & \xrightarrow{n} & H \end{array}$$

where $m(x, y) = xy^{-1}$ (resp. $n(x, y) = xy^{-1}$) for x, y in G (resp. H). Then $n \circ (f \times f)$ is analytic, hence $f \circ m$ is analytic and hence m is analytic since f is an immersion. Therefore G is a Lie group.

The Lie group structure on G is called the *inverse image* of the Lie group structure on H under f.

COROLLARY. *Let* G *be a topological group,* N *a discrete normal subgroup of* G *and* π *the canonical mapping of G onto* G/N. *Suppose that an analytic manifold structure is given on* G/N, *compatible with the topological group structure on* G/N. *Then there exists a unique manifold structure on* G *such that* π *is an immersion; this structure is the inverse image under* π *of the manifold structure on* G/N. *With this structure,* π *is étale,* G *is a Lie group and* G/N *is the quotient Lie group of* G *by* N.

Remark. Let H be a connected real or complex Lie group, $\hat{H}$ its universal covering† and π the canonical mapping of $\hat{H}$ onto H. When we speak of $\hat{H}$

† Cf. *General Topology,* Chapter XI; whilst awaiting the publication of this chapter, see for example L. S. Pontrjagin, *Topological groups,* 2nd edition translated from Russian, Gordon and Breach, 1966; or G. Hochschild, *The structure of Lie groups,* Holden-Day, 1965.

as a Lie group, we shall always mean with the inverse image structure of that on H under π.

10. GROUP GERMS

Definition 5. *A Lie group germ over* K *is a system* (G, e, θ, m) *satisfying the following conditions:*

(i) G *is an analytic manifold over* K;

(ii) $e \in G$;

(iii) θ *is an analytic mapping of* G *into* G;

(iv) m *is an analytic mapping of an open subset* Ω *of* $G \times G$ *into* G;

(v) *for all* $g \in G$, $(e, g) \in \Omega$, $(g, e) \in \Omega$, $m(e, g) = m(g, e) = g$;

(vi) *for all* $g \in G$, $(g, \theta(g)) \in \Omega$, $(\theta(g), g) \in \Omega$, $m(g, \theta(g)) = m(\theta(g), g) = e$;

(vii) *if* g, h, k *are elements of* G *such that* $(g, h) \in \Omega$, $(h, k) \in \Omega$, $(m(g, h), k) \in \Omega$, $(g, m(h, k)) \in \Omega$, then $m(m(g, h), k) = m(g, m(h, k))$.

e is called the identity element of the group germ. We often write gh instead of $m(g, h)$ and (by an abuse of notation) g^{-1} instead of $\theta(g)$.

A Lie group G is a Lie group germ with the obvious choice of e, θ, m.

Let G be a Lie group germ. Then $ee^{-1} = e$, that is

$$e^{-1} = e. \tag{8}$$

For all $g \in G$,

$$g = eg = ((g^{-1})^{-1}g^{-1})g = (g^{-1})^{-1}(g^{-1}g) = (g^{-1})^{-1}e,$$

that is

$$(g^{-1})^{-1} = g. \tag{9}$$

A subset of G invariant under the mapping $g \mapsto g^{-1}$ is called symmetric.

The manifold G, with the point e, the mapping $g \mapsto g^{-1}$ and the mapping $(g, h) \mapsto hg$ is a Lie Group germ $G^{\vee}$ called the opposite of G.

The Lie group germ G is called commutative if, for all $(g, h) \in G \times G$ such that gh is defined, hg is defined and equal to gh.

Let G be a Lie group germ. The set of $(g, h) \in G \times G$ such that gh is defined is a neighbourhood of (e, e). On the other hand, the mappings $(g, h) \mapsto gh$ and $g \mapsto g^{-1}$ are continuous. Hence $(gh)k = g(hk)$ for g, h, k sufficiently close to e. Similarly, $(h^{-1}g^{-1})(gh) = h^{-1}(eh) = h^{-1}h = e$ for g, h sufficiently close to e, whence multiplying on the right by $(gh)^{-1}$,

$$(gh)^{-1} = h^{-1}g^{-1} \quad \text{for } g, h \text{ sufficiently close to } e. \tag{10}$$

Proposition 20. *Let* G *be a Lie group germ and* $g \in G$. *There exist an open neighbourhood* U *of* e *and an open neighbourhood* V *of* g *with the following properties:*

(a) ug *is defined for all* $u \in U$;

(b) vg^{-1} *is defined for all* $v \in V$;

(c) *the mappings* $u \mapsto ug$, $v \mapsto vg^{-1}$ *are inverse analytic isomorphisms of one another of* U *onto* V *and of* V *onto* U.

As the set of definition of the product is open in $G \times G$, there exist an open neighbourhood U of e and an open neighbourhood V of g with properties (a) and (b). Let $\eta(u) = ug$ for $u \in U$, $\eta'(v) = vg^{-1}$ for $v \in V$. By shrinking U and V, it can be assumed that $(ug)g^{-1} = u$ and $(vg^{-1})g = v$ for $u \in U$ and $v \in V$. Then η and η' are injections. By shrinking U further, it can be assumed that $\eta(U) \subset V$. Then $\eta'(V) \supset U$ and $\eta(U)$ is the inverse image of U under η' and hence is an open neighbourhood of g in V. Replacing V by $\eta(U)$, we finally arrive at the situation where η and η' are analytic inverse bijections.

Let G_1, G_2 be two Lie group germs with identity elements e_1, e_2. A mapping f of G_1 into G_2 is called a *morphism* if f satisfies the following conditions:

(i) f is analytic;

(ii) $f(e_1) = e_2$;

(iii) if g, h are elements of G_1 such that gh is defined, then $f(g)f(h)$ is defined and is equal to $f(gh)$.

Let $g \in G_1$. As gg^{-1} is defined and is equal to e_1, $f(g)f(g^{-1})$ is defined and is equal to e_2 and hence

$$f(g)^{-1} = f(g)^{-1}(f(g)f(g^{-1})) = (f(g)^{-1}f(g))f(g^{-1})$$

that is

$$f(g)^{-1} = f(g^{-1}). \tag{11}$$

The composition of two morphisms is a morphism.

If $f: G_1 \to G_2$ and $f': G_2 \to G_1$ are two inverse morphisms of one another, they are isomorphisms (using in particular formula (11)).

Let G_1, G_2 be two Lie group germs, and f a mapping of G_1 into G_2 satisfying conditions (ii) and (iii) above, which is analytic in an open neighbourhood of e_1. Using Proposition 20, it can be proved as in Proposition 4 that f is a morphism.

Let (G, e, θ, m) be a Lie group germ and Ω the set of definition of m. Let H be a submanifold of G containing e, which is stable under θ. Suppose that the set Ω_1 of $(x, y) \in \Omega \cap (H \times H)$ such that $m(x, y) \in H$ is open in $H \times H$. Then $(H, e, \theta|H, m|\Omega_1)$ is a Lie group germ. Such a Lie group germ is called a *Lie subgroup germ of* G. The canonical injection of H into G is a morphism. If $f: L \to G$ is a morphism of Lie group germs such that $f(L) \subset H$, then $f: L \to H$ is a morphism of Lie group germs.

Suppose that K is of characteristic 0. If we replace the hypothesis that H is a submanifold of G by the hypothesis that H is a quasi-submanifold of G, the results of the above paragraph remain true (cf. *Differentiable and Analytic Manifolds*, R, 5.8.5). H is then called a Lie quasi-subgroup germ of G.

If G is a Lie group germ with identity element e, every symmetric open neighbourhood of e in G is a Lie subgroup germ of G. (This applies in particular when G is a Lie group.) Let H be a Lie subgroup germ of G; if H is a neighbourhood of e in G, then H is open in G by Proposition 20.

The *product* Lie group germ of a finite number of Lie group germs is defined in an obvious way.

PROPOSITION 21. *Let* G, H *be two Lie group germs and* ϕ *a morphism of* G *into* H. *The following conditions are equivalent:*

(i) ϕ *is étale at* e;

(ii) *there exist open Lie subgroup germs* G′, H′ *of* G, H *such that* $\phi|G'$ *is an isomorphism of* G′ *onto* H′.

The implication (ii) ⇒ (i) is obvious. Suppose that ϕ is étale at e. There exists an open Lie subgroup germ G_1 of G such that $\phi(G_1)$ is open in H and $\phi|G_1$ is an isomorphism of the manifold G_1 onto the manifold $\phi(G_1)$. Then there exists an open Lie subgroup germ G′ of G_1 such that the product in G of two elements of G′ is always defined and belongs to G_1. If g, g' are elements of G′ such that $gg' \in G'$, then $\phi(g)\phi(g') = \phi(gg') \in \phi(G')$; if g, g' are elements of G′ such that $gg' \in G_1 - G'$, then

$$\phi(g)\phi(g') = \phi(gg') \in \phi(G_1) - \phi(G').$$

Hence $\phi|G'$ is an isomorphism of the Lie group germ G′ onto the open Lie subgroup germ $\phi(G')$ of H.

If the conditions of Proposition 21 hold, G and H are called *locally isomorphic.*

PROPOSITION 22. *Let* H *be a Lie group,* U *a Lie subgroup germ of* H *and* N *the set of* $g \in H$ *such that* U *and* gUg^{-1} *have the same germ at* e (*General Topology*, Chapter I, § 6, no. 10). *Then N is a subgroup of* H *containing* U. *There exists one and only one analytic manifold structure on* N *with the following properties:*

(i) N *with this structure is a Lie group;*

(ii) U *is an open submanifold of* N;

(iii) *the canonical injection of* N *into* H *is an immersion.*

Clearly N is a subgroup of H. If $g \in U$, then $ge \in U$ and $geg^{-1} \in U$, hence $gu \in U$ and $gug^{-1} \in U$ for u sufficiently close to e in U and hence the germ of gUg^{-1} at e is contained in that of U; exchanging g and g^{-1}, we see that the germs of gUg^{-1} and U at e are equal. Hence $U \subset N$.

Let V be an open neighbourhood of e in U such that $V = V^{-1}$, $V^2 \subset U$. Conditions (i), (ii), (iii) of Proposition 18 of no. 9 (where G is replaced by N) are satisfied. Hence there exists an analytic manifold structure on N with the following properties: (α) N with this structure is a Lie group; (β) V is open in N; (γ) the manifold structures on N and U induce the same structure on V. Since V is a submanifold of H, the canonical injection of N into H is an immersion at e and hence at every point of N. Let $u \in U$. There exists an open neighbourhood V′ of e in V such that the mapping $v \mapsto uv$ is an analytic isomorphism of V′ onto an open neighbourhood of u in U (Proposition 20) and at the same time onto an open neighbourhood of u in N. Hence U is open in N and the identity mapping of U is an isomorphism with the given manifold structure on U and the open submanifold structure on N; in other words, U is an open submanifold of N.

Finally, we consider an analytic manifold structure on N with properties (i) and (ii) of the proposition and let N* be the Lie group thus obtained. Then the identity mapping of N into N* is étale at e and hence a Lie group isomorphism. This proves the uniqueness assertion of the proposition.

Let H be a Lie group, U a Lie quasi-subgroup germ of H and N the set of $g \in$ H such that U and gUg^{-1} have the same germ at e. If K is of characteristic 0, there exists on G one and only one manifold structure with properties (i) and (ii) of Proposition 22. The proof is the same as for Proposition 22.

COROLLARY. *Preserving the notation of Proposition* 22, *let* G *be the subgroup of* H *generated by* U. *Then* G *is an open subgroup of* N. *There exists one and only one Lie group structure on* G *such that* U *is an open submanifold of* G *and the canonical injection of* G *into* H *is an immersion.*

Remark. Preserving the notation of Proposition 22 and its corollary, suppose that K is of characteristic 0, that H is finite-dimensional and that the topology on U admits a countable base. Even with all these hypotheses it is possible that G is not closed in H (Exercise 3). But, *if* G *is closed,* G *is a Lie subgroup of* H. For the mapping $(g, h) \mapsto gh$ is a law of analytic left operation of G on H. The orbit of e is G. Our assertion then follows from Propositions 2 (iv) and 14 (iii).

11. LAW CHUNKS OF OPERATION

Let (G, e, θ, m) be a Lie group germ and X a manifold of class C^r.

DEFINITION 6. *A law chunk of left operation of class* C^r *of* G *on* X *is a mapping* ψ *defined on an open subset* Ω *of* $G \times X$ *containing* $\{e\} \times X$, *with values in* X *and with the following properties:*

(i) ψ *is of class* C^r;

(ii) *for all* $x \in X$, $\psi(e, x) = x$;

(iii) *there exists a neighbourhood* Ω_1 *of* $\{e\} \times \{e\} \times X$ *in* $G \times G \times X$ *such that, for* $(g, g', x) \in \Omega_1$, *the elements of* $m(g, g')$, $\psi(m(g, g'), x)$, $\psi(g, \psi(g', x))$ *are defined and* $\psi(g, \psi(g', x)) = \psi(m(g, g'), x)$.

Law chunks of right operation of class C^r are defined similarly.

We often write gx instead of $\psi(g, x)$.

Let G′ be a Lie subgroup germ of G and X′ a submanifold of X. Suppose that the set Ω' of $(g, x) \in \Omega \cap (G' \times X')$ such that $\psi(g, x) \in X'$ is open in $G' \times X'$ (a condition which is always fulfilled if X' is open in X). Then $\psi|\Omega'$ is a law chunk of left operation of class C^r of G' on X', which is said to be derived from ψ by restriction to G' and X'.

PROPOSITION 23. *Let (G, e, θ, m) be a Lie group germ, X a manifold of class C^r, x_0 a point of X, Ω an open neighbourhood of (e, x_0) in $G \times X$ and ψ a mapping of Ω into X with the following properties:*

(i) *ψ is of class C^r;*

(ii) *$\psi(e, x)$ is equal to x for x sufficiently close to x_0;*

(iii) *$\psi(m(g, g'), x) = \psi(g, \psi(g', x))$ for (g, g', x) sufficiently close to (e, e, x_0).*

Then there exist an open neighbourhood X' of x_0 in X and an open subset Ω' of $\Omega \cap (G \times X')$ such that $\psi|\Omega'$ is a law chunk of left operation of class C^r of G on X'.

There exist an open neighbourhood X' of x_0 in X and an open neighbourhood G' of e in G such that $\psi(e, x) = x$ for all $x \in X$, and

$$\psi(g, \psi(g', x)) = \psi(m(g, g'), x)$$

for $(g, g', x) \in G' \times G' \times X'$. Let Ω' be the set of $(g, x) \in \Omega \cap (G' \times X')$ such that $\psi(g, x) \in X'$. Then Ω' is open in $G \times X'$ and X', Ω' have the properties of the proposition.

Lemma 3. Let X be a normal space and $(X_i)_{i \in I}$ a locally finite open covering of X. For all $(i, j) \in I \times I$ and all $x \in X_i \cap X_j$, let $V_{ij}(x)$ be a neighbourhood of x contained in $X_i \cap X_j$. Then we can associate with every $x \in X$ a neighbourhood $V(x)$ of x such that the following conditions are fulfilled:

(a) *the relation $x \in X_i \cap X_j$ implies $V(x) \subset V_{ij}(x)$;*

(b) *if $V(x)$ and $V(y)$ meet, there exists $i \in I$ such that $V(x) \cup V(y) \subset X_i$.*

There exists an open covering $(X'_i)_{i \in I}$ of X such that $\overline{X'_i} \subset X_i$ for all $i \in I$ (*General Topology*, Chapter IX, § 4, Theorem 3). Let $x \in X$. Let $V_1(x)$ be the intersection of the $V_{ij}(x)$ and the X'_k which contain x; this is an open neighbourhood of x. Let $V_2(x)$ be a neighbourhood of x contained in $V_1(x)$ and meeting only a finite number of X_i. Then $V_2(x)$ meets only a finite number of $\overline{X'_i}$ and hence the set

$$V(x) = V_2(x) \cap \bigcap_{i \in I, x \notin \overline{X}'_i} (X - \overline{X'_i})$$

is a neighbourhood of x. If $x \in X_i \cap X_j$, then $V_1(x) \subset X_i \cap X_j$ and hence $V(x) \subset X_i \cap X_j$. Let x, y be in X and suppose that $V(x)$ and $V(y)$ meet. There exists $i \in I$ such that $x \in X'_i$. Then $V_1(x) \subset X'_i$, hence $V(x) \subset X'_i$ and hence $V(y) \cap \overline{X'_i} \neq \emptyset$. Then $y \in \overline{X'_i}$ by definition of $V(y)$, whence $y \in X_i$ and $V(y) \subset X_i$. Thus X_i contains $V(x)$ and $V(y)$.

PROPOSITION 24. *Let* G *be a Lie group germ,* X *a manifold of class* C^r *and* $(X_i)_{i \in I}$ *a locally finite open covering of* X. *For all* $i \in I$, *let* ψ_i *be a law chunk of left operation of class* C^r *of* G *on* X_i. *Suppose that the underlying topological space of* X *is normal and that, for all* $(i, j) \in I \times I$ *and all* $x \in X_i \cap X_j$, ψ_i *and* ψ_j *coincide on a neighbourhood of* (e, x). *There exists a law chunk* ψ *of left operation of class* C^r *of* G *on* X *such that, for all* $i \in I$ *and all* $x \in X_i$, ψ_i and ψ *coincide on a neighbourhood of* (e, x).

For all $(i, j) \in I \times I$ and all $x \in X_i \cap X_j$ choose an open neighbourhood $V_{ij}(x)$ of x in $X_i \cap X_j$ such that ψ_i and ψ_j are defined and equal on a neighbourhood of $\{e\} \times V_{ij}(x)$ in $G \times X$. For all $x \in X$ choose an open neighbourhood $V(x)$ of x in X such that conditions (a) and (b) of Lemma 3 are fulfilled. Let I_x be the set of $i \in I$ such that $x \in X_i$. This is a finite set. Let U_x be the set of $(g, y) \in G \times V(x)$ such that the ψ_i for $i \in I_x$ are defined and coincide on a neighbourhood of (g, y). Then U_x is open and $(e, x) \in U_x$. The ψ_i for $i \in I_x$ all have the same restriction to U_x. Let x, y be in X. If U_x and U_y meet, $V(x)$ and $V(y)$ meet and hence there exists $i \in I$ such that

$$V(x) \cup V(y) \subset X_i.$$

Then $i \in I_x$, $i \in I_y$, $\psi_i|U_x = \psi_x$, $\psi_i|U_y = \psi_y$ and hence

$$\psi_x|(U_x \cap U_y) = \psi_y|(U_x \cap U_y).$$

The ψ_x therefore define a mapping ψ of $U = \bigcup_{x \in X} U_x$ into X and U is an open neighbourhood of $\{e\} \times X$ in $G \times X$. Clearly ψ is of class C^r and $\psi(e, x) = x$ for all $x \in X$. For all $i \in I$ and all $x \in X_i$, ψ coincides with ψ_x and hence with ψ_i in a neighbourhood of (e, x) and hence ψ satisfies condition (iii) of Definition 6.

§ 2. GROUP OF TANGENT VECTORS TO A LIE GROUP

1. TANGENT LAWS OF COMPOSITION

Let X and Y be manifolds of class C^r. We know (*Differentiable and Analytic Manifolds,* R, 8.1.4) that $X \times Y$ is a manifold of class C^r and that the mapping $(T(pr_1), T(pr_2))$, the product of the tangent mappings to the canonical projections, is an isomorphism of class C^{r-1} of $T(X \times Y)$ onto $T(X) \times T(Y)$.† This isomorphism is compatible with the vector bundle structures with base space $X \times Y$ and allows us to identify $T(X \times Y)$ with $T(X) \times T(Y)$. Let

† For $r = 1$, this means that $(T(pr_1), T(pr_2))$ is a homomorphism of $T(X \times Y)$ onto $T(X) \times T(Y)$.

$a \in X$, $b \in Y$, $u \in T_a(X)$, $v \in T_b(Y)$; the above identification allows us to consider (u, v) as an element of $T_{(a,b)}(X \times Y)$; then

$$(u, v) = (u, 0) + (0, v)$$

and $(u, 0)$ (resp. $(0, v)$) is the image of u (resp. v) under the tangent mapping to the immersion $x \mapsto (x, b)$ (resp. $y \mapsto (a, y)$) of X (resp. Y) into $X \times Y$. When it is necessary to be precise, we shall write 0_a for the zero element of $T_a(x)$.

Now let X, Y, Z be manifolds of class C^r and f a mapping of class C^r of $X \times Y$ into Z. The tangent mapping is, using the above identification, a mapping of class C^{r-1} of $T(X) \times T(Y)$ into $T(Z)$. For $u \in T_a(X)$ and $v \in T_b(Y)$,

$$\text{(1)} \qquad T(f)(u, v) = T(f)(u, 0_b) + T(f)(0_a, v),$$

$$\text{(2)} \qquad T(f)(0_a, 0_b) = 0_{f(a,b)}.$$

On the other hand, the mapping $y \mapsto f(a, y)$ is the composition of the immersion $y \mapsto (a, y)$ and f; it follows that

(3) $T(f)(0, v)$ is the image of v under the tangent mapping to $y \mapsto f(a, y)$.

Similarly

(4) $T(f)(u, 0)$ is the image of u under the tangent mapping to $x \mapsto f(x, b)$.

If the mapping f of $X \times Y$ into Z is denoted by $(x, y) \mapsto xy$, uv is often used to denote the element $T(f)(u, v)$ for $u \in T(X)$, $v \in T(Y)$.

Let X be a manifold of class C^r and $m: X \times X \to X$ a law of composition of class C^r on X. Then $T(m)$ is a law of composition of class C^{r-1} on $T(X)$. It is called the *law of composition tangent to m*. The canonical projection p of $T(X)$ onto X is compatible with the laws m and $T(m)$; in other words,

$$\text{(5)} \qquad p \circ T(m) = m \circ (p \times p).$$

It follows from (2) that

$$\text{(6)} \qquad T(m)(0_x, 0_y) = 0_{m(x,y)}$$

for all x, y in X; in other words, the zero section $x \mapsto 0_x$ of $T(X)$ is compatible with the laws m and $T(m)$.

PROPOSITION 1. *Let* X *be a manifold of class* C^r *and m a law of composition of class* C^r *on* X. *If m is associative* (resp. *commutative*), *then* $T(m)$ *is associative* (resp. *commutative*).

If m is associative, then $m \circ (m \times \mathrm{Id}_X) = m \circ (\mathrm{Id}_X \times m)$, whence

$$T(m) \circ (T(m) \times \mathrm{Id}_{T(X)}) = T(m) \circ (\mathrm{Id}_{T(X)} \times T(m))$$

and hence $T(m)$ is associative. Let s be the mapping $(x, y) \mapsto (y, x)$ of $X \times X$ into $X \times X$. If m is commutative, then $m \circ s = m$ and hence

$$T(m) \circ T(s) = T(m).$$

But $T(s)$ is the mapping $(u, v) \mapsto (v, u)$ of $T(X) \times T(X)$ into $T(X) \times T(X)$. Hence $T(m)$ is commutative.

PROPOSITION 2. *Let* X *be a manifold of class* C^r, *m a law of composition of class* C^r *on* X *and e an identity element for m.*

(i) *The vector* 0_e *is an identity element for* $T(m)$.

(ii) $T_e(X)$ *is stable under* $T(m)$ *and the law of composition induced on* $T_e(X)$ *by* $T(m)$ *is the vector space addition on* $T_e(X)$.

(iii) *Let* U *be an open subset of* X *and* α *a mapping of class* C^r *of* U *into* X *such that, for all* $x \in U$, $\alpha(x)$ *is the inverse of x under m. Then, for all* $u \in T(U)$, $T(\alpha)u$ *is the inverse of u under* $T(m)$.

Properties (3) and (4) show that $T(m)(0_e, u) = T(m)(u, 0_e) = u$ for all $u \in T(X)$, whence (i). For u, v in $T_e(X)$,

$$T(m)(u, v) = T(m)(u, 0_e) + T(m)(0_e, v) = u + v,$$

whence (ii). Finally the relations $m(x, \alpha(x)) = m(\alpha(x), x) = e$ for all $x \in U$ imply

$$T(m)(u, T(\alpha)(u)) = T(m)(T(\alpha)u, u) = 0_e$$

for all $u \in T(U)$, whence (iii).

PROPOSITION 3. *Let* $X_1, X_2, \ldots, X_p, Y$ *be manifolds of class* C^r, *i an integer of* $[1, p]$, m_i (resp. *n*) *a law of composition of class* C^r *on* X_i (resp. Y) *and u a mapping of class* C^r *of* $X_1 \times X_2 \times \cdots \times X_p$ *into* Y. *If u is distributive relative to the variable of index i, then* $T(u)$ *is distributive relative to the variable of index i.*

The proof is analogous to that of Proposition 1.

2. GROUP OF TANGENT VECTORS TO A LIE GROUP

PROPOSITION 4. *Let* G *be a Lie group. Then* $T(G)$, *with the law of composition tangent to the multiplication of* G, *is a Lie group. The identity element of* $T(G)$ *is the vector* 0_e.

This follows from Propositions 1 and 2.

PROPOSITION 5. *Let* G *and* H *be Lie groups and f a morphism of* G *into* H. *Then* $T(f)$ *is a morphism of the Lie group* $T(G)$ *into the Lie group* $T(H)$.

We know that $T(f)$ is analytic. On the other hand, let *m* (resp. *n*) denote the multiplication on G (resp. H). Then $f \circ m = n \circ (f \times f)$, whence

$$T(f) \circ T(m) = T(n) \circ (T(f) \times T(f)),$$

which expresses the fact that $T(f)$ is a group homomorphism.

COROLLARY. *Let* $G_1, \ldots, G_n$ *be Lie groups. The canonical isomorphism of the manifold* $T(G_1 \times \cdots \times G_n)$ *onto the manifold* $T(G_1) \times \cdots \times T(G_n)$ *is a Lie group isomorphism.*

pr_i is a morphism of $G_1 \times \cdots \times G_n$ into G_i and hence $T(pr_i)$ is a morphism of $T(G_1 \times \cdots \times G_n)$ into $T(G_i)$.

PROPOSITION 6. *Let* G *be a Lie group.*

(i) *The canonical projection* $p: T(G) \to G$ *is a Lie group morphism.*

(ii) *The kernel of* p *is* $T_e(G)$. *It is a Lie subgroup of* $T(G)$. *The Lie group structure induced on* $T_e(G)$ *by that on* $T(G)$ *is the Lie group structure of the complete normable space* $T_e(G)$.

(iii) *The zero section* s *is an isomorphism of the Lie group* G *onto a Lie subgroup* $s(G)$ *of* $T(G)$ *(which subgroup we identify with* G*).*

(iv) *The Lie group* $T(G)$ *is the semi-direct product of* G *by* $T_e(G)$.

Assertion (i) follows from (5). Assertion (ii) is obvious taking account of Proposition 2 (ii). Assertions (iii) and (iv) follow from (6) and § 1, Proposition 8.

Let $u \in T(G)$ and $g \in G$. By (3) and (4), the products ug, gu calculated in the group $T(G)$ are the images of u under $T(\delta(g^{-1}))$ and $T(\gamma(g))$. It follows from § 1, Corollary 2 to Proposition 17 that the mapping $(g, u) \mapsto gu$ of $G \times T_e(G)$ into $T(G)$ is an isomorphism of the trivial vector bundle $G \times T_e(G)$ with base space G onto the vector bundle $T(G)$. The inverse isomorphism is called the *left trivialization* of $T(G)$. By considering the mapping $(g, u) \mapsto ug$, the *right trivialization* of $T(G)$ is defined similarly.

PROPOSITION 7. *Let* G *be a Lie group,* M *a manifold of class* C^r *and* f *and* g *mappings of class* C^r *of* M *into* G, *so that* fg *is a mapping of class* C^r *of* M *into* G. *Let* $m \in M$, $x = f(m)$, $y = g(m)$, $u \in T_m(M)$. *Then*

$$(T fg)u = T(f)u.y + x.T(g)u.$$

Let m be the multiplication of G. Then $fg = m \circ (f, g)$. Now

$$T(f, g)(u) = (T(f)u, T(g)u),$$

hence $T(fg)u = T(f)u.T(g)u$. It then suffices to apply (1) with f replaced by m.

COROLLARY. *Let* $n \in \mathbf{Z}$. *The tangent mapping at* e *to the mapping* $g \mapsto g^n$ *of* G *into* G *is the mapping* $x \mapsto nx$ *of* $T_e(G)$ *into* $T_e(G)$.

For $n \geqslant 0$, this follows by induction on n from Proposition 7. On the other hand, the tangent mapping at e to the mapping $g \mapsto g^{-1}$ is the mapping $x \mapsto -x$ (no. 1, Proposition 2).

Let G be a Lie group, X a manifold of class C^r and $(g, x) \mapsto gx$ a law of left operation of class C^r of G on X. Arguing as for Proposition 1, we derive a law of left operation of class C^{r-1} of T(G) on T(X), which we shall also denote by $(u, v) \mapsto uv$. Identifying G (resp. X) with the image of the zero section of T(G) (resp. T(X)), we see by (6) that the law of left operation of T(G) on T(X) extends the law of left operation of G on X. For all $u \in T_g(G)$ and $v \in T_x(X)$, by (1),

$$uv = gv + ux. \tag{7}$$

If $g \in G$ and $v \in T_x(X)$, gv is, by (3), the image of v under the tangent mapping at x to the mapping $y \mapsto gy$ of X into X. This tangent mapping is an isomorphism of $T_x(X)$ onto $T_{gx}(X)$. In particular,

$$g(v + v') = gv + gv', \quad g(\lambda v) = \lambda(gv) \quad \text{for } v, v' \text{ in } T_x(X), \lambda \in K. \tag{8}$$

If $x \in X$ and $u \in T_g(G)$, ux is by (4) the image of u under the tangent mapping at g to the mapping $h \mapsto hx$ of G into X. Hence

$$(u + u')x = ux + u'x, \quad (\lambda u)x = \lambda(ux) \quad \text{for } u, u' \text{ in } T_g(G), \lambda \in K. \tag{9}$$

The above can be applied to the case of a Lie group operating on itself by left (resp. right) translation. The corresponding law of operation of T(G) on T(G) is defined by left (resp. right) translation of the Lie group T(G). Formulae (7), (8) and (9) are therefore valid in T(G).

PROPOSITION 8. *Let G_1 and G_2 be Lie groups, X_1 and X_2 manifolds of class C^r and f_i a law of left operation of class C^r of G_i on X_i $(i = 1, 2)$. Let ϕ be a morphism of G_1 into G_2 and ψ a ϕ-morphism of X_1 into X_2. Then $T(\psi)$ is a $T(\phi)$-morphism of $T(X_1)$ into $T(X_2)$.*

$f_2 \circ (\phi \times \psi) = \psi \circ f_1$, whence

$$T(f_2) \circ (T(\phi) \times T(\psi)) = T(\psi) \circ T(f_1).$$

Let G be a Lie group, X a manifold of class C^r and $(g, x) \mapsto gx$ a law of left operation of class C^r of G on X. Let I be an open subset of K containing 0 and $\gamma: I \to G$ a mapping of class C^r such that $\gamma(0) = e$. Let

$$a = T_0(\gamma)1 \in T_e(G).$$

Let $x \in X$. Using (4), ax is the image under the tangent mapping to $\lambda \mapsto \gamma(\lambda)x$ of the tangent vector 1 to I at 0. Hence *the vector field $x \mapsto ax$ on X is the vector field defined by the mapping $(\lambda, x) \mapsto \gamma(\lambda)x$ in the sense of Differentiable and Analytic Manifolds*, R, 8.4.5.

3. CASE OF GROUP GERMS

Let (G, e, θ, m) be a Lie group germ and Ω the set of definition of m. Then $T(\Omega)$ is identified with an open subset of $T(G) \times T(G)$ and $T(m)$ is an analytic mapping of $T(\Omega)$ into $T(G)$. It can be verified as in no. 2 that $(T(G), 0_e, T(\theta), T(m))$ is a Lie group germ. The products of G and T(G) are often written multiplicatively. The canonical projection of T(G) onto G is a morphism of Lie group germs. The restriction of $T_e(m)$ to $T_e(G)$ is the vector space addition of $T_e(G)$. The zero section of T(G) is an isomorphism of the Lie group germ G onto a Lie subgroup germ of T(G) which we identify with G. If f is a morphism of G into a Lie group germ H, $T(f): T(G) \to T(H)$ is a morphism of Lie group germs.

The mapping $\phi: (g, u) \mapsto gu$ of $G \times T_e(G)$ into T(G) is an isomorphism of the trivial vector bundle $G \times T_e(G)$ with base space G onto the vector bundle T(G); for ϕ and ϕ^{-1} are analytic and are vector bundle morphisms, so that it suffices to apply *Differentiable and Analytic Manifolds*, R, 7.2.1. (The proof of no. 2 could also be adapted.) The isomorphism ϕ^{-1} is called the left trivialization of T(G). The inverse isomorphism of the mapping $(g, u) \mapsto ug$ is called the right trivialization.

Let X be a manifold of class C^r and ψ a law chunk of left operation of class C^r of G on X. Then $T(\psi)$ is a law chunk of left operation of class C^{r-1} of T(G) on T(X) extending ψ. Formulae (7), (8) and (9) remain valid if gx is defined. If I is an open subset of K containing 0, if $\gamma: I \to G$ is a mapping of class C^r such that $\gamma(0) = e$ and if $a = T_0(\gamma)1$, the vector field $x \mapsto ax$ defined on X is the vector field defined by the mapping $(\lambda, x) \mapsto \gamma(\lambda)x$ in the sense of *Differentiable and Analytic Manifolds*, R, 8.4.5.

§ 3. PASSAGE FROM A LIE GROUP TO ITS LIE ALGEBRA

1. CONVOLUTION OF POINT DISTRIBUTIONS ON A LIE GROUP

DEFINITION 1. *Let G be a Lie group, g and g′ two points of G and let* $t \in T_g^{(\infty)}(G)$ $t' \in T_{g'}^{(\infty)}(G)$ *be two point distributions at g and g′ on G (Differentiable and Analytic Manifolds, R, 13.2.1). The convolution product of t and t′, denoted by* $t * t'$, *is the image of* $t \otimes t'$ *under the mapping* $(h, h') \mapsto hh'$ *of* $G \times G$ *into G (Differentiable and Analytic Manifolds, R, 13.2.3).*

PROPOSITION 1. (i) *If* $t \in T_g^{(s)}(G)$ *and* $t' \in T_{g'}^{(s')}(G)$, *then* $t * t' \in T_{gg'}^{(s+s')}(G)$.

(ii) *If t or t has no constant term,* $t * t'$ *has no constant term.*

(iii) $\varepsilon_g * \varepsilon_{g'} = \varepsilon_{gg'}$.

(iv) *Let* $t \in T_g^{(s)}(G)$, $t' \in T_{g'}^{(s')}(G)$ *and let f be a function of class* $C^{s+s'}$ *in an open neighbourhood of gg′ with values in a Hausdorff polynormed space. Then*

$$\begin{aligned}\langle t * t', f\rangle &= \langle t', h' \mapsto \langle t, h \mapsto f(hh')\rangle\rangle \\ &= \langle t, h \mapsto \langle t', h' \mapsto f(hh')\rangle\rangle.\end{aligned}$$

This follows from *Differentiable and Analytic Manifolds*, R, 13.4.1, 13.2.3 and 13.4.4.

Suppose that K = **R** or **C** and that G is finite-dimensional. Then G is locally compact. If t, t' are point measures, the definition of $t * t'$ agrees with that of *Integration*, Chapter VIII, § 1. We shall see later that the convolution product of measures and that of point distributions are two special cases of the convolution product of distributions which are not necessarily point distributions.

Let $\mathscr{T}^{(\infty)}(G)$ be the direct sum of the $T_g^{(\infty)}(G)$ for $g \in G$ (cf. *Differentiable and Analytic Manifolds*, R, 13.6.1). We define the convolution product in $\mathscr{T}^{(\infty)}(G)$ as the bilinear mapping of $\mathscr{T}^{(\infty)}(G) \times \mathscr{T}^{(\infty)}(G)$ into $\mathscr{T}^{(\infty)}(G)$ extending the convolution product of Definition 1. We also denote it by $*$. Thus $\mathscr{T}^{(\infty)}(G)$ has an algebra structure filtered by the $\mathscr{T}^{(s)}(G)$. The subalgebra $\mathscr{T}^{(0)}(G) = \bigoplus_{g \in G} T_g^{(0)}(G)$ is identified with the algebra $K^{(G)}$ of the group G over K.

PROPOSITION 2. *The algebra $\mathscr{T}^{(\infty)}(G)$ is associative. It is commutative if and only if G is commutative.*

Let $t \in \mathscr{T}^{(\infty)}(G)$, $t' \in \mathscr{T}^{(\infty)}(G)$, $t'' \in \mathscr{T}^{(\infty)}(G)$. Then $t * (t' * t'')$ is the image of $t \otimes t' \otimes t''$ under the mapping $(g, g', g'') \mapsto g(g'g'')$ of $G \times G \times G$ into G and $(t * t') * t''$ is the image of $t \otimes t' \otimes t''$ under the mapping $(g, g', g'') \mapsto (gg')g''$ of $G \times G \times G$ into G. Hence $(t * t') * t'' = t * (t' * t'')$. It is seen similarly that, if G is commutative, $t * t' = t' * t$. If the convolution product is commutative, G is commutative by Proposition 1 (iii).

PROPOSITION 3. *If $t \in \mathscr{T}^{(\infty)}(G)$ and $g \in G$, then $\gamma(g)_* t = \varepsilon_g * t$, $\delta(g)_* t = t * \varepsilon_{g^{-1}}$, $(\operatorname{Int} g)_* t = \varepsilon_g * t * \varepsilon_{g^{-1}}$. In particular, ε_e is the unit element of $\mathscr{T}^{(\infty)}(G)$.*

Consider the diagram

$$G \xrightarrow{\phi} G \times G \xrightarrow{\psi} G$$

where ϕ is the mapping $h \mapsto (g, h)$ and ψ is the mapping $(h', h) \mapsto h'h$. Then $\gamma(g) = \psi \circ \phi$ and hence $\gamma(g)_* t = \psi_*(\phi_*(t))$. But $\phi_*(t) = \varepsilon_g \otimes t$ and hence $\psi_*(\phi_*(t)) = \varepsilon_g * t$. The argument is similar for $\delta(g)_* t$. Finally,

$$\operatorname{Int} g = \gamma(g) \circ \delta(g)$$

and hence $(\operatorname{Int} g)_* = \gamma(g)_* \circ \delta(g)_*$.

It is therefore seen that, for $t \in T(G)$, $\varepsilon_g * t$ and $t * \varepsilon_g$ are equal to gt and tg calculated in the group T(G) (§ 2, no. 2). But it should be noted that, for t, t' in T(G), the product tt' in the sense of § 2 is in general different from $t * t'$.

DEFINITION 2. *Let* G *be a Lie group. The subalgebra of* $\mathscr{T}^{(\infty)}(G)$ *consisting of the distributions with support contained in* e *is denoted by* U(G).

This algebra is filtered by the subspaces

$$U_s(G) = U(G) \cap \mathscr{T}^{(s)}(G) = T_e^{(s)}(G).$$

We write $U^+(G) = T_e^{(\infty)+}(G)$, $U_s^+(G) = U^+(G) \cap U_s(G)$ (cf. *Differentiable and Analytic Manifolds*, R, 13.2.1). Recall that $U_0(G)$ is identified with K and $U_1^+(G)$ with the tangent space $T_e(G)$. In U(G), $U^+(G)$ is a two-sided ideal supplementary to $U_0(G)$.

Example. Let E be a complete normable space considered as a Lie group. Then the vector space U(E) is canonically identified with the vector space TS(E) (*Differentiable and Analytic Manifolds*, R, 13.2.4). Let $m: E \times E \to E$ be addition on E. Then

$$m_*: \mathsf{TS}(E \times E) \to \mathsf{TS}(E)$$

is equal to $\mathsf{TS}(m)$ (*Differentiable and Analytic Manifolds*, R, 13.2.4). For t, t' in $U(E) = \mathsf{TS}(E)$, the image $t * t'$ of the symmetric tensor product $t \otimes t'$ under m_* is therefore $\mathsf{TS}(m)(t \otimes t')$. By *Algebra*, Chapter IV, § 5, no. 6, Proposition 7, this image is just the product tt' in the algebra TS(E). Thus the algebra U(E) is identified with the algebra TS(E).

PROPOSITION 4. *Consider the bilinear mapping* $(u, v) \mapsto u * v$ (resp. $(u, v) \mapsto v * u$) *of* $U(G) \otimes K^{(G)}$ *into* $\mathscr{T}^{(\infty)}(G)$. *The corresponding linear mapping of* $U(G) \otimes K^{(G)}$ *into* $\mathscr{T}^{(\infty)}(G)$ *is a vector space isomorphism.*

$K^{(G)}$ is the direct sum of the $K\varepsilon_x$ for $x \in G$. On the other hand, the mapping $u \mapsto u * \varepsilon_g$ (resp. $u \mapsto \varepsilon_g * u$) is an isomorphism of the vector space $U(G) = \mathscr{T}_e^{(\infty)}(G)$ onto the vector space $\mathscr{T}_g^{(\infty)}(G)$ by Proposition 3. Finally, $\mathscr{T}^{(\infty)}(G)$ is the direct sum of the $T_g^{(\infty)}(G)$ for $g \in G$.

Let X be a manifold of class C^r ($r \geqslant \infty$) and $x \in X$. We have defined (*Differentiable and Analytic Manifolds*, R, 13.3.1) a canonical filtration on the vector space $\mathscr{T}_x^{(\infty)}(X)$ and a canonical isomorphism $i_{X,x}$ of the associated graded vector space onto the graded vector space $\mathsf{TS}(T_x(X))$. In particular, let $T_e(G) = L$; then $i_{G,e}$ is an isomorphism of the graded vector space gr U(G) onto the graded vector space TS(L). But U(G) is a filtered algebra, from which we obtain a graded algebra structure on gr U(G).

PROPOSITION 5. *The isomorphism* $i_{G,e}$: gr U(G) $\to$ TS(L) *is an algebra isomorphism.*

Let p be the mapping $(t, t') \mapsto t \otimes t'$ of $U(G) \times U(G)$ into $U(G \times G)$. Let c be the mapping $(t, t') \mapsto t * t'$ of $U(G) \times U(G)$ into U(G). Let m be the mapping $(g, g') \mapsto gg'$ of $G \times G$ into G. Then by Definition 1

$$c = m_* \circ p. \tag{1}$$

Consider the diagram

$$\begin{array}{ccccc} \operatorname{gr} U(G) \times \operatorname{gr} U(G) & \xrightarrow{\operatorname{gr}(p)} & \operatorname{gr} U(G \times G) & \xrightarrow{\operatorname{gr}(m_*)} & \operatorname{gr} U(G) \\ \Big\downarrow {\scriptstyle i_{G,e} \times i_{G,e}} & & \Big\downarrow {\scriptstyle i_{G \times G, e}} & & \Big\downarrow {\scriptstyle i_{G,e}} \\ \mathsf{TS}(L) \times \mathsf{TS}(L) & \xrightarrow{q} & \mathsf{TS}(L \times L) & \xrightarrow{\mathsf{TS}(T(m))} & \mathsf{TS}(L) \end{array}$$

where q is the mapping derived from the canonical isomorphism of $\mathsf{TS}(L) \times \mathsf{TS}(L)$ onto $\mathsf{TS}(L \times L)$. By *Differentiable and Analytic Manifolds*, R, 13.4.6 and 13.3.5, the two squares of the diagram are commutative. Hence by (1) the diagram

$$\begin{array}{ccc} \operatorname{gr} U(G) \times \operatorname{gr} U(G) & \xrightarrow{\operatorname{gr}(c)} & \operatorname{gr} U(G) \\ \Big\downarrow {\scriptstyle i_{G,e} \times i_{G,e}} & & \Big\downarrow {\scriptstyle i_{G,e}} \\ \mathsf{TS}(L) \times \mathsf{TS}(L) & \xrightarrow{\mathsf{TS}(T(m)) \circ q} & \mathsf{TS}(L) \end{array}$$

is commutative. Now $T(m) : L \times L \rightarrow L$ maps (x, y) to $x + y$ (§ 2, no. 1, Proposition 2 (ii)). By *Algebra*, Chapter IV, § 5, no. 6, Proposition 7, $\mathsf{TS}(T(m)) \circ q$ is therefore the multiplication of the algebra $\mathsf{TS}(L)$.

2. FUNCTORIAL PROPERTIES

PROPOSITION 6. *Let* G, H *be Lie groups and* ϕ *a morphism of* G *into* H. *For* t, t' *in* $\mathscr{T}^{(\infty)}(G)$, $\phi_*(t * t') = \phi_*(t) * \phi_*(t')$.

Consider the diagram

$$\begin{array}{ccc} G \times G & \xrightarrow{m} & G \\ \Big\downarrow {\scriptstyle \phi \times \phi} & & \Big\downarrow {\scriptstyle \phi} \\ H \times H & \xrightarrow{n} & H \end{array}$$

where $m(g, g') = gg'$, $n(h, h') = hh'$. This diagram is commutative. Hence

$$\begin{aligned} \phi_*(t * t') &= \phi_*(m_*(t \otimes t')) = n_*((\phi \times \phi)_*(t \otimes t')) \\ &= n_*(\phi_*(t) \otimes \phi_*(t')) = \phi_*(t) * \phi_*(t'). \end{aligned}$$

The Lie groups G and $G^{\vee}$ have the same underlying manifold and hence the vector spaces $\mathscr{T}^{(\infty)}(G)$ and $\mathscr{T}^{(\infty)}(G^{\vee})$ are the same. Let θ be the mapping $g \mapsto g^{-1}$, which is an isomorphism of the Lie group G onto the Lie group $G^{\vee}$. Then θ^* is an automorphism of the vector space $\mathscr{T}^{(\infty)}(G)$, which automorphism we denote by $t \mapsto t^{\vee}$. Then $(\varepsilon_g)^{\vee} = \varepsilon_{g^{-1}}$. If $t \in T_e(G)$, then

(2) $$t^{\vee} = -t \quad (\S\ 2, \text{Proposition } 2).$$

Example. Suppose that G is the Lie group defined by a complete normable space E. Then U(G) is identified with $\mathsf{TS}(E)$ and the restriction θ_* to U(G) is identified with $\mathsf{TS}(T_e(\theta))$ (*Differentiable and Analytic Manifolds*, R, 13.2.4). Therefore, if $t \in \mathsf{TS}^s(E)$, $t^{\vee} = (-1)^s t$.

Proposition 7. *Let* G *be a Lie group. Let* t, t' *be in* $\mathscr{T}^{(\infty)}(\mathrm{G})$.

(i) *The product* $t * t'$ *calculated relative to* $\mathrm{G}^{\vee}$ *is equal to the product* $t' * t$ *calculated relative to* G.

(ii) $(t * t')^{\vee} = t'^{\vee} * t^{\vee}$.

Consider the diagram

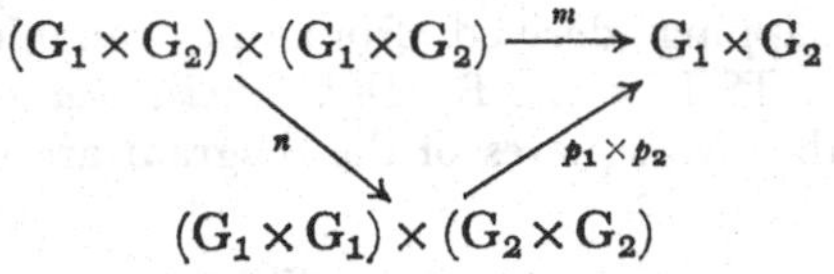

where $s(g, g') = (g', g)$, $m(g, g') = gg'$, $n(g, g') = g'g$ for all g, g' in G. This diagram is commutative. Hence $n_*(t \otimes t') = m_*(s_*(t \otimes t')) = m_*(t' \otimes t)$. This equality is precisely (i). Assertion (ii) follows from (i) and Proposition 6.

Proposition 8. *Let* G, H *be Lie groups and* ϕ *a morphism of* G *into* H. *If* $t \in \mathscr{T}^{(\infty)}(\mathrm{G})$, *then* $\phi_*(t^{\vee}) = (\phi_*(t))^{\vee}$.

Let θ (resp. θ') be the mapping $g \mapsto g^{-1}$ of G into G (resp. of H into H). Then $\phi \circ \theta = \theta' \circ \phi$, whence $\phi_*(\theta_*(t)) = \theta'_*(\phi_*(t))$.

Proposition 9. *Let* $\mathrm{G}_1, \cdots, \mathrm{G}_n$ *be Lie groups and* $\mathrm{G} = \mathrm{G}_1 \times \cdots \times \mathrm{G}_n$. *If the vector spaces* $\mathscr{T}^{(\infty)}(\mathrm{G})$ *and* $\mathscr{T}^{(\infty)}(\mathrm{G}_1) \otimes \cdots \otimes \mathscr{T}^{(\infty)}(\mathrm{G}_n)$ *are canonically identified, the algebra* $\mathscr{T}^{(\infty)}(\mathrm{G})$ *is the tensor product of the algebras* $\mathscr{T}^{(\infty)}(\mathrm{G}_1), \ldots \mathscr{T}^{(\infty)}(\mathrm{G}_n)$. *If* $t_i \in \mathscr{T}^{(\infty)}(\mathrm{G}_i)$ *for* $i = 1, \ldots, n$, *then*

$$(t_1 \otimes \cdots \otimes t_n)^{\vee} = t_1^{\vee} \otimes \cdots \otimes t_n^{\vee}.$$

It suffices to consider the case $n = 2$. Let t_1, t'_1 be in $\mathscr{T}^{(\infty)}(\mathrm{G}_1)$, t_2, t'_2 in $\mathscr{T}^{(\infty)}(\mathrm{G}_2)$. We need to show that $(t_1 \otimes t_2) * (t'_1 \otimes t'_2) = (t_1 * t'_1) \otimes (t_2 * t'_2)$ and that $(t_1 \otimes t_2)^{\vee} = t_1^{\vee} \otimes t_2^{\vee}$. Consider the diagram

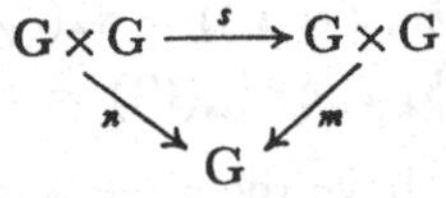

where $m((x_1, x_2), (x'_1, x'_2)) = (x_1x'_1, x_2x'_2)$,

$$n((x_1, x_2), (x'_1, x'_2)) = ((x_1, x'_1), (x_2, x'_2)),$$

$p_1(x_1, x'_1) = x_1x'_1$, $p_2(x_2, x'_2) = x_2x'_2$. This diagram is commutative. Hence

$$m_*((t_1 \otimes t_2) \otimes (t'_1 \otimes t'_2)) = (p_1 \otimes p_2)_*(n_*((t_1 \otimes t_2) \otimes (t'_1 \otimes t'_2))),$$

that is

$$\begin{aligned}(t_1 \otimes t_2) * (t'_1 \otimes t'_2) &= (p_1 \otimes p_2)_*((t_1 \otimes t'_1) \otimes (t_2 \otimes t'_2)) \\ &= p_{1*}(t_1 \otimes t'_1) \otimes p_{2*}(t_2 \otimes t'_2) \\ &= (t_1 * t'_1) \otimes (t_2 * t'_2).\end{aligned}$$

It is seen analogously that $(t_1 \otimes t_2)^\vee = t_1^\vee \otimes t_2^\vee$.

PROPOSITION 10. *Let* H *be a Lie subgroup of* G *and* $i\colon \mathrm{H} \to \mathrm{G}$ *the canonical injection. Then* i_* *is an injective homomorphism of the algebra* $\mathscr{T}^{(\infty)}(\mathrm{H})$ *into the algebra* $\mathscr{T}^{(\infty)}(\mathrm{H})$ *and* $i_*(t^\vee) = (i_*(t))^\vee$ *for all* $t \in \mathscr{T}^{(\infty)}(\mathrm{H})$.

This follows from Propositions 6 and 8 and *Differentiable and Analytic Manifolds*, R, 13.2.3.

$\mathscr{T}^{(\infty)}(\mathrm{H})$ is identified with a subalgebra of $\mathscr{T}^{(\infty)}(\mathrm{G})$ by means of the isomorphism of Proposition 10.

Remark. Proposition 10 remains valid if H is a Lie quasi-subgroup.

We recall (*Differentiable and Analytic Manifolds*, R, 13.5.1) that, if V is an analytic manifold over K, $\mathscr{T}^{(\infty)}(\mathrm{V})$ has canonically a cogebra structure over K with a counit; the counit is the linear mapping of $\mathscr{T}^{(\infty)}(\mathrm{G})$ into K which associates with each element of $\mathrm{T}_x^{(\infty)}(\mathrm{V})$ its constant term.

PROPOSITION 11. *Let* G *be a Lie group.*

(i) *The cogebra* $\mathscr{T}^{(\infty)}(\mathrm{G})$, *with convolution, is a bigebra* (*Algebra*, Chapter III, § 11, no. 4).

(ii) *Let c be the coproduct on* $\mathscr{T}^{(\infty)}(\mathrm{G})$. *Let* $t \in \mathscr{T}^{(\infty)}(\mathrm{G})$ *and write*

$$c(t) = \sum_{i=1}^{n} t_i \otimes t_i'.$$

Then $c(t^\vee) = \sum_{i=1}^{n} t_i^\vee \otimes t_i'^\vee$.

We prove (i). In the definition of bigebras referred to, condition (1) follows from Propositions 2 and 3 and condition (2) follows from *Differentiable and Analytic Manifolds*, R, 13.5.1. Let d be the mapping $g \mapsto (g, g)$ of G into $\mathrm{G} \times \mathrm{G}$. Then $c = d_*$ and hence c is an algebra morphism (Propositions 6 and 9), which is condition (3). Let $t \in \mathrm{T}_g^{(\infty)}(\mathrm{G})$, $t' \in \mathrm{T}_{g'}^{(\infty)}(\mathrm{G})$ have no constant term and λ, λ' be elements of K; then $\varepsilon_g \otimes tt'$, $t \otimes \varepsilon_{g'}$, $t \otimes t'$ are without constant term (*Differentiable and Analytic Manifolds*, R, 13.4.1) and hence the constant term of $(\lambda\varepsilon_g + t) * (\lambda'\varepsilon_{g'} + t')$ is $\lambda\lambda'$; hence condition (4) holds.

We prove (ii). By Propositions 8 and 9,

$$c(t^\vee) = d_*(t^\vee) = (d_*(t))^\vee = \left(\sum_{i=1}^{n} t_i \otimes t_i'\right)^\vee = \sum_{i=1} t_i'^\vee \otimes t_i'^\vee.$$

PROPOSITION 12. *Let* G, H *be two Lie groups and* ϕ *a morphism of* G *into* H. *Then* ϕ_* *is a bigebra morphism of* $\mathscr{T}^{(\infty)}(\mathrm{G})$ *into* $\mathscr{T}^{(\infty)}(\mathrm{H})$.

This follows from Proposition 6 and *Differentiable and Analytic Manifolds*, R, 13.5.1.

Let G be a Lie group. The restrictions of the convolution and the coproduct to U(G) define a bigebra structure on U(G). We have $U(G)^\vee = U(G)$. If $\phi: G \to H$ is a Lie group morphism, we denote by $U(\phi)$ the mapping $t \mapsto \phi_*(t)$ of U(G) into U(H); this is a bigebra morphism. If $\psi: H \to L$ is another Lie group morphism, then $U(\psi \circ \phi) = U(\psi) \circ U(\phi)$. If ϕ is an immersion (resp. a submersion), $U(\phi)$ is injective (resp. surjective) by *Differentiable and Analytic Manifolds*, R, 13.2.3. In particular, if H is a Lie subgroup of G, U(H) is identified with a subalgebra of U(G), the coproduct on U(H) being the restriction of the coproduct on U(G). If H is open in G, then U(H) = U(G). If G_1, G_2 are Lie groups, $U(G_1 \times G_2)$ is identified with $U(G_1) \times U(G_2)$. The primitive elements of U(G) are those of $T_e(G)$ (*Differentiable and Analytic Manifolds*, R, 13.5.3).

Again let $\phi: G \to H$ be a Lie group morphism. If gr U(G) is identified $\mathsf{TS}(T_e(G))$ and gr U(H) with $\mathsf{TS}(T_e(H))$, then gr $U(\phi)$ is identified with $\mathsf{TS}(T_e(\phi))$ (*Differentiable and Analytic Manifolds*, R, 13.3.5). We apply this to the isomorphism $g \mapsto g^{-1}$ of G onto $G^\vee$; then $T_e(\phi) = -1$ and hence

$$(3) \qquad t \in U_s(G) \Rightarrow t^\vee \equiv (-1)^s t \bmod U_{s-1}(G).$$

3. CASE OF A GROUP OPERATING ON A MANIFOLD

Let G be a Lie group, X a manifold of class C^r and f a law of left operation of class C^r of G on X. If $t \in T_g^{(s)}(G)$ and $u \in T_x^{(s')}(X)$ and $s + s' \leqslant r$, we denote by $t * u$ the image of $t \otimes u$ under f_*. We extend the product $*$ to a bilinear mapping also denoted by $*$, of $\mathscr{T}^{(s)}(G) + \mathscr{T}^{(s')}(X)$ into $\mathscr{T}^{(s+s')}(X)$. Proposition 1 of no. 1 can be extended with obvious modifications to the present situation.

When G operates on itself by left translation, we recover Definition 1 of no. 1.

PROPOSITION 13. *Let $t \in \mathscr{T}^{(s)}(G)$, $t' \in \mathscr{T}^{(s')}(G)$, $u \in \mathscr{T}^{(s'')}(X)$, such that*

$$s + s' + s'' \leqslant r.$$

Then $(t * t') * u = t * (t' * u)$.

This can be proved as is Proposition 2 of no. 1.

In particular, if $r \leqslant \infty$, the vector space $\mathscr{T}^{(\infty)}(X)$ is a left module over the algebra $\mathscr{T}^{(\infty)}(G)$ with the product $*$.

PROPOSITION 14. (i) *Let $g_0 \in G$ and $\tau(g_0)$ be the mapping $x \mapsto f(g_0, x)$* of X into X. *If $u \in \mathscr{T}^{(r)}(X)$, then $\tau(g_0)_* u = \varepsilon_{g_0} * u$.*

(ii) *Let $x_0 \in X$ and $\rho(x_0)$ be the mapping $g \mapsto f(g, x_0)$ of G into X. If $t \in T^{(r)}(G)$, then $\rho(x_0)_* t = t * \varepsilon_{x_0}$.*

This can be proved as is Proposition 3 of no. 1.

In particular, if $u \in \mathrm{T}(\mathrm{X})$ and $t \in \mathrm{T}(\mathrm{G})$, $\varepsilon_{g_0} * u$ and $t * \varepsilon_{x_0}$ are equal to the products $g_0 u$ and $t x_0$ defined in § 2, no. 2.

PROPOSITION 15. *Let* G (resp. G′) *be a Lie group and* X (resp. X′) *a manifold of class* C^r. *Suppose that a law of left operation of class* C^r *of* G (resp. G′) *on* X (resp. X′) *is given. Let* ϕ *be a morphism of* G *into* G′ *and* ψ *a* ϕ*-morphism of* X *into* X′. *Let* $t \in \mathscr{T}^{(s)}(\mathrm{G})$, $u \in \mathrm{T}^{(s')}(\mathrm{X})$ *be such that* $s + s' \leqslant r$. *Then*

$$\psi_*(t * u) = \phi_*(t) * \psi_*(u).$$

This can be proved as is Proposition 6 of no. 2.

Remark. Let f be a law of right operation of class C^r of G on X. If $t \in \mathscr{T}^{(s)}(\mathrm{G})$ and $u \in \mathscr{T}^{(s')}(\mathrm{X})$, with $s + s' \leqslant r$, we denote by $u * t$ the image of $u \otimes t$ under f_*. Propositions 13, 14, 15 go over to this situation in an obvious way.

PROPOSITION 16. *Let* G, G′ *be Lie groups,* X *a manifold of class* C^r *and suppose that* G (resp. G′) *operates on* X *on the left* (resp. *right*), *with* $(gx)g' = g(xg')$ *for all* $x \in \mathrm{X}$, $g \in \mathrm{G}$, $g' \in \mathrm{G}'$. *Let* $t \in \mathscr{T}^{(s)}(\mathrm{G})$, $t' \in \mathscr{T}^{(s')}(\mathrm{G}')$, $t'' \in \mathrm{T}^{(s'')}(\mathrm{X})$, *with* $s + s' + s'' \leqslant r$. *Then*

$$(t * t'') * t' = t * (t'' * t').$$

$(t * t'') * t'$ (resp. $t * (t'' * t')$) is the image of $t \otimes t'' \otimes t'$ under the mapping $(g, x, g') \mapsto (gx)g'$ (resp. $g(xg')$) of $\mathrm{G} \times \mathrm{X} \times \mathrm{G}'$ into X.

4. CONVOLUTION OF POINT DISTRIBUTIONS AND FUNCTIONS

Let G be a Lie group, X a manifold of class C^r and $(g, x) \mapsto gx$ a law of left operation of class C^r of G on X. For all $x \in \mathrm{X}$, let $\rho(x)$ denote the orbital mapping of x.

DEFINITION 3. *Let* $t \in \mathscr{T}^{(s)}(\mathrm{G})$ *with* $s \leqslant r$. *Let* $f: \mathrm{X} \to \mathrm{F}$ *be a function of class* C^r *with values in a Hausdorff polynormed space* (*for example* $\mathrm{F} = \mathrm{K}$). *The convolution of* t *and* f, *denoted by* $t * f$, *is the function on* X *with values in* F *defined by*

$$(t * f)(x) = \langle t^{\vee} * \varepsilon_x, f\rangle.$$

Then

$$(4) \qquad \begin{aligned} (t * f)(x) &= \langle \rho(x)_*(t^{\vee}), f\rangle && \text{(no. 3, Proposition 14 (ii))} \\ &= \langle t^{\vee}, f \circ \rho(x)\rangle && (\textit{Diff. \& Anal. Man.}, \text{R}, 13.2.3) \\ &= \langle t, (f \circ \rho(x))^{\vee}\rangle && (\textit{Diff. \& Anal. Man.}, \text{R}, 13.2.3). \end{aligned}$$

Note that Definition 3 can also be written in the more symmetric form

$$(5) \qquad \langle \varepsilon_x, t * f\rangle = \langle t^{\vee} * \varepsilon_x, f\rangle.$$

The function $(g, x) \mapsto f(gx) = (f \circ \rho(x))(g)$ on $\mathrm{G} \times \mathrm{X}$ is of class C^r. By *Differentiable and Analytic Manifolds*, R, 13.4.4, the function $x \mapsto \langle t^{\vee}, f \circ \rho(x)\rangle$ is therefore of class C^{r-s} if $s < \infty$. In other words, if $s < \infty$, *$t * f$ is of class* C^{r-s}.

Clearly $t * f$ depends linearly on t and f.

Formula (4) implies in particular, for $g \in G$,

$$(6) \qquad (\varepsilon_g * f)(x) = f(g^{-1}x)$$

that is

$$(7) \qquad \varepsilon_g * f = \gamma(g)f.$$

Suppose that $K = \mathbf{R}$ or $\mathbf{C}$, that G and X are finite-dimensional and that X has a positive measure invariant under G. The definition of $\varepsilon_g * f$ agrees with that of *Integration*, Chapter VIII, § 4, no. 1 (cf. formula (2), *loc. cit.*).

Proposition 17. *Let* $t \in \mathscr{T}^{(s)}(G)$, $t' \in \mathscr{T}^{(s')}(X)$ *and* $f: X \to F$ *a function of class* C^r *with* $s + s' \leqslant r$. *Then*

$$\langle t', t * f\rangle = \langle t^\vee * t', f\rangle.$$

$$\begin{aligned}\langle t', t * f\rangle &= \langle t', x \mapsto \langle t, g \mapsto f(g^{-1}x)\rangle\rangle && \text{by (4)}\\ &= \langle t \otimes t', (g, x) \mapsto f(g^{-1}x)\rangle && (\textit{Diff. \& Anal. Man.}, \text{R}, 13.4.4(\\ &= \langle t^\vee \otimes t', (g, x) \mapsto f(gx)\rangle && (\textit{Diff. \& Anal. Man.}, \text{R}, 13.2.3)\\ &= \langle t^\vee * t', f\rangle.\end{aligned}$$

Proposition 18. *Let* $t \in \mathscr{T}^{(s)}(G)$, $t' \in \mathscr{T}^{(s')}(G)$ *and* $f: X \to F$ *a function of class* C^r, *with* $s + s' \leqslant r$. *Then*

$$(t * t') * f = t * (t' * f).$$

For all $x \in X$,

$$\begin{aligned}\langle \varepsilon_x, (t * t') * f\rangle &= \langle (t * t')^\vee * \varepsilon_x, f\rangle && \text{by (5)}\\ &= \langle t'^\vee * (t^\vee * \varepsilon_x), f\rangle && \text{(Propositions 2 and 7)}\\ &= \langle t^\vee * \varepsilon_x, t' * f\rangle && \text{(Proposition 17)}\\ &= \langle \varepsilon_x, t * (t' * f)\rangle && \text{(Proposition 17).}\end{aligned}$$

If $r \geqslant \infty$, we see that the set of functions of class C^∞ on X with values in F is a left module over the algebra $\mathscr{T}^{(\infty)}(G)$.

Proposition 19. *Let* $t \in \mathscr{T}^{(s)}(G)$, with $s \leqslant r$. *Let* f (resp. f') *be a function of class* C^r *on* X *with values in a Hausdorff polynormed space* F (resp. F'). *Let* $(u, u') \mapsto uu'$ *be a continuous bilinear mapping of* $F \times F'$ *into a Hausdorff polynormed space* F", *so that* ff' *is a function of class* C^r *on* X *with values in* F". *Let* $\sum_{i=1}^{n} t_i \otimes t'_i$ *be the image of* t *in* $\mathscr{T}^{(s)}(G) \otimes \mathscr{T}^{(s)}(G)$ *under the coproduct. Then*

$$t * (ff') = \sum_{i=1}^{n} (t_i * f)(t'_i * f').$$

Let $x \in X$ and let $\rho(x)$ denote the orbital mapping of x. Then

$$\begin{aligned}\langle \varepsilon_x, t * (ff')\rangle &= \langle t^\vee, (ff') \circ \rho(x)\rangle && \text{by (4)}\\ &= \langle t^\vee, (f \circ \rho(x))(f' \circ \rho(x))\rangle \\ &= \sum_{i=1}^{n} \langle t_i^\vee, f \circ \rho(x)\rangle\langle t_i'^\vee, f' \circ \rho(x)\rangle \\ & && \text{(\textit{Diff. \& Man. Anal.}, R, 13.5.2)}\\ &= \sum_{i=1}^{n} \langle \varepsilon_x, t_i * f\rangle\langle \varepsilon_x, t_i' * f'\rangle && \text{by (4).}\end{aligned}$$

Remark 1. Let G be a Lie group, X a manifold of class C^r and $(x, g) \mapsto xg$ a law of right operation of class C^r of G on X. If $t \in \mathscr{T}^{(s)}(G)$ with $s \leqslant r$ and $f: X \to F$ is a function of class C^r on X, we denote by $f * t$ the function on X defined by

$$\text{(8)} \qquad \begin{aligned}\langle \varepsilon_x, f * t\rangle &= \langle \varepsilon_x * t^\vee, f\rangle \\ &= \langle \rho(x)_*(t^\vee), f\rangle \\ &= \langle t^\vee, f \circ \rho(x)\rangle \\ &= \langle t, (f \circ \rho(x))^\vee\rangle.\end{aligned}$$

In particular

$$\text{(9)} \qquad (f * \varepsilon_g)(x) = f(xg^{-1})$$

that is

$$\text{(10)} \qquad f * \varepsilon_g = \delta(g)^{-1}f.$$

Propositions 17, 18, 19 become, in the obvious notation,

$$\text{(11)} \qquad \langle t', f * t\rangle = \langle t' * t^\vee, f\rangle$$

$$\text{(12)} \qquad f * (t * t') = (f * t) * t'$$

$$\text{(13)} \qquad (ff') * t = \sum_{i=1}^{n} (f * t_i)(f' * t_i').$$

PROPOSITION 20. *Let* G, G' *be Lie groups*, X *a manifold of class* C^r *and* $(g, x) \mapsto gx$ (*resp.* $(x, g') \mapsto xg'$) *a law of left* (resp. *right*) *operation of class* C^r *of* G (resp. G') *on* X. *Suppose that* $(gx)g' = g(xg')$ *for all* $x \in X$, $g \in G$, $g' \in G'$. *Let* $t \in \mathscr{T}^{(s)}(G)$, $t' \in \mathscr{T}^{(s')}(G')$ *and* $f: X \to F$ *be a function of class* C^r *such that* $s + s' \leqslant r$. *Then*

$$(t * f) * t' = t * (f * t').$$

For all $x \in X$,

$$\begin{aligned}\langle \varepsilon_x, (t * f) * t'\rangle &= \langle \varepsilon_x * t'^\vee, t * f\rangle && \text{by (8)}\\ &= \langle t^\vee * (\varepsilon_x * t'^\vee), f\rangle && \text{(Proposition 17)}\\ &= \langle t^\vee * \varepsilon_x, f * t'\rangle && \text{(Proposition 2 and (11))}\\ &= \langle \varepsilon_x, t * (f * t')\rangle && \text{by (5).}\end{aligned}$$

In particular, consider G as operating on itself by left and right translations. If $f\colon G \to F$ is a function of class C^r on G and $t \in \mathscr{T}^{(s)}(G)$ (with $s \leqslant r$), $t * f$ and $f * t$ are, if $s < \infty$, functions of class C^{r-s} on G. Further, let $t' \in \mathscr{T}^{(s')}(G)$, with $s + s' \leqslant r$. Then

$$(t * f) * t' = t * (f * t'). \tag{14}$$

In particular, $\mathscr{C}^\infty(G)$ is a $(\mathscr{T}^{(\infty)}(G), \mathscr{T}^{(\infty)}(G))$-bimodule. Formulae (5) and (8) admit as special cases

$$\langle t, f \rangle = \langle \varepsilon_e, t^\vee * f \rangle = \langle \varepsilon_e, f * t^\vee \rangle. \tag{15}$$

Remark 2. Let $(g, x) \mapsto gx$ be a law of left operation of class C^r of G on X. Let $t \in U_s(G)$ with $s \leqslant r$, Ω be an open subset of X and $f\colon \Omega \to F$ a function of class C^r. $t * f$ can also be defined by formula (4) or (5); it is a function defined on Ω with values in F, of class C^{r-s} if $s < \infty$. The results of this no. extend in an obvious way to this situation.

5. FIELDS OF POINT DISTRIBUTIONS DEFINED BY THE ACTION OF A GROUP ON A MANIFOLD

Let $(g, x) \mapsto \lambda(g, x) = gx$ be a law of left operation of class C^r of G on X. Let $s \leqslant r$ and $t \in U_s(G)$. For all $x \in X$, $t * \varepsilon_x \in T_x^{(s)}(X)$. The mapping $x \mapsto t * \varepsilon_x$ is called *the field of point distributions defined by t and the action of* G *on* X and denoted sometimes by D_t^λ or simply D_t. Let Ω be an open subset of X and F a Hausdorff polynormed space. If $f\colon \Omega \to F$ is of class C^r and $s \leqslant r$, the function $t^\vee * f$ on Ω is also denoted by $D_t f$. Then

$$(D_t f)(x) = \langle t * \varepsilon_x, f \rangle. \tag{16}$$

If $s < \infty$, then $D_t f \in \mathscr{C}^{r-s}(\Omega, F)$ by no. 4. Thus $f \mapsto D_t f$ is a mapping of $\mathscr{C}^r(\Omega, F)$ into $\mathscr{C}^{r-s}(\Omega, F)$ (often denoted by D_t by an abuse of notation).

If $t \in U_s(G)$, $t' \in U_{s'}(G)$ and $s + s' \leqslant r$, then, by Proposition 18 of no. 4,

$$D_{t*t'} f = D_{t'}(D_t f) \tag{17}$$

and hence, using the abuse of notation indicated above,

$$D_{t*t'} = D_{t'} \circ D_t. \tag{18}$$

Suppose that G and X are finite-dimensional. The mapping $(t, x) \mapsto t \otimes \varepsilon_x$ of $T^{(s)}(G) \times X$ into the vector bundle $T^{(s)}(G \times X)$ (cf. *Differentiable and Analytic Manifolds*, R, 13.2.5) is of class C^{r-s}. Hence (*Differentiable and Analytic Manifolds*, R, 13.2.5) the mapping $(t, x) \mapsto t * \varepsilon_x$ of $T^{(s)}(G) \times X$ into the vector bundle $T^{(s)}(X)$ is of class C^{r-s}. In particular, D_t is a differential operator of order $\leqslant s$ and class C^{r-s} in the sense of *Differentiable and Analytic Manifolds*, R, 14.1.6. By formula (16), the function $D_t f$ is then the image of f under this differential operator (*Differentiable and Analytic Manifolds*, R, 14.1.4).

We now no longer suppose that G and X are finite-dimensional. Let ψ be an automorphism of the manifold X and Δ a field of point distributions on X. Conforming with the general definitions, the transform of Δ under ψ is the field of point distributions on X whose value at $\psi(x)$ is $\psi_*(\Delta(x))$; we denote this mapping by $\psi(\Delta)$. If $g \in G$ and $\tau(g)$ denotes the automorphism $x \mapsto gx$ of X, the transform of Δ under $\tau(g)$ is also called the transform of Δ under g.

Proposition 21. *Let ψ be an automorphism of* X *commuting with the operations of* G. *Then D_t is invariant under ψ.*

For all $x \in X$,

$$\begin{aligned}(\psi(D_t))(\psi(x)) &= \psi_*(D_t(x)) = \psi_*(t * \varepsilon_x)\\ &= t * \psi_*(\varepsilon_x) \qquad \text{(Proposition 15)}\\ &= t * \varepsilon_{\psi(x)} = D_t(\psi(x)).\end{aligned}$$

Proposition 22. *If $g \in G$, the transform of D_t under g is $D_{\varepsilon_g * t * \varepsilon_{g^{-1}}}$.*

The value of this transform at gx is

$$\begin{aligned}\tau(g)_*(D_t(x)) &= \tau(g)_*(t * \varepsilon_x)\\ &= \varepsilon_g * (t * \varepsilon_x) \qquad \text{(Proposition 14 (i))}\\ &= (\varepsilon_g * t * \varepsilon_{g^{-1}}) * \varepsilon_{gx} \qquad \text{(Propositions 1 and 2)}\\ &= D_{\varepsilon_g * t * \varepsilon_{g^{-1}}}(gx).\end{aligned}$$

Let $(x, g) \mapsto \mu(x, g) = xg$ be a law of right operation of class C^r of G on X. Let $s \leqslant r$ and $t \in U_s(G)$. For all $x \in X$, $\varepsilon_x * t \in T_x^{(s)}(X)$. The mapping $x \mapsto \varepsilon_x * t$ is called the field of distributions defined by t and the action of G on X and is sometimes denoted by D_t^μ or simply D_t. Let Ω be an open subset of X. If $f: \Omega \to F$ is of class C^r, the function $f * t^\vee$ is denoted by $D_t f$. Then

$$(D_t f)(x) = \langle \varepsilon_x * t, f \rangle \tag{19}$$

and, in the obvious notation,

$$D_{t * t'} f = D_t(D_{t'} f) \tag{20}$$

$$D_{t * t'} = D_t \circ D_{t'}. \tag{21}$$

Proposition 21 remains valid. Let $g \in G$. The transform of D_t under g (that is under the automorphism $x \mapsto xg$ of X) is $D_{\varepsilon_{g^{-1}} * t * \varepsilon_g}$.

6. INVARIANT FIELDS OF POINT DISTRIBUTIONS ON A LIE GROUP

Definition 4. *Let* G *be a Lie group. A field of distributions on* G *is called left* (resp. *right*) *invariant if it is invariant under left* (resp. *right*) *translations of* G.

In other words, a field of distributions $g \mapsto \Delta_g$ on G is left invariant if

$$\Delta_{gg'} = \gamma(g)_* \beta_{g'} \quad \text{for } g, g' \text{ in G},$$

or again if

$$\Delta_{gg'} = \varepsilon_g * \Delta_{g'} \quad \text{for } g, g' \text{ in G}.$$

It is right invariant if

$$\Delta_{gg'} = \delta(g'^{-1})_* \Delta_g \quad \text{for } g, g' \text{ in G},$$

or again if

$$\Delta_{gg'} = \Delta_g * \varepsilon_{g'} \quad \text{for } g, g' \text{ in G}.$$

DEFINITION 5. *Let* G *be a Lie group and* $t \in \mathrm{U}(\mathrm{G})$. *Let* L_t *denote the field of distributions* $g \mapsto \varepsilon_g * t$ *on* G *and* R_t *the field of distributions* $g \mapsto t * \varepsilon_g$ *on* G.

In other words, L_t (resp. R_t) is the field of distributions defined by t and G operating on G on the right (resp. left) by means of the mapping $(g, g') \mapsto gg'$. Let Ω be an open subset of G and F a Hausdorff polynormed space; if $f \in \mathscr{C}^\omega(\Omega, \mathrm{F})$, then $\mathrm{L}_t f = f * t^\vee \in \mathscr{C}^\omega(\Omega, \mathrm{F})$ and

$$\mathrm{R}_t f = t^\vee * f \in \mathscr{C}^\omega(\Omega, \mathrm{F})$$

(no. 5). If G is finite-dimensional, the differential operators L_t and R_t are of class C^ω (no. 5).

PROPOSITION 23. (i) *The mapping* $t \mapsto \mathrm{L}_t$ (resp. $t \mapsto \mathrm{R}_t$) *is an isomorphism of the vector space* U(G) *onto the vector field of left* (resp. *right*) *invariant distributions on* G.

(ii) *For* t, t' *in* U(G), $\mathrm{L}_{t*t'} = \mathrm{L}_t \circ \mathrm{L}_{t'}$, $\mathrm{R}_{t\,t'} = \mathrm{R}_{t'} \circ \mathrm{R}_t$, $\mathrm{L}_t \circ \mathrm{R}_{t'} = \mathrm{R}_{t'} \circ \mathrm{L}_t$ (*with the abuse of notation of no.* 5).

(iii) *If* θ *is the mapping* $g \mapsto g^{-1}$ *of* G *onto* G, *then* $\theta(\mathrm{L}_t) = \mathrm{R}_{t^\vee}$.

(iv) *If* $t \in \mathrm{U}(\mathrm{G})$ *and* $g \in \mathrm{G}$, *then* $(\mathrm{L}_t)_g = (\mathrm{R}_{\varepsilon_g * t * \varepsilon_{g^{-1}}})_g$.

In G every right translation commutes with every left translation. By Proposition 21 of no. 5, L_t is therefore left invariant. As $(\mathrm{L}_t)_e = t$, the mapping $t \mapsto \mathrm{L}_t$ is injective. Let Δ be a field of left invariant distributions on G; let $t = \Delta_e$; then Δ and L_t have the same value at e and are left invariant and hence $\Delta = \mathrm{L}_t$. This proves (i) for L_t and the argument is similar for R_t. The formulae $\mathrm{L}_{t*t'} = \mathrm{L}_t \circ \mathrm{L}_{t'}$, $\mathrm{R}_{t*t'} = \mathrm{R}_{t'} \circ \mathrm{R}_t$ follow from (21) and (18). Let $t \in \mathrm{U}_s(\mathrm{G})$, $t' \in \mathrm{U}_{s'}(\mathrm{G})$, $f \in \mathscr{C}^r(\Omega, \mathrm{F})$, where Ω is open in G and $s + s' \leqslant r$; then

$$\begin{aligned} \mathrm{L}_t \mathrm{R}_{t'} f &= \mathrm{L}_t(t'^\vee * f) = (t'^\vee * f) * t \\ &= t'^\vee * (f * t^\vee) \qquad \text{(Proposition 20)} \\ &= \mathrm{R}_{t'} \mathrm{L}_t f \end{aligned}$$

and hence $\mathrm{L}_t \circ \mathrm{R}_{t'} = \mathrm{R}_{t'} \circ \mathrm{L}_t$. As θ is an isomorphism of G onto $\mathrm{G}^\vee$, $\theta(\mathrm{L}_t)$ is a field of right invariant distributions on G; its value at e is $\theta^*(t) = t^\vee$; hence $\theta(\mathrm{L}_t) = \mathrm{R}_{t^\vee}$. Finally,

$$(\mathrm{L}_t)_g = \varepsilon_g * t = (\varepsilon_g * t * \varepsilon_{g^{-1}}) * \varepsilon_g = (\mathrm{R}_{\varepsilon_g * t * \varepsilon_{g^{-1}}})_g.$$

Remark 1. It is the action of G on itself by *right* translation which defines the fields of *left* invariant distributions.

Remark 2. Suppose that G is finite-dimensional. The mapping

$$(t, g) \mapsto (\mathrm{R}_t)_g = t * \varepsilon_g$$

of $\mathrm{U}_s(\mathrm{G}) \times \mathrm{G}$ into $\mathrm{T}^{(s)}(\mathrm{G})$ is an isomorphism of analytic vector bundles; for this mapping is bijective, linear on each fibre and analytic (no. 5); on the other hand, let $\phi\colon \mathrm{T}^{(s)}(\mathrm{G}) \to \mathrm{U}_s(\mathrm{G}) \times \mathrm{G}$ be the inverse bijection; if $t \in \mathrm{T}_g^{(s)}(\mathrm{G})$, then $\phi(t) = (t * \varepsilon_{g^{-1}}, g)$ and hence ϕ is analytic. The isomorphism ϕ is called the right trivialization of $\mathrm{T}^{(s)}(\mathrm{G})$. Similarly, consider the mapping $(t, g) \mapsto (\mathrm{L}_t)_g = \varepsilon_g * t$ of $\mathrm{U}_s(\mathrm{G}) \times \mathrm{G}$ into $\mathrm{T}^{(s)}(\mathrm{G})$; the inverse isomorphism is called the left trivialization of $\mathrm{T}^{(s)}(\mathrm{G})$. By restriction we recover the right and left trivializations of T(G) (§ 2, no. 2).

7. LIE ALGEBRA OF A LIE GROUP

Let G be a Lie group. In U(G), as in any associative algebra, we write $[t, t'] = t * t' - t' * t$. As $\mathrm{T}_e(\mathrm{G})$ is the set of primitive elements of U(G), $[\mathrm{T}_e(\mathrm{G}), \mathrm{T}_e(\mathrm{G})] \subset \mathrm{T}_e(\mathrm{G})$ (Chapter II, § 1, no. 2, Proposition 4). The restriction of the bracket to $\mathrm{T}_e(\mathrm{G})$ therefore defines on $\mathrm{T}_e(\mathrm{G})$ a Lie algebra structure.

Lemma 1. *Let* X *and* X′ *be complete normable spaces,* X_0 *an open neighbourhood of* 0 *in* X *and* f *an analytic mapping of* X_0 *into* X′ *such that* $f(0) = 0$. *Let* $f = f_1 + f_2 + f_3 + \cdots$ *be the expansion of* f *as an integral series about* 0, *where* f_i *is a homogeneous continuous-polynomial of degree* i *on* X *with values in* X′. *Let* t *be an element of* $\mathrm{TS}^2(\mathrm{X})$, *considered as a point distribution on* X *with support contained in* $\{0\}$. *Let* $t' = f_*(t) \in \mathrm{TS}(\mathrm{X}')$. *The homogeneous component of* t' *of degree* 1 *is* $\langle f_2, t\rangle$.

Let t'_1 be this component. Then, for every continuous linear mapping u of X′ into a polynormed space,

$$\begin{aligned} u(t'_1) &= \langle t', u\rangle && \text{because } u \text{ is continuous and linear} \\ &= \langle t, u \circ f\rangle && (\textit{Diff. \& Anal. Man.}, \text{ R, 13.2.3}) \\ &= \langle t, u \circ f_2\rangle && \text{because } t \in \mathrm{TS}^2(\mathrm{X}) \\ &= u(\langle t, f_2\rangle) && (\textit{Diff. \& Anal. Man.}, \text{ R, 13.2.2}), \end{aligned}$$

whence the lemma.

PROPOSITION 24. *Let* G *be a Lie group and* $(\mathrm{U}, \phi, \mathrm{E})$ *a chart on* G *such that* $\phi(e) = 0$. *Let* V *be an open neighbourhood of* e *such that* $\mathrm{V}^2 \subset \mathrm{U}$. *Let* m *be the analytic mapping* $(a, b) \mapsto \phi(\phi^{-1}(a)\phi^{-1}(b))$ *of* $\phi(\mathrm{V}) \times \phi(\mathrm{V})$ *into* E. *Let*

$$m = \sum_{i, j \geqslant 0} m_{i,j}$$

be the expansion of m *as an integral series about* $(0, 0)$, *where* $m_{i,j}$ *is a bihomogeneous continuous-polynomial of bidegree* (i, j) *on* $E \times E$ *with values in* E.

(i) $m_{i,0} = m_{0,j} = 0$ *for all* $i \neq 1$ and $j \neq 1$.

(ii) $m_{1,0}(a, b) = a$ and $m_{0,1}(a, b) = b$ *for all* $a \in E$, $b \in E$.

(iii) *Let* $\psi\colon T_e(G) \to E$ *be the differential of* ϕ *at* e. *For all* u, v *in* $T_e(G)$,

$$\psi([u, v]) = m_{1,1}(\psi(u), \psi(v)) - m_{1,1}(\psi(v), \psi(u)).$$

$m(a, 0) = a$, $m(0, b) = b$ for all a, b in $\phi(V)$, which proves (i) and (ii). Let u, v be in $T_e(G)$. Let $T_0(E)$ be identified with E and hence ψ with $T_e(\phi)$. The images of u and v under $T_e(\phi)$ are $\psi(u)$ and $\psi(v)$. The tensor product point distribution of these images is the symmetric product of $(\psi(u), 0)$ and $(0, \psi(v))$ in $TS(E \times E) = TS(E) \times TS(E)$, that is

$$(\psi(u), 0) \otimes (0, \psi(v)) + (0, \psi(v)) \otimes (\psi(u), 0).$$

Hence $\phi * (u * v)$ is the image of the above element under the mapping m of $\phi(V) \times \phi(V)$ into E. Its component of degree 1 in TS(E) is, by Lemma 1,

$$x = \langle m_{1,1}, (\psi(u), 0) \otimes (0, \psi(v)) + (0, \psi(v)) \otimes (\psi(u), 0)\rangle.$$

We define a bilinear mapping $n\colon (E \times E)^2 \to E$ by

$$n((a, b), (a', b')) = m_{1,1}(a, b').$$

Then $n((a, b), (a, b)) = m_{1,1}(a, b)$ and hence

$$\begin{aligned} x &= \langle n, (\psi(u), 0) \otimes (0, \psi(v)) + (0, \psi(v)) \otimes (\psi(u), 0)\rangle \\ &= m_{1,1}(\psi(u), \psi(v)) + m_{1,1}(0, 0) = m_{1,1}(\psi(u), \psi(v)). \end{aligned}$$

Similarly, $\phi * (v * u)$ admits $m_{1,1}(\phi(v), \phi(u))$ as component of degree 1 in TS(E). As $\phi([u, v])$ is of degree 1, this proves (iii).

COROLLARY. *The normable space* $T_e(G)$ *together with the bracket, is a normable Lie algebra.*

DEFINITION 6. *The normable space* $T_e(G)$, *together with the bracket, is called the normable Lie algebra of* G, *or simply the Lie algebra of* G, *and is denoted by* L(G).

PROPOSITION 25. *Let* G *be a Lie group and* E(G) *the enveloping algebra of* L(G). *The canonical injection of* L(G) *into* E(G) *defines a homomorphism* θ *of the algebra* E(G) *into the algebra* U(G). *If* K *is of characteristic* 0, η *is a bigebra isomorphism.*

The bigebra U(G) is cocommutative (*Differentiable and Analytic Manifolds*, R, 13.5.1) and the filtration $(U_s(G))$ is compatible with the bigebra structure. The set of primitive elements of U(G) is L(G). It then suffices to apply chapter II, § 1, no. 6, Theorem 1.

When K is of characteristic 0 we shall in future identify U(G) with the enveloping algebra of L(G). By (2) and Proposition 7 (ii), the mapping $t \mapsto t^{\vee}$ of U(G) into U(G) is then identified with the principal antiautomorphism of U(G) (Chapter I, § 2, no. 4).

PROPOSITION 26. *Suppose that* K *is of characteristic* $p > 0$. *For all* $a \in L(G)$, $a^p \in L(G)$ *and* $\operatorname{ad}(a^p) = (\operatorname{ad} a)^p$ *(the power* a^p *being calculated in* U(G)).

If $a \in L(G)$, a is primitive in U(G), hence a^p is primitive in U(G) (Chapter II, § 1, no. 2, *Remark* 1) and hence $a^p \in L(G)$. Let σ_a (resp. τ_a) be the linear mapping $x \mapsto a * x$ (resp. $x \mapsto x * a$) of U(G) into U(G). For all $x \in U(G)$, $(\operatorname{ad} a)(x) = (\sigma_a - \tau_a)(x)$ and hence $(\operatorname{ad} a)^p = (\sigma_a - \tau_a)^p$. But σ_a and τ_a commute and therefore $(\sigma_a - \tau_a)^p = (\sigma_a)^p - (\tau_a)^p = \tau_{a^p} - \sigma_{a^p}$, whence the second assertion.

DEFINITION 7. *Let* X *be a manifold of class* C^r $(r \geqslant 2)$ *and* $\mathfrak{g}$ *a complete normable Lie algebra. A law of infinitesimal left* (resp. *right*) *operation of class* C^{r-1} *of* $\mathfrak{g}$ *on* X *is a mapping* $a \mapsto D_a$ *of* $\mathfrak{g}$ *into the set of vector fields on* X *with the following properties:*

(a) *the mapping* $(a, x) \mapsto D_a(x)$ *is a morphism of class* C^{r-1} *of the trivial vector bundle* $\mathfrak{g} \times X$ *into the vector bundle* T(X);

(b) $[D_a, D_b] = -D_{[a,b]}$ (resp. $[D_a, D_b] = D_{[a,b]}$) *for all* a, b *in* $\mathfrak{g}$.

In particular, each vector field D_a is of class C^{r-1}.

Remark. Let X be a manifold of class C^r, $\mathfrak{g}$ a finite-dimensional Lie algebra and $a \mapsto D_a$ a linear mapping of $\mathfrak{g}$ into the vector space of vector fields of class C^{r-1} on X. Then condition (a) of Definition 7 holds. For, by considering a basis of $\mathfrak{g}$ and applying *Differentiable and Analytic Manifolds*, R, 7.7.1, the problem is reduced to the case $\dim \mathfrak{g} = 1$ and our assertion is then obvious.

PROPOSITION 27. *Let* G *be a Lie group and* X *a manifold of class* C^r. *Suppose that a law of left* (resp. *right*) *operation of class* C^r *of* G *on* X *is given. For all* $a \in L(G)$, *let* D_a *be the field of point distributions defined by* a *on* X.

(i) *The mapping* $(a, x) \mapsto D_a(x)$ *is a morphism of class* C^{r-1} *of the trivial vector bundle* $L(G) \times X$ *into the vector bundle* T(X).

(ii) *Let* I *be an open subset of* K *containing* 0 *and* $\gamma: I \to G$ *a mapping of class* C^r *such that* $\gamma(0) = e$. *Let* $a = T_0(\gamma)1 \in L(G)$. *If* f *is a function of class* C^r *on an open subset of* X, *then*

$$(D_a f)(x) = \lim_{k \in K^*, k \to 0} k^{-1}(f(\gamma(k)x) - f(x)) \quad \textit{if G operates on the left,}$$

$$(D_a f)(x) = \lim_{k \in K^*, k \to 0} k^{-1}(f(x\gamma(k)) - f(x)) \quad \textit{if G operates on the right.}$$

(iii) *If* $r \geqslant 2$, *the mapping* $a \mapsto D_a$ *is a law of infinitesimal left* (resp. *right*) *operation of class* C^{r-1} *of* L(G) *on* X.

Suppose that G operates on X on the left. Let $\phi: G \times X \to X$ be the law of operation. Then $T(\phi)$ is a ϕ-morphism of class C^{r-1} of the vector bundle $T(G) \times T(X)$ into the vector bundle T(X) (*Differentiable and Analytic*

Manifolds, R, 8.1.2). The induced vector bundle $(T(G) \times T(X))|(\{e\} \times X)$ is identified with $E = L(G) \times T(X)$. Hence $T(\phi)|E$ is a vector bundle morphism of class C^{r-1}. For $(a, x) \in L(G) \times X$, $T(\phi)(a, x) = D_a(x)$, whence (i).

The formula giving $(D_a f)(x)$ follows from § 2, the end of no. 2, and *Differentiable and Analytic Manifolds*, R, 8.4.5.

Suppose that $r \geqslant 2$. Let a, b be in L(G) and f be a function of class C^r on an open subset of X. Then

$$D_{[a,b]}f = D_b(D_a f) - D_a(D_b f) \quad \text{by (17)}$$
$$= [D_b, D_a]f \quad (\textit{Diff. \& Anal. Man.}, \text{R}, 8.5.3).$$

Let $x \in X$. By taking f to be a chart on an open neighbourhood of x, it follows that $D_{[a,b]}(x) = [D_b, D_a](x)$, whence (iii). The argument is similar if G operates on X on the right.

When $r \geqslant 2$, the mapping $a \mapsto D_a$ is called the law of infinitesimal operation *associated* with the given law of operation.

8. FUNCTORIAL PROPERTIES OF THE LIE ALGEBRA

Let G and H be Lie groups and ϕ a morphism of G into H. The restriction of $U(\phi)$ to L(G), which is just $T_e(\phi)$, is a continuous morphism of L(G) into L(H), which we denote by $L(\phi)$. If ψ is a morphism of H into a Lie group, then $L(\psi \circ \phi) = L(\psi) \circ L(\phi)$.

For ϕ to be an immersion, it is necessary and sufficient that $L(\phi)$ be an isomorphism of L(G) onto a subalgebra of L(H) admitting a topological supplement. In particular, if G is a Lie subgroup of H and ϕ is the canonical injection, L(G) is identified with a Lie subalgebra of L(H) by means of $L(\phi)$. More particularly, if G is an open subgroup of H, L(G) = L(H).

If G is a Lie quasi-subgroup of H, L(G) is also identified with a closed Lie subgroup of L(H).

For ϕ to be a submersion, it is necessary and sufficient that $L(\phi)$ be surjective and that its kernel admit a topological supplement. In that case, the kernel N of ϕ is a Lie subgroup of G and $L(N) = \operatorname{Ker} L(\phi)$. In particular, if H is the quotient Lie group of G by a normal Lie subgroup P, L(P) is an ideal of L(G) and, if ϕ is the canonical surjection of G onto H, L(G/P) is identified with L(G)/L(P) by means of the morphism derived from $L(\phi)$ when passing to the quotient.

Let I be a finite set, $(G_i)_{i \in I}$ a family of Lie groups, G their product and p_i the canonical morphism of G onto G_i. Then $(L(p_i))_{i \in I}$ is a morphism of the Lie algebra L(G) into the Lie algebra $\prod_{i \in I} L(G_i)$ and is an isomorphism of normable spaces. L(G) is therefore identified with $\prod_{i \in I} L(G_i)$ by means of $(L(p_i))_{i \in I}$.

PROPOSITION 28. *Let* G *and* H *be Lie groups and* ϕ *a morphism of* G *into* H. *Suppose that* K *is of characteristic* 0 *and that* H *is finite-dimensional.*

(i) *The kernel* N *of* ϕ *is a Lie subgroup of* G *and* $L(N) = \text{Ker } L(\phi)$.

(ii) *The morphism* ψ *of* G/N *into* H *derived from* ϕ *when passing to the quotient is an immersion.*

(iii) *If* $\phi(G)$ *is closed in* H *and the topology of* G *has a countable base,* $\phi(G)$ *is a Lie subgroup of* H, ψ *is an isomorphism of the Lie group* G/N *onto the Lie group* $\phi(G)$ *and* $L(\phi(G)) = \text{Im } L(\phi)$.

Let G operate on H on the left by the mapping $(g, h) \mapsto \phi(g)h$. It suffices to apply Proposition 14 of § 1, no. 7, to the orbit of e.

PROPOSITION 29. *Let* G *and* H *be Lie groups and* ϕ *a morphism of* G *into* H. *Suppose that* K *is of characteristic* 0 *and that* H *is finite-dimensional. If* H' *is a Lie subgroup of* H, *then* $G' = \phi^{-1}(H')$ *is a Lie subgroup of* G *and* $L(G') = L(\phi)^{-1}(L(H'))$.

Let π be the canonical mapping of H into the homogeneous space X = H/H'. Let G operate on X on the left by the mapping $(g, x) \mapsto \phi(g)x$. The stabilizer of $\pi(e)$ is G', which is therefore a Lie subgroup of G (§ 1, no. 7, Proposition 14). The orbital mapping of $\pi(e)$ is $\pi \circ \phi$. By Proposition 14 of § 1, L(G') is the kernel of $L(\pi \circ \phi) = T_e(\pi) \circ L(\phi)$. The kernel of $T_e(\pi)$ is L(H') (§ 1, no. 6, Proposition 11 (i)) and hence $\text{Ker } L(\pi \circ \phi) = L(\phi)^{-1}(L(H'))$.

COROLLARY 1. *Let* G, H *be Lie groups and* ϕ_1 *and* ϕ_2 *morphisms of* G *into* H. *Suppose that* K *is of characteristic* 0 *and that* H *is finite-dimensional. The set of* $g \in G$ *such that* $\phi_1(g) = \phi_2(g)$ *is a Lie subgroup* G' *of* G *and* L(G') *is the set of* $x \in L(G)$ *such that* $L(\phi_1)x = L(\phi_2)x$.

We write $\phi(g) = (\phi_1(g), \phi_2(g))$ for all $g \in G$, so that ϕ is a morphism of G into H × H. Let Δ be the diagonal subgroup of H × H. Then $G' = \phi^{-1}(\Delta)$ and $L(\phi)x = (L(\phi_1)x, L(\phi_2)x)$ for all $x \in L(G)$. It now suffices to apply Proposition 29.

COROLLARY 2. *Let* G *be a finite-dimensional Lie group and* G_1 *and* G_2 *two Lie subgroups of* G. *Suppose that* K *is of characteristic* 0. *Then* $G_1 \cap G_2$ *is a Lie subgroup of* G *with Lie algebra* $L(G_1) \cap L(G_2)$.

We apply Proposition 29 to the canonical injection of G_1 into G and the subgroup G_2.

COROLLARY 3. *Let* G, G', H *be Lie groups and* $\phi: G \to H$ *and* $\phi': G' \to H$ *Lie group morphisms. Suppose that* K *is of characteristic* 0 *and that* H *is finite-dimensional. Let* F *be the set of* $(g, g') \in G \times G'$ *such that* $\phi(g) = \phi'(g')$. *Then* F *is a Lie subgroup of* G × G' *and* L(F) *is the set of* $(x, x') \in L(G) \times L(G')$ *such that* $L(\phi)x = L(\phi')x'$.

We apply Corollary 1 to the morphisms $(g, g') \mapsto \phi(g)$ and $(g, g') \mapsto \phi'(g')$ of G × G' into H.

PROPOSITION 30. *Let* G *be a finite-dimensional Lie group with a countable base and*

H *and* H′ *Lie subgroups of* G. *Suppose that* K *is of characteristic* 0 *and that* HH′ *is locally closed in* G.

(i) HH′ *is a submanifold of* G *and* $T_e(HH') = L(H) + L(H')$.

(ii) *Suppose that every element of* H *commutes with every element of* H′. *Then* HH′ *is a Lie subgroup of* G. *Let* ϕ *be the mapping* $(h, h') \mapsto hh'$ *of* $H \times H'$ *onto* HH′. *The kernel of* ϕ *is the set of* (m, m^{-1}) *where* $m \in H \cap H'$ *and the morphism of* $(H \times H')/\mathrm{Ker}\,\phi$ *onto* HH′ *derived from* ϕ *by passing to the quotient is a Lie group isomorphism.*

Let $H \times H'$ operate on G on the right by the mapping $((h, h'), g) \mapsto hgh'^{-1}$. The orbital mapping ρ of e is $(h, h') \mapsto hh'^{-1}$. By Proposition 14 (iii) of § 1, no. 7, HH′ is a submanifold of G and $T_e(HH') = \mathrm{Im}\, T_e(\rho)$. Now

$$T_e(\rho)(L(H) \times \{0\}) = L(H) \quad \text{and} \quad T_e(\rho)(\{0\} \times L(H')) = L(H')$$

and hence $T_e(HH') = L(H) + L(H')$. Suppose that every element of H commutes with every element of H′. Then HH′ is a subgroup of G. By (i), it is a Lie subgroup of G. The rest of the statement follows from Proposition 28.

Proposition 31. *Let* G *be a finite-dimensional Lie group with countable base,* H *a normal Lie subgroup of* G *and* A *a Lie subgroup of* G. *Suppose that* K *is of characteristic* 0 *and that* AH *is closed. Let* ϕ *be the canonical morphism of* G *onto* G/H. *Then the canonical mappings*

$$A/(H \cap A) \to \phi(A), \qquad AH/H \to \phi(A)$$

are Lie group isomorphisms.

By Proposition 30, AH is a Lie subgroup of G. By Corollary 2 to Proposition 29, $H \cap A$ is a Lie subgroup of G. It is therefore meaningful to speak of the groups AH/H and $A/(H \cap A)$. On the other hand, $\phi(A)$, which is the canonical image of AH in G/H, is closed and hence is a Lie subgroup of G/H (Proposition 28 (iii)). Proposition 28, applied to the composite morphisms $A \to G \to G/H$ and $AH \to G \to G/H$, proves that the canonical mappings of the proposition are Lie group isomorphisms.

Proposition 32. *Let* G *and* H *be Lie groups,* k *a non-discrete closed subfield of* K *and* ϕ *a morphism of* G *into* H *for the Lie group structures over* k. *Suppose that* K *is of characteristic* 0. *If* $L(\phi)$ *is* K*-linear,* ϕ *is a morphism for the Lie group structures over* K.

For all $g \in G$,

$$T_g(\phi) = T_e(\gamma(\phi(g))) \circ L(\phi) \circ T_g(\gamma(g)^{-1})$$

and hence $T_g(\phi)$ is K-linear. The proposition then follows from *Differentiable and Analytic Manifolds*, R, 5.14.6.

9. LIE ALGEBRA OF THE GROUP OF INVERTIBLE ELEMENTS OF AN ALGEBRA

Let A be a complete normable associative algebra with unit element e. Let A^* be the group of invertible elements of A. We have seen (§ 1, no. 1) that A^* is an open submanifold of A and is a Lie group. Let G be a Lie group and f a morphism of the Lie group G into the Lie group A^*. f can be considered as an analytic mapping of G into the complete normable space A. Hence, if $t \in \mathscr{T}^{(\infty)}(G)$, we can form $\langle t, f\rangle$, which is an element of A.

PROPOSITION 33. *The mapping* $t \mapsto \langle t, f\rangle$ *is a morphism of the algebra* $\mathscr{T}^{(\infty)}(G)$ *into the algebra* A.

It suffices to verify that, if t and t' are point distributions on G, then $\langle t * t', f\rangle = \langle t, f\rangle\langle t', f\rangle$. But

$$\begin{aligned}\langle t * t', f\rangle &= \langle t \otimes t', (g, g') \mapsto f(gg')\rangle \\ &= \langle t \otimes t', (g, g') \mapsto f(g)f(g')\rangle \\ &= \langle t, f\rangle\langle t', f\rangle\end{aligned}$$

(*Diff. & Anal. Man.*, R, 13.4.3).

The morphism of Proposition 33 is said to be *associated* with f.

Take G to be the group A^* itself and f to be the identity mapping ι of A^*. We obtain a morphism, called *canonical*, of the algebra $\mathscr{T}^{(\infty)}(A^*)$ into the algebra A. The tangent space $T_e(A^*)$ is canonically identified with A; and if $t \in T_e(A^*)$, the definition of this identification is such that $\langle t, \iota\rangle = t$. Then Proposition 33 implies the following corollary:

COROLLARY. *The canonical mapping* ζ *of* $L(A^*)$ *into* A *is an isomorphism of the Lie algebra,* $L(A^*)$ *onto the Lie algebra* A. *In other words,*

$$\zeta([a, b]) = \zeta(a)\zeta(b) - \zeta(b)\zeta(a)$$

for all a, b in $L(A^*)$. *If* K *is of characteristic* $p > 0$, *then* $\zeta(a^p) = \zeta(a)^p$ *for all* $a \in L(A^*)$.

Henceforth $L(A^*)$ and A are identified by means of the isomorphism ζ.

The canonical morphism of $\mathscr{T}^{(\infty)}(A^*)$ into A has been obtained as a special case of the morphism of Proposition 33. But it is possible to argue in the opposite direction:

PROPOSITION 34. *Let* H *be a Lie group,* A *a unital complete normable associative algebra and* $\phi: H \to A^*$ *a Lie group morphism. The associated morphism* ϕ' *of* $\mathscr{T}^{(\infty)}(H)$ *into* A *is obtained by composing* ϕ_* *with the canonical morphism of* $\mathscr{T}^{(\infty)}(A^*)$ *into* A. *In particular,* $\phi'(x) = L(\phi)(x)$ *for all* $x \in L(H)$.

Let i be the identity mapping of A^* into A. Then, for all $t \in \mathscr{T}^{(\infty)}(H)$,

$$\begin{aligned}\phi'(t) = \langle t, \phi\rangle &= \langle t, i \circ \phi\rangle \\ &= \langle \phi_*(t), i\rangle\end{aligned}$$

(*Diff. & Anal. Man.*, R, 13.2.3).

10. LIE ALGEBRAS OF CERTAIN LINEAR GROUPS

Let E be a complete normable space. Then $\mathscr{L}(E)$ is a unital complete normable algebra and $\mathbf{GL}(E)$ is a Lie group. By the Corollary to Proposition 33, no. 9, if $T_1(\mathbf{GL}(E))$ is canonically identified with $\mathscr{L}(E)$, the Lie algebra structure on $L(\mathbf{GL}(E))$ is given by the bracket $(x, y) \mapsto xy - yx$ of two elements of $\mathscr{L}(E)$. In particular, $L(\mathbf{GL}(n, K))$ is canonically identified with $\mathfrak{gl}(n, K)$ (Chapter I, § 1, no. 2).

PROPOSITION 35. *Let E be a finite-dimensional vector space. Let ϕ be the morphism $g \mapsto \det g$ of the Lie group $\mathbf{GL}(E)$ into the Lie group K^*. The mapping $L(\phi)$ of $\mathscr{L}(E)$ into K is the mapping $x \mapsto \operatorname{Tr} x$. The kernel $\mathbf{SL}(E)$ of ϕ is a Lie subgroup of $\mathbf{GL}(E)$ with Lie algebra $\mathfrak{sl}(E)$.*

We choose a norm and a basis of E. The expansion of the determinant proves that

$$\det(1 + u) \in 1 + \operatorname{Tr} u + o(\|u\|)$$

when u tends to 0 in $\mathscr{L}(E)$. Hence, using Proposition 34, no. 9, for $x \in \mathscr{L}(E) = L(\mathbf{GL}(E))$:

$$L(\phi)(x) = \langle x, \phi \rangle = \operatorname{Tr} x.$$

It follows that ϕ is a submersion. Therefore, $\operatorname{Ker} \phi = \mathbf{SL}(E)$ is a Lie subgroup of $\mathbf{GL}(E)$ whose Lie algebra is $\operatorname{Ker} L(\phi) = \mathfrak{sl}(E)$.

Let $E_1, \ldots, E_n$ be complete normable spaces and E their direct sum. Every $x \in \mathscr{L}(E)$ can be represented by a matrix $(x_{ij})_{1 \leqslant i, j \leqslant n}$, where $x_{ij} \in \mathscr{L}(E_i, E_j)$.

PROPOSITION 36. *Let I be a subset of $\{1, 2, \ldots, n\}$ and G the subgroup of $\mathbf{GL}(E)$ consisting of the $g = (g_{ij})_{1 \leqslant i, j \leqslant n} \in \mathbf{GL}(E)$ such that $g_{ij} = 0$ for $i < j$ and $g_{ii} = 1$ for $i \in I$. Then G is a Lie subgroup of $\mathbf{GL}(E)$ and $L(G)$ is the set of $x = (x_{ij})_{1 \leqslant i, j \leqslant n} \in \mathscr{L}(E)$ such that $x_{ij} = 0$ for $i < j$ and $x_{ii} = 0$ for $i \in I$.*

Let S be the set of $(x_{ij}) \in \mathscr{L}(E)$ such that $x_{ij} = 0$ for $i < j$ and $x_{ii} = 0$ for $i \in I$. Then G is the intersection of $\mathbf{GL}(E)$ and the affine subspace $1 + S$ of $\mathscr{L}(E)$. Hence G is a submanifold of $\mathbf{GL}(E)$ and the tangent space to G at 1 is identified with S.

In particular, in $\mathbf{GL}(n, K)$, the total lower triangular subgroup and the lower strict triangular subgroup, defined as in *Integration*, Chapter VII, § 3, no. 3, are Lie subgroups with Lie algebras $\mathfrak{t}(n, K)$ and $\mathfrak{n}(n, K)$ (Chapter I, § 1, no. 2).

PROPOSITION 37. *Let A be a complete normable unital associative algebra and $x \mapsto x^\iota$ a continuous linear mapping of A into A such that $(x^\iota)^\iota = x$ and $(xy)^\iota = y^\iota x^\iota$ for all x, y in A. Suppose that K is of characteristic $\neq 2$. Let G be the subgroup of A^* consisting of the $x \in A$ such that $xx^\iota = x^\iota x = 1$. Then G is a Lie subgroup of A^* and $L(G)$ is the set of $y \in A$ such that $y^\iota = -y$.*

Let S (resp. S′) be the set of $y \in A$ such that $y = y^\iota$ (resp. $y = -y^\iota$). Then S, S′ are closed vector subspaces of A. The formula

$$y = \tfrac{1}{2}(y + y^\iota) + \tfrac{1}{2}(y - y^\iota)$$

proves that A is the topological direct sum of S and S′. Let f be the mapping of A into S defined by $f(x) = xx^\iota$. This mapping is analytic. For all $y \in A$, $f(1 + y) = 1 + y + y^\iota + yy^\iota$; choose a norm on A compatible with its algebra structure; then

$$f(1 + y) \in 1 + y + y^\iota + o(\|y\|) \quad \text{for } y \text{ tending to } 0.$$

Thus, $T_1(f)(y) = y + y^\iota$, so that f is a submersion at 1. Therefore, there exists an open neighbourhood U of 1 in A such that $U \cap G$ is a submanifold of U. Hence (§ 1, no. 3, Proposition 6) G is a Lie subgroup of A^*. Moreover, $L(G) = T_e(G) = \operatorname{Ker} T_1(f)$.

COROLLARY 1. *Suppose that* K *is of characteristic* $\neq 2$. *Let* E *be a finite-dimensional vector space over* K *and* ϕ *a non-degenerate symmetric* (resp. *alternating*) *bilinear form on* E. *For all* $u \in \mathscr{L}(E)$, *let* u^* *be the adjoint of* u *relative to* ϕ. *Let* G *be the orthogonal* (resp. *symplectic*) *group of* ϕ. *Then* G *is a Lie subgroup of* **GL**(E) *and* L(G) *is the set of* $x \in \mathscr{L}(E)$ *such that* $x^* = -x$.

We apply Proposition 37 with $A = \mathscr{L}(E)$ and $x^\iota = x^*$.

Remark. Let B be a basis of E and J the matrix of ϕ with respect to B. Then L(G) is the set of elements of $\mathscr{L}(E)$ whose matrix X with respect to B satisfies the equation

$$^tX = -JXJ^{-1}.$$

This follows from *Algebra*, Chapter IX, § 1, formula (50).

COROLLARY 2. *Let* E *be a complex* (resp. *real*) *Hilbert space and* U *the unitary group of* E. *Then* U *is a real subgroup of* **GL**(E) *and* L(U) *is the set of* $x \in \mathscr{L}(E)$ *such that* $x^* = -x$.

We apply Proposition 37 with $A = \mathscr{L}(E)$ considered as an algebra over **R** and $x^\iota = x^*$.

COROLLARY 3. *Let* E *be a finite-dimensional complex vector space,* ϕ *a non-degenerate Hermitian sesquilinear form on* E *and* U *the unitary group of* ϕ. *Then* U *is a real Lie subgroup of* **GL**(E) *and* L(U) *is the set of* $x \in \mathscr{L}(E)$ *such that* ix *is Hermitian.*

When $E \neq \{0\}$, U is *not* a Lie subgroup of the complex Lie group **GL**(E), for L(U) is not a complex vector subspace of $\mathscr{L}(E)$.

11. LINEAR REPRESENTATIONS

Let G be a Lie group, E a complete normable space and π an analytic linear representation of G on E (§ 1, no. 2). The associated morphism $t \mapsto \langle t, \pi \rangle$ of

$\mathcal{T}^{(\infty)}(G)$ into $\mathcal{L}(E)$ is an algebra morphism (no. 9, Proposition 33) and its restriction to $L(G)$ is $L(\pi)$. Hence $L(\pi)$ is a representation of $L(G)$ on E (Chapter I, § 3, Definition 1).

PROPOSITION 38. *Consider* G *as operating on* E *on the left by the mapping* $(g, x) \mapsto \pi(g)x$. *Let* $b \in E$ *and* $\rho(b)$ *be its orbital mapping. Let* $T_b(E)$ *be canonically identified with* E. *For all* $t \in L(G)$,

$$(L(\pi)t)(b) = \langle t, \rho(b)\rangle = \rho(b)_* t = t * \varepsilon_b.$$

In particular, the vector field defined by t *on* E *is the field* $b \mapsto (L(\pi)t)(b)$.

$L(\pi)t = \langle t, \pi\rangle$ (no. 9, Proposition 34). As the mapping $A \mapsto Ab$ of $\mathcal{L}(E)$ into E is continuous and linear, it follows that

$$\begin{aligned}(L(\pi)t)(b) &= \langle t, g \mapsto \pi(g)b\rangle \\ &= \langle t, \mathrm{Id}_E \circ \rho(b)\rangle \\ &= \langle \rho(b)_* t, \mathrm{Id}_E\rangle \quad \text{(\textit{Diff. \& Anal. Man.}, R, 13.2.3)} \\ &= \rho(b)_* t.\end{aligned}$$

Finally, $\rho(b)_* t = t * \varepsilon_b$ (no. 3, Proposition 14 (ii)).

PROPOSITION 39. *Suppose that* K *is of characteristic* 0. *Let* G *be a Lie group,* E *a finite-dimensional vector space and* π *an analytic linear representation of* G *on* E. *Let* E_1, E_2 *be vector subspaces of* E *such that* $E_2 \subset E_1$. *The set* G_1 *of* $g \in G$ *such that* $\pi(g)x \equiv x \pmod{E_2}$ *for all* $x \in E_1$ *is a Lie subgroup of* G *and* $L(G_1)$ *is the set of* $a \in L(G)$ *such that* $L(\pi)a$ *maps* E_1 *into* E_2.

This follows from Propositions 29 (no. 8) and 36 (no. 10).

COROLLARY 1. *In the notation of Proposition* 39, *the set of* $g \in G$ *such that* $\pi(g)(E_1) \subset E_1$ *is a Lie subgroup of* G *and its Lie algebra is the set of* $a \in L(G)$ *such that* $L(\pi)a$ *maps* E_1 *into* E_1.

We apply Proposition 39 with $E_1 = E_2$.

COROLLARY 2. *Let* G, E, π *be as in Proposition* 39. *Let* F *be a subset of* E. *The set of* $g \in G$ *such that* $\pi(g)x = x$ *for all* $x \in F$ *is a Lie subgroup of* G *and its Lie algebra is the set of* $a \in L(G)$ *such that* $(L(\pi)a)(x) = 0$ *for all* $x \in F$.

We apply Proposition 39 with $E_2 = \{0\}$ and E_1 the vector subspace of E generated by F.

Let $\pi_1, \pi_2, \ldots, \pi_n$ be analytic linear representations of G. Clearly the direct sum π of the π_i (*Algebra*, Chapter VIII, § 13, no. 1) is an analytic linear representation of G and $L(\pi)$ is the direct sum of $L(\pi_1), L(\pi_2), \ldots, L(\pi_n)$ (Chapter I, § 3, no. 1).

PROPOSITION 40. *Let* G *be a Lie group,* E *a complete normable space,* π *an analytic linear representation of* G *on* E *and* F *a closed vector subspace of* E *stable under* $\pi(G)$. *Suppose that* K *is of characteristic* 0, *or that* F *is a direct factor of* E.

(i) *The subrepresentation* π_1 *and the quotient representation* π_2 *of* π *defined by* F *are analytic representations.*

(ii) F *is stable under* $L(\pi)(L(G))$.

(iii) *Let* ρ_1 *and* ρ_2 *be the subrepresentation and quotient representation of* $L(\pi)$ *defined by* F. *Then* $L(\pi_1) = \rho_1$, $L(\pi_2) = \rho_2$.

Let A be the set of $u \in \mathscr{L}(E)$ such that $u(F) \subset F$. Then A is a closed vector subspace of $\mathscr{L}(E)$ and π takes its values in A. By virtue of the hypotheses on K and F, the mapping $\pi': G \to A$ with the same graph as π is analytic (*Differentiable and Analytic Manifolds*, R, 5.8.5). The canonical mappings $\theta_1: A \to \mathscr{L}(F)$ and $\theta_2: A \to \mathscr{L}(E/F)$ are continuous and linear and hence analytic. This proves (i). The mappings $T_e(\pi)$ and $T_e(\pi')$ have the same graph and hence $L(\pi)(L(G)) \subset A$, which proves (ii). We have

$$T_e(\pi_1) = T_e(\theta_1 \circ \pi') = \theta_1 \circ T_e(\pi') = \rho_1$$
$$T_e(\pi_2) = T_e(\theta_2 \circ \pi') = \theta_2 \circ T_e(\pi') = \rho_2.$$

PROPOSITION 41. *Let* G *be a Lie group and* $\pi_1, \pi_2, \ldots, \pi_n, \pi$ *analytic linear representation of* G *on complete normable spaces* $E_1, E_2, \ldots, E_n$, E. *Let*

$$(x_1, x_2, \ldots, x_n) \mapsto x_1 x_2 \ldots x_n$$

be a continuous multilinear mapping of $E_1 \times E_2 \times \cdots \times E_n$ *into* E. *Suppose that*

$$\pi(g)(x_1 x_2 \ldots x_n) = (\pi_1(g)x_1)(\pi_2(g)x_2)\ldots(\pi_n(g)x_n)$$

for all $g \in G$, $x_1 \in E_1, \ldots, x_n \in E_n$. *Then*

$$(L(\pi)a)(x_1 x_2 \ldots x_n) = \sum_{i=1}^{n} x_1 x_2 \ldots x_{i-1}((L(\pi_i)a)x_i)x_{i+1}\ldots x_n$$

for all $a \in L(G)$, $x_1 \in E_1, \ldots, x_n \in E_n$.

As an example we perform the calculation for $n = 2$.

$$\begin{aligned}(L(\pi)a)(x_1 x_2) &= \langle a, g \mapsto \pi(g)(x_1 x_2)\rangle && \text{(Proposition 38)}\\ &= \langle a, (g \mapsto \pi_1(g)x_1)(g \mapsto \pi_2(g)x_2)\rangle \\ &= \langle a, g \mapsto \pi_1(g)x_1\rangle . x_2 + x_1 . \langle a, g \mapsto \pi_2(g)x_2\rangle && \text{(\textit{Diff. \& Anal. Man.}, R, 5.5.6)}\\ &= ((L(\pi_1)a)x_1).x_2 + x_1.((L(\pi_2)a)x_2) && \text{(Proposition 38).}\end{aligned}$$

COROLLARY 1. *Let* G *be a Lie group,* $E_1, \ldots, E_{n+1}$ *complete normable spaces and* $\pi_1, \ldots, \pi_{n+1}$ *analytic linear representations of* G *on* $E_1, \ldots, E_{n+1}$. *Let*

$$E = \mathscr{L}(E_1, \ldots, E_n; E_{n+1})$$

the complete normable space of continuous multilinear mappings of $E_1 \times \cdots \times E_n$ *into* E_{n+1} (*General Topology*, Chapter X, § 3, no. 2). *For all* $g \in G$, *let* $\pi(g)$ *be the automorphism of* E *defined by*

$$(\pi(g)u)(x_1, \ldots, x_n) = \pi_{n+1}(g)(u(\pi_1(g)^{-1}x_1, \ldots, \pi_n(g)^{-1}x_n)).$$

Then π *is an analytic linear representation of* G *on* E *and*

$$((\mathrm{L}(\pi)a)u)(x_1, \ldots, x_n) = -\sum_{i=1}^{n} u(x_1, \ldots, x_{i-1}, (\mathrm{L}(\pi_i)a)x_i, x_{i+1}, \ldots, x_n) + (\mathrm{L}(\pi_{n+1})a)(u(x_1, \ldots, x_n))$$

for all $a \in \mathrm{L(G)}$, $u \in \mathrm{E}$, $x_1 \in \mathrm{E}_1, \ldots, x_n \in \mathrm{E}_n$.

Every element $(A_1, \ldots, A_{n+1})$ of $\mathscr{L}(\mathrm{E}_1) \times \cdots \times \mathscr{L}(\mathrm{E}_{n+1})$ defines a continuous endomorphism $\theta(A_1, \ldots, A_{n+1})$ of E by the formula

$$(\theta(A_1, \ldots, A_{n+1})u)(x_1, \ldots, x_n) = A_{n+1}(u(A_1x_1, \ldots, A_nx_n)).$$

The mapping θ of $\mathscr{L}(\mathrm{E}_1) \times \cdots \times \mathscr{L}(\mathrm{E}_{n+1})$ into $\mathscr{L}(\mathrm{E})$ is continuous and multilinear. Then, for all $g \in \mathrm{G}$,

$$\pi(g) = \theta(\pi_1(g^{-1}), \ldots, \pi_n(g^{-1}), \pi_{n+1}(g))$$

and hence π is analytic. We apply Proposition 41 to the mapping

$$(x_1, \ldots, x_n, u) \mapsto u(x_1, \ldots, x_n)$$

of $\mathrm{E}_1 \times \cdots \times \mathrm{E}_n \times \mathrm{E}$ into E_{n+1}. Then

$$\pi_{n+1}(g)(u(x_1, \ldots, x_n)) = (\pi(g)u)(\pi_1(g)x_1, \ldots, \pi_n(g)x_n)$$

and hence

$$(\mathrm{L}(\pi_{n+1})a)(u(x_1, \ldots, x_n)) = \sum_{i=1}^{n} u(x_1, \ldots, (\mathrm{L}(\pi_i)a)x_i, \ldots, x_n) + ((\mathrm{L}(\pi)a)u)(x_1, \ldots, x_n).$$

When the E_i are finite-dimensional, the representation $\mathrm{L}(\pi)$ of L(G) is derived from the representations $\mathrm{L}(\pi_1), \ldots, \mathrm{L}(\pi_{n+1})$ under the procedure of Chapter I, § 3, Proposition 3.

COROLLARY 2. *Let* G *be a Lie group and* π *an analytic linear representation of* G *on a complete normable space* E. *Then* $g \mapsto {}^t\pi(g)^{-1}$ *is an analytic linear representation* ρ *of* G *on the complete normable space* $\mathscr{L}(\mathrm{E}, \mathrm{K})$† *and* $\mathrm{L}(\rho)a = -{}^t(\mathrm{L}(\pi)a)$ *for all* $a \in \mathrm{L(G)}$.

This is a special case of Corollary 1.

ρ is called the *contragredient* representation of π.

When E is finite-dimensional, $\mathrm{L}(\rho)$ is the dual representation of $\mathrm{L}(\pi)$ in the sense of Chapter I, § 3, no. 3.

COROLLARY 3. *Let* G *be a Lie group and* $\pi_1, \ldots, \pi_n$ *analytic linear representations*

† As when K = **R** or **C**, the transpose ${}^t\pi(g)$ considered here is the restriction to $\mathscr{L}(\mathrm{E}, \mathrm{K})$ of the transpose of $\pi(g)$ in the purely algebraic sense.

of G *on finite-dimensional vector spaces* $E_1, \ldots, E_n$. *Then the representation* $\pi_1 \otimes \cdots \otimes \pi_n$ *of* G (*Appendix*) *is analytic and* $L(\pi_1 \otimes \cdots \otimes \pi_n)$ *is the tensor product of* $L(\pi_1), \ldots, L(\pi_n)$.

The mapping $(A_1, \ldots, A_n) \mapsto A_1 \otimes \cdots \otimes A_n$ of $\mathscr{L}(E_1) \times \cdots \times \mathscr{L}(E_n)$ into $\mathscr{L}(E_1 \otimes \cdots \otimes E_n)$ is multilinear, whence the fact that π is analytic. Consider the mapping $(x_1, \ldots, x_n) \mapsto x_1 \otimes \cdots \otimes x_n$ of $E_1 \times \cdots \times E_n$ into

$$E_1 \otimes \cdots \otimes E_n.$$

By Proposition 41, we see that

$$(L(\pi)a)(x_1 \otimes \cdots \otimes x_n) = \sum_{i=1}^{n} x_1 \otimes \cdots \otimes (L(\pi_i)a)x_i \otimes \cdots \otimes x_n$$

for all $a \in L(G)$, $x_i \in E_i$ for $1 \leqslant i \leqslant n$. Hence $L(\pi)$ is the tensor product of the $L(\pi_i)$.

Corollary 4. *Let* G *be a Lie group and* π *an analytic linear representation of* G *on a finite-dimensional vector space* E. *Then the representations* $T^n(\pi)$, $S^n(\pi)$ *and* $\bigwedge^n(\pi)$ *of* G (*Appendix*) *are analytic and*

$$L(T^n(\pi)) = T^n(L(\pi)), \qquad L(S^n(\pi)) = S^n(L(\pi)), \qquad L(\textstyle\bigwedge^n(\pi)) = \textstyle\bigwedge^n(L(\pi)).$$

This follows from Corollary 3 and Proposition 40.

Corollary 5. *Let* A *be a finite-dimensional algebra. Suppose that* K *is of characteristic* 0. *The automorphism group* Aut(A) *of* A *is a Lie subgroup of* **GL**(A) *and* L(Aut(A)) *is the Lie algebra of derivation of* A.

This follows from Corollary 1 (applied to $E = \mathscr{L}(A, A; A)$) and Corollary 2 of Proposition 39 (applied to the subset of E consisting only of multiplication on A).

Remark. We apply Corollary 1 with $G = \mathbf{GL}(F)$ (F a complete normable spacc), $\pi_1 = \pi_2 = \mathrm{Id}_G$ and π_3 the trivial representation of G on K. We obtain an analytic representation π of **GL**(F) on $\mathscr{L}(F, F; K)$. We assume that F is finite-dimensional and that K is of characteristic 0. Applying Corollary 2 to Proposition 39 to π, we recover part of Corollary 1 to Proposition 37.

Proposition 42. *Let* G *be a Lie group,* X *an analytic manifold,* $(g, x) \mapsto gx$ (resp. xg) *a law of analytic left* (resp. *right*) *operation of* G *on* X *and* x_0 *a point of* X *which is invariant under* G. *For all* $g \in G$, *let* $\tau(g)$ *be the automorphism* $x \mapsto gx$ (resp. xg) *of* X *and let* $\pi(g)$ *be the automorphism of* $T_{x_0}(X)$ *tangent at* x_0 *to* $\tau(g)$.

(i) π *is an analytic representation of* G (resp. $G^\vee$) *on* $T_{x_0}(X)$.

(ii) *For all* $a \in L(G)$ *and all* $\xi_0 \in T_{x_0}(X)$, $L(\pi)a.\xi_0$ *can be calculated as follows: let* D_a *be the vector field defined by* a *on* X *and* ξ *a vector field of class* C^1 *in an open neighbourhood of* x_0 *such that* $\xi(x_0) = \xi_0$; *then*

$$L(\pi)a.\xi_0 = -[D_a, \xi](x_0).$$

$\tau(gg') = \tau(g)\tau(g')$ (resp. $\tau(g')\tau(g)$) and hence $\pi(gg') = \pi(g)\pi(g')$ (resp. $\pi(g')\pi(g)$). On the other hand, since TX is a vector G-bundle of class C^ω (§ 1, no. 8, Proposition 16), π is analytic, whence (i).

To prove (ii), suppose that G operates on the left. There exists an open neighbourhood I of 0 in K and an analytic mapping γ of I into G such that $\gamma(0) = e$, $T_0(\gamma)1 = a$. Then D_a is the vector field on X defined by the mapping $\phi: (\lambda, x) \mapsto \gamma(\lambda)x$ of I $\times$ X into X (§ 2, no. 2). If ϕ_λ denotes the bijection $x \mapsto \gamma(\lambda)x$ of X into X, then

$$[D_a, \xi](x_0) = \left(\frac{d}{d\lambda}(T_{\phi_\lambda(x_0)}(\phi_\lambda^{-1})\xi(\phi_\lambda(x_0)))\right)_{\lambda=0} \quad (\textit{Diff. \& Anal. Man.}, \text{R}, 8.4.5)$$

$$= \left(\frac{d}{d\lambda}(T_{x_0}(\phi_\lambda^{-1})\xi_0)\right)_{\lambda=0}$$

$$= \left(\frac{d}{d}(\pi(\gamma(\lambda))^{-1}\xi_0)\right)_{\lambda=0}.$$

As the mappings $\lambda \mapsto \gamma(\lambda)^{-1}$ and $\lambda \mapsto \gamma(-\lambda)$ are tangent at 0, this is also equal to

$$-\left(\frac{d}{d\lambda}(\pi(\gamma(\lambda))\xi_0)\right)_{\lambda=0}$$

$$= -\left(\frac{d}{d\lambda}(\pi\circ\gamma)(\lambda)\right)_{\lambda=0}\xi_0$$

$$= -L(\pi)a.\xi_0.$$

12. ADJOINT REPRESENTATION

Let G be a Lie group. Consider the law of analytic left operation

$$(g, g') \mapsto gg'g^{-1} = (\operatorname{Int} g)g'$$

of G into G. This law of operation defines, by no. 3, a bilinear mapping of $\mathscr{T}^{(\infty)}(G) \times \mathscr{T}^{(\infty)}(G)$ into $\mathscr{T}^{(\infty)}(G)$, which we shall denote by $\top$ in this no. By Proposition 13 of no. 3,

$$(t * t') \top t'' = t \top (t' \top t'') \tag{22}$$

for all t, t', t'' in $\mathscr{T}^{(\infty)}(G)$. By Proposition 14 (i) of no. 3,

$$\varepsilon_g \top t = (\operatorname{Int} g)_* t \tag{23}$$

for all $g \in G$ and $t \in \mathscr{T}^{(\infty)}(G)$. In particular, the mapping $t \mapsto \varepsilon_g \top t$ of $\mathscr{T}^{(\infty)}(G)$ into $\mathscr{T}^{(\infty)}(G)$ is an automorphism of the bigebra $\mathscr{T}^{(\infty)}(G)$. Its restrictions to U(G), $U_s(G)$, L(G) are denoted by $\mathrm{Ad}_{U(G)}(g)$, $\mathrm{Ad}_{U_s(G)}(g)$, $\mathrm{Ad}_{L(G)}(g)$. We often write Ad(g) instead of $\mathrm{Ad}_{L(G)}(g)$ when no confusion is possible. By (23), Ad(g) *is the tangent mapping at e to* Int(g). It is an automorphism of the normable

Lie algebra L(G). When K is of characteristic 0, $\mathrm{Ad}_{U(G)}(g)$ is the unique automorphism of U(G) which extends Ad(g).

If ϕ is a morphism of the Lie group G into a Lie group H, then

$$\phi_*(t \top t') = \phi_*(t) \top \phi_*(t') \tag{24}$$

for all t, t' in $\mathscr{T}^{(\infty)}(G)$; this follows from Proposition 15 of no. 3.

PROPOSITION 43. *Let t, u be in $\mathscr{T}^{(\infty)}(G)$. Let $\sum_{i=1}^{n} t_i \otimes t_i'$ be the image of t under the coproduct. Then*

$$t \top u = \sum_{i=1}^{n} t_i * u * t_i'^{\vee}.$$

By definition, $t \top u$ is the image of $t \otimes u$ under the mapping $(g, g') \mapsto gg'g^{-1}$ of G $\times$ G into G. Now this mapping is obtained by composing the following mappings:

$\alpha: (g, g') \mapsto (g, g, g')$ of G × G into G × G × G

$\beta: (g, g', g'') \mapsto (g, g'^{-1}, g'')$ of G × G × G into G × G × G

$\gamma: (g, g', g'') \mapsto gg''g'$ of G × G × G into G.

On the other hand:

$$\alpha_*(t \otimes u) = \sum_{i=1}^{n} (t_i \otimes t_i') \otimes u = \sum_{i=1}^{n} t_i \otimes t_i' \otimes u$$

$$\beta_*\left(\sum_{i=1}^{n} t_i \otimes t_i' \otimes u\right) = \sum_{i=1}^{n} t_i \otimes t_i'^{\vee} \otimes u$$

$$\gamma_*\left(\sum_{i=1}^{n} t_i \otimes t_i'^{\vee} \otimes u\right) = \sum_{i=1}^{n} t_i * u * t_i'^{\vee}.$$

COROLLARY 1. *Let $u \in L(G)$ and $u' \in \mathscr{T}^{(\infty)}(G)$. Then $u \top u' = u * u' - u' * u$.*

The image of u under the coproduct is $u \otimes \varepsilon_e + \varepsilon_e \otimes u$, whence

$$u \top u' = u * u' * \varepsilon_e + \varepsilon_e * u' * u^{\vee} = u * u' - u' * u.$$

COROLLARY 2. *Let $t \in \mathscr{T}^{(\infty)}(G)$ and $g \in G$. Then $\varepsilon_g \top t = \varepsilon_g * t * \varepsilon_{g^{-1}}$. If $t \in L(G)$, then $\varepsilon_g \top t = gtg^{-1}$ (where the latter product is evaluated in the group T(G)).*

The image of ε_g under the coproduct is $\varepsilon_g \otimes \varepsilon_g$.

COROLLARY 3. *Let $a \in L(G)$. The vector field defined by a and the left operation $g \mapsto \mathrm{Int}\, g$ of G on G is the field $R_a - L_a$.*

The value of this field at g is

$$\begin{aligned} a \top \varepsilon_g &= a * \varepsilon_g - \varepsilon_g * a && \text{(Corollary 1)} \\ &= (R_a)_g - (L_a)_g && \text{(Definition 5).} \end{aligned}$$

For all $g \in G$ and all $t \in L(G)$,

$$(Ad\, g)(t) = \varepsilon_g \top t = \varepsilon_g * t * \varepsilon_{g^{-1}} = gtg^{-1}. \tag{25}$$

Since $Ad\, g = T_e(Int\, g)$, Proposition 42 of no. 11 proves that Ad is an analytic linear representation of G on the normable space L(G).

DEFINITION 8. *The representation* Ad *of* G *on* L(G) *is called the adjoint representation of* G.

PROPOSITION 44. *For all* $a \in L(G)$,

$$(L(Ad))(a) = ad_{L(G)}a.$$

Let $b \in L(G)$. By Proposition 42 (ii) of no. 11 and Corollary 3 to Proposition 43,

$$(L(Ad))(a).b = -[R_a - L_a, L_b](e).$$

Now $R_a \circ L_b = L_b \circ R_a$ (no. 6, Proposition 23 (ii)), whence $[R_a, L_b] = 0$; then, using Proposition 23 (ii),

$$(L(Ad))(a).b = [L_a, L_b](e) = L_{[a,b]}(e) = [a, b] = (ad_{L(G)}a)b.$$

PROPOSITION 45. *Suppose that* G *is finite-dimensional and that* K *is of characteristic* 0. *Let s be an integer* $\geqslant 0$. *Then the mapping* $\pi: g \mapsto Ad_{U_s(G)}(g)$ *is an analytic linear representation of* G *on* $U_s(G)$ *and* $L(\pi)a = ad_{U_s(G)}\, a$ *for all* $a \in L(G)$.

The linear representation π is a quotient of $\bigoplus_{r=0}^{s} T^r(Ad)$ and is hence analytic. For $a \in L(G)$ and $x_1, x_2, \ldots, x_s$ in $L(G)$,

$$\begin{aligned}(L(\pi)a)(x_1x_2\ldots x_s) &= \sum_{i=1}^{s} x_1\ldots(L(Ad)a.x_i)\ldots x_s \quad \text{(Proposition 41)}\\ &= \sum_{i=1}^{s} x_1\ldots([a, x_i])\ldots x_s \quad \text{(Proposition 44)}\\ &= (Ad_{U_s(G)}\, a)(x_1x_2\ldots x_s).\end{aligned}$$

PROPOSITION 46. *Let* $h \in G$, $x \in T_h(G)$ *and* $a \in L(G)$. *Let* ϕ *be the mapping* $(g, g') \mapsto gg'g^{-1}$ *of* $G \times G$ *into* G. *The image* y *of* $(a, x) \in T_e(G) \times T_h(G)$ *under* $T_{(e,h)}(\phi)$ *is* $y = x + h((Ad\, h^{-1})a - a)$.

We have

$$\begin{aligned}y &= (T_{(e,h)}\phi)(a \otimes \varepsilon_h + \varepsilon_e \otimes x)\\ &= a \top \varepsilon_h + \varepsilon_e \top x\\ &= a * \varepsilon_h - \varepsilon_h * a + x\\ &= h((Ad\, h^{-1})a) - ha + x.\end{aligned}$$

PROPOSITION 47. *Let* G *be a Lie group,* H *and* E *Lie subgroups of* G *and suppose that* $hEh^{-1} = E$ *for all* $h \in H$. *Then* $\mathscr{T}^{(\infty)}(H) \top \mathscr{T}^{(\infty)}(E) \subset \mathscr{T}^{(\infty)}(E)$. *In particular,* $\mathrm{Ad}(H)(L(E)) \subset L(E)$ *and* $[L(H), L(E)] \subset L(E)$.

If $t \in \mathscr{T}^{(\infty)}(H)$ and $t' \in \mathscr{T}^{(\infty)}(E)$, then $t \otimes t' \in \mathscr{T}^{(\infty)}(H \times E)$ and the image of $H \times E$ under the mapping $(g, g') \mapsto gg'g^{-1}$ is contained in E.

PROPOSITION 48. *Let* G *be a Lie group and* H *and* E *Lie subgroups of* G. *Suppose that* G *is, as a Lie group, the semi-direct product of* H *by* E. *Let* ρ *be the linear representation* $g \mapsto (\mathrm{Ad}\, g) \mid L(E)$ *of the Lie group* G *on* L(E) (cf. Proposition 47) *and let* σ *be the restriction of* ρ *to* H. *Then:*

(i) L(G) *is the topological direct sum of* L(H) *and* L(E);

(ii) L(H) *is a subalgebra of* L(G) *and* L(E) *is an ideal of* L(G);

(iii) L(σ) *is a linear representation of* L(H) *on the Lie algebra of derivations of* L(E);

(iv) L(G) *is the semi-direct product of* L(H) *by* L(E) *defined by* L(σ) (Chapter I, § 1, no. 8).

(i) is obvious and (ii) follows from Proposition 47. $L(\sigma) = L(\rho) \mid L(H)$. Now by Propositions 40 (no. 11) and 44 (no. 12), $L(\rho)(t)$ is, for all $t \in L(G)$, the restriction of $\mathrm{ad}_{L(G)}\, t$ to L(E). This proves (iii). Using (i) and (ii), this also proves (iv).

COROLLARY. *Let* G *be a Lie group. Let* $T_e(G)$ *be given its unique commutative Lie algebra structure. Let* τ *be an adjoint representation of* L(G). *Then the Lie algebra of* T(G) *is the semi-direct product of* L(G) *by* $T_e(G)$ *defined by* τ. *In other words, for* x, x' *in* L(G) *and* y, y' *in* $T_e(G)$,

$$[(x, y), (x', y')] = ([x, x'], [x, y'] + [y, x'])$$

(*where the bracket on the left is evaluated in* L(T(G)) *and the brackets on the right in* L(G)).

This follows from Proposition 48 and Proposition 6 of § 2, no. 2.

PROPOSITION 49. *Let* A *be a complete normable unital associative algebra. We identify* A *with* $L(A^*)$. *Then, if* $g \in A^*$ *and* $y \in A$, $(\mathrm{Ad}\, g)y = gyg^{-1}$.

Recall that $\mathrm{Ad}\, g = T_1(\mathrm{Int}\, g)$. Let u_g be the mapping $x \mapsto gxg^{-1}$ of A into A. The identity chart of A^* into A transforms $\mathrm{Int}\, g$ into $u_g \mid A^*$. The tangent mapping at each point of A^* to this mapping is equal to u_g, whence the proposition.

COROLLARY. *For all* $g \in A^*$, *let* $i(g)$ *be the automorphism* $y \mapsto gyg^{-1}$ *of* A, *so that* i *is an analytic linear representation of* A^* *on* A. *For all* $z \in L(A^*) = A$, $L(i)z$ *is the inner derivation* $y \mapsto zy - yz$ *of* A.

This follows from Propositions 49 and 44.

13. TENSORS AND INVARIANT FORMS

Let G be a Lie group. We consider G as operating on itself by left (resp. right) translation. Let λ be a vector functor of class C^ω for isomorphisms. Then $\lambda(TG)$ is an analytic left (resp. right) vector G-bundle (§ 1, no. 8, Corollary to Proposition 16). The mapping $(g, u) \mapsto gu$ (resp. ug) of $G \times \lambda(L(G))$ onto $\lambda(TG)$ is an isomorphism ϕ (resp. ψ) of vector G-bundles (§ 7, no. 8, Corollary 2 to Proposition 17). Every G-invariant section of $\lambda(TG)$ is analytic and determined by its value at e (§ 1, no. 8, Corollary 1 to Proposition 17). Such a section is called *left* (resp. *right*) *invariant*. Let σ be a left invariant section of $\lambda(TG)$; the transform σ' of σ under a right translation $\delta(g)$ is defined by $\sigma'(\delta(g)h) = \lambda(T_h(\delta(g)))\sigma(h)$ for all $h \in G$; it is also left invariant; it is also derived from σ by $\gamma(g) \circ \delta(g) = \text{Int}(g)$ and hence

$$\sigma'(e) = \lambda(\text{Ad}\, g).\sigma(e). \tag{26}$$

Similarly, let τ be a right invariant section of $\lambda(TG)$; the transform τ' of τ under a left translation $\gamma(g)$ is also right variant and

$$\tau'(e) = \lambda(\text{Ad}\, g).\tau(e). \tag{27}$$

We now consider $G \times G$ as operating on G on the left by

$$((g, g'), g'') \mapsto gg''g'^{-1}.$$

Then G is a left Lie homogeneous space of $G \times G$ (§ 1, no. 6, *Example*). Hence $\lambda(TG)$ is an analytic left vector $(G \times G)$-bundle. A section of $\lambda(TG)$ is called *biinvariant* if it is invariant under the action of $G \times G$ on $\lambda(TG)$, in other words if it is invariant under left and right translations. Let $\lambda(L(G))_0$ be the set of elements of $\lambda(L(G))$ invariant under $\lambda(\text{Ad}(G))$. For all $u \in \lambda(L(G))_0$, let σ_u be the mapping of G into $\lambda(TG)$ defined by $\sigma_u(g) = gu = ug$. Then $u \mapsto \sigma_u$ is a bijection of $\lambda(L(G))_0$ onto the set of biinvariant sections of $\lambda(TG)$ (§ 1, no. 8, Corollary 1 to Proposition 17).

PROPOSITION 50. *Let* G *be a Lie Group (assumed to be finite-dimensional if* K *is of characteristic* >0). *Let* E *be the vector space of continuous alternating multilinear forms of degree* k *on* $T_e(G)$. *For all* $u \in E$, *let* ω^u *be the differential form of degree* k *on* G *such that* $(\omega^u)_g$ *is the multilinear form on* $T_g(G)$ *derived from* u *by the translation* $h \mapsto gh$ (resp. $h \mapsto hg$). *Then* ω^u *is analytic and left* (resp. *right*) *invariant on* G. *The mapping* $u \mapsto \omega^u$ *is an isomorphism of* E *onto the vector space of left* (resp. *right*) *invariant differential forms of degree* k *on* G.

This is a special case of what we have said above.

Let F be a complete normable space. Proposition 50 remains true if differential forms on G with values in K are replaced by differential forms on G with values in F. For every continuous linear mapping u of $T_e(G)$ into F, there exists a differential form ω^u of degree 1 on G, with values in F, such that $(\omega^u)_g = u \circ T_g(\gamma(g)^{-1})$. In particular, take $F = T_e(G)$ and $u = \text{Id}_{T_e(G)}$. We then obtain the differential form ω on G such that $\omega_g = T_g(\gamma(g^{-1}))$; this differential form is left invariant and analytic; it is called the *left canonical differential form* of G. $\omega_g(t) = g^{-1}t$ for all $t \in T_g(G)$.

If F is again an arbitrary complete normable space and $u \in \mathscr{L}(T_e(G), F)$, then $\omega^u = u \circ \omega$. In particular (taking F = K), the mapping $v \mapsto v \circ \omega$ is a linear bijection of the dual of $T_e(G)$ onto the vector space of differential forms of degree 1 with values in K which are left invariant under G.

Similarly, the differential form ω' on G such that $\omega'_g = T_g(\delta(g))$ is called the *right canonical differential form* of G. There are analogous properties to those of ω, which we leave to the reader to state. The mapping $g \mapsto g^{-1}$ of G onto G transforms ω into ω'.

14. MAURER–CARTAN FORMULAE

Let X be a manifold of class C^r, of finite dimension if K is of characteristic >0, and let L be a complete normable Lie algebra. Let α be a differential form of degree 1 on X with values in L of class C^{r-1}. Let $x \in X$. The mapping

$$(u_1, u_2) \mapsto [\alpha_x(u_1), \alpha_x(u_2)]$$

of $T_x(X) \times T_x(X)$ into L is a continuous alternating bilinear form on $T_x(X)$ with values in L. We shall denote it by $[\alpha]^2_x$, so that $[\alpha]^2$ is a differential form of degree 2 on X with values in L. Identifying an open neighbourhood of x in X with an open subset of a Banach space, we see immediately that $[\alpha]^2$ is of class C^{r-1}. If X′ is a manifold of class C^r and f: X′ → X is a morphism, then

$$[f^*(\alpha)]^2 = f([\alpha]^2). \tag{28}$$

Let α, β be two differential forms of degree 1 on X with values in L of class C^{r-1}. The exterior product $\alpha \wedge \beta$ of α and β (*Differentiable and Analytic Manifolds*, R, 7.8.2) is a differential form of degree 2 on X with values in L of class C^{r-1}; we have

$$(\alpha \wedge \beta)_x(u_1, u_2) = [\alpha_x(u_1), \beta_x(u_2)] - [\alpha_x(u_2), \beta_x(u_1)] \tag{29}$$

for u_1, u_2 in $T_x(X)$. It is immediate that

$$[\alpha + \beta]^2 = [\alpha]^2 + [\beta]^2 + \alpha \wedge \beta \tag{30}$$

$$\alpha \wedge \alpha = 2[\alpha]^2. \tag{31}$$

PROPOSITION 51. *Let* G *be a Lie group, of finite dimension if* K *is of characteristic* >0, *and let* $a_1, \ldots, a_p$ *be elements of* L(G), F *a complete normable space and* α *a differential form of degree* $p - 1$ *on* G *with values in* F. *If* α *is left invariant, then*

$$(d\alpha)_e(a_1, \ldots, a_p) = \sum_{i<j} (-1)^{i+j}\alpha_e([a_i, a_j], a_1, \ldots, a_{i-1}, a_{i+1}, \ldots, a_{j-1}, a_{j+1}, \ldots, a_p).$$

If α *is right invariant, then*

$$(d\alpha)_e(a_1, \ldots, a_p) = -\sum_{i<j} (-1)^{i+j}\alpha_e([a_i, a_j], a_1, \ldots, a_{i-1}, a_{i+1}, \ldots, a_{j-1}, a_{j+1}, \ldots, a_p).$$

Suppose that α is left invariant. By *Differentiable and Analytic Manifolds*, R, 8.5.7, then

$$(d\alpha)(L_{a_1}, \ldots, L_{a_p}) = \sum_i (-1)^{i-1}L_{a_i}\,\alpha(L_{a_1}, \ldots, L_{a_{i-1}}, L_{a_{i+1}}, \ldots, L_{a_p})$$
$$+ \sum_{i<j} (-1)^{i+j}\alpha([L_{a_i}, L_{a_j}], L_{a_1}, \ldots, L_{a_{i-1}}, L_{a_{i+1}}, \ldots, L_{a_{j-1}}, L_{a_{j+1}}, \ldots, L_{a_p}).$$

But the functions $\alpha(L_{a_1}, \ldots, L_{a_{i-1}}, L_{a_{i+1}}, \ldots, L_{a_p})$ on G are left invariant and therefore constant. Hence

$$L_{a_i}\alpha(L_{a_1}, \ldots, L_{a_{i-1}}, L_{a_{i+1}}, \ldots, L_{a_p}) = 0.$$

Moreover, $[L_{a_i}, L_{a_j}] = L_{[a_i, a_j]}$ (Proposition 23), whence the first formula of Proposition 51. The second can be established analogously, this time using the relation $[R_{a_i}, R_{a_j}] = -R_{[a_i, a_j]}$.

COROLLARY 1. *Let* G *be a Lie group, of finite dimension if* K *is of characteristic* >0, *and* ω *and* ω' *the left and right canonical differential forms of* G. *Then*

$$d\omega + [\omega]^2 = 0, \qquad d\omega' - [\omega']^2 = 0.$$

By Proposition 51,

$$(d\omega)_e(a_1, a_2) = -\omega_e([a_1, a_2]) = -[a_1, a_2] = -[\omega_e(a_1), \omega_e(a_2)]$$
$$= -[\omega]_e^2(a_1, a_2)$$

whence the first formula. The second can be established analogously.

COROLLARY 2. *Suppose that* G *is finite-dimensional. Let* $(e_1, \ldots, e_n)$ *be a basis of* L(G), $(e_1^*, \ldots, e_n^*)$ *the dual basis,* (c_{ijk}) *the constants of structure of* L(G) *relative to the basis* $(e_1, \ldots, e_n)$ *and* ω_i (resp. ω_i') *the left* (resp. *right*) *invariant differential form on* G *with values in* K *such that* $(\omega_i)_e = e_i^*$ (resp. $(\omega_i')_e = e_i^*$). *Then*

$$d\omega_k + \sum_{i<j} c_{ijk}\omega_i \wedge \omega_j = 0 \quad (k = 1, 2, \ldots, n)$$

$$d\omega_k' - \sum_{i<j} c_{ijk}\omega_i' \wedge \omega_j' = 0 \quad (k = 1, 2, \ldots, n).$$

If $r < s$,

$$(d\omega_k)_e(e_r, e_s) = -(\omega_k)_e([e_r, e_s])$$
$$= -\sum_l c_{rsl}(\omega_k)_e(e_l)$$
$$= -c_{rsk}$$
$$= -\sum_{i<j} c_{ijk}(\omega_i \wedge \omega_j)_e(e_r, e_s).$$

The argument is similar for the ω_k'.

15. CONSTRUCTION OF INVARIANT DIFFERENTIAL FORMS

Lemma 2. *Let* G *be a Lie group,* U *a symmetric open neighbourhood of* e *in* G, E *a complete normable space and* $\phi: U^2 \to E$ *an analytic mapping. For all* $g \in U$, *let* ω_g *be the differential at the point* g *of the mapping* $h \mapsto \phi(g^{-1}h)$. *Then* ω *is the restriction to* U *of the left invariant differential form on* G *whose value at* e *is* $d_e\phi$.

Clearly $\omega_e = d_e\phi$. For all $g \in U$ and all $t \in T_e(G)$,

$$\begin{aligned}\langle \omega_g, T_e(\gamma(g))t\rangle &= \langle d_g(\phi \circ \gamma(g)^{-1}), T_e(\gamma(g))t\rangle \\ &= \langle d_e\phi \circ T_g(\gamma(g)^{-1}), T_e(\gamma(g))t\rangle = \langle d_e\phi, t\rangle\end{aligned}$$

and hence ω_g is derived from $d_e\phi$ by $T_e(\gamma(g))$.

PROPOSITION 52. *Let* n *be an integer* >0, G *an* n*-dimensional Lie group,* U *a symmetric open neighbourhood of* e *in* G *and* $\psi: U^2 \to K^n$ *a chart of* G *such that* $\psi(e) = 0$. *If* $(x_1, \ldots, x_n)$ *are the coordinates of* $x \in \psi(U)$ *and* $(y_1, \ldots, y_n)$ *the coordinates of* $y \in \psi(U)$, *we denote by*

$$m_1(x_1, \ldots, x_n, y_1, \ldots, y_n), \ldots, m_n(x_1, \ldots, x_n, y_1, \ldots, y_n)$$

the coordinates of $\psi(\psi^{-1}(x)^{-1}\psi^{-1}(y))$. *Then, if we write, for* $1 \leqslant k \leqslant n$,

$$(32) \quad \varpi_k(x_1, \ldots, x_n) = D_{n+1}m_k(x_1, \ldots, x_n, x_1, \ldots, x_n)\, dx_1 + \cdots + D_{2n}m_k(x_1, \ldots, x_n, x_1, \ldots, x_n)\, dx_n,$$

the differential forms ϖ_k *on* $\psi(U)$ *are derived through* ψ *from left invariant differential forms on* G *and are such that* $\varpi_k(0, \ldots, 0) = dx_k$.

We apply Lemma 2 with E = K, taking $\phi(g)$ to be the coordinate of $\psi(g)$ of index k. We obtain a differential form ω_k; let ϖ_k be its transform under ψ. The value of ϖ_k at $(x_1, \ldots, x_n)$ is the differential at $(x_1, \ldots, x_n)$ of the function $y \mapsto m_k(x_1, \ldots, x_n, y_1, \ldots, y_n)$; this value is therefore given by formula (32). It then suffices to use the conclusion to Lemma 2.

PROPOSITION 53. *Let* G *be a Lie group,* A *a complete normable algebra and* ϕ *a Lie group morphism of* G *into* A^*. *For all* $g \in G$, *let* $\omega_g = \phi(g)^{-1}.d_g\phi$. *Then* ω *is the left invariant differential form on* G *whose value at* e *is* $d_e\phi$.

We apply Lemma 2 with E = A and U = G. The differential at g of the mapping $h \mapsto \phi(g^{-1}h) = \phi(g)^{-1}\phi(h)$ is $\phi(g)^{-1}.d_g\phi$.

16. HAAR MEASURE ON A LIE GROUP

Let G be a Lie group of finite dimension n. Then $\bigwedge^n(T_e(G))$ is of dimension 1. Hence (no. 13) the vector space S of left invariant differential forms of degree n on G is of dimension 1. Let $(\omega_1, \ldots, \omega_n)$ be a basis of the space of left invariant differential forms of degree 1 on G; then $\omega_1 \wedge \omega_2 \wedge \cdots \wedge \omega_n$ is a basis of S.

PROPOSITION 54. *Let* G *be a Lie group of finite dimension n,* ω *a left invariant differential form of degree n on* G *and* ϕ *an endomorphism of* G. *Then*

$$\phi^*(\omega) = (\det L(\phi))\omega.$$

We write $L(\phi) = u$, $\omega_e = f$ and $\phi^*(\omega)_e = g$. For all $x_1, \ldots, x_n$ in L(G),

$$g(x_1, \ldots, x_n) = f(ux_1, \ldots, ux_n) = (\det u)f(x_1, \ldots, x_n)$$

and hence $\phi^*(\omega)_e = \det L(\phi) . \omega_e$. On the other hand, if $g \in G$,

$$\phi \circ \gamma(g) = \gamma(\phi(g)) \circ \phi$$

and hence $\gamma(g)^*\phi^*(\omega) = \phi^*(\omega)$. Thus $\phi^*(\omega)$ is left invariant, whence the proposition.

COROLLARY. *For all* $g \in G$,

$$\delta(g)^*\omega = (\det \operatorname{Ad} g)\omega.$$

$\delta(g)^*\omega = \delta(g)^*\gamma(g)^*\omega = (\operatorname{Int} g)^*\omega$ and $L(\operatorname{Int} g) = \operatorname{Ad} g$.

Let G be a locally compact group and ϕ an endomorphism of G. Suppose that there exist open neighbourhoods V, V′ of e such that $\phi(V) = V'$ and $\phi \mid V$ is a local isomorphism of G into G. Let μ be a left Haar measure on G. By *Integration*, Chapter VII, § 1, Corollary to Proposition 9, there exists a unique number $a > 0$ such that $\phi(\mu \mid V) = a^{-1}\mu \mid V'$. Clearly a is independent of the choice of V, V′ and μ. It is called the modulus of ϕ and is denoted by $\operatorname{mod}_G \phi$ or simply $\operatorname{mod} \phi$. When ϕ is an automorphism of G, we recover Definition 4 of *Integration*, Chapter VII, § 1.

PROPOSITION 55. *Suppose that* K *is locally compact. Let* μ *be a Haar measure on the additive group of* K. *Let* G *be a Lie group of finite dimension n.*

(i) *Let* ω *be a non-zero left invariant differential form of degree n on* G. *Then the measure* $\operatorname{mod}(\omega)_\mu$ (*Differentiable and Analytic Manifolds*, R, 10.1.6) *is a left Haar measure on* G. *If* K = **R** *and* G *has the orientation defined by* ω, *the measure defined by* ω (*Differentiable and Analytic Manifolds*, R, 10.4.3) *is a left Haar measure on* G.

(ii) *Let* ϕ *be an étale endomorphism of* G. *Then* $\operatorname{mod} \phi = \operatorname{mod} \det L(\phi)$.

(i) is obvious. Let V, V′ be open neighbourhoods of e such that $\phi(V) = V'$ and $\phi \mid V$ is a local isomorphism of G into G. Then

$$\begin{aligned}
\phi^{-1}(\operatorname{mod}(\omega)_\mu \mid V') &= \operatorname{mod}(\phi^*(\omega))_\mu \mid V) && \text{by transport of structure} \\
&= \operatorname{mod}(\det L(\phi)\omega \mid V)_\mu && \text{(Proposition 54)} \\
&= \operatorname{mod} \det L(\phi)(\operatorname{mod}(\omega)_\mu \mid V)
\end{aligned}$$

whence $\operatorname{mod} \phi = \operatorname{mod} \det L(\phi)$ by definition of $\operatorname{mod} \phi$.

COROLLARY. *For all* $g \in G$, $\Delta_G(g) = (\text{mod det Ad } g)^{-1}$. *In particular, for* G *to be unimodular, it is necessary and sufficient that* mod det Ad $g = 1$ *for all* $g \in G$.

$$\begin{aligned} \Delta_G(g) &= (\text{mod Int } g)^{-1} && (\textit{Integration}, \text{ Chapter VII, § 1, formula (33)}) \\ &= (\text{mod det L}(\text{Int } g))^{-1} && (\text{Proposition 55}) \\ &= (\text{mod det Ad } g)^{-1}. \end{aligned}$$

Remark. Preserving the hypotheses and notation of Proposition 52, suppose that K is locally compact. Let μ be the measure

$$\text{mod det}(D_{n+i}m_k(x_1, \ldots, x_n, x_1, \ldots, x_n))_{1 \leqslant i, k \leqslant n}\, dx_1 \ldots dx_n$$

on $\psi(U)$. Then $\psi^{-1}(\mu)$ is the restriction to U of a Haar measure on G.

PROPOSITION 56. *Let* G *be a Lie group of finite dimension n,* H *a p-dimensional Lie subgroup and* X *the Lie homogeneous space* G/H. *Suppose that*

$$\det \text{Ad}_{L(G)}\, h = \det \text{Ad}_{L(H)}\, h$$

for all $h \in H$. *Then:*

(i) *The differential forms of degree* $n - p$ *on* X *which are invariant under* G *are analytic.*

(ii) *The vector space of these forms is of dimension* 1.

(iii) *If* ω *is such a non-zero form and* K *is locally compact,* $\text{mod}(\omega)_\mu$ *is a non-zero measure on* X *which is invariant under* G.

By § 1, no. 8, *Examples*, $\text{Alt}^{n-p}(TX, K)$ is an analytic vector G-bundle. Let x_0 be the canonical image of e in X; its stabilizer is H. The fibre of $\text{Alt}^{n-p}(TX, K)$ at x_0 is $\bigwedge^{n-p} T_{x_0}(X)^*$ and $T_{x_0}(X)$ is canonically identified with L(G)/L(H). If $h \in H$, the automorphism τ_h of X defined by h is derived when passing to the quotient from the automorphism $g \mapsto hgh^{-1}$ of G. Hence the automorphism $T_{x_0}(\tau_h)$ is derived when passing to the quotient from $\text{Ad}_{L(G)}(h)$. As

$$\det \text{Ad}_{L(G)}\, h = (\det \text{Ad}_{L(H)}\, h) . (\det T_{x_0}(\tau_h)),$$

the hypothesis implies that $\det T_{x_0}(h) = 1$. Thus, every element of $\bigwedge^{n-p} T_{x_0}(X)^*$ is invariant under H. Then (i) and (ii) follow from § 1, no. 8, Corollary 1 to Proposition 17 and (iii) is obvious.

The existence of a non-zero positive measure on X invariant under G follows from *Integration*, Chapter VII, § 2, Corollary 2 to Theorem 3, for the hypothesis of Proposition 56 implies $\Delta_G | H = \Delta_H$ (Corollary to Proposition 55).

PROPOSITION 57. *Let* G *be a Lie group of finite dimension n. Choose a basis for* $\bigwedge^n T_e(G)^*$; *by means of the right* (resp. *left*) *trivialization of* $\bigwedge^n T(G)^*$, *we can* *identify this vector bundle with the trivial vector bundle* G × K, *so that the transpose of a scalar differential operator is identified with a scalar differential operator.*

Then, if $u \in \mathrm{U(G)}$, *the transpose of* L_u (resp. R_u) is $\mathrm{L}_{\check{u}}$ (resp. $\mathrm{R}_{\check{u}}$).

We shall consider the case where $\bigwedge^n \mathrm{T(G)}^*$ has been trivialized using a right invariant form ω.

Suppose that the proposition has been proved for elements u_1, u_2 of U(G). Then,

$$\begin{aligned}
{}^t(\mathrm{L}_{u_1 * u_2}) &= {}^t(\mathrm{L}_{u_1} \circ \mathrm{L}_{u_2}) && \text{(Proposition 23)}\\
&= {}^t(\mathrm{L}_{u_2}) \circ {}^t(\mathrm{L}_{u_1}) && (\textit{Diff. \& Anal. Man.}, \text{R, 14.3.3})\\
&= \mathrm{L}_{\check{u}_2} \circ \mathrm{L}_{\check{u}_1} && \text{by hypothesis}\\
&= \mathrm{L}_{\check{u}_2 * \check{u}_1} && \text{(Proposition 23)}\\
&= \mathrm{L}_{(u_1 * u_2)^\vee} && \text{(Proposition 7)}
\end{aligned}$$

and hence the proposition is true for $u_1 * u_2$. It therefore suffices to prove the proposition when $u \in \mathrm{T}_e(\mathrm{G})$. Now L_u is defined by G operating on G on the right (no. 6) and hence $\theta_{\mathrm{L}_u}\omega = 0$ since ω is right invariant (*Differentiable and Analytic Manifolds*, R, 8.4.5); therefore, if f is an analytic function in an open neighbourhood of e with values in K, then $\theta_{\mathrm{L}_u}(f\omega) = (\theta_{\mathrm{L}_u}f)\omega$ (*Differentiable and Analytic Manifolds*, R, 8.4.8). Using the identifications made and *Differentiable and Analytic Manifolds*, R, 14.4.1, the transpose of L_u is $-\mathrm{L}_u$, that is $\mathrm{L}_{\check{u}}$.

Corollary. *Let* G *be a finite-dimensional real Lie group,* μ (resp. ν) *a left* (resp. *right*) *Haar measure on* G, *k an integer* $\geqslant 0$, $u \in \mathrm{U}_k(\mathrm{G})$ *and f and g real-valued functions of class* C^k *on* G *with compact support. Then*

$$\int_{\mathrm{G}} (\mathrm{R}_u f)\, g\, d\mu = \int_{\mathrm{G}} f(\mathrm{R}_{\check{u}} g)\, d\mu$$

$$\int_{\mathrm{G}} (\mathrm{L}_u f)\, g\, d\nu = \int_{\mathrm{G}} f(\mathrm{L}_{\check{u}} g)\, d\nu.$$

This follows from Proposition 57 and *Differentiable and Analytic Manifolds*, R, 14.3.8.

17. LEFT DIFFERENTIAL

Definition 9. *Let* G *be a Lie group,* M *a manifold of class* C^r *and f a mapping of class* C^r *of* M *into* G. *The left* (resp. *right*) *differential of f is the differential form of degree* 1 *on* M *with values in* L(G) *which associates with every vector* $u \in \mathrm{T}_m(\mathrm{M})$ *the element* $f(m)^{-1}.(\mathrm{T}_m f)(u)$ (resp. $(\mathrm{T}_m f)(u).f(m)^{-1}$).

In this chapter we shall only consider the left differential, which we shall denote by $f^{-1}.df$, and leave to the reader the task of translating the results for the right differential.

If f is the identity mapping of G, $f^{-1}.df$ is the canonical left differential form ω of G. Returning to the general case of Definition 8,

$$(f^{-1}.df)_m = \omega_{f(m)} \circ \mathrm{T}_m(f)$$

and hence $f^{-1}.df = f^*(\omega)$. This implies that $f^{-1}.df$ is of class C^{r-1}.

Examples. (1) If G is the additive group of a complete normable space and $\mathrm{T}_0(\mathrm{E})$ is canonically identified with E, $f^{-1}.df$ is the differential df defined in *Differentiable and Analytic Manifolds*, R, 8.2.2.

(2) Suppose that G is the multiplicative group A* associated with a complete normable algebra A. Then f can be considered as a mapping of M into A and hence the differential df in the sense of *Differentiable and Analytic Manifolds*, R, 8.2.2 is defined and the product $f^{-1}df$ in the sense of *Differentiable and Analytic Manifolds*, R, 8.3.2 is defined. Clearly the latter form is identical with the left differential of f.

PROPOSITION 58. *Let* G *and* H *be two Lie groups,* M *a manifold of class* C^r, f *a mapping of class* C^r *of* M *into* G *and* h *a morphism of* G *into* H. *Then*

$$(h \circ f)^{-1}.d(h \circ f) = \mathrm{L}(h) \circ (f^{-1}.df) = (h^{-1}.dh) \circ \mathrm{T}(f).$$

For all $x \in \mathrm{M}$ and $u \in \mathrm{T}_x(\mathrm{M})$,

$$(h \circ f)^{-1}.d(h \circ f)(u) = ((h \circ f)(x))^{-1}.\mathrm{T}(h \circ f)(u).$$

The latter expression is equal, on the one hand, to

$$\begin{aligned} &\mathrm{T}(h)(f(x)^{-1}.\mathrm{T}(f)(u)) \quad (\S\, 2, \text{Proposition } 5)\\ &= \mathrm{T}_e(h)((f^{-1}.df)(u)) \end{aligned}$$

and, on the other hand, to

$$\begin{aligned} &h(f(x))^{-1}\mathrm{T}(h)(\mathrm{T}(f)u)\\ &= (h^{-1}.dh)(\mathrm{T}(f)u). \end{aligned}$$

PROPOSITION 59. *Let* G *be a Lie group,* M *a manifold of class* C^r, f *and* g *mappings of class* C^r *of* M *into* G *and* p *the canonical surjection of* TM *onto* M.

(i) $(fg)^{-1}.d(fg) = (\mathrm{Ad} \circ g \circ p)^{-1} \circ (f^{-1}.df) + g^{-1}.dg.$
(ii) *Writing* $h(m) = f(m)^{-1}$ *for all* $m \in \mathrm{M}$,

$$h^{-1}.dh = -(\mathrm{Ad} \circ f \circ p) \circ (f^{-1}.df).$$

Assertion (i) follows from § 2, no. 2, Proposition 7. Assertion (ii) follows from (i) by putting $g = h$.

COROLLARY 1. *Let* $s \in \mathrm{G}$ *and* sg *be the mapping* $x \mapsto sg(x)$ *of* M *into* G. *Then* $(sg)^{-1}.d(sg) = g^{-1}.dg$.

This follows from Proposition 59 (i) taking f to be the constant mapping $x \mapsto s$ of M into G.

COROLLARY 2. *If the mappings f and g of* M *into* G *have the same left differential, the tangent mapping to fg^{-1} is everywhere zero. If further* K *is of characteristic* 0, *then fg^{-1} is locally constant.*

By Proposition 59,

$$(fg^{-1})^{-1}.d(fg^{-1}) = (\mathrm{Ad} \circ g \circ p) \circ (f^{-1}.df) - (\mathrm{Ad} \circ g \circ p) \circ (g^{-1}.dg).$$

If $f^{-1}.df = g^{-1}.dg$, then $(fg^{-1})^{-1}.\, d(fg^{-1}) = 0$, that is $\mathrm{T}_x(fg^{-1}) = 0$ for all $x \in \mathrm{M}$. This proves the first assertion. The second follows from it by *Differentiable and Analytic Manifolds*, R, 5.5.3.

PROPOSITION 60. *Let* G *be a Lie group, of finite dimension if* K *is of characteristic* >0, M *a manifold of class* C^r, f *a mapping of class* C^r *of* M *into* G *and* α *the left differential of* f. *Then* $d\alpha + [\alpha]^2 = 0$.

Let ω be the canonical left differential form of G. Using Corollary 1 to Proposition 51, no. 14, we have

$$\begin{aligned} d\alpha &= d(f^*(\omega)) = f^*(d\omega) = f^*(-[\omega]^2) \\ &= -[f^*(\omega)]^2 = -[\alpha]^2. \end{aligned}$$

18. LIE ALGEBRA OF A LIE GROUP GERM

In this no. $(\mathrm{G}, e, \theta, m)$ denotes a Lie group germ. A large part of the results of the § are still true with the same proof. We shall review those which we shall find useful.

18.1. Let Ω be the set of definition of m. Let $(g, g') \in \Omega$, $t \in \mathrm{T}_g^{(\infty)}(\mathrm{G})$, $t' \in \mathrm{T}_{g'}^{(\infty)}(\mathrm{G})$. As in no. 1, the convolution product of t and t', denoted by $t * t'$, is the image of $t \otimes t'$ under m. We write $\mathrm{U}(\mathrm{G}) = \mathrm{T}_e^{(\infty)}(\mathrm{G})$, $\mathrm{U}_s(\mathrm{G}) = \mathrm{T}_e^{(s)}(\mathrm{G})$, $\mathrm{U}^+(\mathrm{G}) = \mathrm{T}_e^{(\infty)+}(\mathrm{G})$, $\mathrm{U}_s^+(\mathrm{G}) = \mathrm{T}_e^{(s)+}(\mathrm{G})$. For t, t' in $\mathrm{U}(\mathrm{G})$, $t * t'$ is defined and belongs to $\mathrm{U}(\mathrm{G})$. With the convolution product, $\mathrm{U}(\mathrm{G})$ is an associative algebra with unit element ε_e, filtered by the $\mathrm{U}_s(\mathrm{G})$. The canonical isomorphism $i_{\mathrm{G},e}$ of $\mathrm{gr}\,\mathrm{U}(\mathrm{G})$ onto $\mathsf{TS}(\mathrm{T}_e(\mathrm{G}))$ is an algebra isomorphism.

18.2. Let G, H be Lie group germs and $\phi\colon \mathrm{G} \to \mathrm{H}$ a morphism. If $t \in \mathrm{U}(\mathrm{G})$, the image $\mathrm{U}(\phi)(t)$ of t under ϕ_* is an element of $\mathrm{U}(\mathrm{H})$ and $\mathrm{U}(\phi)$ is a morphism of the algebra $\mathrm{U}(\mathrm{G})$ into the algebra $\mathrm{U}(\mathrm{H})$. The mapping $\theta\colon x \mapsto x^{-1}$ of G into G defines a mapping $t \mapsto t^\vee$ of $\mathrm{U}(\mathrm{G})$ into $\mathrm{U}(\mathrm{G})$. For t, t' in $\mathrm{U}(\mathrm{G})$, the product $t * t'$ evaluated relative to $\mathrm{G}^\vee$ is equal to the product $t' * t$ evaluated relative to G and $(t * t')^\vee = t'^\vee * t^\vee$. Then $\mathrm{U}(\phi)(t^\vee) = (\mathrm{U}(\phi)t)^\vee$. If $\mathrm{G}_1, \ldots, \mathrm{G}_n$ are Lie group germs and $\mathrm{G} = \mathrm{G}_1 \times \cdots \times \mathrm{G}_n$, the canonical isomorphism of $\mathrm{U}(\mathrm{G}_1) \otimes \cdots \otimes \mathrm{U}(\mathrm{G}_n)$ onto $\mathrm{U}(\mathrm{G})$ is an algebra isomorphism;

for $t_1, \ldots, t_n$ in U(G), $(t_1 \otimes \cdots \otimes t_n)^\vee = t_1^\vee \otimes \cdots \otimes t_n^\vee$. Let H be a Lie subgroup germ of G and $i: H \to G$ the canonical injection. Then U(i) is an injective homomorphism of the algebra U(H) into U(G) and

$$U(i)(t^\vee) = (U(i)(t))^\vee$$

for all $t \in U(H)$. With the convolution product and the coproduct defined by the manifold structure on G, U(G) is a bigebra and U(ϕ) is a bigebra morphism.

18.3. Let G be a Lie group germ, X a manifold of class C^r and ψ a law chunk of left operation of class C^r of G on X. Let Ω be the set of definition of ψ. If $t \in T_g^{(s)}(G)$, $u \in T_x^{(s')}(X)$, $(g, x) \in \Omega$ and $s + s' \leqslant r$, let $t * u$ denote the image of $t \otimes u$ under ψ_*. Let $t \in T_g^{(s)}(G)$, $t' \in T_x^{(s')}(G)$, $u \in T_x^{(s'')}(X)$; if $s + s' + s'' \leqslant r$ and gg', $(gg')x$, $g'x$, $g(g'x)$ are defined, then

$$(t * t') * u = t * (t' * u).$$

Let $x_0 \in X$ and $\rho(x_0)$ be the mapping $g \mapsto gx_0$, which is defined in an open neighbourhood of e. If $t \in U_r(G)$, then $\rho(x_0)_* t = t * \varepsilon_{x_0}$. Here and in the rest of this no., we shall leave to the reader the task of translating the results for law chunks of right operation.

18.4. Preserving the notation of 18.3, let $t \in U_s(G)$ with $s \leqslant r$. Let f be a function of class C^r on X with values in a Hausdorff polynormed space. Let $t * f$ denote the function on X defined by

$$\begin{aligned}(t * f)(x) &= \langle t, g \mapsto f(\psi(\theta(g), x))\rangle \\ &= \langle t^\vee, f \circ \rho(x)\rangle = \langle \rho(x)_*(t^\vee), f\rangle = \langle t^\vee * \varepsilon_x, f\rangle.\end{aligned}$$

If $t \in U_s(G)$, $t' \in U_{s'}(G)$ and $s + s' \leqslant r$, then $\langle t', t * f\rangle = \langle t^\vee * t', f\rangle$ and $(t * t') * f = t * (t' * f)$. Let $t \in U_s(G)$, f and f' be functions of class C^r on X with values in Hausdorff polynormed spaces F, F' and $(u, u') \mapsto u.u'$ be a continuous bilinear mapping of F $\times$ F' into a Hausdorff polynormed space; let $\sum_{i=1}^{n} t_i \otimes t_i'$ be the image of t under the coproduct; if $s \leqslant r$, then

$$t * (ff') = \sum_{i=1}^{n} (t_i * f)(t_i' * f').$$

18.5. Preserving the notation of 18.3, let $t \in U_s(G)$ with $s \leqslant r$. The mapping $x \mapsto t * \varepsilon_x$ is called the field of point distributions defined by t and the law chunk of operation and is sometimes denoted by D_t^ψ or D_t. If $f: X \to F$ is a function of class C^r, the function $t^\vee * f$ on X is also denoted by $D_t f$; it is of class C^{r-s} if $s < \infty$. If $t \in U_s(G)$, $t' \in U_{s'}(G)$ and $s + s' \leqslant r$, then $D_{t * t'} f = D_{t'}(D_t f)$. If G and X are finite-dimensional, D_t is a differential operator on X of order $\leqslant s$ and of class C^{r-s} (if $s < \infty$). The function $D_t f$ is then the transform of f under this differential operator.

18.6. Let G be a Lie group germ and $t \in U(G)$. L_t denotes the field of point distributions $g \mapsto \varepsilon_g * t$ on G and R_t the field of point distributions $g \mapsto t * \varepsilon_g$ on G. If $f \in \mathscr{C}^\omega(G, F)$, then $L_t f \in \mathscr{C}^\omega(G, F)$ and $R_t f \in \mathscr{C}^\omega(G, F)$. For t, t' in U(G), $L_{t*t'} = L_t \circ L_{t'}$, $R_{t*t'} = R_{t'} \circ R_t$, $L_t \circ R_{t'} = R_{t'} \circ L_{t'}$, $\theta(L_t) = R_t^\vee$.

18.7. As $T_e(G)$ is the set of primitive elements of U(G),

$$[T_e(G), T_e(G)] \subset T_e(G).$$

The normable space $T_e(G)$, together with the bracket, is a normable Lie algebra, called the normable Lie algebra of G (or Lie algebra of G) and denoted by L(G). Let E(G) be the enveloping algebra of L(G). The canonical injection of L(G) into U(G) defines a homomorphism η of the algebra E(G) into the algebra U(G); if K is of characteristic 0, η is a bigebra isomorphism, by means of which U(G) is identified with E(G). Using the notation of 18.3, for all $a \in L(G)$, let D_a be the field of point distributions defined by a on X. The mapping $(a, x) \mapsto D_a(x)$ is a morphism of class C^{r-1} of the trivial vector bundle $L(G) \times X$ into the vector bundle T(X). Let I be an open subset of K containing 0 and $\gamma: I \to G$ a mapping of class C^r such that $\gamma(0) = e$. Let $a = T_0(\gamma)1 \in L(G)$. If $f: X \to F$ is a function of class C^r, then

$$(D_a f)(x) = \lim_{k \in K^*, k \to 0} k^{-1}(f(\gamma(k)x) - f(x)).$$

If $r \geqslant 2$, the mapping $a \mapsto D_a$ is a law of left infinitesimal operation of class C^{r-1} of L(G) on X.

18.8. Let G and H be Lie group germs and ϕ a morphism of G into H. The restriction of $U(\phi)$ to L(G), which is just $T_e(\phi)$, is a continuous morphism of L(G) into L(H), which we denote by $L(\phi)$. If ψ is a morphism of H into a Lie group germ, then $L(\psi \circ \phi) = L(\psi) \circ L(\phi)$. For ϕ to be an immersion, it is necessary and sufficient that $L(\phi)$ be an isomorphism of L(G) onto a Lie subalgebra of L(H) which admits a topological supplement. In particular, if G is a Lie subgroup germ of H and ϕ is the canonical injection, L(G) is identified with a Lie subalgebra of L(H) by means of $L(\phi)$. If $(G_\iota)_{\iota \in I}$ is a finite family of Lie group germs and G is their product, L(G) is canonically identified with $\prod_{\iota \in I} L(G_\iota)$.

18.9. Let G be a Lie group germ, of finite dimension if K is of characteristic > 0. Let F be a complete normable space. Let α be a differential form of degree k on G with values in F. α is called left invariant on G if α_g is derived from α_e by the mapping $h \mapsto gh$ of a neighbourhood of e onto a neighbourhood of g. If α is left invariant, α is analytic. The mapping $\alpha \mapsto \alpha_e$ is a bijection of the set of left invariant differential forms of degree k on G with values in F onto the set of continuous alternating k-linear mappings of $T_e(G)$ into F. If $\alpha_e = \mathrm{Id}_{T(G)}$, α is called the *left canonical differential form of* G. The definitions of right invariant differential forms and the right canonical differential form of G are analogous. If ω is the left canonical differential form of G, then $d\omega + [\omega]^2 = 0$. Let M be a manifold of class C^r and f a mapping of class C^r of M into G. The left differential of f, denoted by $f^{-1}.df$, is the differential form of degree 1 on M with values in L(G) which associates with each vector $u \in T_m(M)$ the element $f(m)^{-1}.(T_m f)(u)$. Then $f^{-1}.df = f^*(\omega)$ and $d\alpha + [\alpha]^2 = 0$. If two mappings f and g of M into G have the same left differential and K is of characteristic 0, then fg^{-1} is locally constant.

§ 4. PASSAGE FROM LIE ALGEBRAS TO LIE GROUPS

Recall that, until the end of the chapter, K is assumed to be of characteristic 0.

1. PASSAGE FROM LIE ALGEBRA MORPHISMS TO LIE GROUP MORPHISMS

Lemma 1. *Let* G *be a Lie group germ and* $\mathfrak{h}$ *a Lie subalgebra of* L(G) *admitting a topological supplement. The union of the* $g\mathfrak{h}$ (resp. $\mathfrak{h}g$) *for* $g \in G$ *is an integrable vector subbundle of* T(G).

By considering the left trivialization of T(G) (§ 2, no. 3), it is seen immediately that the $g\mathfrak{h}$, for $g \in G$, are the fibres of a vector subbundle E of T(G). Let $g \in G$. The set of $(L_a)_g$, where $a \in \mathfrak{h}$, is equal to $g\mathfrak{h}$. Now, if a and b belong to $\mathfrak{h}$, then $[L_a, L_b] = L_{[a,b]}$ and $[a, b] \in \mathfrak{h}$. Hence E is integrable (*Differentiable and Analytic Manifolds*, R, 9.3.3 (iv)). The argument is similar for the $\mathfrak{h}g$.

The integral foliation (*Differentiable and Analytic Manifolds*, R, 9.3.2) of the union of the $g\mathfrak{h}$ (resp. $\mathfrak{h}g$) is called the *left* (resp. *right*) *foliation* of G associated with $\mathfrak{h}$.

THEOREM 1. *Let* G *and* H *be Lie group germs and* f *a continuous morphism of* L(G) *into* L(H).

(i) *There exist an open Lie subgroup germ* G′ *of* G *and a morphism* ϕ *of* G′ *into* H *such that* $f = L(\phi)$.

(ii) *Let* G_1, G_2 *be open Lie subgroup germs of* G *and* ϕ_i *a morphism of* G_i *into* H *such that* $f = L(\phi_i)$ *for* $i = 1, 2$. *Then* ϕ_1 *and* ϕ_2 *coincide on a neighbourhood of* e.

Let $p_1: G \times H \to G$, $p_2: G \times H \to H$ be the canonical projections. For all $(g, h) \in G \times H$, let $f_{g,h}$ be the mapping $ga \mapsto hf(a)$ of $T_g(G) = gL(G)$ into $T_h(H) = hL(H)$. By considering the left trivializations of T(G) and T(H), it is seen immediately that the $f_{g,h}$ define a morphism of $p_1^*T(G)$ into

$p_2^*\mathrm{T}(\mathrm{H})$. Let $\mathfrak{a}$ be the graph of f; it is a closed Lie subalgebra of $\mathrm{L}(\mathrm{G}) \times \mathrm{L}(\mathrm{H})$ which admits $\{0\} \times \mathrm{L}(\mathrm{H})$ as topological supplement. For all $(g, h) \in \mathrm{G} \times \mathrm{H}$, the graph of $f_{g,h}$ is $(g, h) . \mathfrak{a}$. The union of these graphs is an integrable vector subbundle of $\mathrm{T}(\mathrm{G} \times \mathrm{H})$ (Lemma 1). Then there exist (*Differentiable and Analytic Manifolds*, R, 9.3.7) an open neighbourhood U of e_{G} in G and an analytic mapping ϕ of U into H such that $\phi(e_{\mathrm{G}}) = e_{\mathrm{H}}$ and $\mathrm{T}_g(\phi) = f_{g, \phi(g)}$ for all $g \in \mathrm{U}$. In particular, $\mathrm{T}_{e_{\mathrm{G}}}(\phi) = f$.

Let V be an open neighbourhood of e_{G} in G such that, for $(s, t) \in \mathrm{V} \times \mathrm{V}$, the products st and $\phi(s)\phi(t)$ are defined and $st \in \mathrm{U}$. Consider the mappings α_1, α_2 of $\mathrm{V} \times \mathrm{V}$ into H defined by

$$\alpha_1(s, t) = \phi(ts), \qquad \alpha_2(s, t) = \phi(t)\phi(s).$$

Then $\alpha_1(t, e) = \phi(t) = \alpha_2(t, e)$. On the other hand, let t be fixed in V and β_i be the mapping $s \mapsto \alpha_i(s, t)$ of V into H. Then, for all $s \in \mathrm{V}$ and all $a \in \mathrm{L}(\mathrm{G})$,

$$\begin{aligned}
\mathrm{T}_s(\beta_1)(sa) &= \mathrm{T}_{ts}(\phi)(tsa) = f_{ts, \phi(ts)}(tsa) \\
&= \phi(ts) f(a) = f_{s, \beta_1(s)}(sa) \\
\mathrm{T}_s(\beta_2)(sa) &= \phi(t)\mathrm{T}_s(\phi)(sa) = \phi(t) f_{s, \phi(s)}(sa) \\
&= \phi(t)\phi(s) f(a) = f_{s, \beta_2(s)}(sa).
\end{aligned}$$

Hence (*Differentiable and Analytic Manifolds*, R, 9.3.7) α_1 and α_2 coincide on a neighbourhood of $(e_{\mathrm{G}}, e_{\mathrm{G}})$. The restriction of ϕ to a sufficiently small symmetric open neighbourhood of e_{G} is therefore a morphism of Lie group germs, whence (i).

Let $\mathrm{G}_1, \mathrm{G}_2, \phi_1, \phi_2$ be as in (ii) and let us prove that ϕ_1, ϕ_2 coincide on a neighbourhood of e_{G}. There exists an open neighbourhood W of e_{G} such that $\phi_1(ts) = \phi_1(t)\phi_1(s)$, $\phi_2(ts) = \phi_2(t)\phi_2(s)$ for all s, t in W. Then, if $s \in \mathrm{W}$ and $a \in \mathrm{L}(\mathrm{G})$,

$$\mathrm{T}_s(\phi_i)(sa) = \phi_i(s)\mathrm{T}_e(\phi_i)(a) = \phi_i(s) f(a) = f_{s, \phi_i(s)}(sa)$$

for $i = 1, 2$. As $\phi_1(e_{\mathrm{G}}) = e_{\mathrm{H}} = \phi_2(e_{\mathrm{G}})$, it follows from *Differentiable and Analytic Manifolds*, R, 9.3.7 that ϕ_1 and ϕ_2 coincide on a neighbourhood of e_{G}.

COROLLARY 1. *Let* G *and* H *be two Lie group germs. If* L(G) *and* L(H) *are isomorphic,* G *and* H *are locally isomorphic.*

This follows from Theorem 1 and § 1, no. 10, Proposition 21.

COROLLARY 2. *Let* G *be a Lie group germ. If* L(G) *is commutative,* G *is locally isomorphic to the additive Lie group* L(G).

The Lie algebra of the additive group L(G) is isomorphic to L(G). Hence it suffices to apply Corollary 1.

COROLLARY 3. *Let* G *be a Lie group. If* L(G) *is commutative,* G *contains a commutative open subgroup.*

There exists an open Lie subgroup germ U of G which is commutative (Corollary 2). Let V be a neighbourhood of e such that $V^2 \subset U$. Then $xy = yx$ for all x, y in V. Hence the subgroup of G generated by V is commutative; it is obviously open.

2. PASSAGE FROM LIE ALGEBRAS TO LIE GROUPS

We shall denote the Hausdorff series by H(X, Y) (Chapter II, § 6, no. 4, Definition 1).

Lemma 2. *Let* L *be a complete normed Lie algebra over* **R** *or* **C**. *Let* G *be the set of* $x \in L$ *such that* $\|x\| < \frac{1}{3}\log\frac{3}{2}$. *Let* θ *be the mapping* $x \mapsto -x$ *of* G *into* G. *Let* H *be the restriction to* G × G *of the Hausdorff function of* L (Chapter II, § 7, no. 2).

(i) (G, 0, θ, H) *is a Lie group germ.*

(ii) *Let* ϕ *be the identity mapping of* G *into* L. *The differential of* ϕ *at* 0 *is an isomorphism of the normable Lie algebra* L(G) *onto* L.

(i) follows from Chapter II, § 7, no. 2.

As ϕ is a chart on G, the differential ψ of ϕ at 0 is an isomorphism of normable spaces. On the other hand, the expansion as an integral series $H = \sum_{i,j \geqslant 0} H_{ij}$ of the mapping H is such that $H_{11}(x, y) = \frac{1}{2}[x, y]$. By § 3, Proposition 24, for all a, b in L(G),

$$\psi([a, b]) = H_{11}(\psi(a), \psi(b)) - H_{11}(\psi(b), \psi(a)) = [\psi(a), \psi(b)]$$

which proves (ii).

G is called the *Lie group germ defined by* L.

Suppose that K is ultrametric. Let p be the characteristic of the residue field of K. If $p \neq 0$, let $\lambda = |p|^{1/(p-1)}$; if $p = 0$, let $\lambda = 1$.

Lemma 3. *Let* L *be a complete normed Lie algebra over* K. *Let* G *be the set of* $x \in L$ *such that* $\|x\| < \lambda$. *Let* H: G × G → G *be the Hausdorff function of* L (Chapter II, § 8, no. 3).

(i) *With the law of composition* H, G *is a Lie group in which* 0 *is identity element and* $-x$ *the inverse of* x *for all* $x \in G$.

(ii) *Let* ϕ *be the identity mapping of* G *into* L. *The differential of* ϕ *at* 0 *is an isomorphism of the normable Lie algebra* L(G) *onto* L.

(iii) *For all* $\mu \in \mathbf{R}_+^*$, *let* G_μ *be the set of* $x \in L$ *such that* $\|x\| < \mu$. *Then the* G_μ, *for* $\mu < \lambda$, *form a fundamental system of open and closed neighbourhoods of* 0 *and are subgroups of* G.

Assertions (i) and (iii) follow from Chapter II, § 8, no. 3, Proposition 3 and (ii) can be proved as in Lemma 2.

G is called the *Lie group defined by* L.

THEOREM 2. *Let* L *be a complete normable Lie algebra. There exists a Lie group germ G such that* L(G) *is isomorphic to* L. *Two such Lie group germs are locally isomorphic.*

The first assertion follows from Lemmas 2 and 3. The second assertion follows from Corollary 1 to Theorem 1 of no. 1.

COROLLARY 1. *Let* G *be a Lie group. There exists a neighbourhood of e which contains no finite subgroup distinct from $\{e\}$. If* $K = \mathbf{R}$ *or* $\mathbf{C}$, *there exists an open neighbourhood of e which contains no subgroup distinct from $\{e\}$.*

We write $L(G) = L$. Choose a norm on L defining the topology on L and such that $\|[x, y]\| \leqslant \|x\| \|y\|$ for all x, y in L.

Suppose that $K = \mathbf{R}$ or $\mathbf{C}$. Let G′ be the Lie group germ defined by L. There exist an open ball U′ of centre 0 in G′ and an isomorphism ϕ of the Lie group germ U′ onto an open neighbourhood U of e in G. Let $V' = \frac{1}{2}U'$, $V = \phi(V')$, H be a subgroup of G contained in V and $h \in H$. Let $x = \phi^{-1}(h) \in V'$. If $x \neq 0$, there exists an integer $n > 0$ such that $x, 2x, \ldots, nx$ are in V′, $(n+1)x \in U'$, $(n+1)x \notin V'$. Then $h, h^2, \ldots, h^n$ are in V,

$$h^{n+1} \in U, \qquad h^{n+1} \notin V,$$

which is absurd. Hence $H = \{e\}$.

Suppose that K is ultrametric. It suffices to prove the corollary when G is the Lie group associated with L. If $g \in G$, the powers of g evaluated in G are the elements of $\mathbf{Z}g$ evaluated in L. These are distinct if $g \neq e$. Hence G contains no finite subgroup distinct from $\{e\}$.

COROLLARY 2. *Let k be a non-discrete closed subfield of* K, G *a Lie group over k and* $L = L(G)$. *Suppose that* L *has a normable Lie* K*-algebra structure* L′, *compatible with the normable Lie k-algebra structure and invariant under the adjoint representation of* G. *Then there exists on* G *one and only one Lie* K*-group structure compatible with the Lie k-group structure and for which the Lie algebra is* L′.

There exists a Lie group germ G_1 over K such that $L(G_1) = L'$ (Theorem 2). By Corollary 1 to Theorem 1 of no. 1, G and G_1, considered as Lie k-group germs, are locally isomorphic. Hence there exist an open neighbourhood G′ of e in G and a Lie K-group germ structure on G′, with Lie algebra L, which is compatible with the Lie group germ structure over k. Let V be a symmetric open neighbourhood of e in G such that $V^2 \subset G'$. Let $g \in G$. Then $\phi = \operatorname{Int} g$ is a k-isomorphism of a sufficiently small open Lie subgroup germ of G′ onto an open Lie subgroup germ of G′; and $T_e(\phi)$ is K-linear and hence $T_x(\phi)$ is K-linear for x sufficiently close to e; therefore the restriction of Int g to a sufficiently small open neighbourhood of e in V is K-analytic (*Differentiable and Analytic Manifolds*, R, 5.14.6). By § 1, no. 9, Proposition 18, there exists on G an analytic K-manifold structure with which G is a Lie

K-group and V is an open sub-K-manifold of G. By translation, it is seen that the underlying k-manifold structure of G is the given structure. The Lie algebra of the Lie K-group G is the same as that of the open Lie K-subgroup germ V and hence is L′. Finally, the uniqueness stated in the corollary follows from § 3, no. 8, Proposition 32.

THEOREM 3. *Let* G *be a Lie group germ and* $\mathfrak{h}$ *a Lie subalgebra of* L(G) *admitting a topological supplement. There exists a Lie subgroup germ* H *of* G *such that* L(H) = $\mathfrak{h}$. *If* H_1 *and* H_2 *are Lie subgroup germs of* G *such that*

$$L(H_1) = L(H_2) = \mathfrak{h},$$

then $H_1 \cap H_2$ *is open in* H_1 *and* H_2.

There exists a Lie group germ H′ with Lie algebra isomorphic to $\mathfrak{h}$ (Theorem 2). Shrinking H′ if necessary, it can be assumed that there exists a morphism ϕ of H′ into G such that $L(\phi)$ is an isomorphism of L(H′) onto $\mathfrak{h}$ (no. 1, Theorem 1). As $\mathfrak{h}$ admits a topological supplement, ϕ is an immersion at e. Hence, shrinking H′ further, it can be assumed that ϕ is an isomorphism of the manifold H′ onto a submanifold of G. This proves the existence of H. The second assertion follows from the following proposition:

PROPOSITION 1. *Let* G *be a Lie group germ and* H *and* H′ *two Lie subgroup germs. In order that* L(H) ⊃ L(H′), *it is necessary and sufficient that* H ∩ H′ *be open in* H′.

If H ∩ H′ is open in H′, then L(H′) = L(H ∩ H′) ⊂ L(H). Suppose that L(H) ⊃ L(H′). Let i, i' be the canonical injections of H, H′ into G. By shrinking H′ if necessary, it can be assumed that there exists a morphism ψ of H′ into H such that $L(\psi)$ is the canonical injection of L(H′) into L(H) (no. 1, Theorem 1). Then $L(i \circ \psi) = L(i')$ and hence there exists a neighbourhood V of $e_{H'}$ in H′ such that $i \circ \psi$ and i' coincide on V (Theorem 1). Therefore V ⊂ H, hence V ⊂ H ∩ H′ and H ∩ H′ is open in H′ (§ 1, no. 10).

PROPOSITION 2. *Let* G *be a Lie group over* K, k *a non-discrete closed subfield of* K *and* H *a Lie subgroup of the Lie* k*-group* G. *Suppose that* L(H) *is a vector sub-K-space of* L(G) *which admits a topological supplement. Then* H *is a Lie subgroup of the Lie* K*-group* G.

There exists a Lie subgroup germ H′ of the Lie K-group G such that L(H′) = L(H) (Theorem 3). Consider G, H, H′ as Lie k-group germs; Theorem 3 then proves that H ∩ H′ is open in H and H′. Hence there exists an open neighbourhood U of e in G such that U ∩ H is a submanifold of G over K. Therefore, H is a Lie subgroup of the Lie K-group G (§ 1, no. 3, Proposition 6).

3. EXPONENTIAL MAPPINGS

THEOREM 4. *Let* G *be a Lie group germ,* L *its Lie algebra,* V *an open neighbourhood of* 0 *in* L, ϕ *an analytic mapping of* V *into* G *such that* $\phi(0) = 0$ *and* $T_0(\phi) = Id_L$. *The following conditions are equivalent:*

(i) *For all* $b \in L$, $\phi((\lambda + \lambda')b) = \phi(\lambda b)\phi(\lambda' b)$ *for* $|\lambda|$ *and* $|\lambda'|$ *sufficiently small.*

(ii) *For all* $b \in L$ *and every integer* $n > 0$, $\phi_*(b^n)$ *is homogeneous of degree* n *in* U(G) ($T_0^{(\infty)}(L)$ *is identified with* TS(L) *and* b^n *is evaluated in* TS(L)).

(iii) *The mapping* ϕ_* *of* TS(L) *into* U(G) *is compatible with the graduations of* TS(L) *and* U(G).

(iv) *The mapping* ϕ_* *of* TS(L) *into* U(G) *is the canonical mapping of* TS(L) *into the enveloping algebra of* L.

(v) *There exist a norm on* L *defining the topology of* L *and such that*

$$\|[x, y]\| \leqslant \|x\| \, \|y\|$$

for all x, y *in* L *and an open subgroup germ* $W \subset V$ *of the Lie group germ defined by* L (no. 2) *such that* $\phi | W$ *is an isomorphism of* W *onto an open Lie subgroup germ of* G.

(v) $\Rightarrow$ (i): obvious, for $(\lambda b) . (\lambda' b) = (\lambda + \lambda')b$ in W for $|\lambda|$ and $|\lambda'|$ sufficiently small.

(i) $\Rightarrow$ (ii): suppose that condition (i) is satisfied. Let $b \in L$. Let ψ be the restriction of ϕ to $V \cap Kb$. By hypothesis, there exists a symmetric neighbourhood T of 0 in the additive Lie group Kb such that $\psi | T$ is a morphism of the Lie group germ T into G. Hence

$$\phi_*(b^n) = (\psi | T)_*(b^n) = ((\psi | T)_*(b))^n = (\phi_*(b))^n,$$

so that $\phi_*(b^n)$ is homogeneous of degree n in U(G).

(ii) $\Rightarrow$ (iii): this follows from the fact that $TS^n(L)$ is the vector subspace of TS(L) generated by the n-th powers of the elements of L (*Algebra*, Chapter IV, § 5, Proposition 5).

(iii) $\Rightarrow$ (iv): the canonical mapping of TS(L) into the enveloping algebra of L is the unique morphism of graded cogebras mapping 1 to 1 and extending Id_L (Chapter II, § 1, no. 5, *Remark* 3). Now ϕ_* is a cogebra morphism and $\phi_* | L = Id_L$ by hypothesis. If condition (iii) holds, it is seen that condition (iv) also holds.

(iv) $\Rightarrow$ (v): suppose that condition (iv) is satisfied. Choose a norm on L defining the topology of L and such that $\|[x, y]\| \leqslant \|x\| \, \|y\|$ for all x, y in L. Let H be the Lie group germ defined by the normed Lie algebra L. By Theorem 1, there exist an open subgroup germ $S \subset V$ of H and an isomorphism ϕ' of S onto an open subgroup germ of G. As we know already that (v) $\Rightarrow$ (iv), the mapping ϕ'_* of TS(L) into U(G) is the canonical mapping of TS(L) into the enveloping algebra of L. Thus $\phi_*(t) = \phi'_*(t)$ for all $t \in T_0^{(\infty)}(L)$. As ϕ and ϕ' are analytic, ϕ and ϕ' coincide on a neighbourhood of 0.

DEFINITION 1. *Let* G *be a Lie group germ and* L *its Lie algebra. An exponential mapping of* G *is any analytic mapping* ϕ *defined on an open neighbourhood of* 0 *in* L, *with values in* G *and satisfying the conditions of Theorem* 4.

Theorem 4 implies immediately that, for every Lie group germ G, *there exists an exponential mapping of* G and that *two exponential mappings of* G *coincide on a neighbourhood of* 0.

Examples. (1) Let G be the additive group of a complete normable space E. The canonical isomorphism of L(G) onto E satisfies condition (i) of Theorem 4 and is therefore an exponential mapping of G.

(2) Let A be a complete normed unital associative algebra. Let A* be the Lie group consisting of the invertible elements of A. We identify L(A*) with A (§ 3, no. 9, Corollary to Proposition 33). If K = **R** or **C**, we know that the mapping exp of A into A* defined in Chapter II, § 7, no. 3 satisfies condition (i) of Theorem 4 and hence is an exponential mapping. Now let K be ultrametric. Let p be the characteristic of the residue field of K. If $p \neq 0$, let $\lambda = |p|^{1/(p-1)}$; if $p = 0$, let $\lambda = 1$. Let U be the set of $x \in A$ such that $\|x\| < \lambda$. We know (Chapter II, § 8, no. 4) that the mapping exp of U into A* satisfies condition (i) of Theorem 4 and hence is an exponential mapping. Note that U is an additive subgroup of A.

This example explains the terminology adopted in Definition 1.

Let G be a Lie group germ and ϕ an exponential mapping of G. Then ϕ is étale at 0 and hence there exists an open neighbourhood U of 0 in L(G) such that $\phi(U)$ is open in G and $\phi|U$ is an isomorphism of the analytic manifold U onto the analytic manifold $\phi(U)$.

A *canonical chart* (*of the first species*) on G is a chart ψ on the analytic manifold G whose inverse mapping is an exponential mapping. If further G is finite-dimensional and a basis of L(G) is chosen, the coordinate system defined by ψ and this basis in the domain of ψ is called a *canonical coordinate system* (*of the first species*).

PROPOSITION 3. *Let* G *be a Lie group germ,* L *its Lie algebra and* ϕ *an exponential mapping of* G. *Let* $L_1, \dots, L_n$ *be vector subspaces of* L *such that* L *is the topological direct sum of* $L_1, \dots, L_n$. *The mapping*

$$(b_1, b_2, \dots, b_n) \mapsto \theta(b_1, b_2, \dots, b_n) = \phi(b_1)\phi(b_2)\dots\phi(b_n),$$

defined on an open subset of $L_1 \times L_2 \times \dots \times L_n$, *is analytic. The tangent mapping at* $(0, 0, \dots, 0)$ *to* θ *is the canonical mapping of* $L_1 \times \dots \times L_n$ *into* L.

Let k_i be the canonical injection of L_i into $L_1 \times L_2 \times \dots \times L_n$. Then, for all $b \in L_i$, $(T_{(0,\dots,0)}\theta)(T_0 k_i)(b) = (T_0\phi)(b) = b$ and hence $(T_{(0,\dots,0)}\theta)|L_i$ is the canonical injection of L_i into L.

In particular, θ is étale at $(0, 0, \dots, 0)$. Its restriction to a sufficiently small open neighbourhood U of $(0, 0, \ldots, 0)$ has an open image in G and is an isomorphism of the manifold U onto the manifold $\theta(U)$. The inverse mapping η of $\theta(U)$ onto U is called a *canonical chart of the second species* of G, associated with the given decomposition of L as a direct sum. If further G is finite-dimensional and each L_i is generated by a non-zero vector e_i, the coordinate system on $\theta(U)$ defined by η and the e_i is called a *canonical coordinate system of the second species*.

PROPOSITION 4. *Let G be a Lie group germ and ϕ an injective exponential mapping of G. For all x, y in L(G),*

$$(1) \qquad x + y = \lim_{\lambda \in K^*, \lambda \to 0} \lambda^{-1}\phi^{-1}(\phi(\lambda x)\phi(\lambda y))$$

$$(2) \qquad [x, y] = \lim_{\lambda \in K^*, \lambda \to 0} \lambda^{-2}\phi^{-1}(\phi(\lambda x)\phi(\lambda y)\phi(-\lambda x)\phi(-\lambda y))$$

(note that $\phi^{-1}(\phi(\lambda x)\phi(\lambda y))$ and $\phi^{-1}(\phi(\lambda x)\phi(\lambda y)\phi(-\lambda x)\phi(-\lambda y))$ are defined for $|\lambda|$ sufficiently small).

Let $L = L(G)$ be given a norm defining the topology of L and such that $\|[x, y]\| \leqslant \|x\| \, \|y\|$ for all x, y in L. Using Theorems 2 and 4, it can be assumed that G is the Lie group germ defined by L and that $\phi = \mathrm{Id}_G$. Let $(x, y) \mapsto x.y$ denote the product in the group G. The formulae to be proved can then be written

$$(3) \qquad x + y = \lim_{\lambda \in K^*, \lambda \to 0} \lambda^{-1}((\lambda x).(\lambda y))$$

$$(4) \qquad [x, y] = \lim_{\lambda \in K^*, \lambda \to 0} \lambda^{-2}((\lambda x).(\lambda y).(-\lambda x).(-\lambda y)).$$

There exists an open neighbourhood V of 0 in K such that the function

$$\lambda \mapsto f(\lambda) = (\lambda x).(\lambda y)$$

is defined and analytic on V. By Chapter II, § 6, no. 4, *Remark* 2, the expansion of f as an integral series about the origin is

$$\lambda(x + y) + \tfrac{1}{2}\lambda^2[x, y] + \cdots$$

and this proves (3). On the other hand, for u, v in G and $\|u\|$, $\|v\|$ sufficiently small, $u.v$ is an analytic function of (u, v) and the terms of degrees 1 and 2 in the expansion of this function as an integral series about the origin are $u + v + \frac{1}{2}[u, v]$. By *Differentiable and Analytic Manifolds*, R, 3.2.7 and 4.2.3, the terms of degrees 1 and 2 in the expansion of the function $f(\lambda).f(-\lambda)$ as an integral series about the origin are the terms of degrees 1 and 2 in

$$f(\lambda) + f(-\lambda) + \tfrac{1}{2}[f(\lambda), f(-\lambda)]$$

or also in

$$\lambda(x + y) + \tfrac{1}{2}\lambda^2[x, y] - \lambda(x + y) + \tfrac{1}{2}\lambda^2[x, y] + \tfrac{1}{2}[\lambda(x + y), -\lambda(x + y)] = \lambda^2[x, y]$$

and this proves (4).

PROPOSITION 5. *Let* G *be a Lie group, k a non-discrete closed subfield of* K, G′ *the group* G *considered as a Lie group over k and* ϕ (resp. ϕ') *an exponential mapping of* G (resp. G′). *Then* ϕ *and* ϕ' *coincide on a neighbourhood of* 0.

ϕ satisfies hypothesis (i) of Theorem 4 relative to G′ and is therefore an exponential mapping of G′.

PROPOSITION 6. *Let* G *be a Lie group germ,* L *its Lie algebra and* $\phi: V \to G$ *an exponential mapping of* G. *For all* $x \in V$, *let* $T_x(L)$ *be identified with* L, *so that the right differential* $\varpi(x)$ *of* ϕ *at* x *is a linear mapping of* L *into* L. *For* x *sufficiently close to* 0,

$$\varpi(x) = \sum_{n \geqslant 0} \frac{1}{(n+1)!} (\operatorname{ad} x)^n.$$

Let L be given a norm compatible with its topology and such that $\|[x, y]\| \leqslant \|x\| \, \|y\|$ for all $x, y \in L$. It suffices to consider the case where G is the Lie group germ defined by L and $\phi = \mathrm{Id}_G$. By definition, $\varpi(x)$ is then the tangent mapping at x to the mapping $y \mapsto y.x^{-1}$ of G into G. If H(X, Y) denotes the Hausdorff series, $\varpi(x)$ is therefore, for $\|x\|$ sufficiently small, the tangent mapping at 0 to the mapping $y \mapsto H(x, + y, -x)$ of G into G. In $H(X + Y, -X)$, the sum of terms of the first degree in Y is

$$\sum_{m \geqslant 0} \frac{1}{(m+1)!} (\operatorname{ad} X)^m Y$$

(Chapter II, § 6, no. 5, Proposition 5). The proposition then follows from *Differentiable and Analytic Manifolds*, R, 3.2.4 and 4.2.3.

Let G be a Lie group germ and $t \in K$. A *t-th power mapping of* G is any mapping, defined and analytic on an open neighbourhood of e, with values in G and coinciding on a neighbourhood of e with a mapping

$$g \mapsto \phi(t\phi^{-1}(g))$$

where ϕ is an injective exponential mapping of G.

PROPOSITION 7. (i) *If* $t \in \mathbf{Z}$, *a t-th power mapping coincides on a neighbourhood of e with the mapping* $g \mapsto g^t$.

(ii) *The tangent mapping at e to a t-th power mapping is the homothety of ratio t.*

(iii) *If h is a t-th power mapping and h' a t'-th power mapping of* G, $h \circ h'$ *is a (tt')-th power mapping and* $g \mapsto h(g)h'(g)$ *is a $(t + t')$-th power mapping.*

(iv) *If h is a t-th power mapping and* $u \in U^n(G)$, *then* $h_*(u) = t^n u$.

It suffices to prove the proposition when G is the Lie group germ defined by a complete normed Lie algebra and when the t-th power mappings considered are constructed using the exponential mapping $\phi = \mathrm{Id}_G$. But in that case everything is obvious.

4. FUNCTORIALITY OF EXPONENTIAL MAPPINGS

PROPOSITION 8. *Let* G *and* H *be Lie group germs, h a morphism of* G *into* H *and* ϕ_G *and* ϕ_H *exponential mappings of* G *and* H. *There exists a neighbourhood* V *of* 0 *in* L(G) *such that* $h \circ \phi_G$ *and* $\phi_H \circ L(h)$ *coincide on* V.

Let L(G) and L(H) be given norms defining their topologies and such that $\|[x, y]\| \leqslant \|x\| \|y\|$ for all x and y. It can be assumed that G (resp. H) is the Lie group germ defined by L(G) (resp. L(H)), so that ϕ_G (resp. ϕ_H) coincides with Id_G (resp. Id_H) in a neighbourhood of 0. On the other hand, there exists a symmetric open neighbourhood W of 0 in L(G) such that L(h) is a morphism of the Lie group germ W into H. By Theorem 1, L(h) coincides with h on a neighbourhood of 0, whence the proposition.

Loosely speaking, if G and H are identified in a neighbourhood of the identity element with L(G) and L(H) by means of exponential mappings, every morphism of G into H is *linear* in a neighbourhood of 0.

COROLLARY 1. *Let* G *be a Lie group germ,* G′ *a Lie subgroup germ of* G *and* ϕ *an exponential mapping of* G.

(i) *There exists an open neighbourhood* V *of* 0 *in* L(G′) *such that* $\phi | V$ *is an isomorphism of the manifold* V *onto an open neighbourhood of e in* G′.

(ii) *Let* $x \in L(G)$. *The following conditions are equivalent:* (a) $x \in L(G')$; (b) $\phi(\lambda x) \in G'$ *for* $|\lambda|$ *sufficiently small.*

(i) is obtained by applying Proposition 8 to the canonical injection of G′ into G and (ii) follows from (i).

COROLLARY 2. *Let* G *be a Lie group,* ρ *an analytic linear representation of* G *and* ϕ *an exponential mapping of* G. *There exists a neighbourhood* V *of* 0 *in* L(G) *such that*

$$\rho(\phi(x)) = \exp(L(\rho)x)$$

for all $x \in V$.

This follows from Proposition 8 and *Example* 2 of no. 3.

COROLLARY 3. *Let* G *be a Lie group and* ϕ *an exponential mapping of* G.

(i) *There exists a neighbourhood* V *of* 0 *in* L(G) *such that*

$$\mathrm{Ad}(\phi(x)) = \exp \operatorname{ad} x.$$

for all $x \in V$.

(ii) *If* $g \in G$, *there exists a neighbourhood* W *of* 0 *in* L(G) *such that*

$$g\phi(x)g^{-1} = (\mathrm{Ad}\, g . x)$$

for all $x \in W$.

(i) follows from Corollary 2 and § 3, no. 12, Proposition 44.

(ii) follows from Proposition 8 applied to Int g.

5. STRUCTURE INDUCED ON A SUBGROUP

Lemma 4. *Let* G *be a finite-dimensional Lie group,* Ω *a symmetric open neighbourhood of* e *in* G *and* H *a subset of* Ω *containing* e *such that the conditions* $x \in \mathrm{H}$, $y \in \mathrm{H}$, $xy^{-1} \in \Omega$ *imply* $xy^{-1} \in \mathrm{H}$. *Let* $r \in \mathrm{N}_{\mathrm{K}}$. *For all* $x \in \mathrm{H}$, *let* $\mathfrak{h}_x$ *be the set of* $a \in \mathrm{T}_x(\mathrm{G})$ *with the following property: there exist an open neighbourhood* I *of* 0 *in* K *and a mapping* f *of class* C^r *of* I *into* G *such that* $f(0) = x, f(\mathrm{I}) \subset \mathrm{H}$, $(\mathrm{T}_0 f)(1) = a$.

(i) *Let* $\mathfrak{h}_e = \mathfrak{h}$. *Then* $\mathfrak{h}$ *is a Lie subalgebra of* L(G) *which is invariant under* $\mathrm{Ad}_{\mathrm{L(G)}}(\mathrm{H})$.

(ii) $\mathfrak{h}_x = x\mathfrak{h} = \mathfrak{h}x$ *for all* $x \in \mathrm{H}$ (*where* $x\mathfrak{h}$ *and* $\mathfrak{h}x$ *are evaluated in* T(G)).

(iii) *Let* V *be a manifold of class* C^r, v_0 *a point of* V *and* f *a mapping of class* C^r *of* V *into* G *such that* $f(v_0) = e$ *and* $f(\mathrm{V}) \subset \mathrm{H}$. *For every Lie subgroup germ* H′ *of* G *with Lie algebra* $\mathfrak{h}$, $f(v) \in \mathrm{H}'$ *for* v *sufficiently close to* v_0.

(iv) *For every Lie subgroup germ* H′ *of* G *with Lie algebra* $\mathfrak{h}$, $\mathrm{H}' \cap \mathrm{H}$ *is a neighbourhood of* e *in* H′.

(v) *For all* $x \in \mathrm{H}$ *and all* $a \in \mathfrak{h}_x$, *there exist an open neighbourhood* I *of* 0 *in* K *and a mapping* f *of class* C^ω *of* I *into* G *such that* $f(0) = x, f(\mathrm{I}) \subset \mathrm{H}$, $(\mathrm{T}_0 f)(1) = a$.

Clearly $\mathrm{K}\mathfrak{h} = \mathfrak{h}$ and $x\mathfrak{h}_y z = \mathfrak{h}_{xyz}$ for x, y, xy, xyz in H. This implies (ii) and the fact that $\mathfrak{h}$ is invariant under $\mathrm{Ad}_{\mathrm{L(G)}}(\mathrm{H})$.

Let a_1, a_2 be in $\mathfrak{h}$. Let I be an open neighbourhood of 0 in K and f_1, f_2 mappings of class C^r of I into G such that $f_j(0) = e, f_j(\mathrm{I}) \subset \mathrm{H}$, $(\mathrm{T}_0 f_j)(1) = a_j$ $(j = 1, 2)$. We define $f: \mathrm{I} \to \mathrm{G}$ by $f(\lambda) = f_1(\lambda) f_2(\lambda)$. Then f is of class C^r and $f(0) = e$. By shrinking I if necessary, we have $f(\mathrm{I}) \subset \mathrm{H}$. On the other hand, the mapping of $\mathrm{T}_e(\mathrm{G}) \times \mathrm{T}_e(\mathrm{G})$ into $\mathrm{T}_e(\mathrm{G})$ tangent to the mapping $(g, g') \to gg'$ is addition; hence $(\mathrm{T}_0 f) = a_1 + a_2$. Hence $a_1 + a_2 \in \mathfrak{h}$ and $\mathfrak{h}$ is a vector subspace of L(G). Since $x\mathfrak{h}x^{-1} = \mathfrak{h}$ for all $x \in \mathrm{H}$, $(\mathrm{Ad}\, f_1(\lambda)) . a_2 \in \mathfrak{h}$ for all $\lambda \in \mathrm{I}$. The tangent mapping at 0 to the mapping $\lambda \mapsto \mathrm{Ad}\, f_1(\lambda)$ is, by Proposition 44 of § 3, no. 12, the mapping $\lambda \mapsto \mathrm{ad}(\lambda a_1)$; hence

$$[a_1, a_2] = (\mathrm{ad}\, a_1) . a_2 \in \mathfrak{h}$$

since $\mathfrak{h}$ is closed in L(G). Hence we have proved (i). In the rest of the proof we fix a Lie subgroup germ H′ of G with Lie algebra $\mathfrak{h}$.

Let V, v_0, f be as in (iii). Let Y be the left foliation of G associated with $\mathfrak{h}$ (no. 1). For all $y \in \mathrm{H}'$, $\mathrm{T}_y(\mathrm{H}') = y\mathfrak{h}$. On the other hand, for all $v \in \mathrm{V}$, the image of $\mathrm{T}_v(\mathrm{V})$ under $\mathrm{T}_v(f)$ is contained in $\mathfrak{h}_{f(v)} = f(v)\mathfrak{h}$ (by definition of $\mathfrak{h}_{f(v)}$). By *Differentiable and Analytic Manifolds*, R, 9.3.2, f is a morphism of V into Y. As H′ is a leaf of Y (*Differentiable and Analytic Manifolds*, R, 9.2.8), $f(v) \in \mathrm{H}'$ for v sufficiently close to v_0.

Let $(a_1, \ldots, a_s)$ be a basis of $\mathfrak{h}$. There exist an open neighbourhood I of 0 in K and mappings $f_1, \ldots, f_s$ of class C^r of I into G such that $f_j(0) = e$, $f_j(\mathrm{I}) \subset \mathrm{H}$, $(\mathrm{T}f_j)1 = a_j$ for all j. By (iii), $f_j(\lambda) \in \mathrm{H}'$ for $|\lambda|$ sufficiently small. Hence the $f_1(\lambda_1) f_2(\lambda_2) \ldots f_s(\lambda_s)$ constitute, for $|\lambda_1|, |\lambda_2|, \ldots, |\lambda_s|$ sufficiently small, a neighbourhood of e in H'; and this neighbourhood is contained in H. Hence (iv).

If $a \in \mathfrak{h}$, there exist an open neighbourhood I of 0 in K and a mapping f of class C^ω of I into G such that $f(0) = e$, $f(I) \subset H'$, $(T_0 f)1 = a$. This, with (iv), implies (v).

Definition 2. *$\mathfrak{h}$ is called the subalgebra tangent to* H *at* e.

Proposition 9. *Let* G *be a finite-dimensional Lie group and* H *a subgroup of* G.

(i) *There exists on* H *one and only one analytic manifold structure with the following property: for all r between* 1 *and* ω, *for every manifold* V *of class* C^r *and for every mapping f of* V *into* H, *f is of class* C^r *as a mapping of* V *into* H *if and only if f is of class* C^r *as a mapping of* V *into* G.

(ii) *With this structure,* H *is a Lie group, the canonical injection i of* H *into* G *is an immersion and* $L(i)(L(H))$ *is the Lie subalgebra tangent at e to* H.

In (i) the uniqueness is obvious. We prove the existence. Let $\mathfrak{h}$ be the Lie algebra tangent at e to H. Let H' be a Lie subgroup germ of G with Lie algebra $\mathfrak{h}$. By replacing H' by an open subgroup germ of H', it can be assumed that $H' \subset H$ (Lemma 4 (iv)). For all $x \in H$, $xH'x^{-1}$ is a Lie subgroup germ of G with Lie algebra $x\mathfrak{h}x^{-1} = \mathfrak{h}$. Hence $H' \cap (xH'x^{-1})$ is open in H' (no. 2, Theorem 3) and the mapping $y \mapsto xyx^{-1}$ is an isomorphism of $H' \cap x^{-1}H'x$ onto $xH'x^{-1} \cap H'$. Using Proposition 18 of § 1, no. 9, there exist an open Lie subgroup germ W of H' and a Lie group structure on H with the following properties: W is open in H and the manifold structures on H and H' induce the same structure on W. It then follows that the canonical injection i of H into G is an immersion and that $L(i)(L(H)) = L(H') = \mathfrak{h}$. Moreover, let V and f be as in (i). If $f: V \to H$ is of class C^r, $i \circ f: V \to G$ is of class C^r. Suppose that $i \circ f: V \to G$ is of class C^r; then we prove that $f: V \to H$ is of class C^r. By translation, it suffices to consider the case where there exists $v_0 \in V$ such that $f(v_0) = e$ and to prove that $f: V \to H$ is of class C^r in an open neighbourhood of v_0. Now, by Lemma 4 (iii), $f(v) \in H'$ for v sufficiently close to v_0, whence our assertion. Thus we have proved (i) and (ii) has been obtained on the way.

Definition 3. *The Lie group structure on* H *defined in Proposition* 9 *is called the structure induced on* H *by the Lie group structure on* G.

If H is a Lie subgroup of G, its Lie group structure is induced by that on G (*Differentiable and Analytic Manifolds*, R, 5.8.5).

If $G = \mathbf{R}$ and $H = \mathbf{Q}$, then $\mathfrak{h} = \{0\}$ and hence the structure induced on H is the discrete Lie group structure. Similarly if $G = \mathbf{C}$ (considered as a complex Lie group) and $H = \mathbf{R}$.

6. PRIMITIVES OF DIFFERENTIAL FORMS WITH VALUES IN A LIE ALGEBRA

Lemma 5. *Let* X *be a manifold of class* C^r, F *and* F′ *vector bundles of class* C^r *with base space* X *and* ϕ *a morphism of* F *into* F′. *For all* $x \in X$, *let* S_x *be the set of*

$$(a, \phi(a)) \in F_x \oplus F'_x$$

for $a \in F_x$. *Then the union* S *of the* S_x *is a vector subbundle of* $F \oplus F'$.

Let θ and θ' be the mappings of $F \oplus F'$ into itself defined as follows: if $(u, v) \in F_x \oplus F'_x$, then

$$\theta(u, v) = (u, v + \phi(u)), \qquad \theta'(u, v) = (u, v - \phi(u)).$$

By *Differentiable and Analytic Manifolds*, R, 7.7.1, θ and θ' are morphisms of $F \oplus F'$ into itself. Clearly $\theta \circ \theta' = \theta' \circ \theta = \mathrm{Id}_{F \oplus F'}$. Hence θ and θ' are automorphisms of $F \oplus F'$. Therefore, $S = \theta(F \oplus \{0\})$ is a vector subbundle of $F \oplus F'$.

Lemma 6. *Let* G *be a Lie group germ,* ω *the canonical left differential form of* G (§ 3, no. 18.9), M *a manifold of class* C^r $(r \geqslant 2)$ *and* α *a differential form of class* C^{r-1} *and degree* 1 *on* M *with values in* L(G).

(i) *The elements of* T(M × G) *at which the differential form*

$$\theta = \mathrm{pr}_1^*\alpha - \mathrm{pr}_2^*\omega$$

is zero constitute a vector subbundle S *of* T(M × G) *of class* C^{r-1}.

(ii) *For all* $(x, g) \in M \times G$, $T(\mathrm{pr}_1)|S_{(x,g)}$ *is an isomorphism of* $S_{(x,g)}$ *onto* $T_x(M)$.

(iii) *If* $d\alpha + [\alpha]^2 = 0$ (cf. § 3, no. 14) *the vector subbundle* S *is integrable.*

If $(x, g) \in M \times G$ and $(u, v) \in T_x(M) \times T_g(G)$, then

$$\theta_{(x,g)}(u, v) = \alpha(u) - g^{-1}v.$$

Hence the kernel of $\theta_{(x,g)}$ is the set $S_{(x,g)}$ of $(u, g\alpha(u))$ for $u \in T_x(M)$, whence (ii). We consider T(M × G) as the direct sum of two vector bundles F and F′ with $F_{(x,g)} = T_x(M) \times \{0\}$ and $F'_{(x,g)} = \{0\} \times T_g(G)$ for all

$$(x, g) \in M \times G.$$

For $u \in T_x(M) \times \{0\}$, we write $\phi(u) = (0, g\alpha(u))$. Using the left trivialization of T(G), it is seen that ϕ is a morphism of F into F′, whence (i) (Lemma 5). Finally, if $a\alpha + [\alpha]^2 = 0$, then

$$\begin{aligned} d\theta &= \mathrm{pr}_1^*(d\alpha) - \mathrm{pr}_2^*(d\omega) \\ &= -\tfrac{1}{2}(\mathrm{pr}_1^*\alpha \wedge \mathrm{pr}_1^*\alpha - \mathrm{pr}_2^*\omega \wedge \mathrm{pr}_2^*\omega) \\ &= -\tfrac{1}{2}(\mathrm{pr}_1^*\alpha - \mathrm{pr}_2^*\omega) \wedge (\mathrm{pr}_1^*\alpha + \mathrm{pr}_2^*\omega) \\ &= -\tfrac{1}{2}\theta \wedge (\mathrm{pr}_1^*\alpha + \mathrm{pr}_2^*\omega) \end{aligned}$$

and hence S is integrable (*Differentiable and Analytic Manifolds*, R, 9.3.6).

THEOREM 5. *Let* G *be a Lie group germ,* M *a manifold of class* C^r $(r \geqslant 2)$ *and* α *a differential form of class* C^{r-1} *and degree* 1 *on* M *with values in* L(G), *such that* $d\alpha + [\alpha]^2 = 0$. *For all* $x \in M$ *and all* $g \in G$, *there exists a mapping* f, *defined and of class* C^{r-1} *on an open neighbourhood of* x, *with values in* G, *such that* $f(x) = g$ *and* $f^{-1}.df = \alpha$. *Two mappings which satisfy these conditions coincide on a neighbourhood of* x.

Let $x \in M$ and $g \in G$. By Lemma 6 (whose notation we adopt) and *Differentiable and Analytic Manifolds*, R, 9.3.7, there exist an open neighbourhood U of x in M and a mapping $m \mapsto \phi(m) = (m, f(m))$ of class C^{r-1} of U into $M \times G$ such that $f(x) = g$ and $\phi^*(\theta) = 0$. Then

$$\begin{aligned} f^{-1}.df &= f^*(\omega) && (\S\ 3, \text{no. } 18.9)\\ &= (\mathrm{pr}_2 \circ \phi)^*(\omega) && (\text{for } f = \mathrm{pr}_2 \circ \phi)\\ &= \phi^*(\mathrm{pr}_1^* \alpha - \theta) && (\text{Lemma 6})\\ &= \phi^*(\mathrm{pr}_1^* \alpha) && (\text{for } \phi^*(\theta) = 0)\\ &= \alpha && (\text{for } \mathrm{pr}_1 \circ \phi = \mathrm{Id}_U). \end{aligned}$$

Let f' be a mapping of class C^{r-1} of U into G such that $f'(x) = g$ and $f'^{-1}df' = \alpha$. By § 3, 18.9, ff'^{-1} is locally constant and hence $f' = f$ in a neighbourhood of x.

PROPOSITION 10. *Let* M *be an analytic manifold,* $\mathfrak{g}$ *a complete normable Lie algebra and* α *an analytic differential form of degree* 1 *on* M, *with values in* $\mathfrak{g}$, *with the following properties:*

(a) *for all* $m \in M$, α_m *is an isomorphism of* $T_m(M)$ *onto* $\mathfrak{g}$;

(b) $d\alpha + [\alpha]^2 = 0$.

Then, for all $m_0 \in M$, *there exist an open neighbourhood* M′ *of* m_0 *in* M *and a Lie group germ structure on* M′, *compatible with the manifold structure on* M′, *with identity element* m_0 *and with the following properties:*

(i) α_{m_0} *is an isomorphism of* L(M′) *onto* $\mathfrak{g}$;

(ii) *the differential form* $m \mapsto \alpha_{m_0}^{-1} \circ \alpha_m$ *is the left canonical differential form of* M′.

If M'_1 *and* M'_2 *are two such group germs,* M'_1 *and* M'_2 *have a common open subgroup germ.*

There exists a Lie group germ G such that $L(G) = \mathfrak{g}$. Let $m_0 \in M$. By Theorem 5, there exist an open neighbourhood M′ of m_0 in M and an analytic mapping f of M′ into G such that $f(m_0) = e$ and $f^{-1}.df = \alpha$. Then $T_{m_0}(f) = \alpha_{m_0}$ is an isomorphism of $T_{m_0}(M)$ onto $\mathfrak{g}$; hence, shrinking M′ and G, it can be assumed that f is an isomorphism of the manifold M′ onto the manifold G. We transport to M′ the Lie group germ structure on G by means of f^{-1}. Then $T_{m_0}(f)$ becomes an isomorphism of L(M′) onto $L(G) = \mathfrak{g}$, whence (i). On the other hand, if ω denotes the left canonical differential form of G, then

$$\begin{aligned}\alpha_{m_0}^{-1} \circ \alpha_m &= (\mathrm{T}_{m_0} f)^{-1} \circ (f^{-1}.df)(m) \\ &= (\mathrm{T}_m f)^{-1} \circ \omega(f(m)) \circ \mathrm{T}_m f\end{aligned}$$

and hence $m \mapsto \alpha_{m_0}^{-1} \circ \alpha_m$ is the left canonical differential form of M'.

Let M″ be an open neighbourhood of m_0, with a Lie group germ structure, with identity element m_0 and with the analogous properties to properties (i) and (ii). Then α_{m_0} is an isomorphism of L(M') onto $\mathfrak{g}$ and also of L(M″) onto $\mathfrak{g}$ and hence L(M') = L(M″). Therefore, shrinking M' and M″, it can be assumed that there exists an isomorphism ϕ of the group germ M' onto the group germ M″ (no. 1, Corollary 1 to Theorem 1). Then $\phi^{-1}.d\phi$ is the canonical left differential of M'. On the other hand, let ψ be the canonical injection of the manifold $\mathrm{M}' \cap \mathrm{M}''$ into the Lie group germ M″; clearly $\psi^{-1}.d\psi$ is a restriction of the canonical left differential of M″. Hence $(\psi^{-1}.d\psi)(m) = \alpha_{m_0}^{-1} \circ \alpha_m = (\phi^{-1}.d\phi)(m)$ for all $m \in \mathrm{M}' \cap \mathrm{M}''$. Therefore ϕ and ψ coincide on a neighbourhood of m_0 (§ 3, 18.9). This proves the last assertion of the proposition.

COROLLARY. *Let* M *be an analytic manifold of finite dimension n. Let* $\omega_1, \ldots, \omega_n$ *be analytic differential forms of degree* 1 *on* M, *with scalar values, which are linearly independent at each point of* M *and such that, for all* $k = 1, \ldots, n$, $d\omega_k$ *is a linear combination with constant coefficients of the* $\omega_i \wedge \omega_j$. *Then, for all* $m_0 \in \mathrm{M}$, *there exists an open neighbourhood* M' *of* m_0 *in* M *and a Lie group germ structure on* M' *compatible with the manifold structure on* M', *with identity element* m_0 *and such that* $\omega_1|\mathrm{M}', \ldots, \omega_n|\mathrm{M}'$ *form a basis of the space of left invariant differential forms on* M' *of degree* 1 *with scalar values.*

If M'_1 *and* M'_2 *are two such group germs,* M'_1 *and* M'_2 *have a common open subgroup germ.*

Let $\mathrm{X}_1, \ldots, \mathrm{X}_n$ be the vector fields on M such that, at each point m of M, the $(\mathrm{X}_i)_m$ constitute the basis of $\mathrm{T}_m(\mathrm{M})$ dual to $((\omega_1)_m, \ldots, (\omega_n)_m)$. These fields are analytic. By hypothesis, there exist $c_{ijk} \in \mathrm{K}$ $(1 \leqslant i, j, k \leqslant n)$ such that $c_{ijk} = -c_{jik}$ and $d\omega_k = \sum_{i<j} c_{ijk}\omega_i \wedge \omega_j$. By *Differentiable and Analytic Manifolds*, R, 8.5.7, formula (11),

$$\langle[\mathrm{X}_i, \mathrm{X}_j], \omega_k\rangle = -(d\omega_k)(\mathrm{X}_i, \mathrm{X}_j) = -\Big(\sum_{r<s} c_{rsk}\omega_r \wedge \omega_s\Big)(\mathrm{X}_i, \mathrm{X}_j) = -c_{ijk}$$

and hence $[\mathrm{X}_i, \mathrm{X}_j] = -\sum_k c_{ijk}\mathrm{X}_k$. It then follows that the $-c_{ijk}$ are the constants of structure of a Lie algebra $\mathfrak{g}$ relative to a basis $(e_1, \ldots, e_n)$. For all $m \in \mathrm{M}$, let α_m be the linear mapping of $\mathrm{T}_m(\mathrm{M})$ into $\mathfrak{g}$ which maps $(\mathrm{X}_1)_m$ to $e_1, \ldots, (\mathrm{X}_n)_m$ to e_n. Then α is an analytic differential form on M of degree

1 with values in $\mathfrak{g}$ and α_m is an isomorphism of $T_m(M)$ onto $\mathfrak{g}$. On the other hand, $\alpha = \sum_{k=1}^{n} \omega_k e_k$ and hence

$$d\alpha = \sum_{k=1}^{n} (d\omega_k)e_k = \sum_{k=1}^{n} \Big(\sum_{i<j} c_{ijk}\omega_i \wedge \omega_j\Big)e_k$$

and

$$(5) \qquad [\alpha]^2 = \sum_{k=1}^{n} [\omega_k e_k]^2 + \sum_{i<j} (\omega_i e_i) \wedge (\omega_j e_j) \quad (\S\ 3, \text{formula } (30))$$
$$= \sum_{i<j} (\omega_i \wedge \omega_j)[e_i, e_j]$$
$$= -\sum_{k=1}^{n}\sum_{i<j} (c_{ijk}\omega_i \wedge \omega_j)e_k$$
$$= -d\alpha.$$

It then suffices to apply Proposition 10.

7. PASSAGE FROM LAWS OF INFINITESIMAL OPERATION TO LAWS OF OPERATION

PROPOSITION 11. *Let G_1 and G_2 be Lie group germs and X_1 and X_2 manifolds of class C^r ($r \geqslant 2$). For $i = 1, 2$, let ψ_i be a law chunk of left operation of class C^r of G_i on X_i and D_i the associated law of infinitesimal operation. Let $\mu: G_1 \to G_2$ be a morphism and $\phi: X_1 \to X_2$ a mapping of class C^r. Suppose that, for all $a \in L(G)$, the vector fields $(D_1)_a$ and $(D_2)_{L(\mu)a}$ are ϕ-related (Differentiable and Analytic Manifolds, R, 8.2.6). Then there exists a neighbourhood Ω of $\{e\} \times X_1$ in $G_1 \times X_1$ such that $\phi(\psi_1(g, x)) = \psi_2(\mu(g), \phi(x))$ for all $(g, x) \in \Omega$.*

Let $p_1: G_1 \times X_2 \to G_1$, $p_2: G_1 \times X_2 \to X_2$ be the canonical projections. For all $(g_1, x_2) \in G_1 \times X_2$, let f_{g_1,x_2} be the mapping $g_1 a \mapsto (D_2)_{L(\mu)a}(x_2)$ of $T_{g_1}(G_1) = g_1 L(G_1)$ into $T_{x_2}(X_2)$. The f_{g_1,x_2} define a morphism of $p_1^*T(G_1)$ into $p_2^*T(X_2)$.

Let $x_0 \in X_1$. There exist an open neighbourhood G of e in G_1 and an open neighbourhood X of x_0 in X_1 such that $\psi_1(g, x)$ and $\psi_2(\mu(g), \phi(x))$ are defined for $(g, x) \in G \times X$. We write, for $(g, x) \in G \times X$,

$$\alpha(g, x) = \phi(\psi_1(g, x)) \in X_2, \qquad \beta(g, x) = \psi_2(\mu(g), \phi(x)) \in X_2.$$

If G and X are sufficiently small, then, for all $(a, g, x) \in L(G_1) \times G \times X$,

$$\begin{aligned}(T\alpha)(ag, 0_x) &= (T\phi)((D_1)_a(\psi_1(g, x)))\\ &= (D_2)_{L(\mu)a}(\phi(\psi_1(g, x)))\\ &= (D_2)_{L(\mu)a}\alpha(g, x),\\ (T\beta)(ag, 0_x) &= (T\psi_2)(L(\mu)a.\mu(g), \phi(x))\\ &= (D_2)_{L(\mu)a}(\psi_2(\mu(g), \phi(x)))\\ &= (D_2)_{L(\mu)a}\beta(g, x).\end{aligned}$$

Hence for $x \in X$, the morphisms $g \mapsto \alpha(g, x)$ and $g \mapsto \beta(g, x)$ are integrals of f; as

$$\beta(e, x) = \phi(x) = \alpha(e, x)$$

for all $x \in X$, it follows from *Differentiable and Analytic Manifolds*, R, 9.3.7, that α and β coincide on a neighbourhood of (e, x_0). Hence the proposition.

COROLLARY. *Let G be a Lie group germ and X a manifold of class C^r. Consider two law chunks of left operation of class C^r of G on X. Suppose that, for all $a \in L(G)$, the corresponding vector field D_a on X is the same for both law chunks. Then these two law chunks coincide on a neighbourhood of $\{e\} \times X$.*

THEOREM 6. *Let G be a Lie group germ, X a manifold of class C^r $(r \geqslant 2)$ and x_0 a point of X. Let $a \mapsto D_a$ be a law of left infinitesimal operation of class C^{r-1} of L(G) on X.*

(i) *There exists an open neighbourhood X′ of x_0 in X and a law chunk of left operation of class C^{r-1} of G on X′ such that the associated law of infinitesimal operation is $a \mapsto D_a|X'$.*

(ii) *Let there be two law chunks of left operation of class C^{r-1} of G on an open neighbourhood X″ of x_0; if they admit $a \mapsto D_a|X''$ as associated law of infinitesimal operation, they coincide on a neighbourhood of (e, x_0).*

Assertion (ii) follows from the Corollary to Proposition 11. We prove (i). For all $(g, x) \in G \times X$ and all $a \in L(G)$, we write

$$Q_a(g, x) = (ag, D_a(x)) \in T_g(G) \times T_x(X).$$

Let $S_{(g,x)}$ be the set of $Q_a(g, x)$ for $a \in L(G)$. By Lemma 5 of no. 6, the $S_{(g,x)}$ are the fibres of a vector subbundle S of $T(G) \times T(X)$. Let a, b be in L(G); then

$$\begin{aligned}[Q_a, Q_b](g, x) &= ([R_a, R_b](g), [D_a, D_b](x)) \\ &= (-R_{[a,b]}(g), -D_{[a,b]}(x)) \qquad (\S\ 3, 18.6) \\ &= Q_{-[a,b]}(g, x)\end{aligned}$$

and hence S is integrable (*Differentiable and Analytic Manifolds*, R, 9.3.3, (iv)).

By *Differentiable and Analytic Manifolds*, R, 9.3.7, there exist an open neighbourhood G_1 of e in G, an open neighbourhood X_1 of x_0 in X and a mapping $(g, x) \mapsto gx$ of class C^{r-1} of $G_1 \times X_1$ into X such that $ex = x$ for all $x \in X_1$ and

(6) $$(ag)x = D_a(gx) \quad \text{for } a \in L(G), g \in G_1, x \in X_1.$$

In particular

(7) $$ax = D_a(x).$$

Let G_2 be an open neighbourhood of e in G_1 and X_2 an open neighbourhood of x_0 in X_1 such that gg' is defined and belongs to G_1 for g, g' in G_2 and gx

is defined and belongs to X_1 for $(g, x) \in G_2 \times X_2$. Consider the mappings α_1, α_2 of $G_2 \times (G_2 \times X_2)$ into X defined by

$$\alpha_1(g, (h, x)) = g(hx), \qquad \alpha_2(g, (h, x)) = (gh)x.$$

They are of class C^{r-1}. Then

$$\alpha_1(e, (h, x)) = hx = \alpha_2(e, (h, x)).$$

On the other hand

$$\begin{aligned} T(\alpha_1)(ag, 0_{(h,x)}) &= (ag)(hx) \\ &= D_a(g(hx)) \qquad \text{by (6)} \\ &= D_a(\alpha_1(g, (h, x))), \\ T(\alpha_2)(ag, 0_{(h,x)}) &= (agh)x \\ &= D_a((gh)x) \qquad \text{by (6)} \\ &= D_a(\alpha_2(g, (h, x))). \end{aligned}$$

By *Differentiable and Analytic Manifolds*, R, 9.3.7, α_1 and α_2 coincide on a neighbourhood of $(e, (e, x_0))$. Then (i) follows from (7) and Proposition 23 of § 1, no. 11.

COROLLARY 1. *Let G be a Lie group germ and X a paracompact manifold of class C^r $(r \geqslant 2)$. Let $a \mapsto D_a$ be a law of left infinitesimal operation of class C^{r-1} of L(G) on X.*

(i) *There exists a law chunk of left operation of class C^{r-1} of G on X such that the associated law of infinitesimal operation is $a \mapsto D_a$.*

(ii) *Two laws of left operation of class C^{r-1} of G on X which admit $a \mapsto D_a$ as associated law of infinitesimal operation coincide on a neighbourhood of $\{e\} \times X$.*

Assertion (ii) follows from the Corollary to Proposition 11. By Theorem 6 (i), there exist an open covering $(X_i)_{i \in I}$ of X and, for all $i \in I$, a law chunk of left operation ψ_i of Class C^{r-1} of G on X_i such that the associated law of infinitesimal operation is $a \mapsto D_a|X_i$. As X is paracompact, the covering $(X_i)_{i \in I}$ can be assumed to be locally finite. For all $(i, j) \in I \times I$ and all $x \in X_i \cap X_j$, ψ_i and ψ_j coincide on a neighbourhood of (e, x) (Corollary to Proposition 11). As X is normal, we can apply Proposition 24 of § 1, no. 11, which proves (i).

COROLLARY 2. *Let X be a paracompact manifold of class C^r $(r \geqslant 2)$ and ξ a vector field of class C^{r-1} on X. There exists a law chunk of operation ψ of class C^{r-1} of K on X such that for all $x \in X$, $\xi(x)$ is the image under $t \mapsto (t, x)$ of the tangent vector 1 to K at 0. Two law chunks of operation with the above property coincide on a neighbourhood of $\{0\} \times X$.*

This is a special case of Corollary 1.

Remark. Laws of left operation can of course be replaced, throughout this no., by laws of right operation.

§ 5. FORMAL CALCULATIONS IN LIE GROUPS

Let f, g be two formal power series with coefficients in K in the same indeterminates, let f_i (resp. g_i) be the homogeneous component of f (resp. g) of degree i. We shall write

$$f \equiv g \mod \deg p$$

if $f_i = g_i$ for $i < p$.

In this §, G denotes a Lie group germ of finite dimension n; the base field K is assumed to be of characteristic zero. We identify once and for all, by means of a chart, an open neighbourhood of e in G with an open neighbourhood U of 0 in K^n, so that e is identified with 0. For x, y in U and $n \in \mathbf{Z}$, $x.y$ denotes the product of x and y and $x^{[m]}$ the m-th power of x in G (when they are defined). The coordinates of $x \in U$ are denoted by $x_1, x_2, \ldots, x_n$.

1. THE COEFFICIENTS $c_{\alpha\beta\gamma}$

Let Ω be the set of $(x, y) \in U \times U$ such that $x.y$ is defined and belongs to U. Then Ω is open in $U \times U$ and the mapping $(x, y) \mapsto x.y$ of Ω into U is analytic. The coordinates $z_1, \ldots, z_n$ of $z = x.y$ therefore admit expansions as integral series about the origin in the powers of $x_1, \ldots, x_n, y_1, \ldots, y_n$. Therefore, there exist well defined constants $c_{\alpha_1, \ldots, \alpha_n, \beta_1, \ldots, \beta_n, \gamma_1, \ldots, \gamma_n} \in K$, such that

$$(1) \qquad z_1^{\gamma_1} \ldots z_n^{\gamma_n} = \sum_{\alpha_1, \ldots, \beta_n \in \mathbf{N}} c_{\alpha_1, \ldots, \alpha_n, \beta_1, \ldots, \beta_n, \gamma_1, \ldots, \gamma_n} x_1^{\alpha_1} \ldots x_n^{\alpha_n} y_1^{\beta_1} \ldots y_n^{\beta_n}$$

for $\gamma_1, \ldots, \gamma_n$ in $\mathbf{N}$. Adopting the conventions of *Differentiable and Analytic Manifolds*, R, we shall write these formulae more briefly:

$$(2) \qquad (x.y)^\gamma = \sum_{\alpha, \beta \in \mathbf{N}^n} c_{\alpha, \beta, \gamma} x^\alpha y^\beta \quad (\gamma \subset \mathbf{N}^n).$$

Since $x.0 = 0.x = x$ for $x \in U$,

$$(3) \qquad c_{\alpha, 0, \gamma} = c_{0, \alpha, \gamma} = \delta_{\alpha\gamma}$$

where $\delta_{\alpha\gamma}$ is the Kronecker index. In particular, writing henceforth k instead of ε_k for $k = 1, \ldots, n$,

$$(4) \qquad (x.y)_k = x_k + y_k + \sum_{|\alpha| \geqslant 1, |\beta| \geqslant 1} c_{\alpha, \beta, k} x^\alpha y^\beta.$$

Writing $c_{\alpha\beta} = (c_{\alpha\beta 1}, c_{\alpha\beta 2}, \ldots, c_{\alpha\beta n}) \in K^n$, it then follows that

$$(5) \qquad x.y = x + y + \sum_{|\alpha| \geqslant 1, |\beta| \geqslant 1} c_{\alpha\beta} x^\alpha y^\beta.$$

On the right hand side of (5), we consider the homogeneous component of degree 2:

$$B(x, y) = \sum_{i,j=1}^{n} c_{ij} x_i y_j$$

so that $(x, y) \mapsto B(x, y)$ is a bilinear mapping of $K^n \times K^n$ into K^n. Then

$$x.y \equiv x + y + B(x, y) \mod \deg 3. \tag{6}$$

Formula (4) implies on the other hand that

$$c_{\alpha,\beta,\gamma} = 0 \quad \text{if } |\alpha| + |\beta| < |\gamma| \tag{7}$$

and that the terms of total degree $|\gamma|$ in the expansion of z^γ are also those of $(x_1 + y_1)^{\gamma_1}(x_2 + y_2)^{\gamma_2} \dots (x_n + y_n)^{\gamma_n} = \sum_{\alpha+\beta=\gamma} ((\alpha, \beta)) x^\alpha y^\beta$ (cf. *Differentiable and Analytic Manifolds*, R, Notation and conventions). Hence:

$$c_{\alpha,\beta,\gamma} = 0 \quad \text{if} \quad |\alpha| + |\beta| = |\gamma| \quad \text{but} \quad \alpha + \beta \neq \gamma \tag{8}$$

$$c_{\alpha,\beta,\alpha+\beta} = ((\alpha, \beta)). \tag{9}$$

The associativity of the product implies the relation

$$\sum_{\alpha,\xi} c_{\alpha\xi\eta} x^\alpha \Big(\sum_{\beta,\gamma} c_{\beta\gamma\xi} y^\beta z^\gamma\Big) = \sum_{\xi,\gamma} c_{\xi\gamma\eta} \Big(\sum_{\alpha,\beta} c_{\alpha\beta\xi} x^\alpha y^\beta\Big) z^\gamma$$

for all x, y, z sufficiently close to 0, whence

$$\sum_{\xi} c_{\alpha\xi\eta} c_{\beta\gamma\xi} = \sum_{\xi} c_{\xi\gamma\eta} c_{\alpha\beta\xi} \quad (\alpha, \beta, \gamma, \eta \text{ in } \mathbf{N}^n). \tag{10}$$

The group germ G admits a commutative open subgroup germ if and only if $c_{\alpha\beta\gamma} = c_{\beta\alpha\gamma}$ for all α, β, γ in $\mathbf{N}^n$.

2. BRACKET IN THE LIE ALGEBRA

For $\alpha \in \mathbf{N}^n$, let e_α be the point distribution $\frac{1}{\alpha!}\frac{\partial^\alpha}{\partial x^\alpha}$ at the origin. In particular, $e_k = e_{\varepsilon_k} = \frac{\partial}{\partial x_k}$. The e_α form a basis of the vector space U(G). If f is an analytic function on an open neighbourhood of 0 in G and $f(x) = \sum_\alpha \lambda_\alpha x^\alpha$ is its expansion as an integral series about the origin, then

$$\langle e_\alpha, f \rangle = \lambda_\alpha.$$

In particular,

$$\langle e_\alpha, x^\gamma \rangle = \delta_{\alpha\gamma}.$$

Hence

$$\begin{aligned}\langle e_\alpha * e_\beta, x^\gamma\rangle &= \langle e_\alpha \otimes e_\beta, (x.y)^\gamma\rangle \\ &= \langle e_\alpha \otimes e_\beta, \sum_{\alpha', \beta'} c_{\alpha'\beta'\gamma} x^{\alpha'} y^{\beta'}\rangle \\ &= \sum_{\alpha', \beta'} c_{\alpha'\beta'\gamma}\langle e_\alpha, x^{\alpha'}\rangle\langle e_\beta, y^{\beta'}\rangle = c_{\alpha\beta\gamma}\end{aligned}$$

and hence

$$e_\alpha * e_\beta = \sum_\gamma c_{\alpha\beta\gamma} e_\gamma. \tag{11}$$

(Formula (10) then expresses associativity on U(G).)

In particular, since L(G) is stable under the bracket.

$$[e_i, e_j] = \sum_k (c_{ijk} - c_{jik}) e_k. \tag{12}$$

The constants of structure of L(G) relative to the basis $(e_1, \ldots, e_n)$ are therefore the $c_{ijk} - c_{jik}$. In other words, canonically identifying L(G) with K^n, we obtain:

$$[x, y] = B(x, y) - B(y, x). \tag{13}$$

PROPOSITION 1.

(i) $x^{[-1]} \equiv -x + B(x, x)$ mod deg 3

(ii) $x.y.x^{[-1]} \equiv y + [x, y]$ mod deg 3

(iii) $y^{[-1]}.x.y \equiv x + [x, y]$ mod deg 3

(iv) $x^{[-1]}.y^{[-1]}.x.y \equiv [x, y]$ mod deg 3

(v) $x.y.x^{[-1]}.y^{[-1]} \equiv [x, y]$ mod deg 3.

(In (i), $x^{[-1]}$ of course represents the expansion of the function $x \mapsto x^{[-1]}$ as an integral series about the origin; the other formulae can be interpreted analogously.)

Let g_1 and g_2 be the homogeneous components of $x^{[-1]}$ of degrees 1 and 2. Then

$$\begin{aligned}0 &= x.x^{[-1]} \\ &\equiv x + g_1(x) + B(x, g_1(x)) \quad \text{mod deg 2} \quad \text{(by (6))} \\ &\equiv x + g_1(x) \quad \text{mod deg 2}\end{aligned}$$

and hence $g_1(x) = -x$. Then

$$\begin{aligned}0 &= x.x^{[-1]} \\ &\equiv x + (-x + g_2(x)) + B(x, -x + g_2(x)) \quad \text{mod deg 3} \\ &\equiv g_2(x) - B(x, x) \quad \text{mod deg 3}\end{aligned}$$

and hence $g_2(x) = \mathrm{B}(x, x)$. This proves (i). Then, using (i),

$$\begin{aligned} x.y.x^{[-1]} &\equiv (x + y + \mathrm{B}(x,y)).(-x + \mathrm{B}(x,x)) && \text{mod deg } 3\\ &\equiv x + y + \mathrm{B}(x,y) + (-x + \mathrm{B}(x,x)) + \mathrm{B}(x+y, -x) && \text{mod deg } 3\\ &\equiv y + \mathrm{B}(x,y) - \mathrm{B}(y,x) && \text{mod deg } 3\\ &\equiv y + [x,y] && \text{mod deg } 3 \quad (\text{by } (13))\end{aligned}$$

whence (ii). The proof of (iii) is analogous. Combining (i) and (iii), we obtain

$$\begin{aligned} x^{[-1]}.y^{[-1]}.x.y &\equiv (-x + \mathrm{B}(x,x)).(x + [x,y]) && \text{mod deg } 3\\ &\equiv -x + \mathrm{B}(x,x) + x + [x,y] + \mathrm{B}(-x,x) && \text{mod deg } 3\\ &\equiv [x,y] && \text{mod deg } 3\end{aligned}$$

whence (iv). The proof of (v) is analogous.

3. POWERS

Consider j points of G:

$$\begin{aligned} x(1) &= (x(1)_1, x(1)_2, \ldots, x(1)_n)\\ x(2) &= (x(2)_1, x(2)_2, \ldots, x(2)_n)\\ &\ldots\ldots\ldots\ldots\\ x(j) &= (x(j)_1, x(j)_2, \ldots, x(j)_n).\end{aligned}$$

The mapping $(x(1), x(2), \ldots, x(j)) \mapsto x(1).x(2)\ldots x(j)$ admits an expansion as an integral series about the origin:

$$(14)\qquad x(1).x(2)\ldots x(j) = \sum_{\alpha(1),\alpha(2),\ldots,\alpha(j)\in\mathbf{N}^n} a_{\alpha(1),\ldots,\alpha(j)}x(1)^{\alpha(1)}\ldots x(j)^{\alpha(j)}$$

where the $a_{\alpha(1),\ldots,\alpha(j)}$ are elements of K^n. We write, for $j = 0, 1, 2, \ldots,$

$$(15)\qquad \psi_j(x) = \sum_{\alpha(1)\neq 0,\ldots,\alpha(j)\neq 0} a_{\alpha(1),\ldots,\alpha(j)}x^{\alpha(1)+\ldots+\alpha(j)}$$

where the right hand side is a convergent integral series in the variable $x \in \mathrm{K}^n$. This series is obtained by suppressing in (14) the terms in which one of the variables $x(1), \ldots, x(j)$ does not occur explicitly and then writing $x(1) = x(2) = \cdots = x(j) = x$.

If $t \in \mathrm{K}$, all the t-th power mappings of G have the same expansion as an integral series about the origin, since any two of them coincide on a neighbourhood of 0. We denote this expansion as an integral series by $x^{[t]}$.

PROPOSITION 2. (i) $\psi_j \equiv 0 \bmod \deg j$.

(ii) *If* $t \in \mathrm{K}$, *then*

$$(16)\qquad x^{[t]} = \sum_{i=1}^{\infty} \binom{t}{i}\psi_i(x)$$

where the formal power series on the right is meaningful because of (i).

(*We write*

$$\binom{t}{i} = \frac{t(t-1)\dots(t-i+1)}{i!}$$

for all $t \in \mathrm{K}$.)

Assertion (i) is obvious from the definition of the ψ_j.

We prove (ii) for t an integer $\geqslant 0$. By (14),

$$x^{[t]} = \sum_{\alpha(1),\dots,\alpha(t) \in \mathbf{N}^n} a_{\alpha(1),\dots,\alpha(t)}\, x^{\alpha(1)+\dots+\alpha(t)}. \tag{17}$$

For $\alpha = (\alpha(1), \dots, \alpha(t)) \in (\mathbf{N}^n)^t$, let $\sigma(\alpha)$ denote the set of $j \in \{1, 2, \dots, t\}$ such that $\alpha(j) \neq 0$. If, in the sum (17), we group together the terms for which $\sigma(\alpha)$ is the same, then

$$x^{[t]} = \sum_{\sigma \subset (1,t)} h_{t,\sigma}(x) \tag{18}$$

where

$$h_{t,\sigma}(x) = \sum_{\sigma(\alpha)=\sigma} a_{\alpha(1),\dots,\alpha(t)}\, x^{\alpha(1)+\dots+\alpha(t)}. \tag{19}$$

Let $\sigma = \{j_1, j_2, \dots, j_q\}$ with $j_1 < j_2 < \dots < j_q$. In (14) (where j is replaced by t), we substitute 0 for $x(k)$ for $k \notin \sigma$; as 0 is the identity element of G, we obtain the expansion of $x(j_1).x(j_2)\dots x(j_q)$ as an integral series about the origin:

$$x(j_1).x(j_2)\dots x(j_q) = \sum_{\sigma(\alpha)\subset\sigma} a_{\alpha(1),\dots,\alpha(t)}\, x(j_1)^{\alpha(j_1)} x(j_2)^{\alpha(j_2)} \dots x(j_q)^{\alpha(j_q)}$$

and hence, by the definition of ψ_q:

$$\psi_q(x) = \sum_{\sigma(\alpha)=\sigma} a_{\alpha(1),\dots,\alpha(t)}\, x^{\alpha(j_1)+\dots+\alpha(j_q)}. \tag{20}$$

By (19) and (20), we see that $h_{t,\sigma}(x) = \psi_{\mathrm{Card}\,\sigma}(x)$. Then (18) implies

$$x^{[t]} = \sum_{i=0}^{t} \binom{t}{i} \psi_i(x) = \sum_{i=0}^{\infty} \binom{t}{i} \psi_i(x).$$

Then we write $x^{[t]\prime} = \sum_{i=0}^{\infty} \binom{t}{i} \psi_i(x)$ for all $t \in \mathrm{K}$. In the integral series $x^{[t]}$ and $x^{[t]\prime}$, each coefficient is a polynomial function of t. For this is obvious for $x^{[t]\prime}$. As far as $x^{[t]}$ is concerned, it suffices to prove that, for all $u \in \mathrm{U(G)}$, the image of u under $x \mapsto x^{[t]}$ is a polynomial function of t. Now, for $u \in \mathrm{U}^m(\mathrm{G})$, this image is $t^m u$ (§ 4, no. 3, Proposition 7 (iv)).

As $x^{[t]} = x^{[t]\prime}$ for t an integer $\geqslant 0$, it then follows that $x^{[t]} = x^{[t]\prime}$ for all $t \in \mathrm{K}$.

Remarks. (1) We write condition (ii) of Proposition 2 for t an integer $\geqslant 0$:

$$\begin{aligned} 0 &= \psi_0(x) \\ x &= \psi_0(x) + \psi_1(x) \\ x^{[2]} &= \psi_0(x) + 2\psi_1(x) + \psi_2(x) \\ &\cdots\cdots\cdots \end{aligned}$$

These formulae suffice to determine the ψ_i.

(2) We see that $\psi_0(x) = 0$, $\psi_1(x) = x$, $\psi_2(x) = x^{[2]} - 2x$,

$$x^{[-1]} = \sum_{i=1}^{\infty} (-1)^i \psi_i(x).$$

(3) The above expression for ψ_2 and formula (6) prove that

$$\psi_2(x) \equiv \mathrm{B}(x, x) \quad \text{mod deg } 3. \tag{21}$$

Using Proposition 2, (i) and (ii), we see that

$$x^{[t]} \equiv tx + \binom{t}{2}\mathrm{B}(x, x) \quad \text{mod deg } 3. \tag{22}$$

(4) Let $\psi_{p,m}(x)$ and $h_{t,m}(x)$ denote the homogeneous components of $\psi_p(x)$ and $x^{[t]}$ of degree m. Then $\psi_{p,m} = 0$ for $m < p$. On the other hand, Proposition 2 (ii) gives

$$h_{t,m}(x) = \sum_{p \leqslant m} \frac{t(t-1)\dots(t-p+1)}{p!} \psi_{p,m}(x) \tag{23}$$

that is

$$h_{t,m}(x) = \sum_{1 \leqslant r \leqslant m} t^r \phi_{r,m}(x) \tag{24}$$

where the $\phi_{r,m}$ are homogeneous polynomial mappings of degree m of K^n into K^n. In particular, by (23),

$$\phi_{1,m}(x) = \sum_{p \leqslant m} \frac{(-1)^{p-1}}{p} \psi_{p,m}(x) \tag{25}$$

$$\phi_{m,m}(x) = \frac{1}{m!} \psi_{m,m}(x). \tag{26}$$

(5) If K is of characteristic >0, the results of nos. 1 and 2 remain true, provided e_α, in no. 2, is defined by $\langle e_\alpha, \sum_\beta \lambda_\beta x^\beta \rangle = \lambda_\alpha$. In no. 3, if the ψ_i are defined as above, the argument again proves that $x^{[t]} = \sum_{i=1}^{\infty} \binom{t}{i} \psi_i(x)$ if $t \in \mathbf{N}$.

4. EXPONENTIAL

Let $E(x)$ be the expansion of an exponential mapping of G as an integral series about 0. Let $L(x)$ be the expansion of the inverse mapping of an injective exponential mapping of G as an integral series about 0. Since the tangent mapping at 0 to any exponential mapping is the identity of $L(G)$, $E(x) \equiv x \bmod \deg 2$ and $L(x) \equiv x \bmod \deg 2$. Since $E(L(x)) = L(E(x))$ for x sufficiently close to 0, the formal power series E and L are such that $E(L(X)) = L(E(X)) = X$. An analogous argument shows that

$$E(tX) = (E(X))^{[t]}, \qquad L(X^{[t]}) = tL(X)$$

for $t \in K$.

PROPOSITION 3.

(27) $$L = \sum_{p=1}^{\infty} \frac{(-1)^{p-1}}{p} \psi_p$$

(28) $$E = \sum_{p=1}^{\infty} \frac{1}{p!} \psi_{p,p}$$

(Recall that $\psi_{p,p}$ is the homogeneous component of ψ_p of degree p.)

$$E(tx) = (E(x))^{[t]}$$

or, by (24),

$$E(tx) = \sum_{m \geqslant 0} \sum_{1 \leqslant r \leqslant m} t^r \phi_{r,m}(E(x)).$$

The two sides are formal power series in t and x. Equating the terms of first degree in t, we obtain

(29) $$x = \sum_{m \geqslant 0} \phi_{1,m}(E(x)).$$

Now the inversion of a system of formal power series, when it is possible, is possible in only one way (*Algebra*, Chapter IV, § 6, Corollary to Proposition 8). Then

$$\begin{aligned} L(x) &= \sum_{m \geqslant 0} \phi_{1,m}(x) \quad \text{by (29)} \\ &= \sum_{p,m} \frac{(-1)^p}{p} \psi_{p,m}(x) \quad \text{by (25)} \\ &= \sum_{p} \frac{(-1)^p}{p} \psi_p(x), \end{aligned}$$

whence (i). Similarly, for $t \neq 0$,

$$\begin{aligned} L(tx) &= tL((tx)^{[t^{-1}]}) \\ &= tL\Big(\sum_{m \geqslant 0} \sum_{1 \leqslant r \leqslant m} t^{m-r} \phi_{r,m}(x)\Big) \quad \text{by (24).} \end{aligned}$$

Equating the terms of first degree in t, we obtain

$$x = \mathrm{L}\Big(\sum_{m \geqslant 0} \phi_{m,m}(x)\Big)$$

whence

$$\begin{aligned} \mathrm{E}(x) &= \sum_{m \geqslant 0} \phi_{m,m}(x) \\ &= \sum_{m \geqslant 0} \frac{1}{m!} \psi_{m,m}(x) \quad \text{by (26).} \end{aligned}$$

PROPOSITION 4. *For the chart of* G *used to be canonical, it is necessary and sufficient that* $\psi_j = 0$ *for* $j \geqslant 2$.

This is sufficient by Proposition 3. Suppose that the chart is canonical and that $\psi_i = 0$ for $2 \leqslant i < n$. Then $nx = x^{[n]} = \sum_{i=0}^{n} \binom{n}{i} \psi_i(x) = nx + \psi_n(x)$, whence $\psi_n = 0$. Hence $\psi_j = 0$ for $j \geqslant 2$ by induction on j.

§ 6. REAL AND COMPLEX LIE GROUPS

In this paragraph, K is assumed to be equal to **R** or **C**.

1. PASSAGE FROM LIE ALGEBRA MORPHISMS TO LIE GROUP MORPHISMS

Lemma 1. *Let* G *be a simply connected*† *topological group,* W *a symmetric connected open neighbourhood of* e, H *a group and* f *a mapping of* W^3 *into* H *such that*

$$f(xyz) = f(x)f(y)f(z)$$

for x, y, z *in* W. *There exists a morphism* f' *of* G *into* H *such that* $f' \mid \mathrm{W} = f \mid \mathrm{W}$.

For $(g, h) \in \mathrm{G} \times \mathrm{H}$ and U an open neighbourhood of e in W, let $\mathrm{A}(g, h, \mathrm{U})$ be the set of $(gu, hf(u)) \in \mathrm{G} \times \mathrm{H}$ for $u \in \mathrm{U}$. Then $(g, h) \in \mathrm{A}(g, h, \mathrm{U})$ and $\mathrm{A}(g, h, \mathrm{U}_1) \cap \mathrm{A}(g, h, \mathrm{U}_2) = \mathrm{A}(g, h, \mathrm{U}_1 \cap \mathrm{U}_2)$. Let $(s, t) \in \mathrm{A}(g, h, \mathrm{U})$; then $s = gu$ and $t = hf(u)$ for $u \in \mathrm{U}$; there exists an open neighbourhood U′ of e in W such that $u\mathrm{U}' \in \mathrm{U}$; then, for $u' \in \mathrm{U}'$,

$$(su', tf(u')) = (guu', hf(uu')) \in \mathrm{A}(g, h, \mathrm{U})$$

† Cf. Chapter XI of *General Topology* (to appear). It is proved in this chapter that if G_1, G_2 are connected topological groups, ϕ is an open continuous homomorphism of G_1 onto G_2 with discrete kernel and G_2 is simply connected, then ϕ is a homeomorphism. Recall on the other hand that a simply connected space is connected.

and hence $A(s, t, U') \subset A(g, h, U)$. It then follows that the $A(g, h, U)$ form the base of a topology on $G \times H$. We shall denote by Y the set $G \times H$ with this topology and denote by p the canonical projection of Y onto G, which is open. The restriction of p to $A(g, h, U)$ is a homeomorphism of $A(g, h, U)$ onto gU. Hence (Y, p) is a covering space of G. Let Y_0 be the subgroup of Y generated by $A(e, e, W)$ and let $\mathfrak{B}$ be the set of $A(e, e, U)$. Clearly $\mathfrak{B}$ satisfies conditions (GV'_I) and (GV'_{II}) of *General Topology*, Chapter III, § 1, no. 2. The set Y'_0 of $y \in Y_0$ such that the mappings $z \mapsto yzy^{-1}$ and $z \mapsto y^{-1}zy$ of Y_0 into Y_0 are continuous at (e, e) is a subgroup of Y_0. Let $w \in W$. The mapping $w' \mapsto ww'w^{-1}$ of W into G is continuous and hence the mapping

$$(w', f(w')) \mapsto (ww'w^{-1}, f(ww'w^{-1}))$$

of $A(e, e, W)$ into Y is continuous at (e, e). Now $f(ww'w^{-1}) = f(w)f(w')f(w^{-1})$ and therefore $(ww'w^{-1}, f(ww'w^{-1})) = (w, f(w))(w', f(w'))(w, f(w))^{-1}$.

As $w^{-1} \in W$, we see that $(w, f(w)) \in Y'_0$. Thus, $A(e, e, W) \subset Y'_0$, so that $Y'_0 = Y_0$. The group Y_0, with the filter base $\mathfrak{B}$, therefore satisfies condition (GV'_{III}) of *General Topology*, Chapter III, § 1, no. 2. As

$$(g, h).A(e, e, U) = A(g, h, U),$$

Y_0 is a topological group which is connected since $A(e, e, W)$ is connected. Then $p(Y_0)$ is an open subgroup of G, whence $p(Y_0) = G$ since G is connected. The kernel of $p \mid Y_0$ is discrete. As G is simply connected, $p \mid Y_0$ is a homeomorphism of Y_0 onto G. Therefore Y_0 is the graph of a morphism f' of G into H. For $g \in W$, $(g, f(g)) \in A(e, e, W) \subset Y_0$, whence $f(g) = f'(g)$.

THEOREM 1. *Let* G *and* H *be Lie groups and h a continuous morphism of* L(G) *into* L(H). *Suppose that* G *is simply connected. Then there exists one and only one Lie group morphism* ϕ *of* G *into* H *such that* $h = L(\phi)$.

The existence of ϕ follows from Lemma 1 and § 4, no. 1, Theorem 1 (i). The uniqueness of ϕ follows from § 4, no. 1, Theorem 1 (ii) and the fact that G is connected.

COROLLARY. *Let* G *be a finite-dimensional simply connected Lie group. There exists a finite-dimensional analytic linear representation of* G *whose kernel is discrete.*

There exist (Chapter I, § 7, Theorem 2) a finite-dimensional vector spacc E and an injective morphism h of L(G) into the Lie algebra End(E). By Theorem 1, there exists a morphism ϕ of G into **GL**(E) such that $L(\phi) = h$. Hence ϕ is an immersion and therefore its kernel is discrete.

Remarks. (1) There exist finite-dimensional simply connected Lie groups which have no finite-dimensional injective analytic linear representation (Exercise 2).

(2) There exist finite-dimensional connected Lie groups G such that every finite-dimensional analytic linear representation of G has non-discrete kernel (Exercises 3 and 4).

2. INTEGRAL SUBGROUPS

DEFINITION 1. *Let* G *be a Lie group. An integral subgroup of* G *is a subgroup* H *with a* connected *Lie group structure such that the canonical injection of* H *into* G *is an immersion.*

A one-parameter subgroup of G is a 1-dimensional integral subgroup of G.

Let H be an integral subgroup of G and i the canonical injection of H into G. Then $L(i)$ defines an isomorphism of L(H) onto a Lie subalgebra of L(G) admitting a topological supplement. L(H) is identified with its image under $L(i)$.

Examples. (1) A connected Lie subgroup of G is an integral subgroup of G.

(2) Suppose that G is finite-dimensional. Let H be a subgroup of G; we give it the structure induced by the Lie group structure on G (§ 4, no. 5, Definition 3). Then its identity component H_0 is an integral subgroup of G and the subalgebra tangent to H at e is $L(H_0)$ (§ 4, no. 5, Proposition 9 (ii)).

(3) Let G be a complex Lie group, H an integral subgroup of G and G_1 (resp. H_1) the underlying real Lie group of G (resp. H). Then H_1 is an integral subgroup of G_1 and $L(H_1)$ is the underlying real Lie algebra of L(H).

THEOREM 2. *Let* G *be a Lie group.*

(i) *The mapping* $H \mapsto L(H)$ *is a bijection of the set of integral subgroups of* G *onto the set of Lie subalgebras of* L(G) *admitting a topological supplement.*

(ii) *Let* H *be an integral subgroup of* G. *Every connected Lie subgroup germ of* G *with Lie algebra* L(H) *is an open submanifold of* H *which generates* H.

(a) Let $\mathfrak{h}$ be a Lie subalgebra of L(G) admitting a topological supplement. Let H_1 be a Lie subgroup germ of G such that $L(H_1) = \mathfrak{h}$ (§ 4, Theorem 3). H_1 can be chosen so that it is connected. Let H be the subgroup of G generated by H_1. There exists (§ 1, Corollary to Proposition 22) a Lie group structure on H such that H_1 is an open submanifold of H and the canonical injection of H into G is an immersion. As H_1 is connected, H is connected and is hence an integral subgroup of G. Then $L(H) = L(H_1) = \mathfrak{h}$. This proves that the mapping considered in (i) is surjective.

(b) Let H be an integral subgroup of G and N_1 a connected Lie subgroup germ of G with Lie algebra L(H). As the canonical injection of H into G is an immersion, there exists an open subgroup germ H_1 of H which is at the same time a submanifold of G and hence a Lie subgroup germ of G with Lie algebra L(H). On the other hand, let N be the subgroup of G generated by N_1; by part (a) of the proof, it has the structure of an integral subgroup of G such that N_1 is an open submanifold of N. By § 4, Theorem 3, $H_1 \cap N_1$ is open in H_1 and N_1. Hence the subgroup of G generated by $H_1 \cap N_1$ is equal on the one hand to H and on the other to N. Therefore the Lie groups H and N are equal. This proves (ii) and also proves that the mapping considered in (i) is injective.

Remark 1. Let H be an integral subgroup of G. Let Y be the left foliation of G associated with L(H). If $g \in G$, let gH be given the manifold structure derived from that on H by $\gamma(g)$. By *Differentiable and Analytic Manifolds*, R, 9.3.2, the canonical injection of gH into Y is a morphism. This morphism is étale. Hence the maximal connected leaves of Y are the left cosets modulo H.

PROPOSITION 1. *Let* G *and* M *be Lie groups,* H *an integral subgroup of* G *and* ϕ *a morphism of* M *into* G *such that* $L(\phi)(L(M)) \subset L(H)$. *Suppose that* M *is connected. Then* $\phi(M) \subset H$ *and* ϕ, *considered as a mapping of* M *into* H, *is a Lie group morphism.*

In the notation of *Remark* 1, ϕ is a morphism of M into Y (*Differentiable and Analytic Manifolds*, R, 9.3.2) and hence $\phi(M) \subset H$ since M is connected.

COROLLARY 1. *Let* G *and* H *be Lie groups,* ϕ *a Lie group morphism of* G *into* H, N *the kernel of* ϕ *and* $h = L(\phi)$. *Suppose that* G *is connected and that* H *is finite-dimensional.*

(i) N *is a Lie subgroup of* G *and* $L(N) = \operatorname{Ker} h$.

(ii) *Let* H′ *be the integral subgroup of* H *with Lie algebra* Im h. *Then* $\phi(G) = H'$.

(iii) *The mapping of* G/N *into* H′ *derived from* ϕ *when passing to the quotient is a Lie group isomorphism.*

(i) has already been proved (§ 3, no. 8, Proposition 28).

Let ψ be the Lie group morphism of G/N into H derived from ϕ when passing to the quotient; it is an immersion (§ 3, no. 8, Proposition 28). By Proposition 1, ψ is a Lie group morphism of G/N into H′. This morphism is étale and hence $\psi(G/N) = H'$ since H′ is connected; this proves (ii). Then $\psi: G/N \to H'$ is bijective and is a Lie group isomorphism, which proves (iii).

COROLLARY 2. *Let* G *be a Lie group and* H_1 *and* H_2 *integral subgroups of* G. *If* $L(H_2) \subset L(H_1)$, *then* H_2 *is an integral subgroup of* H_1.

Let $i_1: H_1 \to G$, $i_2: H_2 \to G$ be the canonical injections. Then

$$L(i_2)(L(H_2)) = L(H_2) \subset L(H_1).$$

By Proposition 1, i_2 is an analytic mapping of H_2 into H_1 and even an immersion of H_2 into H_1 since $L(i_2)$ is an isomorphism of $L(H_2)$ onto a subalgebra of $L(H_1)$ admitting a topological supplement.

COROLLARY 3. *Let* G *be a finite-dimensional Lie group and* $(H_\iota)_{\iota \in I}$ *a family of Lie subgroups of* G. *Then* $H = \bigcap_{\iota \in I} H_\iota$ *is a Lie subgroup of* G *and*

$$L(H) = \bigcap_{\iota \in I} L(H_\iota).$$

There exists a finite subset J of I such that $\bigcap_{\iota \in J} L(H_\iota)$ is equal to the intersection M of all the $L(H_i)$. We know that $H^* = \bigcap_{i \in J} H_i$ is a Lie subgroup such that $L(H^*) = M$ (§ 3, no. 8, Corollary 2 to Proposition 29). Let H_0 be the identity component of H^*. It is a Lie subgroup of G and $L(H_0) = M$. By Corollary 2, $H_0 \subset H_i$ for all i and hence $H_0 \subset H \subset H^*$, whence the corollary.

COROLLARY 4. *Let* G *be a finite-dimensional connected Lie group. The following conditions are equivalent:*

(i) G *is unimodular* (*Integration*, Chapter VII, § 1, no. 3, Definition 3);

(ii) $\det \operatorname{Ad} g = 1$ *for all* $g \in G$;

(iii) $\operatorname{Tr} \operatorname{ad} a = 0$ *for all* $a \in L(G)$.

The mapping $g \mapsto \det \operatorname{Ad} g$ is a morphism ϕ of G into K^*. By § 3, Propositions 35 (no. 10) and 44 (no. 12), $L(\phi)a = \operatorname{Tr} \operatorname{ad} a$ for all $a \in L(G)$. Clearly $\operatorname{Im} L(\phi) = \{0\}$ or K. In the first (resp. second) case, $\operatorname{Im} \phi = \{1\}$ (resp. $\operatorname{Im} \phi = K^*$) by Corollary 1 and hence G is unimodular (resp. not unimodular) by § 3, no. 16, Corollary to Proposition 55.

PROPOSITION 2. *Let* G *be a finite-dimensional Lie group and* H *an integral subgroup of* G. *The following conditions are equivalent:*

(i) H *is closed*;

(ii) *the topology on* H *is induced by that on* G;

(iii) H *is a Lie subgroup of* G.

(i) ⇒ (iii): this follows from § 1, Propositions 2 (iv) (no. 1) and 14 (iii) (no. 7).

(iii) ⇒ (ii): obvious.

(ii) ⇒ (i): if the topology on H is induced by that on G, H is closed because H is complete (§ 1, no. 1, Proposition 1).

PROPOSITION 3. *Let* G *be a Lie group,* H *an integral subgroup of* G, M *a non-empty connected analytic manifold,* f *a mapping of* M *into* G *and* $r \in N_K$. *Consider the following conditions:*

(i) f *is of class* C^r *and* $f(M) \subset H$;

(ii) $f(M) \subset H$ *and* f, *considered as a mapping of* M *into* H, *is of class* C^r;

(iii) f *is of class* C^r, $f(M)$ *meets* H *and the image of* $T_m(M)$ *is contained in* $f(m) . L(H)$ *for all* $m \in M$.

Then (ii) ⇔ (iii) ⇒ (i). *If the topology on* H *admits a countable base, the three conditions are equivalent.*

(ii) ⇒ (i) and (ii) ⇒ (iii): obvious.

(iii) ⇒ (ii): suppose that condition (iii) holds. By *Differentiable and Analytic Manifolds*, R, 9.2.8, f is a morphism of class C^r of M into the left foliation associated with L(H). As M is connected, $f(M) \subset H$.

If the topology on H admits a countable base, condition (i) implies that f is a mapping of class C^r of M into H (*Differentiable and Analytic Manifolds*, R, 9.2.8); hence (i) ⇒ (ii).

COROLLARY 1. *Let* G *be a finite-dimensional Lie group and* H *an integral subgroup of* G. *Then the Lie subalgebra tangent to* H *at* e (§ 4, no. 5, Definitions 2 and 3) *is* L(H) *and the Lie group structure on* H *is the structure induced by that on* G.

As H is connected and finite-dimensional, its topology admits a countable base.

COROLLARY 2. *Let* G *be a Lie group and* H_1 *and* H_2 *integral subgroups of* G. *Suppose that the topology on* H_1 *admits a countable base. Then*

$$H_2 \subset H_1 \Leftrightarrow L(H_2) \subset L(H_1)$$

and, if these conditions hold, H_2 *is an integral subgroup of* H_1.

The last assertion and the implication $L(H_2) \subset L(H_1) \Rightarrow H_2 \subset H_1$ follow from Corollary 2 to Proposition 1. The converse implication follows from Proposition 3.

COROLLARY 3. *Let* G *be a Lie group and* H_1 *and* H_2 *integral subgroups of* G *whose topology admits a countable base. If* H_1 *and* H_2 *have the same underlying set, the Lie group structures on* H_1 *and* H_2 *are the same.*

This follows from Corollary 2.

Remark 2. Let G be a finite-dimensional Lie group. Let H be a subgroup of G. We shall say, by an abuse of language, that H is an integral subgroup of G if there exists a Lie group structure S on H such that H, together with S, is an integral subgroup of G. By Corollary 3 to Proposition 3, if S exists, S is unique.

Remark 3. Let V be a manifold of class C^r. Let M be a subset of V and x and y elements of M. Consider the following property:

$P_{M,x,y}$: there exist I, $x_0, x_1, \ldots, x_n, f_1, \ldots, f_n$ such that: (a) I is a connected open subset of K; (b) $x_0, \ldots, x_n$ are in M, $x_0 = x$, $x_n = y$; (c) for $1 \leqslant i \leqslant n$, f_i is a mapping of class C^r of I into V which takes the values x_{i-1} and x_i and $f_i(I) \subset M$.

We shall say that M is a C^r-*connected* subset of V if, for all elements x, y of M, the property $P_{M,x,y}$ holds.

PROPOSITION 4. *Let* G *be a finite-dimensional Lie group and* H *a subgroup of* G. *Let* $r \in N_K$. *The following conditions are equivalent:*

(i) H *is an integral subgroup of* G;

(ii) *with the Lie group structure induced by that on* G, H *is connected*;

(iii) H *is* C^r*-connected.*

(ii) ⇒ (i): obvious.

(i) ⇒ (iii): suppose that H has a Lie group structure such that H is an integral subgroup of G. Using the notation of *Remark* 3, the set of $y \in H$ such that property $P_{H,e,y}$ holds is an open subgroup of H. As H is connected, this subgroup is equal to H and hence condition (iii) is satisfied.

(iii) ⇒ (ii): suppose that condition (iii) is satisfied and let H be given the structure induced by the Lie group structure on G. Let $\mathfrak{h}$ be the subalgebra tangent to H at e. The identity component H_0 of H is an integral subgroup of G such that $L(H_0) = \mathfrak{h}$. We show that $H = H_0$. It suffices to prove the following: let I be a connected open subset of K, f a mapping of class C^r of I into G such that $f(I) \subset H$ and λ and μ two points of I; if $f(\lambda) \in H_0$, then $f(\mu) \in H_0$. But, for all $\nu \in I$, $(T_\nu f)(K) \subset f(\nu)\mathfrak{h}$ by definition of $\mathfrak{h}$, so that our assertion follows from Proposition 3.

Remark 4. If $K = \mathbf{R}$, the integral subgroups of G can also be characterized as the subgroups which, with the topology induced by that on G, are *arcwise connected* (§ 8, Exercise 4). However, subgroups may be *connected* but not integral (*Commutative Algebra*, Chapter VI, § 9, Exercise 2).

COROLLARY. *Let* G *be a finite-dimensional Lie group and* H_1 *and* H_2 *two integral subgroups of* G. *The subgroup of* G *generated by* H_1 *and* H_2 *and the subgroup* (H_1, H_2) *of* G *are integral subgroups of* G.

The subgroup (G, G) of G is not always closed (§ 9, Exercise 6).

Recall (§ 3, no. 11, Corollary 5 to Proposition 41) that if $\mathfrak{a}$ is a finite-dimensional algebra, $\mathrm{Aut}(\mathfrak{a})$ is a Lie subalgebra of $\mathbf{GL}(\mathfrak{a})$ and that $L(\mathrm{Aut}(\mathfrak{a}))$ is the Lie algebra of derivations of $\mathfrak{a}$.

DEFINITION 2. *Let* $\mathfrak{a}$ *be a finite-dimensional Lie algebra. Let* $\mathrm{Ad}(\mathfrak{a})$ *or* $\mathrm{Int}(\mathfrak{a})$ *denote the integral subgroup of* $\mathrm{Aut}(\mathfrak{a})$ *with Lie algebra* $\mathrm{ad}(\mathfrak{a})$. *The elements of this group are called inner automorphisms of* $\mathfrak{a}$.

By transport of structure, $\mathrm{ad}(\mathfrak{a})$ is invariant under $\mathrm{Aut}(\mathfrak{a})$ and hence $\mathrm{Int}(\mathfrak{a})$ is normal in $\mathrm{Aut}(\mathfrak{a})$. By § 4, no. 4, Corollary 1 to Proposition 8 and the fact that $\mathrm{Int}(\mathfrak{a})$ is connected, the elements of $\mathrm{Int}(\mathfrak{a})$ are the finite products of automorphisms of the form $\exp \mathrm{ad}\, x$ where $x \in \mathfrak{a}$. In general, $\mathrm{Int}(\mathfrak{a})$ is not a Lie subgroup of $\mathrm{Aut}(\mathfrak{a})$ (Exercise 14).

3. PASSAGE FROM LIE ALGEBRAS TO LIE GROUPS

THEOREM 3. (i) *If* L *is a finite-dimensional Lie algebra, there exists a simply connected Lie group* G *such that* L(G) *is isomorphic to* L.

(ii) *Let* G_1 *and* G_2 *be two connected Lie groups, with* G_1 *simply connected. Let* f *be an isomorphism of* $L(G_1)$ *onto* $L(G_2)$, ϕ *the morphism of* G_1 *into* G_2 *such that* $L(\phi) = f$ *and* N *the kernel of* ϕ. *Then* N *is a discrete subgroup of the centre of* G_1 *and the morphism of* G_1/N *into* G_2 *derived from* ϕ *is a Lie group isomorphism. If* G_2 *is simply connected,* ϕ *is an isomorphism.*

Let L be a finite-dimensional Lie algebra. There exists a finite-dimensional vector space E such that L can be identified with a Lie subalgebra of End(E) (Chapter I, § 7, Theorem 2). Let H be the integral subgroup of $\mathbf{GL}(E)$ with

Lie algebra L. Let $\hat{H}$ be its universal covering (§1, no. 9, *Remark*). Then $L(\hat{H})$ is isomorphic to L, whence (i).

Let G_1, G_2, f, ϕ, N be as in (ii). Then ϕ is étale, hence $\phi(G_1)$ is an open subgroup of G_2 and hence $\phi(G_1) = G_2$. On the other hand, N is discrete and hence contained in the centre of G_1 (*Integration*, Chapter VII, §3, Lemma 4). Clearly the morphism of G_1/N onto G_2 derived from ϕ is a Lie group isomorphism. If G_2 is simply connected, every étale mapping of G_1 onto G_2 is injective and hence $N = \{e\}$.

PROPOSITION 5. *Let* G *be a connected real Lie group. Suppose that* L(G) *has a complex normable Lie algebra structure* L′ *compatible with its real normable Lie algebra structure. There exists on* G *one and only one complex Lie group structure compatible with the real Lie group structure and with Lie algebra* L′.

By §4, no. 2, Corollary 2 to Theorem 2, it suffices to prove that the structure of L′ is invariant under Ad G. Let ϕ be an exponential mapping of G. By §4, no. 4, Corollary 3 (i) to Proposition 8, there exists a neighbourhood V of 0 in L(G) such that the structure of L′ is invariant under Ad $\phi(V)$. But Ad $\phi(V)$ generates G because G is connected.

The conclusion of Proposition 5 is not necessarily true if G is not assumed to be connected (Exercise 7).

PROPOSITION 6. *Let* G *be a connected complex Lie group. If* G *is compact,* G *is commutative.*

The holomorphic mapping $g \mapsto \operatorname{Ad} g$ of G into $\mathscr{L}(L(G))$ is constant (*Differentiable and Analytic Manifolds*, R, 3.3.7) and hence $\operatorname{ad} a = 0$ for all $a \in L(G)$ (§3, no. 12, Proposition 44). Hence G is commutative (§4, Corollary 3 to Theorem 1).

1. EXPONENTIAL MAPPING

THEOREM 4. *Let* G *be a Lie group. There exists one and only one exponential mapping of* G *defined on* L(G).

There exist a convex open neighbourhood U of 0 in L(G) and an exponential mapping of G defined on U. It can be assumed, by choosing U sufficiently small, that

$$\phi((\lambda + \lambda')a) = \phi(\lambda a)\phi(\lambda' a) \tag{1}$$

for $a \in L(G)$, λ, λ' in K, λa, $\lambda' a$, $(\lambda + \lambda')a$ in U.

Let $a \in L(G)$. There exists an integer $n > 0$ such that $\frac{1}{n} a \in U$. If m is another integer such that $\frac{1}{m} a \in U$, then $\frac{1}{nm} a \in U$ and relation (1) implies

$$\phi\left(\frac{1}{n} a\right) = \left(\phi\left(\frac{1}{nm} a\right)\right)^m, \quad \phi\left(\frac{1}{m} a\right) = \left(\phi\left(\frac{1}{nm} a\right)\right)^n;$$

hence $\left(\phi\left(\frac{1}{n}a\right)\right)^n = \left(\phi\left(\frac{1}{m}a\right)\right)^m$. There exists an extension $\psi\colon L(G) \to G$ of ϕ such that $\psi(a) = \left(\phi\left(\frac{1}{n}a\right)\right)^n$ for $a \in L(G)$ and n an integer >0 such that $\frac{1}{n}a \in U$. Clearly ψ is analytic and is an exponential mapping of G. If $\psi'\colon L(G) \to G$ is an exponential mapping of G, ψ and ψ' coincide on a neighbourhood of 0 and hence are equal since L(G) is connected.

Henceforth when we speak of *the exponential mapping of* G, we shall mean the mapping considered in Theorem 4. It will be denoted by $\exp_G$ or exp if there is no risk of confusion.

Example. Let A be a complete normed unital associative algebra. Then $\exp_{A^*}$ is the exponential mapping defined in Chapter II, § 7, no. 3.

PROPOSITION 7. *Let* G *be a Lie group and* a *an element of* L(G). *The mapping* $\lambda \mapsto \exp(\lambda a)$ *of* K *into* G *is the unique morphism* ϕ *of the Lie group* K *into* G *such that* $(T_0\phi)1 = a$.

The mappings $(\lambda, \lambda') \mapsto \exp(\lambda a)\exp(\lambda' a)$ and $(\lambda, \lambda') \mapsto \exp(\lambda + \lambda')a$ of $K \times K$ into G are analytic and coincide on a neighbourhood of (0, 0). As $K \times K$ is connected, these mappings are equal. Hence $\phi\colon \lambda \mapsto \exp(\lambda a)$ is a Lie group morphism of K into G. The tangent mapping at 0 to $\lambda \mapsto \lambda a$ is the mapping $\lambda \mapsto \lambda a$; and $T_e(\exp) = \mathrm{Id}_{L(G)}$; hence $(T_0\phi)1 = a$. The uniqueness assertion of the proposition follows from Theorem 1.

PROPOSITION 8. *Let* G *be a Lie group. For all* x, y *in* L(G) *and* n *an integer,*

$$\exp(x + y) = \lim_{n \to +\infty} \left(\left(\exp\frac{1}{n}x\right)\left(\exp\frac{1}{n}y\right)\right)^n \tag{2}$$

$$\exp[x, y] = \lim_{n \to +\infty} \left(\left(\exp\frac{1}{n}x\right)\left(\exp\frac{1}{n}y\right)\left(\exp\frac{1}{n}x\right)^{-1}\left(\exp\frac{1}{n}y\right)^{-1}\right)^{n^2}. \tag{3}$$

By Proposition 7, this follows from Proposition 4 of § 4, no. 3, taking $\lambda = \frac{1}{n}$.

PROPOSITION 9. *Let* G *be a complex Lie group and* G′ *the underlying real Lie group. Then* $\exp_G = \exp_{G'}$.

This follows from Proposition 5 of § 4, no. 3 and the analyticity of $\exp_G$ and $\exp_{G'}$.

PROPOSITION 10. *Let* G *and* H *be Lie groups and* ϕ *a morphism of* G *into* H.

(i) $\phi \circ \exp_G = \exp_H \circ L(\phi)$.

(ii) *If* G *is an integral subgroup of* H, *then* $\exp_G = \exp_H \mid L(G)$.

The two sides of the equality (i) are analytic mappings of L(G) into H which coincide in a neighbourhood of 0 (§ 4, Proposition 8, no. 4) and hence are equal. Assertion (ii) is a special case of (i).

COROLLARY 1. *Let* G *be a Lie group,* G′ *a Lie subgroup of* G *and* $a \in L(G)$. *The following conditions are equivalent:*

(i) $a \in L(G')$;

(ii) $\exp(\lambda a) \in G'$ *for all* $\lambda \in K$ *and* $|\lambda|$ *sufficiently small;*

(iii) $\exp(\lambda a) \in G$ *for all* $\lambda \in K$.

The argument is as in § 4, no. 4, Corollary 1 to Proposition 8.

COROLLARY 2. *Let* G *be a Lie group,* H *an integral subgroup of* G *and* $a \in L(G)$. *Consider the following conditions:*

(i) $a \in L(H)$;

(ii) $\exp_G(\lambda a) \in H$ *for all* $\lambda \in K$.

Then (i) ⇒ (ii). *If the topology of* H *admits a countable base, then* (i) ⇔ (ii).

Let i be the canonical injection of H into G. If $a \in L(H)$, then

$$\exp_G(\lambda a) = (\exp_G \circ L(i))(\lambda a) = (i \circ \exp_H)(\lambda a) \in H.$$

Hence (i) ⇒ (ii). The converse when H has a countable base follows from Proposition 3.

COROLLARY 3. *Let* G *be a Lie group,* ρ *an analytic linear representation of* G, $x \in L(G)$ *and* $g \in G$.

(i) $\rho(\exp x) = \exp L(\rho)x$;

(ii) $\mathrm{Ad}(\exp x) = \exp \mathrm{ad}\, x$;

(iii) $g(\exp x)g^{-1} = \exp(\mathrm{Ad}\, g.x)$.

The argument is as in § 4, no. 4, Corollaries 2 and 3 to Proposition 8.

COROLLARY 4. *Let* G *be a finite-dimensional connected Lie group.*

(i) $\mathrm{Int}(L(G)) = \mathrm{Ad}(G)$.

(ii) *Let* Z *be the centre of* G. *Then* Z *is a Lie subgroup of* G *whose Lie algebra is the centre of* L(G). *The mapping of* G/Z *into* Int L(G) *derived from* $g \mapsto \mathrm{Ad}\, g$ *when passing to the quotient is a Lie group isomorphism.*

Assertion (i) follows from Corollary 3 (ii) and the remarks after Definition 2. Let $g \in G$. In order that $\mathrm{Ad}\, g = \mathrm{Id}_{L(G)}$, it is necessary and sufficient that Int g coincide with Id_G on a neighbourhood of e (§ 4, no. 1, Theorem 1 (ii)) and hence on the whole of G; in other words, it is necessary and sufficient that $g \in Z$. Then (ii) follows from Corollary 1 to Proposition 1.

DEFINITION 3. *Let* G *be a finite-dimensional connected Lie group. The Lie group* $\mathrm{Int}(L(G)) = \mathrm{Ad}(G)$ *is called the adjoint group of* G.

PROPOSITION 11. *Let* G *be a connected commutative Lie group.*

(i) exp *is an étale morphism of the additive Lie group* L(G) *onto* G.

(ii) *If* $K = \mathbf{R}$ *and* G *is finite-dimensional,* G *is isomorphic to a Lie group of the form* $\mathbf{R}^p \times \mathbf{T}^q$ *(p, q integers* $\geqslant 0$*).*

By the Hausdorff formula, $(\exp x)(\exp y) = \exp(x + y)$ for x, y sufficiently close to 0 and hence for all x, y in L(G) by analytic continuation. Hence exp is a group homomorphism and is étale since

$$T_e(\exp) = \mathrm{Id}_{L(G)}.$$

Hence (i). Assertion (ii) follows from (i) and *General Topology*, Chapter VII, § 1, Proposition 9.

PROPOSITION 12. *Let* G *be a Lie group and* $L = L(G)$. *For all* $x \in L$, *let* $T_x(L)$ *be identified with* L, *so that the right differential* $\varpi(x)$ *of* exp *at* x *is a linear mapping of* L *into* L. *For all* $x \in L$,

$$\varpi(x) = \sum_{n \geqslant 0} \frac{1}{(n+1)!} (\mathrm{ad}\, x)^n.$$

The two sides are analytic functions of x and are equal for x sufficiently close to 0 (§ 4, no. 3, Proposition 6).

Remark. $\varpi(x).(\mathrm{ad}\, x) = \exp \mathrm{ad}\, x - 1$. We write, by an abuse of notation,

$$\varpi(x) = \frac{\exp \mathrm{ad}\, x - 1}{\mathrm{ad}\, x}.$$

COROLLARY. *Let* G *be a complex Lie group and* $x \in L(G)$. *The tangent mapping at* x *to* $\exp_G$ *has kernel* $\bigoplus_{n \in \mathbf{Z} - \{0\}} \mathrm{Ker}(\mathrm{ad}\, x - 2i\pi n)$.

The integral function $z \mapsto \sum_{n \geqslant 0} \frac{1}{(n+1)!} z^n$, equal to $\frac{e^z - 1}{z}$ for $z \neq 0$, admits as zeros the points of $2\pi i\mathbf{Z} - \{0\}$, which are all simple zeros. The corollary then follows from Proposition 12 and the following lemma:

Lemma 2. *Let* E *be a complex Banach space,* u *an element of* $\mathscr{L}(E)$, S *the spectrum of* u *in* $\mathscr{L}(E)$ (*Spectral Theories*, Chapter I, § 1, no. 2) *and* f *a holomorphic complex function on an open neighbourhood* Ω *of* S. *Suppose that* f *admits in* Ω *only a finite number of distinct zeros* $z_1, \ldots, z_n$, *of multiplicities* $h_1, \ldots, h_n$. *Then* $\mathrm{Ker} f(u)$ *is the direct sum of the* $\mathrm{Ker}(u - z_i)^{h_i}$ *for* $1 \leqslant i \leqslant n$.

(For the definition of $f(u)$, see *Spectral Theories*, Chapter I, § 4, no. 8.)

There exists a holomorphic function g on Ω, everywhere non-zero, such that $f(z) = (z - z_1)^{h_1} \ldots (z - z_n)^{h_n} g(z)$. Then $g(u)g^{-1}(u) = g^{-1}(u)g(u) = 1$ and hence $\mathrm{Ker} f(u) = \mathrm{Ker} \prod_{i=1}^{n} (u - z_i)^{h_i}$. Considering $\mathrm{Ker} f(u)$ as a $\mathbf{C}[X]$-module by means of the external law $(h, x) \mapsto h(u)x$ for $h \in \mathbf{C}[X]$, $x \in \mathrm{Ker} f(u)$, we see that $\mathrm{Ker} f(u)$ is the direct sum of the $\mathrm{Ker}(u - z_i)^{h_i}$, using *Algebra*, Chapter VII, § 2, no. 1, Proposition 1.

5. APPLICATION TO LINEAR REPRESENTATIONS

PROPOSITION 13. *Let* G *be a connected Lie group and* ρ *an analytic linear representation of* G *on a complete normable space* E. *Let* E_1, E_2 *be two closed vector subspaces of* E *such that* $E_2 \subset E_1$. *The following conditions are equivalent:*

(i) $\rho(g)x \equiv x \pmod{E_2}$ *for all* $g \in G$ *and all* $x \in E_1$;

(ii) $L(\rho)(L(G))$ *maps* E_1 *into* E_2.

$$\begin{aligned} &\rho(g)x \equiv x && \pmod{E_2} \quad \text{for all } g \in G \text{ and all } x \in E_1 \\ \Leftrightarrow\ &\rho(\exp a)x \equiv x && \pmod{E_2} \quad \text{for all } a \in L(G) \text{ and all } x \in E_1 \\ \Leftrightarrow\ &(\exp L(\rho)a)x \equiv x && \pmod{E_2} \quad \text{for all } a \in L(G) \text{ and all } x \in E_1. \end{aligned}$$

On the other hand, if $u \in \mathscr{L}(E)$, then

$$\begin{aligned} &\exp(\lambda u)x \equiv x && \pmod{E_2} \quad \text{for all } \lambda \in K \text{ and all } x \in E_1 \\ \Leftrightarrow\ &u(E_1) \subset E_2 \end{aligned}$$

whence the proposition.

COROLLARY 1. *For* E_1 *to be stable under* ρ, *it is necessary and sufficient that* E_1 *be stable under* $L(\rho)$.

It suffices to take $E_1 = E_2$ in Proposition 13.

COROLLARY 2. *Suppose that* ρ *is finite-dimensional. For* ρ *to be simple* (resp. *semi-simple*), *it is necessary and sufficient that* $L(\rho)$ *be simple* (resp. *semi-simple*).

This follows from Corollary 1.

COROLLARY 3. *Let* $x \in E$. *For* x *to be invariant under* $\rho(G)$, *it is necessary and sufficient that* x *be annihilated by* $L(\rho)(L(G))$ (that is that x be invariant under $L(\rho)$ in the sense of Chapter I, § 3, Definition 3).

It suffices to take $E_1 = Kx$ and $E_2 = 0$ in Proposition 13.

COROLLARY 4. *Let* ρ' *be another analytic linear representation of* G *on a complete normable space* E'. *Let* $T \in \mathscr{L}(E, E')$. *The following conditions are equivalent:*

(i) $T\rho(g) = \rho'(g)T$ *for all* $g \in G$;

(ii) $TL(\rho)(a) = L(\rho')(a)T$ *for all* $a \in L(G)$.

Let σ be the linear representation of G on $\mathscr{L}(E, E')$ derived from ρ and ρ' (§ 3, no. 11, Corollary 1 to Proposition 41). Condition (i) means that T is invariant under $\sigma(G)$. Condition (ii) means that T is annihilated by $L(\sigma)(L(G))$. It then suffices to apply Corollary 3.

COROLLARY 5. *Suppose that* ρ *and* ρ' *are finite-dimensional. For* ρ *and* ρ' *to be equivalent, it is necessary and sufficient that* $L(\rho)$ *and* $L(\rho')$ *be equivalent.*

This is a special case of Corollary 4.

COROLLARY 6. *Suppose that* G *is finite-dimensional. Let* $t \in U(G)$. *For* L_t (resp. R_t) *to be right* (resp. *left*) *invariant, it is necessary and sufficient that t belong to the centre of* $U(G)$.

For L_t (resp. R_t) to be right (resp. left) invariant, it is necessary and sufficient that $\varepsilon_g * t = t * \varepsilon_g$ for all $g \in G$, that is that $(\operatorname{Int} g)_* t = t$. There exists an integer n such that $t \in U_n(G)$. By Corollary 3 and Proposition 45 of § 3, no. 12, $(\operatorname{Int} g)_* t = t$ for all $g \in G$ if and only if $[a, t] = 0$ for all $a \in L(G)$, that is if and only if t commutes with $U(G)$.

6. NORMAL INTEGRAL SUBGROUPS

Lemma 3. *Let* G *be a Lie group,* H_1 *and* H_2 *integral subgroups whose topology admits a countable base and* $g \in G$. *Then*

$$gH_1g^{-1} = H_2 \Leftrightarrow (\operatorname{Ad} g)(L(H_1)) = L(H_2).$$

$\operatorname{Ad} g = T_e(\operatorname{Int} g)$. Hence, by transport of structure, $(\operatorname{Int} g)(H_1)$ has Lie algebra $(\operatorname{Ad} g)(L(H_1))$. As H_1 and H_2 have countable bases, to say that the sets H_2 and $(\operatorname{Int} g)(H_1)$ are equal amounts to saying that the integral subgroups H_2 and $(\operatorname{Int} g)(H_1)$ are equal (no. 2, Corollary 3 to Proposition 3). Then the lemma follows from Theorem 2 (i).

PROPOSITION 14. *Let* G *be a Lie group and* H *an integral subgroup whose topology admits a countable base. The following conditions are equivalent:*

(i) H *is normal in* G;

(ii) L(H) *is invariant under* Ad(G).

If further G *is connected, these conditions are equivalent to the following:*

(iii) L(H) *is an ideal of* L(G).

If further G *is simply connected and* L(H) *is of finite codimension in* L(G), *these conditions imply that* H *is a Lie subgroup of* G *and that* G/H *is simply connected.*

The equivalence (i) $\Leftrightarrow$ (ii) follows from Lemma 3. If further G is connected, condition (ii) is equivalent to saying that L(H) is stable under ad L(G) (no. 5, Corollary 1 to Proposition 13 and § 3, no. 12, Proposition 44).

Suppose that G is simply connected and that L(H) is an ideal of finite codimension in L(G). By Theorem 3 of no. 3, there exists a simply connected Lie group G′ such that L(G′) is isomorphic to L(G)/L(H). There exists a continuous morphism f of L(G) onto L(G′) with kernel L(H). By Theorem 1 of no. 1, there exists a morphism ϕ of G into G′ such that $L(\phi) = f$. This morphism is a submersion and hence its kernel N is a Lie subgroup of G such that $L(N) = \operatorname{Ker} f = L(H)$. Hence H is the identity component of N and is therefore a Lie subgroup of G. Let ψ be the morphism of G/H into G′ derived from ϕ when passing to the quotient. This morphism is étale; since G is simply connected, ψ is an isomorphism of G/H onto G′.

COROLLARY 1. *Let* G *be a finite-dimensional simply connected Lie group. Let* $\mathfrak{m}$, $\mathfrak{h}$ *be Lie subalgebras of* L(G) *such that* L(G) *is the semi-direct product of* $\mathfrak{m}$ *by* $\mathfrak{h}$. *Let* M, H *be the corresponding integral subalgebras of* G. *Then* M *and* H *are simply connected Lie subgroups of* G *and, as a Lie group,* G *is the semi-direct product of* M *by* H.

By Proposition 14, H is a normal Lie subgroup of G and the Lie group G/H is simply connected. Let π be the canonical morphism of G onto G/H. There exists a morphism θ of G/H into M such that $L(\theta)$ is the canonical isomorphism of $L(G)/L(H) = L(G)/\mathfrak{h}$ onto $L(M) = \mathfrak{m}$. Then

$$L(\pi \circ \theta) = L(\pi) \circ L(\theta) = \mathrm{Id}_{L(G/H)}$$

and hence $\pi \circ \theta = \mathrm{Id}_{G/H}$. By no. 1, Corollary 1 to Proposition 1, $\theta(G/H) = M$. By Proposition 8 of § 1, no. 4, M is a Lie subgroup of G and the Lie group G is the semi-direct product of M by H.

COROLLARY 2. *Let* G *be a finite-dimensional simply connected Lie group,* H *a normal connected Lie subgroup of* G *and* π *the canonical morphism of* G *onto* G/H.

(i) *There exists an analytic mapping* ρ *of* G/H *into* G *such that* $\pi \circ \rho = \mathrm{Id}_{G/H}$.

(ii) *For every mapping* ρ *with the properties of* (i), *the mapping* $(h, m) \mapsto h\rho(m)$ *of* $H \times (G/H)$ *into* G *is an isomorphism of analytic manifolds.*

(iii) H *and* G/H *are simply connected.*

Let $n = \dim G - \dim H$. The corollary is obvious for $n = 0$. We argue by induction on n.

Suppose that there exists an ideal of L(G) containing L(H) distinct from L(G) and L(H). Let H′ be the corresponding connected Lie subgroup of G. Let $\pi_1 : G \to G/H'$ and $\pi_2 : H' \to H'/H$ be the canonical morphisms. By the induction hypothesis, there exist analytic mappings $\rho_1 : G/H' \to G$, $\rho_2 : H'/H \to H'$ such that $\pi_1 \circ \rho_1 = \mathrm{Id}_{G/H'}$, $\pi_2 \circ \rho_2 = \mathrm{Id}_{H'/H}$. Let

$$\pi_3 : G/H \to G/H'$$

be the canonical morphism. If $x \in G/H$ and y is a representative of x in G, y and $\rho_1(\pi_3(x))$ have the same canonical image modulo H′ and hence $x^{-1}\pi(\rho_1(\pi_3(x))) \in H'/H$. Let

$$\rho(x) = \rho_1(\pi_3(x))\rho_2(\pi(\rho_1(\pi_3(x)))^{-1}x) \in G.$$

Clearly ρ is an analytic mapping of G/H into G and

$$\begin{aligned}\pi(\rho(x)) &= \pi(\rho_1(\pi_3(x)))\pi_2(\rho_2(\pi(\rho_1(\pi_3(x)))^{-1}x)) \\ &= \pi(\rho_1(\pi_3(x)))\pi(\rho_1(\pi_3(x)))^{-1}x = x.\end{aligned}$$

If now the only ideals of L(G) containing L(H) are L(G) and L(H), the Lie algebra $L(G)/L(H)$ is either 1-dimensional or simple. In both cases, $L(G)$ is the semi-direct product of a subalgebra by $L(H)$; this is obvious in the first case and in the second this follows from Chapter I, § 6, Corollary 3 to Theorem 5. Assertion (i) then follows from Corollary 1.

Assertion (ii) is obvious. Assertion (iii) follows from (i) and (ii).

The conclusions of Corollary 2 are no longer necessarily true when G is infinite-dimensional or when H is not normal (Exercises 8 and 15).

COROLLARY 3. *Let* G *be a finite-dimensional connected Lie group and* H *a normal connected Lie subgroup of* G. *The canonical morphism of* $\pi_1(H)$ *into* $\pi_1(G)$ *is injective.*

Let G_1 be the universal covering of G and λ the canonical mapping of G_1 onto G. The Lie algebra of G_1 is identified with $L(G)$. The Lie subgroup $\lambda^{-1}(H)$ of G_1 is normal in G_1 and its Lie algebra is $L(H)$. Let H_1 be the identity component of $\lambda^{-1}(H)$ and $\lambda_1 = \lambda \mid H_1$. Then $L(H_1) = L(H)$ and hence λ_1 is an étale morphism of H_1 onto H. On the other hand, H_1 is simply connected (Corollary 2) and hence is identified with the universal covering of H. Then, by *General Topology*, Chapter XI, the canonical morphism of $\pi_1(H)$ into $\pi_1(G)$ is identified with the canonical injection of Ker λ_1 into Ker λ.

7. PRIMITIVES OF DIFFERENTIAL FORMS WITH VALUES IN A LIE ALGEBRA

PROPOSITION 15. *Let* G *be a Lie group,* M *a manifold of class* C^r $(r \geqslant 2)$ *and* α *a differential form of class* C^{r-1} *and degree* 1 *on* M *with values in* $L(G)$, *such that* $d\alpha + [\alpha]^2 = 0$. *Suppose that* M *is simply connected. For all* $x \in M$ *and all* $s \in G$, *there exists one and only one mapping* f *of class* C^{r-1} *of* M *into* G *such that* $f(x) = s$ *and* $f^{-1}.df = \alpha$.

The uniqueness of f follows from § 3, no. 17, Corollary 2 to Proposition 59 and the fact that M is connected. We prove the existence of f. There exist an open covering $(U_i)_{i \in I}$ of M and, for all $i \in I$, a mapping $g_i: U_i \rightarrow G$ of class C^{r-1} such that $g_i^{-1}.dg = \alpha$ on U_i (§ 4, no. 6, Theorem 5). By § 3, no. 17, Corollary 2 to Proposition 59, $g_i g_j^{-1}$ is locally constant on $U_i \cap U_j$. Let $g_i g_j^{-1} = g_{ij}$. Let G_d be the group G with the discrete topology. The $g_{ij}: U_i \cap U_j \rightarrow G_d$ are continuous and $g_{ij} g_{jk} = g_{ik}$ on $U_i \cap U_j \cap U_k$. Since M is simply connected, there exist continuous mappings $\lambda_i: U_i \rightarrow G_d$ such that $g_i g_j^{-1} = \lambda_i \lambda_j^{-1}$ on $U_i \cap U_j$. Let g be the mapping of M into G whose restriction to U_i is $\lambda_i^{-1} g_i$ for all $i \in I$. This mapping is of class C^{r-1} and $g^{-1}dg = \alpha$. The mapping f of M into G defined by $f = s(g(x))^{-1}g$ satisfies the conditions $f^{-1}.df = \alpha$ and $f(x) = s$.

8. PASSAGE FROM LAWS OF INFINITESIMAL OPERATION TO LAWS OF OPERATION

Lemma 4. *Let* G *be a connected topological group,* X *a Hausdorff topological space and* f_1, f_2 *laws of left* (resp. *right*) *operation of* G *on* X *such that, for all* $x \in X$, *the mappings*

$s \mapsto f_1(s, x)$, $s \mapsto f_2(s, x)$ *of* G *into* X *are continuous. Suppose that there exists a neighbourhood* V *of* $\{e\} \times X$ *in* $G \times X$ *such that* f_1 *and* f_2 *coincide on* V. *Then* $f_1 = f_2$.

Let $x \in X$ and A be the set of $g \in G$ such that $f_1(g, x) = f_2(g, x)$. Then A is closed in G. On the other hand, let $g \in A$; we write $y = f_1(g, x) = f_2(g, x)$. There exists a neighbourhood U of e in G such that $f_1(t, y) = f_2(t, y)$ for $t \in U$, in other words such that $f_1(t', x) = f_2(t', x)$ for $t' \in Ug$ (resp. gU). Hence A is open in G and therefore A = G.

PROPOSITION 16. *Let* G *be a connected Lie group,* X *a Hausdorff manifold of class* C^r *and* f_1, f_2 *laws of left* (resp. *right*) *operation of class* C^r *of* G *on* X. *If the laws of infinitesimal operation associated with* f_1 *and* f_2 *are equal, then* $f_1 = f_2$.

By § 4, no. 7, Corollary to Proposition 11, there exists a neighbourhood V of $\{e\} \times X$ in $G \times X$ such that f_1 and f_2 coincide on V. Hence $f_1 = f_2$ (Lemma 4).

Lemma 5. *Let* G *be a simply connected topological group,* X *a Hausdorff topological space,* U *an open neighbourhood of* e *in* G *and* ψ *a continuous mapping of* $U \times X$ *into* X *such that* $\psi(e, x) = x$ *and* $\psi(s, \psi(t, x)) = \psi(st, x)$ *for all* $x \in X$ *and* s, t *in* U *such that* $st \in U$. *Let* W *be a symmetric connected open neighbourhood of* e *such that* $W^3 \subset U$. *There exists one and only one law of continuous left operation* ψ' *of* G *on* X *such that* ψ' *and* ψ *coincide on* $W \times X$. *If* G *is a Lie group and* X *a manifold of class* C^r *and* ψ *is of class* C^r, *then* ψ' *is of class* C^r.

The uniqueness on ψ follows from Lemma 4. Let P be the permutation group of X. For $u \in W^3$, the mapping $x \mapsto \psi(u, x)$ is an element $f(u)$ of P and

$$f(u_1 u_2 u_3) = f(u_1) f(u_2) f(u_3)$$

for u_1, u_2, u_3 in W. Applying Lemma 1 of no. 1, we obtain a morphism f' of G into P which extends $f|W$. Let $\psi'(g, x) = f'(g)(x)$ for all $(g, x) \in G \times X$. Then ψ' is a law of left operation of G on X which coincides with ψ on $W \times X$. As $\psi'(g, \psi'(g', x)) = \psi'(gg', x)$ for $(g, g', x) \in G \times G \times X$, the continuity of ψ on $W \times X$ implies the continuity of ψ' on $gW \times X$ for all $g \in G$. Hence ψ' is continuous. If ψ is of class C^r, it is seen similarly that ψ' is of class C^r.

THEOREM 5. *Let* G *be a simply connected Lie group,* X *a compact manifold of class* C^r $(r \geqslant 2)$ *and* $a \mapsto D_a$ *a law of left* (resp. *right*) *infinitesimal operation of class* C^{r-1} *of* L(G) *on* X. *There exists one and only one law of left* (resp. *right*) *operation of class* C^{r-1} *of* G *on* X *such that the associated law of infinitesimal operation is* $a \mapsto D_a$.

The uniqueness follows from Proposition 16. By § 4, no. 7, Corollary 1 to Theorem 6, there exist a neighbourhood V of $\{e\} \times X$ in $G \times X$ and a law chunk of left (resp. right) operation of class C^{r-1} of G on X, defined on V, such that the associated law of infinitesimal operation is $a \mapsto D_a$. As X is compact, V can be assumed to be of the form $U \times X$, where U is an open neighbourhood of e in G. It then suffices to apply Lemma 5.

9. EXPONENTIAL MAPPING IN THE LINEAR GROUP

PROPOSITION 17. *Let Δ be the set of $z \in \mathbf{C}$ such that $-\pi < \mathscr{I}(z) < \pi$ and Δ' the set of $z \in \mathbf{C}$ which are not real $\leqslant 0$. Let* E *be a complete normable space over* **C** *and* A (resp. A′) *the set of $x \in \mathscr{L}(E)$ whose spectrum* Sp(x) *is contained in* Δ (resp. Δ'). *Then* A (resp. A′) *is an open subset of* $\mathscr{L}(E)$ (resp. **GL**(E)) *and the mappings* $\exp: A \to A'$ *and* $\log: A' \to A$ (*Spectral Theories*, Chapter I, § 4, no. 9) *are inverse isomorphisms of analytic manifolds.*

This follows from *Spectral Theories*, Chapter I, § 4, Proposition 10 and no. 9.

THEOREM 6. *Let* E *be a real or complex Hilbert space and* U *the unitary group of* E.

(i) *The set* H *of Hermitian elements of* $\mathscr{L}(E)$ *is, with the real normed space structure, a closed vector subspace of* $\mathscr{L}(E)$ *admitting a topological supplement.*

(ii) *The set* H′ *of elements* $\geqslant 0$ *of* **GL**(E) *is a real analytic submanifold of* **GL**(E).

(iii) *The restriction to* H *of the mapping* exp *is an isomorphism of real analytic manifolds of* H *onto* H′.

(iv) *The mapping* $(h, u) \mapsto (\exp h)u$ *of* $H \times U$ *into* **GL**(E) *is an isomorphism of real analytic manifolds.*

Recall that, if $x \in \mathscr{L}(E)$, x^* denotes the adjoint of x. Let H_1 be the set of $x \in \mathscr{L}(E)$ such that $x^* = -x$. The formula $x = \frac{1}{2}(x + x^*) + \frac{1}{2}(x - x^*)$ proves that, with its real normed space structure, $\mathscr{L}(E)$ is the topological direct sum of H and H_1, whence (i).

Suppose that $K = \mathbf{C}$. In the notation of Proposition 17, H′ is the set of $h \in H \cap A'$ such that $\operatorname{Sp} h \subset \mathbf{R}_+^*$. As $\exp(\mathbf{R}) = \mathbf{R}_+^*$, (ii) and (iii) follow from Proposition 17 and *Spectral Theories*, Chapter I, § 4, Proposition 8 and § 6, no. 5. The mapping $(h, u) \mapsto y = (\exp h)u$ of $H \times U$ into **GL**(E) is bijective by *Spectral Theories*, Chapter I, § 6, Proposition 15. It is real analytic by the above. The mapping $y \mapsto h = \frac{1}{2}\log(yy^*)$ is real analytic and so also is the mapping $y \mapsto u = (\exp h)^{-1}y$. Hence (iv).

Suppose that $K = \mathbf{R}$. Let $\tilde{E}$ be the complexified Hilbert space of E and J the mapping $\xi + i\eta \mapsto \xi - i\eta$ (for ξ, η in E) of $\tilde{E}$ into $\tilde{E}$. Let $\tilde{H}$, $\tilde{H}'$, $\tilde{U}$ denote the sets defined for $\tilde{E}$ as were H, H′, U for E. Then H (resp. H′, U) is identified with the set of $x \in \tilde{H}$ (resp. $\tilde{H}'$, $\tilde{U}$) such that $JxJ^{-1} = x$. Properties (ii), (iii) and (iv) then follow easily from (i) and the analogous properties in the complex case.

PROPOSITION 18. *Let* E *be a complete normable space over* **C**, $v \in \mathscr{L}(E)$ *and* $g = \exp v$. *Suppose that* Sp(v) *contains none of the points* $2i\pi n$ *with* $n \in \mathbf{Z} - \{0\}$. *Then, for all* $x \in E$, *the conditions* $vx = 0$ *and* $gx = x$ *are equivalent.*

This follows from Lemma 2 of no. 4, applied to the function $z \mapsto e^z - 1$.

COROLLARY 1. *Let* E *be a complete normable space over* **C** *and* F *the space of continuous n-linear mappings of* E^n *into* E. *For all* $v \in \mathscr{L}(E)$, *let* $\sigma(v)$ *be the element of* $\mathscr{L}(F)$ *defined by*

$$(\sigma(v)f)(x_1, \ldots, x_n) = v(f(x_1, \ldots, x_n)) - \sum_{i=1}^{n} f(x_1, \ldots, vx_i, \ldots, x_n).$$

For all $g \in \mathbf{GL}(E)$, *let* $\rho(g)$ *be the element of* $\mathbf{GL}(F)$ *defined by*

$$(\rho(g)f)(x_1, \ldots, x_n) = g(f(g^{-1}x_1, \ldots, g^{-1}x_n)).$$

Let $u \in \mathscr{L}(E)$ *be such that every* $z \in \operatorname{Sp} u$ *satisfies* $|\mathscr{I}(z)| < \dfrac{2\pi}{n+1}$. *Then, for all* $f \in F$, *the conditions* $\sigma(u)f = 0$ *and* $\rho(\exp u)f = f$ *are equivalent.*

$L(\rho) = \sigma$ (§ 3, no. 11, Corollary 1 to Proposition 41) and hence

$$\rho(\exp u) = \exp \sigma(u)$$

(no. 4, Corollary 3 to Proposition 10). By Proposition 18 it then suffices to prove that $\operatorname{Sp} \sigma(u)$ does not meet $2i\pi(\mathbf{Z} - \{0\})$. But this follows from the following lemma:

Lemma 6. *If* $v \in \mathscr{L}(E)$, *then* $\operatorname{Sp} \sigma(v) \subset \operatorname{Sp} v + \operatorname{Sp} v + \cdots + \operatorname{Sp} v$, *where the sum comprises* $n + 1$ *terms.*

We define elements $v_0, v_1, \ldots, v_n$ of $\mathscr{L}(F)$ by writing, for all $f \in F$,

$$(v_0 f)(x_1, \ldots, x_n) = v(f(x_1, \ldots, x_n))$$
$$(v_i f)(x_1, \ldots, v_n) = -f(x_1, \ldots, vx_i, \ldots, x_n) \quad \text{for } 1 \leqslant i \leqslant n.$$

Then $\sigma(v) = \sum_{i=0}^{n} v_i$ and the v_i are pairwise permutable. Let A be the total closed subalgebra of $\mathscr{L}(F)$ generated by the v_i; it is commutative (*Spectral Theories*, Chapter I, § 1, no. 4) and $\operatorname{Sp}_{\mathscr{L}(F)} v' = \operatorname{Sp}_A v' \subset \sum_{i=0}^{n} \operatorname{Sp} v_i$ (*Spectral Theories*, Chapter I, § 3, Proposition 3 (ii)). Now, if $\lambda \in \mathbf{C}$ is such that $v - \lambda$ is invertible, clearly the $v_i - \lambda_i$ are invertible and hence $\operatorname{Sp} v_i \subset \operatorname{Sp} v$ for all i.

COROLLARY 2. *Let* E *be a complete normable algebra over* **C** *and* $w \in \mathscr{L}(E)$. *Suppose that every* $z \in \operatorname{Sp} w$ *satisfies* $|\mathscr{I}(z)| < \dfrac{2\pi}{3}$. *The following conditions are equivalent:*

(i) *w is a derivation of* E;

(ii) exp *w is an automorphism of* E.

This follows from Corollary 1 with $n = 2$ and f the multiplication of E.

PROPOSITION 19. *Let* E *be a complete normable space over* **C**, $v \in \mathscr{L}(\mathrm{E})$ *and* $g = \exp v$. *Suppose that every* $z \in \mathrm{Sp}\, v$ *satisfies* $-\pi < \mathscr{I}(z) < \pi$. *Then, for every closed vector subspace* E′ of E, *the conditions* $v(\mathrm{E}') \subset \mathrm{E}'$ *and* $g(\mathrm{E}') = \mathrm{E}'$ *are equivalent.*

The condition $v(\mathrm{E}') \subset \mathrm{E}'$ implies $g(\mathrm{E}') \subset \mathrm{E}'$ and $g^{-1}(\mathrm{E}') \subset \mathrm{E}'$ and hence $g(\mathrm{E}') = \mathrm{E}'$. Suppose that $g(\mathrm{E}') = \mathrm{E}'$. We use the notation Δ, Δ' of Proposition 17. Since $\mathrm{Sp}\, v$ is a compact subset of Δ, there exists a compact rectangle $\mathrm{Q} = [a, b] \times [a', b']$ such that $\mathrm{Sp}\, v \subset \mathrm{Q} \subset \Delta$. The set $\Delta - \mathrm{Q}$ is connected. Hence $\mathrm{Sp}\, g \subset \exp \mathrm{Q} \subset \Delta'$, the set $\exp \mathrm{Q}$ is compact and the set $\Delta' - \exp \mathrm{Q}$ is connected. The closure of the latter contains $]-\infty, 0]$ and hence

$$(\Delta' - \exp \mathrm{Q}) \cup\,]-\infty, 0] = \mathbf{C} - \exp \mathrm{Q}$$

is connected. Then $\exp \mathrm{Q}$ is polynomially convex (*Spectral Theories*, Chapter I, § 3, Corollary 2 to Proposition 9) and hence the function log, defined on Δ', is a limit in $\mathscr{O}(\exp \mathrm{Q})$ of polynomial functions (*Spectral Theories*, Chapter I, § 4, Proposition 3). Hence $v = \log g$ is the limit in $\mathscr{L}(\mathrm{E})$ of elements of the form $\mathrm{P}(g)$, where P is a polynomial (*Spectral Theories*, Chapter I, § 4, Theorem 3). As $\mathrm{P}(g)(\mathrm{E}') \subset \mathrm{E}'$, it follows that $v(\mathrm{E}') \subset \mathrm{E}'$.

COROLLARY. *Let* E *be a complete normed space over* **C**, $v \in \mathscr{L}(\mathrm{E})$ *and* $g = \exp v$. *Suppose that every* $z \in \mathrm{Sp}\, v$ *satisfies* $-\frac{\pi}{2} < \mathscr{I}(z) < \frac{\pi}{2}$. *Then, for every closed vector subspace* M of $\mathscr{L}(\mathrm{E})$, *the conditions* $g\mathrm{M}g^{-1} = \mathrm{M}$ *and* $[v, \mathrm{M}] \subset \mathrm{M}$ *are equivalent.*

Let $\mathrm{F} = \mathscr{L}(\mathrm{E})$, g' be the mapping $f \mapsto gfg^{-1}$ of F into F and v' be the mapping $f \mapsto [v, f]$ of F into F. Then $g' = \exp v'$ (no. 4, Corollary 3 to Proposition 10 and § 3, no. 11, Corollary 1 to Proposition 41). Lemma 6 proves that $-\pi < \mathscr{I}(z) < \pi$ for all $z \in \mathrm{Sp}\, v'$. It then suffices to apply Proposition 19.

10. COMPLEXIFICATION OF A FINITE-DIMENSIONAL REAL LIE GROUP

Lemma 7. *Let* B *be a group,* A *a normal subgroup of* B, C *the group* B/A *and* $i\colon \mathrm{A} \to \mathrm{B}$ and $p\colon \mathrm{B} \to \mathrm{C}$ *the canonical morphisms. Let* A′ *be a group and* f *a homomorphism of* A *into* A′. *Let* ω *be a morphism of* B *into the automorphism group of* A′. *Suppose that, for* $a \in \mathrm{A}'$, $a' \in \mathrm{A}'$, $b \in \mathrm{B}$,

$$f\,(bab^{-1}) = \omega(b) f(a), \qquad \omega(a)a' = f(a)a'f(a)^{-1}.$$

Let B″ *be the semi-direct product of* B *by* A′ *relative to* ω *and* q *the canonical morphism of* B″ *onto* B.

(i) *The mapping* $a \mapsto (f(a^{-1}), i(a))$ *of* A *into* B″ *is a morphism of* A *onto a normal subgroup* D *of* B″. *Let* $\mathrm{B}' = \mathrm{B}''/\mathrm{D}$ *and* $i'\colon \mathrm{A}' \to \mathrm{B}'$, $g\colon \mathrm{B} \to \mathrm{B}'$ *be the morphisms of* A′ *and* B *into* B′ *derived when taking quotients from the canonical injections of* A′ *and* **B** *into* B″.

(ii) *The morphism* $p \circ q$ *of* B″ *into* C *defines when passing to the quotient a morphism of* B′ *into* C.

(iii) *i' is injective, p' is surjective,* $\mathrm{Ker}(p') = \mathrm{Im}(i')$ *and the following diagram is commutative*

$$\begin{array}{ccccc} A & \xrightarrow{i} & B & \xrightarrow{p} & C \\ {\scriptstyle f}\downarrow & & \downarrow{\scriptstyle g} & & \downarrow{\scriptstyle \mathrm{Id}_C} \\ A' & \xrightarrow{i'} & B' & \xrightarrow{p'} & C. \end{array} \tag{4}$$

(iv) *If $b \in B$ and $a' \in A'$, then*

$$g(b)i'(a')g(b)^{-1} = i'(\omega(b)a').$$

(i) For a_1, a_2 in A, we have, in B″,

$$\begin{aligned}(f(a_1^{-1}), i(a_1))(f(a_2^{-1}), i(a_2)) &= (f(a_1^{-1})(\omega(a_1)f(a_2^{-1})), i(a_1)i(a_2))\\ &= (f(a_1^{-1})f(a_1a_2^{-1}a_1^{-1}), i(a_1a_2))\\ &= (f((a_1a_2)^{-1}, i(a_1a_2))\end{aligned}$$

and hence $a \mapsto (f(a^{-1}), i(a))$ is a homomorphism h of A into B″. Let $a \in A$, $a' \in A'$, $b \in B$; then, in B″,

$$\begin{aligned} bh(a)b^{-1} &= bf(a^{-1})ab^{-1} = (\omega(b)f(a^{-1}))(bab^{-1})\\ &= f(ba^{-1}b^{-1})(bab^{-1}) = h(bab^{-1})\\ a'h(a)a'^{-1} &= a'f(a^{-1})aa'^{-1} = a'f(a^{-1})(\omega(a)a'^{-1})a\\ &= a'f(a^{-1})f(a)a'^{-1}f(a^{-1})a = h(a)\end{aligned}$$

and hence $h(A) = D$ is normal in B″.

(ii) For $a \in A$,

$$(p \circ q)(h(a)) = p(q(f(a^{-1})a)) = p(a) = e$$

and hence $p \circ q$ is trivial on D.

(iii) Let $a' \in A'$ be such that $i'(a') = e$; then $a' \in D$ and hence there exists $a \in A$ such that $a' = f(a^{-1})a$; this implies $a = e$, whence $a' = e$; thus i' is injective. As p and q are surjective, p' is surjective.

Let r denote the canonical morphism of B″ onto B′. Let $a' \in A'$, $b \in B$ and $b' = r(a'b)$. If $b' \in \mathrm{Im}(i')$, there exists $a'_1 \in A'$ such that $b' = r(a'_1)$; then there exists $a \in A$ such that $a'b = a'_1 f(a^{-1})a$, whence $b = a \in A$ and

$$p'(b') = p(q(a'b)) = p(b) = e;$$

thus, $\mathrm{Im}(i') \subset \mathrm{Ker}(p')$. We preserve the notation a', b, b' but assume that $b' \in \mathrm{Ker}(p')$; then $e = p'(b') = p(q(a'b)) = p(b)$ and hence $b \in A$, whence

$$b' = r(a'f(b)f(b^{-1})b) = r(a'f(b)) \in \mathrm{Im}(i');$$

thus $\mathrm{Ker}(p') \subset \mathrm{Im}(i')$.

If $a \in A$, then

$$i'(f(a)) = r(f(a)) = r(f(a)f(a^{-1})a) = r(a) = g(i(a)).$$

If $b \in B$, then

$$p'(g(b)) = p(b)$$

and hence diagram (4) is commutative.

(iv) Let $b \in B$, $a' \in A'$. Then

$$\begin{aligned} g(b)i'(a')g(b)^{-1} &= r(b)r(a')r(b)^{-1} = r(ba'b^{-1}) \\ &= r(\omega(b)a') = i'(\omega(b)a'). \end{aligned}$$

Proposition 20. *Let* G *be a finite-dimensional real Lie group.*

(i) *There exist a complex Lie group* $\tilde{G}$ *and an* **R**-*analytic morphism* γ *of* G *into* $\tilde{G}$ *with the following properties: for every complex Lie group* H *and every* **R**-*analytic morphism* ϕ *of* G *into* H, *there exists one and only one* **C**-*analytic morphism* ψ *of* $\tilde{G}$ *into* H *such that* $\phi = \psi \circ \gamma$.

(ii) *If* $(\tilde{G}', \gamma')$ *has the same properties as* $(\tilde{G}, \gamma)$, *there exists one and only one isomorphism* θ *of* $\tilde{G}$ *onto* $\tilde{G}'$ *such that* $\theta \circ \gamma = \gamma'$.

(iii) *The* **C**-*linear mapping of* $L(G) \otimes \mathbf{C}$ *into* $L(\tilde{G})$ *which extends* $L(\gamma)$ *is surjective; in particular* $\dim_{\mathbf{C}}(\tilde{G}) \leqslant \dim_{\mathbf{R}}(G)$.

Assertion (ii) is obvious. We prove the existence of an ordered pair $(\tilde{G}, \gamma)$ with properties (i) and (iii).

(a) Suppose first that G is connected. Let $\mathfrak{g} = L(G)$, $\mathfrak{g}_{\mathbf{C}} = \mathfrak{g} \otimes_{\mathbf{R}} \mathbf{C}$ be the complexification of $\mathfrak{g}$, S (resp. S′) the simply connected real (resp. complex) Lie group with Lie algebra $\mathfrak{g}$ (resp. $\mathfrak{g}_{\mathbf{C}}$) and σ the unique **R**-analytic morphism of S into S′ such that $L(\sigma)$ is the canonical injection of $\mathfrak{g}$ into $\mathfrak{g}_{\mathbf{C}}$. Let π be the unique **R**-analytic morphism of S onto G such that $L(\pi) = \mathrm{Id}_{L(G)}$ and $F = \mathrm{Ker}\,\pi$.

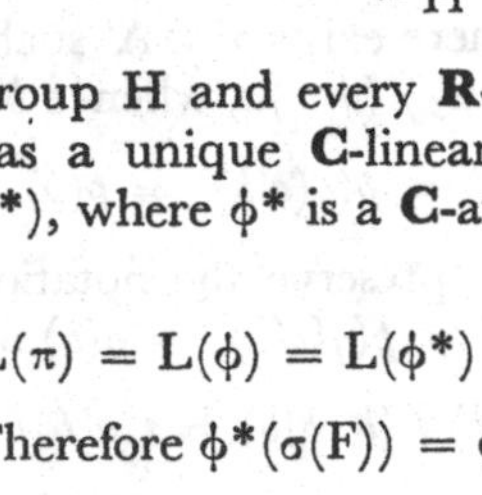

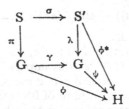

For every complex Lie group H and every **R**-analytic morphism ϕ of G into H, $L(\phi) : \mathfrak{g} \to L(H)$ has a unique **C**-linear extension to $\mathfrak{g}_{\mathbf{C}}$ and this extension is of the form $L(\phi^*)$, where ϕ^* is a **C**-analytic morphism of S′ into H. Then

$$L(\phi \circ \pi) = L(\phi) \circ L(\pi) = L(\phi) = L(\phi^*) \circ L(\sigma) = L(\phi^* \circ \sigma)$$

and hence $\phi \circ \pi = \phi^* \circ \sigma$. Therefore $\phi^*(\sigma(F)) = \phi(\pi(F)) = \{e\}$, whence

$$\sigma(F) \subset \mathrm{Ker}\,\phi^*.$$

Let P be the intersection of the $\mathrm{Ker}\,\phi^*$ for variable ϕ. This is a normal Lie subgroup of S′ (no. 2, Corollary 3 to Proposition 1). Let $\tilde{G} = S'/P$ and

$\lambda: S' \to \tilde{G}$ be the canonical morphism. Then $\sigma(F) \subset P$ and hence there exists one and only one **R**-analytic morphism γ of G into $\tilde{G}$ such that $\gamma \circ \pi = \lambda \circ \sigma$. If $\psi: \tilde{G} \to H$ denotes the morphism derived from ϕ^* when passing to the quotient, then

$$(\phi \circ \gamma) \circ \pi = \psi \circ (\lambda \circ \sigma) = \phi^* \circ \sigma = \phi \circ \pi$$

whence $\psi \circ \gamma = \phi$. Clearly $L(\psi)$, and hence ψ, are determined uniquely by the equality $\psi \circ \gamma = \phi$. Thus we have proved that the ordered pair $(\tilde{G}, \gamma)$ has properties (i) and (iii).

(b) We pass now to the general case. Let F be the identity component of G, $M = G/F$ and $i: F \to G$ and $p: G \to M$ be the canonical morphisms. We apply part (a) of the proof to F. We obtain an ordered pair $(\tilde{F}, \delta)$. For all $g \in G$, $\operatorname{Int} g|F = \omega'(g)$ is an automorphism of F. By the universal property of $\tilde{F}$, there exists one and only one automorphism $\omega(g)$ of the complex Lie group $\tilde{F}$ such that $\delta \circ \omega'(g) = \omega(g) \circ \delta$. Clearly ω is a morphism of G into $\operatorname{Aut}(\tilde{H})$. If $g \in G$ and $f \in F$, then

$$\delta(gfg^{-1}) = (\delta \circ \omega'(g))(f) = (\omega(g) \circ \delta)(f) = \omega(g)(\delta(f)).$$

If $f \in F$, then $\delta \circ (\operatorname{Int}_F f) = (\operatorname{Int}_{\tilde{F}} \delta(f)) \circ \delta$ and $\operatorname{Int}_{\tilde{F}} \delta(f)$ is an automorphism of the complex Lie group $\tilde{F}$; hence $\operatorname{Int}_{\tilde{F}} \delta(f) = \omega(f)$.

Hence Lemma 7 can be applied, which gives a diagram

$$\begin{array}{ccccc} F & \xrightarrow{i} & G & \xrightarrow{p} & M \\ \downarrow{\scriptstyle\delta} & & \downarrow{\scriptstyle\gamma} & & \downarrow{\scriptstyle\mathrm{Id}} \\ \tilde{F} & \xrightarrow{\tilde{i}} & \tilde{G} & \xrightarrow{\tilde{p}} & M. \end{array}$$

Let $\tilde{F}$ be identified with a normal subgroup of $\tilde{G}$ by means of $\tilde{i}$. The group $\tilde{G}$ is generated by $\tilde{F}$ and $\gamma(G)$; hence the automorphisms of $\tilde{F}$ defined by the elements of $\tilde{G}$ are automorphisms of the complex Lie group structure. By § 1, no. 9, Proposition 18, there exists one and only one complex Lie group structure on $\tilde{G}$ such that $\tilde{F}$ is an open Lie subgroup of $\tilde{G}$. Henceforth $\tilde{G}$ will have this structure. As δ is **R**-analytic, γ is **R**-analytic.

The ordered pair $(\tilde{G}, \gamma)$ has property (iii) of the proposition. We show that it has property (i). Let H be a complex Lie group and ψ an **R**-analytic morphism of G into H. There exists a **C**-analytic morphism η of $\tilde{F}$ into H such that $\eta \circ \delta = \phi|F$. Let $g \in G$. The mappings

$$f \mapsto \eta(\omega(g)f), \qquad f \mapsto \phi(g)\eta(f)\phi(g)^{-1}$$

of $\tilde{F}$ into H are **C**-analytic morphisms; they coincide on $\delta(F)$, for, if $f' \in F$, then

$$\phi(g)\eta(\delta(f'))\phi(g)^{-1} = \phi(g)\phi(f')\phi(g)^{-1} = \phi(gf'g^{-1})$$
$$= \eta(\delta(gf'g^{-1})) = \eta(\omega(g)\delta(f'));$$

therefore $\eta(\omega(g)f) = \phi(g)\eta(f)\phi(g)^{-1}$ for all $g \in G$ and all $f \in \tilde{F}$. If G′ denotes the semi-direct product of G by $\tilde{F}$ relative to ω, there then exists a morphism ζ of the group G′ into H which coincides with ϕ on G and with η on $\tilde{F}$. For $f \in F$,

$$\zeta(\delta(f^{-1})f) = \eta(\delta(f^{-1}))\phi(f) = \phi(f^{-1})\phi(f) = e.$$

Hence ζ defines on passing to the quotient a morphism ψ of $\tilde{G}$ into H. Then $\psi \circ \gamma = \phi$ and $\psi \circ \tilde{\imath} = \eta$; the latter inequality implies that ψ is **C**-analytic.

Finally, let ψ' be a **C**-analytic morphism of $\tilde{G}$ into H such that $\phi = \phi' \circ \gamma$. Then

$$\psi' \circ \tilde{\imath} \circ \delta = \psi' \circ \gamma \circ i = \phi \circ i = \psi \circ \tilde{\imath} \circ \delta$$

and hence $\psi' \circ \tilde{\imath} = \psi \circ \tilde{\imath}$. As $\tilde{G}$ is generated by $\tilde{\imath}(\tilde{F})$ and $\gamma(G)$, $\psi' = \psi$.

DEFINITION 4. (G, γ), *or simply* $\tilde{G}$, *is called the universal complexification of* G.

Remarks. (1) Let $(\tilde{G}, \gamma)$ be the universal complexification of G. Let G_0 (resp. $\tilde{G}_0$) be the identity component of G (resp. $\tilde{G}$). By the proof of Proposition 20, $(\tilde{G}_0, \gamma|G_0)$ is the universal complexification of G_0 and the composite morphism

$$G \to \tilde{G} \to \tilde{G}/\tilde{G}_0$$

defines on passing to the quotient an isomorphism of G/G_0 onto $\tilde{G}/\tilde{G}_0$.

(2) Suppose that G is simply connected. Let $\mathfrak{g} = L(G)$, $\mathfrak{g}_{\mathbf{C}}$ be the complexification of $\mathfrak{g}$, S′ the simply connected complex Lie group with Lie algebra $\mathfrak{g}_{\mathbf{C}}$ and σ the morphism of G into S′ such that $L(\sigma)$ is the canonical injection of $\mathfrak{g}$ into $\mathfrak{g}_{\mathbf{C}}$. We again use the notation of the proof of Proposition 20, part (a). If $H = S'$ and $\phi = \sigma$, then $\phi^* = \mathrm{Id}_{S'}$. Hence (S', σ) is the universal complexification of G. Note that σ is not in general injective (Exercise 16); however *its kernel is discrete* since $L(\sigma)$ is injective. On the other hand, let θ be the involution of $\mathfrak{g}_{\mathbf{C}}$ defined by $\mathfrak{g}$ and let η be the corresponding automorphism of the underlying real Lie group of S′; let S'^{η} be the set of points of S′ invariant under η; it is a real Lie subgroup of S′ with Lie algebra $\mathfrak{g}$ (§ 3, no. 8, Corollary 1 to Proposition 29). By no. 1, Corollary 1 to Proposition 1, $\sigma(G)$ is a real integral subgroup of S′ with Lie algebra $\mathfrak{g}$ and hence $\sigma(G)$ *is the identity component of* S'^{η}; in particular $\sigma(G)$ is a real Lie subgroup of S′.

§ 7. LIE GROUPS OVER AN ULTRAMETRIC FIELD

In this paragraph, the valued field K is assumed to be ultrametric and of characteristic 0. Let A denote the valuation ring of K, $\mathfrak{m}$ the maximal ideal of A and p the characteristic of the residue field $A/\mathfrak{m}$. If K is locally compact, then $p \neq 0$ (*Commutative Algebra*, Chapter VI, § 9, Theorem 1).

1. PASSAGE FROM LIE ALGEBRAS TO LIE GROUPS

PROPOSITION 1. *Let* G *be a Lie group germ with identity element e. There exists a fundamental system of open neighbourhoods of e in* G *consisting of Lie subgroups of* G.

Let L(G) be given a norm compatible with its topology and such that $\|[x, y]\| \leqslant \|x\| \|y\|$ for all x, y in L(G). Let G_1 be the Lie group defined by L(G). By § 4, no. 2, Theorem 2, G and G_1 are locally isomorphic. Then it suffices to apply § 4, no. 2, Lemma 3 (iii).

THEOREM 1. *Let* L *be a complete normable Lie algebra. There exists a Lie group* G *such that* L(G) *is isomorphic to* L. *Two such groups are locally isomorphic.*

The first assertion has been proved in § 4, no. 2, Lemma 3. The second is a special case of § 4, no. 2, Theorem 2.

THEOREM 2. *Let* G *be a Lie group and* $\mathfrak{h}$ *a Lie subalgebra of* L(G) *admitting a topological supplement. There exists a Lie subgroup* H *of* G *such that* L(H) = $\mathfrak{h}$. *If* H_1 *and* H_2 *are Lie subgroups of* G *such that* $L(H_1) = L(H_2) = \mathfrak{h}$, *then* $H_1 \cap H_2$ *is open in* H_1 *and* H_2.

The first assertion follows from Proposition 1 and § 4, no. 2, Theorem 3. The second is a special case of § 4, no. 2, Theorem 3.

THEOREM 3. *Let* G *and* H *be Lie groups and h a continuous morphism of* L(G) *into* L(H).

(i) *There exist an open subgroup* G′ *of* G *and a Lie group morphism* ϕ *of* G′ *into* H *such that* $h = L(\phi)$.

(ii) *Let* G_1, G_2 *be open subgroups of* G *and* ϕ_i *a morphism of* G_i *into* H *such that* $h = L(\phi_i)$. *Then* ϕ_1 *and* ϕ_2 *coincide on an open subgroup of* G.

By Proposition 1, this follows from § 4, no. 1, Theorem 1.

PROPOSITION 2. *Let* G *be a Lie group and* $\mathfrak{h}$ *a Lie subalgebra of* L(G) *admitting a topological supplement. The following conditions are equivalent:*

(i) *There exist an open subgroup* G′ *of* G *and a normal Lie subgroup* H *of* G′ *such that* L(H) = $\mathfrak{h}$.

(ii) $\mathfrak{h}$ *is an ideal of* L(H).

If there exist G′ and H with the properties of (i), then L(G′) = L(G) and L(H) is an ideal of L(G′) by § 3, no. 12, Proposition 47.

Suppose that $\mathfrak{h}$ is an ideal of L(G). There exists a Lie group F such that $L(F) = L(G)/\mathfrak{h}$ (Theorem 1). Let h be the canonical morphism of L(G) onto L(F). By Theorem 3 (i), there exist an open subgroup G′ of G and a Lie group morphism ϕ of G′ into F such that $L(\phi) = h$. By § 3, no. 8, the kernel H of ϕ is a Lie subgroup of G′ and $L(H) = \operatorname{Ker} L(\phi) = \operatorname{Ker} h = \mathfrak{h}$. Finally, H is normal in G′ since $H = \operatorname{Ker} \phi$.

2. EXPONENTIAL MAPPINGS

PROPOSITION 3. *Let* G *be a Lie group. There exists an exponential mapping* ϕ *of* G *with the following properties:*

(i) ϕ *is defined on an open subgroup* U *of the additive group* L(G);

(ii) $\phi(\mathrm{U})$ *is an open subgroup of* G *and* ϕ *is an isomorphism of the analytic manifold* U *onto the analytic manifold* $\phi(\mathrm{U})$;

(iii) $\phi(nx) = \phi(x)^n$ *for all* $x \in \mathrm{U}$ *and all* $n \in \mathbf{Z}$.

Let L(G) be given a norm compatible with its topology and such that $\|[x, y]\| \leqslant \|x\| \|y\|$ for x, y in L(G). Let G_1 be the Lie group defined by L(G). Let $\psi = \mathrm{Id}_{\mathrm{G}_1}$, which is an exponential mapping of G_1. For all $\mu > 0$, let L_μ be the set of $x \in \mathrm{L(G)}$ such that $\|x\| < \mu$. Then, for μ sufficiently small, L_μ is an open subgroup of the additive group L(G), $\psi(\mathrm{L}_\mu)$ is an open subgroup of G_1 (§ 4, no. 2, Lemma 3), $\psi|\mathrm{L}_\mu$ is an isomorphism of analytic manifolds of L_μ onto $\psi(\mathrm{L}_\mu)$ and $\psi(nx) = \psi(x)^n$ for all $x \in \mathrm{L}_\mu$ and all $n \in \mathbf{Z}$. The L_μ form a fundamental system of neighbourhoods of 0 in L(G). By Theorem 1, there exist μ and an open subgroup G′ of G such that $\psi(\mathrm{L}_\mu)$ and G′ are isomorphic, whence the proposition.

PROPOSITION 4. *Let* G *be a Lie group and* ϕ *an injective exponential mapping of* G. *Suppose that* $p > 0$. *For all* x, y *in* L(G),

$$x + y = \lim_{n \to +\infty} p^{-n} \phi^{-1}(\phi(p^n x)\phi(p^n y)) \tag{1}$$

$$[x, y] = \lim_{n \to +\infty} p^{-2n} \phi^{-1}(\phi(p^n x)\phi(p^n y)\phi(-p^n x)\phi(-p^n y)). \tag{2}$$

These are special cases of Proposition 4 of § 4, no. 3.

3. STANDARD GROUPS†

If $\mathrm{S}(\mathrm{X}_1, \mathrm{X}_2, \ldots, \mathrm{X}_n)$ is a formal power series with coefficients in A, then, for all $x_1, \ldots, x_r$ in $\mathfrak{m}$, the series $\mathrm{S}(x_1, x_2, \ldots, x_r)$ is convergent. More precisely, $\mathfrak{m} \times \mathfrak{m} \times \cdots \times \mathfrak{m}$ is contained in the domain of absolute convergence of S (*Differentiable and Analytic Manifolds*, R, 4.1.3).

DEFINITION 1. *Let* r *be an integer* $\geqslant 0$. *A standard group of dimension* r *over* K *is a Lie group* G *with the following properties:*

(i) *the underlying analytic manifold of* G *is* $\mathfrak{m} \times \mathfrak{m} \times \cdots \times \mathfrak{m}$ *(r factors);*

(ii) *there exists a formal power series* F *in* $2r$ *variables with coefficients in* A^r, *without constant term, such that* $x.y = \mathrm{F}(x, y)$ *for all* x, y *in* G.

Then $0.0 = 0$ and hence the identity element of G is the origin of $\mathfrak{m} \times \mathfrak{m} \cdots \times \mathfrak{m}$.

---

† The results of nos. 3 and 4 and their proofs remain true when the characteristic of K is >0.

L(G) will be identified with K^r. By § 5, formula (13), the constants of structure of L(G) with respect to the canonical basis belong to A. We shall need, in the same proof to consider the elements of $\mathfrak{m} \times \cdots \times \mathfrak{m}$, now as elements of G, now as elements of L(G).

Example. Let $G = 1 + \mathbf{M}_n(\mathfrak{m})$, which is an open subset of $\mathbf{M}_n(K)$. If $x \in G$, then $\det x \in 1 + \mathfrak{m}$ and hence $G \subset \mathbf{GL}(n, K)$. Clearly $GG \subset G$. If $x = 1 + y$ with $y \in \mathbf{M}_n(\mathfrak{m})$, the calculation of the inverse of a matrix proves first that $x^{-1} \in \mathbf{M}_n(A)$; if we write $x^{-1} = 1 + y'$, then $y + y' + yy' = 0$, hence $y' \in \mathbf{M}_n(\mathfrak{m})$ and therefore $x^{-1} \in G$. Thus G is an open subgroup of $\mathbf{GL}(n, K)$. We identify G with $\mathfrak{m}^{n^2}$ by means of the mapping $(\delta_{ij} + y_{ij}) \mapsto (y_{ij})$. Clearly G is a standard group.

THEOREM 4. *Let G be a finite-dimensional Lie group. There exists an open subgroup of G isomorphic to a standard group.*

By replacing G by a group isomorphic to an open subgroup of G, the problem is reduced to the case where G is an open subset of K^r, with identity element 0 and where the coordinates of the product $x.y$ and the inverse $x^{[-1]}$ are given by formulae

$$(x.y)_i = x_i + y_i + \sum_{|\alpha| \geqslant 1, |\beta| \geqslant 1} c_{\alpha\beta i} x^\alpha y^\beta \quad (i = 1, 2, \ldots, r) \tag{3}$$

$$(x^{[-1]})_i = -x_i + \sum_{|\alpha| > 1} d_{\alpha i} x^\alpha \quad (i = 1, 2, \ldots, r) \tag{4}$$

where the series on the right hand side are convergent for x, y in G (§ 5, no. 1). Let $\lambda \in K^*$ and let the group law be transported from G to $G' = \lambda G$ by the homothety of ratio λ. For x', y' in G', the product $x'.y'$ and the inverse $x'^{[-1]}$ evaluated in G' have coordinates

$$(x'.y')_i = x'_i + y'_i + \sum_{|\alpha| \geqslant 1, |\beta| \geqslant 1} c'_{\alpha\beta i} x'^\alpha y'^\beta \quad (i = 1, 2, \ldots, r)$$

$$(x'^{[-1]})_i = -x'_i + \sum_{|\alpha| > 1} d'_{\alpha i} x'^\alpha \quad (i = 1, 2, \ldots, r)$$

where

$$c'_{\alpha\beta i} = \lambda^{-|\alpha| - |\beta| + 1} c_{\alpha\beta i}, \qquad d'_{\alpha i} = \lambda^{-|\alpha| + 1} d_{\alpha i}.$$

As the series (3) and (4) are convergent, we see that, for $|\lambda|$ sufficiently large, for all α, β, i,

$$|c'_{\alpha\beta i}| \leqslant 1, \qquad |d'_{\alpha i}| \leqslant 1$$

that is $c'_{\alpha\beta i} \in A$ and $d'_{\alpha i} \in A$; and on the other hand $G' \supset \mathfrak{m} \times \mathfrak{m} \times \cdots \times \mathfrak{m}$. Then $\mathfrak{m} \times \mathfrak{m} \times \cdots \times \mathfrak{m}$ is an open subgroup of G' and is a standard group.

4. FILTRATION OF STANDARD GROUPS

We again use the notation of Definition 1. We choose a number $a > 1$ and a real valuation v of K such that $|x| = a^{-v(x)}$ for all $x \in K$ (*Commutative Algebra*, Chapter VI, § 6, Proposition 3). If $\mathfrak{a}$ is a non-zero (and hence open) ideal of A contained in $\mathfrak{m}$, let $G(\mathfrak{a})$ denote the set of elements of G whose coordinates belong to $\mathfrak{a}$. If $\lambda \in \mathbf{R}$, let $\mathfrak{a}_\lambda$ (resp. $\mathfrak{a}_\lambda^+$) denote the set of $x \in K$ such that $v(x) \geqslant \lambda$ (resp. $v(x) > \lambda$); then $\mathfrak{a}_0 = A$, $\mathfrak{a}_0^+ = \mathfrak{m}$. For $x = (x_1, \ldots, x_r) \in G$, we shall write

$$\omega(x) = \inf(v(x_1), \ldots, v(x_r)). \tag{5}$$

PROPOSITION 5. *Let* G *be a standard group.*

(i) *If* $\mathfrak{a}$ *is a non-zero ideal of* A *contained in* $\mathfrak{m}$, $G(\mathfrak{a})$ *is an open normal subgroup of* G.

(ii) *The* $G(\mathfrak{a}_\lambda)$, *for* $\lambda > 0$, *form a fundamental system of neighbourhoods of* e *in* G.

(iii) *Suppose that* $\mathfrak{a}_\lambda \subset \mathfrak{a}$ *for* $\lambda \geqslant \lambda_0$ *and let the* $G(\mathfrak{a})/G(\mathfrak{a}_\lambda)$, *for* $\lambda \geqslant \lambda_0$, *be given the discrete topology. Then the topological group* $G(\mathfrak{a})$ *is the inverse limit of the groups* $G(\mathfrak{a})/G(\mathfrak{a}_\lambda)$.

(iv) *Let* $\mathfrak{a}$, $\mathfrak{b}$ *be non-zero ideals of* A *contained in* $\mathfrak{m}$ *such that* $\mathfrak{a} \supset \mathfrak{b} \supset \mathfrak{a}^2$. *The mapping* $x \mapsto (x_1 \bmod \mathfrak{b}, \ldots, x_r \bmod \mathfrak{b})$ *of* $G(\mathfrak{a})$ *into* $(\mathfrak{a}/\mathfrak{b}) \times \cdots \times (\mathfrak{a}/\mathfrak{b})$ *defines on passing to the quotient an isomorphism of the group* $G(\mathfrak{a})/G(\mathfrak{b})$ *onto the additive group* $(\mathfrak{a}/\mathfrak{b}) \times \cdots \times (\mathfrak{a}/\mathfrak{b})$.

If $x \in G$ and $y \in G(\mathfrak{a})$, the coordinates of x and $x.y$ are equal modulo $\mathfrak{a}$. Hence, for x', x'' in G and y', y'' in $G(\mathfrak{a})$, the coordinates of $x'.x''$ and $(x'.y').(x''.y'')$ are equal modulo $\mathfrak{a}$. This proves (i).

(ii) is obvious.

(iii) follows from the above and *General Topology*, Chapter III, § 7, Proposition 2.

If $x \in G(\mathfrak{a})$ and $y \in G(\mathfrak{a})$, the coordinates of $x.y$ are congruent to those of $x + y$ modulo $G(\mathfrak{a}^2)$ by formula (4) of § 5. This proves (iv).

COROLLARY. *Suppose that* K *is locally compact and let* $q = \operatorname{Card}(A/\mathfrak{m})$.

(i) *If* $\mathfrak{a} = \mathfrak{m}^a$ and $\mathfrak{b} = \mathfrak{m}^b$ *with* $b \geqslant a \geqslant 1$, $G(\mathfrak{a})/G(\mathfrak{b})$ *is a p-group of cardinal* $q^{r(b-a)}$.

(ii) $G(\mathfrak{a})$ *is an inverse limit of p-groups.*

The number of elements in $G(\mathfrak{a})/G(\mathfrak{b})$ is $(\operatorname{Card}(\mathfrak{a}/\mathfrak{b}))^r$; if $b = a + 1$, $\mathfrak{a}/\mathfrak{b}$ is a 1-dimensional vector space over $A/\mathfrak{m}$, whence (i) in this case; the general case follows by induction on $b - a$. Assertion (ii) follows from (i) and Proposition 5 (iii).

PROPOSITION 6. *Let* $\mathfrak{a}$, $\mathfrak{b}$, $\mathfrak{c}$, $\mathfrak{c}'$ *be non-zero ideals of* A *contained in* $\mathfrak{m}$ *such that*

$$\mathfrak{c}' \subset \mathfrak{c}, \qquad \mathfrak{a}\mathfrak{b} \subset \mathfrak{c}, \qquad \mathfrak{a}\mathfrak{b}^2 \subset \mathfrak{c}', \qquad \mathfrak{a}^2\mathfrak{b} \subset \mathfrak{c}'.$$

If $x \in G(\mathfrak{a})$ *and* $y \in G(\mathfrak{b})$, *then* $x^{[-1]}.y^{[-1]}.x.y$, $x.y.x^{[-1]}y^{[-1]}$ *and* $[x, y]$ *belong to* $G(\mathfrak{c})$ *and are congruent modulo* $G(\mathfrak{c}')$.

By § 5, no. 2, Proposition 1, there exist $c_{\alpha\beta} \in A^r$ such that

$$x^{[-1]}.y^{[-1]}.x.y - [x, y] = \sum_{|\alpha|+|\beta|\geqslant 3} c_{\alpha\beta}x^{\alpha}y^{\beta}.$$

If $x = 0$ or $y = 0$, then $x^{[-1]}.y^{[-1]}.x.y - [x, y] = 0$; hence $c_{0\beta} = c_{\alpha 0} = 0$. On the other hand, the conditions

$$x \in G(\mathfrak{a}), \qquad y \in G(\mathfrak{b}), \qquad |\alpha| \geqslant 1, \qquad |\beta| \geqslant 1, \qquad |\alpha| + |\beta| \geqslant 3$$

imply

$$c_{\alpha\beta}x^{\alpha}y^{\beta} \in G(\mathfrak{a}^2\mathfrak{b} + \mathfrak{a}\mathfrak{b}^2) \subset G(\mathfrak{c}')$$

and hence $x^{[-1]}.y^{[-1]}x.y - [x, y] \in G(\mathfrak{c}')$. We see similarly that

$$x.y.x^{[-1]}.y^{[-1]} - [x, y] \in G(\mathfrak{c}').$$

Finally, by § 5, formula (13), $[x, y] \in G(\mathfrak{a}\mathfrak{b}) \subset G(\mathfrak{c})$.

PROPOSITION 7. (i) *The family* $(G(\mathfrak{a}_\lambda))$ *is a central filtration on* G (Chapter II, § 4, no. 4, Definition 2).

(ii) *For* $\lambda \in \mathbf{R}_+^*$, $G(\mathfrak{a}_\lambda) = \{x \in G \mid \omega(x) \geqslant \lambda\}$, $G(\mathfrak{a}_\lambda^+) = \{x \in G \mid \omega(x) > \lambda\}$.

(ii) is obvious. We prove (i). Clearly $G(\mathfrak{a}_\lambda) = \bigcap_{\mu<\lambda} G(\mathfrak{a}_\mu)$ and $G = \bigcup_{\lambda>0} G(\mathfrak{a}_\lambda)$. On the other hand, if $x \in G(\mathfrak{a}_\lambda)$ and $y \in G(\mathfrak{a}_\mu)$, then

$$x^{[-1]}.y^{[-1]}.x.y \in G(\mathfrak{a}_{\lambda+\mu})$$

by Proposition 6 applied with $\mathfrak{a} = \mathfrak{a}_\lambda$, $\mathfrak{b} = \mathfrak{a}_\mu$, $\mathfrak{c} = \mathfrak{c}' = \mathfrak{a}_{\lambda+\mu}$.

By Chapter II, § 4, no. 4, we can form the group gr(G) associated with the group G, with the central filtration $(G(\mathfrak{a}_\lambda))$. Writing $G_\lambda = G(\mathfrak{a}_\lambda)/G(\mathfrak{a}_\lambda^+)$ for all $\lambda > 0$, we obtain $\mathrm{gr}(G) = \bigoplus_{\lambda>0} G_\lambda$. Recall (*loc. cit.*, Proposition 1) that the commutator in G allows us to define a bracket in gr(G) with which gr(G) is a Lie algebra, as follows: if $\bar{x} \in G_\lambda$ and $\bar{y} \in G_\mu$, choose a representative x of $\bar{x}$ in $G(\mathfrak{a}_\lambda)$ and a representative y of $\bar{y}$ in $G(\mathfrak{a}_\mu)$; then $[\bar{x}, \bar{y}]$ is the class of $x^{[-1]}.y^{[-1]}.x.y \in G(\mathfrak{a}_{\lambda+\mu})$ in $G_{\lambda+\mu}$. By Proposition 6, applied with $\mathfrak{a} = \mathfrak{a}_\lambda$, $\mathfrak{b} = \mathfrak{a}_\mu$, $\mathfrak{c} = \mathfrak{a}_{\lambda+\mu}$, $\mathfrak{c}' = \mathfrak{a}_{\lambda+\mu}^+$, we see that $[\bar{x}, \bar{y}]$ is also the class of $[x, y]$ in $G_{\lambda+\mu}$. Thus, when G is considered as a Lie subalgebra of $L(G) = K^r$, filtered by the $G(\mathfrak{a}_\lambda)$, the associated graded Lie algebra (Chapter II, § 4, no. 3) is equal to gr(G).

5. POWERS IN STANDARD GROUPS

We preserve the notation of no. 4.

PROPOSITION 8. *Let* $n \in \mathbf{Z}$ *and* h_n *be the mapping* $x \mapsto x^n$ *of* G *into* G. *Let* $\mathfrak{a}$ *be a non-zero ideal of* A *contained in* $\mathfrak{m}$, *such that* $n \notin \mathfrak{a}$. *Then* $h_n | G(\mathfrak{a})$ *is an isomorphism of the analytic manifold* $G(\mathfrak{a})$ *onto the analytic manifold* $G(n\mathfrak{a})$.

By definition of standard groups, h_n is equal on the whole of G to the sum of an integral series with coefficients in A^r. By § 5, formula (4), this series is of the form

$$h_n(x) = nx + \sum_{|\alpha| \geq 2} a_\alpha x^\alpha.$$

Hence, for $x \in G$,

$$h_n(nx) = n^2\Big(x + \sum_{|\alpha| \geq 2} a_\alpha n^{|\alpha|-2} x^\alpha\Big)$$
$$= n^2 S(x)$$

where we write $S(x) = x + \sum_{|\alpha| \geq 2} a_\alpha n^{|\alpha|-2} x^\alpha$. This series $S(x)$ defines an analytic mapping, also denoted by S, of G into G. By *Algebra*, Chapter IV, § 6, Proposition 8, there exists an integral series S′ in r variables with coefficients in A^r such that $S'(S(X)) = S(S'(X)) = X$. Hence S is an isomorphism of the analytic manifold G onto itself and, for every non-zero ideal $\mathfrak{b}$ of A contained in $\mathfrak{m}$, $S(G(\mathfrak{b})) \subset G(\mathfrak{b})$, $S'(G(\mathfrak{b})) \subset G(\mathfrak{b})$ and therefore

$$S(G(\mathfrak{b})) = G(\mathfrak{b}).$$

As $h_n(y) = n^2 S\left(\frac{1}{n} y\right)$ for $y \in nG$, we see that $h_n | nG(\mathfrak{b})$ is an isomorphism of the analytic manifold $nG(\mathfrak{b})$ onto the analytic manifold $n^2G(\mathfrak{b})$. But, as $n \notin \mathfrak{a}$, $|n| > |\lambda|$ for all $\lambda \in \mathfrak{a}$, hence $n^{-1}\mathfrak{a} \subset \mathfrak{m}$ and hence $\mathfrak{a}$ is of the form $n\mathfrak{b}$ where $\mathfrak{b}$ is a non-zero ideal of A contained in $\mathfrak{m}$.

COROLLARY. *If n is invertible in* A, *h_n is an isomorphism of the analytic manifold* G *onto itself. For every non-zero ideal $\mathfrak{a}$ of* A *contained in $\mathfrak{m}$, $h_n(G(\mathfrak{a})) = G(\mathfrak{a})$. For all $x \in G$, $\omega(x^n) = \omega(x)$.*

This follows immediately from Proposition 8.

PROPOSITION 9. *Suppose that $p \neq 0$.*

(i) *Let $\mathfrak{a}$, $\mathfrak{b}$ be non-zero ideals of* A *such that $\mathfrak{b} \subset \mathfrak{a} \subset \mathfrak{m}$. In the group $G(\mathfrak{a})/G(\mathfrak{b})$, every element has order a power of p.*

(ii) *Suppose that $v(p) = 1$. If $x \in G$ is such that $\omega(x) > \frac{1}{p-1}$, then*

$$\omega(x^p) = \omega(x) + 1.$$

By § 5, formula (4), for all $x \in G$,

$$x^p = px + \sum_{|\alpha| \geq 2} c_\alpha x^\alpha$$

where $c_\alpha \in A^r$ for all α. Even for proving (i) it can be assumed that $v(p) = 1$. Then if $\omega(x) \geq 1$, it follows that $\omega(x^p) \geq \omega(x) + 1$ and hence $\omega(x^{p^n})$ tends to $+\infty$ as n tends to $+\infty$; this proves (i). As $\binom{p}{i}$ is divisible by p for

$1 \leqslant i \leqslant p-1$, Proposition 2 of § 5, no. 3, proves that $c_\alpha \in p\mathrm{A}^r$ for

$$2 \leqslant |\alpha| \leqslant p-1$$

and hence

$$\omega(c_\alpha x^\alpha) > \omega(px) = \omega(x) + 1 \quad \text{for } 2 \leqslant |\alpha| \leqslant p-1.$$

On the other hand, if $|\alpha| \geqslant p$, $\omega(c_\alpha x^\alpha) \geqslant p\omega(x)$ and $p\omega(x) > \omega(x) + 1$ if $\omega(x) > \frac{1}{p-1}$. This proves (ii).

6. LOGARITHMIC MAPPING

Lemma 1. *Suppose that* $p \neq 0$. *Let* G *be a Lie group,* G_1 *an open subgroup of* G *which is isomorphic to a standard group and* $x \in \mathrm{G}$. *The following conditions are equivalent:*

(i) *there exists a power of* x *which belongs to* G_1;

(ii) *there exists a strictly increasing sequence* (n_i) *of integers such that* x^{n_i} *tends to* e *as* i *tends to* $+\infty$.

(ii) $\Rightarrow$ (i): obvious.

(i) $\Rightarrow$ (ii): suppose that $y = x^m \in \mathrm{G}_1$. By Proposition 9 (i) of no. 5, y^{p^n} tends to e as n tends to $+\infty$, in other words x^{mp^n} tends to e as n tends to $+\infty$.

PROPOSITION 10. *Suppose that* $p \neq 0$. *Let* G *be a finite-dimensional Lie group. Let* G_f *be the set of* $x \in \mathrm{G}$ *for which there exists a strictly increasing sequence* (n_i) *of integers such that* x^{n_i} *tends to* e *as* i *tends to* $+\infty$.

(i) G_f *is open in* G.

(ii) *There exists one and only one mapping* ψ *of* G_f *into* L(G) *with the following properties:*

(a) $\psi(x^n) = n\psi(x)$ *for all* $x \in \mathrm{G}_f$ *and all* $n \in \mathbf{Z}$;

(b) *there exists an open neighbourhood* V *of* e *in* G *such that* $\psi|\mathrm{V}$ *is the inverse mapping of an injective exponential mapping.*

(iii) *The mapping* ψ *is analytic.*

There exists an open subgroup of G which is isomorphic to a standard group (no. 3, Theorem 4). Assertion (i) then follows from Lemma 1.

Let U be an open subgroup of L(G) and $\phi\colon \mathrm{U} \to \phi(\mathrm{U})$ an exponential mapping of G with the properties of Proposition 3 of no. 2. U can be assumed to be small enough for $\phi(\mathrm{U}) \subset \mathrm{G}_f$. Let $x \in \mathrm{G}_f$. There exists $m \in \mathbf{Z} - \{0\}$ such that $x^m \in \phi(\mathrm{U})$. The element $\frac{1}{m}\phi^{-1}(x^m)$ does not depend on the choice of m. For let $m' \in \mathbf{Z}$ be such that $x^{m'} \in \phi(\mathrm{U})$. Then $x^{mm'} \in \phi(\mathrm{U})$ and

$$m'\phi^{-1}(x^m) = \phi^{-1}(x^{mm'}) = m\phi^{-1}(x^{m'}),$$

whence our assertion. Let $\psi(x) = \frac{1}{m}\phi^{-1}(x^m)$. Then $\psi|\phi(\mathrm{U}) = \phi^{-1}$. On the other hand, if $n \in \mathbf{Z}$, then

$$\psi(x^n) = \frac{1}{m}\phi^{-1}(x^{nm}) = \frac{n}{m}\phi^{-1}(x^m) = n\psi(x).$$

Hence ψ has properties (a) and (b) of the proposition. In a neighbourhood of x, ψ is composed of the mappings $x \mapsto x^m$, $y \mapsto \phi^{-1}(y)$ and $z \mapsto \frac{1}{m} z$; hence ψ is analytic on G_f.

Finally, let ψ' be a mapping of G_f into L(G) and V′ a neighbourhood of e in G_f such that $\psi'(x^n) = n\psi'(x)$ for $x \in G_f$ and $n \in \mathbf{Z}$ and such that $\psi'|V'$ is the inverse mapping of an injective exponential mapping. Then ψ and ψ' coincide on a neighbourhood W of e. If $x \in G_f$, there exists $n \in \mathbf{Z}$ such that $x^n \in W$. Then

$$n\psi'(x) = \psi'(x^n) = \psi(x^n) = n\psi(x)$$

and hence $\psi = \psi'$.

Definition 2. *The mapping* ψ *of Proposition* 10 *is called the logarithmic mapping of of* G *and is denoted by* $\log_G$ *or simply* log.

Proposition 11. *Suppose that* $p \neq 0$. *Let* x, y *be two permutable elements of* G_f. *Then* $xy \in G_f$ *and* $\log(xy) = \log x + \log y$.

The fact that $xy \in G_f$ follows from Lemma 1. Let U be an open subgroup of the additive group L(G) and $\phi\colon U \to \phi(U)$ an exponential mapping of G with the properties of Proposition 3 of no. 2; U can be assumed to be small enough for $\log|\psi(U)$ to be the inverse mapping of ϕ. For $n \in \mathbf{Z} - \{0\}$ suitably chosen, $x^n \in \phi(U)$, $y^n \in \phi(U)$. Let $u = \log x^n$, $v = \log y^n$, whence $x^n = \phi(u)$, $y^n = \phi(v)$. By formula (2), $[u, v] = 0$. The Hausdorff formula proves then that $\phi(\lambda(u+v)) = \phi(\lambda u)\phi(\lambda v)$ for $|\lambda|$ sufficiently small; hence, for every sufficiently large integer i,

$$\phi(p^i(u+v)) = \phi(p^i u)\phi(p^i v)$$

that is

$$p^i(\log x^n + \log y^n) = \log(x^{np^i} y^{np^i})$$

or

$$np^i(\log x + \log y) = np^i \log(xy).$$

Proposition 12. *Suppose that* $p \neq 0$. *Let* $x \in G_f$. *The following conditions are equivalent:*

(i) $\log x = 0$;

(ii) x *is of finite order in* G.

If there exists an integer $n > 0$ such that $x^n = e$, it follows that

$$n \log x = \log x^n = 0,$$

whence $\log x = 0$. If $\log x = 0$, let V be a neighbourhood of e in G_f such that $\log|V$ is the inverse mapping of an injective exponential mapping. There exists an integer $n > 0$ such that $x^n \in V$; the equality $\log x^n = 0$ implies $x^n = e$.

Proposition 13. *Suppose that* $p \neq 0$. *If* G *is compact or standard, then* $G_f = G$.

If G is standard, it suffices to use Lemma 1. Suppose that G is compact. Let $x \in G$ and V be a neighbourhood of e in G. Let y be a limit point of the sequence $(x^n)_{n \geqslant 0}$. For all $n > 0$, there exist two integers n_1, n_2 such that $n_1 \geqslant 2n_2 \geqslant n$ and $x^{n_1} \in yV$, $x^{n_2} \in yV$, whence $x^{n_1 - n_2} \in V^{-1}V$ and $n_1 - n_2 \geqslant n$. Hence $x \in G_f$.

COROLLARY. *Suppose that* K *is locally compact. Then* G_f *is the union of the compact subgroups of* G.

Let $x \in G$. If x belongs to a compact subgroup of G, then $x \in G_f$ (Proposition 13). Suppose that $x \in G_f$. As K is locally compact, there exists an open subgroup G_1 of G which is compact. Then there exists an integer $m > 0$ such that $x^m \in G_1$. The closed subgroup G_2 generated by x^m is contained in G_1 and is therefore compact. Then x commutes with the elements of G_2 and hence $G_2 \cup xG_2 \cup \cdots \cup x^{m-1}G_2$ is a compact subgroup of G which contains x.

Example. Suppose that K is locally compact. Let U be the set of invertible elements of A; it is a compact open subgroup of the Lie group K^*. Then $U \subset (K^*)_f$ by Proposition 13; on the other hand, if $x \in K^*$ is such that $x \notin U$, either x^n tends to 0 as n tends to $+\infty$, or x^n tends to 0 as n tends to $-\infty$; hence $U = (K^*)_f$. The function $\log_{K^*}$ is defined and analytic on U, with values in $L(K^*) = K$, and such that $\log_{K^*}(xy) = \log_{K^*}(x) + \log_{K^*}(y)$ for all x, y in U; the elements x of U such that $\log_{K^*}(x) = 0$ are the roots of unity of K.

We again use the notation of nos. 3, 4 and 5.

PROPOSITION 14. *Suppose that* $p \neq 0$ *and that* v *is chosen such that* $v(p) = 1$. *Let* G *be a standard group and* E(X) (*resp.* L(X)) *the expansion of the exponential function of* G (*resp. the logarithmic function of* G) *as an integral series about* 0.

(i) *The domain of absolute convergence* (*Differentiable and Analytic Manifolds*, R, 4.1.3) *of* E *contains the set* Δ *of* $x \in G$ *such that* $\omega(x) > \frac{1}{p-1}$. *Let* E′ *denote the mapping defined on* Δ *by this series. Then* E′ *is an exponential mapping of* G *and is an isomorphism of the manifold* Δ *onto itself.*

(ii) *The domain of absolute convergence of* L *contains* G. *Let* L′ *denote the mapping on* G *defined by this series. Then* L′ *is the logarithmic mapping of* G *and the restriction of* L′ *to* Δ *is the inverse mapping of* E′.

(iii) *The mapping* E′ *is an isomorphism of* Δ, *with the Hausdorff law, onto the subgroup* Δ *of* G.

Using the notation of § 5, nos. 3 and 4, $E = \sum_{m \geqslant 1} \frac{\psi_{m,m}}{m!}$ (§ 5, no. 4, Proposition 3). As the coefficients $c_{\alpha\beta\gamma}$ belong to A, $\|\psi_{m,m}\| \leqslant 1$ (*Differentiable and Analytic Manifolds*, R, Appendix) K^r is assumed to have the norm

$$\|(\lambda_1, \ldots, \lambda_r)\| = \sup(|\lambda_1|, \ldots, |\lambda_r|).$$

By Chapter II, § 8, no. 1, Lemma 1, $v(m!) \leqslant \frac{m-1}{p-1}$. If $\omega(x) > \frac{1}{p-1}$, we see that $m\omega(x) - v(m!)$ tends to $+\infty$ with m, whence

$$\left\|\frac{\psi_{m,m}}{m!}\right\| \|x\|^m \leqslant \frac{1}{|m!|} \|x\|^m$$

which tends to 0 as m tends to $+\infty$ and

$$\omega\left(\frac{\psi_{m,m}(x)}{m!}\right) > \frac{m}{p-1} - \frac{m-1}{p-1} = \frac{1}{p-1} \quad \text{for } m \geqslant 1.$$

Therefore Δ is contained in the domain of absolute convergence of E and $E'(\Delta) \subset \Delta$. Clearly E′ is an exponential mapping.

If L_m denotes the homogeneous component of L of degree m, Proposition 3 of § 5, no. 4, proves that each coefficient of L_m is of the form

$$a_1 + \frac{1}{2}a_2 + \cdots + \frac{1}{m}a_m$$

with $a_1, a_2, \ldots, a_m$ in A; but

$$\inf\left(v(1), v\left(\frac{1}{2}\right), \ldots, v\left(\frac{1}{m}\right)\right) = o(\log m) \quad \text{as } m \text{ tends to } +\infty$$

and

$$\inf\left(v(1), v\left(\frac{1}{2}\right), \ldots, v\left(\frac{1}{m}\right)\right) \geqslant v\left(\frac{1}{m!}\right) \geqslant -\frac{m-1}{p-1}.$$

Therefore, if $\omega(x) > 0$, $\|L_m\| \cdot \|x\|^m$ tends to 0 as m tends to $+\infty$, so that G is contained in the domain of absolute convergence of L. On the other hand, if $\omega(x) > \frac{m}{p-1}$, then $\omega(L_m(x)) > \frac{m}{p-1} - \frac{m-1}{p-1} = \frac{1}{p-1}$ for $m \geqslant 1$ and hence $L'(\Delta) \subset \Delta$.

As the formal power series $L(E(X))$ and $E(L(X))$ are equal to X, no. 4.1.5 of *Differentiable and Analytic Manifolds*, R, proves that

$$L'(E'(x)) = E'(L'(x)) = x$$

for $x \in \Delta$. Hence E′ is an isomorphism of the manifold Δ onto itself and the inverse isomorphism is the restriction of L′ to Δ.

$L(X^{[n]}) = nL(X)$ for n an integer >0 (cf. § 5, no. 4). As G is contained in the domain of absolute convergence of L and $X^{[n]}$, therefore $L'(x^n) = nL'(x)$ for all $x \in G$. The relation $L'|\Delta = E'^{-1}$ implies that $L'(x^n) = \log x^n$ for n sufficiently large. Hence $L'(x) = \log x$. We have thus proved (i) and (ii).

Let $H = \sum_{r,s \geqslant 0} H_{r,s}$ be the Hausdorff formal power series and h the Hausdorff function relative to L(G). The domain of absolute convergence of $\tilde{H}$

contains $\Delta \times \Delta$ and h is defined on $\Delta \times \Delta$ (Chapter II, § 8, Proposition 2). Then

$$E'(x)E'(y) = E'(h(x,y))$$

for x, y sufficiently close to 0 (§ 4, Theorem 4 (v)). Hence, in the notation of no. 3, Definition 1, the formal power series F(E(X), E(Y)) and E(H(X, Y)) are equal. Let x, y be elements of Δ. Then

$$\sup_m \left\|\frac{\psi_{m,m}}{m!}\right\| (\sup \|x\|, \|y\|)^m < 1$$

$$\sup_{r,s} \|H_{r,s}\| \, \|x\|^r \|y\|^s < |p|^{1/(p-1)}$$

by Chapter II, § 8, formula (14). By *Differentiable and Analytic Manifolds*, R, 4.1.5, $E'(x)E'(y)$ is obtained by substituting x for X and y for Y in

$$F(E(X), E(Y))$$

and $E'(h(x,y))$ is obtained by substituting x for X and y for Y in E(H(X, Y)). Hence $E'(x)E'(y) = E'(h(x,y))$.

§ 8. LIE GROUPS OVER R AND $\mathbf{Q}_p$

1. CONTINUOUS MORPHISMS

THEOREM 1. *Let* G *and* H *be two Lie group germs over* **R** *or* $\mathbf{Q}_p$. *Let* f *be a continuous morphism of* G *into* H. *Then* f *is analytic.*

We give L(G) and L(H) norms which define their topologies and such that $\|[x,y]\| \leqslant \|x\| \, \|y\|$ for all x, y. There exist an open ball V of centre 0 in L(G) and an exponential mapping ϕ of G defined on V such that: (1) $\phi(V)$ is an open neighbourhood of e in G; (2) ϕ is an isomorphism of the analytic manifold V onto the analytic manifold $\phi(V)$; (3) $\phi(nx) = \phi(x)^n$ for all $x \in V$ and all $n \in \mathbf{Z}$ such that $nx \in V$. We define similarly W and ψ for H. By shrinking V if necessary, it can be assumed that $f(\phi(V)) \subset \psi(W)$. Then $g = \psi^{-1} \circ f \circ \phi$ is a continuous mapping of V into W.

We show that

$$(1) \qquad (x \in V, \lambda \in \mathbf{Q} \text{ and } \lambda x \in V) \Rightarrow g(\lambda x) = \lambda g(x).$$

It can be assumed that $\lambda \neq 0$. Let $\lambda = \frac{r}{q}$ with r, q in $\mathbf{Z} - \{0\}$. Let $y = \frac{r}{q} x$.

If K = **R**, we write $z = \frac{x}{q} = \frac{y}{r} \in V$. Then $x = qz$, $y = rz$, whence

$$g(x) = \psi^{-1}(f(\phi(qz))) = \psi^{-1}(f(\phi(z)^q)) = \psi^{-1}(f(\phi(z))^q).$$

We show that $\psi^{-1}(f(\phi(z))^q) = q\psi^{-1}(f(\phi(z))) = qg(z)$. It suffices to verify that, if $u \in \psi(W)$ is such that $u^q \in \psi(W)$, then $\psi^{-1}(u^q) = q\psi^{-1}(u)$; but if $u = \psi(v)$, $u^q = \psi(v^1)$, then $v^1/q \in W$ and $(\psi(v^1/q))^q = u^q$, hence $\psi(v^1/q) = u = \psi(v)$ and therefore $v^1 = qv$.

Similarly, $g(y) = rg(z)$, whence (1).

If $K = \mathbf{Q}_p$, we write $z = rx = qy \in V$, whence $g(z) = rg(x) = qg(y)$, whence again (1).

As $\mathbf{Q}$ is dense in K, (1) implies that

$$(2) \qquad (x \in V, \lambda \in K \text{ and } \lambda x \in V) \Rightarrow g(\lambda x) = \lambda g(x).$$

Let $x \in L(G)$ and λ, λ' be elements of K^* such that $\lambda x \in V$, $\lambda' x \in V$. Then

$$g(\lambda' x) = g\left(\frac{\lambda'}{\lambda}\lambda x\right) = \frac{\lambda'}{\lambda} g(\lambda x)$$

by (2) and hence $\frac{1}{\lambda} g(\lambda x) = \frac{1}{\lambda'} g(\lambda' x)$. Thus an extension h of g to L(G) is defined by writing $h(x) = \frac{1}{\lambda} g(\lambda x)$ for all λ such that $\lambda x \in V$. Clearly h is continuous. We show that

$$(3) \qquad (x \in L(G) \text{ and } \lambda \in K) \Rightarrow h(\lambda x) = \lambda h(x).$$

Let $\lambda' \in K^*$ be such that $\lambda' x \in V$ and $\lambda' \lambda x \in V$. Then

$$h(\lambda x) = \frac{1}{\lambda'} g(\lambda' \lambda x) = \frac{1}{\lambda'} \lambda g(\lambda' x) = \lambda \frac{1}{\lambda'} g(\lambda' x) = \lambda h(x).$$

Let x, y be in L(G). Then, by Proposition 4 of § 4, no. 3,

$$\begin{aligned} h(x) + h(y) &= \lim_{\lambda \in K^*, \lambda \to 0} \lambda^{-1}\psi^{-1}(\psi(\lambda h(x))\psi(\lambda h(y))) \\ &= \lim_{\lambda \in K^*, \lambda \to 0} \lambda^{-1}\psi^{-1}(\psi(h(\lambda x))\psi(h(\lambda y))). \end{aligned}$$

For $|\lambda|$ sufficiently small, $\lambda x \in V$ and $\lambda y \in V$ and hence the above expression is equal to

$$\begin{aligned} &\lim_{\lambda \in K^*, \lambda \to 0} \lambda^{-1}\psi^{-1}(f(\phi(\lambda x))f(\phi(\lambda y))) \\ &= \lim_{\lambda \in K^*, \lambda \to 0} \lambda^{-1}(\psi^{-1} \circ f)(\phi(\lambda x)\phi(\lambda y)) \\ &= \lim_{\lambda \in K^*, \lambda \to 0} \lambda^{-1} g(\phi^{-1}(\phi(\lambda x)\phi(\lambda y))) \\ &= \lim_{\lambda \in K^*, \lambda \to 0} h(\lambda^{-1}\phi^{-1}(\phi(\lambda x)\phi(\lambda y))) \\ &= h(\lim_{\lambda \in K^*, \lambda \to 0} \lambda^{-1}\phi^{-1}(\phi(\lambda x)\phi(\lambda y))) \\ &= h(x + y). \end{aligned}$$

Thus h is continuous and linear, hence $g = h|V$ is analytic, hence f is analytic on $\phi(V)$ and hence f is analytic (§ 1, no. 10).

COROLLARY 1. *Let* G *be a topological group. There exists on* G *at most one analytic manifold structure over* $\mathbf{R}$ (resp. $\mathbf{Q}_p$) *compatible with the topological group structure on* G.

This follows immediately from Theorem 1.

Definition 1. *A topological group* G *is called a real* (resp. *p-adic*) *Lie group if there exists on* G *a real* (resp. *p-adic*) *Lie group structure compatible with its topology.*

That structure is then unique and we can speak of the *dimension* of such a group. If G and H are two such groups, every continuous morphism of G into H is analytic.

Corollary 2. *Let* G *be a topological group and* V *an open neighbourhood of e. Suppose that* V *has an analytic manifold structure which makes it into a real* (resp. *p-adic*) *Lie group germ. Then* G *is a real* (resp. *p-adic*) *Lie group.*

Let $g \in G$. There exists an open neighbourhood V′ of e in G such that $V' \cup gV'g^{-1} \subset V$. The mapping $v \mapsto gvg^{-1}$ of V′ into V is a continuous and hence analytic morphism of the Lie group germ V′ into the Lie group germ V. It then suffices to apply Proposition 18 of § 1, no. 9.

Remarks. (1) Theorem 1 and its corollaries are no longer true if $\mathbf{R}$ (resp. $\mathbf{Q}_p$) is replaced by $\mathbf{C}$ (Exercise 1).

(2) Let G be a topological group. It can be shown† that the following conditions are equivalent: (a) G is a finite-dimensional real Lie group; (b) G is locally compact and there exists a neighbourhood of e containing no subgroup distinct from $\{e\}$; (c) there exists an open neighbourhood of e homeomorphic to an open ball of a space $\mathbf{R}^n$. (For a much less difficult result, cf. Exercise 6.)

Proposition 1. *Let* G, G′ *be topological groups and f a continuous morphism of* G *into* G′. *Suppose that one of the following three cases holds:*

(a) G *is a real Lie group and* G′ *is a p-adic Lie group;*

(b) G *is a p-adic Lie group and* G′ *is a real Lie group;*

(c) G *is a p-adic Lie group and* G′ *is a p′-adic Lie group with* $p \neq p'$.

Then f is locally constant.

Case (a). Let G_0 be the identity component of G. Then $f(G_0)$ is a connected subgroup of G′ and hence $f(G_0) = \{e\}$ and G_0 is open in G.

Case (b). Let V′ be a neighbourhood of e in G′ such that every subgroup of G′ contained in V′ reduces to $\{e\}$ (§ 4, no. 2, Corollary 1 to Theorem 2). There exists a neighbourhood V of e in G such that $f(V) \subset V'$. Then there exists an open subgroup G_1 of G such that $G_1 \subset V$ (§ 7, no. 1, Proposition 1). Then $f(G_1) = \{e\}$.

Case (c). By § 7, Theorem 4 and the Corollary to Proposition 8, there exists a neighbourhood V′ of e in G′ such that, for all $x' \in V' - \{e\}$, x'^{p^n} does not tend

† See for example D. Montgomery and L. Zippin, *Topological transformation groups,* Interscience tracts in pure and applied mathematics, no. 1, Interscience publishers, New York 1955 (in particular pp. 169 and 184).

to e as n tends to $+\infty$. There exists a neighbourhood V of e in G such that $f(\mathrm{V}) \subset \mathrm{V}'$. By § 7, Theorem 4 and Proposition 9, there exists an open subgroup G_1 of G such that $\mathrm{G}_1 \subset \mathrm{V}$ and such that, for all $x \in \mathrm{G}_1$, x^{p^n} tends to e as n tends to $+\infty$. Then $f(\mathrm{G}_1) = \{e\}$.

2. CLOSED SUBGROUPS

THEOREM 2. *Let* G *be a finite-dimensional Lie group over* $\mathbf{R}$ *or* $\mathbf{Q}_p$. *Every closed subgroup of* G *is a Lie subgroup of* G. *More generally, let* U *be a symmetric open neighbourhood of* e *in* G *and* H *a non-empty closed subspace of* U *such that the conditions* $x \in \mathrm{H}$, $y \in \mathrm{H}$ *and* $xy^{-1} \in \mathrm{U}$ *imply* $xy^{-1} \in \mathrm{H}$. *Then* H *is a Lie subgroup germ of* G.

Let $\mathfrak{h}$ be the Lie subalgebra tangent to H at e (§ 4, no. 5, Definition 2). There exists a Lie subgroup germ H_0 of G with Lie algebra $\mathfrak{h}$ and contained in H. We prove that H_0 is open in H with the topology induced by that on G. This will prove that H is an analytic submanifold of G and will establish the theorem.

There exist a vector subspace $\mathfrak{k}$ supplementary to $\mathfrak{h}$ in L(G), symmetric open neighbourhoods V_1, V_2 of zero in $\mathfrak{h}$ and $\mathfrak{k}$ respectively and an exponential mapping ϕ of G defined on $\mathrm{V}_1 + \mathrm{V}_2$ and possessing the following properties:

(a) the mapping $(a_1, a_2) \mapsto \phi(a_1)\phi(a_2)$ is an analytic isomorphism of $\mathrm{V}_1 \times \mathrm{V}_2$ onto an open subset V of G;

(b) $\phi(\mathrm{V}_1) \subset \mathrm{H}_0$;

(c) $\mathrm{V}^2 \subset \mathrm{U}$.

We shall prove (and this will achieve the proof) that there exists an open neighbourhood V_2' of 0 in V_2 such that $\mathrm{H} \cap (\phi(\mathrm{V}_1)\phi(\mathrm{V}_2')) = \phi(\mathrm{V}_1)$.

Suppose that the assertion is false. Then we can find a sequence (x_n) in V_1 and a sequence (y_n) in $\mathrm{V}_2 - \{0\}$ tending to 0, such that $\phi(x_n)\phi(y_n) \in \mathrm{H}$ for all n. Then $\phi(y_n) \in \mathrm{H}$ by (c).

If $\mathrm{K} = \mathbf{Q}_p$, it can be assumed that V_2 is an additive subgroup of $\mathfrak{k}$ and that $\phi(pa) = \phi(a)^p$ for all $a \in \mathrm{V}_2$ and all $p \in \mathbf{Z}$. Then $\phi(\lambda y_1) \in \mathrm{H}$ for all $\lambda \in \mathbf{Z}$ and hence by continuity for all $\lambda \in \mathbf{Z}_p$. The mapping $f\colon \lambda \mapsto \phi(\lambda y_1)$ of $\mathbf{Z}_p$ into G is analytic and takes its values in H and $(\mathrm{T}_0 f)(1) = y_1$. Hence $y_1 \in \mathfrak{h}$, which is absurd. The theorem is therefore established in the case of $\mathbf{Q}_p$.

If $\mathrm{K} = \mathbf{R}$, it can be assumed that V_2 is connected and that y_n belongs to $\frac{1}{4}\mathrm{V}_2 - \{0\}$. By taking a subsequence of (y_n) if necessary, we can find a sequence (λ_n) of non-zero scalars such that $\lambda_n^{-1} y_n$ tends to an element y of $\mathrm{V}_2 - \{0\}$. The sequence (λ_n) tends to 0. Let $\lambda \in \mathbf{R}$ be such that $\lambda y \in \frac{1}{4}\mathrm{V}_2$; we prove that $\exp(\lambda y) \in \mathrm{H}$. It can be assumed that $\lambda\lambda_n^{-1} y_n \in \frac{1}{4}\mathrm{V}_2$ for all n. Let $k_n \in \mathbf{Z}$ be such that $|\lambda - k_n\lambda_n|$ tends to 0. For n sufficiently large, $(\lambda - k_n\lambda_n)\lambda_n^{-1} y_n \in \frac{1}{4}\mathrm{V}_2$ and hence $k_n y_n \in \frac{1}{2}\mathrm{V}_2$. Hence $\exp(h y_n) \in \mathrm{H}$ for h an integer and $0 \leqslant |h| \leqslant |k_n|$ (as is seen by induction on $|h|$). Then

$$\begin{aligned}\exp(\lambda y) &= \lim_{n\to\infty} \exp(\lambda\lambda_n^{-1} y_n) = \lim_{n\to\infty} (\exp((\lambda - k_n\lambda_n)\lambda_n^{-1} y_n)\exp(k_n y_n)) \\ &= \lim_{n\to\infty} \exp k_n y_n \in \mathrm{H}.\end{aligned}$$

Hence the mapping $f: \lambda \mapsto \exp \lambda y$, where $\lambda y \in \frac{1}{4}V_2$, takes its values in H and $(T_0 f)(1) = y$. Hence $y \in \mathfrak{h}$, which is absurd. The theorem is thus established in the case of **R**.

Theorem 2 is no longer true if G is not assumed to be finite-dimensional (Exercise 12).

COROLLARY 1. *Let G′ be a locally compact group, G a finite-dimensional Lie group over* **R** (resp. $\mathbf{Q}_p$) *and f a continuous morphism of G′ into G. If the kernel of f is discrete, G′ is a finite-dimensional real* (resp. *p*-adic) *Lie group.*

There exists a compact neighbourhood V of e in G′ such that $f \mid V$ is a homeomorphism of V onto a compact subspace of G. If U is a sufficiently small open neighbourhood of e in G, the hypotheses of Theorem 2 are satisfied with $H = f(V) \cap U$. Hence H is a Lie subgroup germ of G. Let W be the inverse image of H under $f \mid V$. Then W is a neighbourhood of e in G′. Let W be given the analytic manifold structure transported from that on H by $(f \mid W)^{-1}$. For all $z \in G'$, the mapping $x \mapsto f(z)xf(z)^{-1}$ of G into G is analytic; hence there exists an open neighbourhood W′ of e in W such that the mapping $x' \mapsto zx'z^{-1}$ of W′ into W is analytic. By Proposition 18 of § 1, no. 9, there exists on G′ a Lie group structure which induces on a sufficiently small open neighbourhood of e the same analytic structure as W and hence the same topology as the initial topology on G′.

COROLLARY 2. *Let G be a finite-dimensional Lie group over* K, H *a subgroup of* G, V *an open neighbourhood of e in* G *and* $(M_i)_{i \in I}$ *a family of analytic manifolds over* K; *for all* $i \in I$, *let* f_i *be a* K*-analytic mapping of* V *into* M_i *such that*

$$H \cap V = \{x \in V \mid f_i(x) = f_i(e) \text{ for all } i \in I\}.$$

(i) *If* K = **C**, H *is a Lie subgroup of* G.

(ii) *If* K *is a finite extension of* $\mathbf{Q}_p$ *and* I *is finite,* H *is a Lie subgroup of* G.

(i) Suppose that K = **C**. We consider G as a real Lie group. Then H is a real Lie subgroup of G (Theorem 2). Let $a \in L(H)$. There exists a connected open neighbourhood W of 0 in **C** such that $\exp \lambda a \in V$ for all $\lambda \in W$. Let $i \in I$. Then $f_i(\exp \lambda a) = f_i(e)$ if $\lambda \in \mathbf{R} \cap W$. Hence $f_i(\exp \lambda a) = f_i(e)$ if $\lambda \in W$ by analytic continuation. Thus, $\exp \lambda a \in H$ for $\lambda \in W$ and therefore $\mu a \in L(H)$ for all $\mu \in \mathbf{C}$. Therefore H is a Lie subgroup of the complex Lie group G (§ 4, no. 2, Proposition 2).

(ii) Suppose that K is a finite extension of $\mathbf{Q}_p$. We consider G as a Lie group over $\mathbf{Q}_p$. It is finite-dimensional and Theorem 2 implies that H is a *p*-adic Lie subgroup of G. Since I is finite, $\prod_{i \in I} M_i$ is a manifold and it can be assumed that the family (f_i) reduces to a single mapping f. Let $a \in L(G)$. Let ϕ be an exponential mapping of G. Then $f(\phi(\lambda a)) = f(e)$ for $\lambda \in \mathbf{Q}_p$ and $|\lambda|$ sufficiently small. Since f is K-analytic, it follows that $f(\phi(\lambda a)) = f(e)$ for $\lambda \in K$ and $|\lambda|$ sufficiently small. Hence $\phi(\lambda a) \in H$ for $\lambda \in K$ and $|\lambda|$ suffiicently small and therefore $\mu a \in L(H)$ for all $\mu \in K$. The proof is completed as in (i).

Corollary 2 (ii) is no longer true if we omit the hypothesis that I is finite.

§ 9. COMMUTATORS, CENTRALIZERS AND NORMALIZERS IN A LIE GROUP

In this paragraph, K is assumed to be of characteristic zero.

1. COMMUTATORS IN A TOPOLOGICAL GROUP

Let G be a topological group. We define the groups $\overline{D^0}G, \overline{D^1}G, \overline{D^2}G, \ldots$ and $\overline{C^1}G, \overline{C^2}G, \overline{C^3}G, \ldots$ by the formulae

$$\overline{D^0}G = G, \qquad \overline{D^{i+1}}G = \overline{(\overline{D^i}G, \overline{D^i}G)}$$

$$\overline{C^1}G = G, \qquad \overline{C^{i+1}}G = \overline{(G, \overline{C^i}G)}.$$

Proposition 1. *Let* G *be a topological group and* A *and* B *subgroups of* G. *Then* $\overline{(\overline{A}, \overline{B})} = \overline{(A, B)}$, $\overline{D^i}\overline{A} = \overline{D^i A}$, $\overline{C^i}\overline{A} = \overline{C^i A}$.

Let ϕ be the continuous mapping $(x, y) \mapsto x^{-1}y^{-1}xy$ of $G \times G$ into G. Then $\phi(A \times B) \subset (A, B)$, hence $\phi(\overline{A} \times \overline{B}) \subset \overline{(A, B)}$ and hence

$$\overline{(\overline{A}, \overline{B})} \subset \overline{(A, B)};$$

the opposite inclusion is obvious and therefore $\overline{(A, B)} = \overline{(\overline{A}, \overline{B})}$. Clearly $\overline{D^0}\overline{A} = \overline{D^0 A}$; assuming the equality $\overline{D^i}\overline{A} = \overline{D^i A}$, it follows that

$$\overline{D^{i+1}}\overline{A} = \overline{(\overline{D^i}\overline{A}, \overline{D^i}\overline{A})} = \overline{(\overline{D^i A}, \overline{D^i A})} = \overline{(D^i A, D^i A)} = \overline{D^{i+1} A}$$

and hence $\overline{D^i}\overline{A} = \overline{D^i A}$ for all i. The proof of the formula $\overline{C^i}\overline{A} = \overline{C^i A}$ is analogous.

Corollary 1. *If* G *is Hausdorff, the following conditions are equivalent:*

(i) G *is solvable* (resp. *nilpotent*)*;*

(ii) $\overline{D^i}G = \{e\}$ (resp. $\overline{C^i}G = \{e\}$) *for sufficiently large* i.

$D^iG \subset \overline{D^i}G$, $C^iG \subset \overline{C^i}G$ and hence (ii) ⇒ (i). $\{e\} = \overline{\{e\}}$ and hence (i) ⇒ (ii) by Proposition 1.

Corollary 2. *Let* G *be a Hausdorff topological group and* A *a subgroup of* G. *For* A *to be solvable* (resp. *nilpotent, commutative*), *it is necessary and sufficient that* $\overline{A}$ *be so.*

This follows immediately from Proposition 1.

PROPOSITION 2. *Let* G *be a topological group and* A *and* B *subgroups of* G. *If* A *is connected,* (A, B) *is connected.*

For fixed y in B, the set M_y of (x, y) with $x \in A$ is connected (for the mapping $x \mapsto (x, y)$ of A into G is continuous). $e \in M_y$ and hence the union R of the M_y with $y \in B$ is connected. But (A, B) is the subgroup of G generated by R, whence the proposition.

2. COMMUTATORS IN A LIE GROUP

PROPOSITION 3. *Let* G *be a finite-dimensional Lie group and* H_1 *and* H_2 *subgroups of* G. *Let* $\mathfrak{h}_1$, $\mathfrak{h}_2$ *and* $\mathfrak{h}$ *be the Lie subalgebras tangent at* e *to* H_1 H_2 *and* (H_1, H_2) *respectively. Then* $[\mathfrak{h}_1, \mathfrak{h}_2] \subset \mathfrak{h}$.

Let $a \in \mathfrak{h}_1$, $b \in \mathfrak{h}_2$. There exist an open neighbourhood I of 0 in K and analytic mappings f_1, f_2 of I into G such that

$$f_1(0) = f_2(0) = e, \quad f_1(I) \subset H_1, \quad f_2(I) \subset H_2,$$
$$(T_0 f_1)1 = a, \quad (T_0 f_2)1 = b.$$

We write

$$f(\lambda, \mu) = (f_1(\lambda), f_2(\mu)) \in (H_1, H_2) \quad \text{for } \lambda, \mu \text{ in I}.$$

We identify an open neighbourhood of e in G with an open subset of K^r by means of a chart which maps e to 0. Then L(G) is identified with K^r. By § 5, no. 2, Proposition 1, the expansion of $f(\lambda, \mu)$ as an integral series about the origin is

$$f(\lambda, \mu) = \lambda\mu[a, b] + \sum_{i \geqslant 1, j \geqslant 1, i+j \geqslant 2} \lambda^i \mu^j a_{ij}$$

where $a_{ij} \in K^r$ (the terms in λ^i and μ^j in the expansion of $f(\lambda, \mu)$ are zero because $f(\lambda, 0) = f(0, \mu) = 0$). We fix μ in I. Letting λ tend to 0, we see that

$$\mu[a, b] + \sum_{j \geqslant 2} \mu^j a_{1j} \in \mathfrak{h}.$$

Since this is true for all $\mu \in I$, it follows that $[a, b] \in \mathfrak{h}$.

Remark. Even if H_1 and H_2 are connected Lie subgroups of G, the Lie subalgebra of L(G) generated by $[\mathfrak{h}_1, \mathfrak{h}_2]$ is in general distinct from $\mathfrak{h}$.

PROPOSITION 4. *Let* G *be a finite-dimensional real or complex Lie group. Let* A, B, C *be integral subgroups of* G *such that* $[L(A), L(C)] \subset L(C)$ *and*

$$[L(B), L(C)] \subset L(C).$$

If $[L(A), L(B)] \subset L(C)$, *then* $(A, B) \subset C$. *If* $[L(A), L(B)] = L(C)$, *then* $(A, B) = C$.

Suppose that $[L(A), L(B)] \subset L(C)$. The sum $L(A) + L(B) + L(C)$ is a Lie subalgebra of L(G). By considering an integral subgroup of G with Lie algebra $L(A) + L(B) + L(C)$, the problem is reduced to the case where

$$L(A) + L(B) + L(C) = L(G)$$

and G is connected. Then L(C) is an ideal of L(G). Suppose first that G is simply connected. Then C is a normal Lie subgroup of G (§ 6, no. 6, Proposition 14). Let ϕ be the canonical morphism of G onto G/C. Then

$$[L(\phi)(L(A)), L(\phi)(L(B))] = \{0\}$$

and hence $\phi(A)$ and $\phi(B)$ commute by the Hausdorff formula; therefore $(A, B) \subset C$. In the general case, let G′ be the universal covering of G and A′, B′, C′ the integral subgroups of G′ such that $L(A') = L(A)$, $L(B') = L(B)$, $L(C') = L(C)$. Then $(A', B') \subset C'$ and A, B, C are the canonical images of A′, B′, C′ in G, whence $(A, B) \subset C$. On the other hand, (A, B) is the underlying set of an integral subgroup of G (§ 6, no. 2, Corollary to Proposition 4) and its Lie algebra contains $[L(A), L(B)]$ (Proposition 3). If

$$[L(A), L(B)] = L(C),$$

then $(A, B) \supset C$, whence $(A, B) = C$.

Corollary. *Let* G *be a finite-dimensional connected real or complex Lie group with Lie algebra* $\mathfrak{g}$. *The subgroups* D^iG (resp. C^iG) *are integral subgroups with Lie algebras* $\mathscr{D}^i\mathfrak{g}$ (resp. $\mathscr{C}^i\mathfrak{g}$). *If* G *is simply connected, they are Lie subgroups.*

The first assertion follows from Proposition 4 by induction on i. The second follows from the first and § 6, no. 6, Proposition 14.

Proposition 5. *Let* G *be a finite-dimensional real or complex Lie group and* A *an integral subgroup of* G. *Then* $D\overline{A} = DA$. *In particular,* A *is a normal subgroup of* $\overline{A}$ *and* $\overline{A}/A$ *is commutative.*

Let $\mathfrak{a} = L(A)$. Let G_1 be the set of $g \in G$ such that

$$(\mathrm{Ad}\, g)x \equiv x \pmod{\mathscr{D}\mathfrak{a}} \quad \text{for all } x \in \mathfrak{a}.$$

Then G_1 is a closed subgroup of G. If $y \in \mathfrak{a}$, then $\exp y \in G_1$, by § 6, no. 4, Corollary 3 (ii) to Proposition 10. Hence G_1 contains A and therefore $\overline{A}$. Thus, for $g \in \overline{A}$, $L(\mathrm{Int}\, g)$ leaves $\mathfrak{a}$ stable and therefore $\mathrm{Int}\, g$ leaves A stable; more precisely, $L(\mathrm{Int}\, g)$ defines the identity automorphism of $\mathfrak{a}/\mathscr{D}\mathfrak{a}$ and hence $\mathrm{Int}\, g$ defines the identity automorphism of A/DA. This proves that $(\overline{A}, A) \subset DA$. With the real Lie group structure on G, $\overline{A}$ is a Lie subgroup (§ 8, no. 2, Theorem 2); let $\mathfrak{b}$ be its Lie algebra. Let G_2 be the set of $g \in G$ such that

$$(\mathrm{Ad}\, g)x \equiv x \pmod{\mathscr{D}\mathfrak{a}} \quad \text{for all } x \in \mathfrak{b}.$$

By the above, $G_2 \supset A$ and hence $G_2 \supset \overline{A}$. Therefore, for $g \in \overline{A}$, Int g leaves DA stable and defines the identity automorphism of $\overline{A}/DA$. Hence $DA \supset D\overline{A}$.

PROPOSITION 6. *Suppose that* K *is ultrametric. Let* G *be a finite-dimensional Lie group. Let* A, B, C *be Lie subgroups of* G *such that* $[L(A), L(C)] \subset L(C)$, $[L(B), L(C)] \subset L(C)$. *If* $[L(A), L(B)] \subset L(C)$, *there exist open subgroups* A′, B′ *of* A, B *such that* $(A', B') \subset C$. *If* $[L(A), L(B)] \subset L(C)$, *there exist open subgroups* A′, B′, C′ *of* A, B, C *such that* $(A', B') = C'$.

Suppose that $[L(A), L(B)] \subset L(C)$. As in the proof of Proposition 4, the problem reduces to the case where L(C) is an ideal of L(G). Then, by replacing G by an open subgroup, it reduces to the case where C is normal in G (§ 7, no. 1, Proposition 2). Let ϕ be the canonical morphism of G onto G/C. Then

$$[L(\phi)(L(A)), L(\phi)(L(B))] = \{0\}.$$

By the Hausdorff formula, there exist open subgroups A′, B′ of A, B such that $\phi(A')$ and $\phi(B')$ commute, whence $(A', B') \subset C$. Suppose further that

$$[L(A), L(B)] = L(C).$$

By Proposition 3, the Lie subalgebra tangent to (A′, B′) at e contains L(C). Hence (A′, B′) contains a Lie subgroup germ of G with Lie algebra L(C). Therefore (A′, B′) is an open subgroup of C.

COROLLARY. *Suppose that* K *is ultrametric. Let* G *be a finite-dimensional Lie group with Lie algebra* $\mathfrak{g}$. *There exists an open subgroup* G_0 *of* G *such that, for all* i, D^iG_0 (resp. C^iG_0) *is a Lie subgroup of* G *with Lie algebra* $\mathscr{D}^i\mathfrak{g}$ (resp. $\mathscr{C}^i\mathfrak{g}$).

(a) By Proposition 3 applied inductively, for every open subgroup G_1 of G and for all i, D^iG_1 contains a Lie subgroup germ of G with Lie algebra $\mathscr{D}^i\mathfrak{g}$.

(b) Let G′ be an open subgroup of G such that, for $i \leqslant n$, D^iG' is a Lie subgroup of G with Lie algebra $\mathscr{D}^i\mathfrak{g}$. By Proposition 6, there exist open subgroups H_1, H_2 of D^nG', such that (H_1, H_2) is a Lie subgroup with Lie algebra $\mathscr{D}^{n+1}\mathfrak{g}$. Let G″ be an open subgroup of G′ small enough for $D^nG'' \subset H_1 \cap H_2$. Then $D^{n+1}G'' \subset (H_1, H_2)$. The relations

$$D^0G'' \subset D^0G', D^1G'' \subset D^1G', \ldots, D^nG'' \subset D^nG', D^{n+1}G'' \subset (H_1, H_2)$$

prove, using (a), that D^iG'' is, for $i \leqslant n + 1$, a Lie subgroup of G with Lie algebra $\mathscr{D}^i\mathfrak{g}$.

(c) There exists an integer p such that $\mathscr{D}^p\mathfrak{g} = \mathscr{D}^{p+1}\mathfrak{g} = \cdots$. By the above, there exists an open subgroup G_0 of G such that D^iG_0 is, for $i \leqslant p$, a Lie subgroup of G with Lie algebra $\mathscr{D}^i\mathfrak{g}$. But, by (a), the same assertion remains true for $i > p$ since $D^pG_0 \supset D^iG_0$ for $i > p$.

(d) The argument is similar for the C^i.

3. CENTRALIZERS

Recall that two elements x, y of a group are called permutable if $(x, y) = e$, or $(\operatorname{Int} x)y = y$, or $(\operatorname{Int} y)x = x$; and that two elements a, b of a Lie algebra are called permutable if $[a, b] = 0$, or $(\operatorname{ad} a).b = 0$, or $(\operatorname{ad} b).a = 0$. Let G be a Lie group, $x \in \mathrm{G}$, $a \in \mathrm{L(G)}$; x and a are called permutable if $(\operatorname{Ad} x).a = a$, that is if $xa = ax$ in T(G).

Let G be a Lie group, $\mathfrak{g}$ its Lie algebra, A a subset of G and $\mathfrak{a}$ a subset of $\mathfrak{g}$. Let $\mathrm{Z_G(A)}$ (resp. $\mathrm{Z_G}(\mathfrak{a})$) denote the set of elements of G which are permutable with all the elements of A (resp. $\mathfrak{a}$). It is a closed subgroup of G. Let $\mathfrak{z}_\mathfrak{g}(\mathrm{A})$ (resp. $\mathfrak{z}_\mathfrak{g}(\mathfrak{a})$) denote the set of elements of $\mathfrak{g}$ which are permutable with all the elements of A (resp. $\mathfrak{a}$). It is a closed Lie subalgebra of $\mathfrak{g}$.

PROPOSITION 7. *Let* G *be a finite-dimensional Lie group,* $\mathfrak{g}$ *its Lie algebra and* $\mathfrak{a}$ *a subset of* $\mathfrak{g}$. *Then* $\mathrm{Z_G}(\mathfrak{a})$ *is a Lie subgroup of* G *with Lie algebra* $\mathfrak{z}_\mathfrak{g}(\mathfrak{a})$.

This follows from § 3, Proposition 44 and Corollary 2 to Proposition 39.

PROPOSITION 8. *Let* G *be a finite-dimensional real or complex Lie group,* $\mathfrak{g}$ *its Lie algebra and* A *a subset of* G. *Then* $\mathrm{Z_G(A)}$ *is a Lie subgroup of* G *with Lie algebra* $\mathfrak{z}_\mathfrak{g}(\mathrm{A})$.

Suppose that A consists of a single point a. Then $\mathrm{Z_G(A)}$ is the set of fixed points of $\operatorname{Int} a$; hence $\mathrm{Z_G(A)}$ is a Lie subgroup of G and $\mathrm{L(Z_G(A))}$ is the set of fixed points of $\operatorname{Ad} a$, that is $\mathfrak{z}_\mathfrak{g}(\mathrm{A})$ (§ 3, no. 8, Corollary 1 to Proposition 29). The general case follows by means of § 6, no. 2, Corollary 3 to Proposition 1.

PROPOSITION 9. *Let* G *be a finite-dimensional real or complex Lie group,* $\mathfrak{g}$ *its Lie algebra,* A *an integral subgroup of* G *and* $\mathfrak{a} = \mathrm{L(A)}$. *Then* $\mathrm{Z_G(A)} = \mathrm{Z_G}(\mathfrak{a})$, $\mathfrak{z}_\mathfrak{g}(\mathrm{A}) = \mathfrak{z}_\mathfrak{g}(\mathfrak{a})$ *and* $\mathrm{Z_G(A)}$ *is a Lie subgroup of* G *with Lie algebra* $\mathfrak{z}_\mathfrak{g}(\mathfrak{a})$.

Let $x \in \mathrm{G}$. Then

$$\begin{aligned} x \in \mathrm{Z_G(A)} &\Leftrightarrow \mathrm{A} \subset \mathrm{Z_G}(\{x\}) \\ &\Leftrightarrow \mathfrak{a} \subset \mathrm{L}(\mathrm{Z_G}(\{x\})) && \text{(§ 6, Corollary 2 to Proposition 3)} \\ &\Leftrightarrow \mathfrak{a} \subset \mathfrak{z}_\mathfrak{g}(\{x\}) && \text{(Proposition 8)} \\ &\Leftrightarrow x \in \mathrm{Z_G}(\mathfrak{a}) \end{aligned}$$

and hence $\mathrm{Z_G(A)} = \mathrm{Z_G}(\mathfrak{a})$. Let $u \in \mathfrak{g}$. Then

$$\begin{aligned} u \in \mathfrak{z}_\mathfrak{g}(\mathrm{A}) &\Leftrightarrow \mathrm{A} \subset \mathrm{Z_G}(\{u\}) \\ &\Leftrightarrow \mathfrak{a} \subset \mathrm{L}(\mathrm{Z_G}(\{u\})) && \text{(§ 6, Corollary 2 to Proposition 3)} \\ &\Leftrightarrow \mathfrak{a} \subset \mathfrak{z}_\mathfrak{g}(\{u\}) && \text{(Proposition 7)} \\ &\Leftrightarrow u \in \mathfrak{z}_\mathfrak{g}(\mathfrak{a}) \end{aligned}$$

and hence $\mathfrak{z}_\mathfrak{g}(\mathrm{A}) = \mathfrak{z}_\mathfrak{g}(\mathfrak{a})$. The last assertion then follows from Proposition 7 or Proposition 8.

4. NORMALIZERS

Let G be a Lie group, $\mathfrak{g}$ its Lie algebra, A a subset of G and $\mathfrak{a}$ a subset of $\mathfrak{g}$. In this section, $N_G(A)$ will denote the set of $g \in G$ such that $gAg^{-1} = A$. It is a subgroup of G, which is closed if A is closed. $\mathfrak{n}_\mathfrak{g}(\mathfrak{a})$ will denote the set of $x \in \mathfrak{g}$ such that $[x, \mathfrak{a}] \subset \mathfrak{a}$ (cf. Chapter I, § 1, no. 4). It is a subalgebra of $\mathfrak{g}$, which is closed if $\mathfrak{a}$ is closed. $N_G(\mathfrak{a})$ will denote the set of $g \in G$ such that $g\mathfrak{a}g^{-1} = \mathfrak{a}$.

PROPOSITION 10. *Let* G *be a finite-dimensional Lie group,* $\mathfrak{g}$ *its Lie algebra and* $\mathfrak{a}$ *a vector subspace of* $\mathfrak{g}$. *Then* $N_G(\mathfrak{a})$ *is a Lie subgroup of* G *with Lie algebra* $\mathfrak{n}_\mathfrak{g}(\mathfrak{a})$.

This follows from § 3, Proposition 44 and Corollary 1 to Proposition 39.

PROPOSITION 11. *Let* G *be a finite-dimensional real or complex Lie group,* $\mathfrak{g}$ *its Lie algebra,* A *an integral subgroup of* G *and* $\mathfrak{a} = L(A)$. *Then* $N_G(A) = N_G(\mathfrak{a})$ *and* $N_G(A)$ *is a Lie subgroup of* G *containing* $\overline{A}$, *with Lie algebra* $\mathfrak{n}_\mathfrak{g}(\mathfrak{a})$.

The equality $N_G(A) = N_G(\mathfrak{a})$ follows from § 6, no. 2, Corollary 2 to Proposition 3. By Proposition 10, $N_G(A)$ is then a Lie subgroup of G with Lie algebra $\mathfrak{n}_\mathfrak{g}(\mathfrak{a})$. Hence $N_G(A)$ is closed. As $N_G(A) \supset A$, $N_G(A) \supset \overline{A}$.

COROLLARY. *If* $\mathfrak{a} = \mathfrak{n}_\mathfrak{g}(\mathfrak{a})$, A *is a Lie subgroup of* G *and is the identity component of* $N_G(A)$.

This identity component is a Lie subgroup with Lie algebra $\mathfrak{n}_\mathfrak{g}(\mathfrak{a})$ (Proposition 11) and hence is equal to A by § 6, no. 2, Theorem 2 (i).

5. NILPOTENT LIE GROUPS

PROPOSITION 12. *Let* G *be a finite-dimensional Lie group. For* L(G) *to be nilpotent, it is necessary and sufficient that* G *have a nilpotent open subgroup.*

Suppose that G has a nilpotent open subgroup G_0. By the Corollaries to Propositions 4 and 6, no. 2, $\mathscr{C}^i L(G_0) = \{0\}$ for sufficiently large i. Hence $L(G_0) = L(G)$ is nilpotent.

Suppose that L(G) is nilpotent. If K = **R** or **C**, the identity component G_0 of G is nilpotent by the Corollary to Proposition 4, no. 2, and G_0 is open in G. If K is ultrametric, the Corollary to Proposition 6, no. 2, proves that there exist an open subgroup G_1 of G, an integer $i > 0$ and a neighbourhood V of e in G such that $C^i G_1 \cap V = \{e\}$. Then, if G_0 is a sufficiently small subgroup of G_1, $C^i G_0 \subset V$, hence $C^i G_0 = \{e\}$ and G_0 is nilpotent.

Let $\mathfrak{g}$ be a nilpotent Lie algebra. The Hausdorff series H(X, Y) corresponding to $\mathfrak{g}$ has only a finite number of non-zero terms and we know (Chapter II, § 6, no. 5, *Remark* 3) that the law of composition $(x, y) \mapsto H(x, y)$ defines a group structure on $\mathfrak{g}$. Suppose further that $\mathfrak{g}$ is complete and normable. Clearly the law H is a continuous-polynomial (*Differentiable and Analytic Manifolds*, R, Appendix). Hence $\mathfrak{g}$, together with the law H, is a Lie group G, said to be *associated with* $\mathfrak{g}$. By § 4, no. 2, Lemmas 2 and 3, $L(G) = \mathfrak{g}$. The identity mapping ϕ of $\mathfrak{g}$ into G is an exponential mapping of G such that $\phi(\lambda x)\phi(\lambda' x) = \phi((\lambda + \lambda')x)$ for all $x \in \mathfrak{g}$, $\lambda \in K$, $\lambda' \in K$. Every Lie subalgebra $\mathfrak{h}$ of $\mathfrak{g}$ admitting a topological supplement is a Lie subgroup H of G and $L(H) = \mathfrak{h}$.

PROPOSITION 13. *Let* G *be a finite-dimensional simply connected nilpotent Lie group over* **R** *or* **C**.

(i) $\exp_G$ *is an isomorphism of the Lie group associated with* L(G) *onto* G.

(ii) *Every integral subgroup of* G *is a simply connected Lie subgroup of* G.

Let $\mathfrak{g} = L(G)$, which is nilpotent (Proposition 12). As two simply connected Lie groups over **R** or **C** which have the same Lie algebra are isomorphic (§ 6, no. 3, Theorem 3 (ii)), it suffices to prove the proposition when G is the group associated with $\mathfrak{g}$. Then (i) and (ii) follow from what was said before the proposition.

PROPOSITION 14. *Let* G *be a finite-dimensional connected Lie group over* **R** *or* **C**.

(i) *If* G *is nilpotent,* $\exp_G$ *is étale and surjective.*

(ii) *If* K = **C** *and* $\exp_G$ *is étale, then* G *is nilpotent.*

Let G′ be the universal covering space of G. Let ϕ be the canonical morphism of G′ onto G. Then $\exp_G = \phi \circ \exp_{G'}$ (§ 6, no. 4, Proposition 10) and hence (i) follows from Proposition 13 (i).

If K = **C** and exp is étale, then, for all $x \in L(G)$, and x has no eigenvalue belonging to $2i\pi(\mathbf{Z} - \{0\})$ (§ 6, no. 4, Corollary to Proposition 12). Applying this to λx, where λ varies through **C**, it follows that all the eigenvalues of ad x are zero and hence that ad x is nilpotent. Therefore L(G) is nilpotent (Chapter I, § 4, Corollary 1 to Theorem 1) and hence G is nilpotent (Proposition 12).

PROPOSITION 15. *Let* G *be a finite-dimensional connected nilpotent Lie group over* **R** *or* **C** *and* A *an integral subgroup of* G. *Then* $Z_G(A)$ *is the* connected *Lie subgroup of* G *with Lie algebra* $\mathfrak{z}_G(L(A))$.

By Proposition 9 of no. 3 it suffices to prove that $Z_G(A)$ is connected. Let $g \in Z_G(A)$. There exists $x \in L(G)$ such that $g = \exp x$ (Proposition 14). Then $\mathrm{Ad}\, g|L(A) = 1$ (no. 3, Proposition 9), hence $\mathrm{Ad}\, g^n|L(A) = 1$ for all $n \in \mathbf{Z}$ and hence $\exp(\mathrm{ad}\, nx)|L(A) = 1$ for all $n \in \mathbf{Z}$. As the mapping

$$\lambda \mapsto \exp(\mathrm{ad}\, \lambda x)|L(A)$$

of K into $\mathscr{L}(L(A), L(G))$ is polynomial, $\exp(\mathrm{ad}\, \lambda x)|L(A) = 1$ for all $\lambda \in K$, that is $\exp(\lambda x) \in Z_G(A)$ for all $\lambda \in K$.

PROPOSITION 16. *Let* G *be a finite-dimensional nilpotent Lie group over* **R** *or* **C** *and* A *an integral subgroup of* G *distinct from* G. *Then* $N_G(A)$ *is a* connected *Lie subgroup of* G *distinct from* A.

$N_G(A) \neq A$ (*Algebra*, Chapter I, § 6, Corollary 1 to Proposition 8). By Proposition 11 of no. 4, we need only prove that $N_G(A)$ is connected. Let $g \in N_G(A)$. There exists $x \in L(G)$ such that $g = \exp x$ (Proposition 14). Let E be the vector subspace of $\mathscr{L}(L(G))$ consisting of the $u \in \mathscr{L}(L(G))$ such that $u(L(A)) \subset L(A)$. Then $\operatorname{Ad} g^n \in E$ and hence $\exp(\operatorname{ad} nx) \in E$ for all $n \in \mathbf{Z}$. Hence $\exp(\operatorname{ad} \lambda x) \in E$ for all $\lambda \in K$, that is $\exp(\lambda x) \in N_G(A)$ for all $\lambda \in K$.

PROPOSITION 17. *Let $\mathfrak{g}$ be a finite-dimensional nilpotent Lie algebra over* K *and $(\mathfrak{g}_0, \mathfrak{g}_1, \ldots, \mathfrak{g}_n)$ a decreasing sequence of ideals of $\mathfrak{g}$ such that $\mathfrak{g}_0 = \mathfrak{g}$, $\mathfrak{g}_n = \{0\}$ and $[\mathfrak{g}, \mathfrak{g}_i] \subset \mathfrak{g}_{i+1}$ for $0 \leqslant i < n$. Let $\mathfrak{a}_1, \mathfrak{a}_2, \ldots, \mathfrak{a}_p$ be vector subspaces of $\mathfrak{g}$ such that each $\mathfrak{g}_i$ is the direct sum of its intersections with the $\mathfrak{a}_j$. Let $\mathfrak{g}$ be given the Hausdorff law of composition* H. *Let ϕ be the mapping*

$$(x_1, x_2, \ldots, x_p) \mapsto x_1 \,\mathsf{H}\, x_2 \,\mathsf{H} \cdots \mathsf{H}\, x_p$$

of $\mathfrak{a}_1 \times \mathfrak{a}_2 \times \cdots \times \mathfrak{a}_p$ into $\mathfrak{g}$.

(i) *ϕ is a bijection of $\mathfrak{a}_1 \times \mathfrak{a}_2 \times \cdots \times \mathfrak{a}_p$ onto $\mathfrak{g}$;*

(ii) *ϕ and ϕ^{-1} are polynomial mappings;*

(iii) *the mapping $(x, y) \mapsto \phi^{-1}(\phi(x).\phi(y)^{-1})$ of $(\mathfrak{a}_1 \times \mathfrak{a}_2 \times \cdots \times \mathfrak{a}_p)^2$ into $\mathfrak{a}_1 \times \mathfrak{a}_2 \times \cdots \times \mathfrak{a}_p$ is polynomial.*

The proposition is obvious for $\dim \mathfrak{g} = 0$. Suppose that $\dim \mathfrak{g} > 0$ and that the proposition has been established for dimensions $< \dim \mathfrak{g}$. It can be assumed that $\mathfrak{g}_{n-1} \neq \{0\}$ and $\mathfrak{g}_{n-1}$ is then a non-zero central ideal of $\mathfrak{g}$. There exists an index j such that $\mathfrak{h} = \mathfrak{g}_{n-1} \cap \mathfrak{a}_j \neq \{0\}$. Let $\mathfrak{g}' = \mathfrak{g}/\mathfrak{h}$, θ be the canonical morphism of $\mathfrak{g}$ onto $\mathfrak{g}'$, $\mathfrak{g}'_i = \theta(\mathfrak{g}_i)$ and $\mathfrak{a}'_i = \theta(\mathfrak{a}_i)$. Then $(\mathfrak{g}'_0, \mathfrak{g}'_1, \ldots, \mathfrak{g}'_n)$ is a decreasing sequence of ideals of $\mathfrak{g}'$ such that $\mathfrak{g}'_0 = \mathfrak{g}'$, $\mathfrak{g}'_n = \{0\}$,

$$[\mathfrak{g}', \mathfrak{g}'_i] \subset \mathfrak{g}'_{i+1}$$

for $0 \leqslant i < n$ and each $\mathfrak{g}'_i$ is the direct sum of its intersections with the $\mathfrak{a}'_j$. Let ϕ' be the mapping

$$(x'_1, x'_2, \ldots, x'_p) \mapsto x'_1 \,\mathsf{H}\, x'_2 \,\mathsf{H} \cdots \mathsf{H}\, x'_p$$

of $\mathfrak{a}'_1 \times \mathfrak{a}'_2 \times \cdots \times \mathfrak{a}'_p$ into $\mathfrak{g}'$. By the induction hypothesis, ϕ' is bijective and ϕ', ϕ'^{-1} are polynomial mappings.

Let $x \in \mathfrak{g}$. We write

$$\phi'^{-1}(\theta(x)) = (x'_1(x), x'_2(x), \ldots, x'_p(x)). \tag{1}$$

Then

$$\theta(x) = x'_1(x) \,\mathsf{H}\, x'_2(x) \,\mathsf{H} \cdots \mathsf{H}\, x'_p(x). \tag{2}$$

Let $\mathfrak{h}_1$ be a vector subspace supplementary to $\mathfrak{h}$ in $\mathfrak{g}$, the sum of the $\mathfrak{a}_k$ with $k \neq j$ and a supplement of $\mathfrak{h}$ in $\mathfrak{a}_j$. There exists a bijection η of $\mathfrak{g}'$ onto $\mathfrak{h}_1$ such that $\theta \circ \eta = \mathrm{Id}_{\mathfrak{g}'}$. For $x \in \mathfrak{g}$, we write

$$\zeta(x) = \eta(x'_1(x)) \,\mathsf{H}\, \eta(x'_2(x)) \,\mathsf{H} \cdots \mathsf{H}\, \eta(x'_p(x)) \in \mathfrak{g}. \tag{3}$$

$$y(x) = \zeta(x)^{-1} \,\mathsf{H}\, x = (-\zeta(x)).x. \tag{4}$$

By (2) and (3), $\theta(\zeta(x)) = \theta(x)$ and hence $y(x) \in \mathfrak{h}$. Finally we write

$$(5) \quad \psi(x) = (\eta(x'_1(x)), \ldots, \eta(x'_j(x)) + y(x), \ldots, \eta(x'_p(x))) \in \mathfrak{a}_1 \times \cdots \times \mathfrak{a}_p.$$

As $y(x)$ is central in $\mathfrak{g}$,

$$\begin{aligned} \phi(\psi(x)) &= \eta(x'_1(x)) \,\mathrm{H} \cdots \mathrm{H}\, \eta(x'_j(x)) \,\mathrm{H} \cdots \mathrm{H}\, \eta(x'_p(x)) \,\mathrm{H}\, y(x) \\ &= \zeta(x) \,\mathrm{H}\, y(x) && \text{by (3)} \\ &= x && \text{by (4).} \end{aligned}$$

Hence $\phi \circ \psi = \mathrm{Id}_{\mathfrak{g}}$. Now let $(x_1, x_2, \ldots, x_p) \in \mathfrak{a}_1 \times \mathfrak{a}_2 \times \cdots \times \mathfrak{a}_p$ and write $x = \phi(x_1, x_2, \ldots, x_p) = x_1 \,\mathrm{H}\, x_2 \,\mathrm{H} \cdots \mathrm{H}\, x_p$. Then

$$\theta(x) = \theta(x_1) \,\mathrm{H}\, \theta(x_2) \,\mathrm{H} \cdots \mathrm{H}\, \theta(x_p),$$

hence $x'_i(x) = \theta(x_i)$ for $1 \leqslant i \leqslant p$ and therefore

$$\zeta(x) = x_1 \,\mathrm{H}\, x_2 \,\mathrm{H} \cdots \mathrm{H}\, (\eta\theta(x_j)) \,\mathrm{H} \cdots \mathrm{H}\, x_p$$
$$y(x) = x_j - \eta\theta(x_j).$$

Then by (5)

$$\psi(x) = (x_1, \ldots, \eta\theta(x_j) + x_j - \eta\theta(x_j), \ldots, x_p) = (x_1, x_2, \ldots, x_p).$$

Hence $\psi \circ \phi = \mathrm{Id}_{\mathfrak{a}_1 \times \ldots \times \mathfrak{a}_p}$. This proves (i). As the Hausdorff law is polynomial, ϕ is polynomial. By the induction hypothesis, ϕ'^{-1} is polynomial; by formula (1), the functions x'_j are polynomial, hence ζ is polynomial (formula (3)), y is polynomial (formula (4)) and ψ is polynomial (formula (5)). This proves (ii). Assertion (iii) follows from (i) and (ii) and the fact that the Hausdorff law is polynomial.

Example of a nilpotent Lie group. Let G be the lower strict triangular subgroup of $\mathbf{GL}(n, \mathrm{K})$. It is a Lie subgroup of $\mathbf{GL}(n, \mathrm{K})$ and $\mathrm{L(G)} \subset \mathfrak{gl}(n, \mathrm{K})$ is the Lie algebra of lower triangular matrices with zero diagonal (§ 3, no. 10, Proposition 36). By Chapter II, § 4, no. 6, *Remark*, G is nilpotent. Suppose henceforth that $\mathrm{K} = \mathbf{R}$ or $\mathbf{C}$. As G is homeomorphic to $\mathrm{K}^{n(n-1)/2}$, G is simply connected. The exponential mapping of L(G) into G is just the mapping

$$u \mapsto \exp u = \sum_{k \geqslant 0} \frac{u^k}{k!} = \sum_{k=0}^{n-1} \frac{u^k}{k!}$$

(§ 6, no. 4, *Example*). By Proposition 13, the exponential is an isomorphism of the manifold L(G) onto the manifold G. Proposition 17 of § 6, no. 9 gives the inverse bijection log. We give K^n a norm. By *Spectral Theories*, Chapter I, § 4, no. 9, for $g \in \mathrm{G}$ and $\|g - 1\| < 1$,

$$\log g = \sum_{k \geqslant 1} \frac{(-1)^{k-1}}{k} (g-1)^k$$

that is

$$(6) \qquad \log g = \sum_{k=1}^{n-1} \frac{(-1)^{k-1}}{k} (g-1)^k.$$

But the two sides of (6) are analytic functions of g for $g \in \mathrm{G}$ and are therefore equal for all $g \in \mathrm{G}$.

Proposition 18. *Let k be a commutative field.* V *a vector space of finite dimension* >0 *over k and* G *a subgroup of* $\mathbf{GL}(\mathrm{V})$ *whose elements are unipotent.*

(i) *There exists a non-zero element v of* V *such that* $gv = v$ *for all* $g \in \mathrm{G}$.

(ii) *There exists a basis* B *of* V *such that, for all* $g \in \mathrm{G}$, *the matrix of g with respect to* B *is lower triangular and has all its diagonal elements equal to* 1.

(iii) *The group* G *is nilpotent.*

(a) Suppose first that k is algebraically closed and that the identity representation of G is simple. Let a, b be in G. Then

$$\mathrm{Tr}(a(b-1)) = \mathrm{Tr}(ab-1) - \mathrm{Tr}(a-1) = 0 - 0 = 0$$

for $ab - 1$ and $a - 1$ are nilpotent. As the vector subspace of $\mathscr{L}(\mathrm{V})$ generated by G is $\mathscr{L}(\mathrm{V})$ (*Algebra*, Chapter VIII, § 4, Corollary 1 to Proposition 2), $\mathrm{Tr}(u(b-1)) = 0$ for all $u \in \mathscr{L}(\mathrm{V})$ and hence $b = 1$. Thus $\mathrm{G} = \{1\}$.

(b) We pass now to the general case. Let $\bar{k}$ be an algebraic closure of k, $\overline{\mathrm{V}} = \mathrm{V} \otimes_k \bar{k}$ and $\overline{\mathrm{G}} \subset \mathbf{GL}(\overline{\mathrm{V}})$ the set of $a \otimes 1$ for $a \in \mathrm{G}$. Let W (resp. W′) be the set of elements of V (resp. $\overline{\mathrm{V}}$) invariant under G (resp. $\overline{\mathrm{G}}$). Then $\mathrm{W}' = \mathrm{W} \otimes_k \bar{k}$ for $\mathrm{W} = \bigcap_{g \in \mathrm{G}} \mathrm{Ker}(g-1)$ and $\mathrm{W}' = \bigcap_{g \in \mathrm{G}} \mathrm{Ker}(g-1) \otimes 1$. If V_1 denotes a minimal element in the set of non-zero vector subspaces of $\overline{\mathrm{V}}$ which are stable under $\overline{\mathrm{G}}$, then $\mathrm{V}_1 \subset \mathrm{W}'$ by part (a) of the proof; hence $\mathrm{W} \neq \{0\}$, which proves (i).

(c) By induction on dim V, it follows from (i) that there exists an increasing sequence $(\mathrm{V}_1, \mathrm{V}_2, \ldots, \mathrm{V}_n)$ of vector subspaces of V which are stable under G such that $\mathrm{V}_n = \mathrm{V}$ and the automorphism group of $\mathrm{V}_i/\mathrm{V}_{i-1}$ canonically derived from G reduces to $\{1\}$ for all i (we make the convention that $\mathrm{V}_r = \{0\}$ for $r \leqslant 0$), This implies first (ii) and consequently (iii) (Chapter II, § 4, no. 6, *Remark*).

Corollary 1. *Let* G *be a finite-dimensional connected real or complex Lie group. For* G *to be nilpotent, it is necessary and sufficient that every element of* Ad G *be nilpotent.*

If every element of Ad G is unipotent, Ad G is nilpotent (Proposition 18) and hence G, which is a central extension of Ad G, is nilpotent. If G is nilpotent, L(G) is nilpotent, hence ad x is nilpotent for all $x \in \mathrm{L(G)}$ and hence $\mathrm{Ad}(\exp x) = \exp \mathrm{ad}\, x$ is unipotent; but every element of G is of the form $\exp x$ for some $x \in \mathrm{L(G)}$ (Proposition 14).

COROLLARY 2. *Every analytic subgroup of* $\mathbf{GL}(n, K)$ *consisting of unipotent elements is a simply connected Lie subgroup.*

This follows from Propositions 13 (ii) and 18 (ii) and the fact that the lower strict triangular group is simply connected.

6. SOLVABLE LIE GROUPS

PROPOSITION 19. *Let G be a finite-dimensional Lie group. For L(G) to be solvable, it is necessary and sufficient that G possess a solvable open subgroup.*

The proof is analogous to that of Proposition 12 of no. 5.

PROPOSITION 20. *Let G be a simply connected solvable Lie group of finite-dimension n over* $\mathbf{R}$ *or* $\mathbf{C}$ *and* $\mathfrak{g} = L(G)$. *Let* $(\mathfrak{g}_n, \mathfrak{g}_{n-1}, \ldots, \mathfrak{g}_0)$ *be a sequence of subalgebras of* $\mathfrak{g}$ *of dimensions* $n, n-1, \ldots, 0$, *such that* $\mathfrak{g}_{i-1}$ *is an ideal of* $\mathfrak{g}_i$ *for* $i = n, n-1, \ldots, 1$.† *Let* G_i *be the integral subgroup of G corresponding to* $\mathfrak{g}_i$. *Let* x_i *be a vector of* $\mathfrak{g}_i$ *not belonging to* $\mathfrak{g}_{i-1}$. *Let* ϕ_i *be the mapping*

$$(\lambda_1, \lambda_2, \ldots, \lambda_i) \mapsto (\exp \lambda_1 x_1)(\exp \lambda_2 x_2) \ldots (\exp \lambda_i x_i)$$

of K^i *into G. Then* ϕ_n *is an isomorphism of analytic manifolds and* $\phi_i(K^i) = G_i$ *for all i.*

For $n = 0$ the proposition is obvious. We argue by induction on n. Let H be the integral subgroup of G such that $L(H) = Kx_n$. By § 6, no. 6, Corollary 1 to Proposition 14, H and G_{n-1} are simply connected Lie subgroups of G and, as a Lie group, G is the semi-direct product of H by G_{n-1}. Hence $\lambda \mapsto \exp(\lambda x_n)$ is an isomorphism of K onto H and by the induction hypothesis the mapping

$$(\lambda_1, \lambda_2, \ldots, \lambda_{n-1}) \mapsto (\exp \lambda_1 x_1)(\exp \lambda_2 x_2) \ldots (\exp \lambda_{n-1} x_{n-1})$$

is an isomorphism of the analytic manifold K^{n-1} onto the analytic manifold G_{n-1} which maps $K^i \times \{0\}$ to G_i for $i = 1, 2, \ldots, n-1$. Hence the proposition.

PROPOSITION 21. *Let G be a simply connected finite-dimensional solvable Lie group over* $\mathbf{R}$ *or* $\mathbf{C}$ *and M an integral subgroup of G. Then M is a Lie subgroup of G and is simply connected.*

We continue to use the notation n, $\mathfrak{g}$, $\mathfrak{g}_i$, x_i, ϕ of Proposition 20 but impose on the x_i the following supplementary condition: let $i_p > i_{p-1} > \cdots > i_1$ be the integers i such that $L(M) \cap \mathfrak{g}_i \neq L(M) \cap \mathfrak{g}_{i-1}$; then we take

$$x_{i_k} \in L(M) \cap \mathfrak{g}_{i_k}$$

† Such a sequence exists by Chapter I, § 5, Proposition 2.

for $k = 1, 2, \ldots, p$. By induction on n, it is easily seen that $(x_{i_p}, x_{i_{p-1}}, \ldots, x_{i_1})$ is a basis of L(M). Let N be a simply connected Lie group such that there exists an isomorphism h of L(N) onto L(M). Let

$$y_p = h^{-1}(x_{i_p}), \ldots, y_1 = h^{-1}(x_{i_1}).$$

By Proposition 20, the mapping

$$(\lambda_1, \lambda_2, \ldots, \lambda_p) \mapsto (\exp \lambda_1 y_1)(\exp \lambda_2 y_2) \ldots (\exp \lambda_p y_p)$$

is an isomorphism of the manifold K^p onto the manifold N. There exists a Lie group morphism τ of N into G such that $h = L(\tau)$ and $\tau(N) = M$ (§ 6, no. 2, Corollary 1 to Proposition 1). Hence M is the set of elements of G of the form

$$\begin{aligned}\tau((\exp \lambda_1 y_1) \ldots (\exp \lambda_p y_p)) &= \exp(\lambda_1 L(\tau) y_1) \ldots \exp(\lambda_p L(\tau) y_p) \\ &= \exp(\lambda_1 x_{i_1}) \ldots \exp(\lambda_p x_{i_p}).\end{aligned}$$

Thus $M = \phi(T)$ where T is a vector subspace of K^n.

PROPOSITION 22. *Suppose that* K = **R** *or* **C**. *Let* V *be a finite-dimensional vector space and* G *a connected solvable subgroup of* **GL**(V). *Suppose that the identity representation of* G *is simple.*

(i) *If* K = **R**, *then* dim V $\leqslant$ 2 *and* G *is commutative.*

(ii) *If* K = **C**, *then* dim V = 1.

(i) Suppose that K = **R**. Then the closure H of G in **GL**(V) is a solvable connected Lie subgroup of **GL**(V) (no. 1, Corollary 2 to Proposition 1). Hence L(H) is solvable (Proposition 19). The identity representation of L(G) is simple (§ 6, no. 5, Corollary 2 to Proposition 13). Hence dim V $\leqslant$ 2 and L(G) is commutative (Chapter I, § 5, Corollaries 1 and 4 to Theorem 1). Hence G is commutative.

(ii) Suppose that K = **C**. Let W be a minimal element among the non-zero real vector subspaces of V which are stable under G. The complex vector subspace of V generated by W is equal to V since the identity representation of G is simple. By (i), G|W is commutative. Hence G is commutative. Therefore every element of G is a homothety (*Algebra*, Chapter VIII, § 4, Corollary 1 to Proposition 2), so that dim V = 1.

COROLLARY. *Let* V *be a complex vector space of finite dimension* >0 *and* G *a connected solvable subgroup of* **GL**(V).

(i) *There exists a non-zero element* v *of* V *such that* $gv \in \mathbf{C}v$ *for all* $g \in G$.

(ii) *There exists a basis* B *of* V *such that, for all* $g \in G$, *the matrix of* g *with respect to* B *is lower triangular.*

Let V_1 be a minimal element among the non-zero vector subspaces of V which are stable under G. By Proposition 22 (ii), dim $V_1 = 1$. This proves (i). By induction on dim V, it then follows that there exists an increasing sequence $(V_1, V_2, \ldots, V_n)$ of vector subspaces of V which are stable under G such that $\dim V_{i+1}/V_i = 1$ for $i < n$ and $V_n = V$; hence (ii).

7. RADICAL OF A LIE GROUP

PROPOSITION 23. *Let* G *be a finite-dimensional rear or complex Lie group,* $\mathfrak{r}$ *the radical of* L(G) (Chapter I, § 5, Definition 2) *and* $\mathfrak{n}$ *the largest nilpotent ideal of* L(G) (Chapter I, § 4, no. 4). *Let* R (resp. N) *be the integral subgroup of* G *with Lie algebra* $\mathfrak{r}$ (resp. $\mathfrak{n}$). *Then* R (resp. N) *is a solvable* (resp. *nilpotent*) *Lie subgroup of* G, *which is invariant under every continuous automorphism of* G. *Every solvable* (resp. *nilpotent*) *connected normal subgroup of* G *is contained in* R (resp. N).

The group R is solvable (§ 6, Proposition 19). Suppose that $K = \mathbf{R}$. Let G′ be a connected solvable normal subgroup of G. Then $\overline{G'}$ is a connected solvable (no. 1, Corollary 2 to Proposition 1) normal Lie subgroup of G (§ 8, no. 2, Theorem 2). Hence $L(\overline{G'})$ is a solvable ideal of L(G), whence $L(\overline{G'}) \subset \mathfrak{r}$ and $\overline{G'} \subset R$. In particular, $\overline{R} \subset R$, hence R is closed and therefore is a Lie subgroup of G. Suppose that $K = \mathbf{C}$. Let H be the underlying real Lie subgroup of G. If $\mathfrak{r}'$ is the radical of L(H), $i\mathfrak{r}'$ is a solvable ideal of L(H), whence $\mathfrak{r}' = i\mathfrak{r}'$; hence $\mathfrak{r} \subset \mathfrak{r}' \subset \mathfrak{r}$ and, by the above, R is closed in H and therefore in G; thus R is a Lie subgroup of G. Every connected solvable normal subgroup of G is a connected solvable normal subgroup of H and therefore contained in R. Hence we have proved for $K = \mathbf{C}$ and for $K = \mathbf{R}$ that R is the largest connected solvable normal subgroup of G; therefore R is invariant under every continuous automorphism of G. The proof for N is completely analogous.

DEFINITION 1. Let G *be a finite-dimensional real or complex Lie group. The radical of* G *is the largest connected solvable normal subgroup of* G.

Remark. Even if G is connected there may be solvable normal subgroups of G not contained in the radical of G.

PROPOSITION 24. *Suppose that* $K = \mathbf{R}$ or $\mathbf{C}$. *Let* G_1, G_2 *be two finite-dimensional connected Lie groups,* R_1 *and* R_2 *their radicals and* ϕ *a surjective morphism of* G_1 *into* G_2. *Then* $\phi(R_1) = R_2$.

By § 3, no. 8, Proposition 28, $L(\phi)$ is surjective. Hence $L(\phi)(L(R_1)) = L(R_2)$ (Chapter I, § 6, Corollary 3 to Proposition 2). Let i be the canonical injection of R_1 into G_1. Then the image of $\phi \circ i$ is R_2 (§ 6, no. 2, Corollary 1 to Proposition 1).

PROPOSITION 25. *Suppose that* $K = \mathbf{R}$ *or* $\mathbf{C}$. *Let* G_1, G_2 *be finite-dimensional connected Lie groups and* R_1 *and* R_2 *their radicals. The radical of* $G_1 \times G_2$ *is* $R_1 \times R_2$.

This follows from Chapter I, § 5, Proposition 4.

8. SEMI-SIMPLE LIE GROUPS

PROPOSITION 26. *Let* G *be a finite-dimensional connected real or complex Lie group. The following conditions are equivalent:*

(i) L(G) *is semi-simple;*

(ii) *the radical of* G *is* $\{e\}$;

(iii) *every normal commutative integral subgroup of* G *is equal to* $\{e\}$.

Condition (ii) means that the radical of L(G) is $\{0\}$ and hence (i) $\Leftrightarrow$ (ii) (Chapter I, § 6, Theorem 1). The equivalence of (i) and (iii) follows from § 6, no. 6, Proposition 14.

DEFINITION 2. *A connected real or complex Lie group is called semi-simple if it is finite-dimensional and satisfies the conditions of Proposition* 26.

Remark 1. Let G be a finite-dimensional connected real or complex Lie group. If G is not semi-simple, G has a commutative connected Lie subgroup G′ which is invariant under every continuous automorphism, such that $G' \neq \{e\}$. For let $\mathfrak{n}$ be the largest nilpotent ideal of L(G); then $\mathfrak{n} \neq \{0\}$ and the corresponding analytic subgroup N is a Lie subgroup which is invariant under every continuous automorphism of G (no. 7, Proposition 23); the centre G′ of N has the desired properties.

PROPOSITION 27. *Let* G *be a finite-dimensional connected real or complex Lie group. The following conditions are equivalent:*

(i) L(G) *is simple;*

(ii) *the only normal integral subgroups of* G *are* $\{e\}$ *and* G *and further* G *is not commutative.*

This follows from § 6, no. 6, Proposition 14.

DEFINITION 3. *A connected real or complex Lie group is called almost simple if it is finite-dimensional and satisfies the conditions of Proposition* 27.

PROPOSITION 28. *Let* G *be a simply connected real or complex Lie group. The following conditions are equivalent:*

(i) G *is semi-simple;*

(ii) G *is isomorphic to the product of a finite number of almost simple groups.*

If G is a finite product of almost simple Lie groups, L(G) is a finite product of simple Lie algebras and is hence semi-simple. If G is semi-simple, L(G) is isomorphic to a product of simple Lie algebras $L_1, \ldots, L_n$. Let G_i be a simply connected Lie group with Lie algebra L_i, which is therefore almost simple. Then G and $G_1 \times \cdots \times G_n$ are simply connected and have isomorphic Lie algebras and are therefore isomorphic.

Lemma 1. *Let* G *be a connected topological group,* Z *its centre and* Z′ *a discrete subgroup of* Z. *Then the centre of* G/Z′ *is* Z/Z′.

Let y be an element of G whose class modulo Z′ is a central element of

G/Z′. Let ϕ be the mapping $g \mapsto gyg^{-1}y^{-1}$ of G into G. Then $\phi(G)$ is connected and contained in Z′ and hence $\phi(G) = \phi(\{e\}) = \{e\}$. Therefore $y \in Z$.

PROPOSITION 29. *Let* G *be a semi-simple connected real or complex Lie group.*

(i) G = (G, G).

(ii) *The centre* Z *of* G *is discrete.*

(iii) *The centre of* G/Z *is* $\{e\}$.

Assertion (i) follows from the Corollary to Proposition 4, no. 2 and Chapter I, § 6, Theorem 1.

Assertion (ii) follows from § 6, no. 4, Corollary 4 to Proposition 10 and Chapter I, § 6, no. 1, *Remark* 2.

Assertion (iii) follows from (ii) and Lemma 1.

PROPOSITION 30. (i) *Let* $\mathfrak{g}$ *be a semi-simple real or complex Lie algebra. Then* Int $\mathfrak{g}$ *is the identity component of* Aut $\mathfrak{g}$.

(ii) *Let* G *be a semi-simple connected real or complex Lie group. The adjoint group of* G *is the identity component of* Aut L(G). *Its centre reduces to the identity element.*

Every derivation of $\mathfrak{g}$ is inner (Chapter I, § 6, Corollary 3 to Proposition 1) and hence L(Int $\mathfrak{g}$) = L(Aut $\mathfrak{g}$), which proves (i). The first assertion of (ii) follows from (i). The second follows from Proposition 29 (iii) and § 6, no. 4, Corollary 4 (ii) to Proposition 10.

Remark 2. Let $\mathfrak{g}$ be a complex semi-simple Lie algebra and $\mathfrak{g}_0$ its underlying real Lie algebra. Then Aut($\mathfrak{g}$) is open in Aut($\mathfrak{g}_0$), for Int($\mathfrak{g}_0$) $\subset$ Aut($\mathfrak{g}$).

PROPOSITION 31. *Let* G *be a simply connected finite-dimensional real or complex Lie group and* R *its radical. There exists a semi-simple simply connected Lie subgroup* S *of* G *such that* G, *as a Lie group, is the semi-direct product of* S *by* R. *If* S′ *is a semi-simple integral subgroup of* G, *there exists* x *in the nilpotent radical of* L(G) *such that*

$$(\mathrm{Ad}\exp x)(S') \subset S.$$

This follows from § 6, no. 6, Corollary 1 to Proposition 14 and Chapter I, § 6, Theorem 5 and Corollary 1.

Lemma 2. *Let* G *be a group* (resp. *a topological group*), G′ *a normal subgroup of* G, V *a finite-dimensional vector space over a commutative field* k (resp. *over* K), ρ *a linear representation* (resp. *a continuous linear representation*) *of* G *on* V *and* $\rho' = \rho|G'$.

(i) *If* ρ *is semi-simple,* ρ' *is semi-simple.*

(ii) *If* ρ' *is semi-simple and every finite-dimensional linear* k*-representation* (resp. *continuous linear* K*-representation*) *of* G/G′ (resp. $G/\overline{G'}$) *is semi-simple, then* ρ *is semi-simple.*

Suppose that ρ is semi-simple; we prove that ρ' is semi-simple. It suffices to consider the case where ρ is simple. Let V′ be a minimal non-zero sub-G′-module of V. For all $g \in G$, $\rho(G')\rho(g)V' = \rho(g)\rho(G')V' = \rho(g)V'$, in other words $\rho(g)V'$ is stable under $\rho(G')$; if V'' is a sub-G'-module of $\rho(g)V'$, then $\rho(g)^{-1}V''$ is a sub-G'-module of V' and hence V'' is equal to $\{0\}$ or $p(g)V'$. Thus, for all $g \in G$, $\rho(g)V'$ is a simple G'-module. But $\sum_{g \in G} \rho(g)V'$ is a non-zero sub-G-module of V, whence $V = \sum_{g \in G} \rho(g)V'$. Hence ρ' is semi-simple.

Suppose that ρ' is semi-simple. Let W be a non-zero sub-G-module of V. As ρ' is semi-simple, there exists a projector f_0 of V onto W which commutes with $\rho'(G)$. Let E be the set of $f \in \mathscr{L}(V, V)$ which commute with $\rho(G')$, which map V into W and whose restriction to W is a homothety; for $f \in E$, let $\alpha(f)$ denote the ratio of the homothety $f|W$. Then $f_0 \in E$ and $\alpha(f_0) = 1$. Clearly α is a linear form on E. Let $F = \operatorname{Ker} \alpha$, which is a hyperplane of E. For $f \in E$ and $g \in G$, we write $\sigma(g)f = \rho(g) \circ f \circ \rho(g)^{-1}$; then $\sigma(g)f$ maps V into W and its restriction to W is the homothety of ratio $\alpha(f)$; if $g' \in G'$, then

$$\begin{aligned}\sigma(g)f \circ \rho(g') &= \rho(g) \circ f \circ \rho(g)^{-1} \circ \rho(g') \\ &= \rho(g) \circ f \circ \rho(g^{-1}g'g) \circ \rho(g^{-1}) \\ &= \rho(g) \circ \rho(g^{-1}g'g) \circ f \circ \rho(g^{-1}) \\ &= \rho(g') \circ \rho(g) \circ f \circ \rho(g^{-1}) \\ &= \rho(g') \circ \sigma(g)f.\end{aligned}$$

Hence $\sigma(g)f \in E$. Therefore σ is a linear k-representation (resp. continuous linear K-representation) of G on E leaving F stable. Then $\sigma(g) = \mathrm{Id}_E$ for $g \in G'$ and hence for $g \in \overline{G'}$ in the topological case. Suppose that every finite-dimensional k-linear representation (resp. continuous K-linear representation) of G/G' (resp. $G/\overline{G'}$) is semi-simple. Then there exists in E a supplement of F which is stable under G. In other words, there exists $f \in E$ such that $\alpha(f) = 1$, which is invariant under G. Then f is a projector of V onto W, and, for $g \in G$, $\rho(g) \circ f \circ \rho(g^{-1}) = f$, that is f commutes with $\rho(G)$. Thus ρ is semi-simple.

THEOREM 1. *Let* G *be a finite-dimensional real or complex Lie group,* G_0 *its identity component,* R *its radical and* $\mathfrak{r}$ *the radical of* L(G); *suppose that* G/G_0 *is finite. Let* ρ *be a finite-dimensional analytic linear representation of* G. *The following conditions are equivalent:*

(i) ρ *is semi-simple;*
(ii) $\rho|G_0$ *is semi-simple;*
(iii) $\rho|R$ *is semi-simple;*
(iv) $L(\rho)$ *is semi-simple;*
(v) $L(\rho)|\mathfrak{r}$ *is semi-simple.*

(i) ⇔ (ii) by Lemma 2 and *Integration*, Chapter VII, § 3, Proposition 1. (ii) ⇔ (iv) and (iii) ⇔ (v) by § 6, no. 5, Corollary 2 to Proposition 13. (iv) ⇔ (v) by Chapter I, § 6, Theorem 4.

COROLLARY 1. *Let* ρ, ρ_1, ρ_2 *be finite-dimensional semi-simple analytic linear repre-*

sentations of G *and n an integer* $\geqslant 0$. *Then* $\rho_1 \otimes \rho_2$, $T^n\rho$, $S^n\rho$, $\bigwedge^n\rho$ (*Appendix*) *are semi-simple.*

The semi-simplicity of $\rho_1 \otimes \rho_2$ follows from Theorem 1 and Chapter I, § 6, Corollary 1 to Theorem 4. The semi-simplicity of $T^n\rho$, $S^n\rho$, $\bigwedge^n\rho$ follows from the semi-simplicity of $\rho_1 \otimes \rho_2$.

We shall see later that, if k is a commutative field of characteristic 0, Γ a group and ρ_1 and ρ_2 finite-dimensional semi-simple linear k-representations of Γ, then $\rho_1 \otimes \rho_2$ is semi-simple.

Corollary 2. *Let* ρ *be a finite-dimensional semi-simple analytic linear representation of* G *on a vector space* V, S *the symmetric algebra of* V *and* S^G *the subalgebra of* S *consisting of the elements invariant under* $(S\rho)(G)$. *Then* S^G *is a finitely generated algebra.*

This follows from Theorem 1, Chapter I, § 6, Theorem 6 (a) and *Commutative Algebra*, Chapter V, § 1, Theorem 2.

Corollary 3. *Let* G *be a real or complex Lie group and* G_0 *its identity component. Suppose that* G_0 *is semi-simple and that* G/G_0 *is finite. Then every finite-dimensional analytic linear representation of* G *is semi-simple.*

Proposition 32. *Let* G *be a finite-dimensional connected real Lie group. Suppose that* L(G) *is reductive. The following conditions are equivalent:*

(i) $G/\overline{D^1G}$ *is compact;*

(ii) (resp. (ii')) *every finite-dimensional analytic linear representation of* G *on a complex* (resp. *real*) *vector space is semi-simple.*

(i) $\Rightarrow$ (ii'): Suppose that $G/\overline{D^1G}$ is compact. Then every continuous linear representation of $G/\overline{D^1G}$ on a finite-dimensional real vector space is semi-simple (*Integration*, Chapter VII, § 3, Proposition 1). Let ρ be a finite-dimensional analytic linear representation of G on a real vector space. Then $\rho|D^1G$ is analytic, D^1G is semi-simple (Chapter I, § 6, Proposition 5) and hence $\rho|D^1G$ is semi-simple (Corollary 3 to Theorem 1). Hence ρ is semi-simple (Lemma 2).

It can be seen similarly that (i) $\Rightarrow$ (ii).

(i) $\Rightarrow$ (ii): Suppose that $G/\overline{D^1G}$ is not compact and hence isomorphic to a group of the form $\mathbf{R}^p \times \mathbf{T}^q$ with $p > 0$ (§ 6, no. 4, Proposition 11 (ii)). Then there exists a surjective morphism of $G/\overline{D^1G}$ onto $\mathbf{R}$ and hence a surjective morphism of G onto $\mathbf{R}$. The mapping

$$g \mapsto \sigma(g) = \begin{pmatrix} 1 & 0 \\ \rho(g) & 1 \end{pmatrix}$$

is an analytic linear representation of G on $\mathbf{R}^2$ which is not semi-simple, for the only 1-dimensional vector subspace of $\mathbf{R}^2$ which is stable under $\sigma(G)$ is $\mathbf{R}(0, 1)$.

It can be seen similarly that (ii) $\Rightarrow$ (i).

PROPOSITION 33. *Let* G *be a finite-dimensional complex Lie group whose number of connected components is finite,* ρ *a finite-dimensional analytic linear representation of* G *and* G′ *an integral subgroup of the real Lie group* G *such that* L(G′) *generates* L(G) *over* **C**. *Then, for* ρ *to be semi-simple, it is necessary and sufficient that* $\rho|G'$ *be semi-simple.*

Let $\rho' = \rho|G'$. For ρ (resp. ρ') to be semi-simple, it is necessary and sufficient that $L(\rho)$ (resp. $L(\rho')$) be semi-simple (Theorem 1). Let V be the space of ρ. For a vector subspace of V to be stable under $L(\rho)(L(G))$, it is necessary and sufficient that it be stable under $L(\rho')(L(G'))$. Hence the proposition.

§10. THE AUTOMORPHISM GROUP OF A LIE GROUP

In this paragraph, K is assumed to be of characteristic zero.

1. INFINITESIMAL AUTOMORPHISMS

Lemma 1. *Let* G *be a Lie group and* α *a vector field on* G. *For all* $g \in G$, *let*

$$\beta(g) = \alpha(g)g^{-1} \in L(G).$$

The following conditions are equivalent:

(i) α *is a homomorphism of the group* G *into the group* T(G);

(ii) *for all* g, g' *in* G, $\alpha(gg') = \alpha(g)g' + g\alpha(g')$;

(iii) *for all* g, g' *in* G, $\beta(gg_1) = \beta(g) + (\operatorname{Ad} g)\beta(g')$.

Condition (i) means that, for all g, g' in G, we have in the group T(G):

$$\beta(g)g\beta(g')g' = \beta(gg')gg'$$

or

$$\beta(g)((\operatorname{Ad} g)\beta(g'))gg' = \beta(gg')gg'.$$

But the product of $\beta(g)$ and $(\operatorname{Ad} g)\beta(g')$ in T(G) is just the sum of $\beta(g)$ and $(\operatorname{Ad} g)\beta(g')$ in L(G) (§ 2, no. 1, Proposition 2). Hence (i) $\Leftrightarrow$ (iii). On the other hand, condition (ii) may be written $\beta(gg')gg' = \beta(g)gg' + g\beta(g')g'$, or

$$\beta(gg' = \beta(g) + (\operatorname{Ad} g)\beta(g')$$

and hence (ii) $\Leftrightarrow$ (iii).

DEFINITION 1. *Let* G *be a Lie group. An infinitesimal automorphism of* G *is any analytic vector field on* G *satisfying the conditions of Lemma* 1.

Lemma 2. *Let* K′ *be a non-discrete closed subfield of* K, A *a* K′*-manifold,* B *and* C K*-manifolds and* f *a* K′*-analytic mapping of* A × B *into* C. *Suppose that, for all*

$a \in A$, *the mapping* $b \mapsto f(a, b)$ *of* B *into* C *is* K-*analytic. Then, for all* $t \in TA$, *the mapping* $u \mapsto (Tf)(t, u)$ *of* TB *into* TC *is* K-*analytic.*

We fix $t \in TA$ and write $g(u) = (Tf)(t, u)$. Clearly g is K′-analytic. By *Differentiable and Analytic Manifolds*, R, 5.14.6, it suffices to prove that the tangent mappings to g are K-linear. It can be assumed that A, B, C are open neighbourhoods of 0 in complete normable spaces E, F, G over K′, K, K and that t is tangent to A at 0. We identify TA, TB, TC, with $A \times E$, $B \times F$, $C \times G$ and t with an element of E. Then for all $(x, y) \in TB = B \times F$.

$$g(x, y) = (f(0, x), (D_1 f)(0, x)(t) + (D_2 f)(0, x)(y)).$$

We identify $T(B \times F)$ with $(B \times F) \times (F \times F)$ and $T(C \times G)$ with $(C \times G) \times (G \times G)$. Then, for all

$$((x, y), (h, k)) \in T(B \times F) = (B \times F) \times (F \times F),$$

$(Tg)((x, y), (h, k)) = ((a, b), (c, d))$, where

$$a = f(0, x),$$
$$b = (D_1 f)(0, x)(t) + (D_2 f)(0, x)(y), \qquad c = (D_2 f)(0, x)(h),$$
$$d = (D_2 D_1 f)(0, x)(t, h) + (D_2 D_2 f)(0, x)(y, h) + (D_2 f)(0, x)(k).$$

We now fix $(x, y) \in B \times F$. We need to prove that the mapping $(h, k) \mapsto (c, d)$ of $F \times F$ into $G \times G$ is K-linear. As the mapping $x \mapsto f(0, x)$ of B into C is K-analytic, the mappings

$$(h, k) \mapsto (D_2 f)(0, x)(h), \qquad (h, k) \mapsto (D_2 D_2 f)(0, x)(y, h),$$
$$(h, k) \mapsto (D_2 f)(0, x)(k)$$

are K-linear. On the other hand

$$(D_2 D_1 f)(0, x)(t, h) = \lim_{\lambda \in K'^*, \lambda \to 0} \lambda^{-1}((D_2 f)(\lambda t, x)(h) - (D_2 f)(0, x)(h))$$

and, for fixed λ, the mapping $x \mapsto f(\lambda t, x)$ is K-analytic, so that the mapping $h \mapsto (D_2 f)(\lambda t, x)(h)$ is K-linear.

Proposition 1. *Let* K′ *be a non-discrete closed subfield of* K, G *a Lie group over* K, V *a manifold over* K′ *and* $(v, g) \mapsto vg$ *a* K′-*analytic mapping of* $V \times G$ *into* G. *Suppose that, for all* $v \in V$, *the mapping* $g \mapsto vg$ *of* G *into* G *is an automorphism of* G. *Let* ε *be an element of* V *such that* $\varepsilon g = g$ *for all* $g \in G$ *and* $a \in T_\varepsilon(V)$. *Then the vector field* $g \mapsto ag$ *on* G *is an infinitesimal automorphism of* G.

For $v \in V$, $g_1 \in G$, $g_2 \in G$, $v(g_1 g_2) = (vg_1)(vg_2)$. Hence, for $u_1 \in TG$, $u_2 \in TG$, $a(u_1 u_2) = (au_1)(au_2)$ (§ 2, no. 1, Proposition 3). In particular, the mapping $g \mapsto ag$ of G into TG is a group homomorphism. On the other hand, this mapping is analytic by Lemma 2.

Proposition 2. *Let* G *be a real or complex Lie group and* α *an infinitesimal auto-*

morphism of G. *There exists a law of analytic operation* $(\lambda, g) \mapsto \phi_\lambda(g)$ *of* K *on* G *with the following properties:*

(1) *if* D *is the associated law of infinitesimal operation, then* $D(1) = \alpha$;

(2) *for all* $\lambda \in K$, $\phi_\lambda \in \operatorname{Aut} G$.

(a) For all $\mu > 0$, let K_μ be the open ball of centre 0 and radius μ in K. For all $g \in G$, let $\mathscr{F}_g$ be the set of analytic integral curves f of α defined in a ball K_μ and such that $f(0) = g$. By *Differentiable and Analytic Manifolds*, R, 9.1.3 and 9.1.5, $\mathscr{F}_g$ is non-empty and two elements of $\mathscr{F}_g$ coincide on the intersection of their domains of definition; let $\mu(g)$ be the least upper bound of the numbers μ such that there exists an element of $\mathscr{F}_g$ defined in K_μ; there exists a unique element of $\mathscr{F}_g$ defined in $K_{\mu(g)}$; we denote it by f_g.

(b) Let g_1, g_2 be in G, $f_1 \in \mathscr{F}_{g_1}$, $f_2 \in \mathscr{F}_{g_2}$ with f_1 and f_2 defined on the same ball K_μ. Then $f_1 f_2 \colon K_\mu \to G$ is analytic and $(f_1 f_2)(0) = g_1 g_2$. On the other hand, for all $\lambda \in K_\mu$,

$$\begin{aligned}(T_\lambda(f_1 f_2))1 &= (T_\lambda f_1)1 . f_2(\lambda) + f_1(\lambda) . (T_\lambda f_2)1 && (\text{§ 2, Proposition 7})\\ &= \alpha(f_1(\lambda)) f_2(\lambda) + f_1(\lambda)\alpha(f_2(\lambda))\\ &= \alpha((f_1 f_2)(\lambda)) && (\text{Lemma 1})\end{aligned}$$

and hence $f_1 f_2 \in \mathscr{F}_{g_1 g_2}$. This proves that $\mu(g_1 g_2) \geqslant \inf(\mu(g_1), \mu(g_2))$.

(c) By *Differentiable and Analytic Manifolds*, R, 9.1.4 and 9.1.5, there exists a neighbourhood V of e in G such that $\sigma = \inf_{g \in V} \mu(g) > 0$. Let $h \in G$ and C be its connected component. For all $h' \in C$, $\mu(h') \geqslant \inf(\sigma, \mu(h)) > 0$ by (b). On the other hand, the functions $f_{h'}$, where $h' \in C$, take their values in C. By *Differentiable and Analytic Manifolds*, R, 9.1.4 and 9.1.5, $\mu = +\infty$ in C and finally $\mu = +\infty$ in G. Then let $f_g(\lambda) = \phi_\lambda(g)$ for all $g \in G$ and all $\lambda \in K$. By *Differentiable and Analytic Manifolds*, R, 9.1.4 and 9.1.5, the mapping $(\lambda, g) \mapsto \phi_\lambda(g)$ is a law of analytic operation of K on G. Clearly, if D is the associated law of infinitesimal operation, $D(1) = \alpha$. By (b),

$$\phi_\lambda(g_1 g_2) = \phi_\lambda(g_1)\phi_\lambda(g_2)$$

for all $\lambda \in K$, $g_1 \in G$, $g_1 \in G$.

PROPOSITION 3. *Suppose that* K *is ultrametric. Let* G *be a compact Lie group and* α *an infinitesimal automorphism of* G. *There exist an open subgroup* I *of* K *and a law of analytic operation* $(\lambda, g) \mapsto \phi_\lambda(g)$ *of* I *on* G *with the following properties:*

(1) *if* D *is the associated law of infinitesimal operation, then* $D(1) = \alpha$;

(2) *for all* $\lambda \in I$, $\phi_\lambda \in \operatorname{Aut} G$.

As G is compact, there exist an open subgroup I′ of K and a law of analytic operation $(\lambda, g) \mapsto \phi_\lambda(g)$ of I′ on G with property (1) of the proposition

(§ 4, no. 7, Corollary 2 to Theorem 6). We write $\phi_\lambda(g) = f_g(\lambda)$ for $\lambda \in I'$ and $g \in G$. Then, for g_1, g_2 in G and $\lambda \in I'$,

$$\begin{aligned}(T_\lambda(f_{g_1} f_{g_2}))1 &= (T_\lambda f_{g_1})1 . f_{g_2}(\lambda) + f_{g_1}(\lambda).(T_\lambda f_{g_2})1 \\ &= \alpha(f_{g_1}(\lambda)) f_{g_2}(\lambda) + f_{g_1}(\lambda)\alpha(f_{g_2}(\lambda)) \\ &= \alpha(f_{g_1}(\lambda) f_{g_2}(\lambda))\end{aligned}$$

and $(f_{g_1} f_{g_2})(0) = g_1 g_2 = f_{g_1 g_2}(0)$. Hence $f_{g'_1 g'_2}(\lambda) = f_{g'_1}(\lambda) f_{g'_2}(\lambda)$ for

$$(g'_1, g'_2, \lambda)$$

in a neighbourhood of $(g_1, g_2, 0)$ (*Differentiable and Analytic Manifolds*, R, 9.1.8). As G is compact, there exists an open subgroup I of I′ such that $f_{g_1 g_2}(\lambda) = f_{g_1}(\lambda) f_{g_2}(\lambda)$ for all $g_1 \in G$, $g_2 \in G$, $\lambda \in I$. In other words, $\phi_\lambda \in \operatorname{Aut} G$ for $\lambda \in I$.

Lemma 3. *Let* G *and* G′ *be Lie groups and* ϕ *a homomorphism of* G *into* Aut(G′). *Let* $f(g, g') = (\phi(g))(g')$ *for* $g \in G$, $g' \in G'$. *Consider the following conditions:*

(i) f *is analytic;*

(ii) f *is analytic in a neighbourhood of* $(e_G, e_{G'})$;

(iii) *for all* $g' \in G'$, *the mapping* $g \mapsto f(g, g')$ *is analytic.*

Then (i) ⇔ ((ii) and (iii)). *If* G *is connected,* (i) ⇔ (ii).

Clearly (i) implies (ii) and (iii). Let $g_0 \in G$, $g'_0 \in G'$. For all $g \in G$, $g' \in G'$,

$$f(gg_0, g'g'_0) = (\phi(g)\phi(g_0))(g'g'_0) = \phi(g)(\phi(g_0)g').\phi(g)(\phi(g_0)g'_0).$$

This proves the implication ((ii) and (iii)) ⇒ (i). Finally, if G′ is connected, G′ is generated by every neighbourhood of $e_{G'}$ and hence (ii) ⇒ (iii).

2. THE AUTOMORPHISM GROUP OF A LIE GROUP (REAL OR COMPLEX CASE)

In this no., we assume that K = **R** or **C**.

Lemma 4. *Let* H *be a finite-dimensional simply connected Lie group.*

(i) *For all* $u \in \operatorname{Aut} L(H)$, *let* $\theta(u)$ *be the unique automorphism of* H *such that* $L(\theta(u)) = u$. *Then the mapping* $(u, g) \mapsto \theta(u)g$ *of* (Aut L(H)) × H *into* H *is analytic.*

(ii) *Let* N *be a Lie subgroup of* H *and* Aut(H, N) *the set of* $v \in \operatorname{Aut} H$ *such that* $v(N) = N$. *Then* $\theta^{-1}(\operatorname{Aut}(H, N))$ *is a Lie subgroup of* Aut L(H).

(iii) *Suppose that* N *is discrete and normal, so that the Lie algebra of* G = H/N *is identified with* L(H). *For all* $w \in \operatorname{Aut} G$, *let* $\eta(w)$ *be the unique automorphism of* H *such that* $L(\eta(w)) = L(w)$. *Then the mapping* η *is an isomorphism of the group* Aut G *onto the group* Aut(H, N).

To prove (i), it suffices, by Lemma 3 of no. 1, to verify that the mapping $(u, g) \mapsto \theta(u)g$ is analytic in a neighbourhood of $(\mathrm{Id}_{L(H)}, e)$. There exists an open neighbourhood B of 0 in L(H) such that $\psi = \exp_H|B$ is an analytic isomorphism of B onto an open neighbourhood of e in H. There exist an open neighbourhood U of $\mathrm{Id}_{L(H)}$ in Aut L(H) and an open neighbourhood B′ of 0 in L(H) such that U(B′) ⊂ B. Then the mapping $(u, g) \mapsto \theta(u)g$ of $U \times \psi(B')$ into H is composed of the following mappings:

the mapping $(u, g) \mapsto (u, \psi^{-1}(g))$	of $U \times \psi(B')$	into	$U \times B'$;
the mapping $(u, x) \mapsto u(x)$	of $U \times B'$	into	B;
the mapping $y \mapsto \psi(y)$	of B	into	G.

Hence this mapping is analytic.

Let p be the canonical mapping of H into the homogeneous space H/N. Then $\theta^{-1}(\mathrm{Aut}(H, N))$ is the set of $u \in \mathrm{Aut}\, L(H)$ such that

$$p(\theta(u)g) = p(e), \qquad p(\theta(u^{-1})g) = p(e)$$

for all $g \in N$. By § 8, no. 2, Theorem 2 and Corollary 2 to Theorem 2, this proves (ii).

Suppose that N is discrete and normal. Let $w \in \mathrm{Aut}\, G$. Then

$$L(p \circ \eta(w)) = L(\eta(w)) = L(w) = L(w \circ p)$$

hence $p \circ \eta(w) = w \circ p$ and therefore $\eta(w) \in \mathrm{Aut}(H, N)$. Clearly the mapping η of Aut G into Aut(H, N) is an injective homomorphism. This homomorphism is surjective because $p: H \to G$ is a submersion.

Let G be a locally compact group and Γ the automorphism group of G. Recall that a topology $\mathcal{T}_\beta$ has been defined on Γ (*General Topology*, Chapter X, § 3, no. 5). It is the coarsest topology for which the mappings $v \mapsto v$ and $v \mapsto v^{-1}$ of Γ into $\mathscr{C}_c(G; G)$ (space of continuous mappings of G into G with the compact convergence topology) are continuous. The topology $\mathcal{T}_\beta$ is compatible with the group structure on Γ (*loc. cit.*). For every compact subset L of G and every neighbourhood U of e_G in G, let N(L, U) be the set of $\phi \in \Gamma$ such that $\phi(g) \in gU$ and $\phi^{-1}(g) \in gU$ for all $g \in L$; then the N(L, U) form a fundamental system of neighbourhoods of e_Γ. If G is generated by a compact subset C, the topology $\mathcal{T}_\beta$ is also the coarsest topology for which the mappings $v \mapsto v|C$ and $v \mapsto v^{-1}|C$ of Γ into $\mathscr{C}_u(C; G)$ are continuous (for every compact subset of G is contained in $(C \cup C^{-1})^n$ for sufficiently large n). If K is locally compact and V is a finite-dimensional vector space over K, the topology $\mathcal{T}_\beta$ on **GL**(V) is just the usual topology.

THEOREM 1. *Let* G *be a finite-dimensional Lie group and* G_0 *its identity component. Suppose that* G *is generated by* G_0 *and a finite number of elements.*

(i) *There exists on* Aut G *one and only one analytic manifold structure satisfying the following condition:*

(AUT) *for every analytic manifold* M *and every mapping* f *of* M *into* Aut G, f *is analytic if and only if the mapping* $(m, g) \mapsto f(m)g$ *of* $M \times G$ *into* G *is analytic.*

Suppose in the rest of the statement that Aut G *has this structure.*

(ii) Aut G *is a finite-dimensional Lie group.*

(iii) *The morphism* $\phi: u \mapsto \mathrm{L}(u)$ *of* Aut G *into* Aut L(G) *is analytic.*

(iv) *If* G *is connected,* ϕ *is an isomorphism of the Lie group* Aut G *onto a Lie subgroup of* Aut L(G); *this Lie subgroup is equal to* Aut L(G) *if* G *is simply connected.*

(v) *Let* $\mathfrak{a}$ *be the set of infinitesimal automorphisms of* G. *Then* $\mathfrak{a}$ *is a Lie algebra of vector fields and the law of infinitesimal operation associated with the mapping* $(u, g) \mapsto u(g)$ *of* (Aut G) $\times$ G *into* G *is an isomorphism of* L(Aut G) *onto* $\mathfrak{a}$.

(vi) *The topology of the Lie group* Aut G *is the topology* $\mathcal{T}_\beta$.

(a) The uniqueness of the analytic structure considered in (i) is obvious.

(b) Suppose that G is connected. Let H be the universal covering space of G, p the canonical morphism of H onto G and $\mathrm{N} = \mathrm{Ker}\, p$. We introduce the notation θ, η and Aut(H, N) of Lemma 4. We transport the Lie group structure of Aut L(G) to Aut H by means of θ. Then Aut H becomes a finite-dimensional Lie group and Aut(H, N) a Lie subgroup of Aut H (Lemma 4 (ii)). We transport the Lie group structure of Aut(H, N) to Aut G by means of η^{-1}. Then Aut G becomes a finite-dimensional Lie group. Properties (ii), (iii) and (iv) of the theorem are satisfied and the mapping $(u, g) \mapsto u(g)$ of (Aut G) $\times$ G into G is analytic (Lemma 4 (i)). Let M be an analytic manifold, f a mapping of M into Aut G and ϕ the mapping $(m, g) \mapsto f(m)g$ of $\mathrm{M} \times \mathrm{G}$ into G. Clearly, if f is analytic, ϕ is analytic. Suppose that ϕ is analytic. Then the $\mathrm{T}\phi: \mathrm{TM} \times \mathrm{TG} \to \mathrm{TG}$ is analytic; its restriction to $\mathrm{M} \times \mathrm{L(G)}$, that is the mapping $(m, x) \mapsto \mathrm{L}(f(m))x$ of $\mathrm{M} \times \mathrm{L(G)}$ into L(G) is therefore analytic; as L(G) is finite-dimensional, it follows that the mapping $m \mapsto \mathrm{L}(f(m))$ of M into Aut L(G) is analytic and hence that f is analytic. Thus (i) holds.

Let L(G) be given a norm. For all $\lambda > 0$, let B_λ be the open ball of centre 0 and radius λ in L(G). We choose $\lambda > 0$ sufficiently small for $\psi = \exp_{\mathrm{G}}|\mathrm{B}_\lambda$ to be an isomorphism of the analytic manifold B_λ onto the open submanifold $\psi(\mathrm{B}_\lambda)$ of G. Let Φ be a filter on Aut G. For Φ to converge to Id_{G} in Aut G, it is necessary and sufficient that $\mathrm{L}(\Phi)$ converge to $\mathrm{Id}_{\mathrm{L(G)}}$ in Aut L(G) and hence that $\mathrm{L}(\Phi)|\mathrm{B}_{\lambda/2}$ and $\mathrm{L}(\Phi)^{-1}\mathrm{B}_{\lambda/2}$ converge uniformly to $\mathrm{Id}_{\mathrm{B}_{\lambda/2}}$. This condition implies that $\Phi|\psi(\mathrm{B}_{\lambda/2})$ and $\Phi^{-1}|\psi(\mathrm{B}_{\lambda/2})$ converge uniformly to $\mathrm{Id}_{\psi(\mathrm{B}_{\lambda/2})}$. Conversely, suppose that $\Phi|\psi(\mathrm{B}_{\lambda/2})$ converges uniformly to $\mathrm{Id}_{\psi(\mathrm{B}_{\lambda/2})}$. There exists $\mathrm{M} \in \Phi$ such that, if $u \in \mathrm{M}$, then $u(\psi(\mathrm{B}_{\lambda/2})) \subset \psi(\mathrm{B}_{2\lambda/3})$; then $\mathrm{L}(u)(\mathrm{B}_{\lambda/2})$ is a connected subset of L(G) whose image under $\exp_{\mathrm{G}}$ is contained in $\psi(\mathrm{B}_{2\lambda/3})$, hence $\mathrm{L}(u)(\mathrm{B}_{\lambda/2})$ does not meet $\mathrm{B}_\lambda - \mathrm{B}_{2\lambda/3}$ and therefore $\mathrm{L}(u)(\mathrm{B}_{\lambda/2}) \subset \mathrm{B}_\lambda$; then the hypothesis that $\Phi|\psi(\mathrm{B}_{\lambda/2})$ converges uniformly to $\mathrm{Id}_{\psi(\mathrm{B}_{\lambda/2})}$ implies that $\mathrm{L}(\Phi)|\mathrm{B}_{\lambda/2}$ converges uniformly to $\mathrm{Id}_{\mathrm{B}_{\lambda/2}}$. It then follows that:

$$(\Phi \text{ converges to } \mathrm{Id}_{\mathrm{G}} \text{ in Aut G}) \Leftrightarrow (\Phi \text{ converges to } \mathrm{Id}_{\mathrm{G}} \text{ under } \mathcal{T}_\beta).$$

This proves (vi).

Let D be the law of infinitesimal operation associated with the law of left operation of Aut(G) on G. By Propositions 1 and 2 of no. 1, $D(L(\text{Aut } G)) = \mathfrak{a}$. Hence $\mathfrak{a}$ is a Lie algebra of vector fields and D is a morphism of L(Aut G) onto $\mathfrak{a}$. Let x_1 and x_2 be elements of L(Aut G) such that $D(x_1) = D(x_2)$. Then the laws of operation $(\lambda, g) \mapsto (\exp \lambda x_1) g$ and $(\lambda, g) \mapsto (\exp \lambda x_2) g$ of K on G have the same associated law of infinitesimal operation; hence, for $|\lambda|$ sufficiently small, $\exp \lambda x_1$ and $\exp \lambda x_2$ coincide on a neighbourhood of e (§ 4, no. 7, Theorem 6), whence $\exp \lambda x_1 = \exp \lambda x_2$. It follows that $x_1 = x_2$ and hence D is an isomorphism of L(Aut G) onto $\mathfrak{a}$.

The theorem has thus been completely proved for G connected.

(c) We pass to the general case. By hypothesis, G is generated by G_0 and a finite number of elements $x_1, x_2, \ldots, x_n$. Every $u \in \text{Aut } G$ leaves G_0 stable. Let $\text{Aut}_1 G$ be the set of $u \in \text{Aut } G$ which, on passing to the quotient, give the identity automorphism of G/G_0. This is a normal subgroup of Aut G. By part (b) of the proof, Aut G_0 has a canonical Lie group structure and the mapping $(g_1, g_2, \ldots, g_n, u) \mapsto (ug_1, ug_2, \ldots, ug_n)$ of $G_0^n \times \text{Aut } G_0$ into G_0^n is analytic. Let P be the corresponding semidirect product of Aut G_0 by G_0^n; it is a finite-dimensional Lie group (§ 1, no. 4, Proposition 7).

If $w \in \text{Aut}_1 G$, we write

$$\begin{aligned} w_0 &= w|G_0 \in \text{Aut } G_0 \\ w_i &= x_i^{-1} w(x_i) \in G_0 \qquad (1 \leqslant i \leqslant n) \\ \zeta(w) &= ((w_1, \ldots, w_n), w_0) \in P. \end{aligned}$$

For all w, w' in $\text{Aut}_1 G$,

$$\begin{aligned} \zeta(w)\zeta(w') &= ((w_1, \ldots, w_n)(w_0(w'_1), \ldots, w_0(w'_n)), w_0 w'_0) \\ &= ((w_1 w_0(w'_1), \ldots, w_n w_0(w'_n)), w_0 w'_0) \\ &= ((x^{-1} w(x_1) w(x_1^{-1} w'(x_1)), \ldots, x_n^{-1} w(x_n) w(x_n^{-1} w'(x_n))), w_0 w'_0) \\ &= (((ww')_1, \ldots, (ww')_n), (ww')_0) \\ &= \zeta(ww') \end{aligned}$$

and hence ζ is a homomorphism of $\text{Aut}_1 G$ into P. This homomorphism is obviously injective.

We show that $\zeta(\text{Aut}_1 G)$ is closed in P. Let Φ be a filter on $\text{Aut}_1 G$ such that $\zeta(\Phi)$ converges to a point $((w_1, \ldots, w_n), w_0)$ of P. Then Φ converges pointwise to a mapping v of G into G. Clearly v is an endomorphism of the group G. Moreover, v leaves each coset modulo G_0 stable and $v|G_0 = w_0$. It follows that $v \in \text{Aut}_1 G$. As $\zeta(v) = ((w_1, \ldots, w_n), w_0)$, we have shown that $\zeta(\text{Aut}_1 G)$ is closed in P.

(d) In part (d) of the proof we assume that $K = \mathbf{R}$. By § 8, no. 2, Theorem 2, $\zeta(\text{Aut}_1 G)$ is a Lie subgroup of P. We transport the real Lie group structure on $\zeta(\text{Aut}_1 G)$ to $\text{Aut}_1 G$ by means of ζ^{-1}. Thus $\text{Aut}_1 G$ becomes a finite-dimensional Lie group.

Let M be an analytic manifold, f a mapping of M into $\mathrm{Aut}_1 G$ and ϕ the mapping $(m, g) \mapsto f(m)g$ of $M \times G$ into G. We have the following equivalences:

f analytic
$\Leftrightarrow$ the mappings $m: \mapsto (f(m))_i$, where $0 \leqslant i \leqslant n$, are analytic

$$\Leftrightarrow \begin{cases} \text{the mappings } m \mapsto f(m)x_i \text{ of M into G, for } 1 \leqslant i \leqslant n, \text{ are analytic} \\ \text{and} \\ \text{the mapping } (m, g) \mapsto f(m)g \text{ of } M \times G_0 \text{ into G is analytic} \end{cases}$$

$\Leftrightarrow \phi$ is analytic.

For $w \in \mathrm{Aut}_1 G$, $L(w) = L(w_0)$ and hence the morphism $w \mapsto L(w)$ of $\mathrm{Aut}_1 G$ into Aut L(G) is analytic. We see as in (b) that the law of infinitesimal operation associated with the law of operation of $\mathrm{Aut}_1 G$ on G is an isomorphism of $L(\mathrm{Aut}_1 G)$ onto $\mathfrak{a}$.

Let C be a compact subset of G_0 generating G_0. For a filter Φ to converge to Id_G on $\mathrm{Aut}_1 G$, it is necessary and sufficient that $\Phi|(C \cup \{x_1\} \cup \cdots \cup \{x_n\})$ and $\Phi^{-1}|(C \cup \{x_1\} \cup \cdots \cup \{x_n\})$ converge uniformly to

$$\mathrm{Id}_G|(C \cup \{x_1\} \cup \cdots \cup \{x_n\}).$$

The topology of $\mathrm{Aut}_1 G$ is therefore the topology $\mathscr{T}_\beta$.

Clearly $\mathrm{Aut}_1 G$ is open in Aut G with the topology $\mathscr{T}_\beta$. There exists on Aut G a Lie group structure compatible with this topology and inducing on $\mathrm{Aut}_1 G$ the structure constructed above (§ 8, no. 1, Corollary 2 to Theorem 1). The fact that the Lie group Aut G has the properties of the theorem follows from the corresponding properties for $\mathrm{Aut}_1 G$.

(e) In part (e) of the proof we assume that $K = \mathbf{C}$. By (c) and Theorem 2 of § 8, no. 2, there exists on $\mathrm{Aut}_1 G$ a real Lie group structure such that ζ is an isomorphism of $\mathrm{Aut}_1 G$ onto a real Lie subgroup of P.

The law of operation $(w, g) \mapsto wg$ of $(\mathrm{Aut}_1 G) \times G$ on G is real analytic. Let D be the associated law of infinitesimal operation. By Propositions 1 and 2 of no. 1, $D(L(\mathrm{Aut}_1 G)) = \mathfrak{a}$.

For all $\alpha \in \mathfrak{a}$, let α_0 denote the restriction of α to G_0; it is an infinitesimal automorphism of G_0 which we identify, because of part (b) of the proof, with an element of $L(\mathrm{Aut}\, G_0)$. For $1 \leqslant i \leqslant n$, we write

$$\alpha_i = x_i^{-1}\alpha(x_i) \in L(G) = L(G_0).$$

Finally, we write $f(\alpha) = ((\alpha_1, \ldots, \alpha_n), \alpha_0) \in L(P)$. Then f is a $\mathbf{C}$-linear mapping of $\mathfrak{a}$ into L(P).

On the other hand, clearly $L(\zeta) = f \circ D$. Hence $L(\zeta)(L(\mathrm{Aut}_1 G)) = f(\mathfrak{a})$ is a complex vector subspace of L(P). By Proposition 2 of § 4, no. 2, $\zeta(\mathrm{Aut}_1 G)$ is a complex Lie subgroup of P and we can proceed exactly as in (d): we transport the complex Lie group structure on $\zeta(\mathrm{Aut}_1 G)$ to $\mathrm{Aut}_1 G$ by means of ζ^{-1} and we see as in (d) that $\mathrm{Aut}_1 G$ has the properties analogous to properties (i), (ii), (iii), (v) and (vi) of the theorem.

Clearly $\mathrm{Aut}_1 G$ is open in Aut G with the topology $\mathscr{T}_\beta$. Let $w \in$ Aut G. Let σ be the automorphism $v \mapsto wvw^{-1}$ of $\mathrm{Aut}_1 G$. It is real analytic (§ 8, no. 1, Theorem 1), $L(\sigma)$ is an **R**-automorphism of $L(\mathrm{Aut}_1 G)$ and

$$D \circ L(\mathrm{Aut}_1 G) \circ D^{-1}$$

is an **R**-automorphism of $\mathfrak{a}$. This automorphism is also the automorphism of $\mathfrak{a}$ derived from w by transport of structure; as w is K-analytic, we see that $L(\sigma)$ is K-linear. Hence σ is K-analytic (§ 3, no. 8, Proposition 32). By § 1, no. 9, Proposition 18, there exists on Aut G one and only one Lie K-group structure such that $\mathrm{Aut}_1 G$ is an open Lie subgroup of Aut G. The fact that this structure has the properties of the theorem follows from the corresponding properties for $\mathrm{Aut}_1 G$.

COROLLARY 1. *Let* G *be a finite-dimensional real Lie group and* G_0 *its identity component. Suppose that* G *is generated by* G_0 *and a finite number of elements. Then* Aut G *has the topology* $\mathscr{T}_\beta$ *and is a finite-dimensional real Lie group.*

COROLLARY 2. *Let* G *be a semi-simple connected real or complex Lie group. The group* Int G *is the identity component of* Aut G.

The mapping $u \mapsto L(u)$ is an isomorphism of Aut G onto a Lie subgroup of Aut L(G) (Theorem 1). The image of Int G under this isomorphism is Ad G. But Ad G is the identity component of Aut L(G) (§ 9, no. 8, Proposition 30 (ii)).

3. THE AUTOMORPHISM GROUP OF A LIE GROUP (ULTRAMETRIC CASE)

THEOREM 2. *When* K *is ultrametric and locally compact and* G *is a compact Lie group, assertions* (i), (ii), (iii), (v) *and* (vi) *of Theorem* 1 *are true.*

(a) The uniqueness of the analytic structure considered in (i) is obvious.

(b) Suppose that G is the Lie group defined by the normable Lie algebra L. Then G is an open and closed ball in L. Let $w \in$ Aut G. Then $L(w)$ coincides with w in a neighbourhood of 0. Let $x \in G$. Let p be the characteristic of the residue field. Then $p^n x$ tends to 0 as n tends to $+\infty$. There therefore exists n such that $w(p^n x) = L(w)(p^n x)$. Therefore

$$p^n w(x) = w(x)^{p^n} = w(x^{p^n}) = w(p^n x)$$
$$= L(w)(p^n x) = p^n L(w)(x)$$

whence $w(x) = L(w)(x)$. Thus, $w = L(w)|G$.

Let Γ be the set of $\gamma \in$ Aut L(G) such that $\gamma(G) = G$. As G is open and compact in L(G), Γ is an open subgroup of Aut L(G). By the above, Aut G is identified with Γ, whence there is a Lie group structure on Aut G, with which properties (i), (ii), (iii) and (vi) of Theorem 1 are obvious. Property (v) follows from Propositions 1 and 3 of no. 1.

(c) We pass to the general case. By § 7, no. 1, Proposition 1, there exists an open compact subgroup G_0 of G which is of the type considered in (b). Then G is generated by G_0 and a finite number of elements $x_1, x_2, \ldots, x_n$. Let $\mathrm{Aut}_1 G$ be the set of $u \in \mathrm{Aut}\, G$ such that $u(G_0) = G_0$ and $u(x_i G_0) = x_i G_0$ for $1 \leqslant i \leqslant n$. We define as in the proof of Theorem 1, part (c), a semi-direct product P of Aut G_0 by G_0^n and an injective homomorphism ζ of $\mathrm{Aut}_1 G$ into P, whose image is closed in P.

(d), (e): the argument is exactly as in parts (d), (e) of the proof of Theorem 1 with $\mathbf{R}$ replaced by $\mathbf{Q}_p$ and using Proposition 3 instead of Proposition 2.

Remark. If $K = \mathbf{Q}_p$ and the Lie group G is *generated* by a compact subset (cf. Exercise 2), assertions (i), (ii), (iii) and (vi) of Theorem 1 are still true, but not (v) (Exercise 3).

APPENDIX

OPERATIONS ON LINEAR REPRESENTATIONS

Let G be a group, k a commutative field, $E_1, E_2, \ldots, E_n$ vector spaces over k and π_i a linear representation of G on E_i ($1 \leqslant i \leqslant n$). The mapping

$$g \mapsto \pi_1(g) \otimes \cdots \otimes \pi_n(g)$$

is a linear mapping of G into the vector space $E_1 \otimes \cdots \otimes E_n$, called the *tensor product of* $\pi_1, \ldots, \pi_n$ and denoted by $\pi_1 \otimes \cdots \otimes \pi_n$.

Let E be a vector space over k and π a linear representation of G on E. For all $g \in G$, let $\tau(g)$ (resp. $\sigma(g)$, $\varepsilon(g)$) be the unique automorphism of the algebra T(E) (resp. S(E), $\bigwedge$(E)) which extends $\pi(g)$ (*Algebra*, Chapter III, § 5, no. 2, § 6, no. 2 and § 7, no. 2). Then τ (resp. σ, ε) is a linear representation of G on T(E) (resp. S(E), $\bigwedge$(E)) denoted by $\mathsf{T}(\pi)$ (resp. $\mathsf{S}(\pi)$, $\bigwedge(\pi)$). The subrepresentation of $\mathsf{T}(\pi)$ (resp. $\mathsf{S}(\pi)$, $\bigwedge(\pi)$) defined by $T^n(E)$ (resp. $S^n(E)$, $\bigwedge^n(E)$) is called the n-th tensor (resp. symmetric, exterior) power of π and is denoted by $\mathsf{T}^n(\pi)$ (resp. $\mathsf{S}^n(\pi)$, $\bigwedge^n(\pi)$). Then

$$\mathsf{T}^n(\pi) = \pi \otimes \pi \otimes \cdots \otimes \pi$$

(n factors). The representations $\mathsf{S}(\pi)$, $\bigwedge(\pi)$ are quotient representations of $\mathsf{T}(\pi)$ and hence $\mathsf{S}^n(\pi)$, $\bigwedge^n(\pi)$ are quotient representations of $\mathsf{T}^n(\pi)$.

Let $\mathfrak{g}$ be a Lie algebra over k. The tensor product of a finite number of representations of $\mathfrak{g}$ has already been defined in Chapter I, § 3, no. 2; it is denoted by $\pi_1 \otimes \cdots \otimes \pi_n$. Let E be a vector space over k and π a representation of $\mathfrak{g}$ on E. For all $x \in \mathfrak{g}$, let $\tau'(x)$ (resp. $\sigma'(x)$, $\varepsilon'(x)$) be the unique derivation of the algebra T(E) (resp. S(E), $\bigwedge$(E)) which extends $\pi(x)$ (*Algebra*, Chapter III, § 10, no. 9, *Example* 1). Then τ' (resp. σ', ε') is a linear representation of $\mathfrak{g}$ on T(E) (resp. S(E), $\bigwedge$(E)) by *Algebra*, *loc. cit.*, formula (35), denoted by $\mathsf{T}(\pi)$ (resp. $\mathsf{S}(\pi)$, $\bigwedge(\pi)$). The subrepresentation of $\mathsf{T}(\pi)$ (resp. $\mathsf{S}(\pi)$, $\bigwedge(\pi)$) defined by $\mathsf{T}^n(E)$ (resp. $\mathsf{S}^n(E)$, $\bigwedge^n(E)$) is denoted by $\mathsf{T}^n(\pi)$ (resp. $\mathsf{S}^n(\pi)$, $\bigwedge^n(\pi)$). The representation $\mathsf{T}^n(\pi)$ is the tensor product on n representations identical with π. The representations $\mathsf{S}(\pi)$, $\Lambda(\pi)$ are quotient representations of $\mathsf{T}(\pi)$ and hence $\mathsf{S}^n(\pi)$, $\bigwedge^n(\pi)$ are quotient representations of $\mathsf{T}^n(\pi)$.

EXERCISES

§ 1

1. Let G be a finite-dimensional connected real or complex Lie group. Let H and L be Lie subgroups of G such that HL = G. Show that the canonical mappings $G \to G/H$ and $G \to G/L$ define an isomorphism of analytic manifolds

$$G/(H \cap L) \to (G/H) \times (G/L).$$

2. Let $P = \{z \in \mathbf{C} | \mathscr{I}(z) > 0\}$. Let G be the group of bijections $z \mapsto az + b$ of P ($a > 0$, $b \in \mathbf{R}$). Then G operates on P simply transitively, which allows us to transport the complex analytic manifold structure on P to G. Show that the structure obtained is invariant under left translations of G but not under right translations.

3. Let $\mathbf{R}_d$ be the additive group of real numbers with the discrete manifold structure. Let $\mathbf{R}_d$ operate on the analytic manifold $\mathbf{R}$ by the law of operation $(x, y) \mapsto x + y$. Then $\mathbf{R}_d$ operates transitively on $\mathbf{R}$ but $\mathbf{R}$ is not a Lie homogeneous space of $\mathbf{R}_d$.

4. Let H be a compact real Lie group operating on a finite-dimensional real manifold V; suppose that V is of class C^r, where $r \in N_{\mathbf{R}}$, and that the action of H on V is of class C^r. Let $p \in V$ be invariant under H; the group H operates linearly on the tangent space T to V at p. Show that there exist an open neighbourhood U of p which is stable under H and a C^r-morphism $f: U \to T$ such that:

(*a*) $f(p) = 0$ and the tangent mapping to f at p is the identity;

(*b*) f commutes with the action of H on U and T.

(Choose first an f_0 satisfying (*a*) and then define f by the formula

$$f(x) = \int_H h.f_0(h^{-1}x)dh,$$

where dh is the Haar measure on H with total mass 1.)

Deduce that the H-spaces V and T are locally isomorphic to a neighbourhood of p, in other words that there exist systems of coordinates on V at p with respect to which H operates linearly ("Bochner's Theorem").

5. Perform Exercise 4 for manifolds over an ultrametric field K, assuming that H is a finite group whose order n is such that $n.1 \neq 0$ in K.

¶ 6. Let G be a real Lie group operating properly on a finite-dimensional Hausdorff real analytic manifold V. Let $p \in V$, let H be the stabilizer of p and let Gp be the orbit of p. The group H is compact and the manifold Gp is identified with the Lie homogeneous space G/H.

(*a*) Show that there exists a submanifold S of V, passing through p, stable under H and such that $T_p(V)$ is the direct sum of $T_p(Gp)$ and $T_p(S)$. (Choose a supplement of $T_p(Gp)$ in $T_p(V)$ which is stable under H and apply Exercise 4.)

(*b*) Suppose that S is chosen as above. The group H operates properly and freely on $G \times S$ by $h.(g, s) = (gh, h^{-1}s)$; let $E = (G \times S)/H$ be the corresponding quotient manifold (cf. *Differentiable and Analytic Manifolds*, R, 6.5.1). The mapping $(g, s) \mapsto gs$ defines when passing to the quotient a morphism $\rho: E \to V$. Show that ρ commutes with the action of G on E and V and that there exists an open submanifold S' of S containing p, stable under H and such that ρ induces a manifold isomorphism of $E' = (G \times S')/H$ onto a saturated neighbourhood of the orbit Gp.

(*c*) Let $N_p = T_p(V)/T_p(Gp)$ be the transversal space at p to Gp (*Differentiable and Analytic Manifolds*, R, 5.8.8); the group H operates on N_p. Deduce from (*b*) that knowing H and the linear representation of H on N_p determines the G-space V in a neighbourhood of the orbit of p.

(*d*) If $x \in N_p$, let H_x be the stabilizer of x in H. Show that there exists a saturated neighbourhood V' of Gp with the following property:

For all $p' \in V'$, there exists $x \in N_p$ such that the stabilizer of p' in G is conjugate (in G) to H_x.

(*e*) Show that the H_x are finite in number, up to conjugation in H. (Argue by induction on $\dim N_p = n$, using the action of H on a sphere of dimension $n - 1$ which is stable under H.)†

¶ 7. Let E be a complete normable space over **R** and F a closed vector subspace of E such that there exists no bicontinuous linear bijection of E onto $F \times (E/F)$.‡ Let X be the analytic manifold $\mathbf{R} \times E \times F$. Let G be the

† For more details, see R. S. Palais, On the existence of slices for actions of non-compact Lie groups, *Ann. of Math.*, **72** (1961), pp. 295–323.

‡ For an example of such an ordered pair (E, F), cf. for example A. Douady, Le problème des modules pour les sous-espaces analytiques compacts d'un espace analytique donné, *Ann. Inst. Fourier*, **16** (1966), p. 16.

additive group of F, which operates on X by the mapping

$$(g, (\lambda, e, f)) \mapsto (\lambda, e + g, f + \lambda g)$$

of G × X into X. For all $x \in X$, let R_x be the G-orbit of x, which is a quasi-submanifold of X. Then condition (*a*) of Proposition 10 is satisfied, but there exists no ordered pair (Y, π) with the following properties: Y is an analytic manifold, π is a morphism of X into Y and, for all $x \in X$,

$$T_x(\pi): T_x(X) \to T_{\pi(x)}(Y)$$

is surjective with kernel $T_x(R_x)$.

(Suppose that (Y, π) exists. Let H be the tangent space to Y at $\pi(0, 0, 0)$. Then H is isomorphic to $\mathbf{R} \times F \times (E/F)$. On the other hand H is isomorphic to $T_{\pi(x)}Y$ for x sufficiently close to (0, 0, 0) and is hence isomorphic to $\mathbf{R} \times E$.)

8. Let $\mathbf{T}$ be given the normalized Haar and $\mathscr{L}(L^2(\mathbf{T}))$ the normic topology. Show that the regular representation of $\mathbf{T}$ on $L^2(\mathbf{T})$ is not continuous. (For all $g \in \mathbf{T}$ distinct from e, construct a function $f \in L^2(\mathbf{T})$ such that $\|f\| = 1$, $\|\gamma(g)f - f\| = \sqrt{2}$.)

9. Let I be the set of $(x, \sqrt{2}x) \in \mathbf{R}^2$ for $-\frac{1}{2} < x < \frac{1}{2}$. Let U be the canonical image of I in the real Lie group $H = \mathbf{T}^2$. Show that U is a Lie subgroup germ of H but that the subgroup H′ of H generated by U is dense in H and not closed in H.

10. Let $\mathbf{R}_d$ be the group $\mathbf{R}$ with the discrete manifold structure. Let H be the real Lie group $\mathbf{R} \times \mathbf{R}_d$. Let U be the set of elements of H of the form (x, x) or of the form $(x, -x)$ where $x \in \mathbf{R}$. Then U is a Lie subgroup germ of H, U is discrete, the subgroup G of H generated by U is equal to H and the Lie group structure on G defined by the Corollary to Proposition 22 is the discrete group structure.

§ 3

1. Let G be a Lie group and G_1 and G_2 Lie subgroups of G. Suppose that G_2 is of finite codimension in G and that K is of characteristic 0. Then $G_1 \cap G_2$ is a Lie subgroup of G with Lie algebra $L(G_1) \cap L(G_2)$ (argue as in Proposition 29 and its Corollary 2).

2. Suppose that K is locally compact. Let G be a Lie group of finite dimension n.

(*a*) Let ϕ be an étale endomorphism of G with finite kernel. Let μ be a left Haar measure of G. Then ϕ is proper and

$$\phi(\mu) = \alpha . \mu | \phi(G) \quad \text{where} \quad \alpha = \frac{\text{Card}(\text{Ker}\,\phi)}{\text{mod}\det L(\phi)}.$$

(Use Proposition 55.)

(*b*) Suppose that G is compact. Let ϕ be an étale endomorphism of G. Then Ker ϕ is finite, $\phi(G)$ is of finite index in G and

$$\frac{\mathrm{Card}(\mathrm{Ker}\,\phi)}{\mathrm{Card}(G/\phi(G))} = \mathrm{mod}\,\det L(\phi).$$

(Use (*a*) to transform $\mu(G)$.)

(*c*) Suppose that G is compact and commutative. Let $r \in \mathbf{Z}$ be such that $r.1 \neq 0$ in K and let ϕ be the endomorphism $x \mapsto x^r$ of G. Then

$$\frac{\mathrm{Card}(\mathrm{Ker}\,\phi)}{\mathrm{Card}(G/\phi(G))} = (\mathrm{mod}\, r.1)^n.$$

(Observe that $L(\phi)$ is the homothety of ratio r by § 2, Corollary to Proposition 7.)

3. Let E be the complex vector space of symmetric complex matrices with 2 rows and 2 columns. For all $s \in \mathbf{SL}(2, \mathbf{C})$, we define an automorphism $\rho(s)$ of E by $\rho(s).m = s.m.{}^t s$.

(*a*) Show that ρ is an analytic linear representation of $\mathbf{SL}(2, \mathbf{C})$. Identifying $\begin{pmatrix} a & b \\ b & c \end{pmatrix} \in E$ with $(a, b, c) \in \mathbf{C}^3$, the matrix of $\rho\begin{pmatrix} \alpha & \beta \\ \gamma & \delta \end{pmatrix}$ is

$$\begin{pmatrix} \alpha^2 & 2\alpha\beta & \beta^2 \\ \alpha\gamma & \alpha\delta + \beta\gamma & \beta\delta \\ \gamma^2 & 2\gamma\delta & \delta^2 \end{pmatrix}.$$

(*b*) Show that $m \mapsto \det(m)$ is the non-degenerate quadratic form

$$q(a, b, c) = ac - b^2$$

on E and that ρ is a morphism of $\mathbf{SL}(2, \mathbf{C})$ onto $\mathbf{SO}(q)$ with kernel $\{I, -I\}$. (For the surjectivity, observe that, in the matrix of an element of $\mathbf{SO}(q)$, the 1st column can be expressed in the form $(\alpha^2, \alpha\gamma, \gamma^2)$ and the 3rd column in the form $(\beta^2, \beta\delta, \delta^2)$, where $\alpha^2\delta^2 + \beta^2\gamma^2 - 2\alpha\gamma\beta\delta = 1$. Then observe that 2 elements of $\mathbf{SO}(q)$ which coincide on a non-isotropic plane are equal.)

(*c*) Deduce that the complex Lie group $\mathbf{SO}(3, \mathbf{C})$ is isomorphic to the complex Lie group $\mathbf{SL}(2, \mathbf{C})/\{I, -I\}$.

4. (*a*) Let q be a quadratic form of signature $(1, 2)$ on $\mathbf{R}^3$. Let P be the set of $m \in \mathbf{R}^3$ such that $q(m) > 0$. Then P is the union of two disjoint convex cones C and $-$C. If $s \in \mathbf{SO}(q)$, then either $s(C) = C$ and $s(-C) = -C$ or $s(C) = -C$ and $s(-C) = C$. Let $\mathbf{SO}^+(q)$ be the set of $s \in \mathbf{SO}(q)$ such that $s(C) = C$. Then $\mathbf{SO}^+(q)$ is an open normal subgroup of $\mathbf{SO}(q)$ of index 2 in $\mathbf{SO}(q)$.

(*b*) By imitating the method of Exercise 3, define a morphism of $\mathbf{SL}(2, \mathbf{R})$ onto $\mathbf{SO}^+(q)$ with kernel $\{I, -I\}$. Deduce that the real Lie group $\mathbf{SO}^+(q)$ is isomorphic to the real Lie group $\mathbf{SL}(2, \mathbf{R})/\{I, -I\}$.

(c) For $(\xi, \eta) \in \mathbf{C}^2$ and $(\xi', \eta') \in \mathbf{C}^2$, let $f((\xi, \eta), (\xi', \eta')) = \xi\bar{\xi}' - \eta\bar{\eta}'$. Show that the inner automorphism of $\mathbf{SL}(2, \mathbf{C})$ defined by $\frac{1}{\sqrt{2}}\begin{pmatrix} 1 & -i \\ -i & 1 \end{pmatrix}$ maps $\mathbf{SU}(f)$ to $\mathbf{SL}(2, \mathbf{R})$.

5. Let E be the real vector space of Hermitian complex matrices with 2 rows and 2 columns of zero trace. For all $s \in \mathbf{SU}(2, \mathbf{C})$, we define an automorphism $\rho(s)$ of E by $\rho(s)(m) = sms^*$. We identify the element

$$\begin{pmatrix} x & y + iz \\ y - iz & -x \end{pmatrix}$$

of E with the element (x, y, z) of $\mathbf{R}^3$. By imitating the method of Exercise 3, show that ρ is a morphism of the real Lie group $\mathbf{SU}(2, \mathbf{C})$ onto the real Lie group $\mathbf{SO}(3, \mathbf{R})$ with kernel $\{I, -I\}$ and that $\mathbf{SO}(3, \mathbf{R})$ is isomorphic to $\mathbf{SU}(2, \mathbf{C})/\{I, -I\}$

¶ 6. Let F be the vector space of Hermitian complex matrices with 2 rows and 2 columns. For all $s \in \mathbf{SL}(2, \mathbf{C})$, we define an automorphism $\sigma(s)$ of F by $\sigma(s)(m) = sms^*$. We identify the element $\begin{pmatrix} t + x & y + iz \\ y - iz & t - x \end{pmatrix}$ of F with the element (t, x, y, z) of $\mathbf{R}^4$ and write

$$q(t, x, y, z) = t^2 - x^2 - y^2 - z^2.$$

Let C be the set of $(t, x, y, z) \in \mathbf{R}^4$ such that $t^2 - x^2 - y^2 - z^2 > 0$, $t > 0$. Let $\mathbf{SO}^+(q)$ be the set of $g \in \mathbf{SO}(q)$ such that $g(\mathrm{C}) = \mathrm{C}$; it is an open normal subgroup of $\mathbf{SO}(q)$ of index 2 in $\mathbf{SO}(q)$ (cf. Exercise 4). Show that σ is a morphism of the underlying real Lie group of $\mathbf{SL}(2, \mathbf{C})$ onto the real Lie group $\mathbf{SO}^+(q)$ with kernel $\{I, -I\}$. (For the surjectivity, use Exercise 5.) Deduce that the real Lie group $\mathbf{SO}^+(q)$ is isomorphic to the underlying real Lie group of $\mathbf{SL}(2, \mathbf{C})/\{I, -I\}$, that is (Exercise 3) the underlying real Lie group of $\mathbf{SO}(3, \mathbf{C})$.

7. (a) Let G be the set of quaternions of norm 1. This is a Lie subgroup of the real Lie group $\mathbf{H}^*$, homeomorphic to $\mathbf{S}_3$ and hence simply connected, cf. *General Topology*, Chapter XI.

(b) Let E be the real vector space of pure quaternions, identified with $\mathbf{R}^3$ by

$$(x, y, z) \mapsto xi + yj + zk.$$

For all $g \in \mathrm{G}$, we define an automorphism $\rho(g)$ of E by $\rho(g)q = gqg^{-1}$. Show that ρ is a morphism of the Lie group G onto the real Lie group $\mathbf{SO}(3, \mathbf{R})$ with kernel $\{1, -1\}$. Deduce that the Lie group $\mathbf{SO}(3, \mathbf{R})$ is isomorphic to the Lie group $\mathrm{G}/\{1, -1\}$.

(c) Let $\mathbf{C}$ be identified with the subfield $\mathbf{R} + \mathbf{R}i$ of $\mathbf{H}$. The mapping

$(\lambda, q) \mapsto q\lambda$ of $\mathbf{C} \times \mathbf{H}$ into $\mathbf{H}$ makes $\mathbf{H}$ into a vector space over $\mathbf{C}$ which is identified with $\mathbf{C}^2$ by the choice of basis $(1, j)$. For all $g \in G$, let $\sigma(g)$ be the automorphism of this vector space defined by $\sigma(g)q = gq$. Show that σ is an isomorphism of the real Lie group G onto the real Lie group $\mathbf{SU}(2, \mathbf{C})$. The latter is therefore simply connected by (*a*).

(*d*) For $(q_1, q_2) \in G \times G$, let $\tau(q_1, q_2)$ be the mapping $q \mapsto q_1 q \bar{q}_2$ of $\mathbf{H} = \mathbf{R}^4$ into itself. Show that τ is a morphism of the real Lie group $G \times G$ onto the real Lie group $\mathbf{SO}(4, \mathbf{R})$ with kernel $N = \{(1, 1), (-1, -1)\}$ and hence that $\mathbf{SO}(4, \mathbf{R})$ is isomorphic to $(G \times G)/N$.

8. Let G be a finite-dimensional connected real Lie group and M a finite-dimensional connected manifold of class C^∞; let there be given a law of right operation of class C^∞ of G on M. For all $x \in M$, let $\rho(x)$ be the corresponding orbital mapping.

(*a*) For G to operate transitively on M, it is necessary and sufficient that, for all $x \in M$, the mapping $T_e(\rho(x))$ of L(G) into $T_x(M)$ be surjective. (To see that the condition is sufficient, observe that then every G-orbit in M is open.)

(*b*) Suppose that G operates transitively on M. We identify M with a homogeneous space H\G, where H is the stabilizer of a point x_0 of M. The action of G on M is called imprimitive if there exists a closed submanifold V of M of class C^∞ such that $0 < \dim V < \dim M$ and for which every transform $V.s$ (where $s \in G$) is either equal to V or disjoint from V. For this to be so, it is necessary and sufficient that there exist a Lie subgroup L of G such that $H \subset L \subset G$ and $\dim(H) < \dim(L) < \dim(G)$. If this is not so, the action of G on M is called primitive; for this it suffices that there exist no subalgebra of L(G) lying between L(G) and L(H) and distinct from L(G) and L(H).

(*c*) Under the hypothesis of (*b*), let $\mathscr{L}$ be the image of L(G) under the law of infinitesimal operation associated with the action of G. Let $\mathscr{E}$ be the algebra of functions of class C^∞ on M and $\mathfrak{m}$ the maximal ideal of $\mathscr{E}$ consisting of the functions in $\mathscr{E}$ which are zero at x_0; we denote by $\mathscr{L}_p$ for $p = -1, 0, 1, 2, \ldots$ the set of $x \in \mathscr{L}$ such that $\theta(X)f \in \mathfrak{m}^{p+1}$ for all $f \in \mathscr{E}$; then $\mathscr{L}_{-1} = \mathscr{L}$ and $\mathscr{L}_0$ is the image of L(H) under the law of infinitesimal operation. If $(U, \phi, \mathbf{R}^n)$ is a chart of M centered on x_0 and $X_1, \ldots, X_n$ are the vector fields on U defined by the chart, the elements of $\mathscr{L}_p$ are the $a_1X_1 + \cdots + a_nX_n$ with $a_1, \ldots, a_n$ in $\mathfrak{m}^p|U$. Show that $[\mathscr{L}_p, \mathscr{L}_q] \subset \mathscr{L}_{p+q}$ (with the convention $\mathscr{L}_{-2} = \mathscr{L}$). If there exists $Y \in \mathscr{L}_p$ (for $p \geqslant 0$) such that $Y \notin \mathscr{L}_{p+1}$, there exists $x \in L$ such that $[Y, X] \notin \mathscr{L}_p$. The $\mathscr{L}_p$ are Lie subalgebras of $\mathscr{L}$. For $p \geqslant 0$, $\mathscr{L}_p$ is an ideal of $\mathscr{L}_0$. Let ρ be the canonical linear representation of H on $T_{x_0}(M)$. The Lie algebra of H/(Ker ρ) is isomorphic to $\mathscr{L}_0/\mathscr{L}_1$.

(*d*) Suppose further that the intersection of the $\mathscr{L}_p$ is $\{0\}$. Then there exists a greatest index r such that $\mathscr{L}_r \neq \{0\}$ and all the $\mathscr{L}_j$ of indices $\leqslant r$ are distinct. For $p \geqslant 0$,

$$\dim(\mathscr{L}_p/\mathscr{L}_{p+1}) \leqslant \binom{n+p}{p+1}.$$

(*e*) The hypothesis of (*d*) is satisfied if M is analytic and G operates analytically on M.

9. In the notation of Proposition 30, HH′ may be an open subset of G dense in G and distinct from G (take $G = \mathbf{GL}(2, \mathbf{R})$ with H the upper triangular subgroup and H′ the lower strict triangular subgroup).

§ 4

1. Let G be a Lie group, H a normal Lie quasi-subgroup of G and π the canonical mapping of G onto G/H. There exists on G/H one and only one Lie group structure with the following property: for a homomorphism θ of G/H into a Lie group G′ to be a Lie group morphism, it is necessary and sufficient that $\theta \circ \pi$ be a Lie group morphism. Moreover, L(G/H) is canonically isomorphic to L(G)/L(H). (Let Q be a Lie group germ such that L(Q) = L(G)/L(H). Show that, by shrinking Q if necessary, Q can be identified with an open neighbourhood of 0 in G/H and that Proposition 18 of § 1 can then be applied to G/H.)

¶ 2. Let G and H be two Lie groups and f a Lie group morphism of G into H.

(i) The kernel N of f is a normal Lie quasi-subgroup of G and $L(N) = \operatorname{Ker} L(f)$. (Use exponential mappings of G and H).

(ii) Let $g: G/N \to f(G)$ be the mapping derived from f when passing to the quotient. If f and $L(f)$ have closed images, and the topology of G admits a countable base, then $f(G)$ is a Lie quasi-subgroup of H with Lie algebra $\operatorname{Im} L(f)$ and g is a Lie group isomorphism, where G/N has the structure defined in Exercise 1. (Reduce it by means of (i) to the case where $N = \{e\}$.)

3. Let G be a Lie group, U an open neighbourhood of 0 in L(G) and ϕ an analytic mapping of U into G such that $\phi(0) = e$ and $T_0\phi = \mathrm{Id}_{L(G)}$. The following conditions are equivalent:

(*a*) ϕ is an exponential mapping;

(*b*) There exists $m \in \mathbf{Z}$ distinct from 0, 1, -1 such that $\phi(mx) = \phi(x)^m$ in a neighbourhood of 0.

(If condition (*b*) holds, show that ϕ^* satisfies condition (iii) of Theorem 4.)

¶ 4. (*a*) Let $K = \mathbf{R}$ or $\mathbf{C}$ or an ultrametric field with residue field of characteristic >0. Let G be a Lie group over K, U an open neighbourhood of 0 in L(G) and $\phi: U \to G$ a mapping which is differentiable at 0 and such that $\phi(0) = e$, $T_0(\phi) = \mathrm{Id}_{L(G)}$ and $\phi((\lambda + \lambda')b) = \phi(\lambda b)\phi(\lambda' b)$ for λb, $\lambda' b$, $(\lambda + \lambda')b$ in U. Then ϕ coincides on a neighbourhood of 0 with an exponential mapping.

(*b*) Let k be a field of characteristic 0. Let K be the valued field $k((X))$, which is of characteristic 0. Let G be the additive Lie group $k[[X]]$. Let ϕ be the continuous k-linear mapping of G into G such that $\phi(X^n) = X^n + X^{2n}$ for every integer $n \geqslant 0$. Then ϕ satisfies the conditions of (*a*), but coincides on no neighbourhood of 0 with an exponential mapping.

5. Let (e_1, e_2, e_3) be the canonical basis of $\mathbf{R}^3$. We give $\mathbf{R}^3$ the nilpotent Lie algebra structure $\mathfrak{g}$ such that $[e_1, e_2] = e_3$, $[e_1, e_3] = [e_2, e_3] = 0$. Let c_{ijk} denote the constants of structure of $\mathfrak{g}$ relative to the basis (e_1, e_2, e_3).

(*a*) Show that on $\mathbf{R}^3$ the differential forms

$$\omega_1 = dx_1, \qquad \omega_2 = dx_2, \qquad \omega_3 = -x_1 dx_2 + dx_3$$

satisfy the relations

$$d\omega_k = -\sum_{i<j} c_{ijk}\omega_i \wedge \omega_j \qquad (k = 1, 2, 3)$$

and are linearly independent at each point of $\mathbf{R}^3$.

(*b*) Deduce that there exists, on an open neighbourhood of 0 in $\mathbf{R}^3$, a Lie group germ structure G such that $L(G) = \mathfrak{g}$ and

$$(1) \qquad (a_1, a_2, a_3)(x_1, x_2, x_3) = (x_1 + a_1, x_2 + a_2, x_3 + a_3 + a_1 x_2)$$

for (a_1, a_2, a_3), (x_1, x_2, x_3) sufficiently close to 0. Show that in fact formulae (1) define a nilpotent Lie group structure on $\mathbf{R}^3$.

¶ 6. Let G be a finite-dimensional Lie group, $\mathfrak{g}$ its Lie algebra, $\mathfrak{g}^*$ the dual vector space of $\mathfrak{g}$, σ the adjoint representation of G on $\mathfrak{g}$ and ρ the representation of G on $\mathfrak{g}^*$ defined by $\rho(g) = {}^t\sigma(g)^{-1}$ for all $g \in G$.

(*a*) $L(\rho)(x) = -{}^t(\mathrm{ad}\, x)$ for all $x \in \mathfrak{g}$.

(*b*) Let $f \in \mathfrak{g}^*$. Let G_f denote the stabilizer of f in G; it is a Lie subgroup of G and $\mathfrak{g}_f = L(G_f)$ is the set of $x \in G$ such that ${}^t(\mathrm{ad}\, x)f = 0$.

(*c*) For x, y in $\mathfrak{g}$, we write $B_f(x, y) = f([x, y])$. Show that B_f is an alternating bilinear form on $\mathfrak{g}$, that the mapping s_f of $\mathfrak{g}$ into $\mathfrak{g}^*$ left associated with B_f is the mapping $x \mapsto {}^t(\mathrm{ad}\, x)f$ and that the orthogonal of $\mathfrak{g}$ with respect to B_f is $\mathfrak{g}_f$. Let β_f denote the non-degenerate alternating bilinear form on $\mathfrak{g}/\mathfrak{g}_f$ (which is therefore of even dimension) derived from B_f when passing to the quotient. We denote by β'_f the inverse form of B_f on $(\mathfrak{g}/\mathfrak{g}_f)^*$.

(*d*) Let Ω denote an orbit of G on $\mathfrak{g}^*$; we give it the manifold structure derived from that on G/G_f by transport of structure, where f is an arbitrary point of Ω. Then G operates analytically on Ω on the left. Let D be the associated law of infinitesimal operation. For all $a \in \mathfrak{g}$ and all $f \in \Omega$,

$$D_a(f) = -({}^t\mathrm{ad}\, a)f.$$

The tangent subspace $T_f(\Omega)$ is the image of s_f, that is the orthogonal of $\mathfrak{g}_f$ in $\mathfrak{g}^*$, and is canonically identified with $(\mathfrak{g}/\mathfrak{g}_f)^*$, so that s_f defines a canonical isomorphism r_f of $T_f(\Omega)$ onto its dual. The field $f \mapsto \beta'_f$ is an analytic differential form ω of degree 2 on Ω, which is invariant under G. For $f \in \Omega$, $a \in \mathfrak{g}$, $b \in \mathfrak{g}$, $\omega(D_a(f), D_b(f)) = f([a, b])$.

(*e*) Show that $d\omega = 0$.

(*f*) If α is a differential form of degree 1 on Ω, the isomorphisms r_f allow us to identify α with a vector field. Let A be the set of analytic functions on Ω with values in K. For ϕ, ψ in A, we write $[\phi, \psi] = \omega(d\phi, d\psi) \in A$. Show that A has thus a Lie algebra structure.

(*g*) For all $a \in \mathfrak{g}$, let ψ_a be the function $f \mapsto f(a)$ on Ω. Show that $a \mapsto \psi_a$ is a homomorphism of $\mathfrak{g}$ into the Lie algebra A and that the form $d\psi_a$ is identified, by means of the isomorphisms r_f, with the vector field D_a.

(*h*) Let U be an open subset of $\mathfrak{g}^*$ and ϕ an analytic function on U. For all $f \in U$, the differential $d_f\phi$ of ϕ at f is identified with an element of $\mathfrak{g}$. Suppose that $X\phi = 0$ for every vector field X defined by the action of G on $\mathfrak{g}^*$. Show that then, for all $f \in U$, $d_f\phi$ belongs to the centre of $\mathfrak{g}_f$.

(*i*) For all $f \in \mathfrak{g}^*$, we write $r_f = \dim \mathfrak{g}_f$. Let $r = \inf_{f \in \mathfrak{g}^*} r_f$. Show that the set V of $f \in \mathfrak{g}^*$ such that $r_f = r$ is open and dense in $\mathfrak{g}^*$. Show that, for all $f \in V$, $\mathfrak{g}_f$ is commutative. (Construct r functions ϕ satisfying the conditions of (*h*) in a neighbourhood of f and such that their differentials at f are linearly independent.)

¶ 7. (*a*) Let $(\lambda, x) \mapsto \lambda . x$ be a law of continuous left operation of $\mathbf{R}$ on $\mathbf{T}$. Then one of the two following holds:

(1) either there exists a fixed point and the stabilizer of every point is either $\mathbf{R}$ or $\{0\}$;

(2) or there exists a point whose stabilizer is an infinite discrete subgroup of $\mathbf{R}$ and $\mathbf{R}$ operates transitively on $\mathbf{T}$. (If $\{\lambda \in \mathbf{R} | \lambda . x = x\} = \{0\}$, the orbit of x is homeomorphic to $\mathbf{R}$ and has frontier points $\lim_{\lambda \to +\infty} \lambda . x$ which are fixed. If $\{\lambda \in \mathbf{R} | \lambda . x = x\} = \mathbf{Z}a$ with $a \neq 0$, the orbit of x is homeomorphic to $\mathbf{T}$ and $\mathbf{R}$ is transitive.)

(*b*) If f is a homeomorphism of $\mathbf{T}$ onto itself and k an integer $\geqslant 1$, a point $x \in \mathbf{T}$ is called periodic of period k for f if $f^k(x) = x$ and $f^h(x) \neq x$ for $1 \leqslant h < k$. In the notation of (*a*), let f_λ be the homeomorphism $x \mapsto \lambda . x$. Then the set of periodic points of period $k > 1$ for f_λ is empty or equal to $\mathbf{T}$. (If there exists a periodic point of period $k > 1$ for f_λ, its stabilizer is distinct from $\{0\}$ and $\mathbf{R}$, hence $\mathbf{R}$ operates transitively on $\mathbf{T}$ and all the points of $\mathbf{T}$ are periodic of period k for f_λ.)

(*c*) Let Ω be a neighbourhood of e in $\mathrm{Diff}^\infty(\mathbf{T})$ (*Differentiable and Analytic Manifolds*, R, 15.3.8, Note (¹)). There exist an integer $k > 1$ and $f \in \Omega$ without fixed point such that the set of periodic points of period k for f is neither empty nor equal to $\mathbf{T}$. (Let $p: \mathbf{R} \to \mathbf{T}$ be the canonical mapping. Let ρ be the mapping $p(x) \mapsto p\left(x + \frac{2\pi}{k}\right)$ of $\mathbf{T}$ into $\mathbf{T}$. Let Ω' be a neighbourhood of e in $\mathrm{Diff}^{\infty}(\mathbf{T})$ such that $\Omega'^2 \subset \Omega$. For sufficiently large k, $\rho \in \Omega'$. On the other hand, let σ be a mapping of $\mathbf{T}$ into $\mathbf{T}$ such that $\sigma(y) = y$ for

$$y \notin p\left(\left]0, \frac{2\pi}{k}\right[\right)$$

and $\sigma(p(x)) = p(x + \lambda(x))$ for $x \in \left]0, \frac{2\pi}{k}\right[$, where

$$0 < \lambda(x) < \frac{2\pi}{k} - x;$$

can be chosen such that $\sigma \in \Omega'$. Then $f = \rho \circ \sigma$ has the required properties.)

Such an f cannot, by (*b*), be in the image of a continuous morphism of $\mathbf{R}$ into $\mathrm{Diff}^{\infty}(\mathbf{T})$.

(*d*) Let Y be a compact differentiable manifold of dimension $\geqslant 1$. Every neighbourhood of e in $\mathrm{Diff}^{\infty}(\mathrm{Y})$ contains an element h with the following property: h belongs to the image of no continuous morphism of $\mathbf{R}$ into $\mathrm{Diff}^{0}(\mathrm{Y})$. (The manifold Y contains an open submanifold of the form $\mathbf{T} \times \mathrm{D}$, where D is the open Eulcidean ball of radius 1 and centre 0 in $\mathbf{R}^n$ ($n = \dim \mathrm{Y} - 1$). Let $g \in \mathrm{Diff}^{\infty}(\mathrm{D})$ be such that $g(0) = 0$, $\|g(x)\| > \|x\|$ for $0 < \|x\| < \frac{1}{2}$ and $g(x) = x$ for $\|x\| \geqslant \frac{1}{2}$ and g very close to e in $\mathrm{Diff}^{\infty}(\mathrm{D})$. Let f be as in (*c*), and, for $0 < t < 1$, let $f_t \colon \mathbf{T} \to \mathbf{T}$ be defined as follows: for $x \in \mathbf{T}$, $y \in p^{-1}(x)$, $z \in p^{-1}(f(x))$ and $|z - y| < \pi$, we write $f_t(x) = p(ty + (1 - t)z)$. Then there exists $h \in \mathrm{Diff}^{\infty}(\mathrm{Y})$ such that $h(x, y) = (f_{2\|y\|}(x), g(y))$ for $x \in \mathbf{T}$, $y \in \mathrm{D}$ and $h(u) = u$ for $u \in \mathrm{Y} - (\mathbf{T} \times \mathrm{D})$; moreover, we can choose f and g such that h is arbitrarily close to e in $\mathrm{Diff}^{\infty}(\mathrm{Y})$. The points of $\mathbf{T} \times \{0\}$ are the $y \in \mathrm{Y}$ such that when n tends to $+\infty$, $h^{kn}(x)$ tends to a point not fixed under h (in fact it is periodic of period k). Therefore, every homeomorphism of Y onto Y which commutes with h leaves $\mathbf{T} \times \{0\}$ invariant. Then apply (*c*).)

(*e*) Let Y be a compact differentiable manifold of dimension $\geqslant 1$. There do not exist a Lie group G and continuous morphisms $\mathrm{Diff}^{\infty}(\mathrm{Y}) \to \mathrm{G}$, $\mathrm{G} \to \mathrm{Diff}^{0}(\mathrm{Y})$ whose composition is the canonical injection of $\mathrm{Diff}^{\infty}(\mathrm{Y})$ into $\mathrm{Diff}^{0}(\mathrm{Y})$. (Use (*d*).)

8. Let G be a finite-dimensional Lie group. Suppose that L(G) is simple. Let A be a normal subgroup of G. If A is not open, A is discrete and its centralizer in G is open. (Let $\mathfrak{a}$ be the subalgebra tangent to A at e. Show that $\mathfrak{a} = \{0\}$. Let x be an element of A not belonging to the centre of G. Consider the mapping $y \mapsto yxy^{-1}$ of G into A. Deduce from the relation $\mathfrak{a} = \{0\}$ that the centralizer of x in G is open. Then, using an exponential mapping, show that the centralizer of x in G contains a neighbourhood of e independent of x.)

¶ 9. Let G be a compact Lie group over K of dimension n. Suppose that the Lie algebra L(G) is simple. For all $g \in G$, every integer $m \geqslant 1$ and every neighbourhood V of e in G, let $M(g, m, V)$ denote the set of elements of G of the form $\prod_{i=1} x_i(y_i, g)x_i^{-1}$ with $x_1, \ldots, x_m, y_1, \ldots, y_m$ in V.

(*a*) Show that if g is an element of G whose centralizer is not open, if m is an integer $\geqslant n$ and if V is a neighbourhood of e, then $M(g, m, V)$ is a neighbourhood of e. (There exists $a \in L(G)$ such that $(\mathrm{Ad}\, g)(a) \neq a$. Let ϕ be an exponential mapping of G. Let y be the image of 1 under the tangent mapping at 0 to $\lambda \mapsto (\phi(\lambda a), g)$ (where $\lambda \in K$). Then $b = (\mathrm{Ad}\, g - 1)(a) \neq 0$. There exist $x_1, \ldots, x_m$ in V such that $\{(\mathrm{Ad}\, x_1)(b), \ldots, (\mathrm{Ad}\, x_m)(b)\}$ contains a basis of L(G). Consider the mapping

$$(x'_1, \ldots, x'_m, y_1, \ldots, y_m) \mapsto \prod_{i=1}^{n} x'_i(y_i, g)x_i'^{-1}$$

of G^{2m} into G. Show that the tangent mapping at $(x_1, \ldots, x_m, e, \ldots, e)$ is surjective.)

(*b*) Let U be a neighbourhood of e in G. Let m be an integer $\geqslant 1$. There exists an element g of G whose centralizer is not open, such that $M(g, m, G) \subset U$. (Argue by *reductio ad absurdum* using the compactness of G.)

(*c*) Let G′ be a compact Lie group over K whose Lie algebra is simple. Let $\phi: G \to G'$ be a bijective homomorphism of abstract groups. Then ϕ is continuous. (Apply (*b*) to G′ and (*a*) to G.)

10. Let G be a Lie group over K. Show that there exists a neighbourhood V of e with the following property: for every sequence (x_n) of elements of V, if we define inductively $y_1 = x_1, y_n = (x_n, y_{n-1})$, the sequence (y_n) tends to e.

11. (*a*) For x, y in $\mathbf{S}_{2n-1} \subset \mathbf{C}^n$, we define $\alpha(x, y) \in [0, \pi]$ by

$$\cos \alpha(x, y) = \mathscr{R}((x|y)).$$

For s, t in $\mathbf{U}(n)$, we write $d(s, t) = \sup_{x \in \mathbf{S}_{2n-1}} \alpha(sx, tx)$. Show that d is a left and right invariant distance on $\mathbf{U}(n)$.

(*b*) Let $s \in \mathbf{U}(n)$. Let $\theta_1, \ldots, \theta_m$ be the numbers in $]-\pi, \pi]$ such that the $e^{i\theta_j}$ are the distinct eigenvalues of s. Let $\theta(s) = \sup_{1 \leqslant j \leqslant m} |\theta_j|$. Then $d(e, s) = \theta(s)$.. (Let V_j be the eigenspace of s corresponding to $e^{i\theta_j}$. For $x \in \mathbf{S}_{2n-1}$, find a lower bound to $\mathscr{R}((x|sx))$ by decomposing with respect to the V_j.)

(*c*) Let s, t be in $\mathbf{U}(n)$ be such that $\theta(t) < \frac{\pi}{2}$ and s commutes with (s, t). Then s and t commute. (In the notation of (*b*), let V'_j be the orthogonal supplement of V_j and $W_j = t(V_j)$; show that $W_j = (W_j \cap V_j) + (W_j \cap V'_j)$ and then that $W_j \cap V'_j = \{0\}$.)

(*d*) Show that there exists a compact neighbourhood V of e in $\mathbf{U}(n)$ which has the property of Exercise 10, which is stable under inner automorphisms of $\mathbf{U}(n)$ and which is such that if x, y are non-permutable elements of V then x and (x, y) are not permutable. (Use (*c*).)

(*e*) Let β be the normalized Haar measure of $\mathbf{U}(n)$. Let V_1 be a symmetric compact neighbourhood of e in $\mathbf{U}(n)$ such that $V_1^2 \subset V$. Let p be an integer such that $\beta(V_1) > 1/p$. Show that, for every finite subgroup F of $\mathbf{GL}(n, \mathbf{C})$, there exists a commutative normal subgroup A of F such that $\mathrm{Card}(F/A) \leqslant p$ (Jordan's theorem). (Reduce it to the case where $F \subset \mathbf{U}(n)$. Take A to be the subgroup of F generated by $F \cap V$ and use (*d*).)

12. Let G be a finite-dimensional connected real or complex Lie group and α a left invariant differential form of degree p on G. For α to be right invariant too, it is necessary and sufficient that $d\alpha = 0$. (Observe that the condition for right invariance can be written

$$\wedge^p({}^t\mathrm{Ad}(s)).\alpha(e) = \alpha(e)$$

for all $s \in G$ and hence it is equivalent to the condition

$$\sum_{j=1}^{f} \langle \alpha(e), u_1 \wedge \cdots \wedge u_{j-1} \wedge [u, u_j] \wedge u_{j+1} \wedge \cdots \wedge u_p \rangle = 0$$

for all $u, u_1, \ldots, u_p$ in L(G).) In particular, the left and right invariant differential forms of degree 1 on G form a vector space of dimension

$$\dim L(G) - \dim[L(G), L(G)].$$

13. Let $\mathbf{R}^n$ be given the usual scalar product $((\xi_i), (\eta_i)) \mapsto \sum_{i=1} \xi_i \eta_i$. Let I$(n)$ be the group of isometric affine transformations of $\mathbf{R}^n$. It is canonically isomorphic to the Lie subgroup of $\mathbf{GL}(n+1, \mathbf{R})$ consisting of the matrices

$$S = \begin{pmatrix} U & x \\ 0 & 1 \end{pmatrix}$$

where $U \in \mathbf{O}(n)$ and x is an arbitrary element of $\mathbf{R}^n$ (matrix of type $(n, 1)$). I(n) can be identified with the semi-direct product of $\mathbf{O}(n)$ by $\mathbf{R}^n$ defined by the canonical injection of $\mathbf{O}(n)$ into $\mathbf{GL}(n, \mathbf{R})$ (the matrix S is then denoted by (U, x)). Let $p\colon I(n) \to \mathbf{O}(n)$ be the canonical surjection, with kernel $\mathbf{R}^n$.

(*a*) Let Γ be a discrete subgroup of I(n). Consider the closure $\overline{p(\Gamma)}$ of $p(\Gamma)$ in $\mathbf{O}(n)$. Show that its identity component is commutative. (Let V be a neighbourhood of I in $\mathbf{O}(n)$ with the properties of Exercise 11 (*d*) and such that further $\|U - I\| \leqslant \frac{1}{4}$ for all $U \in V$ (the norm on $\mathrm{End}(\mathbf{R}^n)$ being derived from the Euclidean norm on $\mathbf{R}^n$). Argue by *reductio ad absurdum*, assuming that there exist $S_1 = (U_1, x_1)$, $S_2 = (U_2, x_2)$ such that U_1 and U_2 belong to V and do not commute. Show that $(S_1, S_2) = ((U_1, U_2), y)$ with

$$\|y\| \leqslant \tfrac{1}{4}(\|x_1\| + \|x_2\|).$$

Then define the S_k inductively by $S_k = (S_1, S_{k-1})$ for $k \geqslant 3$. Using the choice of V, show that this sequence has an infinity of distinct terms and is bounded in $I(n)$, which is absurd.)

(*b*) Γ is called *crystallographic* if $I(n)/\Gamma$ is compact. Show that if this is so, for all $x \in \mathbf{R}^n$, the affine subspace L of $\mathbf{R}^n$ generated by Γx is equal to $\mathbf{R}^n$. (Otherwise, for all $y \in \mathbf{R}^n$, all the points of Γy are the same distance from L and, as this distance can be arbitrarily large $I(n)/\Gamma$ is not compact.)

(*c*) If Γ is commutative and crystallographic, then $\Gamma \subset \mathbf{R}^n$. (If $S = (U, x) \in \Gamma$ is such that the vector subspace $V \subset \mathbf{R}^n$ consisting of the points invariant under U is not equal to $\mathbf{R}^n$ and $V^\perp$ denotes the orthogonal subspace of V, then, for all $S' = (U', x') \in \Gamma$, U' leaves V and $V^\perp$ stable. On the other hand, since $U|V^\perp$ has no non-zero invariant point, show that this allows us to assume, by changing the origin, that $x \in V$. Using (*b*), there exists $S' = (U', x') \in \Gamma$ such that the orthogonal projection y' of x' onto $V^\perp$ is $\neq 0$. By evaluating Sx' by the formula $S = S'SS'^{-1}$, deduce that $U.y' = y'$ contrary to the definition of V.)

(*d*) If Γ is crystallographic, $\Gamma \cap \mathbf{R}^n$ is a free commutative group of rank n and $\Gamma/(\Gamma \cap \mathbf{R}^m)$ is a finite group (Bieberbach's Theorem). (Let W be the vector subspace of $\mathbf{R}^n$ generated by $\Gamma \cap \mathbf{R}^n$. The compact group $\overline{p(\Gamma)}$ has only a finite number of connected components. If $W = \{0\}$, Γ contains, by (*a*), a commutative normal subgroup of finite index Γ_1; Γ_1 is then crystallographic, which implies a contradiction with (*c*). Hence $W \neq \{0\}$. Show that $p(\Gamma)$ leaves W stable and that $p(\Gamma)|W$ is finite; otherwise there would exist a sequence (S_m) in Γ such that if $S_m = (U_m, x_m)$, the U_m are distinct and tend to I; then form the $(I, a_j)S_m(I, a_j)^{-1}S_m^{-1}$, where (a_j) is a basis of $\Gamma \cap \mathbf{R}^n$, and obtain a contradiction with the hypothesis that Γ is discrete. Finally, to see that $W = \mathbf{R}^n$, show that otherwise the action of Γ on $\mathbf{R}^n/W$ would be that of a crystallographic group containing no translation $\neq 0$.)

§ 5

1. Suppose that $K = \mathbf{R}$. Let r be an integer $\geqslant 1$ or ∞. A *group of class* C^r is a set G with a group structure and a manifold structure of class C^r such that the mapping $(x, y) \mapsto xy^{-1}$ of $G \times G$ into G is of class C^r.

(*a*) Suppose henceforth that $r \geqslant 2$. A neighbourhood of e in G is identified with a neighbourhood of 0 in a Banach space. Let $xy = P(x, y)$, where P is a mapping of class C^2 on an open neighbourhood of (0, 0). Let

$$(D_1D_2P)(0, 0) = B.$$

Show that

$$xy = x + y + B(x, y) + |x|\,|y|\,o(1)$$

as $(x, y) \to (0, 0)$. (Use an expansion of P of order 2 with integral remainder.)

(*b*) Show that

$$x^{-1} = -x + \mathrm{B}(x, x) + |x|^2 o(1)$$
$$xyx^{-1} = y + \mathrm{B}(x, y) - \mathrm{B}(y, x) + |x|\,|y| o(1)$$
$$x^{-1}y^{-1}xy = \mathrm{B}(x, y) - \mathrm{B}(y, x) + |x|\,|y| o(1)$$

as (x, y) tends to $(0, 0)$.

2. Let G be a group of class C^r with $r \geqslant 2$.

(*a*) Let $t \in \mathscr{T}^{(s)}(\mathrm{G})$, $t' \in \mathscr{T}^{(s')}(\mathrm{G})$. If $s + s' \leqslant r$, define $t * t' \in \mathscr{T}^{(s+s')}(\mathrm{G})$ as in § 3, no. 1. If t and t' are without constant term, $t * t'$ is without constant term. The image of t under the mapping $x \mapsto x^{-1}$ of G into G is denoted by $t^\vee$; then $t^\vee \in \mathscr{T}^{(s)}(\mathrm{G})$. If f is a function of class C^r on G with values in a Hausdorff polynormed space, $f * t$ denotes the function on G defined by $(f * t)(x) = \langle \varepsilon_x * t^\vee, f \rangle$ for all $x \in \mathrm{G}$; this function is of class C^{r-s} if $s < \infty$. Show, as in § 3, no. 4, that $f * (t * t') = (f * t) * t'$.

(*b*) If $t \in \mathrm{T}_e(\mathrm{G})$, the vector field $x \mapsto \varepsilon_x * t$ on G is denoted by L_t. Show, as in § 3, no. 6, that, for t, t' in $\mathrm{T}_e(\mathrm{G})$, $\mathrm{L}_{t*t'} = \mathrm{L}_t \circ \mathrm{L}_{t'}$ and hence that

$$\mathrm{L}_{t*t' - t'*t} = [\mathrm{L}_t, \mathrm{L}_{t'}].$$

Therefore, $t * t' - t' * t$ is an element of $\mathrm{T}_e(\mathrm{G})$ which we denote by $[t, t']$.

(*c*) Using the notation of Exercise 1 (*a*), if $t \in \mathrm{T}_e(\mathrm{G})$, then

$$\mathrm{L}_t(x) = (\mathrm{D}_2\mathrm{P}(x, 0))(t)$$

for x sufficiently close to 0. Deduce that $[t, t'] = \mathrm{B}(t, t') - \mathrm{B}(t', t)$.

(*d*) Show that $\mathrm{T}_e(\mathrm{G})$, with the bracket $(t, t') \mapsto [t, t']$, is a normable Lie algebra. (To prove the Jacobi identity, use (*c*), Exercise 1 (*b*) and identity (5) of *Algebra*, Chapter I, § 6, no. 2.)

(For a sequel to this exercise, cf. § 8, Exercise 6.)

§ 6

1. Let D be a set of elements of $\mathbf{SL}(2, \mathbf{C})$ of the form $\begin{pmatrix} a & b \\ 0 & a^{-1} \end{pmatrix}$ where $a > 0$ and $b \in \mathbf{C}$. Show that D is a Lie subgroup of the underlying real Lie group of $\mathbf{SL}(2, \mathbf{C})$ and that the mapping $(u, d) \mapsto ud$ of $\mathbf{SU}(2, \mathbf{C}) \times \mathrm{D}$ into $\mathbf{SL}(2, \mathbf{C})$ is an isomorphism of real analytic manifolds. Deduce from this and § 3, Exercise 7 (*c*) that $\mathbf{SL}(2, \mathbf{C})$ is simply connected.

2. Let G be the universal covering of $\mathbf{SL}(2, \mathbf{R})$, π the canoncial morphism of G onto $\mathbf{SL}(2, \mathbf{R})$ and N its kernel.

(*a*) Arguing as in Exercise 1, show that there exists an isomorphism of the real analytic manifold $\mathbf{U} \times \mathbf{R}^2$ onto the real analytic manifold $\mathbf{SL}(2, \mathbf{R})$. Deduce that N is isomorphic to $\mathbf{Z}$.

(*b*) If ρ is an analytic linear representation of G on a complex vector space, then Ker $\rho \supset$ N. (The representation $L(\rho)$ of $\mathfrak{sl}(2, \mathbf{R})$ defines, by complexification, a representation of $\mathfrak{sl}(2, \mathbf{C})$ and the latter is, by Exercise 1, of the form $L(\sigma)$, where σ is an analytic linear representation of $\mathbf{SL}(2, \mathbf{C})$. Then $L(\sigma \circ \pi) = L(\rho)$ and hence $\sigma \circ \pi = \rho$.)

3. Let G_1 be the simply connected nilpotent real Lie group defined in Exercise 5 of § 4. Let Z be the centre of G_1, N a non-trivial discrete subgroup of Z and $G = G_1/N$. If ρ is a finite-dimensional analytic linear representation of G, $\rho(Z/N)$ is a semi-simple set of automorphisms, for Z/N is compact; but $Z = (G, G)$ and hence $\rho(Z/N)$ is unipotent (Chapter I, § 6, Proposition 6). Therefore ρ is trivial on Z/N.

4. Let G be a compact connected complex Lie group. Show that every analytic linear representation of G is trivial.

5. In the notation of Exercise 9 of § 1, show that H′ is an integral subgroup of H. Deduce that in the simply connected group $\mathbf{SU}(2, \mathbf{C}) \times \mathbf{SU}(2, \mathbf{C})$ (§ 3, Exercise 7 (c)), there exist one-parameter subgroups which are not closed.

¶ 6. Let I be the interval $(1, 2)$ with the discrete topology. Let E be the complete normed space of functions $f: I \to \mathbf{R}$ such that $\|f\| = \sum_{i \in I} |f(i)| < +\infty$. For all $x \in I$, let ε_x be the element of E such that $\varepsilon_x(t) = 0$ for $t \neq x$ and $\varepsilon_x(x) = 1$. Let P be the subgroup of E generated by the $x\varepsilon_x$ with $x \in I$; it is a discrete subgroup of E. Let G be the Lie group E/P. Let F be the hyperplane of E consisting of the $f \in E$ such that $\sum_{i \in I} f(i) = 0$. Let H be the Lie group $F/(F \cap P)$. Let ϕ be the canonical morphism of H into G. Show that ϕ is bijective and is an immersion but is not a Lie group isomorphism. (Let f be a linear form on E with kernel F; show that $f(P) = \mathbf{R}$ and hence that $F + P = E$.) Deduce that in Proposition 3 the countability hypothesis cannot be omitted.

7. Let ϕ be the morphism of $\mathbf{Z}/2\mathbf{Z}$ into the permutation group of $\mathbf{C}$ which maps the non-identity element to $z \mapsto \bar{z}$. Let G be the semi-direct product of $\mathbf{Z}/2\mathbf{Z}$ by the real Lie group $\mathbf{C}$ corresponding to ϕ. It is a real Lie group with lie algebra $\mathbf{R}^2$. Show that there exists on G no complex Lie group structure compatible with the real Lie group structure.

¶ 8. Let H be a complex Hilbert space of dimension $\aleph_0$ and G the unitary group of H considered as a real Lie group. The group G is simply connected.†

(*a*) Let Z be the centre of G, which is isomorphic to $\mathbf{T}$. Let a be an irrational

† Cf. N. H. Kuiper, The homotopy type of the unitary group of Hilbert space, *Topology*, **3** (1965), pp. 19–30. In this article it is even proved that G is contractible.

number. Let $\mathfrak{s}$ be the Lie subalgebra of $L(G) \times L(G)$ consisting of the (x, ax) where $x \in L(Z)$. Let S be the corresponding integral subgroup of $G \times G$. Then $\mathfrak{s}$ is an ideal of $L(G) \times L(G)$; however S is not closed and is dense in $Z \times Z$.

(*b*) Let $\mathfrak{g} = (L(G) \times L(G))/\mathfrak{s}$. There exists no Lie group with Lie algebra $\mathfrak{g}$. (Let H be such a group. As G is simply connected, there exists a morphism $\phi: G \times G \to H$ such that $L(\phi)$ is the canonical mapping of $L(G) \times L(G)$ onto $\mathfrak{g}$. Let $N = \operatorname{Ker} \phi$. Then $L(N) \supset \mathfrak{s}$, hence $N \supset S$ and therefore $N \supset Z \times Z$ by (*a*). Then $L(N) \supset L(Z) \times L(Z)$, which is absurd.)

9. Let X be a non-empty connected compact complex manifold of dimension n. Suppose that there exist holomorphic vector fields $\xi_1, \ldots, \xi_n$ on X which are linearly independent at each point of X. Show that there exist a complex Lie group G and a discrete subgroup D of G such that X is diffeomorphic to G/D. (Let c_{ijk} be the holomorphic functions on X such that $[\xi_i, \xi_j] = \sum_k c_{ijk}\xi_k$. The c_{ijk} are constants because X is compact. Take G to be the simply connected complex Lie group whose Lie algebra admits the c_{ijk} as constants of structure and use Theorem 5.)

10. Let G be a Lie group with a finite number of connected components. For G to be unimodular, it is necessary and sufficient that $\operatorname{Tr} \operatorname{ad} a = 0$ for all $a \in L(G)$.

11. Let E be a complete normable space over $\mathbf{C}$, $v \in \mathscr{L}(E)$ and $g = \exp(v)$. Suppose that $\operatorname{Sp}(v) \cap 2i\pi(\mathbf{Z} - \{0\}) = \emptyset$. Let E_1, E_2 be closed vector subspaces of E which are stable under v, such that $E_2 \subset E_1$. Suppose that the automorphism of E_1/E_2 derived from g is the identity. Then $v(E_1) \subset E_2$.

¶ 12. Consider a real or complex complete normable space E and a closed vector subspace F of E such that F^0 admits no topological supplement in E′.†

(*a*) Let A (resp. B) be the complete normable space of continuous endomorphisms of E (resp. F). Let C be the set of $u \in A$ such that $u(F) \subset F$. Let α be the mapping $u \mapsto (u, u \mid F)$ of C into $A \times B$. Then α is an isomorphism of the complete normable space C onto a closed vector subspace of $A \times B$ and $\alpha(C)$ admits no topological supplement in $A \times B$. (Let $x \in E$ be such that $x \notin F$. Let

† Let c_0 (resp. l^1, l^∞) be the Banach space of real or complex sequences $(x, x_2, \ldots)$ satisfying $\lim_{n\to\infty} x_n = 0$ $\left(\text{resp. } \sum_n |x_n| < +\infty, \text{ resp. } \sup_n |x_n| < +\infty\right)$ with the norm $\|x\| = \sup_n |x_n|$ $\left(\text{resp. } \sum_n |x_n|, \text{ resp. } \sup_n |x_n|\right)$. There exists a continuous morphism π of l^1 onto c_0; let F be its kernel. Then $(c_0)' = l^1$ is identified with F^0. But a vector subspace of countable type of l^∞ cannot be a topological direct factor in l^∞ (cf. A. Grothendieck, Sur les applications linéaires faiblement compactes d'espaces du type C(K), *Can. J. Math.*, **5** (1953), p. 169).

$\xi \in F^0$ be such that $\langle x, \xi \rangle = 1$. If $\eta \in E'$, let $\tilde{\eta}$ denote the element $y \mapsto \langle y, \eta \rangle x$ of A. Suppose that there exists a projector π of $A \times B$ onto $\alpha(C)$. We define $\tilde{\pi}: E' \to E'$ by $\tilde{\pi}(\eta) = {}^t(\alpha^{-1}(\pi(\tilde{\eta}, 0)))(\xi)$. Then $\tilde{\pi}$ is a projector of E' onto F^0, which is absurd.)

(*b*) A unital real or complex complete normable algebra M is constructed as follows: $M = \bigoplus_{i=0}^{\infty} M_i$ is graded by the M_i; M_0 is the set of scalar multiples of 1; M_1 is the set of scalar multiples of a non-zero element u; $M_2 = \{0\}$; $M_3 = F$; $M_4 = E$; $M_i = \{0\}$ for $i \geqslant 5$; $ux = x$ for all $x \in M_3$. Let N be the complete normable space of continuous endomorphisms of the complete normable space M. Let N_1 be the set of continuous derivations of M. Show, using (*a*), that N_1 has no topological supplement in N.

(*c*) Show that the group of bicontinuous automorphisms of M is a Lie quasi-subgroup of **GL**(M), but not a Lie subgroup of **GL**(M). (Use Corollary 2 to Proposition 18 and (*b*).)

13. Let G be a real Lie group, $L = L(G)$, and $\phi: L \to G$ a mapping which is differentiable at 0, such that $T_0(\phi) = \mathrm{Id}_L$ and such that $\phi(nx) = \phi(x)^n$ for all $x \in L$ and $n \in \mathbf{Z}$. Then $\phi = \exp_G$. (Let V be a neighbourhood of 0 in L and W a neighbourhood of e in G such that $\theta = \exp_G | V$ is an analytic isomorphism of V onto W. Let $\psi = \theta^{-1} \circ (\phi | \phi^{-1}(W))$. Then $T_0(\psi) = \mathrm{Id}_L$, whence $\psi = \mathrm{Id}_L$ using the equality $\psi\left(\frac{1}{n} x\right) = \frac{1}{n} \psi(x)$ for x sufficiently close to 0 and $n \in \mathbf{Z} - \{0\}$.

14. Let $\mathfrak{a}_1$ be the commutative Lie algebra $\mathbf{R}^4$, identified with $\mathbf{C}^2$. Let $\mathfrak{a}_2$ be the commutative Lie algebra **R**. Let ϕ be the homomorphism of $\mathfrak{a}_2$ into $\mathrm{Der}(\mathfrak{a}_1)$ which maps 1 to the derivation $(z_1, z_2) \mapsto (iz_1, i\sqrt{2}z_2)$. Let $\mathfrak{a}$ be the semi-direct product of $\mathfrak{a}_2$ by $\mathfrak{a}_1$ corresponding to ϕ.

(*a*) Show that $(\mathrm{Int}\ \mathfrak{a}) | \mathfrak{a}_1$ contains, for all $\phi \in \mathbf{R}$, the automorphism

$$(z_1, z_2) \mapsto (e^{i\phi} z_1, e^{i\sqrt{2}\phi} z_2).$$

(*b*) Show that the closure G of $(\mathrm{Int}\ \mathfrak{a}) | \mathfrak{a}_1$ in $\mathbf{GL}(\mathfrak{a}_1)$ contains, for all $(\phi, \phi') \in \mathbf{R}^2$, the automorphism $(z_1, z_2) \mapsto (e^{i\phi} z_1, e^{i\phi'} z_2)$.

(*c*) Show that L(G) contains the endomorphism $u: (z_1, z_2) \mapsto (z_1, 0)$ of $\mathfrak{a}_1$ and that there exists no $x \in \mathfrak{a}$ such that $u = (\mathrm{ad}\ x) | \mathfrak{a}_1$.

(*d*) Deduce that $\mathrm{Int}(\mathfrak{a})$ is not closed in $\mathrm{Aut}(\mathfrak{a})$.

15. Show that the finite-dimensional simply connected real Lie group of § 3, Exercise 7 (*a*) contains non-simply connected Lie subgroups (it contains in fact Lie groups isomorphic to **U**).

¶ 16. (*a*) Let $\mathfrak{g}$ be a complex Lie algebra. Let $\bar{\mathfrak{g}}$ be the complex Lie algebra derived from $\mathfrak{g}$ using the automorphism $\lambda \mapsto \bar{\lambda}$ of **C**. Let $\mathfrak{g}_0$ be the real Lie algebra derived from $\mathfrak{g}$ by restriction of scalars. Let $\mathfrak{g}' \supset \mathfrak{g}_0$ be the complex Lie algebra $\mathfrak{g}_0 \otimes_{\mathbf{R}} \mathbf{C}$. For all $x \in \mathfrak{g}$, we write

$$f(x) = \tfrac{1}{2}(x \otimes 1 - (ix) \otimes i) \in \mathfrak{g}', \qquad g(x) = \tfrac{1}{2}(x \otimes 1 + (ix) \otimes i) \in \mathfrak{g}'.$$

Show that f (resp. g) is an isomorphism of $\mathfrak{g}$ (resp. $\overline{\mathfrak{g}}$) onto an ideal $\mathfrak{m}$ (resp. $\mathfrak{n}$) of $\mathfrak{g}'$ and that the ideals $\mathfrak{m}$, $\mathfrak{n}$ are supplementary in $\mathfrak{g}'$. This defines projectors of $\mathfrak{g}'$ onto $\mathfrak{m}$, $\mathfrak{n}$. Show that, for all $x \in \mathfrak{g}$, $f(x) = p(x)$, $g(x) = q(x)$.

(*b*) Suppose that $\mathfrak{g}$ is finite-dimensional. Let G be the simply connected complex Lie group with Lie algebra $\mathfrak{g}$. Let $\overline{G}$ be the conjugate complex Lie group and G_0 the underlying real Lie group. Then $L(\overline{G}) = \overline{\mathfrak{g}}$ and $L(G_0) = \mathfrak{g}_0$. Let M (resp. N) be the simply connected complex Lie group with Lie algebra $\mathfrak{m}$ (resp. $\mathfrak{n}$). Then f defines an isomorphism ϕ of G onto M, g defines an isomorphism γ of $\overline{G}$ onto N and the simply connected complex Lie group G′ with with Lie algebra $\mathfrak{g}'$ is identified with M × N. Show that the real integral subgroup of G′ with Lie algebra $\mathfrak{g}_0$ is closed and simply connected and is identified with G_0. The restriction to $G \subset M \times N$ of pr_1 (resp. pr_2) is the isomorphism ϕ (resp. γ) of G (resp. $\overline{G}$) onto M (resp. N).

(*c*) Deduce from (*b*) and *Remark* 2 of no. 10 that G′, with the canonical injection of G_0 into G′, is the universal complexification of G_0.

(*d*) Let H be a connected complex Lie group with Lie algebra $\mathfrak{g}$, $\overline{H}$ the conjugate complex Lie group and H_0 the underlying real Lie group, such that G is the universal covering space of H. Let K be the (discrete) kernel of the canonical morphism of G onto H. Show that K is a central subgroup of G′. The canonical injection of G_0 into G′ therefore defines an injection i of H_0 into G′/N. Show that (G′/N, i) is the universal complexification of H_0. (Note that this ordered pair has the universal property of Proposition 20.)

(*e*) We write $G'/N = (H_0)_{\mathbf{C}}$. Deduce from (*d*) a canonical morphism ψ of $(H_0)_{\mathbf{C}}$ onto $H \times \overline{H}$, which is a covering mapping. Show, by the example $H = \mathbf{C}^*$, that ψ is not in general an isomorphism.

17. (*a*) Let G, $\overline{G}$, G_0 be as in Exercise 16 (*b*). Let V be a finite-dimensional complex vector space and ρ an irreducible linear representation of G_0 on V. Show that there exist finite-dimensional complex vector spaces X, Y, an analytic irreducible linear representation σ (resp. τ) of G (resp. $\overline{G}$) on X (resp. Y) and an isomorphism of V onto $X \otimes Y$ which transforms ρ into $\sigma \otimes \tau$. (Use Exercise 16 (*a*), Proposition 2 of Chapter I, § 2, and the Corollary to Proposition 8 of *Algebra*, Chapter VIII, § 7.)

(*b*) Show that the conclusion of (*a*) is not necessarily true if we omit the hypothesis that G is simply connected. (Consider the complex Lie group $\mathbf{C}^*$.)

18. Let A be a finite-dimensional connected commutative complex Lie group with Lie algebra $\mathfrak{a}$; let Λ be the kernel of $\exp_A$, so that A is identified with $\mathfrak{a}/\Lambda$.

(*a*) The following conditions are equivalent:

(a1) The canonical mapping $\mathbf{C} \otimes_{\mathbf{Z}} \Lambda \to \mathfrak{a}$ is injective.

(a2) A is isomorphic to a Lie subgroup of some $(\mathbf{C}^*)^n$.

(a3) A is isomorphic to a group $(\mathbf{C}^*)^p \times \mathbf{C}^q$.

(a4) A has a finite-dimensional faithful complex linear representation.

(a5) A has a finite-dimensional semi-simple faithful complex linear representation with closed image.

(*b*) The following conditions are equivalent:

(b1) The canonical mapping $\mathbf{C} \otimes_{\mathbf{Z}} \Lambda \to \mathfrak{a}$ is surjective.

(b2) A is isomorphic to a quotient of a group $(\mathbf{C}^*)^n$.

(b3) No direct factor of A is isomorphic to $\mathbf{C}$.

(b4) Every complex linear representation of A is semi-simple.

(c) The following conditions are equivalent:

(c1) The canonical mapping $\mathbf{C} \otimes_{\mathbf{Z}} \Lambda \to \mathfrak{a}$ is bijective.

(c2) A is isomorphic to a group $(\mathbf{C}^*)^n$.

(*d*) Let F be a finite subgroup of A and let $A' = A/F$. Show that A satisfies conditions (a_i) (resp. (b_i), (c_i)) if and only if A′ does.

19. Let G be a real Lie group. Show that there exists a neighbourhood V of 0 in L(G) such that, for x, y in V:

$$\exp(t(x+y)) = \lim_{n\to+\infty} \left(\exp\frac{tx}{n}.\exp\frac{ty}{n}\right)^n$$

$$\exp(t^2[x,y]) = \lim_{n\to+\infty} \left(\exp\frac{tx}{n}.\exp\frac{ty}{n}.\exp\frac{-tx}{n}.\exp\frac{-ty}{n}\right)^{n^2}$$

uniformly for $t \in [0, 1]$.

20. Let G be a Lie group, H a Lie subgroup of G and A an integral subgroup of G such that $L(H) \cap L(A) = \{0\}$. Show that $H \cap A$ is discrete in the Lie group A.

21. Let G be a real Lie group and H a normal 1-dimensional integral subgroup.

(*a*) If H is not closed, $\overline{H}$ is compact (*Spectral Theories*, Chapter II, § 2, Lemma 1), hence isomorphic to T^n and hence central in G if G is connected (use *General Topology*, Chapter VII, § 2, Proposition 5).

(*b*) Suppose that H is closed. Let a be an element of G and $C(a)$ the set of elements of G which commute with a. Suppose that $H \not\subset C(a)$.

If H is isomorphic to $\mathbf{R}$, then $C(a) \cap H = \{e\}$. (Consider the automorphism $\alpha: h \mapsto a^{-1}ha$ of H.) If H is isomorphic to $\mathbf{T}$, $C(a) \cap H$ has two elements. (Consider α again and use *General Topology*, Chapter VII, § 2, Proposition 6.) The second situation is impossible if G is connected (use (*a*)).

(*c*) With the hypotheses of (*b*), suppose further that G/H is commutative.

Then $G = C(a).H$. (Observe that the mapping $h \mapsto h^{-1}\alpha(h)$ of H into H is surjective.)

22. Let G be a real Lie group, H a normal 1-dimensional integral subgroup and A a closed subgroup of G such that AH is not closed in G. Then H is central in the identity component of $\overline{AH}$. (Reduce it to the case where $G = \overline{AH}$ and G is connected. Then the set B of elements of A which commute with H is a normal subgroup of G. Passing to the quotient by B, reduce it to the case $B = \{e\}$. As every commutator in G commutes with H, A is then commutative. By assuming that H is closed and using Exercise 21 (*c*), show that AH would be closed, whence a contradiction. Hence H is not closed and Exercise 21 (*a*) can be used.)

¶ 23. Let G be a real Lie group and $c = (U, \phi, E)$ a chart on the manifold G centred on the identity element e. If $x \in U$, let $|x|_c$ denote the norm of $\phi(x)$ in the Banach space E.

(*a*) Show that, if $c' = (U', \phi', E')$ is another chart centred on e, there exist constants $\lambda > 0$ and $\mu > 0$ such that

$$\mu|x|_c \leqslant |x|_{c'} \leqslant \lambda|x|_c$$

for all $x \in G$ sufficiently close to e.

(*b*) Show that, for all $\rho > 0$, there exists a neighbourhood U_ρ of e contained in U such that, for x, y in U_ρ, $(x, y) \in U_\rho$ and

$$|(x, y)|_c \leqslant \rho . \inf(|x|_c, |y|_c).$$

(*c*) Suppose that G is finite-dimensional. Let Γ be a discrete subgroup of G. Apply (*b*) with $0 < \rho < 1$ and choose U_ρ to be relatively compact. The set $U_\rho \cap \Gamma$ is finite. Let

$$\{e = \gamma_0, \gamma_1, \ldots, \gamma_m\}$$

be its elements, numbered so that

$$|\gamma_0|_c \leqslant |\gamma_1|_c \leqslant \cdots \leqslant |\gamma_m|_c.$$

Show that, if $i \leqslant m$ and $j \leqslant m$, the commutator (γ_i, γ_j) is equal to one of the γ_k with $k < \inf(i, j)$. Deduce that the subgroup generated by $U_\rho \cap \Gamma$ is nilpotent of class $\leqslant m$.

Deduce from the above the existence of a neighbourhood V of e such that, for every discrete subgroup Γ of G, there exists a nilpotent integral subgroup N of G containing $V \cap \Gamma$.

(*d*) Suppose that G is connected and finite-dimensional and contains an increasing sequence of discrete subgroups D_n whose union is dense in G. Then G is nilpotent. (Using (*c*), prove the existence of a central one-parameter subgroup H which meets D_n for sufficiently large n at a point distinct from e. Argue by induction on dim G, distinguishing two cases according to whether H is closed or relatively compact (Exercise 21 (*a*)).)

24. Let G, G′ be finite-dimensional real Lie groups, f a surjective morphism of G into G′, N its kernel, H an integral subgroup of G and $H' = f(H)$.

(*a*) Suppose that N is finite. For H′ to be closed, it is necessary and sufficient that H be closed.

(*b*) Suppose that N is compact. If H is closed, H′ is closed.

¶ 25. Let G be a finite-dimensional real or complex Lie group. Let $S \subset L(G)$. Suppose that L(G) is the Lie algebra generated by S and that S is stable under homotheties.

(*a*) Let H be the subgroup of G generated by exp S. Show that H is open in G. (Let A be the set of $x \in L(G)$ such that $\exp(Kx) \subset H$ and B the vector subspace of L(G) generated by A. Show that $[A, S] \subset A$ and then that $[A, A] \subset B$. Deduce that B is a subalgebra of L(G) and then use Proposition 3 of § 4.)

(*b*) Suppose further that

$$(x \in S \text{ and } y \in S) \Rightarrow ((\mathrm{Ad}\exp x)(y) \in S).$$

Then the vector subspace V of L(G) generated by S is equal to L(G). (Using (*a*), show that $[L(G), V] \subset V$.)

26. Let G be an n-dimensional connected compact complex Lie group, X a finite-dimensional connected complex analytic manifold and $(g, x) \mapsto gx$ a law of analytic left operation of G on X. For all $g \in G$, let $\rho(g)$ be the mapping $x \mapsto gx$ of X into X. Suppose, that, for all $g \in G$ distinct from e, $\rho(g) \neq \mathrm{Id}_X$. Then every orbit of G in X is an n-dimensional closed submanifold of X. (Let $x \in X$, H be the stabilizer of x and H′ the identity component of H. For all $g \in H'$, let $u(g)$ be the tangent mapping at x to $\rho(g)$. Arguing as in Proposition 6 of no. 3, show that $u(g)$ is the identity for all $g \in H'$. Using § 1, Exercise 4 and the connectedness of X, deduce that $H' = \{e\}$.)

¶ 27. Let G be a compact real Lie group, M a compact manifold of class C^2 and J an open interval of **R** containing 0. Let $(s, (x, \xi)) \mapsto (m_\xi(s, x), \xi)$ be a mapping of class C^2 of $G \times (M \times J)$ into $M \times J$ under which G operates on $M \times J$ on the left.

(*a*) Let X be a vector field of class C^1 on $M \times J$ such that, for all $(x, \xi) \in M \times J$, the second projection of $X_{(x,\xi)}$ is the tangent vector 1 to J. Translating X by G and integrating over G, deduce the existence of a field X′ with the same properties as X, which is moreover invariant under G.

(*b*) Show that there exists a diffeomorphism $(x, \xi) \mapsto (h_\xi(x), \xi)$ of $M \times J$ onto itself such that

(i) for all $\xi \in J$, h_ξ is a diffeomorphism of M onto M;

(ii) $m_\xi(s, x) = h_\xi(m_0(s, h_\xi^{-1}(x)))$ for $s \in G$, $x \in M$, $\xi \in J$.

(Use (*a*) and Theorem 5 of no. 8.)

28. Give an example of a finite-dimensional real Lie group G and two integral subgroups A, B of G such that $L(G) = L(A) + L(B)$ but AB is not a subgroup of G (cf. *Integration*, Chapter VII, § 3, no. 3, Proposition 6).

29. Let G be a finite-dimensional complex connected Lie group, G_0 the underlying real Lie group and H an integral subgroup of G_0. Show that there exists a smallest integral subgroup H^* of G containing H. Give an example where H is closed in G_0 but H^* is not closed in G (take $G = \mathbf{C}^2/\mathbf{Z}^2$).

30. Let G be a compact connected complex Lie group and hence of the form $\mathbf{C}^n/D$, where D is a discrete subgroup of $\mathbf{C}^n$ of rank $2n$.

(*a*) Show that every holomorphic differential 1-form ω on G is invariant. (Let $\pi: \mathbf{C}^n \to G$ be the canonical morphism. Let $\zeta_1, \ldots, \zeta_n$ be the coordinate functions on $\mathbf{C}^n$. Then $\pi^*(\omega)$ is of the form $\sum_j a_j d\zeta_j$, where the a_j are holomorphic functions on $\mathbf{C}^n$ which are invariant under D and hence constant.)

(*b*) Let $G' = \mathbf{C}^n/D'$, where D' is a discrete subgroup of $\mathbf{C}^n$ of rank $2n$. Show that every *analytic manifold* isomorphism $u: G \to G'$ is of the form $s \mapsto v(s) + a'$, where v is a *Lie group* isomorphism and $a' \in G'$. (Let $\pi': \mathbf{C}^n \to G'$ be the canonical morphism. There exists an isomorphism of analytic manifolds $\tilde{u}: \mathbf{C}^n \to \mathbf{C}^n$ such that $u \circ \pi = \pi' \circ \tilde{u}$. For every holomorphic differential 1-form ω' on G', $u^*(\pi'^*(\omega'))$ is a 1-form on $\mathbf{C}^n$ which is invariant under translations by (*a*). Deduce that $\tilde{u}$ is an affine mapping.)

§ 7

1. Let G be a Lie group and $\phi: U \to G$ an exponential mapping such that $\mathbf{Z}U \subset U$ and $\phi(rx) = \phi(x)^r$ for all $x \in U$ and all $r \in \mathbf{Z}$. If $p > 0$, ϕ is an analytic isomorphism of U onto $\phi(U)$. (If $\phi(x) = \phi(y)$, then $\phi(p^n x) = \phi(p^n y)$ for all $n \in \mathbf{N}$ and hence $x = y$. Let W be an open neighbourhood of 0 in L(G) such that ϕ^{-1} is analytic on $\phi(W)$. For all $s \in \phi(U)$, there exist $n \in \mathbf{N}$ and a neighbourhood V of s in $\phi(U)$ such that $t \in V \Rightarrow t^{p^n} \in \phi(W)$.)

2. Let G be a Lie group, $\phi: U \to G$ an exponential mapping with the properties of Proposition 3, V a neighbourhood of 0 in U such that $\mathbf{Z}V \subset V$ and $\psi: V \to G$ a tangent mapping at 0 to ϕ such that $\psi(nx) = \psi(x)^n$ for all $n \in V$ and all $n \in \mathbf{Z}$. If $p > 0$, then $\psi = \phi \mid V$. (Give L(G) a norm. Let $x \subset V$. Then $p^n x$ tends to 0 as n tends to $+\infty$, hence there exists $\alpha_n > 0$ such that α_n tends to 0 and $\|(\phi^{-1} \circ \psi)(p^n x) - p^n x\| \leqslant \alpha_n \|p^n x\|$. But $\psi(p^n x) = \psi(x)^{p^n}$, whence $\|(\phi^{-1} \circ \psi)(x) - x\| \leqslant \alpha_n \|x\|$.)

3. Let U be the set of invertible elements of A and $U' = 1 + \mathfrak{m} \subset U$.

(*a*) Show that $U' \subset U_f$. (If $x = 1 + y$ with $y \in \mathfrak{m}$, then x^{p^n} tends to 1 as n tends to $+\infty$ by the binomial theorem.)

(*b*) Show that U_f is the set of elements of U whose image in $A/\mathfrak{m}$ is a root of unity. (Use (*a*).) Hence recover the fact that $U = U_f$ when K is locally compact ($A/\mathfrak{m}$ is then finite).

4. Let $n \in \mathbf{N}^*$, p be a prime number and G the set of matrices belonging to $\mathbf{GL}(n, \mathbf{Z}_p)$ all of whose elements are congruent to 1 modulo p if $p \neq 2$ (resp. modulo 4 if $p = 2$). Then G is an open subgroup of $\mathbf{GL}(n, \mathbf{Z}_p)$.

(*a*) Show that G has no element of finite order $\neq 1$.

(*b*) Show that every finite subgroup of $\mathbf{GL}(n, \mathbf{Z}_p)$ is isomorphic to a subgroup of $\mathbf{GL}(n, \mathbf{Z}/p\mathbf{Z})$ if $p \neq 2$ (resp. $\mathbf{GL}(n, \mathbf{Z}/4\mathbf{Z})$ if $p = 2$).

¶ 5. (*a*) Let Γ be a compact subgroup of $G = \mathbf{GL}(n, \mathbf{Q}_p)$. Show that there exists a conjugate of Γ which is contained in $\mathbf{GL}(n, \mathbf{Z}_p)$. (If T is a lattice of $\mathbf{Q}_p^n$ with respect to $\mathbf{Z}_p$ (*Commutative Algebra*, Chapter VII, § 4, Definition 1), show that the stabilizer of T in Γ is open in Γ and hence of finite index and $\sum_{\gamma \in \Gamma} \gamma T$ is a lattice which is stable under Γ.)

(*b*) Deduce that G_f is the union of the conjugates of $\mathbf{GL}(n, \mathbf{Z}_p)$.

(*c*) Deduce that every finite subgroup Φ of $\mathbf{GL}(n, \mathbf{Q}_p)$ has order a divisor of $a_n(p)$, where $a_n(p)$ is defined by

$$a_n(p) = (p^n - 1)(p^n - p)\dots(p^n - p^{n-1}) \quad \text{if } p \neq 2$$
$$a_n(p) = 2^{n^2}(2^n - 1)(2^n - 2)\dots(2^n - 2^{n-1}) \quad \text{if } p = 2.$$

(Using (*a*), it can be assumed that $\Phi \subset \mathbf{GL}(n, \mathbf{Z}_p)$. Then use Exercise 4.)

(*d*) Show that every finite subgroup of $\mathbf{SL}(n, \mathbf{Q}_p)$ has order a divisor of $s_n(p)$, where $s_n(p) = \dfrac{a_n(p)}{p-1}$ for $p \neq 2$ and $s_n(2) = \dfrac{a_n(2)}{2}$.

¶ 6. In this exercise, l denotes a prime number $\neq 2$. If a is an integer $\neq 0$, let $v_l(a)$ denote the l-adic valuation of a, i.e. the largest integer e such that $a \equiv 0 \pmod{l^e}$.

(*a*) Let m be an integer $\geqslant 1$. We write

$$\varepsilon(l, m) = 0 \qquad \text{if } m \not\equiv 0 \pmod{(l-1)}$$

$$\varepsilon(l, m) = v_l\left(\frac{m}{l-1}\right) + 1 \qquad \text{if } m \equiv 0 \pmod{(l-1)}.$$

Show that, if x is an integer prime to l, then

$$v_l(x^m - 1) \geqslant \varepsilon(l, m)$$

and that there is equality if the image of x in the cyclic group $(\mathbf{Z}/l^2\mathbf{Z})^*$ is a generator of this group.

(*b*) Let n be an integer $\geqslant 1$. We write

$$r(l, n) = \sum_{m=1}^{n} \varepsilon(l, m).$$

Show that

$$r(l, n) = \left[\frac{n}{l-1}\right] + \left[\frac{n}{l(l-1)}\right] + \left[\frac{n}{l^2(l-1)}\right] + \cdots$$

where the symbol $[\alpha]$ denotes the integral part of the real number α. Show that, if x is an integer prime to l, then

$$v_l((x^n - 1)(x^{n-1} - 1)\dots(x - 1)) \geqslant r(l, n)$$

and that there is equality if the image of x in $(\mathbf{Z}/l^2\mathbf{Z})^*$ is a generator of this group.

(*c*) In the notation of Exercise 5 (*c*), show that $l^{r(l,n)}$ is the greatest power of l which divides all the $a_n(p)$, for p prime $\neq l$ (or for all sufficiently large p, which amounts to the same). (Use (*b*) applied to $x = p$; then choose p such that its image in $(\mathbf{Z}/l^2\mathbf{Z})^*$ is a generator of this group, which is possible by the arithmetic progression theorem.†)

(*d*) Let Γ be a finite subgroup of $\mathbf{GL}(n, \mathbf{Q})$ and let l^e be the greatest power of l dividing the order of Γ. Show that $e \leqslant r(l, n)$. (Use Exercise 5 to prove that l^e divides all the $a_n(p)$ and then apply (*c*) above.)

(*e*) Conversely, show that there exists a finite l-subgroup $\Gamma_{l,n}$ of $\mathbf{GL}(n, \mathbf{Q})$ whose order is $l^{r(l,n)}$. (Reduce it to the case where n is of the form $l^a(l-1)$, with $a \geqslant 0$. Decompose $\mathbf{Q}^n$ as a sum of l^a copies of $\mathbf{Q}^{l-1}$ and use this decomposition to give an operation on $\mathbf{Q}^n$ of the semi-direct product $\mathrm{H}_{l,n}$ of the symmetric group $\mathfrak{S}_{l^a}$ and the group $(\mathbf{Z}/l\mathbf{Z})^{l^a}$. Take $\Gamma_{l,n}$ to be a Sylow l-group of $\mathrm{H}_{l,n}$.)

Show that, if n is even, $\Gamma_{l,n}$ is contained in a conjugate of the symplectic group $\mathbf{Sp}(n, \mathbf{Z})$.

(*f*) *Let Γ be a finite subgroup of $\mathbf{GL}(n, \mathbf{Q})$ which is an l-group. Show that Γ is conjugate to a subgroup of $\Gamma_{l,n}$. (Show first, using a suitable reduction modulo p, that the reduction of Γ is a finite subgroup of the reduction of a conjugate of $\Gamma_{l,n}$; then use the character of the representation of Γ on $\mathbf{Q}^n$.) In particular, every subgroup of $\mathbf{GL}(n, \mathbf{Q})$ of order $l^{r(l,n)}$ is conjugate to $\Gamma_{l,n}$.*

¶ 7. In this exercise, let $v_2(a)$ denote the 2-adic valuation of an integer a.

(*a*) Let Γ be a finite subgroup of $\mathbf{GL}(n, \mathbf{Q})$. Show that there exists a positive definite quadratic form with coefficients in $\mathbf{Z}$, which is invariant under Γ. Deduce (by the same argument as in Exercises 4 and 5) that, for all sufficiently large p, Γ is isomorphic to a subgroup of an orthogonal group $\mathbf{O}(n)$ over the field $\mathbf{F}_p$ (resp. a subgroup of $\mathbf{SO}(n)$ if Γ is contained in $\mathbf{SL}(n, \mathbf{Q})$).

(*b*) Suppose that Γ is contained in $\mathbf{SL}(n, \mathbf{Q})$ and let 2^e denote the greatest power of 2 dividing the order of Γ. Show that, if n is odd, 2^e divides all the integers

$$b_n(p) = (p^{n-1} - 1)(p^{n-3} - 1)\dots(p^2 - 1)$$

for sufficiently large prime p. (Use (*a*) and Exercise 13 of *Algebra*, Chapter IX, § 6.)

---

† For a proof of this theorem, see for example A. Selberg, *Ann. of Math.*, **50** (1949), pp. 297–304.

Show that, if n is even and p is a sufficiently large prime, 2^e divides the l.c.m. $b_n(p)$ of the integers

$$(p^n - 1)(p^{n-2} - 1)\dots(p^2 - 1)/(p^{n/2} + 1)$$
$$(p^n - 1)(p^{n-2} - 1)\dots(p^2 - 1)/(p^{n/2} - 1).$$

(Same method as for n odd.)

(*c*) Under the hypotheses of (*b*), we write

$$r(2, n) = n + \left[\frac{n}{2}\right] + \left[\frac{n}{4}\right] + \cdots = n + v_2(n!).$$

Let A be an integer $\geqslant 3$. Show that $2^{r(2,n)-1}$ is the greatest power of 2 which divides all the $b_n(p)$ for $p \geqslant A$. (Same method† as in Exercise 6; use the existence of a prime number $p \geqslant A$ such that $p \equiv 5 \pmod 8$.) Deduce the inequality $e \leqslant r(2, n) - 1$.

(*d*) Conversely, let C_n be the subgroup of $\mathbf{GL}(n, \mathbf{Z})$ generated by the permutation matrices and the diagonal matrices with coefficients ± 1. The order of C_n is $2^n n!$ and $v_2(2^n n!) = r(2, n)$. If $\Gamma_{2,n}$ denotes the intersection of a Sylow 2-group of C_n with $\mathbf{SL}(n, \mathbf{Z})$, deduce that the order of $\Gamma_{2,n}$ is $2^{r(2,n)-1}$.

8. Let n be an integer $\geqslant 1$. We write

$$M(n) = \prod_l l^{r(l,n)},$$

where the product is taken over all prime numbers l and the $r(l, n)$ are defined as in Exercises 6 and 7.

Then $M(1) = 2$, $M(2) = 2^3.3 = 24$, $M(3) = 2^4.3 = 48$,

$$M(4) = 2^7.3^2.5 = 5760.$$

Deduce from Exercises 6 and 7 that the l.c.m. of the orders of the finite subgroups of $\mathbf{GL}(n, \mathbf{Q})$ (or $\mathbf{GL}(n, \mathbf{Z})$, which amounts to the same) is equal to $M(n)$. Consider the same question for $\mathbf{SL}(n, \mathbf{Q})$ with $M(n)$ replaced by $\frac{1}{2}M(n)$.

9. Suppose that K is locally compact. Let $G = \mathbf{GL}(n, K)$. Let G_1 be the set of $g \in G$ which leaves stable a lattice of K^n with respect to A. Let G_2 be the set of $g \in G$ which generate a relatively compact subgroup in G. Let G_3 be the set of $g \in G$ whose eigenvalues in an algebraic closure of K are of

† For more details on this exercise and on the preceding one, see H. Minkowski, *Gesamm. Abh.*, Leipzig-Berlin, Teubner, 1911 (Bd I, S. 212–218) and W. Burnside, *Theory of groups of finite order* (2nd ed.), Cambridge Univ. Press, 1911 (pp. 479–484).

absolute value 1. Then $G_1 = G_2 = G_3 = G_f$. (Use the argument of Exercise 5 (*a*).)

10. Suppose that K is locally compact. Let G be a standard group of dimension n over K. Let μ be a Haar measure on the additive group K^n. Show that $\mu|G$ is a left and right Haar measure on G. (Use the fact that G is the inverse limit of the $G(\mathfrak{a}_\lambda)$ and *Integration*, Chapter VII, § 1, Proposition 7.)

§ 8

1. The mapping $z \mapsto \bar{z}$ of **C** into **C** is a non-analytic continuous automorphism of the complex Lie group **C**.

2. Let G be the Hilbert space of sequences $(\lambda_1, \lambda_2, \ldots)$ of real numbers such that $\sum_i \lambda_i^2 < +\infty$. Consider G as a real Lie group. Let G_n be the set of $(\lambda_1, \lambda_2, \ldots) \in G$ such that $\lambda_m \in \frac{1}{m}\mathbf{Z}$ for $1 \leqslant m \leqslant n$. The G_n are closed Lie subgroups of G and hence $H = \bigcap_n G_n$ is a closed subgroup of G. But this group is totally disconnected and not discrete and hence is not a Lie subgroup of G.

¶ 3. In $\mathbf{Q}_p \times \mathbf{Q}_p$, every closed subset can be defined by a family of analytic equations. Deduce that Corollary 2 (ii) of Theorem 2 is no longer true if the hypothesis that I is finite is omitted.

¶ 4. Let G be a finite-dimensional real Lie group and A a subgroup of G. We say that an element x of L(G) is A-accessible if, for every neighbourhood U of e, there exists a continuous mapping α of $[0, 1]$ into A such that $\alpha(0) = e$ and $\alpha(t) \in \exp(tx).U$ for $0 \leqslant t \leqslant 1$. Let $\mathfrak{h}$ be the set of A-accessible elements of L(G).

(*a*) Show that $\mathfrak{h}$ is a Lie subalgebra of L(G). (Use Exercise 19 of § 6.)

(*b*) Let H be an integral subgroup of G such that $L(H) = \mathfrak{h}$. Show that $H \subset A$. (Let $I = [-1, 1]$ and let $\mathbf{R}^r$ be given the Euclidean norm. Let $(x_1, \ldots, x_r)$ be a basis of $\mathfrak{h}$. Construct continuous mappings $\alpha_1, \ldots, \alpha_r$ of I into A and continuous mappings $f_1, \ldots, f_r$ from I^r to **R**, such that, for $t = (t_1, \ldots, t_r) \in I^r$,

$$\alpha_1(t_1) \ldots \alpha_r(t_r) = \exp(f_1(t)x_1) \ldots \exp(f_r(t)x_r)$$
$$\|t - (f_1(t), \ldots, f_r(t))\| \leqslant \tfrac{1}{2}.$$

Then apply the following theorem: let f be a continuous mapping of I^r into $\mathbf{R}^r$ such that $\|f(x) - x\| \leqslant \frac{1}{2}$ for all $x \in I^r$; then $f(I^r)$ contains a neighbourhood of 0 in $\mathbf{R}^r$.†)

† This follows from the Brouwer fixed point theorem, which can be consulted, for example, in N. Dunford and J. T. Schwartz, *Linear operators*, part I (Interscience publishers, 1958), pp. 467–470.

(*c*) Deduce that $\mathfrak{h}$ is the subalgebra tangent to A at e.
(*d*) Show that if A is arcwise connected, then A = H.†

¶ 5. Let G be a Hausdorff topological group, H a closed subgroup of G and π the canonical mapping of G onto G/H. Suppose that H is a finite-dimensional real Lie group. There exist a neighbourhood U of $\pi(e)$ in G/H and a continuous mapping σ of U into G such that $\pi \circ \sigma = \mathrm{Id}_U$. (Let ρ be an analytic linear representation of H on $\mathbf{GL}(n, \mathbf{R})$, which is locally a homeomorphism (§ 6, Corollary to Theorem 1). Let f be a continuous function $\geqslant 0$ on G, equal to 1 at e and zero outside a sufficiently small neighbourhood of e. Let ds be a left Haar measure of H. For $x \in G$, we write

$$g(x) = \int f(xs)\rho(s)^{-1}ds \in \mathbf{M}_n(\mathbf{R}).$$

Then $g(xt) = g(x)\rho(t)$ for $x \in G$ and $t \in H$. If V is sufficiently small,

$$g(x) \in \mathbf{GL}(n, \mathbf{R})$$

for x sufficiently close to e. Finally use the fact that the theorem to be proved is true locally for $\mathbf{GL}(n, \mathbf{R})$ and $\rho(H)$.

6. Let G be a group of class C^r (§ 5, Exercise 1) with $r \geqslant 2$. There exists on G one and only one real Lie group structure S such that the underlying manifold structure of class C^r of S is the given structure. (The uniqueness of S follows from Corollary 1 to Theorem 1. Let L(G) be the normable Lie algebra associated with G in Exercise 2 of § 5. There exist a real Lie group germ G′ and an isomorphism h of L(G′) onto L(G) (§ 4, Theorem 3). Verify as in § 4, no. 1 that there exist a symmetric open neighbourhood G″ of e_G in G′ and a mapping ϕ of class C^r of G″ into G such that $T_e(\phi) = h$ and $\phi(g_1 g_2) = \phi(g_1)\phi(g_2)$ for g_1, g_2 in G″. By shrinking G″, it can be assumed that $V = \phi(G'')$ is open in G and that ϕ is an isomorphism of class C^r of the manifold G″ onto the manifold V. There therefore exists on V a real Lie group germ structure such that the underlying manifold structure of class C^r is the given structure. For all $g \in G$, Int g defines as analytic mapping of $V \cap (g^{-1}Vg)$ onto $(gVg^{-1}) \cap V$ (Theorem 1). By Proposition 18 of § 1, there exists on G a real Lie group structure S inducing the same analytic structure as V on an open neighbourhood of e. By translation, the underlying manifold structure of class C^r of S on G is the given structure.)

7. Let G be a real Lie group and H a closed subgroup of G.
(*a*) Let $\mathfrak{h}$ be the set of $x \in L(G)$ such that $\exp(tx) \in H$ for all $t \in \mathbf{R}$. Then $\mathfrak{h}$ is a Lie subalgebra of L(G). (Use Proposition 8 of § 6.)

---

† For more details, cf. M. Goto, On an arcwise connected subgroup of a Lie group, *Proc. Amer. Math. Soc.*, **20** (1969), pp. 157–162.

(*b*) Suppose that H is locally compact. Show that H is a Lie subgroup of G. (Show first that $\mathfrak{h}$ is finite-dimensional by proving the existence in $\mathfrak{h}$ of a precompact neighbourhood of 0. Then imitate the proof of Theorem 2, choosing V_2 such that $(\exp V_2) \cap H$ is relatively compact.)

§ 9

1. Let $G = \mathbf{SO}(3, \mathbf{R})$. Let Δ_1, Δ_2 be two orthogonal lines in $\mathbf{R}^3$ and H_i the subgroup of G consisting of the rotations about Δ_i $(i = 1, 2)$. Then $[L(H_1), L(H_2)]$ is a 1-dimensional Lie subalgebra of G distinct from the Lie subalgebra tangent to (H_1, H_2) at *e*.

2. Suppose that K is ultrametric. Let G be a finite-dimensional Lie group, $\mathfrak{g}$ its Lie algebra and A a finite subset of G. Then $Z_G(A)$ is a Lie subgroup of G with Lie algebra $\mathfrak{z}_{\mathfrak{g}}(A)$. (Argue as in Proposition 8.)

3. Let G be a connected real or complex Lie group. The centre Z of G is a Lie quasi-subgroup of G and L(Z) is the centre of L(G).

4. Suppose that K is ultrametric and $p > 0$ (in the notation of § 7). Let G be a finite-dimensional Lie group, A a group of automorphisms of G and B the corresponding group of automorphisms of L(G). Let G^A (resp. $L(G)^B$) be the set of elements of G (resp. L(G)) which are fixed under A (resp. B). Then G^A is a Lie subgroup of G with Lie algebra $L(G)^B$. (Use the logarithmic mapping.)

5. Let G be a finite-dimensional connected real or complex Lie group. Let $(G_0, G_1, \ldots)$ be the upper central series of G (Chapter II, § 4, Exercise 18) and let $(\mathfrak{g}_0, \mathfrak{g}_1, \ldots)$ be the upper central series of the Lie algebra L(G) (Chapter I, § 1, no. 6). Then, for all *i*, G_i is a Lie subgroup of G such that $L(G_i) = \mathfrak{g}_i$.

6. Let Γ be the 3-dimensional simply connected nilpotent real Lie group defined in Exercise 5 (*b*) of § 4. Let $\alpha \in \mathbf{R}$ be an irrational number. Let P be the discrete subgroup of $\Gamma \times \mathbf{R}$ consisting of the $((0, 0, x), \alpha x)$, where $x \in \mathbf{Z}$. Let $G = (\Gamma \times \mathbf{R})/P$. Show that (G, G) is not closed in G.

7. (*a*) Let G be a semi-simple connected real Lie group, Z its centre and ρ a continuous linear representation of G on a finite-dimensional complex vector space. There exists an integer p such that, for all $z \in Z$, $\rho(z)$ is diagonalizable and all the eigenvalues of $\rho(z)$ are p-th roots of unity. (It can be assumed that ρ is irreducible. Then $\rho(z)$ is scalar by Schur's Lemma. On the other hand, $\det \rho(g) = 1$ for all $g \in G$ because $G = \mathscr{D}G$.)

(*b*) Deduce that, if G admits an injective finite-dimensional continuous linear representation, Z is finite.

¶ 8. Let G be a finite-dimensional connected real Lie group, $\mathfrak{g}$ its Lie algebra, $\mathfrak{n}$ the largest nilpotent ideal of $\mathfrak{g}$ and $\mathfrak{h} = [\mathfrak{g}, \mathfrak{g}] + \mathfrak{n}$.

(*a*) $\mathfrak{h}$ is a characteristic ideal of $\mathfrak{g}$; the radical of $\mathfrak{h}$ is $\mathfrak{n}$; for all $x \in \mathfrak{h}$, $\mathrm{Tr}\, \mathrm{ad}_{\mathfrak{h}} x = 0$; for every Levi section $\mathfrak{l}$ of $\mathfrak{g}$, $\mathfrak{h} = \mathfrak{l} + \mathfrak{n}$.

(*b*) Suppose that G is simply connected. Let Z be its centre, H the Lie subgroup of G with Lie algebra $\mathfrak{h}$ and ϕ the canonical morphism of G onto G/H. Then $\phi(Z)$ is discrete in G/H. (The group G is the semi-direct product of a semi-simple Lie group S and its radical R. Let $z \in Z$. Then $z = y^{-1}x$, where $x \in R$ and y belongs to the centre of S. There exists an integer p such that the eigenvalues of $\mathrm{Ad}_{\mathfrak{g}} y$, and hence also of $\mathrm{Ad}_{\mathfrak{g}} x$, are p-th roots of unity (Exercise 7). Deduce that, if N is the Lie subgroup of G with Lie algebra $\mathfrak{n}$, there exists a neighbourhood U of e in R satisfying $U = UN = NU$,

$$Z \cap (SU) \subset SN = H.)$$

(*c*) We no longer assume that G is simply connected. Let H be the integral subgroup of G with Lie algebra $\mathfrak{h}$. Show that H is a unimodular normal Lie subgroup of G with nilpotent radical, such that G/H is commutative. (Use (*a*) and (*b*).)

(*d*) Deduce that, if (G, G) is dense in G, the radical of G is nilpotent.

9. Let G be the universal covering of $\mathbf{SL}(2, \mathbf{R})$. We identify the kernel of the canonical morphism $G \to \mathbf{SL}(2, \mathbf{R})$ with $\mathbf{Z}$ (§ 6, Exercise 2). Let a be an element of $\mathbf{T}^n$ whose powers are everywhere dense in $\mathbf{T}^n$ (*General Topology*, Chapter VII, § 1, Corollary 2 to Proposition 7). Let D be the discrete subgroup of $G \times \mathbf{T}^n$ generated by $(1, a)$. Let $H = (G \times \mathbf{T}^n)/D$. Then

$$L(H) = \mathfrak{sl}(2, \mathbf{R}) \times \mathbf{R}^n$$

and the integral subgroup H′ of H with Lie algebra $\mathfrak{sl}(2, \mathbf{R})$ is isomorphic to G and dense in H. $D^n H' = H'$ for all $n \geqslant 0$.

¶ 10. Let G be a finite-dimensional real or complex Lie group. Let

$$p = \dim[L(G), L(G)].$$

Let (G, G) be given its structure as an integral subgroup of G. There exists a neighbourhood V of e in (G, G) such that every element of V is the product of p commutators of elements of G. (Let $x_1, y_1, \ldots, x_p, y_p$ be elements of L(G) such that the $[x_i, y_i]$ form a basis of L(G). For s, t in $\mathbf{R}$, we write

$$\rho_i(s, t) = (\exp sy_i)^{-1}(\exp tx_i)^{-1}(\exp sy_i)(\exp tx_i).$$

Apply the implicit function theorem to the mapping

$$(s_1, t_1, \ldots, s_p, t_p) \mapsto \rho_1(s_1, t_1) \ldots \rho_p(s_p, t_p)$$

of $\mathbf{R}^{2p}$ into (G, G).)

11. Let G be the semi-direct product of $B = \mathbf{Z}/2\mathbf{Z}$ by K corresponding to the automorphism $x \mapsto -x$ of the Lie group K. Then L(K) is an ideal of L(G) and $L(B) = \{0\}$, but $(K, B) = K$.

12. Let (e_1, e_2, e_3) be the canonical basis of K^3. Consider the nilpotent Lie algebra structure on K^3 such that $[e_1, e_2] = e_3$, $[e_1, e_3] = [e_2, e_3] = 0$. With the associated group law on K^3 (no. 5),

$$(x, y, z)(x', y', z') = (x + x', y + y', z + z' + \tfrac{1}{2}(xy' - yx')).$$

The mapping

$$(\lambda, \mu, \nu) \mapsto (\exp \lambda e_1)(\exp \mu e_2)(\exp \nu(e_1 + e_3))$$

of K^3 into G is neither surjective (show that (0, 1, 1) is not in the image) nor injective (show that (0, 1, 0) and (1, 1, −1) have the same image).

13. (*a*) A multiplication is defined on $\mathbf{R}^3$ as follows:

$$(x, y, z) . (x', y', z') = (x + x' \cos z - y' \sin z, y + x' \sin z + y' \cos z, z + z').$$

Show that a solvable real Lie group G is thus obtained such that $DG = \mathbf{R}^2 \times \{0\}$. The centre Z of G is $\{0\} \times \{0\} \times 2\pi\mathbf{Z}$.

(*b*) For $(x, y, z) \in G$, let $\pi(x, y, z)$ be the mapping

$$(\lambda, \mu) \mapsto (\lambda \cos z - \mu \sin z + x, \lambda \sin z + \mu \cos z + y)$$

of $\mathbf{R}^2$ into $\mathbf{R}^2$. Show that π is a morphism of G onto a Lie subgroup G′ of the affine group of $\mathbf{R}^2$, generated by the translations and rotations of $\mathbf{R}^2$. Show that $\operatorname{Ker} \pi = \mathbf{Z}$.

(*c*) We canonically identify $L(G) = T_{(0,0,0)}G$ with $\mathbf{R}^3$. Let (e_1, e_2, e_3) be the canonical basis of $\mathbf{R}^3$. Show that $[e_1, e_2] = 0$, $[e_3, e_1] = e_2$, $[e_3, e_2] = -e_1$.

(*d*) Show that, for $c \neq 0$,

$$\exp_G(a, b, c) = \left(\frac{1}{c}(a \sin c + b \cos c - b), \frac{1}{c}(-a \cos c + b \sin c + a), c\right).$$

Deduce that, for all $u \in L(G)$, $Z \subset \exp(\mathbf{R}u)$. Show that $\exp_G : L(G) \to G$ is neither injective nor surjective.

14. (*a*) Let $G = \mathbf{GL}(n, \mathbf{C})$. Show that $\exp_G$ is surjective. (Use the holomorphic functional calculus of *Spectral Theories*, Chapter I, § 4, no. 8.)

(*b*) Let $G' = \mathbf{SL}(2, \mathbf{C})$. Show that, if $\sigma \in \mathbf{C}^*$, the element $\begin{pmatrix} -1 & \sigma \\ 0 & -1 \end{pmatrix}$ of G′ does not belong to the image of $\exp_{G'}$.

15. (*a*) Let $x \in \mathfrak{sl}(2, \mathbf{R})$ and $\Delta = \det x$. Show that

$$e^x = \operatorname{ch} \sqrt{-\Delta}.\mathrm{I} + \frac{\operatorname{sh} \sqrt{-\Delta}}{\sqrt{-\Delta}} \cdot x \quad \text{if } \Delta < 0$$

$$e^x = \cos \sqrt{\Delta}.\mathrm{I} + \frac{\sin \sqrt{\Delta}}{\sqrt{\Delta}}.x \quad \text{if } \Delta > 0.$$

(*b*) Let $\mathrm{H} = \mathbf{SL}(2, \mathbf{R})$. Show that $g = \begin{pmatrix} \lambda & 0 \\ 0 & \lambda^{-1} \end{pmatrix} \in \mathrm{H}$ is the image of $\exp_{\mathrm{H}}$ if and only if $\lambda > 0$ or $\lambda = -1$. If $\lambda = 0$, all the one-parameter subgroups containing g are equal.

16. Let G be a finite-dimensional Lie group. Show that there exists a basis $(x_1, \ldots, x_n)$ of L(G) such that, for all i, $\exp(\mathbf{R}x_i)$ is a Lie subgroup of G. (Use *Spectral Theories*, Chapter II, § 2, Lemma 1.)

¶ 17. (*a*) Let $\mathfrak{c}$ denote a real solvable Lie algebra with a basis (a, b, c) such that $[a, b] = c$, $[a, c] = -b$, $[b, c] = 0$. Let $\mathfrak{d}$ denote a real solvable Lie algebra with a basis (a, b, c, d) such that $[a, b] = c$, $[a, c] = -b$, $[b, c] = d$, $[a, d] = [b, d] = [c, d] = 0$. Let $\mathfrak{g}$ be a real solvable Lie algebra. If $\mathfrak{g}$ contains non-zero elements x, y, z such that $[x, y] = z$ and $[x, z] = -y$, then $\mathfrak{g}$ contains a subalgebra isomorphic to $\mathfrak{c}$ or $\mathfrak{d}$. (Write $a_1 = y$, $b_1 = z$, $c_1 = [y, z]$; define a_i, b_i, c_i inductively by $a_i = [a_{i-1}, c_{i-1}]$, $b_i = [b_{i-1}, c_{i-1}]$, $c_i = [a_i, b_i]$. Consider the smallest integer k such that $c_k = 0$.)

(*b*) Let $\mathfrak{g}$ and $\mathfrak{g}^*$ be real solvable Lie algebras and ϕ a homomorphism of $\mathfrak{g}$ onto $\mathfrak{g}^*$. If $\mathfrak{g}^*$ contains a subalgebra isomorphic to $\mathfrak{c}$ or $\mathfrak{d}$, $\mathfrak{g}$ has the same property.

(*c*) Let $\mathfrak{h}$ be a complex solvable Lie algebra. Consider a Jordan–Hölder series for the adjoint representation of $\mathfrak{h}$; the quotients of this series define 1-dimensional representations of $\mathfrak{h}$ and hence linear forms on $\mathfrak{h}$. These linear forms, which depend only on $\mathfrak{h}$, are called the *roots* of $\mathfrak{h}$. If $\mathfrak{h}'$ is a real solvable Lie algebra, the roots of $\mathfrak{h}'$ are the restrictions to $\mathfrak{h}'$ of the roots of $\mathfrak{h}' \otimes_{\mathbf{R}} \mathbf{C}$.

Then let G be a finite-dimensional simply connected solvable real Lie group and $\mathfrak{g}$ its Lie algebra. Show that the following conditions are equivalent: (α) $\exp_{\mathrm{G}}$ is injective; (β) $\exp_{\mathrm{G}}$ is surjective; (γ) $\exp_{\mathrm{G}}$ is bijective; (δ) $\exp_{\mathrm{G}}$ is an isomorphism of the analytic manifold L(G) onto the analytic manifold G; (ε) L(G) contains no subalgebra isomorphic to $\mathfrak{c}$ or $\mathfrak{d}$; (ζ) there exists no quotient algebra of L(G) admitting a subalgebra isomorphic to $\mathfrak{c}$; (η) every root of $\mathfrak{g}$ is of the form $\phi + i\phi'$ where ϕ, ϕ' are in $\mathfrak{g}^*$ and ϕ' is proportional to ϕ; (θ) for all $x \in \mathfrak{g}$, the only pure imaginary eigenvalue (in $\mathbf{C}$) of $\operatorname{ad} x$ is 0.†

† For more details, cf. M. Saito, Sur certains groupes de Lie résolubles, *Sci. Papers of the College of General Education*, Univ. of Tokyo, **7** (1957), pp. 1–11 and 157–168.

18. Let G be a connected real Lie group, Z its centre, Z_0 the identity component of Z, $\mathfrak{g}$ the Lie algebra of G and $\mathfrak{z}$ the centre of $\mathfrak{g}$. Then Z and Z_0 are Lie quasi-subgroups of G with Lie algebra $\mathfrak{z}$ (Exercise 3). An acceptable norm on $\mathfrak{g}$ is a norm defining the topology of $\mathfrak{g}$ and making $\mathfrak{g}$ into a normed Lie algebra.

(*a*) For all $r > 0$ and $z \in \mathfrak{z}$, there exists an acceptable norm on $\mathfrak{g}$ whose value at z is $<r$. (Let q be an acceptable norm on $\mathfrak{g}$. Show that, for suitably chosen $\lambda > 0$, the function $x \mapsto \|x\| = \lambda q(x) + \inf_{y \in \mathfrak{z}} q(x + y)$ has the required properties.)

(*b*) Show that the following conditions are equivalent:

(i) Z_0 is simply connected;

(ii) for every acceptable norm on $\mathfrak{g}$, the restriction of $\exp_G$ to the open ball of centre 0 and radius π is injective;

(iii) there exists $r > 0$ such that, for every acceptable norm on $\mathfrak{g}$, the restriction of $\exp_G$ to the open ball of centre 0 and radius r is injective.

(To prove (iii) $\Rightarrow$ (i), use (*a*). To prove (i) $\Rightarrow$ (ii), suppose that x, y are in $\mathfrak{g}$, $\|x\| < \pi$, $\|y\| < \pi$, $x \neq y$, $\exp_G x = \exp_G y$. Then $\exp \operatorname{ad} x = \exp \operatorname{ad} y$ and hence $\operatorname{ad} x = \operatorname{ad} y$ by Proposition 17 of § 6 applied to a complexification of $\mathfrak{g}$. Hence there exists a non-zero z in $\mathfrak{z}$ such that $\exp_G z = e$; it follows that (i) is false.)

(*c*) By considering the group G′ of Exercise 13 (*b*), show that the conclusion of (*b*) is no longer true if π is replaced by a number $> \pi$. (In the notation of Exercise 13, use the norm $ae_1 + be_2 + ce_3 \mapsto (a^2 + b^2 + c^2)^{\frac{1}{2}}$ on L(G′).)

19. Let G be a locally compact group and H a closed subgroup of G. Suppose that H is a finite-dimensional simply connected solvable Lie group and that G/H is compact. Show that there exists a compact subgroup L of G such that G is the semi-direct product of L and H. (Argue by induction on dim H. Consider the last non-trivial derived group of H and use *Integration*, Chapter VII, § 3, Proposition 3.)

20. Let G be a finite-dimensional solvable connected real Lie group. We introduce the notation S, S′, F, σ of § 6, no. 10, proof of Proposition 20.

(*a*) Using Proposition 21, show that σ is an isomorphism of S onto a Lie subgroup of the underlying real Lie group of S′.

(*b*) Deduce that the universal complexification $\tilde{G}$ of G is identified with $S'/\sigma(F)$ and that the canonical mapping of G into $\tilde{G}$ is an isomorphism of G onto a Lie subgroup of the underlying real Lie group of $\tilde{G}$.

¶ 21. (*a*) Let G be a simply connected solvable real Lie group with the following properties: (α) L(G), which is n-dimensional, has a commutative ideal of dimension $n - 1$, corresponding to a subgroup A of G; (β) there exists an element σ of the centre of G which does not belong to A. Show that there exists an element x of L(G) such that $\exp x = \sigma$. Show that L(G) is the product of a commutative ideal and an ideal with a basis

$$(x, a_1, b_1, \ldots, a_k, b_k)$$

such that $a_1, b_1, \ldots, a_k, b_k$ belong to L(A), $[x, a_i] = 2\pi n_i b_i$, $[x, b_i] = -2\pi n_i a_i$ ($n_i \in \mathbf{Z} - \{0\}$) for all i. Generalize the results of Exercise 13 (*d*) to G.

(*b*) Let G be a finite-dimensional simply connected solvable real Lie group. Let D be a discrete subgroup of the centre of G. There exist a basis $(x_1, x_2, \ldots, x_n)$ of L(G) and an integer $r \leqslant n$ with the following properties: (α) every element of G can be written uniquely in the form

$$(\exp t_1 x_1) \ldots (\exp t_n x_n)$$

where $t_1, \ldots, t_n$ are in $\mathbf{R}$; (β) $x_1, \ldots, x_r$ are pairwise permutable and

$$(\exp x_1, \ldots, \exp x_r)$$

is a basis of the commutative group D. (Argue by induction on the dimension of G. Let $\mathfrak{a}$ be maximal commutative ideal of L(G). Let A be the corresponding integral subgroup. Then A is closed in G and DA/A is a discrete subgroup of the centre of G/A, to which the induction hypothesis can be applied, whence there are elements $x_1^*, \ldots, x_m^*$ of L(G/A) and an integer s. For $1 \leqslant i \leqslant s$, let σ_i be an element of D whose class modulo A is $\exp x_i^*$. Construct one by one, using (*a*), representatives $x_1, \ldots, x_m$ of $x_1^*, \ldots, x_m^*$ such that $\exp x_i = \sigma_i$ and $[x_i, x_j] = 0$ for $1 \leqslant i, j \leqslant s$.)†

(*c*) Deduce from (*b*) that every finite-dimensional connected solvable real Lie group is homomorphic to a space $\mathbf{R}^n \times \mathbf{T}^m$ (m, n integers $\geqslant 0$).

22. Let G be a finite-dimensional connected real Lie group such that L(G) is reductive. Then $G = (V \times S)/N$ where V is a finite-dimensional real vector space, where S is a simply connected semi-simple real Lie group and N is a central discrete subgroup of $V \times S$. Then $G/\overline{D^1 G}$ is isomorphic to $V/\overline{\mathrm{pr}_1 N}$. Hence $G/\overline{D^1 N}$ is compact if and only if $\mathrm{pr}_1 N$ generates the vector space V.

23. (*a*) Let G be a finite-dimensional commutative real Lie group with only a finite number of connected components. For G to be compact, it is necessary and sufficient that every finite-dimensional analytic linear representation of G on a complex vector space be semi-simple. (Use the proof of Proposition 32.)

(*b*) Let G be a finite-dimensional commutative complex Lie group with only a finite number of connected components. Let N be the kernel of $\exp_G$.

---

† For more details, cf. C. Chevalley, Topological structure of solvable groups, *Ann. of Math.*, **42** (1941), pp. 668–675.

For N to generate over **C** the vector space L(G), it is necessary and sufficient that every finite-dimensional analytic linear representation of G be semi-simple. (Use (*a*), Lemma 1 and Proposition 32.)

24. The Lie group **SL**(2, **R**) is connected and almost simple, but $\{I, -I\}$ is a normal commutative subgroup of **SL**(2, **R**).

25. Let G be an almost simple connected real or complex Lie group. Let A be a normal subgroup of G. If A $\neq$ G, A is discrete and central. (Use Exercise 8 of § 4.) Therefore the quotient of G by its centre is simple as an abstract group.

¶ 26. Let G be a finite-dimensional connected real Lie group. Suppose that G admits a finite-dimensional injective continuous linear representation ρ. Then (G, G) is closed in G. (Let R be the radical of G and S a maximal semi-simple integral subgroup of G. Using Exercise 7 (*b*), reduce it to the case where G is the semi-direct product of S and R. By Chapter I, § 6, Proposition 6, ρ is unipotent on (G, R). Hence ρ((G. R)) is closed in the linear group and therefore (G, R) is closed in G.)

¶ 27. Let G be a finite-dimensional solvable simply connected real Lie group and N the largest connected nilpotent normal subgroup of G. Then G admits a finite-dimensional injective continuous linear representation which is unipotent on N. (Argue by induction on the dimension of G. Use Proposition 20 and Chapter I, § 7, Theorem 1.)†

¶ 28. (*a*) Let k be a commutative field of characteristic 0, L an n-dimensional Lie algebra over k, L_1 an $(n-1)$-dimensional subalgebra containing no non-zero ideal of L and a_0 an element of L not belonging to L_1. For $i = 2, 3, \ldots$, we define L_i inductively to be the set of $x \in L_{i-1}$ such that $[x, a_0] \in L_{i-1}$. We write $L_i = L$ for $i \leqslant 0$. Show that $[L_i, L_j] \subset L_{i+j}$ (by induction on $i + j$) and then that, for $0 \leqslant i \leqslant n$, L_i is a subalgebra of L of codimension i (by induction on i).

(*b*) For $0 < i < n$, choose in L_i an element a_i not belonging to L_{i+1} such that $[a_0, a_i] \equiv i a_{i-1}$ (mod. L_i). Show, by induction on the ordered pairs (i, j) ordered lexicographically, that, for $0 \leqslant i < j < n$, $i + j - 1 < n$ and

$$[a_i, a_j] \equiv (j - i) a_{i+j-1} (\text{mod}. L_{i+j}).$$

(*c*) Deduce that L is either of dimension 1, or of dimension 2 and non-commutative, or isomorphic to $\mathfrak{sl}(2, k)$.

(*d*) Let A be a finite-dimensional real Lie algebra. A subalgebra B of A is called extendable if there exists a subalgebra $B_1 \supset B$ such that

† For more details concerning injective linear representations of Lie groups, cf. G. Hochschild, *The structure of Lie groups*, Holden-Day, 1965, and in particular Chapter XVIII.

$\dim B_1 = \dim B + 1$. Let R′(A) denote the intersection of all the non-extendable subalgebras of A. Let R(A) denote the largest ideal of A with a composition series as an A-module all of whose quotients are of dimension 1.

Show that R′(A) is a characteristic ideal of A. (Observe that R′(A) is stable under Aut(A).)

(*e*) Show that $R'(A) \supset R(A)$ and that $R'(A/R(A)) = R'(A)/R(A)$.

(*f*) Show that, if B is a subalgebra of A, then $R'(B) \supset R'(A) \cap B$.

(*g*) If A is solvable, $R'(A) = R(A)$. (It can be assumed that $R(A) = \{0\}$, by (*e*). Suppose that $R'(A) \neq \{0\}$. By (*d*), there exists a minimal non-zero ideal I of A contained in R′(A). Using (*f*) and Chapter I, § 5, Corollary 1 to Theorem 1, show that $\dim I = 1$, whence a contradiction.)

(*h*) Let P be the radical of R′(A), B a semi-simple subalgebra of A and $C = P + B$ which is a subalgebra of A by (*d*). Using (*f*), show that $\mathrm{ad}_P B$ can be expressed in a triangular form with respect to a suitable basis of P. Conclude that $[P, B] = \{0\}$.

(*k*) R(A) is the radical of R′(A). (Use (*e*), (*f*), (*g*), (*h*) and a Levi decomposition of A.) Let $R'(A) = R(A) \oplus T$ be a Levi decomposition of R′(A). By (*h*), $R'(A) = R(A) \times T$. Then $R(T) = \{0\}$. Applying (*c*) to the simple factors of T, conclude that $T = \{0\}$. It has therefore been shown that $R(A) = R'(A)$.

(*l*) Let G be a connected real Lie group with Lie algebra A. Let $\mathscr{S}(G)$ be the set of subgroups N of G such that there exists a decreasing sequence $(N_m, N_{m-1}, \ldots, N_0)$ of subgroups with the following properties: $N_m = N$, $N_0 = \{e\}$, each N_i is a normal connected Lie subgroup of G and $\dim N_i/N_{i-1} = 1$ for all $i > 0$. Show that the integral subgroup R(G) of G with Lie algebra R(A) is the largest element of $\mathscr{S}(G)$. (Argue by induction on dim R(A). Let I be a 1-dimensional ideal of A and N the corresponding integral subgroup of G. Pass to the quotient by $\overline{N}$. If N is not closed, use Exercise 21 (*a*) of § 6.)

(*m*) A Lie subgroup H of G is called extendable if there exists a Lie subgroup $H_1 \supset H$ such that $\dim H_1 = \dim H + 1$. Let R′(G) denote the intersection of all the non-extendable connected Lie subgroups of G.

Let B be a non-extendable subalgebra of A. Show that the corresponding integral subgroup of G is closed. (Use Proposition 5.) Deduce that

$$L(R'(G)) \subset R(A).$$

whence $R'(G) \subset R(G)$.

(*n*) Show that $R(G) = R'(G)$. (Reduce it to the case where $R'(G) = \{e\}$. Then use Exercise 22 of § 6.)†

¶ 29. A real Lie group G is said to be *of type* (N) if it is finite-dimensional,

† For more details, cf. J. Tits, Sur une classe de groupes de Lie résolubles, *Bull. Soc. Math. Belg.*, **11** (1959), pp. 100–115 and **14** (1962), pp. 196–209.

nilpotent, connected and simply connected. If G is such a group and $\mathfrak{g}$ is its Lie algebra, the mapping $\exp: \mathfrak{g} \to G$ is an isomorphism when $\mathfrak{g}$ is given the group structure defined by the Hausdorff law (cf. Chapter II, § 6, no. 5, *Remark* 3). Let $\log: G \to \mathfrak{g}$ denote the inverse isomorphism.

(*a*) Let V be a vector **Q**-subspace of $\mathfrak{g}$. Show the equivalence of the following conditions:

(i) V is a Lie **Q**-subalgebra of $\mathfrak{g}$;

(ii) exp(V) is a subgroup of G.

(Use Exercise 5 of § 6 of Chapter II.)

For a subgroup H of G to be of the form exp(V), where V is a vector **Q**-subspace of $\mathfrak{g}$, it is necessary and sufficient that H be saturated in G (Chapter II, § 4, Exercise 14), i.e. that the relations $x \in G$, $x^n \in H$, $n \neq 0$ imply $x \in H$ (*loc. cit.*). If this is so, show that H is an integral subgroup of G if and only if log(H) is a Lie **R**-subalgebra of $\mathfrak{g}$.

(*b*) Let V be a Lie **Q**-subalgebra of $\mathfrak{g}$ of finite dimension m, let $(e_1, \ldots, e_m)$ be a basis of V over **Q** and let Λ be the subgroup of V generated by $e_1, \ldots, e_m$. Using the fact that the Hausdorff law is polynomial, show that there exists an integer $d \geqslant 1$ such that, for every integer r which is a non-zero multiple of d, $\exp(r\Lambda)$ is a subgroup of G. Let d be such an integer and r a multiple of d. Show that $\exp(r\Lambda)$ is *discrete* if and only if $(e_1, \ldots, e_m)$ is a free family over **R**, i.e. if the canonical mapping of $V \otimes_{\mathbf{Q}} \mathbf{R}$ into $\mathfrak{g}$ is *injective*. Suppose that this is the case; show that $G/\exp(r\Lambda)$ is *compact* if and only if $V \otimes_{\mathbf{Q}} \mathbf{R} \to \mathfrak{g}$ is *bijective*, i.e. *if* V *is a* **Q**-*form of* $\mathfrak{g}$. (To show that the condition is sufficient, argue by induction on the nilpotency class of $\mathfrak{g}$.)

(*c*) Conversely, let Γ be a discrete subgroup of G, let $\bar{\Gamma}$ be its saturation in G (Chapter II, *loc. cit.*) and let $\mathfrak{g}_\Gamma = \log(\bar{\Gamma}) = \mathbf{Q}.\log(\Gamma)$ be the corresponding Lie **Q**-subalgebra. Suppose that $\mathbf{R}.\mathfrak{g} = \mathfrak{g}$, i.e. that Γ is contained in no distinct integral subgroup of G. Let z be an element $\neq 1$ of the centre of Γ and let $x = \log(z)$. Show that x belongs to the centre of $\mathfrak{g}$. If $X = \exp(\mathbf{R}x)$, show that $X/(\Gamma \cap X)$ is compact and that the image of Γ in G/X is discrete. Deduce, arguing by induction on dim G, that G/Γ is *compact* and that $\mathfrak{g}_\Gamma$ is a G-*form of* $\mathfrak{g}$. If Λ is a lattice of $\mathfrak{g}_\Gamma$, show that there exists an integer $d \neq 0$ such that Γ contains $\exp(d\Lambda)$ and that the index of $\exp(d\Lambda)$ in Γ is then finite.

Let H be an integral subgroup of G with Lie algebra $\mathfrak{h}$. Show that $H/(H \cap \Gamma)$ is compact if and only if $\mathfrak{h}$ is rational with respect to the **Q**-structure $\mathfrak{g}_\Gamma$ (in particular if $\mathfrak{h}$ is one of the terms of the lower—or upper—central series of $\mathfrak{g}$).

(*d*) Let Γ be a discrete subgroup of G, let H be the smallest integral subgroup of G containing Γ and let $\mathfrak{h}$ be its Lie algebra. Show that H/Γ is compact and that $\mathfrak{g} \otimes_{\mathbf{Q}} \mathbf{R} \to \mathfrak{g}$ is injective and has image $\mathfrak{h}$. (Apply (*c*) to the nilpotent group H.)

(*e*) Let Γ be a discrete subgroup of G. Show that there exists a basis $(x_1, \ldots, x_q)$ of L(G) with the following properties:

(i) for all $i \in [1, q]$, $\mathbf{R}x_i + \cdots + \mathbf{R}x_q$ is an ideal $\mathfrak{n}_i$ of $\mathbf{R}x_{i-1} + \cdots + \mathbf{R}x_q$;

(ii) there exists $p \in [1, q]$ such that Γ is the set of products

$$\exp(m_p x_p) \exp(m_{p+1} x_{p+1}) \ldots \exp(m_q x_q)$$

where $m_p, \ldots, m_q$ are in $\mathbf{Z}$;

(iii) if G/Γ is compact, $\{\mathfrak{n}_1, \mathfrak{n}_2, \ldots, \mathfrak{n}_q\}$ contains the lower central series of $\mathfrak{g}$. (Using (*d*) and Proposition 16, reduce it to the case where G/Γ is compact. Then argue by induction on dim G, using (*c*).)

(30) Give an example of a 7-dimensional real nilpotent Lie algebra with no basis with respect to which the constants of structure are rational (cf. Chapter I, § 4, Exercise 18). Deduce that the corresponding group of type (N) (Exercise 29) has no discrete subgroup with compact subgroup. (Use Exercise 29.)

31. Let G and G′ be two Lie groups of type (N) (Exercise 29) and let Γ be a discrete subgroup of G such that G/Γ is compact. Show that every homomorphism $f \colon \Gamma \to G'$ can be extended uniquely to a Lie group morphism of G into G′ (begin by extending f to the saturation $\hat{\Gamma}$ of Γ and derive a homomorphism of the Lie $\mathbf{Q}$-algebra $\log(\hat{\Gamma})$ into the Lie algebra of G′, cf. Exercise 29).

32. Let G be a Lie group of type (N) (Exercise 29) and let Γ be a discrete subgroup of G. Prove the equivalence of the following conditions:

(*a*) G/Γ is compact;

(*b*) the measure of G/Γ is finite (relative to a non-zero G-invariant positive measure);

(*c*) every integral subgroup of G containing Γ is equal to G.

(Use Exercise 29.)

¶ 33. Let Γ be a group. Prove the equivalence of the following conditions:

(*a*) Γ is nilpotent, torsion-free and finitely generated;

(*b*) there exists a Lie group of type (N) (Exercise 29) containing Γ as a discrete subgroup;

(*c*) there exists a Lie group G of type (N) (Exercise 29) containing Γ as a discrete subgroup and such that G/Γ is compact.

(The equivalence of (*b*) and (*c*) follows from Exercise 29. The implication (*c*) ⇒ (*a*) is proved by induction on dim(G) by the method of Exercise 29 (*c*). To prove that (*a*) ⇒ (*c*), show first that the Lie $\mathbf{Q}$-algebra attached to the saturation of Γ (cf. Chapter II, § 6, Exercise 4) is finite-dimensional; then take the tensor product of this algebra with $\mathbf{R}$ and the corresponding group of type (N).)

34. Let G be a finite-dimensional connected real or complex Lie group and ρ an analytic linear representation of G on a finite-dimensional complex vector space V. Suppose that ρ is semi-simple. The group G operates by automorphisms on the algebra $S(V^*)$ of polynomial functions on V.

(*a*) Let $S(V^*)^G$ be the set of G-invariant elements of $S(V^*)$. There exists a projector p of $S(V^*)$ onto $S(V^*)^G$ which commutes with the operations of G and which leaves stable every G-stable vector subspace of $S(V^*)$.

(*b*) Let I, J be ideals of $S(V^*)$ which are stable under G. Let A (resp. B) be the set of zeros of I (resp. J) in V. Suppose that $A \cap B = \varnothing$. There then exists $u \in I$ such that u is G-invariant and that $u = 1$ in B. (By *Commutative Algebra*, Chapter 7, § 3, no. 3, Proposition 2, there exists $v \in I$, $w \in J$ such that $v + w = 1$ Let $u = pv$. Show that u has the required properties.)

35. Let G be a compact subgroup of $\mathbf{GL}(n, \mathbf{R})$.

(*a*) Let A, B be disjoint compact G-invariant subsets of $\mathbf{R}^n$. There exists a G-invariant polynomial function u on $\mathbf{R}^n$, such that $u = 1$ on A and $u = 0$ on B. (There exists a continuous real-valued function v on $\mathbf{R}^n$ such that $v = 1$ on A and $v = 0$ on B. By the Stone-Weierstrass Theorem. there exists a polynomial function w on $\mathbf{R}^n$ such that $|v - w| \leqslant \frac{1}{3}$ on $A \cup B$. Derive u by an integration process with respect to the normalized Haar measure of G.)

(*b*) Let $x \in \mathbf{R}^n$. Then Gx is the set of zeros in $\mathbf{R}^n$ of a finite number of G-invariant polynomials. (Use (*a*) and Corollary 2 to Theorem 1, no. 8.)

36. In $\mathbf{SL}(2, \mathbf{R})$, the elements which can be expressed in the form (a, b), where a, b in $\mathbf{SL}(2, \mathbf{R})$, are the elements $\neq -I$.

¶ 37. Let G be a compact real Lie group and G′ a finite-dimensional connected real Lie group. Suppose that L(G) is simple. Let $\rho: G \to G'$ be a homomorphism of abstract groups. Suppose that there exists a neighbourhood V of e_G in G such that $\rho(V)$ is relatively compact. Then ρ is continuous. (Let V′ be a neighbourhood of $e_{G'}$ in G′. Let $V'' \subset V'$ be a neighbourhood of $e_{G'}$ such that

$$(x \in V'', x_j \in \rho(V), y_j \in \rho(V) \text{ for } 1 \leqslant j \leqslant n) \Rightarrow \left(\prod_{i=1}^{n} x_i(y_i, x)x_i^{-1} \in V'\right)$$

where $n = \dim G$. It can be assumed that ρ is non-trivial. Then Ker ρ is finite by Exercise 25. Therefore $\rho(G)$ is not countable and hence not discrete. Hence there exists $g \in G$ with non-open centralizer such that $\rho(g) \in V''$. Using Exercise 9 of § 4 and its notation, $\rho(M(g, n, V)) \subset V'$ and $M(g, n, V)$ is a neighbourhood of e_G in G.)

38. Let G be a finite-dimensional real Lie group and G_0 its identity component. Consider the following property:

(F) Ad(G) is closed in Aut L(G).

Show that G has property (F) in each of the following cases:

(i) G is connected and nilpotent;

(ii) every derivation of L(G) is inner;

(iii) G_0 has property (F) and G/G_0 is finite;

(iv) G is an upper triangular group;

(v) G_0 is semi-simple.

39. Let G be a finite-dimensional connected real Lie Group, H an integral subgroup of G and $\overline{H}$ its closure in G. Suppose that H has property (F) (Exercise 38).

(*a*) Let $x \in \overline{H}$. Then $\mathrm{Ad}_{L(G)}x$ leaves L(H) stable (no. 2, Proposition 5); let $u(x)$ be its restriction to L(H). Then $u(\overline{H}) = \mathrm{Ad}_{L(H)}(H)$. (Observe that $\mathrm{Ad}_{L(H)}(H)$ is dense in $u(\overline{H})$ and apply property (F).)

(*b*) Let C be the centre of $\overline{H}$. Then $\overline{H} = C.H$ and C is the closure of the centre of H. (By (*a*), if $x \in \overline{H}$, there exists $y \in H$ such that $\mathrm{Ad}_{L(H)}x = \mathrm{Ad}_{L(H)}y$. Then $\mathrm{Ad}(x^{-1}y)$ is the identity on L(H), hence $x^{-1}y \in Z_{\overline{H}}(H)$ and $x \in Z_{\overline{H}}(H).H$. By continuity, $Z_{\overline{H}}(H)$ is equal to C. The group C is closed in $\overline{H}$ and hence in G; therefore $\overline{C \cap H} \subset C$. Let $x \in C$. There exists a sequence (x_n) in H such that x_n tends to x. Then $\mathrm{Ad}_{L(H)}x_n$ tends to 1 and hence there exists a sequence (y_n) in H such that y_n tends to e and $\mathrm{Ad}_{L(H)}y_n = \mathrm{Ad}_{L(H)}x_n$. Then $y_n^{-1}x_n \in C \cap H$ and $y_n^{-1}x_n$ tends to x.)

(*c*) $H = \overline{H}$ if and only if the centre of H is closed in G. (Use (*b*).)

40. Let H be a finite-dimensional real Lie group and H_0 its identity component.

(*a*) Suppose that H_0 satisfies condition (F) of Exercise 38, that H/H_0 is finite and that the centre of H_0 is compact. Let G be a finite-dimensional real Lie group and $f: H \to G$ a continuous homomorphism with discrete kernel. Then $f(H)$ is closed in G. (Reduce it to the case where H is connected. As the kernel of f is discrete, the centre of $f(H)$ is the image under f of the centre of H (Lemma 1) and hence is closed in G. Apply Exercise 39 (*c*).)

(*b*) Suppose that H_0 is semi-simple and that H/H_0 is finite. Then the image of H under a finite-dimensional continuous linear representation is closed in the linear group. (Apply (*a*) and Exercise 7 (*b*).)

41. Let G be a finite-dimensional connected real Lie group and $\rho: G \to \mathbf{GL}(n, \mathbf{C})$ a continuous homomorphism with finite kernel. Then $\rho((G, G))$ is closed in $\mathbf{GL}(n,\mathbf{C})$. (Using Exercises 7 (*b*) and 40 (*a*), reduce it to the case where G is the semi-direct product of a semi-simple group S and its radical R. Every element of $\rho((G, R))$ is unipotent and hence $\rho((G, R))$ is closed. The vector subspace V_1 of fixed points of $\rho((G, R))$ is $\neq \{0\}$. It is stable under $\rho(G)$. By considering $\mathbf{C}^n/V_1$ arguing by induction on n and using the complete reducibility of $\rho(S)$, it is seen that it is possible to write

$$\mathbf{C}^n = V_1 \oplus \cdots \oplus V_q$$

where each V_i is stable under $\rho(S)$ and $\rho((G, R))(V_i) \subset V_1 \oplus \cdots \oplus V_{i-1}$ for for all i. On the other hand, $\rho(S)$ is closed (Exercise 40 (*b*)). Finally, $\rho((G, G)) = \rho(S).\rho((G, R))$.)

42. Let G be a finite-dimensional connected real Lie group. Suppose that

G admits an injective continuous linear representation $\rho: G \to \mathbf{GL}(n, \mathbf{C})$. Then G admits a linear representation $\sigma: G \to \mathbf{GL}(n', \mathbf{C})$ which is a homomorphism of G onto a closed subgroup of $\mathbf{GL}(n', \mathbf{C})$. (By Exercise 41, $\rho((G, G))$ is closed in $\mathbf{GL}(n, \mathbf{C})$ and hence (G, G) is closed in G. Let $\rho: G \to (G, G)$ be the canonical morphism and τ an injective linear representation with closed image $G/(G, G)$ which is connected and commutative. Take $\sigma = \rho \oplus (\tau \circ p)$.)

§ 10

1. Let G be a finite-dimensional connected real Lie group. Show that the canonical mapping of Aut(G) into Aut(L(G)) is not in general surjective (take $G = \mathbf{T}$).

2. Suppose that $K = \mathbf{Q}_p$. Let G be a finite-dimensional Lie group. Show that the following conditions are equivalent:

(*a*) there exist $x_1, x_2, \ldots, x_n$ in G such that the subgroup of G generated by $\{x_1, \ldots, x_n\}$ is dense in G;

(*b*) G is generated by a compact subset.

(To prove that (*b*) implies (*a*), observe that, if $(e_1, \ldots, e_n)$ is a basis of L(G), $(\exp \mathbf{Z}_p e_1)(\exp \mathbf{Z}_p e_2) \ldots (\exp \mathbf{Z}_p e_n)$ is a neighbourhood of e.)

3. Let G be the set of $(x, y) \in \mathbf{Q}_p \times \mathbf{Q}_p$ such that $|y| \leqslant 1$. This is an open subgroup of $\mathbf{Q}_p \times \mathbf{Q}_p$. Then $L(G) = \mathbf{Q}_p \times \mathbf{Q}_p$. Let α be the infinitesimal automorphism $(x, y) \mapsto (0, x)$. Then the conclusion of Proposition 3 is false.

(4) Let K be a quadratic extension of $\mathbf{Q}_p$ and let $\omega \in K - \mathbf{Q}_p$. Let $G = \mathbf{Q}_p + \mathbf{Z}_p\omega$, considered as a Lie group over K. The only automorphisms of G are the $(x, y) \mapsto (\lambda x, y)$ where λ is invertible in $\mathbf{Z}_p$. The set of these automorphisms cannot be given a Lie group structure over K with the properties of Theorem 1.

HISTORICAL NOTE

(Chapters I to III)

I. GENESIS

The theory, called for nearly a century "theory of Lie groups", was essentially developed by one mathematician: Sophus Lie.

Before embarking on the history, we summarize briefly various earlier research which prepared the way.

(a) *Transformation groups* (Klein-Lie, 1869–1872)

About 1860 the theory of permutation groups of a *finite* set was developing and beginning to be used (Serret, Kronecker, Mathieu, Jordan). On the other hand, the theory of invariants, which was in full flight, was familiarizing mathematicians with some infinite sets of geometric transformations stable under composition (notably linear or projective transformations). But before the work of Jordan [7] in 1868 on "groups of movements" (closed subgroups of the group of displacements of 3-dimensional Euclidean space), no conscious link seems to have been established between these two currents of ideas.

In 1869, the young Felix Klein (1849–1925), a pupil of Plücker, established a friendship in Berlin with the Norwegian Sophus Lie (1842–1899), who was a few years older, brought together by their common interest in Plücker's "line geometry" and in particular the theory of line complexes. It was about this time that Lie conceived one of his most original ideas, the introduction of the notion of invariant into Analysis and Differential Geometry; one of the sources was his observation that the classical methods of integration "by quadratures" of differential equations depended entirely on the fact that the equation is invariant under a "continuous" family of transformations. 1869 is the date of the first work (edited by Klein) where Lie used this idea; there he studied the "Reye complex" (set of four lines cutting the faces of a tetrahedron in four points with a given cross-ratio) and the curves and surfaces which admit as tangents lines of this complex [3 a]: his method depends on the invariance of the Reye complex under the 3-parameter commutative group (maximal torus of **PGL**(4, **C**)) which leaves the vertices of the tetrahedron invariant. The same idea dominated the work Klein and Lie wrote together when they were in Paris in the spring of 1870 [1, a]; there they essentially determined the commutative connected subgroups of the projective group of the plane **PGL**(3, **C**) and studied the geometric properties of their orbits (under the name of curves or surfaces V); this gave them, by a uniform procedure, properties of various curves, algebraic or transcendental, such as $y = cx^m$ or the logarithmic spirals. Both works served to underline the profound impression made on them by the theories of Galois and Jordan (Jordan's commentary on Galois had appeared in *Math. Annalen* in 1869; on the other hand, Lie had heard talk of Galois theory as early as 1863). Klein, who in 1871 became interested in non-Euclidean geometries, saw there the start of his research on a classification principle for all known geometries which was to lead him in 1872 to the "Erlangen Programme". For his part, Lie, in a letter of 1873 to A. Mayer ([3], vol. V, p. 584), dated his ideas on transformation groups from his sojourn in Paris and in a work of 1871 ([3 *b*] p. 208) he was already using the term "transformation group" and explicitly posed the problem of the determination of all subgroups ("*continuous or discontinuous*") of **GL**(n, **C**). To be truthful, Klein and Lie had both experienced some difficulty in entering this new mathematical universe and Klein spoke of Jordan's newly appeared "Treatise" as a "*book sealed with seven seals*" ([2], p. 51); he wrote moreover concerning [1 a] and [1 b]: "*To Lie belongs all the credit for the heuristic idea of a continuous group of operators, in particular everything concerning integration of differential equations and partial derivatives. All the notions he developed later in his theory of continuous groups were already there in embryo, but were however so little elaborated, that only after long conversations could I convince him of many details, for example to begin with the very existence of the curves* V" ([2], p. 415).

(b) *Infinitesimal transformations*

The concept of an "infinitely small" transformation goes back at least to the beginnings of Infinitesimal Calculus; we know that Descartes discovered the instantaneous centre of rotation when admitting that "in the infinitely small" every plane movement can be likened to a rotation: the elaboration of Analytical Mechanics in the 18th century is entirely founded on similar ideas. In 1851, Sylvester, seeking to form invariants of the linear group **GL**(3, **C**) and some of its subgroups, gave the parameters appearing in his matrices "infinitely small" increases of the forms $\alpha_j dt$ and expressed the fact that a function $f((z_j))$ was invariant by writing the equation $f((z_j + \alpha_j(dt)) = f((z_j))$; this gave him for f the linear partial differential equation $Xf = 0$, where

$$Xf = \sum_j \alpha_j \frac{\partial f}{\partial z_j}, \tag{1}$$

where X is therefore a *differential operator*, the "derivative in the direction of the direction parameters α_i" ([5], vol. 3, pp. 326 and 327); Sylvester seemed to think that here was a general principle of considerable importance but appears never to have returned to the question. A little later, Cayley ([6], vol. II, pp. 164–178) proceeded similarly for the invariants of $\mathbf{SL}(2, \mathbf{C})$ under certain representations of this group and showed that they are the solutions of two first order partial differential equations $\mathrm{X}f = 0$, $\mathrm{Y}f = 0$, where X and Y are obtained as above from "infinitely small" transformations

$$\begin{pmatrix} 0 & 0 \\ dt & 0 \end{pmatrix} \text{ and } \begin{pmatrix} 0 & dt \\ 0 & 0 \end{pmatrix}.$$

In modern terms, this is expressed by the fact that X and Y generate the Lie algebra $\mathfrak{sl}(2, \mathbf{C})$; moreover Cayley calculated the bracket XY – YX explicitly and showed that it was also derived from an "infinitely small" transformation.

In his memoir of 1868 on groups of movements [7], Jordan used from beginning to end the concept of "infinitely small transformation", but exclusively from a geometric point of view. He is no doubt responsible for the idea of a one-parameter group "generated" by an infinitely small transformation: for Jordan, it was the set of transformations obtained by "*suitably repeating*" the infinitely small transformation (*loc. cit.*, p. 243). Klein and Lie, in their memoir, used the same expression "*repeated infinitely small transformation*" [1 b], but the context shows that they understood by that an integration of a differential system. If the one-parameter group they considered consists of transformations $x' = f(x, y, t)$, $y' = g(x, y, t)$, the corresponding "infinitely small transformation" is given by

$$dx = p(x, y)dt, \qquad dy = q(x, y)dt$$

where $p(x, y) = \frac{\partial f}{\partial t}(x, y, t_0)$, $q(x, y) = \frac{\partial g}{\partial t}(x, y, t_0)$ and t_0 corresponds to the identity transformation of the group. As Klein and Lie knew the functions f and g explicitly, they had no difficulty in verifying that the functions

$$t \mapsto f(x, y, t) \quad \text{and} \quad t \mapsto g(x, y, t)$$

give in parametric form the integral curve of the differential equation

$$q(\xi, \eta)\, d\xi = p(\xi, \eta)\, d\eta$$

passing through the point (x, y), but they give no general argument; moreover they nowhere use this fact in the remainder of their memoir.

(c) *Contact transformations*

In the next two years, Lie seemed to abandon the theory of transformation groups (although he remained in very close contact with Klein, who published his "Programme" in 1872) to study contact transformations, integration of first order partial differential equations and the relations between these two theories. We are not concerned here with the history of these questions and we shall confine ourselves to mentioning a few points which seem to have played an important role in the genesis of the theory of transformation groups.

The notion of contact transformation generalizes both point transformations and inverse polar transformations. Roughly a contact transformation† on $\mathbf{C}^n$ is an isomorphism of an open subset Ω of the manifold $T'(\mathbf{C}^n)$ of cotangent vectors to $\mathbf{C}^n$ onto another open subset Ω' of $T'(\mathbf{C}^n)$ mapping the canonical 1-form of Ω to that of Ω'. In other words, if $(x_1, \ldots, x_n, p_1, \ldots, p_n)$ denote the canonical coordinates of $T'(\mathbf{C}^n)$, a contact transformation is an isomorphism $(x_i, p_i) \mapsto (X_i, P_i)$ satisfying the relation $\sum_{i=1}^{n} P_i dX_i = \sum_{i=1}^{n} p_i dx_i$. Such transformations occur in the study of the integration of partial differential equations of the form

$$(2) \qquad F\left(x_1, x_2, \ldots, x_n, \frac{\partial z}{\partial x_1}, \ldots, \frac{\partial z}{\partial x_n}\right) = 0.$$

Lie became familiar during his research on these questions with the manipulation of Poisson brackets

$$(3) \qquad (f, g) = \sum_{i=1}^{n} \left(\frac{\partial f}{\partial x_i}\frac{\partial g}{\partial p_i} - \frac{\partial g}{\partial x_i}\frac{\partial f}{\partial p_i}\right)$$

and the brackets‡ $[X, Y] = XY - YX$ of differential operators of type (1); he interprets the Poisson bracket (3) as the effect on f of a transformation of type (1) associated with g and observes on this occasion that the Jacobi identity for Poisson brackets means that the bracket of the differential operators corresponding to f and g is associated with the bracket (g, h). Research into functions g such that $(F, g) = 0$, which occurs in Jacobi's method for integrating the

---

† Here we are concerned with "homogeneous" contact transformations. Earlier, studying equations of type (2) but with z occurring in F had led Lie to consider contact transformations in $2n + 1$ variables $z, x_1, \ldots, x_n, p_1, \ldots, p_n$, where it is required to find $2n + 2$ functions Z, P_i, X_i $(1 \leqslant i \leqslant n)$ and ρ (the latter $\neq 0$ at every point) such that $dZ - \sum_i P_i dX_i = \rho(dz - \sum_i p_i dx_i)$. This case which appears to be more general can be easily reduced to the "homogeneous" case ([4], vol. 2, pp. 135–146).

‡ These had already occurred in the Jacobi–Clebsch theory of "complete systems" of first order partial differential equations $X_j f = 0$ $(1 \leqslant j \leqslant r)$, a notion which is equivalent to Frobenius's "completely integrable system"; the fundamental theorem (equivalent to "Frobenius's theorem") which characterizes these systems is that the brackets $[X_i, X_j]$ must be linear combinations (with variable coefficients) of the X_k.

partial differential equation (2), became for Lie the study of infinitesimal contact transformations which leave the given equation invariant. Finally, Lie was led to study sets of functions $(u_j)_{1 \leqslant j \leqslant m}$ of the x_i and p_i such that the brackets (u_j, u_k) are functions of the u_h and called these sets "groups" (they had already essentially been studied by Jacobi) [3 *c*].

II. CONTINUOUS GROUPS AND INFINITESIMAL TRANSFORMATIONS

Suddenly, in the autumn of 1873, Lie again took up the study of transformation groups and obtained decisive results. Insofar as it is possible to follow his line of thought from a few letters to A. Mayer written in 1873–1874 ([3], vol. 5, pp. 584–608), he started from a "continuous group" of transformations on n variables

$$x'_i = f_i(x_1, \ldots, x_n, a_1, \ldots, a_r) \quad (1 \leqslant i \leqslant n) \tag{4}$$

depending effectively† on r parameters $a_1, \ldots, a_r$; he observed that, if the transformation (4) is the identity for the values $a_1^0, \ldots, a_r^0$ of the parameters,‡ then the first order Taylor expansions of the x_i:

$$f_i(x_1, \ldots, x_n, a_1^0 + z_1, \ldots, a_r^0 + z_r) = x_i + \sum_{k=1}^{r} z_k X_{ki}(x_1, \ldots, x_n) + \cdots \quad (1 \leqslant i \leqslant n) \tag{5}$$

give a "generic" infinitely small transformation depending linearly on the r parameters z_j

$$dx_i = \left(\sum_{k=1}^{r} z_k X_{ki}(x_1, \ldots, x_n)\right) dt \quad (1 \leqslant i \leqslant n). \tag{6}$$

Proceeding as in his memoir with Klein, Lie integrated the differential system

$$\frac{d\xi_1}{\sum_k z_k X_{k1}(\xi_1, \ldots, \xi_n)} = \cdots = \frac{d\xi_n}{\sum_k z_k X_{kn}(\xi_1, \ldots, \xi_n)} = dt, \tag{7}$$

† By this Lie meant that the f_i could not be expressed in terms of less than r functions of the a_j, or also that the Jacobian matrix $(\partial f_i/\partial a_j)$ is "in general" of rank r.

‡ In his first notes, Lie thought that he could prove *a priori* the existence of the identity and the inverse in the whole set of transformations (4) stable under composition; he recognized later that his proof was incorrect and Engel provided him with a counter-example reproduced in [4], vol. 1, § 44. However, Lie showed how "continuous" systems (4) stable under composition could be reduced to group germs of transformations: such a system is of the form $G \circ h$, where G is a group germ of transformations and h a transformation of the system ([4], vol. 1, Theorem 26, p. 163 and vol. 3, Theorem 46, p. 572).

which gave him, for each point $(z_1, \ldots, z_r)$ a one-parameter group

(8) $$t \mapsto x'_i = g_i(x_1, \ldots, x_n, z_1, \ldots, z_r, t) \quad (1 \leqslant i \leqslant n)$$

such that $g_i(x_1, \ldots, x_n, z_1, \ldots, z_r, 0) = x_i$ for all i. He showed by an ingenious method, using the fact that the transformations (4) form a set which is stable under composition, that the one-parameter group (8) is a subgroup of the given group [3 d]. The new idea, the key to the whole theory, is to take the Taylor expansions of the functions (4) to the *second order*. The progress of his argument was quite confused and heuristic ([3 d] and [3], vol. 5, pp. 600–601); it can be presented as follows. For sufficiently small z_j, let $t = 1$ in (8); thus new parameters $z_1, \ldots, z_r$ are obtained for the transformations of the group (this is in fact the first appearance of the "canonical parameters"). Then by definition, using (7),

$$\frac{\partial g_i}{\partial t} = \sum_k z_k X_{ki}(x'_1, \ldots, x'_n)$$

whence

$$\frac{\partial^2 g_i}{\partial t^2} = \sum_{k,j} z_k \frac{\partial X_{ki}}{\partial x_j}(x'_1, \ldots, x'_n)\frac{\partial x'_j}{\partial t}$$

$$= \sum_{k,j} z_k \frac{\partial X_{ki}}{\partial x_j}(x'_1, \ldots, x'_n)\Big(\sum_h z_h X_{hj}(x'_1, \ldots, x'_n)\Big)$$

which gives

$$x'_i = x_i + \Big(\sum_k z_k X_{ki}(x_1, \ldots, x_n)\Big)t + \tfrac{1}{2}\Big(\sum_{k,h,j} z_k z_h \frac{\partial X_{ki}}{\partial x_j}(x_1, \ldots, x_n) X_{hj}(x_1, \ldots, x_n)\Big)t^2 + \cdots,$$

whence, for $t = 1$, the Taylor expansions in the parameters z_j

(9) $$x'_i = x_i + \Big(\sum_j z_k X_{ki}\Big) + \tfrac{1}{2}\Big(\sum_{k,h,j} z_k z_h X_{hj} \frac{\partial X_{ki}}{\partial x_j}\Big) + \cdots \quad (1 \leqslant i \leqslant n).$$

We abridge these relations to $x' = G(x, z)$ relating the vectors

$$x = (x_1, \ldots, x_n), \qquad x' = (x'_1, \ldots, x'_n), \qquad z = (z_1, \ldots, z_r);$$

the fundamental stability property of the set of these transformations under composition can be written

(10) $$G(G(x, u), v) = G(x, H(u, v))$$

where $H = (H_1, \ldots, H_r)$ is independent of x; it is immediate that $H(u, 0) = u$ and $H(0, v) = v$, whence the expansions

(11) $$H_i(u, v) = u_i + v_i + \tfrac{1}{2}\sum_{h,k} c_{ikh} u_h v_k + \cdots,$$

where the terms omitted are not linear in u or v. Transforming (10), using (9) and (11) and then comparing the terms in $u_h v_k$ of the two sides. Lie obtained the relations

$$(12)\quad \sum_{j=1}^{n}\left(X_{hj}\frac{\partial X_{ki}}{\partial x_j} - X_{kj}\frac{\partial X_{hi}}{\partial x_j}\right) = \sum_{l=1}^{r} c_{lhk}X_{li} \quad (1 \leqslant h, k \leqslant r, 1 \leqslant i \leqslant n).$$

His experience with the theory of partial differential equations led him to write these conditions in a simpler form: following the pattern of (1), he associated with each of the infinitely small transformations obtained by setting $z_k = 1$, $z_h = 0$ for $h \neq k$ in (6), the differential operator

$$(13)\qquad A_k(f) = \sum_{i=1}^{n} X_{ki}\frac{\partial f}{\partial x_i},$$

and rewrote conditions (12) in the form

$$(14)\qquad [A_h, A_k] = \sum_{l} c_{lhk}A_l,$$

the corner stone of his theory. Until then he used the terms "infinitely small transformation" and "infinitesimal transformation" indifferently (e.g. [3 *e*]); the simplicity of relations (14) led him to call the operator (13) the "symbol" of the infinitesimal transformation $dx_i = X_{ki}dt$ $(1 \leqslant i \leqslant n)$ [3 *d*] and very soon he called the operator (13) itself the "*infinitesimal transformation*" ([3 *d*] and [3], vol. 5, p. 589).

He then became aware of the close links which united the theory of "continuous groups" with his earlier research on contact transformations and partial differential equations. This bringing together filled him with enthusiasm: "*My earlier works were as it were all ready there waiting to found the new theory of transformation groups*" he wrote to Mayer in 1874 ([3], vol. 5, p. 586).

In the following years, Lie pursued the study of transformation groups. Besides the general theorems summarized below (§ III), he obtained some more special results: the determination of the transformation groups of the line and plane, the subgroups of low codimension in the projective groups, the groups in at most 6 parameters, etc. He did not abandon differential equations for long. In fact, it seems that, for him, the theory of transformation groups was a tool for integrating differential equations, where the transformation group played a role analogous to that of the Galois group of an algebraic equation.†

† This research had little influence on the general theory of differential equations, since the automorphism group of such an equation is usually trivial. As a compensation, for certain types of equations (for example linear equations), interesting results have been obtained later by Picard, Vessiot and, more recently, Ritt and Kolchin.

We note that this research led him equally to the introduction of certain transformation sets in an infinity of parameters which he called "infinite continuous groups"‡; he reserved the name "finite continuous groups" for transformation groups in a finite number of parameters of type (4) above.

THE LIE GROUPS-LIE ALGEBRAS "DICTIONARY"

The theory of "finite continuous" groups, developed by Lie in numerous memoirs beginning in 1874, is expounded systematically in the imposing treatise "*Theorie der Transformationsgruppen*" ([4], 1888–1893]), written in collaboration with F. Engel§; it is the object of study in the first volume and the last five chapters of the third, the second being devoted to contact transformations.

As the title indicates, this work is only concerned with transformation groups in the sense of equations (4), where the space of "variables" x_i and the space of "parameters" a_j play initially such important roles. On the other hand, the concept of "abstract" group had not been clearly isolated at that period; when in 1885 ([3*f*], §5) Lie noted that in the notation of (10) the equation $w = \mathrm{H}(u, v)$ which gives the parameters of the composition of two transformations of the group defines a new group, he considered it as a *transformation group* on the space of parameters, thus obtaining what he called the "parameter group" (he even obtained two, which are just the group of left translations and the group of right translations†).

The variables x_i and the parameters a_j in equations (4) were in principle assumed to be complex (except in Chapters XIX–XXIV of volume 3) and the functions f_i to be analytic; Lie and Engel were of course aware of the fact that these functions are not in general defined for all complex values of the x_i and a_j and that therefore the composition of such functions raises serious difficulties ([4], vol. 1, pp. 15–17, pp. 33–40 and *passim*); and although throughout they almost always wrote as if the composition of the transformations they were studying was possible without restriction, this was no doubt for the convenience

‡ Today they are called "Lie pseudo-groups"; care should be taken not to confuse them with the "Banach" Lie groups defined in this volume.

§ From 1886 to 1898, Lie occupied the chair at Leipzig left vacant by Klein and had Engel as his assistant; this circumstance favoured the production of an active mathematical school as well as the diffusion of Lie's ideas, so little known until then (notably because of the fact that his first memoirs were usually written in Norwegian and published in the Comptes Rendus de l'Académie de Christiania, which was little used elsewhere). Thus at a time when it was unusual for young French mathematicians to go to Germany for instruction, E. Vessiot and A. Tresse spent a year studying at Leipzig with Sophus Lie.

† The analogous notion for permutation groups had been introduced and studied by Jordan in his "*Treatise*".

of the results and they explicitly reaffirmed the "local" point of view whenever necessary (cf. *loc. cit.*, p. 168 or 189 for example or *ibid.*, vol. 3, p. 2, note at the bottom of the page); in other words, the mathematical object they studied is close to what we call in this treatise a law chunk of operation. They did not refrain, on occasions, from considering global groups, for example the 4 series of classical groups ([4], vol. 3, p. 682), but do not appear to have asked themselves the question of what in general constitutes a "global group"; they were content to obtain, for the "parameters" of the classical groups (the "variables" of these groups introduced no difficulty, since the transformations in question are linear transformations of $\mathbf{C}^n$) "local" systems of parameters in a neighbourhood of the identity transformation, without worrying about the domain of validity of the formulae they were writing down. They however set themselves a problem which arises neatly out of the local theory‡: the study of "mixed" groups, that is groups with a finite number of connected components, such as the orthogonal group ([4], vol. 1, p. 7). They presented this study as a study of a set of transformations stable under composition and passage to the inverse which is the union of sets H_j each of which is described by systems of functions $(f_i^{(j)})$ as in (4); the number of (essential) parameters of each H_j is even *a priori* assumed to depend on j, but they showed that in fact this number is the same for all the H_j. Their principal result was then the existence of a finite continuous group G such that $\mathrm{H}_j = \mathrm{G} \circ h_j$ for some $h_j \in \mathrm{H}_j$ and for all j; they also established that G is normal in the mixed group and noted that the determination of the invariants of the latter reduces to that of the invariants of G and a discontinuous group ([4], vol. 1, Chapter 18).

The general theory developed in [4] ended (without the authors saying so very systematically) by achieving a "dictionary" translating the properties of "finite continuous" groups into those of the set of their infinitesimal transformations. It is based on the "three theorems of Lie", each of which consists of an assertion and its converse.

The *first theorem* ([4], vol. 1, pp. 33 and 72 and vol. 3, p. 563) affirms in the first place that if in (4) the parameters are effective, the functions f_i satisfy a system of partial differential equations of the form

$$(15) \qquad \frac{\partial f_i}{\partial a_j} = \sum_{k=1}^{r} \xi_{ki}(f(x, a))\psi_{kj}(a) \quad (1 \leqslant i \leqslant n)$$

where the matrix (ξ_{ki}) is of maximum rank and $\det(\psi_{kj}) \neq 0$; conversely, if the

‡ Recall (Historical Note to *Algebra*, Chapter VIII) that following a Note of H. Poincaré ([14], vol. V, pp. 77–79) several authors studied the group of invertible elements of a finite-dimensional associative algebra. It is interesting to note on this subject that E. Study, in his works on this subject, introduced a symbolism which essentially amounted to considering the abstract group defined by the parameter group.

functions f_i have this property, formulae (4) define a group germ of transformations.

The *second theorem* ([4], vol. 1, pp. 149 and 158, and vol. 3, p. 590) gives relations between the ξ_{ki} on the one hand and the ψ_{ij} on the other: the conditions on the ξ_{ki} can be written in the form

$$(16)\quad \sum_{k=1}^{n}\left(\xi_{ik}\frac{\partial \xi_{jl}}{\partial x_k} - \xi_{jk}\frac{\partial \xi_{il}}{\partial x_k}\right) = \sum_{k=1}^{r} c_{ij}^{k}\xi_{kl} \quad (1 \leqslant i, j \leqslant r, 1 \leqslant l \leqslant n)$$

where the c_{ij}^k are constants $(1 \leqslant i, j, k \leqslant r)$ skew-symmetric in i, j. The conditions on the ψ_{ij}, in the form given by Maurer [10], are:

$$(17)\quad \frac{\partial \psi_{kl}}{\partial a_m} - \frac{\partial \psi_{km}}{\partial a_l} = \tfrac{1}{2}\sum_{1\leqslant i,j\leqslant r} c_{ij}^{k}(\psi_{il}\psi_{jm} - \psi_{jl}\psi_{im}) \quad (1 \leqslant k, l, m \leqslant r).$$

By introducing the contragradient matrix (α_{ij}) of (ψ_{ij}) and the infinitesimal transformations

$$(18)\quad X_k = \sum_{i=1}^{n} \xi_{ki}\frac{\partial}{\partial x_i}, \qquad A_k = \sum_{j=1}^{r} \alpha_{kj}\frac{\partial}{\partial a_j} \quad (1 \leqslant k \leqslant r)$$

(16) and (17) can be written respectively:

$$(19)\quad [X_i, X_j] = \sum_{k=1}^{r} c_{ij}^{k} X_k \quad (1 \leqslant i, j \leqslant r).$$

$$(20)\quad [A_i, A_j] = \sum_{k=1}^{r} c_{ij}^{k} A_k$$

Conversely, if r infinitesimal transformations X_k $(1 \leqslant k \leqslant r)$ are given which are linearly independent and satisfy conditions (19), the one-parameter subgroups generated by these transformations generate a transformation group in r essential parameters.

Finally, the *third theorem* ([4], vol. 1, pp. 170 and 297 and vol. 3, p. 597) reduces the determination of the systems of infinitesimal transformations $(X_k)_{1\leqslant k\leqslant r}$ satisfying (19) to a purely algebraic problem: the following must hold:

$$(21)\quad c_{ij}^{k} + c_{ji}^{k} = 0$$

$$(22)\quad \sum_{l=1}^{r} (c_{il}^{m}c_{jk}^{l} + c_{kl}^{m}c_{ij}^{l} + c_{jl}^{m}c_{ki}^{l}) = 0 \quad (1 \leqslant i, j, k, m \leqslant r).$$

Conversely,† if (21) and (22) are satisfied, there exists a system of infinitesimal

† This converse was not obtained without difficulty. The first proof given by Lie [3 *d*] consisted in passing to the adjoint group and was in fact only valid if the centre of the given Lie algebra was {0}. He then gave two general proofs ([4], vol. 2, Chapter 17 and vol. 3, pp. 599–604); it is very significant that the first was based on contact transformations and that Lie found it more natural than the second.

transformations satisfying relations (19), whence a transformation group with r parameters (in other words, the linear combinations of the X_k with constant coefficients form a Lie algebra and conversely every finite-dimensional Lie algebra can be obtained in this way).

These results are completed by studying the questions of isomorphisms. Two transformation groups are called *similar* if it is possible to pass from one to the other by an invertible coordinate transformation on the variables and an invertible coordinate transformation on the parameters: from the very beginning of his research, Lie had been concerned naturally with this notion when defining the "canonical parameters". He showed that two groups are similar if, by a transformation on the "variables", it is possible to transform the infinitesimal transformations of the one into those of the other ([4], vol. 1, p. 329). A necessary condition for this to be the case, is that the Lie algebras of these two groups be isomorphic, which Lie expressed by saying that the groups are "*gleichzusammengesetzt*"; but this condition is not sufficient and a whole chapter ([4], vol. 1, Chapter 19) is devoted to obtaining supplementary conditions assuring that the groups are "similar". The theory of permutation groups on the other hand provided the notion of "holohedric isomorphism" of two such groups (isomorphism of the underlying "abstract" groups); Lie transposed this notion to transformation groups and showed that two such groups are "holohedrically isomorphic" if and only if their Lie algebras are isomorphic ([4], vol. 1, p. 418). In particular, every transformation group is holohedrically isomorphic to each of its parameter groups and this shows that, when we wish to study the structure of the group, the "variables" on which it operates are of little importance and that in fact all that matters is the Lie algebra.†

Always by analogy with the theory of permutation groups, Lie introduced the notions of subgroups, normal subgroups and "merihedric isomorphisms" (surjective homomorphisms) and showed that they correspond to those of subalgebras, ideals and surjective Lie algebra homomorphisms; he had already previously come across an especially important example of a "merihedric isomorphism", the adjoint representation, and had recognized its relations with the centre of the group ([3 *f*], § 3, no. 9). For these results, as for the fundamental theorems, the essential tool is the Jacobi-Clebsch Theorem giving the complete integrability of a differential system (one of the forms of the theorem called "Frobenius's"); he however gave a new proof using one-parameter groups ([4], vol. 1, Chapter 6).

---

† A similar evolution can be pointed out in the theory of "abstract" groups, in particular finite groups. They were first defined as transformation groups, but Cayley had already noted that what is essential is the way the transformations are composed with one another and not the nature of the concrete representation of the group as a permutation group of particular objects.

The notions of transitivity and primitivity, so important for permutation groups, occurred just as naturally for "finite continuous" transformation groups and the Lie–Engel treatise made a detailed study of this ([4], vol. 1, Chapter 13 and *passim*); the relations with the stabilizer subgroups of a point and the notion of homogeneous space were perceived (insofar as was possible without taking a global point of view) ([4], vol. 1, p. 425).

Finally the "dictionary" was completed, in [4], by the introduction of the notions of derived group and solvable group (called "integrable group" by Lie; this terminology, suggested by the theory of differential equations, remained in use until the works of H. Weyl) ([4], vol. 1, p. 261 and vol. 3, pp. 678–679); the relation between commutators and brackets had elsewhere been perceived by Lie in 1883 ([3], vol. 5, p. 358).

Other proofs of the fundamental theorems

In [8] F. Schur showed that in canonical coordinates the ψ_{ik} of (15) satisfy the differential equations

$$(23)\qquad \frac{d}{dt}(t\psi_{ik}(ta)) = \delta_{ik} + \sum_{j,l} c_{jl}^{k} ta_l \psi_{ij}(ta).$$

These are integrated and give a formula equivalent to the formula

$$(24)\qquad \varpi(X) = \sum_{n>0} \frac{1}{(n+1)!}(\operatorname{ad}(X))^n$$

of our Chapter III, § 6, no. 4, Proposition 12; in particular, in canonical coordinates, the ψ_{ij} can be extended to integral functions of the a_k. F. Schur deduced a result which made precise an earlier remark of Lie: if, in definition (4) of transformation groups, it is only assumed that the f_i are of class C^2, then the group is holohedrically isomorphic to an analytic group.† Following his

† Lie had already stated without proof a result of this type ([3 *g*], no. 7). He had been led there by his research on the foundations of geometry ("Helmholz's problem"), where he had remarked that the analyticity hypotheses are not natural.

The result of F. Schur was to lead Hilbert in 1900 to ask if the same conclusion remained true if it was only assumed that the f_i were continuous ("Hilbert's 5th problem"). This problem has stimulated much research. The most complete result along this line is the following theorem, proved by A. Gleason, D. Montgomery and L. Zippin; every locally compact topological group has an open subgroup which is an inverse limit of Lie groups; it implies that every locally Euclidean group is a Lie group. For more details on this question, cf. D. Montgomery and L. Zippin [41].

research on the integration of differential systems, E. Cartan ([12], vol. II_2, p. 371) introduced in 1904 Pfaffian forms

$$(25) \qquad \omega_k = \sum_{i=1}^{r} \psi_{ki}\, da_i \quad (1 \leqslant i \leqslant r)$$

(in the notation of (15)), called later *Maurer-Cartan forms*. The Maurer conditions (17) could be written

$$d\omega_k = -\tfrac{1}{2}\sum_{i,j} c_{ij}^k \omega_i \wedge \omega_j;$$

E. Cartan showed that the theory of finite continuous groups could be developed starting from the ω_k and established the equivalence of this point of view and that of Lie. But, for him, the interest of this method was above all that it could be adapted to "infinite continuous groups" the theory of which he pushed much further than Lie had done and which allowed him to develop his theory of the generalized "moving frame".

IV. THE THEORY OF LIE ALGEBRAS

Having once acquired the correspondence between transformation groups and Lie algebras, the theory took a considerably more algebraic turn and became centred on a deep study of Lie algebras.†

A first, short period, from 1888 to 1894, marked by the works of Engel and his pupil Umlauf and especially Killing and E. Cartan, achieved a series of spectacular results on complex Lie algebras. We have seen earlier that the notion of solvable Lie algebra was due to Lie himself, who had shown (in the complex case) the theorem on the reduction of solvable linear Lie algebras to triangular form ([4], vol. 1, p. 270).‡ Killing observed [11] that there exists in a Lie algebra a largest solvable ideal (which we now call the radical) and that the quotient of the Lie algebra by its radical has zero radical; he called Lie algebras with zero radical *semi-simple* and proved that they are products of simple algebras (the latter notion having already been introduced by Lie, who had proved the simplicity of the "classical" algebras ([4], vol. 3, p. 682)).

On the other hand, Killing introduced in a Lie algebra the characteristic equation $\det(\mathrm{ad}(x) - \omega.1) = 0$, already encountered by Lie when studying 2-dimensional Lie subalgebras containing a given element of a Lie algebra. We shall return in other Historical Notes to this Book to the analysis of the methods by which Killing, whilst engaged on a penetrating study of the pro-

† The term "Lie algebra" was introduced by H. Weyl in 1934: in his work of 1925, he had used the expression "infinitesimal group". Earlier, mathematicians had spoken simply of the "infinitesimal transformations $X_1 f, \ldots, X_r f$" of the group, which Lie and Engel frequently abbreviated by saying "the group $X_1 f, \ldots, X_r f$"!

‡ Almost at the beginning of his research, Lie had encountered solvable linear groups and even in fact nilpotent linear groups [3 *h*].

perties of the roots of the "generic" characteristic equation for a semi-simple algebra, achieved the most remarkable of his results, the *complete* determination of the (complex) simple Lie algebras.§

Killing proved that the derived algebra of a solvable algebra is "of rank 0" (which means that $\operatorname{ad} x$ is nilpotent for every element x of the algebra). Almost immediately, Engel showed that algebras "of rank 0" are solvable (this statement is essentially what we called Engel's Theorem in Chapter I, § 4, no. 2). In his thesis, E. Cartan introduced, on the other hand, what we now call the "Killing form" and established the two fundamental criteria which characterize by means of this form solvable Lie algebras and semi simple Lie algebras.

Killing had affirmed ([11], IV) that the derived algebra of a Lie algebra is the sum of a semi-simple algebra and its radical, which is nilpotent, but his proof was incomplete. A little later, E. Cartan announced without proof ([12], vol. I_1, p. 104) that more generally every Lie algebra is the sum of its radical and a semi-simple subalgebra; the only result in this direction established indisputably at this period was a theorem of Engel affirming the existence, in every non-solvable Lie algebra, of a simple Lie subalgebra of dimension 3. The first published proof (for complex Lie algebras) of Cartan's statement is due to E. E. Levi [18]; another proof (equally valid in the real case) was given by J. H. C. Whitehead in 1936 [26 a]. In 1942, A. Malcev completed this result by the uniqueness theorem for "Levi sections" up to conjugation.

From his very first works, Lie was concerned with the problem of the isomorphism between any Lie algebra and a linear Lie algebra. He had believed that he had solved it affirmatively by considering the adjoint representation (and deduced from it a proof of his "third theorem") [3 *d*]; he realized that his proof was correct only for Lie algebras with zero centre, gave another erroneous proof thereof ([3 *f*], § 3, no. 9) and then recognized ([3 *g*], p. 231) that the question remained open. In fact, it remained so for a long time and was only finally resolved affirmatively in 1935 by Ado [27]. On the other hand, Lie had essentially posed the problem of determining linear representations of simple Lie algebras of minimal dimension and had solved it for the classical groups; in his Thesis, Cartan also solved this problem for the exceptional simple algebras†; the methods he used for this were to be generalized by him twenty years later to obtain all the irreducible representations of real or complex simple Lie algebras.

---

§ Up to the fact that he found two exceptional groups of dimension 52 but did not notice that they were isomorphic. (He only considered complex simple Lie algebras for the more general problem had not been considered at that time; Killing's methods are in fact valid for any algebraically closed field of characteristic 0).

† Cartan's approach consisted of studying the non-trivial extension Lie algebras of a simple Lie algebra and a (commutative) radical of minimal dimension.

The property of complete reducibility of a linear representation seems to have been encountered for the first time (in a geometric form) by Study. In an unpublished manuscript, but quoted in [4], vol. 3, pp. 785–788, he proved this property for linear representations of the Lie algebra **SL**(2, **C**) and obtained partial results for **SL**(3, **C**) and **SL**(4, **C**). Lie and Engel conjectured on this occasion that the complete reducibility theorem was true for **SL**(n, **C**) for all n. The complete reducibility of linear representations of semi-simple Lie algebras was established by H. Weyl in 1925† by a global type of argument (see later). The first algebraic proof was obtained in 1935 by Casimir and van der Waerden [32]; other algebraic proofs have since been given by R. Brauer [31] (this is the one we have reproduced) and J. H. C. Whitehead [26 b].

Finally, in the course of his research on the exponential mapping (cf. *infra*), H. Poincaré ([14], vol. 3) considered the associative algebra of differential operators of all orders, generated by the operators of a Lie algebra; he showed essentially that, if $(X_i)_{1 \leqslant i \leqslant n}$ is a basis of the Lie algebra, the associative algebra generated by the X_i has as basis certain symmetric functions of the X_i (sum of the non-commutative "monomials" derived from a given monomial by all permutations of factors). The essence of his proof was algebraic in nature and allowed him to obtain the enveloping algebra structure which we define abstractly in Chapter I. Analogous proofs were given in 1937 by G. Birkhoff [29 b] and E. Witt [30].‡

Most of the results cited above are limited to real or complex Lie algebras which alone correspond to Lie groups in the usual sense. The study of Lie algebras over a field other than **R** or **C** was embarked upon by Jacobson [28 a] who showed that the great majority of the classical results (i.e. those of Chapter I) remain true for any field of characteristic zero.

V. EXPONENTIAL AND HAUSDORFF FORMULA

The first research on the exponential mapping was due to E. Study and F. Engel; Engel [9 b] remarked that the exponential is not surjective for **SL**(2, **C**) (for example $\begin{pmatrix} -1 & a \\ 0 & -1 \end{pmatrix}$ is not an exponential if $a \neq 0$), but that it is for **GL**(n, **C**) and hence also for **PGL**(n, **C**) (the latter property had already been noted by Study for $n = 2$); thus **SL**(2, **C**) and **PGL**(2, **C**) give an example of two locally isomorphic groups, which are however very different from a global

† H. Weyl remarked on this occasion that the construction given by E. Cartan for irreducible representations implicitly uses this property.

‡ The first use of differential operators of higher order generated by the X_i was no doubt the use of the "Casimir operator" for the proof of the complete reducibility theorem. After 1950, the research of Gelfand and his school and Harish-Chandra, on infinite-dimensional linear representations, has brought these operators into the forefront.

point of view. Engel also showed that the exponential is surjective for the other classical groups augmented by homotheties; this work was taken up and pursued by Maurer, Study and others, without producing substantial new results.

In 1899, H. Poincaré ([14], vol. 3, pp. 169–172 and 173–212) embarked upon the study of the exponential mapping from a different point of view. His memoirs seem to have been hastily edited, for in several places he affirmed that every element of a connected group is an exponential, whereas he gave examples to the contrary elsewhere. His results were mainly concerned with the adjoint representation: he showed that a semi-simple element of such a group G may be the exponential of an infinity of elements of the Lie algebra L(G), whereas a non-semi-simple element may not be an exponential at all. If ad(X) has no eigenvalue which is a non-zero multiple of $2\pi i$, then exp is étale at X. He also proved that, if U and V describe loops in L(G) and W is defined by continuity such that $e^{U}.e^{V} = e^{W}$, it is possible not to return to the original value of W. He used a residue formula which amounts essentially to

$$\Phi(\mathrm{ad}\, X) = \frac{1}{2\pi i}\int \frac{\Phi(\xi)\, d\xi}{\xi - \mathrm{ad}\, X}$$

where ad(X) is a semi-simple element whose non-zero eigenvalues are of multiplicity 1, Φ is an integral series with sufficiently large radius of convergence and the integral is taken over a contour surrounding the eigenvalues of ad X; he also studied what happens as X tends to a transformation with multiple eigenvalues.

Research into expressions for W as a function of U and V in the formula $e^{U}.e^{V} = e^{W}$ had already, just before Poincaré's work, been the object of two memoirs by Campbell [13]. As Baker wrote a little later "... *Lie theory suggests in an obvious way that the product* $e^{U}e^{V}$ *is the form* e^{W} *where* W *is a series of alternants in* U *and* V ...". The later works on this subject aimed at making this assertion precise and giving an explicit formula (or a method of construction) for W ("Hausdorff formula"). After Campbell and Poincaré, Pascal, Baker [15] and Hausdorff [16] returned to the question; each considered that the proofs of his predecessors were not convincing; the principal difficulty resided in what is meant by "alternants": are they special elements of the Lie algebra in question, or universal "symbolic" expressions? Neither Campbell, nor Poincaré, nor Baker expressed himself clearly on this point. Hausdorff's memoir, on the other hand, is perfectly precise; he worked first on the algebra of (non-commutative) associative formal power series in a finite number of indeterminates and considered U, V, W as elements of this algebra. He proved the existence of W by a differential equation argument analogous to that of his predecessors. He used the same argument to prove the convergence of the series when the indeterminates are replaced by elements of a finite-dimensional Lie algebra. As Baker had remarked, and Poincaré independently, this result could be used to give a proof of Lie's third theorem; he clarified the correspondence between Lie groups and Lie algebras, for example where the commutator subgroup is concerned.

In 1947, Dynkin [39] again took up the question and obtained the explicit coefficients of the Hausdorff formula, by considering from the outset a normed Lie algebra (of finite dimension or otherwise, over **R**, **C** or an ultrametric field).†

VI. LINEAR REPRESENTATIVES AND GLOBAL LIE GROUPS

None of the works which we have just mentioned made any honest attempt to define and study global Lie groups. It was H. Weyl who made the first move in this direction. He was inspired by two theories, which until then had developed independently: that of linear representations of complex semi-simple Lie algebras, due to E. Cartan, and that of linear representations of finite groups, due to Frobenius and which had just been transposed to the orthogonal group by I. Schur, using an idea of Hurwitz. The later had shown [17] how to form invariants for the orthogonal group or unitary group by replacing the operation of the mean on a finite group by an integration with respect to an invariant measure. He had also noted that, by applying this method to the unitary group, invariants are obtained for the general linear group, the first example of the "unitarian trick". In 1924, I. Schur [20] used this procedure to show the complete reducibility of the representations of the orthogonal group $\mathbf{O}(n)$ and the unitary group $\mathbf{U}(n)$ by constructing an invariant non-degenerate positive definite Hermitian form; he deduced, by the "unitarian trick", the complete reducibility of the holomorphic representations of $\mathbf{O}(n, \mathbf{C})$ and $\mathbf{SL}(n, \mathbf{C})$, established orthogonality relations for the characters of $\mathbf{O}(n)$ and $\mathbf{U}(n)$ and determined the characters of $\mathbf{O}(n)$. H. Weyl immediately extended this method to complex semi-simple Lie algebras [21]. Given such an algebra $\mathfrak{g}$, he showed that it has a "compact real form" (which amounts to saying that is obtained by extension of scalars from **R** to **C** from an algebra $\mathfrak{g}_0$ over **R** whose adjoint group G_0 is compact). Further, he showed that the fundamental group of G_0 is finite and hence that the universal covering‡ of G_0 is compact. He deduced, by a suitable adaptation of Schur's procedure, the complete reducibility of the representations of $\mathfrak{g}$ and also gave, by a global method, the determination of the

† In the ultrametric case, the classical method of upper bounds cannot be extended without precautions because of the asymptotic behaviour of the p-adic absolute value of $1/n$ as n tends to infinity.

‡ H. Weyl did not define explicitly this notion, with which he had been familiar since the editing of his course on Riemann surfaces (1913). It was O. Schreier [22] who in 1926–1927 gave, for the first time, the definition of a topological group and that of a "continuous" group (i.e. locally homeomorphic to a Euclidean space), and the construction of the universal covering of such a group.

characters of the representations of $\mathfrak{g}$. In a letter to I. Schur (*Sitzungsber. Berlin*, 1924, pp. 338–343), H. Weyl summarized the results of Cartan not known to Schur (cf. [20], p. 299, note at the bottom of the page) and compared the two approaches: Cartan's method gave all the holomorphic representations of the simply connected group with Lie algebra $\mathfrak{g}$; in the case of the orthogonal group, representations were thus obtained of a double covering (called later the spinor group), which had escaped Schur; on the other hand, Schur's method had the advantage of showing the complete reducibility and giving the characters explicitly.

After the work of H. Weyl, E. Cartan adopted an openly global approach in his research on symmetric spaces and Lie groups. This approach was the basis of his exposition of 1930 ([12], vol. I_2, pp. 1165–1225) of the theory of "finite continuous" groups. In particular we find here the first proof of the global version of the 3rd fundamental theorem (the existence of a Lie group with given Lie algebra); Cartan also showed that every closed subgroup of a real Lie group is a Lie group (Chapter III, § 8, no. 2, Theorem 2) which generalized a result of J. von Neumann on closed subgroups of the linear group [23]. In this Memoir, von Neumann showed also that every continuous representation of a semi-simple group is real analytic.

After these works, the theory of Lie groups in the "classical" sense (that is in finite dimension over **R** or **C**) was more or less complete in its essentials. The first detailed exposition of it was given by Pontrjagin in his book on topological groups [36]; there he followed an approach quite close to that of Lie, but carefully distinguished between the local and the global. This was followed by Chevalley's book [38] which also contains the first systematic discussion of the theory of analytic manifolds and exterior differential calculus; the "infinitesimal transformations" of Lie appear there as vector fields and the Lie algebra of a Lie group G is identified with the space of left invariant fields on G. He leaves out any discussion on "group germs" and "transformation groups".

VII. EXTENSIONS OF THE NOTION OF LIE GROUP

Nowadays the vitality of Lie theory is manifested by the diversity of its applications (in topology, differential geometry, arithmetic, etc.) and by the creation of parallel theories where the underlying manifold structure is replaced by a similar structure (p-adic manifold, algebraic variety, scheme, formal scheme, ...). We are not concerned here with the history of these developments and we shall confine ourselves to those touched on in Chapter III: Banach Lie groups and p-adic Lie groups.

(a) *Banach Lie groups*

Here we are concerned with "infinite-dimensional" Lie groups. From the local point of view, a neighbourhood of 0 in Euclidean space is replaced by a neighbourhood of 0 in a Banach space. This is what G. Birkhoff did in 1936 [29 a], thus achieving the notion of a *complete normed Lie algebra* and its correspondence with a "group germ" defined on an open set of a Banach space. About 1950, Dynkin completed these results by extending the Hausdorff formula to this case (cf. *supra*).

The definitions and results of Birkhoff and Dynkin are local. Until recently, it does not seem that any one has sought to make the corresponding global theory explicit, no doubt because of the lack of applications.†

(b) *p-adic Lie groups*

Such groups were encountered for the first time in 1907 in the works of Hensel [19] on *p*-adic analytic functions (defined by expansions as integral series). He studied in particular the exponential and the logarithm; in spite of the *a priori* surprising behaviour of the series which define them (for example the exponential series does not converge everywhere), their fundamental functional properties remain valid, which provides a *local isomorphism* between the additive group and the multiplicative group of $\mathbf{Q}_p$ (or, more generally, of any complete ultrametric field of characteristic zero).

A. Weil [33] and E. Lutz [34] were equally concerned with commutative (but this time non-linear) groups in their work on *p*-adic elliptic curves (1936). Besides arithmetic applications, a local isomorphism is constructed here between the group and the additive group, based on the integration of an invariant differential form. This method applies equally to Abelian varieties, as was noted by C. Chabauty soon afterwards, who used it without further explanation to prove a particular case of the "Mordell conjecture" [35].

From then on, it was clear that the *local* theory of Lie groups could be applied with scarcely any change to the *p*-adic case. The fundamental theorems of the Lie groups–Lie algebras "dictionary" were established in 1942 in the thesis of R. Hooke [37], a pupil of Chevalley; this work also contained the *p*-adic analogue of E. Cartan's theorem on closed subgroups of real Lie groups.

More recently, M. Lazard [42 b] developed a more precise form of the "dictionary" for compact analytic groups over $\mathbf{Q}_p$. He showed that the existence of a *p*-adic analytic structure on a compact group G is closely related to that of certain filtrations on G and gives various applications of this (for example to the cohomology of G). One of Lazard's tools is an improvement of Dynkin's results on the convergence of the *p*-adic Hausdorff series [42 a].

---

† If, in spite of this lack of applications, we have mentioned "Banach" groups in Chapter III, it is because Banach manifolds are being used more and more in Analysis (and even for the study of finite-dimensional manifolds) and because, moreover, this generalization offers no extra difficulty.

VIII. FREE LIE ALGEBRAS

It remains to speak of a series of works on *Lie algebras* where the connection with the theory of *Lie groups* is very tenuous; this research has on the other hand important applications to the theory of "abstract" groups and more especially nilpotent groups.

Its origin is the work of P. Hall [24], which appeared in 1932. However Lie algebras are not discussed here: P. Hall had in mind the study of a certain class of p-groups, those which he called "regular". But this led him to examine in detail iterated commutators and the lower central series of a group; on this subject he established a version of the Jacobi identity (cf. Chapter II, § 4, no. 4, formula (20)) and the "Hall formula"

$$(xy)^n = x^n y^n (x, y)^{n(1-n)/2} \ldots \quad \text{(cf. Chapter II, § 5, Exercise 9).}$$

Almost immediately (in 1935–1937) appeared the fundamental works of W. Magnus ([25 a] and [25 b]) and E. Witt [30]. In [25 a] Magnus used the same algebra of formal power series $\hat{\mathrm{A}}$ as Hausdorff (since called "Magnus algebra"); he embedded the free group F in it and used the natural filtration of $\hat{\mathrm{A}}$ to obtain a decreasing sequence (F_n) of subgroups of F; it is one of the first examples of a *filtration*. He conjectured that the F_n coincide with the terms of the lower central series of F. This conjecture was proved in his second memoir [25 b]; also in this work he showed explicitly the close relation between his ideas and those of P. Hall and defined the free Lie algebra L (as a subalgebra of $\hat{\mathrm{A}}$) which he showed essentially to be identified with the graded algebra of F. In [30], Witt completed this result on various points. He showed notably that the enveloping algebra of L is a free associative algebra and deduced immediately the rank of the homogeneous components of L ("Witt formulae").

As for the proof of the basis of L known as the "Hall basis" (cf. Chapter II, § 2, no. 11), it seems that it only appeared in 1950 in a note of M. Hall [40], although it was implicit in the works of P. Hall and W. Magnus quoted above.

BIBLIOGRAPHY

1. F. Klein and S. Lie: (*a*) Sur une certaine famille de courbes et surfaces, *C. R. Acad. Sci.*, **70** (1870), pp. 1222–1226 and 1275–1279 (=[2], pp. 416–420 and [3], vol. 1, pp. 78–85); (*b*) Über diejenigen ebenen Kurven, welche durch ein geschlossenes System von einfach unendlich vielen vertauschbaren linearen Transformationen in sich übergehen, *Math. Ann.*, **4** (1871), pp. 50–84 (=[2], pp. 424–459 and [3], vol. 1, Abh. XIV, pp. 229–266).
2. F. Klein, *Gesammelte mathematische Abhandlungen*, Bd. I, Berlin (Springer), 1921.
3. S. Lie, *Gesammelte Abhandlungen*, 7 vol., Leipzig (Teubner): (*a*) Über die Reziprozitätsverhältnisse des Reyeschen Komplexes, vol. I, Abh. V, pp. 68–77 (=*Gött. Nach.* (1870), pp. 53–66); (*b*) Über eine Klasse geometrischer Transformationen, vol. I, Abh. XII, pp. 153–214 (=*Christiana For.* (1871), pp. 182–245); (*c*) Über partielle Differentialgleichungen erster Ordnung, vol. III, Abh. VII, pp. 32–63 (=*Christiana For.* (1873), pp. 16–51); (*d*) Theorie der Transformationsgruppen II, vol. V, Abh. III, pp. 42–75 (=*Archiv f. Math.*, **1** (1876), pp. 152–193); (*e*) Ueber Gruppen von Transformationen, vol. V, Abh. I, pp. 1–8 (=*Gött. Nachr.* (1874), pp. 529–542); (*f*) Allgemeine Untersuchungen über Differentialgleichungen, die eine kontinuierliche endliche Gruppe gestatten, vol. VI, Abh. III, pp. 139–223 (=*Math. Ann.*, **25** (1885), pp. 71–151); (*g*) Beiträge zur allgemeinen Transformationenstheorie, vol. VI, Abh. V, pp. 230–236 (=*Leipziger Ber.* (1888), pp. 14–21); (*h*) Theorie der Transformationsgruppen III, vol. V, Abh. IV, pp. 78–133 (=*Archiv f. Math.*, vol. III, (1887), pp. 93–165).
4. S. Lie and F. Engel, *Theorie der Transformationsgruppen*, 3 vol., Leipzig (Teubner), 1888–1893.
5. J. J. Sylvester, *Collected Mathematical Papers*, 4 vols., Cambridge, 1904–1911.
6. A. Cayley, *Collected Mathematical Papers*, 13 vols., Cambridge, 1889–1898.

7. C. Jordan, Mémoire sur les groupes de mouvements, *Annali di Math.*, **11** (1868–1869), pp. 167–215 and 332–345 (= *Oeuvres*, vol. IV, pp. 231–302).
8. F. Schur: (*a*) Zur Theorie der aus Haupteinheiten gebildeten Komplexen, *Math. Ann.*, **33** (1889), pp. 49–60; (*b*) Neue Begründung der Theorie der endlichen Transformationsgruppen, *Math. Ann.*, **35** (1890), pp. 161–197; (*c*) Zur Theorie der endlichen Transformationsgruppen, *Math. Ann.*, **38** (1891), pp. 273–286; (*d*) Über den analytischen Character der eine endliche continuierliche Transformationsgruppe darstellende Funktionen, *Math. Ann.*, **41** (1893), pp. 509–538.
9. F. Engel: (*a*) Über die Definitionsgleichung der continuierlichen Transformationsgruppen, *Math. Ann.*, **27** (1886), pp. 1–57; (*b*) Die Erzeugung der endlichen Transformationen einer projektiven Gruppe durch die infinitesimalen Transformationen der Gruppe, I, *Leipziger Ber.*, **44** (1892), pp. 279–296, II (mit Beiträgen von E. Study), *ibid.*, **45** (1893), pp. 659–696.
10. L. Maurer, Über allgemeinere Invarianten-Systeme, *Sitzungsber. München*, **18** (1888), pp. 103–150.
11. W. Killing, Die Zusammensetzung der stetigen endlichen Transformationsgruppen: (I) *Math. Ann.*, **31** (1888), pp. 252–290; (II) *ibid.*, **33** (1889), pp. 1–48; (III) *ibid.*, **34** (1889), pp. 57–122; (IV) *ibid.*, **36** (1890), pp. 161–189.
12. E. Cartan, *Oeuvres complètes*, 6 vol., Paris (Gauthier-Villars), 1952–54.
13. J. E. Campbell: (*a*) On a law of combination of operators bearing on the theory of continuous transformation groups, *Proc. London Math. Soc.*, (1) **28** (1897), pp. 381–390; (*b*) On a law of combination of operators (second paper), *ibid.*, **29** (1898), pp. 14–32.
14. H. Poincaré, *Oeuvres*, 11 vol., Paris (Gauthier-Villars), 1916–1956.
15. H. F. Baker, Alternants and continuous groups, *Proc. London Math. Soc.*, (2) **3** (1905), pp. 24–47.
16. F. Hausdorff, Die symbolische Exponentialformel in der Gruppentheorie, *Leipziger Ber.*, **58** (1906), pp. 19–48.
17. A. Hurwitz, Über die Erzeugung der Invarianten durch Integration, *Gött, Nachr.* (1897), pp. 71–90 (= *Math. Werke*, vol. II, pp. 546–564).
18. E. E. Levi, Sulla struttura dei gruppi finiti e continui, *Atti Acc. Sci. Torino*, **40** (1905), pp. 551–565 (= *Opere*, vol. I, pp. 101–115).
19. K. Hensel, Über die arithmetischen Eigenschaften der Zahlen, *Jahresber. der D.M.V.*, **16** (1907), pp. 299–319, 388–393, 474–496.
20. I. Schur, Neue Anwendungen der Integralrechnung auf Probleme der Invariantentheorie, *Sitzungsber. Berlin*, 1924, pp. 189–208, 297–321, 346–355.
21. H. Weyl, Theorie der Darstellung kontinuierlicher halb-einfacher Gruppen durch lineare Transformationen, I, *Math. Zeitschr.*, **23**, (1925),

pp. 271–309; II, *ibid.*, **24** (1926), pp. 328–376; III, *ibid.*, **24** (1926), pp. 377–395 (=*Werke*, vol. II, pp. 543–647).

22. O. SCHREIER: (*a*) Abstrakte kontinuierliche Gruppen, *Abh. math. Sem. Hamburg*, **4** (1926), pp. 15–32; (*b*) Die Verwandschaft stetiger Gruppen in grossen, *ibid.*, **5** (1927), pp. 233–244.
23. J. VON NEUMANN, Zur Theorie der Darstellung kontinuierlicher Gruppen, *Sitzungsber. Berlin*, 1927, pp. 76–90 (=*Collected Works*, vol. I, pp. 134–148).
24. P. HALL, A contribution to the theory of groups of prime power order, *Proc. London Math. Soc.*, (3) **4** (1932), pp. 29–95.
25. W. MAGNUS: (*a*) Beziehungen zwischen Gruppen und Idealen in einem speziellen Ring, *Math. Ann.*, **111** (1935), pp. 259–280; (*b*) Über Beziehungen zwischen höheren Kommutatoren, *J. Crelle*, **177** (1937), pp. 105–115.
26. J. H. C. WHITEHEAD: (*a*) On the decomposition of an infinitesimal group, *Proc. Camb. Phil. Soc.*, **32** (1936), pp. 229–237 (=*Mathematical Works*, I, pp. 281–289); (*b*) Certain equations in the algebra of a semi-simple infinitesimal group, *Quart. Journ. of Math.*, (2) **8** (1937), pp. 220–237 (=*Mathematical Works*, I, pp. 291–308).
27. I. ADO: (*a*) Note on the representation of finite continuous groups by means of linear substitutions (in Russian), *Bull. Phys. Math. Soc. Kazan*, **7** (1935), pp. 3–43; (*b*) The representation of Lie algebras by matrices (in Russian), *Uspehi Mat. Nauk*, **2** (1947), pp. 159–173 (English translation: *Amer. Math. Soc. Transl.*, (1) **9**, pp. 308–327).
28. N. JACOBSON: (*a*) Rational methods in the theory of Lie algebras, *Ann. of Math.*, **36** (1935), pp. 875–881; (*b*) Classes of restricted Lie algebras of characteristic p, II, *Duke Math. Journal*, **10** (1943), pp. 107–121.
29. G. BIRKHOFF: (*a*) Continuous groups and linear spaces; *Rec. Math. Moscou*, **1** (1936), pp. 635–642; (*b*) Representability of Lie algebras and Lie groups by matrices, *Ann. of Math.*, **38** (1937), pp. 526–532.
30. E. WITT, Treue Darstellung Lieschen Ringe, *J. Crelle*, **177** (1937), pp. 152–160.
31. R. BRAUER, Eine Bedingung für vollständige Reduzibilität von Darstellungen gewöhnlicher und infinitesimaler Gruppen, *Math. Zeitschr.*, **41** (1936), pp. 330–339.
32. H. CASIMIR-B. L. VAN DER WAERDEN, Algebraischer Beweis der vollständigen Reduzibilität der Darstellungen halbeinfacher Liescher Gruppen, *Math. Ann.*, **111** (1935), pp. 1–12.
33. A. WEIL, Sur les fonctions elliptiques p-adiques, *C. R. Acad. Sci.*, **203** (1935), p. 22.
34. E. LUTZ, Sur l'équation $y^2 = x^3 - Ax - B$ dans les corps p-adiques, *J. Crelle*, **177** (1937), pp. 237–247.
35. C. CHABAUTY, Sur les points rationels des courbes algébriques de genre supérieur à l'unité, *C. R. Acad. Sci.*, **212** (1941), pp. 882–884.

36. L. S. PONTRJAGIN, *Topological groups*, Princeton Univ. Press, 1939.
37. R. HOOKE, Linear p-adic groups and their Lie algebras, *Ann. of Math.*, **43** (1942), pp. 641–655.
38. C. CHEVALLEY, *Theory of Lie groups*, Princeton Univ. Press, 1946.
39. E. DYNKIN: (*a*) Evaluation of the coefficients of the Campbell–Hausdorff formula (in Russian), *Dokl. Akad. Nauk*, **57** (1947), pp. 323–326; (*b*) Normed Lie algebras and analytic groups (in Russian), *Uspehi Mat. Nauk*, **5** (1950), pp. 135–186 (English translation: *Amer. Math. Soc. Transl.*, (1) **9**, pp. 470–534).
40. M. HALL, A basis for free Lie rings and higher commutators in free groups, *Proc. Amer. Math. Soc.*, **1** (1950), pp. 575–581.
41. D. MONTGOMERY-L. ZIPPIN, *Topological Transformation Groups*, New York (Interscience), 1955.
42. M. LAZARD: (*a*) Quelques calculs concernant la formule de Haudorff, *Bull. Soc. Math. France*, **91** (1963), pp. 435–451; (*b*) Groupes analytiques p-adiques, *Publ. Math. I.H.E.S.*, no. 26 (1965), pp. 389–603.

INDEX OF NOTATION

The reference numbers indicate respectively the chapter, paragraph and number (or, occasionally, exercise).

INDEX OF TERMINOLOGY

The reference numbers indicate respectively the chapter, paragraph and number (or, occasionally, exercise)

SUMMARY

of certain properties of finite-dimensional Lie algebras over a field of characteristic 0.

Let $\mathfrak{g}$ be a Lie algebra, $\mathfrak{r}$ its radical, $\mathfrak{n}$ its largest nilpotent ideal, $\mathfrak{s}$ its nilpotent radical and $\mathfrak{k}$ the orthogonal of $\mathfrak{g}$ relative to the Killing form. Then $\mathfrak{r}$, $\mathfrak{n}$, $\mathfrak{s}$, $\mathfrak{k}$ are characteristic ideals and $\mathfrak{r} \supset \mathfrak{k} \supset \mathfrak{n} \supset \mathfrak{s}$.

(I) *Any one of the following properties characterizes semi-simple Lie algebras:*

(1) $\mathfrak{r} = \{0\}$; (2) $\mathfrak{n} = \{0\}$; (3) $\mathfrak{k} = \{0\}$; (4) every commutative ideal of $\mathfrak{g}$ is zero; (5) the algebra $\mathfrak{g}$ is isomorphic to a product of simple Lie algebras; (6) every finite-dimensional representation of $\mathfrak{g}$ is semi-simple.

(II) *Any one of the following properties characterizes reductive Lie algebras:*

(1) $\mathfrak{s} = \{0\}$; (2) $\mathfrak{r}$ is the centre of $\mathfrak{g}$; (3) $\mathscr{D}\mathfrak{g}$ is semi-simple; (4) $\mathfrak{g}$ is the product of a semi-simple algebra and a commutative algebra; (5) the adjoint representation of $\mathfrak{g}$ is semi-simple; (6) $\mathfrak{g}$ has a finite-dimensional representation such that the associated bilinear form is non-degenerate; (7) $\mathfrak{g}$ has a finite-dimensional semi-simple faithful representation.

(III) *Any one of the following properties characterizes solvable Lie algebras:*

(1) $\mathscr{D}^p\mathfrak{g} = \{0\}$ for sufficiently large p; (2) there exists a decreasing sequence $\mathfrak{g} = \mathfrak{g}_0 \supset \mathfrak{g}_1 \supset \cdots \supset \mathfrak{g}_n = \{0\}$ of ideals of $\mathfrak{g}$ such that the algebras $\mathfrak{g}_i/\mathfrak{g}_{i+1}$ are commutative; (3) there exists a decreasing sequence

$$\mathfrak{g} = \mathfrak{g}'_0 \supset \mathfrak{g}'_1 \supset \cdots \supset \mathfrak{g}'_{n'} = \{0\}$$

of subalgebras of $\mathfrak{g}$ such that $\mathfrak{g}'_{i+1}$ is an ideal of $\mathfrak{g}'_i$ and $\mathfrak{g}'_i/\mathfrak{g}'_{i+1}$ is commutative; (4) there exists a decreasing sequence $\mathfrak{g} = \mathfrak{g}''_0 \supset \mathfrak{g}''_1 \supset \cdots \supset \mathfrak{g}''_{n''} = \{0\}$ of subalgebras of $\mathfrak{g}$ such that $\mathfrak{g}''_{i+1}$ is an ideal of codimension 1 in $\mathfrak{g}''_i$; (5) $\mathfrak{k} \supset \mathscr{D}\mathfrak{g}$; (6) $\mathscr{D}\mathfrak{g}$ is nilpotent.

(IV) *Any one of the following properties characterizes nilpotent Lie algebras:*
(1) $\mathscr{C}^p\mathfrak{g} = \{0\}$ for sufficiently large p; (2) $\mathscr{C}_p\mathfrak{g} = \mathfrak{g}$ for sufficiently large p; (3) there exists a decreasing sequence $\mathfrak{g} = \mathfrak{g}_0 \supset \mathfrak{g}_1 \supset \cdots \supset \mathfrak{g}_p = \{0\}$ of ideals of $\mathfrak{g}$ such that $[\mathfrak{g}, \mathfrak{g}_i] \subset \mathfrak{g}_{i+1}$; (4) there exists a decreasing sequence $\mathfrak{g} = \mathfrak{g}'_0 \supset \mathfrak{g}'_1 \supset \cdots \supset \mathfrak{g}'_{p'} = \{0\}$ of ideals of $\mathfrak{g}$ such that $[\mathfrak{g}, \mathfrak{g}'_i] \subset \mathfrak{g}'_{i+1}$ and the $\mathfrak{g}'_i/\mathfrak{g}'_{i+1}$ are 1-dimensional; (5) there exists an integer i such that

$$(\operatorname{ad} x_1) \circ (\operatorname{ad} x_2) \circ \cdots \circ (\operatorname{ad} x_i) = 0$$

for all $x_1, \ldots, x_i$ in $\mathfrak{g}$; (6) for all $x \in \mathfrak{g}$, $\operatorname{ad} x$ is nilpotent.

(V) $\mathfrak{g}$ commutative $\Rightarrow$ $\mathfrak{g}$ nilpotent $\Rightarrow$ the Killing form of $\mathfrak{g}$ is zero $\Rightarrow$ $\mathfrak{g}$ solvable.
$\mathfrak{g}$ commutative $\Rightarrow$ $\mathfrak{g}$ reductive.
$\mathfrak{g}$ semi-simple $\Rightarrow$ $\mathfrak{g}$ reductive.

(VI) *Characterizations of* $\mathfrak{r}$:
(1) $\mathfrak{r}$ is the largest solvable ideal of $\mathfrak{g}$; (2) $\mathfrak{r}$ is the smallest ideal such that $\mathfrak{g}/\mathfrak{r}$ is semi-simple; (3) $\mathfrak{r}$ is the only solvable ideal such that $\mathfrak{g}/\mathfrak{r}$ is semi-simple; (4) $\mathfrak{r}$ is the orthogonal of $\mathscr{D}\mathfrak{g}$ relative to the Killing form.

(VII) *Characterizations of* $\mathfrak{n}$:
(1) $\mathfrak{n}$ is the largest nilpotent ideal of $\mathfrak{g}$; (2) $\mathfrak{n}$ is the largest nilpotent ideal of $\mathfrak{r}$; (3) $\mathfrak{n}$ is the set of $x \in \mathfrak{r}$ such that $\operatorname{ad}_{\mathfrak{g}} x$ is nilpotent; (4) $\mathfrak{n}$ is the set of $x \in \mathfrak{r}$ such that $\operatorname{ad}_{\mathfrak{r}} x$ is nilpotent; (5) $\mathfrak{n}$ is the largest ideal of $\mathfrak{g}$ such that, for all $x \in \mathfrak{n}$, $\operatorname{ad}_{\mathfrak{g}} x$ is nilpotent; (6) $\mathfrak{n}$ is the set of $x \in \mathfrak{g}$ such that $\operatorname{ad}_{\mathfrak{r}} x$ belongs to the radical of the associative algebra generated by 1 and the $\operatorname{ad}_{\mathfrak{g}} y$ ($y \in \mathfrak{g}$).

(VIII) *Characterizations of* $\mathfrak{s}$:
(1) $\mathfrak{s}$ is the intersection of the kernels of the finite-dimensional simple representations of $\mathfrak{g}$; (2) $\mathfrak{s}$ is the smallest of the kernels of the finite-dimensional semi-simple representations of $\mathfrak{g}$; (3) $\mathfrak{s}$ is the intersection of the largest nilpotency ideals of the finite-dimensional representations of $\mathfrak{g}$; (4) $\mathfrak{s}$ is the smallest ideal of $\mathfrak{g}$ such that $\mathfrak{g}/\mathfrak{s}$ is reductive; (5) $\mathfrak{s} = \mathfrak{r} \cap \mathscr{D}\mathfrak{g}$; (6) $\mathfrak{s} = [\mathfrak{r}, \mathfrak{g}]$; (7) $\mathfrak{s}$ is the intersection of the orthogonals of $\mathfrak{g}$ relative to the bilinear forms associated with the finite-dimensional representations of $\mathfrak{g}$.

NICOLAS BOURBAKI

Elements of Mathematics

English translations published:

General Topology, Part I

General Topology, Part II

Theory of Sets

Commutative Algebra

Algebra, Part I

Lie Groups and Lie Algebras, Part I

General Topology, Part I

General Topology, Part II

Theory of Sets

Commutative Algebra

Algebra, Part I

Printed in the United States
By Bookmasters